THE NEW ATLAS OF BREEDING BIRDS IN

BRITAIN AND IRELAND: 1988–1991

Dedicated to the memory of Brian Pashby,
Rob Robertson and Alan Walker

The New Atlas of Breeding Birds in Britain and Ireland: 1988–1991

Compiled by

DAVID WINGFIELD GIBBONS

JAMES B. REID

ROBERT A. CHAPMAN

British Trust for Ornithology

Scottish Ornithologists' Club

Irish Wildbird Conservancy

PUBLISHED BY

T & AD POYSER

ISBN 0 85661 075 5

First published in 1993 by T & A D Poyser Ltd
24–28 Oval Road, London NW1 7DX

Text set in 10/12pt Bembo by
Selwood Systems, Midsomer Norton.
Printed and bound in Great Britain by
Butler & Tanner Ltd., Frome, Somerset.

NEW ATLAS WORKING GROUP

(Note that those members of the group not listed as being affiliated to a particular organisation represented the BTO)

H.P. Sitters (Chairman) (1986–92)
Dr E. Bignal (NCC) (1987–90)
A.W. Brown (SOC) (1988–92)
R.A. Chapman (IWC: Organiser for Irish Republic) (1989–92)
J.C. Davies (SOC) (1986–88)
I.J. Ferguson-Lees (1986–92)
Dr R.J. Fuller (1987–92)
Dr C. Galbraith (JNCC) (1991–92)
Dr D.W. Gibbons (1987–92) Atlas Organiser
A. Goodall (1986–92)
Dr J.J.D. Greenwood (1987–92)
I. Herbert (IWC: Organiser for Irish Republic) (1988)
Dr S.D. Hill (CEGB and The National Grid Company plc) (1987–92)
C. D. Hutchinson (IWC) (1987–88)
Dr P. C. Lack (1986–92)
R. Nairn (IWC) (1986)
Dr R.J. O'Connor (1986–87)
Dr M. Pienkowski (NCC) (1987–90)
T. Poyser (1990–92)
Dr T. Reed (NCC) (1990–91)
Dr J.B. Reid (SOC and SNH: Organiser for Scotland) (1987–92)
Dr A. Richford (T & A D Poyser/Academic Press) (1991–92)
R. Spencer (1990–92)

S. Foulger (Secretary) (1991–92)
J. Mewis (Secretary) (1986–87)
E. Murray (Secretary) (1987–91)

INVITED MEMBERS OF THE NEW ATLAS WORKING GROUP

Dr C. J. Bibby (RSPB) (1991)
Dr D. A. Hill (1991)
Dr M. Tasker (NCC and Seabird Group) (1987)

EDITING

R. Spencer (principal editor), T. Poyser and D.W. Gibbons

DATA PROCESSING

Host Data Services Ltd, Watford and St Annes

MAPPING CONSULTANTS

Dot distribution mapping (DMAP): Dr A. Morton (Imperial College at Silwood Park)
Colour mapping (UNIMAP): I. Woiwod (Institute of Arable Crops Research, Rothamsted)

ILLUSTRATIONS

Richard Allen	Steve Carter	Andrew Hutchinson	Chris Rose
Norman Arlott	John Davis	Rodney Ingram	Brian Small
Kevin Baker	Kim Franklin	Ernest Leahy	Thelma K. Sykes
Bryan Bland	Robert Gillmor	Ian Lewington	David Thelwell
Nik Borrow	Alan Harris	David Nurney	Gordon Trunkfield
George Brown	John Hollyer	Dan Powell	Donald Watson
John Busby	Mark Hulme	Darren Rees	Martin Woodcock

The New Atlas was organised by the BTO, SOC and IWC, and was supported by JNCC.
It was sponsored by The National Grid Company plc, PowerGen plc, National Power
PLC and Nuclear Electric plc, under the DoE's conservation and business sponsorship
initiative and was negotiated by WWF UK (World Wide Fund for Nature).

Contents

Foreword *by HRH The Duke of Edinburgh*   ix

Preface *by Humphrey P. Sitters*   xi

Abbreviations used in the text   xiii

Frequently cited references   xiv

Introduction and methods *by David W. Gibbons*   1

Interpreting the species accounts *by David W. Gibbons*   16

Main species accounts   20

Brief species accounts   440

Bias and the key squares survey *by David W. Gibbons*   448

Population estimates for breeding birds in Britain and Ireland *by Simon Gates, David W. Gibbons and John H. Marchant*   462

The breeding birds of Britain and Ireland : changing status and species richness *by David W. Gibbons*   476

Appendices
  A   Data used for the Abundance maps   484
  B   Names of plants mentioned in the text   487
  C   Names of birds mentioned in the text   488
  D   Names of other animals mentioned in the text   491
  E   Additional maps   492
  F   List of other species recorded   495

References   499

Acknowledgements   514

Index of birds mentioned in the text   516

Foreword

by HRH The Duke of Edinburgh

A great many people in this country are concerned about the status of wildlife - birds, mammals, reptiles, insects and plants - in these islands. The increasing human population and the level of its activities in industry, housing, agriculture and leisure is known to be putting great pressure on many of the less abundant wild species.

People interested in birds in particular, are anxious about the effects of development, commercial forestry, draining and reclamation of wetlands, grubbing-up of hedgerows and chemical agriculture on the habitats of our resident species and on the migratory patterns of visiting birds.

Quite apart from the influence of human activities, nature is a dynamic process and the success of species has always depended on a number of natural variables, including climatic conditions throughout the year and the relative abundance of prey and predators.

Before the process of distinguishing between the effects of human activities and natural fluctuations can begin, there is a vital need for accurate and reliable information. This 'New Atlas' provides just such information and on a scale never previously available. This great work has been achieved by successful collaboration between voluntary bodies and industry, between amateurs and professionals and between Britain and Ireland. I am sure it will be warmly welcomed by birdwatchers of every standard.

[signature]

1993

Preface

by HUMPHREY P. SITTERS, *Chairman of the New Atlas Working Group*

This book breaks new ground in our knowledge of bird distributions in these islands. For 1968–72, *The Atlas of Breeding Birds in Britain and Ireland* published range maps of unprecedented precision, but we still had only the vaguest idea of how numbers varied from region to region. Were there more Skylarks in eastern England or Scotland, more Robins in Wales or northern England, or more Greenfinches in eastern or western Ireland? Now, for the first time, using innovative colour cartography techniques, we have the answers to these and many other questions.

This atlas also breaks new ground as it is the first repeated national survey to allow distributional change to be assessed for all species. Admittedly, when the project was first mooted, this was not regarded as an important objective, as it was thought to be too soon after the 1968–72 atlas for any significant changes to have occurred. With hindsight, that assessment was wrong. As fieldwork progressed, it became ever more apparent that many species had changed their distributions. Quite apart from such dramatic examples as the near extinctions of the Wryneck and Red-backed Shrike, the severe reduction in range of the Corncrake, the recovery and expansion of the Peregrine population and the even more marked spread of the Hobby, the results show that many other species have altered their distributions in a variety of interesting ways. These include the Goldfinch's extension of range into north-east Scotland, the Nuthatch's colonisation of Cumbria and Northumberland, the Goosander's southward spread into Wales and south-west England, and the expansion of the Siskin and Common Crossbill through conifer afforestation, as well as worrying contractions in the ranges of the Black Grouse, Nightingale, Grasshopper Warbler and Corn Bunting. There is a lesson here: bird distribution patterns are dynamic, changing more rapidly than was previously thought. Therefore they should be monitored more closely in future. This may involve a repeat breeding atlas sooner rather than later, or perhaps some kind of rolling sample survey with a selection of grid squares being monitored more often.

This atlas not only addresses new questions but also uses new methods. With cursory consideration, the assessment of bird densities might seem to be simple. Unfortunately that is not the case. Developing a method which would be straightforward, which would cover all species, and which could be adapted to a national scale, exercised the minds of the New Atlas Working Group for a considerable time. Eventually, on the basis of earlier BTO research and a year of pilot fieldwork, it was decided to assess densities of commoner species by reference to frequency of occurrence and those of scarcer species by counts.

Two aspects of fundamental concern to the Working Group were that the results should be free of bias caused by the uneven distribution of birdwatchers, and that the entire project should be precisely repeatable so that any future breeding atlas could properly assess changes in density. For these reasons, the principal method of assessing abundance involved visiting tetrads (2-km squares) for exactly two hours – no more and no less. The Working Group became only too well aware that many observers found the two hour limit extremely irksome, complaining that they had missed species which they were sure were present and that they felt the results would present a false picture. They also thought that elusive species would be under-represented. Neither of these concerns is valid, but they were widely held and to some extent affected the support that the project received. They are invalid because the purpose of the timed visits was not to produce a comprehensive bird list, but a means of assessing relative abundance within each 10-km square. Thus, if a species was missed during a timed visit to a tetrad, the inference was not that it was absent, but that it was less abundant in that 10-km square. Moreover, it does not matter that elusive species were missed more than conspicuous ones because the project did not attempt to show how density varied between species, only within species. It was unimportant that, for example, Nuthatches are more conspicuous than Treecreepers, just so long as a Treecreeper in Surrey is not more or less obvious than one in Yorkshire. On the other hand, the fact that some birds are more elusive than others was taken into account in establishing distribution. For this purpose, observers were encouraged to spend whatever additional time it took to find every species present in each 10-km square.

The results of the *New Atlas* are certainly very interesting and stimulating, but it is in their value to science and conservation that their true importance lies. Like its predecessor, the *New Atlas* will be in the public domain, so it will always be difficult to appreciate the full extent of its use. We do know, however, that atlas data are regarded as of seminal importance by government conservation agencies, such as the Joint Nature Conservation Committee, and major nature conservation charities, such as the Royal Society for the Protection of Birds. More specifically, we know that atlas data have been used in a variety of ways in the cause of conservation: for example, to measure avian diversity for the purposes of identifying sites for special protection; to assess the conservation status of species, as in the designation of 'Red Data Birds' (in turn, conservation status influences how much money is spent on each species); to identify areas where more detailed ecological studies should be undertaken; to identify species under threat (as shown by contractions in range); and to influence land-use strategies. The last of these may have important implications when atlas data are used, as they have been recently, to assess sites for designation as Environmentally Sensitive Areas or to influence regional forestry strategy in Scotland. Another way in which *New Atlas* data will be of value to both science and conservation is in relating bird populations to habitat. This will help us to reach a better understanding of why our birds are distributed in the way they are.

In addition to the use of *New Atlas* data by conservation bodies, they will be valuable to industry and developers as part of the environmental assessment process, identifying and quantifying potential impacts of proposals. The *New Atlas* will be an important reference source for all those involved in such work. Its use by industry extends beyond that for new developments to management of their land, in particular understanding the importance of the habitats within their ownership for different bird species.

Perhaps it is in education that atlas studies have their greatest value for conservation. All who study this book will be led to a deeper understanding of the rich and varied avifauna of these islands. This must have profound effects which, though quite unmeasurable, will influence the way birds are valued by society.

One of the main reasons for the success of the *New Atlas* is surely that it was a co-operative project with a wide variety of people and organisations pulling together to accomplish something which none could have achieved individually. The first breeding atlas was a partnership between the British Trust for Ornithology and the Irish Wildbird Conservancy. This time the BTO and IWC were delighted that the Scottish Ornithologists' Club was able to join in, even providing its own full-time organiser. Of crucial importance, too, was the financial support so generously provided by the *New Atlas*'s principal sponsors, the Central Electricity Generating Board and its successor companies, The National Grid Company, PowerGen, National Power and Nuclear Electric. That was supplemented by assistance for specific items from the Nature Conservancy Council, the Joint Nature Conservation Committee, and WWF UK (World Wide Fund for Nature). The whole project was organised by David Gibbons from the BTO's headquarters, in collaboration with Jim Reid in Scotland and Bob Chapman in Ireland. They in turn were assisted by a network of regional organisers who saw to it that every one of 3,858 10-km squares in Britain and Ireland was surveyed by one or more of an army of several thousand amateur ornithologists. These are the real heroes of this enterprise. For many, the completion of the fieldwork was a crusade carried out with dedication and determination, some of it in remote and dangerous countryside and in difficult conditions. They all must surely be delighted with the fruits of their labours set out in this book. If nothing else, the *New Atlas* is a prime example of the vital role that amateurs continue to play in nature conservation and biological science.

I am confident that *The New Atlas of Breeding Birds in Britain and Ireland: 1988–1991* will be regarded both at home and around the world as a work of innovation and enterprise. In great measure this is due to the inventiveness, imagination and skill of the leader of the team, David Gibbons. He and his colleagues, Jim Reid and Bob Chapman, are each to be warmly congratulated for a task well done.

The *New Atlas* was designed to be repeatable. Our findings have revealed that bird distributions change more rapidly than previously thought which suggests that the interval between the commencement of the present atlas and the next should not be more than twenty years. I am sure that all those who participated in the present atlas will look forward to the next with warm anticipation and in the knowledge that it is all very worthwhile.

Yelverton, Devon, 30th March 1993

ABBREVIATIONS USED IN THE TEXT

The following abbreviations have been used in the text:

BSBI	Botanical Society of the British Isles
BTO	British Trust for Ornithology
CBC	Common Birds Census (BTO)
CCW	Countryside Council for Wales
CEGB	Central Electricity Generating Board
DoE	Department of the Environment
EC	European Community
EN	English Nature
ha	hectare
ICBP	International Council for Bird Preservation
ITE	Institute of Terrestrial Ecology
IWC	Irish Wildbird Conservancy
JNCC	Joint Nature Conservation Committee
km	kilometre
km^2	square kilometre
KSS	Key Squares Survey
m	metre
NCC	Nature Conservancy Council
NERC	Natural Environment Research Council
NNR	National Nature Reserve
NRS	Nest Record Scheme (BTO)
RBBP	Rare Breeding Birds Panel
RO	Atlas Regional Organiser
RSPB	Royal Society for the Protection of Birds
SABAP	Southern African Bird Atlas Project
SCR	Seabird Colony Register (Seabird Group/JNCC)
SNH	Scottish Natural Heritage
SOC	Scottish Ornithologists' Club
SOVON	Samenwerkende Organisaties Vogelonderzoek Nederland (Cooperating Organisations Bird Census Work in The Netherlands)
tetrad	2km × 2km square of the British or Irish National Grids
WBS	Waterways Bird Survey (BTO)
WWT	Wildfowl and Wetlands Trust
10-km square	10km × 10km square of the British or Irish National Grids

FREQUENTLY CITED REFERENCES

The following works of reference are cited frequently and are always referred to by an abbreviated title (shown in parentheses):

BATTEN, L.A., C.J. BIBBY, P. CLEMENT, G.D. ELLIOTT and R.F. PORTER. 1990. *Red Data Birds in Britain*. T & AD Poyser, London. (*Red Data Birds*).

CRAMP, S. (ed) 1977–93. *Handbook of the Birds of Europe, the Middle East and North Africa: the birds of the Western Palearctic*. Oxford University Press, Oxford. 7 Vols (*BWP*).

HUTCHINSON, C.D. 1989. *Birds in Ireland*. T & AD Poyser, Calton. (*Birds in Ireland*).

LACK, P. 1986. *The Atlas of Wintering Birds in Britain and Ireland*. T & AD Poyser, Calton. (*Winter Atlas*).

LLOYD, C., M.L. TASKER and K. PARTRIDGE. 1991. *The Status of Seabirds in Britain and Ireland*. T & AD Poyser, London. (*Status of Seabirds*).

MARCHANT, J.H., R. HUDSON, S.P. CARTER and P. WHITTINGTON. 1990. *Population Trends in British Breeding Birds*. BTO, Tring. (*Trends Guide*).

SHARROCK, J.T.R. 1976. *The Atlas of Breeding Birds in Britain and Ireland*. T & AD Poyser, Berkhamsted. (*68–72 Atlas*).

THOM, V.M. 1986. *Birds in Scotland*. T & AD Poyser, Calton. (*Birds in Scotland*).

WITHERBY, H.F., F.C.R. JOURDAIN, N.F. TICEHURST and B.W. TUCKER. 1938–41. *The Handbook of British Birds*. London. 5 Vols (*The Handbook*).

For clarity, the current volume is referred to as the *88–91 Atlas*. With the exception of titles of species accounts, scientific names of species are not given within the body of the text but in appendices. Information extracted from the reports of the RBBP are referenced simply as RBBP. These reports are published annually in *British Birds* (e.g. SPENCER, R. and the RARE BREEDING BIRDS PANEL. 1990. Rare breeding birds in the United Kingdom in 1988. *Brit. Birds* **83:** 353–390; SPENCER, R. and the RARE BREEDING BIRDS PANEL. 1991. Rare breeding birds in the United Kingdom in 1989. *Brit. Birds* **84:** 349–370, 379–392).

Introduction and Methods

Historical Background

The publication of the *68–72 Atlas* by the BTO and the IWC heralded a new era in ornithology. This was the first attempt at mapping the distributions of an entire breeding avifauna at a national level in an objective manner. This field survey, which was modelled on the BSBI's *Atlas of the British Flora* (Perring and Walters 1962), was undertaken over five breeding seasons from 1968 to 1972. The planning of the survey was overseen by an Atlas Working Group. An appointed organiser (J.T.R. Sharrock) recruited a network of more than 120 volunteer regional organisers, and the fieldwork was carried out by an estimated 10,000–15,000 volunteer observers. The aim of the survey was to record the presence or absence of all breeding species within each 10-km square of the British and Irish National Grids, and in addition to determine the highest level of proof of breeding (categorised as 'possible', 'probable' or 'confirmed') of each species in each square. At the end of the five year period a staggering 282,000 10-km square records had been received, all of which had to be processed, exhaustively checked and edited, and eventually presented as a series of dot distribution maps in the final publication. This volume has become an invaluable conservation tool, and is probably one of the most quoted books in British ornithology. The enthusiasm for this type of survey was such that numerous regions and countries throughout the world have undertaken atlases of their own, using methods based on those developed by the Atlas Working Group. The bulk of the fieldwork for these atlases has been undertaken by volunteer birdwatchers, and it is this partnership between a small group of professionals overseeing the surveys, and the large number of volunteers, that has been responsible for the success of these projects.

Very few of these surveys, however, have been quantitative and few have attempted to map regional variation in bird density, as opposed to simple distribution. Thus, for example, the distribution map of the Robin shown in the *68–72 Atlas* could, if misunderstood, give the impression that Robins were equally abundant throughout their breeding range. That this was not the case has been demonstrated by the results of the CBC, a long-term monitoring scheme organised by the BTO, which uses the territory mapping method. From CBC data approximate estimates of breeding density can be obtained, which show that Robins are more abundant in the south and west than elsewhere in Britain (O'Connor and Shrubb 1986a), yet the simple dot distribution map gave no indication of this.

A small number of surveys, for example the *Atlas des Oiseaux Nicheurs de Belgique* (Devillers *et al.* 1988), the *Atlas van de Nederlandse Vogels* (SOVON 1987) and the European Ornithological Atlas Project (Bekhuis 1991), have used subjective order of magnitude estimates (1–9, 10–99, 100–999 etc.) to determine regional variation in abundance. Whilst these are probably reasonable for species with great variation in breeding density, they are probably less useful for those with little variation in density. In addition, they are prone to an unknown degree of error. Nevertheless, these are attempts at obtaining *absolute* levels of abundance, i.e. an estimate of the total number of birds or pairs of birds within each grid square. All other atlases based on quantitative methods have developed techniques for estimating *relative* abundance only (but see Buckland *et al.* 1990). Such relative estimates may be correlated with absolute density, but often in an unknown and frequently non-linear manner. They can be used to determine areas in which a species is more or less common, but not by precisely how much. Furthermore, relative estimates are generally species specific and cannot be used to make comparisons between species. The method of assessing relative abundance for the *Winter Atlas* (again a BTO/IWC survey and a companion volume to the *68–72 Atlas*) was to ask observers to count all birds seen or heard during a day's fieldwork and to note the total time spent in the field. The counts were then corrected to take into account variation in time in field. The method was both simple and effective. An alternative approach used in the *Atlas of the Birds of the Southern Cape* (Hockey *et al.* 1989), and subsequently by SABAP, was to use an index of frequency, measured as the proportion of cards submitted for each quarter degree grid square, for a given month, which contained a record of the species in question.

During pilot fieldwork for the *68–72 Atlas* a quantitative method was tested. This involved observers visiting all 25 tetrads within their chosen 10-km square and noting the species found in each. A measure of relative abundance of each species, in each of the pilot 10-km squares, was then calculated as the number of tetrads in which it was recorded, divided by 25. Had this method been adopted for the full survey it would have been a quantitative atlas, using frequency of occurrence as its measure of relative abundance. In the event, however, it was believed that such a high level of coverage could be achieved only in well populated areas (Ferguson-Lees and Sharrock 1970) and the method was shelved. This frequency of occurrence method was not forgotten, however, and a revised version was adopted for the *88–91 Atlas*.

In summer 1986, BTO Council decided that a new breeding atlas should commence in 1988, 20 years on from the start of the *68–72 Atlas*. A new Atlas Working Group was appointed in September of that year and its brief was to develop the methods that would ensure, firstly, that the distributions of all breeding species were re-mapped, and secondly, that regional variation in density of each species was measured. The importance of ensuring adequate coverage in Scotland and Ireland was appreciated at an early stage, and the SOC and IWC were asked to help co-organise the survey. The New Atlas Working Group sat on 14 separate occasions between 1986 and 1992, and this volume is, in part, the result of their decisions.

Methods and rationale

Fieldwork for the *88–91 Atlas* was conducted between April 1st and July 31st in each of the four years 1988–91. Initially, it had been hoped that coverage would be complete in three years, but this proved impossible and the survey eventually occupied four seasons. Fieldwork was coordinated by a network of volunteer regional organisers and was undertaken by members of the BTO, SOC, IWC and others (again mostly volunteers) who followed a set of instructions issued prior to the beginning of each breeding season. In brief, the instructions were as follows. Observers visited, either alone or as a coordinated team, a minimum of eight tetrads of their own choice within each 10-km square. Two hours were spent in each tetrad, and a species list was compiled for each. From these timed visits an index of abundance of each species in each square was calculated.

Additional supplementary (non-timed) observations were also requested, to ensure that the species lists for each 10-km square were as complete as possible.

It was recommended that the two-hour period be split into two one-hour visits, one early in the season (April to May) and one late (June to July), although it was stressed that in remote areas a single visit (after mid May to ensure that summer migrants had arrived) would be a more efficient use of time. This recommendation was made for two reasons. Firstly, some species are more detectable early in the season and others late, and splitting the timed period in this manner would help to maximise the number of species recorded. Secondly, it increased the time period during which Atlas fieldwork could be carried out. With the exception of a few early breeding species, such as Mistle Thrush or Common Crossbill, the optimal time for recording breeding birds is probably mid May to mid June. Had fieldwork been restricted to this short time period, however, it would have been impossible to ensure complete coverage, hence the decision to have a four month recording period. If observers had carried out their two-hour tetrad visits only in the first half of the season, they might have missed late arriving migrants (such as the Spotted Flycatcher), hence the split into two one-hour visits, one early and one late in the season.

The observer(s) noted how many and which tetrads had been visited, each tetrad being denoted by a single letter of the alphabet (A to Z, with the exception of O; Fig. 1). If an observer visited a 10-km square that had been visited in a previous year, they were asked to cover tetrads which had not already been surveyed. This was to help ensure that the indices of abundance obtained were representative of the entire 10-km square and not simply of the preferred tetrads within that square.

The frequency of occurrence (= frequency index) of each species in each 10-km square was expressed as the proportion of tetrads visited in which that species was recorded. Thus, if a total of 20 tetrads was surveyed over the four year period in a particular 10-km square, and a species was recorded in 10 of these, then its frequency of occurrence was 0.5 (10÷20). The results of a pilot survey undertaken in 1987 had shown that such indices were correlated with a measure of absolute density (Gibbons 1987), although strictly they reflect how widespread a species is in a given 10-km square. With the exception of species that are either very common or highly clumped in distribution, there are theoretical reasons for expecting, and empirical observations that support, a relationship between frequency and abundance within and between species (Blondel 1975, Blondel et al. 1981, Dawson 1981, Fuller 1982, Verner 1985, Yapp 1956).

For species with highly clumped distributions, however, such a measure of frequency would have greatly reduced the true variation in relative density. For example, a large colony of Sand Martins may have been present in a tetrad in one 10-km square, and a single nest of the same species in a neighbouring 10-km square, yet the frequency of occurrence for the species in the two squares would have been the same (assuming the same number of tetrads was visited in each). To overcome this problem, observers counted the number of apparently occupied nests of colonial nesting species (some seabirds, Grey Heron and Sand Martin) and the number of individual birds, excluding dependent young, of other highly clumped species, such as the birds of open water (e.g. ducks, geese and swans). For some very common species with clumped distributions (e.g. Rook, Swift and House Martin) it was thought to be impractical to ask observers to count either nests or individuals, so these were censused using the frequency index method instead.

An added complication was that of counts of coastal seabird colonies. When the Atlas survey was commencing, the Seabird Group and the NCC were just completing the Seabird Colony Register (SCR), for which counts of coastal seabird colonies in Britain and Ireland had been undertaken between 1985 and 1987. The results of this survey were subsequently published in Status of Seabirds. The Atlas Working Group decided that it would be a waste of resources to repeat this survey as part of the Atlas so instead, following discussions with the Seabird Group/NCC, it was decided to incorporate SCR data directly into the Atlas, despite the different survey dates. SCR coverage of inland seabird colonies was poor, however, so Atlas observers were asked to count nests in seabird colonies that were out of sight of the sea. Such counts were required for the following species: Fulmar, Cormorant, Black-headed Gull, Common Gull, Lesser Black-backed Gull, Herring Gull, Great Black-backed Gull, Common Tern, and Arctic Tern. All other seabird species either had inland counts as part of the SCR, or did not breed inland. Atlas observers were encouraged to provide counts of inland seabird colonies from those parts of the 10-km square in which they did not make timed tetrad visits as well, and these counts were entered (along with a grid reference) in a 'Comments' box on the recording forms. The precise manner in which SCR and Atlas data were merged is discussed later.

Counts were required for one further group of birds: rare species. The between squares variation for these species was expected to be low (little better than presence or absence), so observers were asked to count individuals of all species with less than an estimated 10,000 breeding pairs in Britain. Dependent young were not included in any counts.

All counts, whether of colonial, open-water or rare species, were carried out within the two-hour period, although the clock could be stopped during counting. If observers made two one-hour visits to a tetrad they were asked to count all relevant species on both visits, and to enter the higher of the two counts on the recording form – unless they considered it was significantly inflated by the presence of wintering or passage birds, in which case, the lower count was entered.

The adoption of a set time period within each tetrad was one of the most controversial aspects of the methods, but there were compelling reasons for adhering to it. Parts of NW Scotland and the W coast of Ireland are remote or sparsely populated with birdwatchers, whereas in SE England quite the opposite is the case. If no time period had been set then clearly more time would have been spent in tetrads in well covered areas. As the probability of detecting a given species is likely to increase with time spent observing, this would have led to a tendency to high indices in well covered areas and low indices in poorly covered areas. Thus the measures of abundance would have reflected observer, and not bird, density. Although in poorly covered areas, fewer tetrads would have been visited, the two-hour time period ensured there would be no systematic bias due to variation in effort. Of course, the estimates of abundance would be more prone to error in remote areas, simply because the index would be based on a smaller number of visited tetrads. Thus, rather than spending more than two hours in a tetrad, observers were asked to cover more tetrads, thereby increasing the precision of the estimate and more truly reflecting the content of the entire 10-km square.

The choice of eight tetrads per 10-km square and two hours per tetrad was essentially an arbitrary decision, based on the opinion of the Atlas Working Group that the minimum amount of fieldwork in any 10-km square should be equivalent to two or three man-days (15–16 hours). The combination of eight tetrads and two hours allowed for reasonably high precision and ensured that observers did not have to rush through tetrads. Keeping the time spent per tetrad reasonably short also increased the between-squares variation for very common species, which would otherwise have been low (see Bias and the Key Squares Survey).

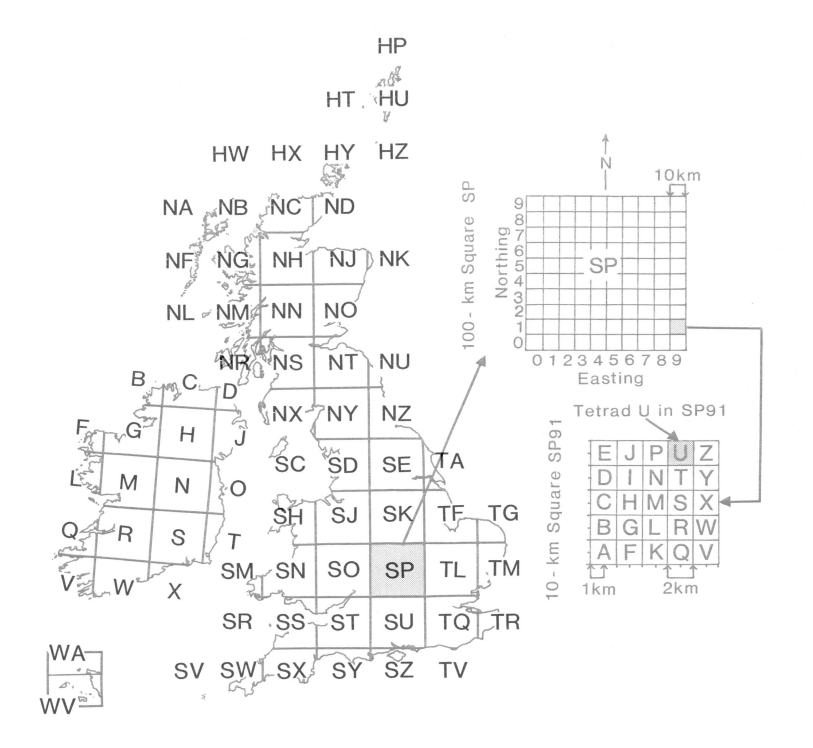

Fig. 1. The letter designations of the 100-km squares of the British and Irish grids, and the method of determining 10-km square and tetrad grid references. Tetrad U in 10-km square SP91 is given as an example.

In some coastal squares entire tetrads are in the sea. The Atlas Working Group ruled that only tetrads whose centres fell on land could be included. 'Land' was defined as ground (or freshwater) above the low water mark as shown on Ordnance Survey Maps. Had the choice of tetrads with only tiny amounts of land been allowed, the indices for inland species would have been decreased, due to the smaller area of land in which to search, whereas the indices for coastal species would have been increased. Conversely, excluding all tetrads which contained any amount of sea would have seriously reduced indices of coastal species. The 'centre on land' rule was a sensible compromise. If a 10-km square contained eight or fewer surveyable tetrads, all were visited. The frequency of occurrence estimates for species in coastal 10-km squares became increasingly imprecise as the number of tetrads on land decreased. In extreme cases, frequencies were either one or zero (essentially presence or absence) for coastal 10-km squares with only a single surveyable tetrad.

Although most records from the April to July survey period were required for the Atlas, some were not. Observers recorded only species considered to be 'using' the tetrad. Hence, gulls flying over were omitted, but feeding Swallows or Kestrels hovering were included. Records of summering non-breeding birds were included, but birds which were, in the observer's opinion, late winterers or passage migrants were omitted. Inevitably this introduced a degree of subjectivity, but where observers were uncertain they were asked to consult their RO.

In addition to the timed tetrad visits, observers recorded whether or not each species was seen or heard, or whether there was evidence that it was breeding in the 10-km square(s). Evidence of breeding could come from the two-hour timed tetrad visits or from any other reliable source, such as local bird reports and atlases, or further supplementary (non-timed) observations. To ensure that the species lists were as complete as possible for each 10-km square, observers were strongly encouraged to return to their 10-km squares after the timed tetrad visits in order to locate species which may have been missed; this was particularly important for elusive and nocturnal species.

A species was considered to be **breeding** if any of the following activities were observed:

Bird apparently holding territory.
Courtship and display; or anxiety call/agitated behaviour of adult indicating presence of young or nest.
Brood patch on trapped bird.
Adult visiting probable nest site.
Nest building (including excavating nest hole).
Distraction display or injury feigning.
Used nest found.
Recently fledged young.
Adult carrying faecal sac or food.
Adult entering or leaving nest site in circumstances indicating occupied nest (including colonies).
Nest with eggs found, or bird sitting but not disturbed, or eggshells found near nest.
Nest with young; or downy young of ducks, gamebirds, waders and other nidifugous species.

'Breeding' records thus referred to birds proved to be breeding and to those that were probably breeding although proof was lacking. Records of birds that were observed (or heard) while using the 10-km square, but with no evidence of breeding, were classified as **seen**. Once again, as for tetrad recording, birds which were late winterers or passage migrants were omitted.

The 'breeding' category of the *88–91 Atlas* thus combined the 'probable' and 'confirmed' categories of the *68–72 Atlas*, and the 'seen' category was comparable to 'possible', with two important exceptions. Firstly, singing males present in the same place on more than one date were included as probably breeding in the *68–72 Atlas*. For the *88–91 Atlas*, singing males were included in the 'seen' category, because song does not necessarily imply breeding (some species sing on passage). As observers were given the option of making single visits to tetrads (particularly in remote areas) the chance of recording singing males on two separate occasions was reduced. Secondly, the *68–72 Atlas* included in the 'possible' category only birds in suitable nesting habitat during the breeding season. The *88–91 Atlas* instructions made no mention of nesting habitat in the 'seen' category. Summering, non-breeding birds were included in the *68–72 Atlas* provided there was suitable breeding habitat in the 10-km square. Summering, non-breeding birds were included in the *88–91 Atlas* regardless of whether or not there was any suitable breeding habitat in the square.

These subtle, but important, changes were made for a number of reasons. 'Probable' and 'confirmed' breeding records were combined to ensure that observers did not spend time searching for nests or other high levels of proof of breeding. It would have been easy for an observer to spend much of a timed tetrad visit nest-finding, rather than species-finding, and this would have affected the indices obtained. Absolute proof of breeding is not essential for mapping the breeding distribution of most species. In most cases it is sufficient to know that a species is holding territory and probably attempting to breed. Although there are some species that will defend a territory but will not breed (e.g. Red Kite), for the great majority of species this is not the case.

The inclusion of summering birds in areas with no suitable breeding habitat was perhaps more controversial, but was done for a number of reasons. Firstly, from a conservation point of view, knowledge of presence of a species during the breeding season is as important as knowledge of breeding. Secondly, defining suitable habitat is not straightforward, as this may vary spatially, temporally and also with observer knowledge. For example, at the time of the *68–72 Atlas*, Marsh Harriers were almost exclusively confined to extensive *Phragmites* reed beds, but this species now breeds regularly in crops. To be certain that there is no suitable breeding habitat for a particular species in a particular 10-km square requires a degree of omniscience. Thirdly, observers for the *88–91 Atlas* were able to refer to the *68–72 Atlas* and could easily have transposed the term 'breeding range' for 'breeding habitat' and thus excluded useful records because of preconceived notions of a species' breeding range. Finally, some species summer in areas in which they subsequently breed, so these records may provide an insight into future changes in breeding range.

During the early years of the 1988–91 survey, it became clear that some observers and ROs were uncertain of the definition of the term 'summering'. To clarify this, during the final stages of editing of records (see below), ROs were asked to include records of species that did not normally breed in their area only if they were from the period mid May to mid June. Unfortunately this instruction was not included in earlier versions of the fieldwork instructions.

In the *68–72* and *88–91 Atlases* the inclusion or exclusion of a small proportion of records led to a degree of observer and RO subjectivity. The 1968–72 survey required a subjective assessment of whether or not suitable breeding habitat was present. The 1988–91 survey required an assessment of whether or not the record was of a passage or wintering bird. Neither of these is satisfactory, but birds recorded during the breeding season may be from wintering, breeding, passage or summering populations, and it is often possible to tease these apart only by subjective assessments. Year-round atlases such as the *Atlas van de Nederlandse Vogels* (SOVON 1987) are an alternative approach to this problem, but require more extensive fieldwork. In addition, in the *68–72 Atlas* a record of a species 'present in suitable breeding habitat' could have been excluded if the observer (or RO) was certain it had not bred. This was not true of the *88–91 Atlas*.

In practice these changes in the rules of acceptance or rejection of a record make very little difference for most species. It is mainly the distributions of seabirds, which nest in a restricted number of localities, but range over a much greater area, which would be affected by the inclusion of summering birds out of breeding habitat. However, as breeding is relatively easy to prove for these species, and as much of the seabird data came from the very comprehensive SCR survey of breeding colonies, all maps of seabirds presented here are based on breeding records *only*. A number of northerly breeding species such as Common Scoter and Dunlin are also affected by the inclusion of summering records, as the Distribution maps show that these have small summering populations on more southerly coasts.

The recording forms

Two separate recording forms were used. The first, a 10-km square worksheet, was for observations during the timed tetrad visits from a single 10-km square. This contained a series of horizontal rows, one for each of the species that was likely to be recorded, and vertical columns, one for each tetrad, thus 25 columns in total. The rows for species which had to be counted were shaded; those for which simple presence or absence in a tetrad was sufficient were not. The observer entered ticks for non-count species, or counts in the relevant row and column, and at the end of the season added up the number of ticks (i.e. number of tetrads in which recorded) for each species and summed all the individual counts (Fig. 2). The observer entered the 10-km square designation (two letters and two digits in Britain and one letter and two digits in Ireland, as shown on Ordnance Survey maps), the number of tetrads visited, which tetrads had been visited, and other information on the header section of the worksheet (Fig. 3).

The second recording form, a supplementary record sheet, was used for 10-km square (not tetrad) records from any part of Britain or Ireland collected outside of the timed visits (Fig. 4). These were circulated widely in an attempt to gain as many additional records as possible. To ensure that very early and very late breeding species were not missed, supplementary records of breeding were welcomed from outside of the April to July period.

Advantages and disadvantages of the method, and comparisons with other quantitative atlases

The methods of data collection enabled two different maps to be produced for each species; one of Abundance and one of Distribution. The former were based on data collected during timed tetrad visits *only*, and were thus corrected for variation in fieldwork effort, whilst the latter were based on these and on further supplementary, non-timed data, and thus were not corrected for effort. The decision to distinguish between data collected during timed tetrad visits and data collected outside of these periods was taken early in the planning stages of the *88–91 Atlas*. The advantage of this has already been stressed – standardisation of effort for the abundance data. There was, however, a disadvantage. Rare and elusive species, particularly nocturnal ones, could be missed during two-hour visits, and only

subsequently found by hard searching outside of the timed periods. Thus for some of these species the Abundance maps yielded little more information than the Distribution maps, so they are not presented in this Atlas (see *Interpreting the Species Accounts*). A possible way around this would have been to allow observers to spend as long as they wished in each tetrad, to count all birds and to record the time spent. This would have increased the chances of detecting any elusive species possibly present, and the counts could have been corrected to a standard time period. This method would have been similar to that of the *Winter Atlas*, but at a finer geographical resolution. It would, however, have led to observers spending more time in fewer tetrads than under the adopted methodology, and this would have led to lower precision and a poorer representation of the nature of the entire 10-km square. In addition, it would have added to the complexity of both the fieldwork and the analysis.

One further problem with the methods of obtaining abundance estimates was that of prior knowledge of individual observers. Observers who knew of the location of a rarity might ensure they included that site in their timed visit. In less familiar areas this would not have occurred. Such bias could be problematical for a few species, and it is thought that, in particular, the Abundance map for the Peregrine might have been affected in this manner.

The *Winter Atlas* methodology allowed untimed supplementary counts and, for the maps, both timed and untimed data were combined, even though untimed counts could not be corrected for time spent in the field. The incorporation of such uncorrected data forced the decision to use the maximum count of a species seen in a day, where more than one count was available for a square. As maxima are sample size dependent this could have led to systematic bias towards higher counts in well covered areas. In practice, however, this problem was largely unfounded, as it was shown that only for 8% of species could variation in coverage have accounted for more than 10% of the variation in bird abundances; and in the cases of Great Tit and Blue Tit, only, did it account for more than 20% (see p16 in the *Winter Atlas*). Within the 8%, most species were commonest in the well covered areas for other reasons.

The 'proportion of cards' technique used by SABAP in southern Africa was not corrected for time in the field, so there may have been some systematic bias, caused by less effort per card in poorly covered areas, and vice versa. In addition, as there was no requirement for a minimum number of cards for each square before the data were acceptable (equivalent to the minimum number of eight tetrads in the *88–91 Atlas*), the technique was prone to error due to imprecision in areas from which very few cards were received. Despite these problems, however, the simplicity of the 'proportion of cards technique' makes it an attractive method.

There were two further advantages of the *88–91 Atlas* methods. Firstly, not only did the timed tetrad visits provide measures of abundance for each species in each 10-km square, they also yielded presence/absence information at the tetrad level. It must, however, be stressed that two hours in each tetrad did not provide a comprehensive species list for that tetrad, just an idea of the species assemblages that occurred within it. Secondly, and most importantly, the *88–91 Atlas* methods are highly repeatable. It was clear that during the 1988–91 survey there was wide regional variation in fieldwork effort (Fig. 5), although this can be quantified only for the timed tetrad visit data. It is likely that there was regional variation during the *68–72 Atlas*, but it was measured only crudely. Unfortunately, variation in effort between the two Atlas periods may lead to apparent changes in distributional range that are not real. To take an extreme example, if a particular county was well populated with enthusiastic birdwatchers during 1968–72, but not so during 1988–91, then any apparent contraction of

25 inland BH nests counted. Do not count individual birds for IN or N species.

A tick for an IN species means that you recorded it but found no inland nests, or recorded it (or its nests) at the coast.

Total number of tetrads in which BH was recorded.

Total number of inland nests counted (i.e. 25 + 40 = 65).

BH must breed in the 10-km as nests were found.

Note that some seabirds only require a tick and not a count of nests; this is because they never breed inland.

Enter a count of 0 in this column if no inland nests were found.

No HG nests recorded during timed visits, but breeding recorded during a casual visit.

2 LO seen in this tetrad during timed visit. Note that this is 2 individuals not 2 pairs. Do not include dependent young in any counts.

Breeding recorded during a casual visit (in this case a nocturnal visit).

A tick for an N species means that it was recorded but no nests were found.

The value in the COUNT column must always equal the sum of the individual tetrad counts (i.e. 2 + 1 + 1 + 2 = 6).

SM recorded in 4 tetrads, and 40 nests found (all in tetrad F). If no nests were found, a 0 should be entered here.

40 SM nests recorded in tetrad F.

Do not give counts for those species that are unshaded; simply tick these species.

S and B should only be entered in these columns — do not enter them elsewhere on the Worksheet, nor enter ticks in these columns.

RT was Seen during tetrad visits, but Breeding was subsequently proven during a casual visit.

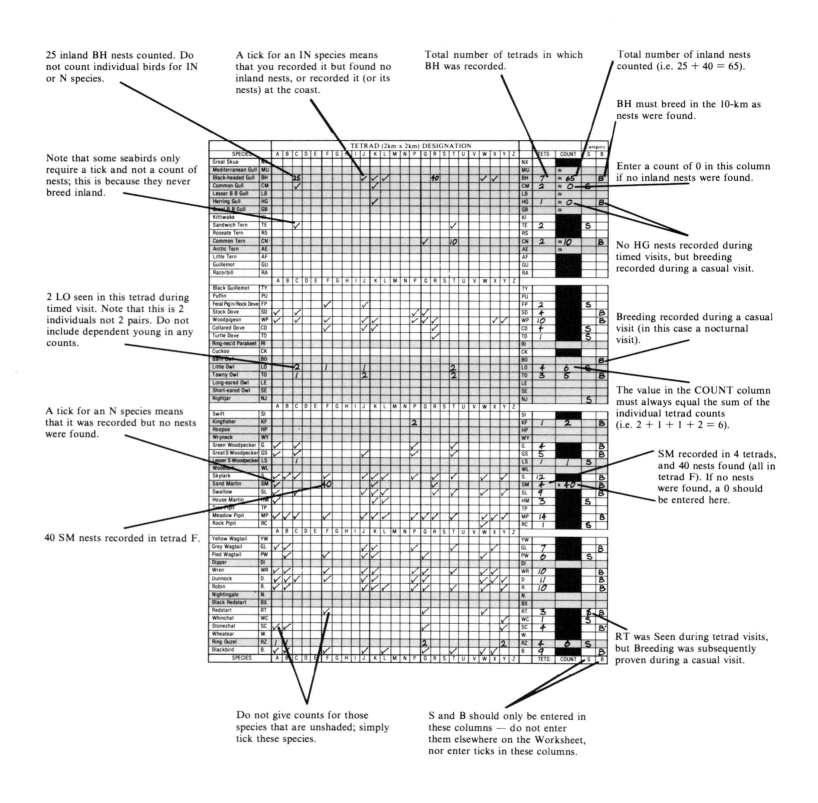

Fig. 2.   A page of a 10-km square worksheet (reduced) taken from the fieldwork instructions, indicating how it was to be filled in by the observer.

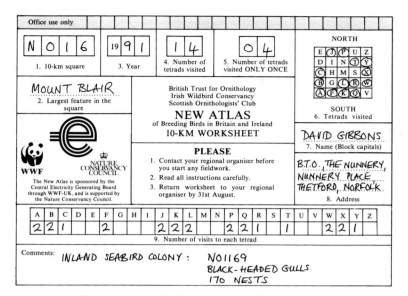

Fig. 3. A completed worksheet heading (reduced).

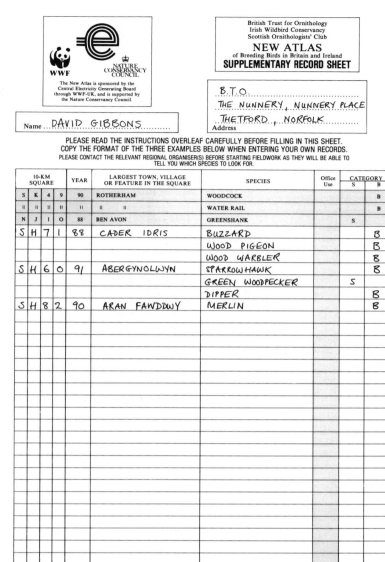

Fig. 4. A completed supplementary record sheet (reduced).

species range after the 20 year interval may simply reflect difference in observer enthusiasm, rather than real changes in distribution. This problem was addressed by the *88–91 Atlas* methods though, obviously, nothing could be done to correct the potential biases arising from uneven coverage in the *68–72 Atlas*. Because a set amount of time and effort was spent in each tetrad during 1988–91, any future atlas using the same methods will be able to determine precise distributional changes, which are not influenced by differential levels of enthusiasm, by comparing distributions tetrad on tetrad (rather than 10-km square on 10-km square).

The *Winter Atlas* and SABAP are both quantitative atlases relying upon fieldwork rather than guesswork to provide measures of abundance. Neither, however, collected information on such a fine geographical scale as the *88–91 Atlas*, nor in such a repeatable manner. In their favour, however, this meant that their methods were more straightforward, were less prone to being misunderstood, and were probably better received by amateur fieldworkers. Treading the fine line between utility of the data and enjoyment of the fieldwork is never easy, and great care has to be taken in the early planning stages of any such survey to ensure that this line is not crossed. It is sincerely hoped that those who contributed to the fieldwork for the *88–91 Atlas* will feel that the relative complexity of the methods was justified when they see the results of their work in this book.

National and regional organisation

Fieldwork within each 10-km square was mainly undertaken by volunteer observers each of whom received instructions from, and returned records to, a volunteer RO. In most parts of the UK, the RO was the BTO Regional Representative (RR), although in some areas the RR had nominated a separate RO for part, or all, of their region. In highly populated areas, fieldwork within a particular 10-km square was sometimes coordinated by a 10-km square 'steward', who reported to the local RO. ROs, in their turn, reported to one of three National Organisers, one for England, Wales and Northern Ireland, one for Scotland and one for the Republic of Ireland. The RO network in the Republic was less complete than in the UK, and in many instances individual observers dealt directly with the National Organiser.

Without this regional network the survey would have been impossible. ROs drummed up enthusiasm in their region, ensured adequate coverage in each of their squares, coordinated fieldwork to reduce duplication of effort, kept tallies of the species recorded in each square, and dealt with mountains of paperwork.

The ROs for each area were as follows and, unless otherwise stated, ROs were responsible for the entire data gathering and checking in their area, for 1988–92:

Regional Organisers

England

Avon	R.L. Bland
Bedfordshire	A.J. Livett (1988), E. Newman (1989–92)
Berkshire	I. Collins
Birmingham and W. Midlands	J.R. Winsper
Buckinghamshire (N)	N.H.F. Stone
Buckinghamshire (S)	A.F. Brown

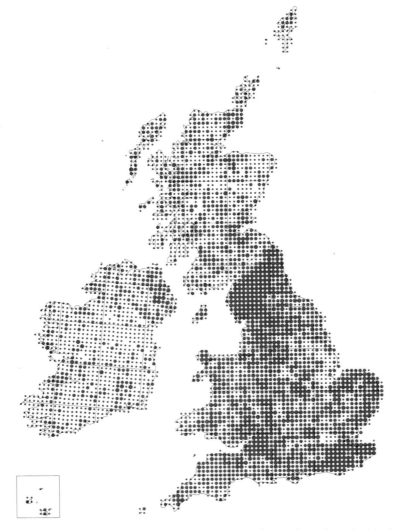

Essex (NE)	G.E. Edwards (1988), P. Dwyer (1989–92)
Essex (S)	G.E. Edwards
Gloucestershire	T.A. Jones assisted by T. J. Davis
Hampshire	G.C. Evans
Herefordshire	K. Mason
Hertfordshire	Dr K.W. Smith
Huntingdonshire and Peterborough	G. Walthew (1988–90), M.R. Coates (1991–92)
Isles of Scilly	J.W. Hale
Isle of Wight	J. Stafford
Kent	Major G.F.A. Munns
Lancashire (E)	A.A. Cooper
Lancashire (NW)	D.J. Sharpe
Lancashire (S)	D. Jackson
Leicestershire and Rutland	C. Measures (1988–90), J. Graham (1991–92)
Lincolnshire (E)	R.K. Watson
Lincolnshire (N)	I. Shepherd
Lincolnshire (S)	R. and K. Heath
Lincolnshire (W)	A. Goodall
London and Middlesex	K. Betton
Manchester	R.G. Williams
Merseyside	A.S. Duckels
Merseyside (Wirral)	A.S. Duckels and S.J. Woolfall (1988), S.D. Jowitt (1989–92)
Norfolk (NE)	Dr M.P. Taylor (1988–89), A. Lowe (1990–92)
Norfolk (NW)	Dr M.P. Taylor (1988–89), M. Barrett (1990–92)
Norfolk (SE)	Dr M.P. Taylor (1988–89), K. Johns (1990–92)
Norfolk (SW)	Dr M.P. Taylor (1988–89), A. Waterman (1990-92)
Northamptonshire	P.W. Richardson
Northumberland (N)	B.N. Rossiter
Northumberland (SE)	T. and M. Cadwallender
Northumberland (SW)	M. Macfarlane
Nottinghamshire	E. Cowley
Oxfordshire (N)	M.F. Oliver
Oxfordshire (S)	C. Ross
Rugby	R.B. Ratcliffe (1988–91)
Shropshire	C.E. Wright
Somerset	W.J. Webber
Staffordshire (N)	F.C. Gribble
Staffordshire (S)	P.K. Dedicoat
Suffolk (E)	M.T. Wright
Suffolk (W)	R. Waters
Surrey	Dr E.F.J. Garcia
Sussex (E)	M. Scott-Ham
Sussex (W)	V. Bentley
Warwickshire	J.A. Hardman
Wiltshire (E)	S.B. Edwards
Wiltshire (W)	R. Turner
Worcestershire	G.H. Green
Yorkshire (Bradford)	M.L. Denton
Yorkshire (E)	The late B.S. Pashby (1988–91), S. Pashby

Fig. 5. Variation in observer effort measured as the number of tetrads visited in each 10-km square. The three different dot sizes represent differing levels of coverage. These are, from the smallest to the largest, 1–8, 9–16 and 17–25 tetrads visited per 10-km square. In 10-km squares which are blank no timed visits were carried out as part of the main survey, but see *Bias and the Key Squares Survey*.

Cambridgeshire	M.J. Allen (1988), R. Clarke (1989–92) assisted by A. Dobson (1989)
Cheshire (N and E)	B. Martin (1988), C. Richards (1989–92)
Cheshire (Mid)	R.S. Leigh
Cheshire (S)	R.S. Leigh (1988), C. Lythgoe (1989–92)
Cleveland	R.T. McAndrew
Cornwall	S.F. Jackson
Cumbria (N)	J.C. Callion
Cumbria (S)	I. Kinley
Derbyshire	G.P. Mawson
Devon	H.P. Sitters
Dorset	R. Gleason, assisted by J. Morgan (1989) and R. Peart (1990)
Durham	D. Sowerbutts
Essex (NW)	G.E. Edwards (1988), G. Smith (1989–92)

Yorkshire (Harrogate)	The late A.F.G. Walker (1988–90), M.F. Brown (1991–92)
Yorkshire (Leeds and Wakefield)	S.P. Singleton (1988–89), A.D. Mitchell (1990), T.Dolan (1991–92)
Yorkshire (N)	C.M. Hind
Yorkshire (NE)	S. Cochrane
Yorkshire (N. Humberside)	P. Scanlan (1988–90), F. Moffatt (1991–92)
Yorkshire (NW)	M.M. Priestley
Yorkshire (SE)	K. Hayhow
Yorkshire (SW)	Dr A.H.V. Smith
Yorkshire (York)	P. Hutchinson

Wales

Anglesey	J. Clark
Brecon	M. Peers
Caernarfon	J. Barnes
Carmarthen	D.H.V. Roberts
Cardigan (N)	R. Squires
Cardigan (S)	P. Davis
Clwyd (E)	J.C. Peters
Clwyd (W)	R.D. Corran
Glamorgan (Mid)	S.J. Moon (1988), R. Poole (1989–92)
Glamorgan (S)	P. Bristow (1988), B. Edge (1989), R. Poole (1990–92)
Glamorgan (W)	Dr D.K. Thomas
Gwent	Dr S.J. Tyler
Merioneth	A. Seddon and M. Garnett assisted by J.B. Grasse (1990)
Montgomery	K.E. Stott
Pembrokeshire	G. Rees
Radnor	E. Morgan (1988), P. Jennings (1989–92)

Northern Ireland

Antrim and Belfast	G.A. Acheson
Armagh	D.W.A. Knight
Down	G. McElwaine
Fermanagh	B.H. Nelson
Londonderry	H. Dick
Tyrone	P.S. Grosse

Scotland

Aberdeen and Banff	A. Webb (1988–90), P. Doyle (1991–92)
Angus	B.M. Lynch
Argyll (N. and Mull)	M. Madders assisted by D.C. Jardine (1988)
Argyll (S. and Gigha)	D.C. Jardine
Arran, Bute and Cumbrae	D. Warner
Ayrshire	B.D. Kerr
Benbecula and the Uists	P.R. Boyer
Borders (Roxburgh and Berwickshire)	The late R.J. Robertson (1988–90), R.D. Murray (1991–92)
Borders (Tweeddale and Ettrick/Lauderdale)	R.D. Murray

Caithness	E.W.E. Maughan
Central Scotland	M.E. Phillips
Dumfries	R. Mearns
Fife and Kinross	N. Elkins
Inverness (Badenoch and Strathspey)	R. Dennis
Inverness (E)	R.L. Swann
Inverness (W)	D.S. Whitaker
Islay, Jura and Colonsay	D.C. Jardine assisted by M. Ogilvie (Islay only)
Isle of May	B. Zonfrillo
Kincardine	R.M. Laing assisted by P. Doyle (1992)
Kirkcudbright	G. Shaw
Lanarkshire, Renfrewshire and Dunbartonshire	I.P. Gibson
Lewis and Harris	C.M. Reynolds assisted by S. Angus (1989)
Lothians	A.W. and L. Brown (1988–89), L. Souter (1990), G.D. Smith (1991–92)
Moray and Nairn	M.J.H. Cook
Orkney	C.J. Corse
Perthshire	R.E. Youngman
Ross-shire	A.D.K. Ramsay assisted by R.L. Swann (1990–92)
Rum, Eigg, Canna and Muck	R.L. Swann
Shetland	J.D. Okill
Skye	A. Currie
Sutherland	G. Bates
Wigtown	G. Sheppard

Isle of Man

	Dr J.P. Cullen

Channel Islands

	I.J. Buxton

Republic of Ireland

Carlow	S. Corry
Cavan	T. Cooney
Clare	P. Brennan
Cork	P. Smiddy
Donegal	R. Sheppard
Dublin	T. Cooney and H. Brazier
Galway	N. Sharkey
Kilkenny	R. Goodwillie and I. Logan
Limerick	E. Jones
Longford	M. Ryan
Louth	L. Lenehan
Meath	L. Lenehan
Monaghan	T. Cooney
Tipperary	C. Wilson and J. Coman
Wexford	D. Daly
Wicklow	P. Farrelly

Wherever possible fieldwork was carried out by ornithologists with local knowledge. As the survey progressed, however, it became clear that in some regions (particularly Ireland and NW Scotland) there was a paucity of observers. To fill in these gaps expeditions were mounted involving 86 volunteers, and a total of 13 field staff were employed, during May and June 1990 and 1991. The help of two other organisations was crucial here. Staff from the National Parks and Wildlife Service (Ireland) – formerly the Wildlife Service of the Office of Public Works – helped cover many squares in Ireland, and members of the Royal Air Force Ornithological Society covered some of the most remote terrain in Scotland. In addition, all National Organisers undertook a great deal of fieldwork in poorly covered areas.

At the start of the 1990 season, all ROs, and ultimately all observers, were provided with a list of the species that had been recorded in each of their 10-km squares; and at the beginning of the final (1991) season, each was provided with, not only a list of all species recorded during 1988–90, but also lists from the *68–72 Atlas* for comparison, and a tabulation of the tetrads that had been visited in each square. With this information ROs were able to ensure that the minimum amount of fieldwork (eight tetrads for non-coastal squares) had been undertaken in every 10-km square, and that the species list for each was as complete as possible. Throughout the survey, ROs attempted to coordinate fieldworkers to ensure that the same tetrad was not visited twice. Thus, observers who visited eight tetrads in one year, were asked to visit different tetrads in the following year.

Additional data

Although not submitted directly to the Atlas, records submitted to a number of other schemes have been included by the kind permission of those named in parentheses. These are:

1. JNCC/Seabird Group SCR (P. Walsh and M. Tasker): this is expanded on below.
2. Data from the BTO's Integrated Population Monitoring Programme carried out under contract to JNCC:
   (a) NRS (H. Crick and C. Dudley): records for 87 species from 1988–90.
   (b) CBC (J. Marchant and S. Carter): records of 107 species from CBC and WBS plots during 1988–90.
   (c) Heronries Census (S. Carter): locations of breeding colonies during 1989 in England and Wales.
3. RSPB 1988–89 Hen Harrier survey of Scotland (B. Etheridge and C. Bibby).
4. RSPB/IWC 1988 Corncrake survey (T. Stowe, A. Hudson and E. Mayes).
5. RSPB 1989 and 1991 Cirl Bunting surveys (K. Smith and A. Evans).
6. RSPB Stone Curlew survey data (R. Green).
7. RSPB Black-throated Diver survey data (R. Summers and L. Campbell).
8. RSPB/University of Dundee Crested Tit study data (H. Young)
9. NCC 1988 Dotterel survey (D. Thompson).
10. NCC surveys of low intensity agricultural land and moorland (A. Brown).
11. WWT/BTO 1990 Mute Swan survey (S. Delaney and J. Greenwood).
12. Data from the Kite Committee (P. Davis).
13. Data from the Sea Eagle Project Team (J. Love).
14. Data from the Golden Oriole Study Group (J. Allsop, P. Mason and M. Raines).
15. 1989 Quail survey data for Britain (S. Dudley) and Scotland (R. Murray, *Scottish Bird Report* **22**: 45–50).

16. BTO/RSPB/JNCC/Regional Raptor Groups 1991 Peregrine Survey (D. Ratcliffe).
17. Irish Wildlife Service/IWC 1991 Peregrine Survey (J. Wilson and D. Norris).
18. Data from RSPB reserves (G. Hirons).
19. Data from NNRs (T. Reed).
20. Local atlases; data were received from the following local atlases: Avon (R. Bland and J. Tully), Bedfordshire (E. Newman), Berkshire (A. Swash and I. Collins), Dorset (R. Gleason), Essex (M. Dennis), Hampshire (G. Evans), Hertfordshire (K. Smith), London (K. Betton), Northumberland (N. Rossiter), Shropshire (C. Wright), Suffolk (M. Wright and R. Waters), Surrey (E. Garcia), Sussex (V. Bentley and M. Scott-Ham), Borders (R. Murray), Clyde (I. Gibson), Lothians (M. Leven and M. Holling) and Co. Wicklow. This list is by no means exhaustive, as other areas submitted data without necessarily making it clear whether or not the data were collected as part of the local or national surveys. In addition, individuals other than those listed were involved in extracting the data prior to its submission.

All these data were incorporated as supplementary records and are thus included on the Distribution maps. Data from 1, 4 and 5 were also used in the production of Abundance maps.

Counts of seabird colonies

Seabird colony counts came from the following sources:

(a) The JNCC/Seabird Group SCR. These data included the results of the near complete coverage of seabird colonies on British and Irish coasts, published in *Status of Seabirds*, as well as data from inland colonies from a variety of sources such as local bird reports. Many other organisations and individuals contributed data to the SCR – most notably the RSPB, the National Parks and Wildlife Service (Ireland) and the Shetland Oil Terminal Environmental Advisory Group (SOTEAG). Most data for the SCR are from the period 1985–87, although for some species they may date from 1981–91 (see below).

(b) Counts of inland seabird colonies from timed tetrad visits and supplementary records of other inland, and some Irish coastal, colonies (from the 'Comments' box of the worksheet) by Atlas fieldworkers during 1988–91.

(c) A complete survey of breeding Arctic Terns in Orkney and Shetland by the RSPB in 1989.

The data from all these sources were combined and edited to provide 10-km square totals of breeding seabirds for inclusion in the *88–91 Atlas*. Coverage of inland 10-km squares was incomplete, as often only a proportion of tetrads were visited in each 10-km square, and the time spent in each was limited: thus the totals for inland 10-km squares are necessarily minimal. There was, however, no simple solution to this problem.

The following seabirds were dealt with in this manner. Fulmar, Manx Shearwater, Storm Petrel, Leach's Petrel, Gannet, Cormorant, Shag, Arctic Skua, Great Skua, Black-headed Gull, Common Gull, Lesser Black-backed Gull, Herring Gull, Great Black-backed Gull, Kittiwake, Sandwich Tern, Roseate Tern, Common Tern, Arctic Tern, Little Tern, Guillemot, Razorbill, Black Guillemot and Puffin.

The following rules were adopted by Paul Walsh and Mark Tasker of JNCC/Seabird Group when selecting the most appropriate counts for inclusion in the *88–91 Atlas*, from the 25,000 records which were available:

1. The quality of individual counts was assessed on the basis of apparent accuracy, count units used, date of count and other criteria (see *Status of Seabirds*).

2. Where complete and accurate counts were available for a colony within the period 1988–91, these were used in preference to older counts. 1988–91 timed tetrad counts were treated as minimal counts.

3. If several counts were available from 1988–91, the highest was generally selected.

4. For certain species (in particular Cormorant, Black-headed Gull, and all the terns), colony locations may shift from year to year (*Status of Seabirds*). To minimise the chance of double counting these species within a 10-km square, the year with most complete coverage was generally used. Movements between colonies within a season, or between 10-km squares, can also occur, and colonies of other species (e.g. Kittiwake in some regions) may also shift between years. Full allowance for the latter points would have over complicated the analysis, and so were ignored.

5. If no suitable count was available from 1988–91, then the most recent good-quality count from the period 1984–87 was used, taking 4 into account when necessary, but colonies which were known to have become extinct or to have been abandoned more recently, were not used.

6. Counts from the period 1981–83 were used in only a few cases. Full censuses of Arctic Skuas were made in Orkney in 1982, and of Black Guillemots in Shetland and Orkney in 1982–84. More recent coverage of these islands has been very incomplete, so the older counts have mostly been used. For other species and regions, 1981–83 counts were used only if the colony was known to exist in any year during 1988–91.

7. Pre-1988 breeding records without counts were not used. 1988–91 records of this type were used only in the Distribution maps.

8. Where there was sufficient doubt that a count referred to actual breeding birds, it was ignored.

9. Counts from colonies which crossed 10-km square boundaries were allocated to only one of their constituent 10-km squares; that which occurred at the colony centre. In the Channel Islands, all counts of colonies on Guernsey, Alderney, Herm and Sark are shown in their correct geographical position, whilst those from Jersey and Les Minquiers have been allocated to a single 10-km square (WV65).

10. For most species of seabird, the count units used for mapping were 'pairs': that is, apparently occupied nests, sites or territories, depending on the species – see *Status of Seabirds*. In the case of the four auks, however, it is difficult to count breeding pairs. Most auk counts were necessarily of individual adult-plumaged birds at colonies – on land in the case of Guillemot and Razorbill, and on land, on nearby sea or flying above the colony in the case of the Puffin. For Black Guillemots, the best counts are of adults in likely breeding habitat, on land or nearby sea, in the early mornings of April or early May (Ewins and Tasker 1985). Details of conversion factors between different count units are given in the species accounts in *Status of Seabirds*.

A comparison of the Distribution and Abundance maps for seabirds will show that for a small proportion of 10-km squares, although there is evidence of breeding, no count has been provided. This may be for one of two reasons. First, as mentioned above, the colony may have extended across a 10-km square boundary and although breeding evidence was provided for both squares, the count was allocated to only one of the two squares. Second, the colony(ies) may not have been counted. Although it would have been straightforward to allocate a mean count to such squares, this has not been done, because the reason why a count was not performed was generally unknown. If, for example, this was because a colony was very small and considered unimportant, then the allocation of a mean value would have overestimated the colony size within the square.

Data processing and checking

ROs checked all data received from observers and, in particular, queried unlikely records prior to submitting them to their National Organiser and, ultimately, to BTO headquarters. A total of 5,820 worksheets and approximately 11,000 supplementary sheets was submitted and all were checked, manually, prior to processing. The worksheets contained a line for each species which included its two-letter computer code (e.g. CH= Chaffinch); supplementary record sheets did not, and had to be coded manually, so the processing of these was more prone to error.

The data were sent to a commercial data processing firm (Host Data Services) at the end of each season. They were processed in a previously agreed format, were written to tape and transferred on to the BTO's Prime Mini Computer. During the early years of the survey the data were only encoded – that is typed into the computer once. As it became apparent that the rapid typing of the processors introduced unwanted typographic errors, and as it took a disproportionate amount of time to find and correct them, data from the later years were encoded and verified. This procedure ensured that all data were input twice by the processors, the second round of inputting verified against the first, and any inconsistencies found and corrected. This procedure, though expensive, removed all typographic errors. In total, data processing required about 26,000,000 key depressions.

All data were processed at the grid definition at which they were collected. Thus data collected during timed tetrad visits and submitted on worksheets were processed in a format which ensured that the data for each tetrad were computerised. Although most maps in this volume are based on 10-km square data, the tetrad information is on computer. The NCC (and subsequently JNCC) considered that access to the data at this fine grid level was so important that they generously funded the data processing.

Although data verification revealed all typographic errors, there was still a large number of observer errors to be detected and corrected. For this to be undertaken automatically, a suite of checking programs was developed which identified all records with errors; the data files were then edited. Thus, for example, the program checked that species were not recorded for tetrads that had not been visited, that the total number of tetrads in which each species was recorded in a 10-km square agreed with the value entered by the observer, and that the 10-km square entered on the worksheet actually existed. Another advantage of the worksheet over the supplementary record sheet in terms of data processing was that data from worksheets were always entered in taxonomic order, thus errors in processing of the species code (e.g. typing GO = Goldfinch for GD = Goosander or vice versa) could easily be found and corrected. The same was not true of the supplementary record sheets, and many of the errors on the draft maps produced throughout the project were of this kind.

Some sorts of check were, however, impossible by computer. If, for example, an observer had recorded a Little Grebe but unintentionally filled in the row for an adjacent species (e.g. Great Crested Grebe) then this error could be spotted only by visual inspection of the data, and local knowledge. For this reason all original data, plus a computer printout of the same, were returned to ROs after the final field season for line by line checking. This was extremely laborious, but proved invaluable as many errors were discovered.

Even this, however, was not the final check. Immediately prior to the production of final maps, all ROs were sent a list of the species recorded in each of their squares for a final check. Fortunately, this stage found few new errors, most having been discovered at an earlier stage.

For the duration of the project, all data handling was undertaken on the Prime Mini Computer using the SAS (SAS Institute Inc.) software program. This extremely powerful software enabled customised programs to be developed that not only allowed statistical analyses of the data but, perhaps more importantly, allowed the data to be manipulated into a variety of formats (e.g. for checking by ROs and for map production).

Confidential records

In the *68–72 Atlas*, 1,800 confidential records (of rarities or observations of species well outside their normal range) were submitted. These records were published (in accordance with the advice of the RBBP) only when the original observer had given written consent. This inevitably required a great deal of correspondence and resulted in a delay in the date of publication. Learning from this experience, and benefiting from a change in attitude towards these types of record, the Atlas Working Group took care to ensure that the BTO was not constrained by the need to obtain observer consent prior to publication. The following wording was used in the fieldwork instructions:

> Rare species should be dealt with, if possible, in the same manner as other species. However, you may send the record directly to the National Organiser if you prefer.... All confidential records will be held under lock and key and will only be published in accordance with the advice of the Rare Breeding Birds Panel.

Observers were not obliged to provide the tetrad location of confidential records but it was stressed that the conservation and scientific value of the record would be limited if they did not do so. Subsequently an article in *BTO News* (1991, **173**:1) during the last year of fieldwork expanded on this:

> Prior to publication, the Atlas maps of rare species will be shown to members of the Rare Breeding Birds Panel. These maps will include those records that observers have asked to be treated in confidence. The RBBP will then decide how best the data can be presented to safeguard confidentiality of the record, whilst ensuring that the final maps convey as much information as possible. If any observer is concerned over a particular record, they are welcome to communicate this to the panel.

In the event, about 800 confidential records were received. Maps containing these records and others judged by the Atlas Organiser to be sensitive, were produced and presented to a special meeting of the RBBP in January 1992. The following were present at the meeting: Leo Batten (EN, Chairman RBBP), Robert Spencer (Secretary RBBP), Colin Bibby (ICBP), Jeremy Greenwood (BTO), Tim Sharrock (*British Birds*), Ken Smith (RSPB), David Stroud (JNCC) and Ron Summers (RSPB). Bob Chapman (representing the IWC) and David Gibbons (the Atlas Organiser) were also present. The RBBP laid down a number of general criteria which were used to decide whether or not a particular record was to be treated confidentially, and not located accurately on the published Atlas map. These were:

1. The strength of local opinion.
2. The susceptibility of the species to disturbance (e.g. egg collecting).
3. Whether or not the record was from a protected site.
4. Whether or not the record was already well known.
5. The probability that the species would return to the same site in subsequent years.
6. Whether or not the record was extralimital (i.e. well outside the species' normal range).
7. Records of species with specialised local habitats that might be threatened.
8. How records of the species had been treated in the *68–72 Atlas*.

Ultimately a total of five separate methods was used to deal with the records the RBBP considered confidential. These were (a) record omitted from map, (b) record downgraded from 'breeding' to 'seen', (c) record shifted by up to two 10-km squares, (d) records centralised within a 50-km or 100-km square and (e) records centralised within a larger area. The method used depended largely on the distribution of the records of the species in question.

It must be stressed that some records which were submitted in confidence to the Atlas have not been published as such here, and vice versa. Publication of these records has always followed the advice of the RBBP or, for the Golden Eagle and Dotterel, that of local specialists nominated by the RBBP. Paul Hillis (IWC) gave further advice on the treatment of confidential records from the Republic of Ireland.

Map Production

Although more complete guidance on how to interpret the maps is given later (see *Interpreting the Species Accounts*), this section explains the data that were used, and the method of map production. Much of the latter part of this section has been provided for the technical reader interested in mapping species' distribution and abundance, and is not intended for the general reader.

Generally, three maps are presented for each species: a Distribution map, an Abundance map and a Change map.

The Distribution maps are based on all records submitted whether from timed or non-timed fieldwork. For each species in each square the higher level of presence – 'breeding' taking precedence over 'seen' – recorded over the four year period is plotted.

With the exception of seabirds, Corncrake and Cirl Bunting (for which see later) the Abundance maps are based on timed tetrad visits only and are thus corrected for effort. The value used for map production is either the frequency of occurrence, or, for count species, the mean count per tetrad, for each species in each square.

The Change maps document the change in distribution that has occurred over the 20 years between the two Atlas periods (1968–72 to 1988–91). They show 10-km squares in which the species has either appeared or disappeared but do not show those in which it was present during both surveys. The best way of documenting distributional change for each species would have been to present maps of the 1968–72 distribution, the 1988–91 distribution, and the change in distribution (1968–72 to 1988–91) side by side. Unfortunately, space is limited and only two of these maps could be presented. As the 1968–72 distribution maps had already been published, it was decided to present only the 1988–91 distribution and the distributional Change map. Attempts to produce satisfactory single maps combining all of this information failed because they required too many types of dots and were difficult to interpret.

The maps presented in the *68–72 Atlas* were produced on an adapted electric typewriter, and those of the *Winter Atlas* commercially by laser projection on to microfiche. All maps presented here were produced by the Atlas Organiser using one of two different systems.

The Distribution and Change maps were produced on the 'distribution and coincidence' mapping program DMAP, run on a Personal Computer. This program has many unique features, two of which were of overwhelming importance. Firstly, the program had been specially modified to handle the Irish Grid. This is rotated at approximately 5° to the British Grid, and all Irish data were automatically rotated within the program prior to map production. Secondly, it can produce PostScript output files. When reproduced on a printer with a resolution of 1,200 dots per inch, these files produced very high quality maps. All dot maps presented in this volume were generated in this manner, the map data being read from disk.

Traditionally, maps showing regional variations in abundance have used increasing dot sizes to represent increasing numbers. In this volume they are mostly represented by a colour scale from light blue (= low density) to dark red (= high density). These maps were produced within the UNIMAP interactive mapping system, run on a VAX computer at the Institute of Arable Crops Research at Rothamsted. Unlike the dot distribution maps, they were by no means straightforward to produce. UNIMAP cannot work on more than one grid at a time, so the data for Ireland had to be rotated before running the program. To make them consistent with the dot maps, the rotation was exactly the same, using a translation of the relevant source code of the DMAP program. For non-coastal squares the abundance values for each species were allocated to the centre of the square, following the convention of placing distributional records in the centre of a grid square. For coastal squares, however, this represented a potential bias, as the data may have been collected in only a small proportion of the square, the rest being in the sea. To overcome this, the data for coastal 10-km squares were given amended grid references to ensure that the data point fell on the approximate centre of the land in the square.

The abundance data for each species were run through a customised UNIMAP program using the GINTP1 interpolation routine (see Morgan and Thorp 1990 for a review of UNIMAP interpolation routines). UNIMAP works by taking grid referenced data and interpreting from those data on to its own internal grid, the definition of which is user-specified. For all the colour abundance maps presented here, an interpolation grid of 5-km was used, and the program printed colour at the nodes (intercepts) of the grid. The value (i.e. colour) given to each node depended on where the data fell. If this was at the centre of a non-coastal square then it was allocated the value actually entered for that 10-km square (the true value). If, however, it fell at the corner or mid edge of the 10-km square it was allocated an interpolated value, calculated as the mean of the nearest true values. Interpolating values near the coast was more complex because the true data did not necessarily fall at a grid node. Values for the few 10-km squares with no abundance data were interpolated in the same manner.

To ensure consistency between the dot and Abundance maps the same digitised coastline file was used for each, although that for the Abundance maps had to be modified slightly. Some islands were too small to accept colour within their coastlines. By and large, these were also islands which had no tetrads whose centres fell on land, and thus no abundance data were available. These islands were omitted from the map. There was, however, a second group of islands and peninsulas sufficiently large to accept colour, but which contained none of the interpolated grid nodes. They were sufficiently small that all the 5-km grid nodes fell in the sea around them, and thus colour was not printed (colour only being printed within the coastline). To overcome this problem, the coastline file was artificially modified around such islands and peninsulas so that at least one grid node fell within each.

It was the problem of islands and peninsulas that made the use of a 5-km interpolation grid necessary. If a 10-km grid had been used, many larger islands would have lacked colour. A 1-km grid would have overcome most of these problems, but would have exponentially increased both the computer processing time and the expense. The 5-km grid was a sensible compromise.

The colour scale chosen was designed to ensure that the colours ran from light to dark, as well as from blue to red. This had two advantages. Had the colour scale run through the full spectrum of blue to red, at a uniform colour density, it would not have been apparent which way the scale ran; hence the light to dark, as well as the blue to red scale. The other reason is that red-green colour blindness affects about 1 in 12 of the male population of Britain. The use of different colour densities for red and green should enable most readers with this visual handicap to discriminate between the colours on the basis of colour density.

Varying the degree of lightness/darkness, however, presented its own problems, as the shading produced on the draft maps could not be reproduced in print because of interference patterns (moiré patterns) produced when the colour separations were overlaid. Instead, the output file from the UNIMAP program was converted into a Colour PostScript file and reproduced electronically at the printers as for the dot maps.

Colour Abundance maps have not been provided for one group of species – the seabirds. This is because large seabird colonies frequently occur on small islands and cannot be represented realistically using UNIMAP. For all seabirds covered by the SCR, simple dot abundance maps are given. These use increasing symbol sizes to represent increasing levels of abundance.

Coverage

A total of 551,370 10-km square records were submitted to the Atlas. These were based on 320,595 records collected during timed tetrad visits, and 230,775 supplementary records. The 10-km square timed visit records were themselves based on 1,262,231 tetrad records from 42,736 tetrads (34,601 in Britain, the Channel Islands and the Isle of Man, and 8,135 in Ireland) in 3,672 10-km squares (2,720 in Britain, the Channel Islands and the Isle of Man, and 952 in Ireland).

Many records were duplicates of the same species in the same 10-km square, and the total number of non-duplicate 10-km square records (i.e. a single record of each species in each square, and thus equivalent to the total number of dots on the Distribution maps) was 275,732.

Whilst it is difficult to compare coverage between the *68–72* and *88–91 Atlases* directly, one simple comparison is the total number of records.

This does not, however, take into consideration the true underlying changes. Nevertheless the results of this comparison are shown in Table 1.

Table 1. *A comparison of the number of non-duplicate 10-km square records submitted during 1968–72 and 1988–91.*

	No. of 10-km square records		
	1968–72	1988–91	Ratio of 1988–91 : 1968–72
Britain	217,615	219,129	1.01 (0.99)
Ireland	64,106	56,603	0.88 (0.86)
Both	281,721	275,732	0.98 (0.96)

For convenience, Britain includes the Channel Islands and Isle of Man. The ratios in parentheses are after exclusion of 'possible' (1968–72) and 'seen' (1988–91) seabird records.

Even taking into account the small increase in the number of records through the inclusion of summering birds (mostly seabirds), coverage in Britain during the two Atlas periods was clearly comparable. The same was not true of Ireland. This is a problem which complicates interpretation of the Change maps for Ireland, for a species may have apparently disappeared from a 10-km square simply because of lower search effort in 1988–91, compared with 1968–72, rather than a real range change. Extending the survey into a fifth year in Ireland would have helped solve this problem, but unfortunately the vagaries of funding, sponsorship and publication deadlines did not allow this. There were many reasons for the lower coverage in Ireland, two of them outstanding. These are, firstly, as discussed by Hutchinson (*Birds in Ireland*), that complicated methods are rarely well received by ornithologists in Ireland, and the apparently complex methods of the *88–91 Atlas* did not help in this regard. Secondly, and probably more importantly, the lack of a coherent RO network in Ireland was significant. Not all counties in Ireland had committed ROs, and in many areas observers dealt directly with the Irish Organiser. This meant that in parts of the Republic there was no individual responsible for enthusing the local birdwatchers to help with the survey. Nevertheless, the coverage of Ireland was still a remarkable achievement, with records being submitted from all except one mainland 10-km square (C53). What is astonishing, however, is the coverage achieved in Ireland during the *68–72 Atlas*.

As a lower emphasis was put on finding evidence of breeding in the *88–91* compared with the *68–72 Atlas* (to give priority to the quantitative methodology), it is perhaps not surprising that the percentage of records with evidence of breeding was lower, again especially in Ireland (Table 2).

Table 2. *A comparison between 1968–72 and 1988–91 of the percentage of records with evidence of breeding.*

	Number of 10-km square records with:					
	No breeding evidence ('possible' or 'seen')			Breeding evidence ('probable' & 'confirmed' or 'breeding')		
	1968–72	1988–91	Ratio of 1988–91 : 1968–72	1968–72	1988–91	Ratio of 1988–91 : 1968–72
Britain	23,228	48,052	2.07	194,387	171,077	0.88
Ireland	6,176	16,519	2.67	57,930	40,084	0.69
Both	29,404	64,571	2.20	252,317	211,161	0.84

For convenience, Britain includes the Channel Islands and Isle of Man. The figures are of non-duplicate records.

Records were received from 3,858 10-km squares during 1988–91 (2,830 in Britain and the Isle of Man, 14 in the Channel Islands, and 1,014 in Ireland). Remarkably, during 1968–72, records were received from exactly this same number of 10-km squares (and not 3,862 as reported in

the *68–72 Atlas*). However, the 10-km squares covered in the two Atlases were slightly different, and the total covered by the two different Atlases was 3,875. In the *68–72 Atlas*, Fair Isle and each of the Channel Islands were all treated as single 10-km squares, although they in fact include 18 10-km squares. In the *88–91 Atlas* all 10-km squares have been treated separately. In addition, records were received from 17 10-km squares (C53, L53, T37, V63, V81, NL97, NS00, NS22, NS23, NY05, SD35, SH59, SS77, SW81, SX03, SX14 and TR08) during 1968–72 but not 1988–91, and from five 10-km squares (HP62, NA93, NF61, NG07 and TR46) in 1988–91 but not 1968–72. Although V81 (Fastnet Rock) and V63 were visited during 1988–91, no birds were found. Overall, ten fewer 10-km squares were covered in 1988–91 than in 1968–72. With the exception of Irish 10-km square C53, all of the 10-km squares omitted in 1988–91 contain only tiny amounts of land.

Because a great deal of non-timed fieldwork was undertaken whilst searching for supplementary records, it is not possible to provide a total for the number of hours spent in the field between 1988 and 1991, nor an idea of how this varied regionally. However, in Table 3 and Fig. 5 the amount of fieldwork devoted to timed visits is shown, and it is clear that

Fig. 6. The number of species recorded in each 10-km square. The symbols of increasing size refer to 1–25, 26–50, 51–75 and 76+ species per 10-km square.

Table 3. *Annual variation in fieldwork effort and total number of records (duplicates included) submitted for Britain and Ireland combined.*

Year	Number of hours spent in field (timed visits only)	Number of 10-km square records from timed visits	Number of non-timed (supplementary) 10-km square records
1988	39,728	133,442	39,959
1989	27,762	96,088	45,777
1990	21,042	77,306	59,632
1991	3,798	13,709	32,795
year not specified	16	50	52,612
TOTAL	92,346	320,595	230,775

there was enormous regional variation in fieldwork effort, and that this frequently followed county boundaries, showing how enthusiasm varied at the local level. This striking regional variation in effort shows how important it was to develop a method that would take this into account, reduce bias and hence not affect the Abundance maps. It is also clear that most of the timed visits were undertaken during the first two seasons, whilst supplementary observations were more evenly spread over all four seasons.

The total number of species recorded in each 10-km square during 1988–91 is shown in Fig. 6, although this should not be taken to reflect the true species richness in each square, as variation in effort played a part in determining the pattern. This map is included to allow comparison with Fig. 8 on p25 of the *68–72 Atlas*, which shows a similar analysis for 1968–72. The same scale was used for both maps.

Text and artwork

Texts were written by a wide variety of authors, experts on the particular species wherever possible, following a clear set of guidelines. Draft texts were written in 1991 following the third (1990) field season. They were edited and returned to authors in 1992 with near final maps after inclusion of data from the fourth season, and were amended by the authors in the light of the new data.

The species vignettes were commissioned by the publishers and follow the advice of Robert Gillmor. All artwork was vetted by the New Atlas Working Group prior to its inclusion in the Atlas.

Interpreting the Species Accounts

The species accounts are divided into three sections. Species for which four or more breeding records were submitted to the Atlas are included within the main species accounts, and each is given a double-page spread. Those for which between one and four breeding records were submitted are included in the brief species accounts. Those for which only records of occurrence, not of breeding, were submitted are merely listed in Appendix F. There are exceptions, however. For example, Red-backed Shrike and Wryneck have each less than four breeding records, yet are treated in the main accounts. This is because at the time of the *68–72 Atlas* they had a much wider distribution, and their declines warrant full accounts.

Each species in the main accounts has a full text accompanied by up to three maps and a table, with additional material when relevant. These three maps are, as outlined in the *Introduction and Methods*, the Distribution, Abundance and Change maps. For clarity these are always placed in the same position on the double-paged spread, the Abundance map (in full colour for all species except seabirds) being at the bottom right of the left-hand page, and the Distribution map at the bottom left and the Change map at the top right of the right-hand page.

Species in the brief accounts have only a short text, a Distribution map and an accompanying table.

The texts

Where possible the texts cover the following topics: (a) an interpretation of the pattern of distribution and abundance of each species, including possible causes of the observed pattern; (b) a comparison of the 1968–72 and 1988–91 distributions, and an interpretation of the possible causes of any distributional change; (c) a review of recent applied research, particularly that relevant to conservation and ecology; (d) a review of recent pure research, e.g. breeding biology and behaviour; finally, (e) an estimate of the size of the species' populations in Britain and Ireland, separately, and their importance in a European or world context.

It is important to realise that interpretation of the patterns of distribution and abundance are often subjective, and based on the experience of the author of each text. Unless supportive evidence is provided, such as reference to other research, then these interpretations must be considered only hypothetical. As part of the RSPB/NERC funded Bird Distribution Modelling Project, the BTO is currently testing some of the hypotheses that have been proposed to account for the observed patterns of bird distribution, abundance and distributional change. The results of this work will be published elsewhere.

Wherever possible, population estimates for each species were calculated centrally using a suite of methods, though in some cases the estimate was provided by the text author. A full account of the methods used for generating population estimates is given later in *Population Estimates for Breeding Birds in Britain and Ireland*.

The maps

A comprehensive account of the method of map production is given in the *Introduction and Methods*. The present section is provided to enable readers to interpret the maps readily without recourse to the *Introduction and Methods*.

The Distribution map

This shows the distribution of each species recorded during 1988–91. Two different dot sizes are used on the map. A large dot refers to a 10-km square with evidence of breeding (the 'breeding' category; see *Introduction and Methods*). A small dot refers to presence during the breeding season, but with no stronger evidence of breeding (the 'seen' category). These data came from both timed and non-timed observations, and thus do not take into account variation in observer effort. Only 10-km squares with evidence of breeding are presented for seabirds (see *Introduction and Methods*). All dots are placed centrally within each 10-km square and for coastal squares may occasionally fall in the sea.

The Abundance map

For all species with an Abundance map (with the exception of seabirds) colour is used to represent regional variation in relative abundance. The same ten colours are used for each species, running from blue through to red, and simultaneously from light to dark, with light blues referring to a low level of abundance and dark red to a high level. However, the class limits and intervals to which these colours refer varies from species to species, and are shown alongside the map.

The class limits and intervals were calculated by dividing the abundance data into ten quantiles (i.e. deciles, see Robinson *et al.* 1984). These were determined by arranging the observed values in the order of their magnitude, from the lowest to the highest. The first decile value was calculated by counting one-tenth of the number of observations from the bottom of the ordered array; the limits of the decile being the first and last value in this one-tenth of the data, and the interval being the difference between them. Higher order deciles (second, third etc) were calculated in the same manner. The quantile method of presentation was chosen instead of, for example, fixed class intervals, because it maximised the within-species variation. The outcome of this manner of presentation is that each of the colours covers approximately one-tenth of the coloured area on the final map. For 22 ubiquitous species, however, it was impossible to divide the data into deciles. This was because they were recorded in every tetrad in numerous 10-km squares, and thus had a frequency index of 1 in more than a tenth of the 10-km squares in which they occurred (see Table 6 in *Bias and the Key Squares Survey*). For these species, all 10-km squares with a value of 0·96 (or occasionally 0·98) or more were allocated to the highest class (i.e. top of the scale), and the remaining data divided into nine quantiles.

A complete listing of the data used for the production of each Abundance map is given in Appendix A. For species marked 'Frequency' in Appendix A, the data are frequency of occurrence (the proportion of tetrads visited in each 10-km square in which the species was recorded). For those marked 'Count', the data are, in general, the mean number of individuals counted per tetrad in each 10-km square. There are, however, a number of exceptions to this rule. For Grey Heron and Sand Martin, Count refers to the mean number of apparently occupied nests recorded per tetrad in a 10-km square; for Corncrake it refers to the total number of calling males; and for Cirl Bunting the total number of pairs in each 10-km square, the data for the latter two species being provided by the RSPB. For seabirds (except auks) the data are the number of pairs in each 10-km square, and for auks they are the number of individuals (see *Introduction and Methods*). For all species with the exception of the seabirds, Corncrake and Cirl Bunting, all the data presented on the Abundance maps were collected during timed visits and are thus corrected for variation in observer effort. For the seabirds, Corncrake and Cirl Bunting, it is thought the counts reflect total numbers in each 10-km square fairly well.

Because species vary in their detectability, because the measures of abundance used are mostly only relative indices, and because the maps for each species are presented on species-specific scales, it is generally not possible to make between-species comparisons from the Abundance maps.

Some species in the main accounts do not have an Abundance map, simply because too few records were collected to present one. This was because these species were either too rare, or too elusive and thus not recorded during timed visits, or both. Abundance maps are generally presented only for those species for which 20 or more abundance records were received and for which at least 35% as many abundance as distribution records were submitted. These thresholds were entirely arbitrary and were decided upon by inspection of the maps.

Because of the problems of portraying, accurately, large seabird colonies on small islands using colour mapping, all seabird Abundance maps in the main species accounts use four increasing symbol sizes to represent increasing colony sizes. The class limits (and intervals) were calculated as for the colour Abundance maps, except that the data were divided into four (quartiles) rather than ten (deciles). There are thus similar numbers of colonies within each of the four classes for each species. A key showing the class limits is given alongside each seabird Abundance map.

The Change map

This map documents the recorded change in distribution between 1968–72 and 1988–91. It uses four different types of dot, the definitions of which are as follows:

Small open dot = possible breeding in 1968–72, not recorded in 1988–91.

Large open dot = probable or confirmed breeding in 1968–72, not recorded in 1988–91.

Small filled dot = not recorded in 1968–72 but present, although no evidence of breeding, in 1988–91.

Large filled dot = not recorded in 1968–72, but evidence of breeding in 1988–91.

10-km squares in which the species was recorded during *both* time periods are not shown on the map, even if the level of proof of breeding had changed. For seabirds, only 10-km squares with evidence of breeding (i.e. probable or confirmed for 1968–72, breeding for 1988–91) were used in the generation of Change maps. There are thus no small dots on seabird Change maps.

It is tempting to conclude that the Change map is a precise measure of the real distributional change. Unfortunately, this may not be the case everywhere due to differing levels of coverage during the two Atlases.

Although coverage in Britain during 1988–91 was comparable to that in 1968–72, coverage in Ireland was only about 90% (see Table 1 in the *Introduction and Methods*). Thus the Change maps probably represent Britain well, but need to be interpreted more carefully for Ireland. Coverage in the Republic of Ireland was, however, fairly even, so large scale changes in geographical range can still be determined.

Even within Britain, coverage at a regional level probably differed between the two Atlases, some counties (or BTO regions) being better surveyed in 1968–72 than 1988–91, or vice versa. In areas that were less well covered in 1988–91, the extent of the decline may well be exaggerated for two particular groups of species. These are, firstly, elusive species such as the Water Rail, and, secondly, nocturnal species. Nearly 60% of 10-km square records were collected during timed visits, the rest being from supplementary observations, and it is likely that the vast majority of timed visits was undertaken during daylight hours. If observers were unable to return at dusk (even though they were asked to do so whenever possible), then nocturnal species may have been underrecorded. In the methodology for the *68–72 Atlas* there were no timed visits, and it is possible that a higher proportion of fieldwork was undertaken at dusk, although whether or not this was the case is unknown. The two species most characteristic of this group are probably Woodcock and Barn Owl.

Because of the problems inherent in visiting seabird colonies (e.g. inaccessibility) the Change maps for these species should be treated with particular caution. The measured change will, partly, be dependent on whether or not a special effort was made to visit remote colonies during the two Atlas periods. In addition an apparent marked decline in distribution of a colonial seabird may not infer a similar decline in numbers. All major colonies were covered for the SCR, and losses on the Change maps may be only of small numbers of birds.

Changes in the rules of acceptance or rejection of a record for a 10-km square also affect the Change map. In particular, the inclusion of records of summering non-breeding birds away from suitable nesting habitat, influences the numbers of small filled dots on the Change map, although for only a small number of species. In addition, a record of a species in suitable breeding habitat could have been excluded from the *68–72 Atlas* if it was thought not to have bred; such a record would have been included in the *88–91 Atlas*.

It is because of these complications in interpreting distributional change that no statistic of range change (*sensu* Buckland *et al.* 1990) has been given.

Despite these difficulties, however, the Change maps do reflect the real underlying distributional changes, and the balance of filled to open dots provides a clue to the fortunes of each species in the 20 year period between the two Atlases. Indeed, many of the apparent changes revealed by comparison of the two Atlases are so striking that they cannot be explained away as consequences of coverage. For many species the Change maps are the best evidence we have of recent changes in status.

Presentation of confidential records on the maps

As outlined in the *Introduction and Methods*, five separate methods have been used to deal with confidential records. These are: (a) record omitted from map, (b) record downgraded from 'breeding' to 'seen' (large dot to small dot), (c) record shifted by up to two 10-km squares, (d) record centralised within a 50-km or 100-km square, and (e) record centralised within a larger area.

On the Distribution and Change maps, treatment of confidential records using methods (d) and (e) is obvious, as the area within which the records have been centralised is shaded. Where this occurs, no further comment is made in the adjoining text. Where methods (a), (b) and (c) have been used, however, mention is made in the text.

The true grid references of 1968–72 confidential records, and not those published in the *68–72 Atlas*, were used in the production of the Change maps. In general, only 1988–91 records have been moved on these maps so, occasionally, previously unpublished 10-km grid references from the *68–72 Atlas* are shown. Given the passage of time since 1968–72, it is thought that this is unlikely to be detrimental to these species. Using methods (d) and (e), however, it was necessary to centralise 1968–72 records (open dots) in those shaded areas in which 1988–91 records (filled dots) were centralised. This was because the centralised 1988–91 records sometimes fell in the same 10-km squares as non-centralised 1968–72 records, and thus masked them.

For a handful of species, a comparison of the Change map with the distribution map presented in the *68–72 Atlas* would have divulged the 10-km square grid reference of some 1988–91 confidential records. To overcome this, inaccurate records have been placed on the Change map. Mention has been made in the text whenever this was done.

For only six species (Red-throated Diver, Golden Eagle, Buzzard, Peregrine, Little Tern and Chough) was it necessary to move or omit records from the Abundance maps to maintain confidentiality. Again, the text states where this was done.

Tabular information

A table, listing Britain and Ireland separately, shows the numbers of squares occupied by each species during the two Atlas periods. For convenience, the Isle of Man is included in the British figures. The Channel Islands are included in the 'Both' column only. The number of 10-km squares with evidence of breeding ('probable' and 'confirmed' for the *68–72 Atlas*, and 'breeding' for the *88–91 Atlas*) and of presence, with no evidence of breeding ('possible' for the *68–72 Atlas*, and 'seen' for the *88–91 Atlas*) is given for Britain and Ireland, and for each time period separately, as is the total number of records for Britain, Ireland and both combined. Percentage change values between the two Atlas periods are given only for the summary columns: all records in Britain, Ireland and both combined. All percentage change values should be interpreted carefully, bearing in mind the problems that have been outlined for the Change maps (above). For seabirds, only records with evidence of breeding ('probable' and 'confirmed' from the *68–72 Atlas*, and 'breeding' from the *88–91 Atlas*) are tabulated.

For a small number of species the change in number of records with breeding evidence, rather than the total number of records, may provide a more realistic measure of change (e.g. Goldeneye and Dunlin). To avoid a plethora of statistics, however, these values are not given, although the data included in the tables allow them to be calculated approximately. Occasionally no percentage change value is shown, because the species status has changed from presence to absence, or vice versa. For example, the Ruddy Duck was not recorded in Ireland during 1968–72, but was present in eight 10-km squares during 1988–91. The percentage increase of the Ruddy Duck in Ireland is thus infinite. Furthermore, large percentage change values are particularly prevalent among species which were recorded in only a few squares. For example, the Red-necked Phalarope has undergone a 75% reduction in range in Ireland between the two Atlas periods, even though it was lost from only three 10-km squares; it was recorded in four Irish 10-km squares in 1968–72, and only one in 1988–91.

Channel Islands, Fair Isle, uncovered squares and errors

In the *68–72 Atlas*, Fair Isle, Guernsey, Jersey, Alderney, Sark and Herm were all treated as single 10-km squares, whereas actually they cover eighteen 10-km squares. In the *88–91 Atlas* these squares have all been treated separately, although this change in method has been taken into consideration when calculating distributional changes. Thus, on the Distribution maps, all eighteen 10-km squares are plotted separately and show the true distribution. On the Change maps, however, a false picture would be given if these squares were considered separately. Instead, the 1988–91 data for each island has been 'merged' into a single square (as in the *68–72 Atlas*) prior to the generation of the Change map data. Furthermore, the Change maps do not include records from 10-km squares which were covered (i.e. from which at least one record was received) in only one of the two Atlas periods; records must have been received from both surveys for it to be included.

The percentage change figures take these problems into account in the same manner as the Change maps. This can lead to apparent anomalies. An extreme example of this is shown by the Short-toed Treecreeper, which is restricted to the Channel Islands and was recorded in five 10-km squares during 1968–72, and ten during 1988–91, yet the percentage change value instead of being 100% is −40%. This is because it was recorded on all five of the Channel Islands in 1968–72, and on only three in 1988–91, hence a decline in distribution of 40%.

The Channel Islands are conventionally placed in a box in the SW corner of each map.

The *68–72 Atlas* data were processed, analysed and mapped some years before the BTO acquired computing facilities, and thus simple tasks – such as counting up the total number of records for each species, or editing the data files after changes were made to the maps – were tedious to do. Because of this, the number of records for each species included on the *68–72 Atlas* computer files sometimes differs slightly from those published in the tables accompanying each species account, which in turn may differ from the number of dots on the 1968–72 published map. Unfortunately, it is often not clear whether it is the 1968–72 data on the computer files or on the maps which are correct. The Change map and tabular data in this book have been generated from these computer files, thus there may be small discrepancies between the 1968–72 data shown here and that in the *68–72 Atlas*; this will be most noticeable in the tables accompanying each species account.

Methods of referencing

Frequently cited works of reference are referred to by an abbreviated title (see p xiii). All other references are included in a single list on pp 499–513.

The Species Accounts

Red-throated Diver

Gavia stellata

The Red-throated Diver is one of the more scarce but characteristic breeding species of boggy, moorland areas in the far N and W of Scotland. Whilst occasionally nesting on the shores of lochs, it favours the smaller lochans and pools, which are particularly widespread in areas such as Shetland, the Outer Hebrides and the Caithness flows. These pools are rarely sufficiently large or productive to support breeding adults or their chicks, and Red-throated Divers normally feed in nearby coastal waters or, if breeding far inland, on larger lochs. Divers calling as they commute between nesting and feeding areas are amongst the most evocative features of northern moors in summer.

The major strongholds for this species are in Shetland, Orkney, the Outer Hebrides and N Scotland, particularly Caithness, Sutherland and Wester Ross. They are also widespread, but at low density, along the whole W of Scotland south to the Mull of Kintyre, and a few pairs breed in the N of Ireland. Although breeding does occur well inland, distribution away from the coast is dependent on the availability of small breeding lochans with larger lochs nearby for feeding, but the pattern is much more patchy. Breeding density in parts of Shetland exceeds one pair per km².

At first sight the overall pattern of distribution in the *88–91 Atlas* is similar to that shown by the *68–72 Atlas*, the most obvious change in distribution being an eastwards extension in E Highland and W Grampian. Parallel evidence of an eastward expansion is provided by various recent records of breeding in Grampian, Badenoch and Strathspey (e.g. Dennis 1984).

Detailed work in Shetland (Gomersall *et al.* 1984) has shown that no evidence of breeding could be found for a third of all lochs and pools on which divers were actually seen. They regularly commute distances of several kilometres between their nests and feeding areas, and the only reliable criterion for determining whether divers are actually breeding in a particular area are the sighting of an incubating adult, eggs or young. Despite this, and the fact that a lower proportion of birds was recorded as breeding during 1988–91, the number of 10-km squares with breeding evidence was very similar during the two Atlas periods. There were, however, twice as many non-breeding records in 1988–91 as in 1968–72, and although some of these may have referred to commuting rather than breeding birds, this, taken in conjunction with the Change map, suggests a small expansion in overall range since 1968–72.

The data on diver abundance need to be treated with some caution. Unless counts are confined to birds known to be breeding in a specific tetrad, there is a danger, especially in coastal squares, that a commuting bird might be recorded in several tetrads. Notwithstanding this caveat, the Abundance map does broadly confirm the known pattern of relative abundance, with the greatest numbers in Shetland and Orkney, and generally lower numbers elsewhere, especially away from the coast.

In Shetland, Okill and Wanless (1990) have shown that breeding success varies considerably from year to year. Gomersall (1986) reviewed available data back to 1918 and was unable to detect any trend in annual productivity, which averaged 0.45 fledged young per pair. Although in the majority of cases the causes of failure were unknown, water level fluctuations, in contrast to the Black-throated Diver, were not found to be a major factor and only 4.5% of eggs were lost to flooding. Disturbance and predation are known to cause breeding failure and there is more general concern that loss of breeding sites to forestry and declining fish-stocks may threaten the populations as a whole.

Although there has been no systematic attempt to survey the breeding population throughout its range in Britain and Ireland, Shetland was fully surveyed in 1983 (700 pairs: Gomersall *et al.* 1984) and Orkney holds 90–95 pairs (Booth *et al.* 1984). Estimates have been made for several other areas (e.g. Argyll 68–88 pairs, Broad *et al.* 1986). *Red Data Birds* suggests

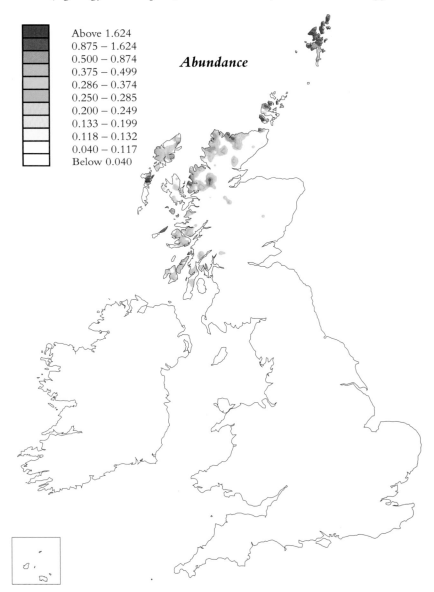

| Above 1.624 |
| 0.875 – 1.624 |
| 0.500 – 0.874 |
| 0.375 – 0.499 |
| 0.286 – 0.374 |
| 0.250 – 0.285 |
| 0.200 – 0.249 |
| 0.133 – 0.199 |
| 0.118 – 0.132 |
| 0.040 – 0.117 |
| Below 0.040 |

Abundance

a total British population of 1,200–1,500 pairs. Extrapolation from tetrad counts during April to July in 1988–91, yields a total of 2,850 birds in Britain, which would suggest a population of perhaps 1,425 pairs (although some may have been non-breeding birds), towards the higher end of the *Red Data Birds* estimate. The Irish population has remained stable at less than ten pairs for the last 50 years (*Birds in Ireland*). These estimates are substantially more than those given in *68–72 Atlas*, the difference largely being explained by the availability, since then, of comprehensive data for Shetland.

LEN CAMPBELL

Only two of the Irish records were obtained during timed visits. To maintain confidentiality these have been omitted from the Abundance map.

68–72 Atlas p 30

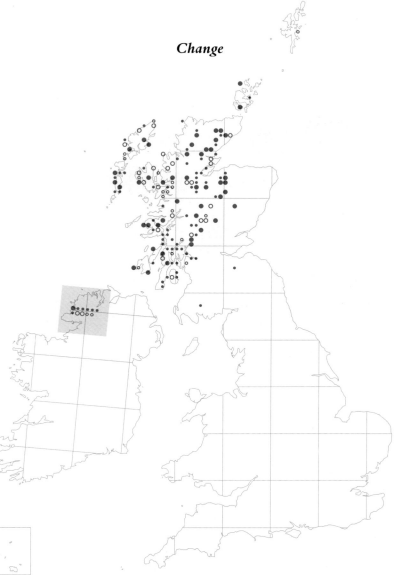

Change

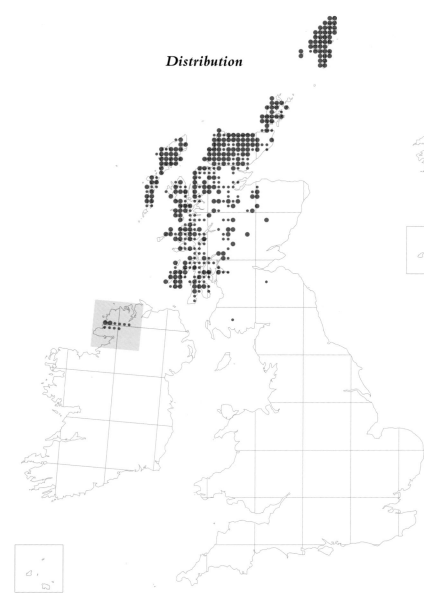

Distribution

Years	Present, no breeding evidence		Breeding evidence		All records		
	Br	Ir	Br	Ir	Br	Ir	Both (+CI)
1968–72	67	3	243	4	310	7	317
1988–91	133	8	246	2	379	10	389
				% change	22.3	42.9	22.7

Black-throated Diver

Gavia arctica

Larger, and visually more spectacular than the commoner and typically moorland-breeding Red-throated Diver, the Black-throated Diver breeds almost exclusively on larger lochs in the Outer Hebrides and the mainland of N and W Scotland, where it ranks alongside the Golden Eagle as a flagship species of the Highland avifauna. It is absent from Orkney and Shetland.

Although small lochs or lochans may occasionally be used, nests are usually found on the shores of the bigger lochs, where islands are particularly favoured (in a study area in NW Scotland 76% of nests were on islands; Campbell and Mudge 1989). In contrast to Red-throated Divers, many of which regularly commute to the sea to feed, breeding Black-throated Divers feed almost entirely on freshwater, even when the nesting loch is close to the coast. Some pairs feed mainly on their breeding loch, whilst others may make regular use of other lochs nearby.

The major strongholds are in N and W Sutherland, Wester Ross and Lewis, with small numbers in Caithness, Inverness-shire and Tayside, and a scattering of pairs elsewhere in Strathclyde and SW Scotland. Although most lochs hold only a single breeding pair, one regularly has up to 10 pairs, one holds three, while four others hold at least two pairs each (Campbell 1988; RSPB unpubl.). The Irish breeding record was of a pair with young on the sea; unfortunately, although it is possible that they bred nearby, this could not be confirmed.

Whilst the general pattern of distribution in 1988–91 was similar to that given in the *68–72 Atlas*, the number of occupied 10-km squares has slightly decreased. However, the difficulties inherent in surveying divers, and differences in the survey approach of the two Atlases (causing, in particular, an overall lower level of proof of breeding in 1988–91) are sufficiently large to prevent a reliable assessment of the scale of any decline.

Black-throated Divers nest and feed on large lochs where waves, bad light and distance may make them very difficult to detect, and a carefully targeted survey procedure is needed which allows relatively prolonged observation from good vantage points (Campbell and Talbot 1987). Thus even in intensive studies (Campbell and Mudge 1989), confirmation of breeding may require several visits to a loch, since the only acceptable evidence of breeding is the sighting of an incubating adult, eggs or young. Records from these recent intensive surveys by the RSPB and the former NCC have been included on the Distribution map.

Despite these limitations the Change map provides evidence of a slight decline in the population which is broadly consistent with the conclusions of a recent comprehensive review of the current status of Black-throated Divers in Scotland (Mudge *et al.* 1991). They identified at least 22 sites which had ceased to be used by breeding Black-throated Divers, mainly in the inter-atlas years, and concluded that, although the available data were inadequate to establish its scale, a decline in the population had taken place in the last 20 years. No significant decline could be detected in the stronghold areas in NW Scotland, and the changes appeared to be occurring mainly at the edge of the range, where notable losses include the

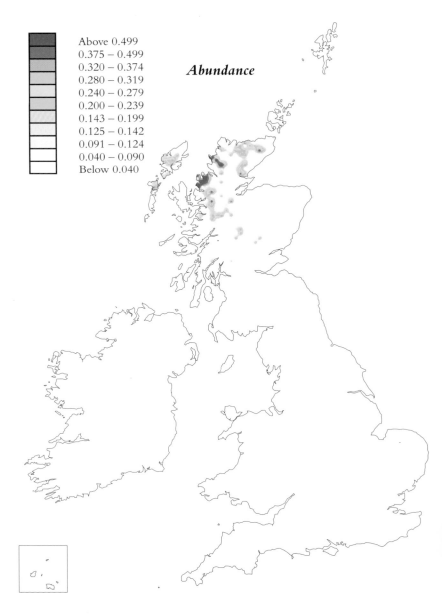

	Above 0.499
	0.375 – 0.499
	0.320 – 0.374
	0.280 – 0.319
	0.240 – 0.279
	0.200 – 0.239
	0.143 – 0.199
	0.125 – 0.142
	0.091 – 0.124
	0.040 – 0.090
	Below 0.040

Abundance

comparatively well documented pairs on Arran and Mull.

Although most lochs have only a single pair of Black-throated Divers, a big loch may include several tetrads, possibly in two or more 10-km squares. There is a risk that relative abundance will be significantly influenced by chance factors, such as a pairs of divers using several tetrads on the same loch. Despite this the Abundance map highlights the importance of some of the known concentrations in the northwest, with lower relative abundance in a variety of mainly eastern 10-km squares, which agrees with the figures available from more detailed surveys.

Campbell and Mudge (1989) have shown that breeding success in Scotland over six years averaged 0.25 fledged chicks per summering pair; sufficiently low to indicate that any decline in the population might be the result of poor recruitment. Flooding and predation were the two major causes of clutch failure, and disturbance was relatively unimportant. At least 40% of hatched chicks died before fledging, raising fears that inadequate food, perhaps as a result of water catchment acidification, might be having an impact on the population.

A comprehensive survey of breeding Black-throated Divers in Scotland was carried out in 1985 from which it was estimated that there were 151

Change

Distribution

summering pairs (Campbell and Talbot 1987). This estimate has been substantiated by the recent status review (Mudge *et al.* 1991) and is similar to the estimate given in the *68–72 Atlas*, but higher than that quoted in *Birds in Scotland*. Extrapolation from tetrad counts during April to July in 1988–91 yields a figure of 310 birds, equivalent to perhaps 155 pairs, although some may have been non-breeding birds.

LEN CAMPBELL

Two records have been moved by up to two 10-km squares on the Distribution and Change maps.

68–72 Atlas p 28

Years	Present, no breeding evidence		Breeding evidence		All records		
	Br	Ir	Br	Ir	Br	Ir	Both (+CI)
1968–72	67	0	145	0	212	0	212
1988–91	87	1	112	1	199	2	201
				% change	− 6.1		− 5.2

Little Grebe

Tachybaptus ruficollis

The Little Grebe prefers seclusion and is a secretive bird during the breeding season. It frequents still or slow-moving waters, and for a breeding site needs submerged and emergent vegetation. In these conditions its presence is more likely to be detected by its whinnying call than by sight, and when silent it may easily be overlooked by an Atlas observer with many other species to distract attention.

It is widespread throughout lowland Britain, being absent only from areas lacking suitable waters (such as chalk downlands) and from the fenlands of East Anglia, where the waterways lack the necessary cover. It is sparsely distributed in the West Country and in W Wales, but beyond the Welsh mountains, reappears on Anglesey. In N England and in Scotland it similarly keeps to lower ground. The Central Lowlands are its stronghold in Scotland, but it also reaches the Highland fringes north of Perth, and penetrates the Highlands from Loch Ness and the Spey valley. In Ireland it is found throughout low-lying wetland areas.

The Abundance map, based on actual counts, shows that the highest numbers in England are found along major river valleys, e.g. Test, Hampshire Avon and Trent. In Ireland the species occurs at greater densities per occupied 10-km square than in Britain.

Changes between the 1968–72 map and that for 1988–91 are difficult to evaluate. The Distribution map indicates the loss of the species from parts of its earlier range, but this finding is at odds with other surveys. For example, the *Trends Guide* concludes that there is little evidence of change in numbers over the period 1974–88, and the breeding range may even have expanded with the species' greater use of rivers and canals. The Change map shows that the species has indeed expanded its range in some areas, such as Wales and N Scotland, notably in Caithness. However, other areas have lost their Little Grebes, especially in the E Midlands, W Scotland and much of the Irish Republic.

Dobinson and Richards (1964) showed that Little Grebes are very susceptible to prolonged frost, and that they virtually disappeared from N Britain after the severe winter of 1962/63. Moss and Moss (1993) also detected cold weather losses. Their NRS analysis compared breeding performance of pairs which used different types of water body. Pairs which nested on canals, dykes and lakes laid earlier than those on rivers and streams; and nests on ponds were the most likely to succeed, with rivers and streams again the least favourable waters. The differences were attributed to the significantly greater incidence of flooding suffered by nests on the margins of running waters.

Although many nest recorders had difficulty in making a complete record of breeding attempts, Moss and Moss found that the peak laying period is early May, and that very few clutches were started outside April – July. The most frequently recorded clutch size was four, but brood size was difficult to record reliably due to the likelihood that some young were hidden in aquatic vegetation during a recorder's visit. Eggs were hatched in only 53% of nests. In view of the high failure rate and the difficulty of observing repeat breeding with certainty, there is little firm evidence of double brooding, although it is generally assumed to be the norm (e.g. *68–72 Atlas*). Glue (1990a) gave a confirmed example of a triple brood.

Little Grebes breed over much of the Old World (*BWP*), occupying most of Africa outside the desert regions, and across Eurasia as far east as Japan and parts of Indonesia. Populations in central Europe are migratory, being unable to survive the severe winter conditions there. Those breeding in Caithness are amongst the most northerly in the world.

Following the assumption of 5–10 pairs per occupied 10-km square

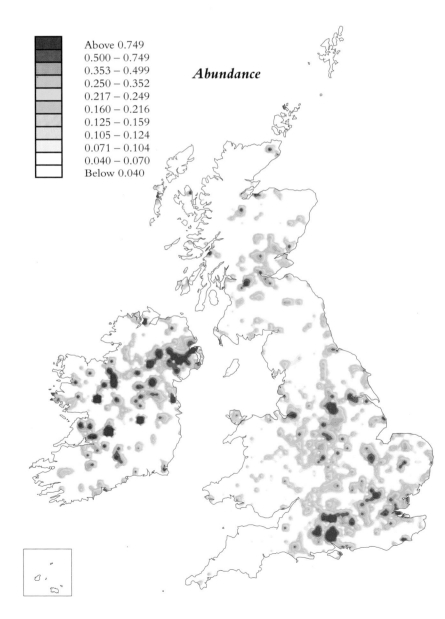

Above 0.749
0.500 – 0.749
0.353 – 0.499
0.250 – 0.352
0.217 – 0.249
0.160 – 0.216
0.125 – 0.159
0.105 – 0.124
0.071 – 0.104
0.040 – 0.070
Below 0.040

Abundance

used in the *68–72 Atlas*, but taking into account the different relative densities in Britain and Ireland shown by the timed counts, the Little Grebe population in Britain is estimated at 5,000–10,000 pairs, and that in Ireland at 3,000–6,000.

DORIAN MOSS

68–72 Atlas p 38

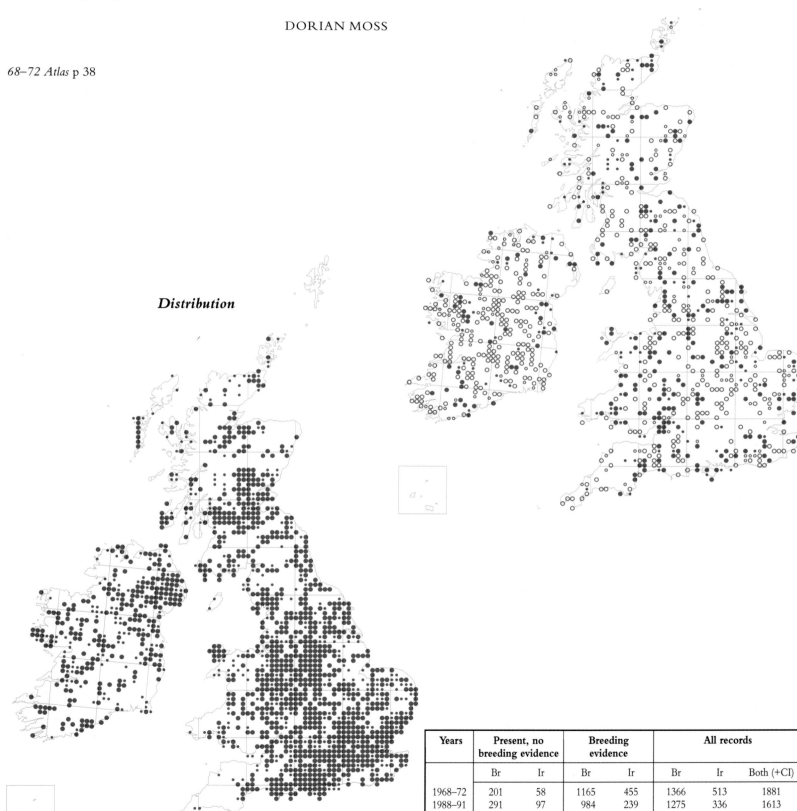

Change

Distribution

Years	Present, no breeding evidence		Breeding evidence		All records		
	Br	Ir	Br	Ir	Br	Ir	Both (+CI)
1968–72	201	58	1165	455	1366	513	1881
1988–91	291	97	984	239	1275	336	1613
% change					−6.6	−34.4	−14.2

Great Crested Grebe

Podiceps cristatus

Great Crested Grebes are perhaps our most elegant waterbirds, with beautiful breeding plumage, fascinating courtship display and the engaging habit of carrying their young on their backs. In less enlightened times, their head plumes were much sought after by the trade in 'grebe fur', and the species was all but exterminated in Britain by 1860. Fortunately, protection was given, attitudes changed and numbers recovered, helped by the creation of new habitats. The recovery from persecution has been well documented by surveys over almost half a century (Harrisson and Hollom 1932, Hollom 1936, 1951, 1959, Prestt and Mills 1966, Hughes *et al.* 1979).

Great Crested Grebes breed on large, shallow waters with some fringing vegetation in which the nest is sited, usually concealed in emergent vegetation, or protected from predators by its inaccessibility (Simmons 1974). Still waters – lakes, reservoirs and gravel pits – are favoured, but they can use reedy fringes of slow-flowing rivers. The Distribution map shows a clear preference for lowland Britain and Ireland, with the species being found throughout SE England and the Midlands, most of East Anglia except parts of the fens, the Forth-Clyde corridor of Scotland, and the low wetland belt of Ireland. Within this range, the species is absent only from areas without large water bodies, especially the chalk ridges such as the South Downs. In S and SW England, S Wales, N England and S Scotland, the distribution is patchy; and in S Ireland and N Scotland, it is virtually absent. In the more upland areas, water levels are subject to rapid variation, which threatens the success of grebes during the incubation period.

The Abundance map indicates that numbers are greatest at the centres of the species' distribution in England, for example in the Midlands, and the northeastern half of the range in Ireland.

There have been changes between the distribution map shown in the *68–72 Atlas* and that shown here. Great Crested Grebes have colonised many areas peripheral to their former range, for example in E England, and along a belt close to the England/Scotland border. There is no clear pattern in the distribution of squares where the species has been lost. The *Trends Guide* reports that the long-term increase has continued over 1974–88, and this is reflected in the increased number of squares occupied in Britain. Hutchinson (*Birds in Ireland*) reported increases in Ireland continuing into the 1980s, but the close similarity in the numbers of squares occupied in 1968–72 and 1988–91 suggests this may not have led to any significant range expansion there.

Hughes *et al.* (1979) reported a complete census of all known Great Crested Grebe breeding waters made in 1975, in a comparison with a similar census made 10 years earlier (Prestt and Mills 1966). Their best estimate was that the population had increased over that period by 46% in England and Wales, but only 9% in Scotland. The former increase was in line with the trend observed between 1931 and 1965. Contributing factors had included the increase in availability and suitability of gravel pits, especially in S England, where the proportion of pairs in this habitat had increased from 31% to 43%, replacing reservoirs as the water type most preferred for breeding. A new tendency to breed on lowland rivers was observed in the 1960s, as preferred habitats became saturated (Harvey 1979).

In Scotland, by comparison, natural lakes were used by the majority of birds. The lack of a rise in numbers comparable to that observed in England and Wales is discussed by Thom (*Birds in Scotland*), who suggests that intensive management for angling or excessive disturbance by water sports might have been contributory factors at the relatively small number of lakes holding the majority of the Scottish population in 1965. A study in Denmark (Asbirk and Dybbro 1978) found that densities of Great Crested Grebes were correlated with eutrophication of the water, but

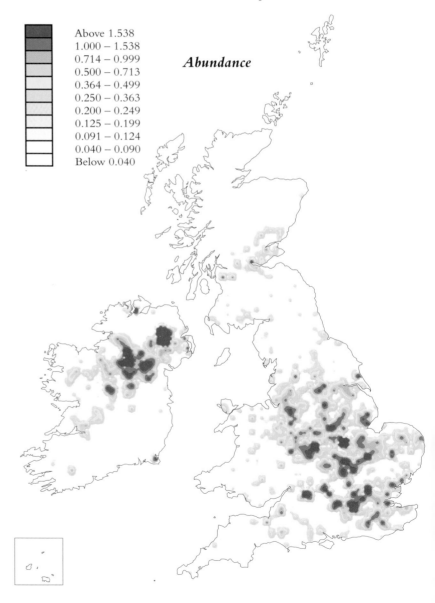

Above 1.538	
1.000 – 1.538	
0.714 – 0.999	*Abundance*
0.500 – 0.713	
0.364 – 0.499	
0.250 – 0.363	
0.200 – 0.249	
0.125 – 0.199	
0.091 – 0.124	
0.040 – 0.090	
Below 0.040	

decreased with increasing disturbance by boats.

Great Crested Grebes are opportunistic breeders, with an extended breeding season, with May–June the peak laying period (*BWP*, Simmons 1974). The majority of clutches are of 3–4 eggs. The incubation period is about 28 days, and the young are usually fed by their parents for 13 weeks, so only 10% of pairs have time to attempt second broods (*BWP*).

The world range of the Great Crested Grebe extends across temperate Eurasia as far as China, and over much of this range it is strictly a summer visitor in its breeding areas (*BWP*). There are other races in a few discrete populations in Africa, and also in SE Australia.

The most recent national census, in 1975, found almost 7,000 birds in England, Wales and Scotland (Hughes *et al.* 1979), while contemporaneous censuses in Northern Ireland and the Irish Republic counted 661 and 758 birds respectively (*Birds in Ireland*). In view of the organisational complexity of the 1975 survey, this has not been repeated more recently. Extrapolation from tetrad counts during April to July in 1988–91 yields higher estimates of 8,000 birds in Britain, and 4,150 in Ireland.

DORIAN MOSS

68–72 Atlas p 32

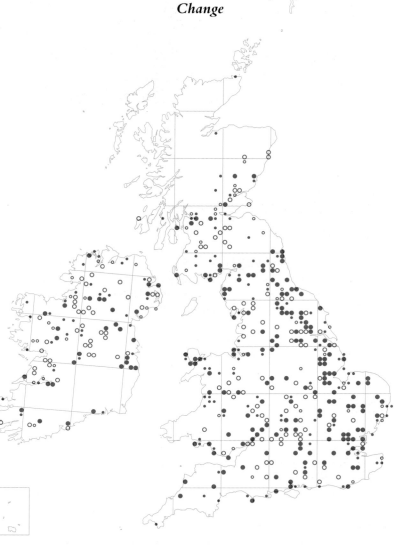

Change

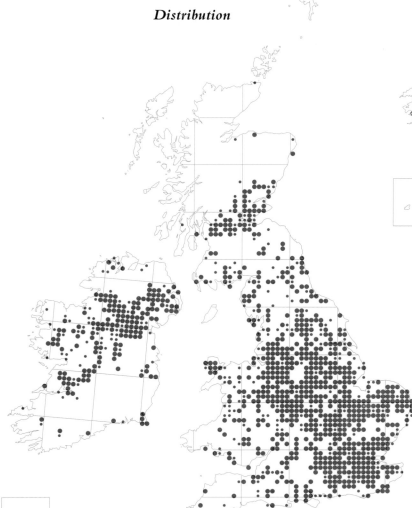

Distribution

Years	Present, no breeding evidence		Breeding evidence		All records		
	Br	Ir	Br	Ir	Br	Ir	Both (+CI)
1968–72	97	30	665	193	762	223	985
1988–91	166	64	726	161	892	225	1117
				% change	17.1	0.9	13.4

Slavonian Grebe

Podiceps auritus

In the breeding season this is a most beautiful bird, with golden ear tufts and chestnut neck which it shows off to perfection during its elaborate displays. At the same time, the trilling calls add to the magic of spring time at a few select Highland lochs where the birds nest.

The Distribution map shows that the headquarters of this bird in Britain are in the Scottish Highlands, centred in Inverness district, where it was first recorded nesting in 1908. Breeding was proved in ten 10-km squares in that county, with a further two squares in Badenoch and Strathspey and four in Moray and Nairn. Further sightings came from S and E Scotland, but the only breeding record there during the Atlas period came to light in 1991, when one pair, which had been present for several summers, was located breeding in E Tayside.

Comparison with the *68–72 Atlas* map shows that the distribution of the Slavonian Grebe has changed little, although there is the suggestion of a slight expansion around the periphery of its range, particularly north and east. Unfortunately, the Caithness population recorded during 1968–72 became extinct. However, these small scale shifts in distribution underestimate larger changes that have occurred to the population in the inter-Atlas years.

Since 1971, the population has been monitored annually by the RSPB and by local ornithologists, especially by M.I. Harvey working to the north of the Great Glen. During the 1968–72 survey, the population in this area numbered 11–13 pairs. It increased to a peak of 25–26 pairs in 1977 and 1980, but has now declined to 17–23 pairs. South of the Great Glen, the comparable figures were 27–29 pairs, increasing to a peak of 48–50 pairs in 1978–79 and now reduced to 30–34 pairs.

Outside the main area, the Strathspey population, which started with one pair in 1971, increased to a peak of nine pairs at five lochs in 1984 but has now dropped to 5–7 pairs at two lochs. The Moray birds which peaked at seven pairs in 1971, are now steady at 4–5 pairs at two lochs. Slavonian Grebes nested in Perthshire in 1973, reaching a maximum of three pairs in 1982, but failed to nest after 1983. One pair nested in Aberdeenshire in 1974 and nesting occurred recently in Angus.

In Scotland, breeding tends to be on mesotrophic lochs with small bays containing bottle sedge, which is the favoured nesting habitat. In recent years, some pairs have bred on more productive lochs and, at one of these, sedge beds are absent and the grebes nest in loch-side bushes and overhanging branches. This has allowed earlier nesting, with good fledging success and even cases of double broods. Overall breeding success for the population is generally poor, ranging from 0.4 chicks per pair in poor years to 0.8 chicks per pair in good years. Failures have been caused by disturbance by fishermen, birdwatchers, photographers and tourists as well as by illegal egg taking. Predation of eggs, young and even adults by mammals, birds and pike has been recorded, and nests have been flooded or left high and dry by water level fluctuations.

Several lochs have been affected by conifer afforestation including complete desertion of the site; one site was deserted following the building of a waterside house and one loch suffered a sharp drop in numbers following stocking with rainbow trout. Nevertheless, other sites, where no obvious changes have taken place, are suddenly deserted. Part of one of the best sites, and historically the loch where Slavonian Grebes were first recorded nesting in Britain, is now a reserve owned by the RSPB. There, at Loch Ruthven, the nesting grebes may be watched at close-quarters from a public hide without risk of disturbance. This loch is also an SSSI, but protection for many of the other sites is still required.

The NW Europe population is centred in Finland (3,000 pairs), Sweden (1,000 pairs), Norway (500 pairs) and Iceland (500–750 pairs). The species also breeds in Russia and North America. The Scottish population, despite increasing from very low numbers, is still relatively small and vulnerable. It rose steadily from 52 pairs in 1971 to a peak of 80 pairs in 1978–80 but has now levelled out at about 60 pairs.

ROY DENNIS

68–72 Atlas p 34

Change

Distribution

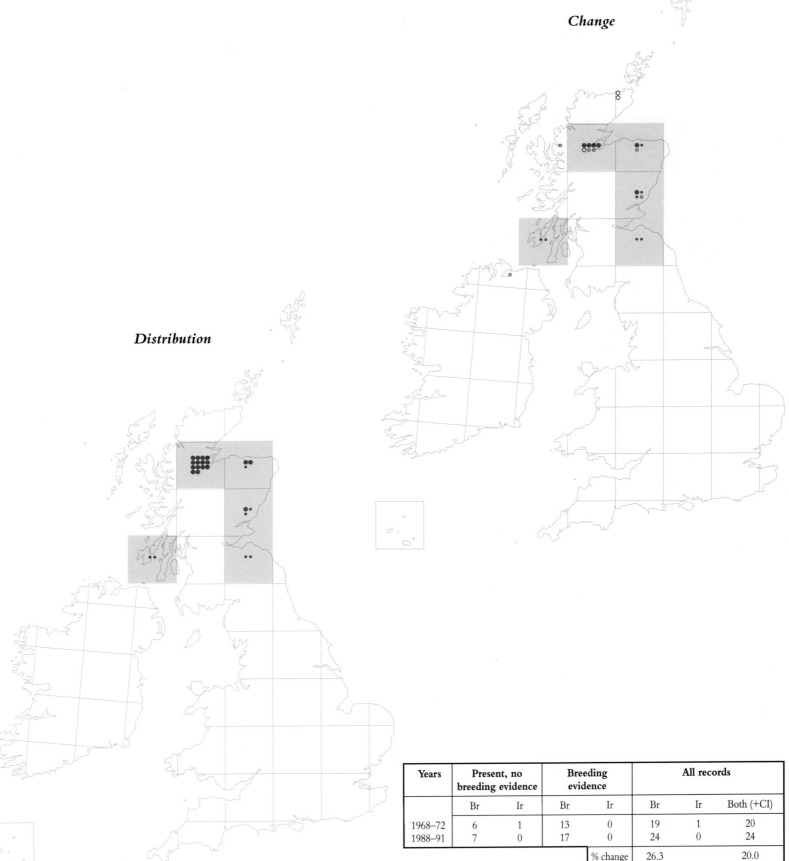

Years	Present, no breeding evidence		Breeding evidence		All records		
	Br	Ir	Br	Ir	Br	Ir	Both (+CI)
1968–72	6	1	13	0	19	1	20
1988–91	7	0	17	0	24	0	24
				% change	26.3		20.0

Black-necked Grebe

Podiceps nigricollis

This elusive small grebe remains a scarce breeder in Britain and Ireland, restricted to a handful of confidential sites but increasingly likely to appear at quiet lowland waters. It is adept at vanishing into dense reeds, and breeding can be difficult to prove until family parties are seen feeding amongst the emergent vegetation.

The Distribution map shows a fairly widespread pattern from central Scotland, south to the English Midlands, but regular breeding is concentrated at three or four main sites, with an increasing tendency for sporadic breeding elsewhere. The ecological requirements of the Black-necked Grebe in Britain have not been studied, but a preliminary examination of the factors common to established sites shows an apparent liking for shallow eutrophic lakes, with an abundance of emergent and submerged vegetation. Such sites are typically quiet and undisturbed, with extensive fringing vegetation such as willow or reeds, and often with patches of amphibious bistort or water lilies covering the surface of the open water. Regularly occupied sites often hold breeding Black-headed Gulls which may provide protection from aerial predators.

In Britain and Ireland few natural sites conform to the apparent requirements of Black-necked Grebes. Breeding has, however, been recorded from established water-filled gravel pits and from reservoirs, suggesting that with careful management birds could be encouraged to colonise artificial waters.

It is not yet clear which factors may be limiting the population within Britain and Ireland, but disturbance, particularly from boats, water sports and shore based pursuits, has been held responsible for birds failing to breed at otherwise apparently suitable sites. Predatory fish such as pike have also been implicated in breeding failures, and may well cause brood depletion. Nevertheless, the most obvious cause recorded for reduction of established breeding pairs has been habitat loss or change brought about by natural succession or drainage.

The most striking feature of the Change map is the increase in breeding records from the English Midlands, perhaps an unexpected event in a species so long confined to a handful of sites, largely in Scotland.

A remarkable event was the discovery in 1929 of a colony at Lough Funshinagh, Co. Roscommon, which numbered over 250 pairs in 1930, before eventual drainage in 1934. Interestingly, few birds apparently settled elsewhere in Ireland, and the population dwindled rapidly, only a handful of pairs persisting until 1959. Since then there have been only three published records of breeding, the latest in 1982, although occasional breeding probably still occurs (*Birds in Ireland*, Hillis and Cotton 1989, Kennedy *et al.* 1954).

In Britain, many early colonies failed to persist, apparently due to habitat change. Thus, in Anglesey, the colony present since breeding was first recorded in 1904 had disappeared by 1929, and a colony that survived in Strathclyde from 1938 to 1951 was finally lost when the site was drained. Since 1904, throughout England and Wales, breeding has undoubtedly occurred intermittently away from known sites, but many attempts will have gone unrecorded. Only Scotland has published records of regular site occupancy, and seems to have been the species' stronghold since breeding was first confirmed, in Midlothian, in 1930 (*Birds in Scotland*).

Breeding birds normally arrive at suitable sites during April. The nest, located in shallow water, usually consists of a floating heap of water weed anchored in the cover of reed or sedges. Eggs are laid from late April until July, the usual clutch numbering 3–5 (*BWP*). The incidence of second broods in Britain and Ireland is unclear, but in some years some pairs are clearly double brooded.

The European population fluctuates from year to year but, since at least 1980, has tended to increase on the northern and western edges of the range. Populations in the Netherlands and Belgium have both shown recent growth: the former having increased dramatically from about 100 pairs in 1984 to 230–250 pairs in 1989 (Hustings 1991). The Belgian population has been more erratic, with breeding years interspersed with total absences; but a peak of 28 pairs was recorded in 1983 (Devillers *et al.* 1988).

There is no evidence that before 1970 the British population ever exceeded 10 pairs in any one year, and although huge colonies of many hundreds of pairs occur in North America, the large Irish colony of 250 pairs appears to be without precedent in W Europe. With the demise of that colony, the presence of Black-necked Grebes in Britain and Ireland has been precarious. Since about 1970, however, despite fluctuations, the population has sustained a slow but steady increase up to the present time. During 1988, 1989 and 1990 a minimum of 15 and maximum of 40 pairs were recorded breeding in Britain, and if the European increase continues we may expect further population growth in Britain and Ireland.

GRAHAM ELLIOTT

Two records have been moved by up to two 10-km squares on the Distribution and Change maps.

68–72 Atlas p 36

Two-year averages for Black-necked Grebes in Britain 1973–90

	73–74	75–76	77–78	79–80	81–82	83–84	85–86	87–88	89–90
Confirmed	17	6	12	12	8	14	10	20	23
Possible	19	11	9	6	10	17	18	15	16
Max total	35	17	21	18	18	31	28	35	39

Based on RBBP reports: number of pairs breeding

Change

Distribution

Years	Present, no breeding evidence		Breeding evidence		All records		
	Br	Ir	Br	Ir	Br	Ir	Both (+CI)
1968–72	5	0	6	1	11	1	12
1988–91	24	0	11	0	35	0	35
			% change		218.2		191.7

Fulmar

Fulmarus glacialis

With its distinctive stiff-winged, gliding flight, its noisy cackling when ashore, and its virtually ubiquitous distribution around British and Irish coasts, the Fulmar is nowadays one of our most familiar seabirds. Yet until late last century, St Kilda, off NW Scotland, was the only colony known in these islands.

As evident from the Abundance map, population density is still highest in N Scotland, and Scotland as a whole held 93% of the British and Irish population during 1985–87 counts (*Status of Seabirds*). The largest colony (63,000 occupied nest sites in 1987) has always been St Kilda, where Fulmars have bred since the 12th century or earlier. Foula, in Shetland, now the second largest colony, with 47,000 sites in 1987, was colonised in about 1878, possibly by Faeroese/Icelandic birds (Fisher 1952). Subsequent expansion, particularly early this century, was rapid.

Counts at ten year intervals from 1939 to 1969 confirmed that, for colonies other than St Kilda, rapid population growth was continuing at an average rate of 7–8% per annum (Fisher 1952, Cramp *et al.* 1974). St Kildan and other populations continued to grow between 1969 and the 1985–87 survey, by 3–4% per annum (*Status of Seabirds*).

The formation of new colonies, since the *68–72 Atlas*, is particularly evident in S England and Wales, around the Irish Sea and in W Scotland (Change map). Suffolk, Jersey and Guernsey have all been colonised. Counts for the SCR indicate that the greatest population changes since 1969 have been in SE England, SE Ireland, Wales and NW England (increases of 170–290%). All other regions showed some increase, mainly in the range 80–100% (*Status of Seabirds*).

The bulk of the Fulmar population, in N Scotland, occupies high sea-cliffs, but atypical nest situations, such as on flat ground or among ruined buildings, are also used. Further south, very low cliffs or banks, sometimes with no other seabird company, are increasingly being occupied as the population expands into more marginal habitat. Castle ramparts and even occupied buildings are sometimes used, particularly in E Scotland and NE England. Inland quarries or crags (for instance in the latter regions and Northern Ireland) are also occupied, although most are within 20 km of the sea (SCR).

The original factors behind the Fulmar's spectacular expansion around these islands are still unclear. Theories (summarised in Cramp *et al.* 1974 and *Status of Seabirds*) include Fisher's (1952) suggestion that increased food availability, in the form of waste from the whaling and trawling industries, triggered a huge increase in the Icelandic population. Another suggestion was the appearance within that population of a genotype favouring range expansion.

Growth in numbers, on the scale which occurred in Britain and Ireland up to 1939, probably involved continuing immigration. Evidence for this comes from a long-term study on Eynhallow in Orkney. Survival rates of immature and, especially, breeding age Fulmars are high there, but Dunnet *et al.* (1979) calculated that the breeding output of 0.16–0.52 chicks fledged per pair, annually, would not have allowed population increase at pre-1939 rates, without immigration. On a smaller geographical scale, immigration has undoubtedly been a major contributor to the continued rapid growth of Fulmar numbers in, for instance, SE England.

The breeding output of Fulmars has been assessed annually since the mid 1980s at up to 30 British colonies (Walsh *et al.* 1991). Productivity has generally been within the range 0.3–0.6 chicks fledged per pair, with variations likely to be reflecting such factors as nest predation and food availability. Geographical patterns have been unclear (most study colonies are in Scotland), but some regions have shown striking changes from year to year. A marked fall in breeding success of Shetland Fulmars in 1988 may have been related to reduced availability of sandeels, though several other seabird species were more severely affected.

Fulmars have a broad dietary range and this has almost certainly contributed to their success. A study on Foula found that the diet during the

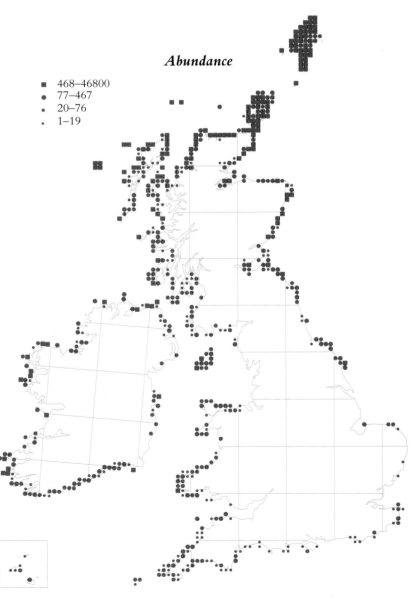

Abundance

- ■ 468–46800
- ● 77–467
- ● 20–76
- · 1–19

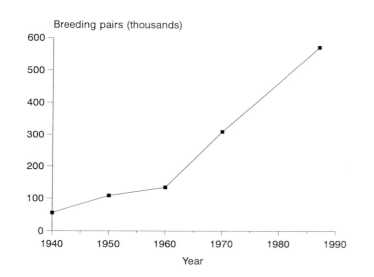

Breeding pairs (thousands)

Population growth of British and Irish Fulmars since 1939 (from data summarised in *Status of Seabirds*).

Change

Distribution

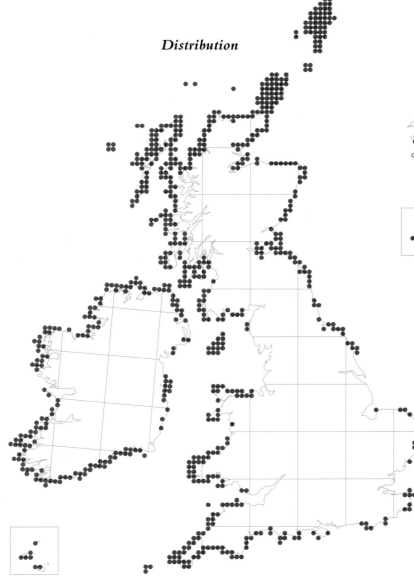

chick rearing period was mainly sandeels, while St Kildan birds at the same time fed predominantly on zooplankton (Furness and Todd 1984).

About 571,000 pairs of Fulmars breed on British and Irish coasts, 31,000 in Ireland and 540,000 in Britain (*Status of Seabirds*). Perhaps another 500 pairs occupy inland colonies (SCR). These figures inevitably include some non-breeding or 'prospecting' birds. The total represents about 7–8% of the North Atlantic breeding population, and 3–4% of the world population.

PAUL M. WALSH and MARK L. TASKER

68–72 Atlas p 40

Years	Breeding evidence		
	Br	Ir	Both (+CI)
1968–72	506	155	662
1988–91	550	159	716
% change	7.7	2.6	7.1

20 years. A few new colonies have been 'discovered', but it is often difficult to determine if this is due to increase in observer effort or genuine colonisation. Encouraging efforts have been made to rid some islands of rats; Cardigan Island and the Calf of Man have each had extermination programmes, but colonisation (or recolonisation) has been slow at both sites. Numbers on the islands of SW Wales and on Rum appear to have increased (*Status of Seabirds*). Evidence for population change from the Kerry islands is more confusing.

The distribution of Manx Shearwaters is affected by the presence or absence of rats. All of the colonies holding more than a few pairs are on rat-free islands or on cliffs inaccessible to rats. The colonies on Rum are interesting exceptions to this. Thompson (1987) found that rats were present in coastal habitats throughout the year and that the inland hills where the Manx Shearwaters nest, at an altitude of *ca* 600m, did not support any rats during the winter. Thus, when the Shearwaters arrive at their colonies during March and April, they are returning to rat-free habitat. Rat numbers build up on the island during the summer, but only reach the shearwater colonies at the time of chick fledging. The rats then eat dead chicks and eggs, before running out of food again. The Shear-

Manx Shearwater

Puffinus puffinus

Most birdwatchers will associate Manx Shearwaters with large scale passage movements off headlands: on some days many thousands of these elegant seabirds can pass W coast headlands of Britain and Ireland. Fewer bird-watchers will have seen the very large rafts that gather at feeding sites at sea or those that assemble off colonies, ready to visit land by night. Yet fewer will have had the privilege of being in a large colony by night, accompanied by the birds' eerie cries.

The maps illustrate the degree to which this species is concentrated into rather few sites. Three areas or colonies hold the vast majority of breeding birds. Off the W coast of Scotland, Rum supports the largest Manx Shearwater colony in world; two islands off SW Wales (Skomer and Skokholm) support similar numbers in total, while lower but still very important numbers are present on islands off Kerry. Away from these areas Manx Shearwaters breed on at least a further 32 islands (*Status of Seabirds*). For breeding, Manx Shearwaters need locations free of mammalian pred-ators; conditions which are primarily provided by offshore islands. There is, however, a notable lack of colonies in Orkney and Shetland and further south in the North Sea, so presumably other factors are involved in determining distribution. The three main breeding areas are in waters close to the productive zone where the Atlantic mixes with coastal currents. There is thus an indication that oceanography affects breeding distribution. Although the presence of suitable feeding conditions is important for Manx Shearwaters, these need not be close as they can forage at some distance from their colonies. For example, large concentrations of these birds occur in the N Minch during the breeding season, at least 170 km from the Rum colony (Webb *et al.* 1990), although they need not all be breeding birds.

There have been relatively few changes in distribution during the past

Abundance

- ■ 1100–100000
- ● 100–1099
- · 27–99
- · 1–26

waters thus, in effect, nest on a rat-free 'island' within Rum.

Brooke (1990) has reviewed much of the recent work on the breeding biology of this species, and those wishing to know more are referred there.

There is some dispute over the precise taxonomic relationship of the Manx Shearwater to other similar forms that are distributed around the world. Although numbers are not precisely known, over 90% of the population of this shearwater breeds in Britain and Ireland. Much of the remainder of the population nests on the Faeroe Islands and in Iceland (*Status of Seabirds*). Britain and Ireland thus have a particular responsibility for this bird and hence it is included in *Red Data Birds*.

The *Status of Seabirds* estimates that between 220,000 and 250,000 pairs breed in Britain and between 30,000 and 50,000 in Ireland.

MARK L. TASKER and PAUL M. WALSH

68–72 Atlas p 42

Change

Distribution

Years	Breeding evidence		
	Br	Ir	Both (+CI)
1968–72	32	16	48
1988–91	22	13	38
% change	−31.3	−18.8	−20.8

Storm Petrel

Hydrobates pelagicus

Only a lucky few have seen this little seabird at its breeding sites. These are located on small offshore islands to the north and west of Britain and Ireland, and by day the birds are either in their nesting burrows or are far out to sea. Change-over on the nests occurs by night, so that most birdwatchers' views of these oceanic wanderers are through a telescope while seawatching.

Due to its nesting habits, this species is almost impossible to census, and indeed it is likely that quite a few colonies await discovery. It is plainly a bird of the Atlantic Ocean margins, with none known to nest south of Orkney within the North Sea, and the largest concentration of birds being in SW Ireland. Storm Petrels are predominantly plankton-eaters, though many also feed on scraps of offal from trawlers. It is likely that their breeding distribution is linked to the availability of food, but this association has not been investigated. Away from their colonies, Storm Petrels are rarely seen from land in the breeding season, but, using taped calls, relatively large numbers have been lured to mist-nets set by night on the coast, often in areas distant from known colonies. This might indicate that the birds are coming into nearshore waters to feed at night, when avian predators such as gulls are not a threat and plankton may be closer to the surface of the water. Virtually all colonies (and certainly all large ones) are on islands free of mammalian predators.

The difficulty of censusing Storm Petrel colonies means that it is virtually impossible to assess any recent changes. Apparent changes since the *68–72 Atlas* are likely to reflect whether or not a colony was visited in the relevant years. However, one colony, Skea Skerries in Orkney, was washed away in heavy winter gales in the 1970s, and two other Scottish

colonies (Noss and Canna) may now be extinct due to the activities of cats and rats. Numbers on Annet in the Isles of Scilly have also fallen in recent years, possibly due to rising numbers of large gulls (Harvey 1983).

The RSPB has recently been experimenting with the use of endoscopes in nesting cavities, and taped calls to examine Storm Petrel colonies. Further development of these techniques should in due course lead to enhanced conservation of this species. Knowledge of its status is essential before the success or otherwise of conservation action can be assessed.

Between about two-thirds and three-quarters of the world population of this species breeds in Britain and Ireland (*Status of Seabirds*). Its breeding distribution is virtually confined to Europe. Numbers in Britain probably lie between 20,000 and 150,000 pairs, and in Ireland between 50,000 and 100,000 pairs. A speculative total of 160,000 pairs for Britain and Ireland was used in *Status of Seabirds*.

MARK L. TASKER and PAUL M. WALSH

68–72 Atlas p 44

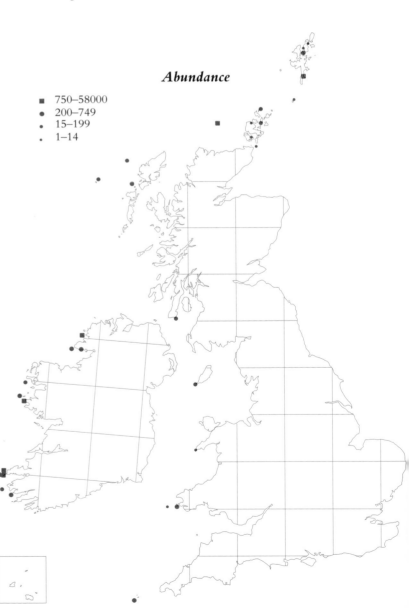

Abundance

■ 750–58000
● 200–749
· 15–199
· 1–14

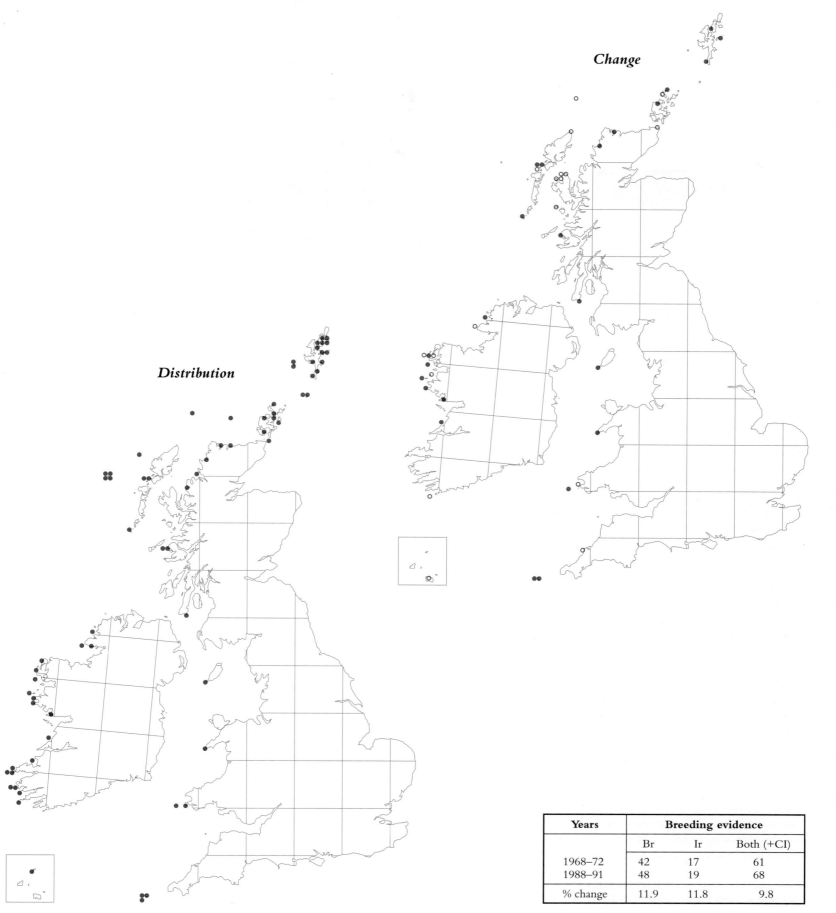

Change

Distribution

Years	Breeding evidence		
	Br	Ir	Both (+CI)
1968–72	42	17	61
1988–91	48	19	68
% change	11.9	11.8	9.8

Leach's Petrel

Oceanodroma leucorhoa

This truly oceanic bird must be one of the least encountered species on the British and Irish breeding list. It is seen most frequently during autumn gales, particularly at W coast sites. Those lucky enough to go to sea in search of Leach's Petrels have most often encountered them beyond the edge of the continental shelf to the west of Scotland (Webb *et al.* 1990). A few birds are attracted at night to taped calls played on the Atlantic and North Sea coasts of Britain and Ireland.

As with Storm Petrels, this species is almost impossible to census. It is known to nest only on Ramna Stacks, Shetland; Foula; Sule Skerry; North Rona; Sula Sgeir; Flannan Islands; St Kilda; and the Stags of Broadhaven, Co. Mayo. These are all islands, mostly in the far northwest, and are free of mammalian predators. This distribution presumably relates to their need to be close to feeding areas over or beyond the edge of the continental shelf, and to the need for safe nesting burrows. The largest concentration is on the St Kilda group, where Tasker *et al* (1988) estimated that Boreray alone held over 3,000 occupied burrows.

There have been very few estimates of Leach's Petrel colony sizes, and most of these are based on informed guesswork. It is thus impossible to assess any changes in colony size. The total of eight colonies currently known in British and Irish waters is three more than in 1968–72. Two of these were 'rediscoveries'; breeding on Sule Skerry was last noted in 1933 while breeding on the Stags of Broadhaven had been suspected in the 1940s. The colony on the Ramna Stacks was found in 1981 (Fowler 1982); that on Foula was strongly suspected at the time of the *68–72 Atlas* and confirmed in 1974. It is impossible to tell whether these colonies have been present continuously and not recognised as such, are recolonisations, or are quite new. Several other islands have probable or confirmed breeding records from earlier in this century. Bearasay and Haskeir in the Outer Hebrides, Mingulay and Rum in Scotland, Duvillaun Beg off Co. Mayo, and Inishtearaght, Inishnabro and Blackrock off Ireland were all reported as breeding sites last century.

Very little is known of status changes in British and Irish colonies. In New England, numbers are known to have declined due to predation by rats, domestic cats and dogs, to destruction of nesting habitat by sheep and to disturbance by tourists, bombs and research workers. Adult mortality may have increased through ingestion of plastic particles at sea (Buckley and Buckley 1984). Ainslie and Atkinson (1937) summarised knowledge of breeding biology and behaviour at British colonies. There has been little research since then.

The species nests in both the North Pacific and North Atlantic. The British and Irish population probably lies between 10,000 and 100,000 pairs (*Status of Seabirds*). This is a very large proportion of those nesting in Europe, but only a small fraction of the world population.

MARK L. TASKER and PAUL M. WALSH

68–72 Atlas p 46

Change

Distribution

Years	Breeding evidence		
	Br	Ir	Both (+CI)
1968–72	9	0	9
1988–91	10	1	11
% change	11.1		22.2

Gannet

Morus bassanus

The Gannet breeds in colonies which are usually situated on the cliffs or upper slopes of islands. Few ornithological experiences can rival the sight and sound of a gannetry at the height of the breeding season, with thousands of brilliant white birds packed densely on their breeding ledges, and the clamour of their calls filling the air.

All gannetries are situated close to areas where the birds' main prey, herring, mackerel and sandeels *Ammodytes* sp are abundant (Nelson 1978). Besides good fishing, the species needs islands and usually cliffs, partly for protection against large, mammalian predators but also for air currents to aid take offs and landings at the nest. The right combination of nesting and feeding habitat occurs mainly in the N and W of Britain, and the species' strongholds coincide with some of the wildest and most spectacular coastal scenery in Britain and Ireland.

There is a long tradition of Gannet counting and the species is one of the easier seabirds to census. Many colonies are counted from photographs, in which case the counting unit is usually the 'apparently occupied site' (one or two Gannets occupying a site suitable for breeding, irrespective of whether nest material can be distinguished). Some of the smaller colonies are, however, counted in the field and these counts are normally given as occupied nests.

The 16 gannetries occupied at the time of the 1969–70 survey were all extant in 1988–91 but during the past 20 years new colonies have been founded on Foula and Fair Isle (Shetland), Troup Head (Grampian), Clare Island (Co. Mayo) and Ireland's Eye (Co. Dublin). Two sites on the Shiant Islands (Western Isles) were occupied periodically between 1975 and 1986 but were abandoned in 1987 (Murray and Wanless 1986, Wanless 1987, *Status of Seabirds*).

The St Kilda (Western Isles) gannetry on the island of Boreray and its attendant stacks, Stac Lee and Stac an Armin, is the largest colony with 50,050 occupied sites, but numbers at four other colonies – the Bass Rock (Lothian), Ailsa Craig (Strathclyde), Grassholm (Dyfed) and Little Skellig (Co. Kerry) all exceed 20,000 occupied sites. Together these five gannetries accounted for 77% of the total British and Irish population in 1984–85 (Wanless 1987).

Of the 16 colonies occupied in 1969–70, numbers at 12 have increased while those at the other four have remained more or less stable. In general, rates of increase at the larger colonies have been *ca* 2–3% per annum, close to the intrinsic rate of increase calculated for the species (Nelson 1978). Numbers at many of the smaller, newly colonised gannetries have increased at a much faster rate, indicating that they are attracting recruits from other colonies. This pattern of change is remarkably similar to that recorded throughout this century. The reasons behind the steady increase seem, at least in part, to be due to the Gannets' recovery from heavy persecution in the 19th century and favourable feeding conditions.

As a breeding bird the Gannet is confined to the North Atlantic. The current British and Irish totals of 162,100 and 24,500 nests (see Table) respectively together represent *ca* 70% of the world population.

SARAH WANLESS

68–72 Atlas p 48

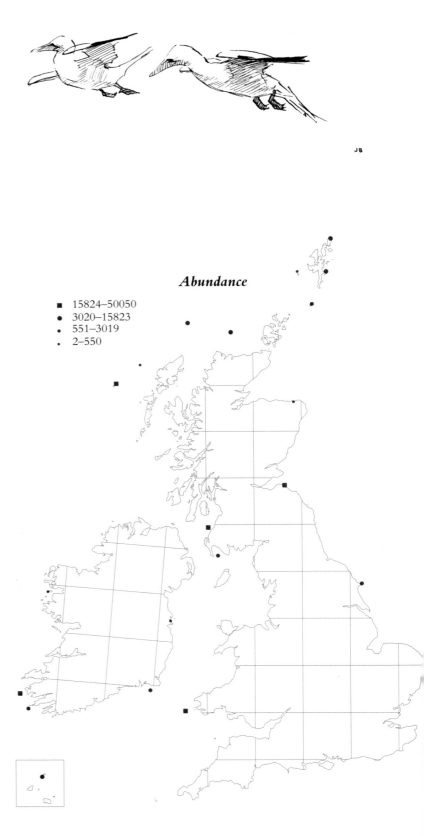

Abundance

- ■ 15824–50050
- ● 3020–15823
- ● 551–3019
- · 2–550

	Numbers of occupied nests or sites			
	1969–70	1984–85	Most recent count	(year)
Hermaness, Shetland	5,894	8,063	10,057	(1991)
Noss, Shetland	4,300	5,231	5,904	(1991)
Foula, Shetland	0	140	280	(1991)
Fair Isle, Shetland	0	138	687	(1991)
Sule Stack, Orkney	4,018	5,900*	N.U.	
Sula Sgeir, Western Isles	8,964	9,140*	N.U.	
St Kilda, Western Isles	52,000	50,050*	N.U.	
Flannan Islands, Western Isles	16	223	414	(1988)
Shiant Islands, Western Isles	0	1	0	(1987)
Troup Head, Grampian	0	0	94	(1991)
Bass Rock, Lothian	8,977	21,600*	N.U.	
Ailsa Craig, Strathclyde	13,054	22,811	N.U.	
Scar Rocks, Dumfries and Galloway	450	770	N.U.	
Bempton, Humberside	18	529*	1,252	(1991)
Ortac, Alderney	1,000	1,897*	1,868	(1989)*
Les Etacs, Alderney	2,000	1,950*	2,660	(1989)*
Grassholm, Dyfed	16,128	28,600*	N.U.	
British Total	16,819	157,043	162,087	
Clare Island, Co. Mayo	0	2*	2	(1986)
Little Skellig, Co. Kerry	20,000	21,919*	N.U.	
Bull Rock, Co. Cork	1,500	1,511*	N.U.	
Great Saltee, Co. Wexford	155	541*	1.050	(1990)*
Ireland's Eye, Co. Dublin	0	0	27	(1990)*
Irish Total	21,655	23,973	24,509	
Grand Total	138,474	181,016	186,596	

Sources: Cramp *et al* (1974), Murray and Wanless (1986), Wanless (1987), *Status of Seabirds*, SCR

N.U. No update since 1984–85.

Count units: 1969–70 pairs occupying nests; 1984–85 and
most recent counts given as occupied nests with
counts of apparently occupied sites indicated by asterisks.

Change

Distribution

Years	Breeding evidence		
	Br	Ir	Both (+CI)
1968–72	15	3	19
1988–91	18	5	24
% change	28.6	66.7	33.3

Cormorant

Phalacrocorax carbo

The Cormorant is a large and conspicuous seabird, heraldic in posture when drying its wings, and of almost reptilian appearance when seen at close quarters. Identification poses few problems, and breeding is usually easy to confirm, although smaller colonies, especially when isolated, are easily overlooked. The nest is a bulky tangle of twigs and seaweed, and in Britain and Ireland most colonies are of 10–300 pairs.

The Distribution map emphasises the essentially coastal nature of the species as a breeding bird. A bias towards W Britain and the west of Ireland is evident, these being the areas best supplied with cliffs, stacks and rocky islets, the preferred breeding habitat. It is absent from a large part of NE Scotland and, apart from one colony in East Sussex, breeds nowhere on the SE coast of England from the Humber to the Solent. The present distribution is very similar to that of the *68–72 Atlas*, but careful comparison reveals an apparent net decrease in the number of colonies, especially in NW Scotland, the Western Isles and Cornwall. By contrast the number of inland colonies in Ireland has increased and, in what is probably the most exciting development since the *68–72 Atlas*, a number of inland breeding attempts, mostly in trees, have been made in England and Wales (one colony only). Some colonies have failed to become established, however, and most are still quite small (< 40 pairs), but one, that at Abberton Reservoir, Essex, has expanded rapidly. Nesting first took place there in 1981, when nine pairs bred, and by 1990 there were no fewer than 356 pairs, making it one of the largest colonies in Britain. The origins of these birds are not known for certain but their proximity to the rapidly expanding and predominantly tree-nesting Dutch and Danish populations has prompted speculation that they may come from the Continent. The plumages of the two populations differ slightly and an investigation of the Abberton birds suggests that those currently breeding there are mainly, but not exclusively, of the N Atlantic race *P.c.carbo* rather than of the Continental race *P.c.sinensis*, implying that most originate from within Britain and Ireland (Sellers 1993).

Recent colour-ringing studies in Wales have shown that some Cormorants breed in their second summers and that the majority have begun to breed by the time they are three years old (R.M. Sellers and S.J. Sutcliffe). Many non-breeding second-year birds attend the natal colonies so most of the records away from the breeding colonies shown on the map in Appendix E are likely to be first-year birds.

The population biology of British and Irish Cormorants is still poorly known, but based on the 'Operation Seafarer' (1969–70) and SCR (1985–87) results, the current trend appears to be for a decrease in breeding numbers in N and NW Scotland, a small increase in Wales, somewhat larger increases in SW Scotland, the Forth-Humber area and SW England, and substantial growth in Ireland (*Status of Seabirds*, Sellers 1991). Elsewhere in W Europe the past 30 years have seen a rapid and sustained growth.

Cormorants have been persecuted by man in the past because of the adverse effect they are thought to have on fish stocks. The problem is

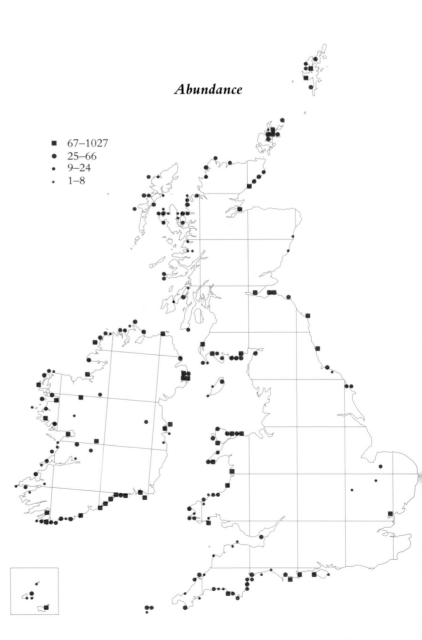

Abundance

■ 67–1027
● 25–66
• 9–24
· 1–8

probably nowhere very severe in Britain but there are instances of Cormorants being serious pests in other parts of the world, for instance in the Netherlands (Moerbeek *et al.* 1987). In Ireland, Cormorants are afforded full protection under the Wildlife Act 1976, whilst in the UK they are protected under the Wildlife and Countryside Act 1981. The latter allows the species to be shot under licence 'to prevent serious damage' but only after 'other methods have been tried and failed'. There are no published figures on how many licences have been issued but the total is probably small. Despite this, substantial numbers of ringed Cormorants are still reported as having been shot. Mortality due to shooting and tangling in nets appears to be particularly high in NW Scotland and this may account in part for the recent declines in that area. Fish predation by Cormorants at fish farms and put-and-take lakes is usually thought of as a winter 'problem' in Britain and Ireland, but the growing numbers summering, and now breeding, inland increases the risk of conflict between conservation and fisheries interests and the position needs to be monitored closely.

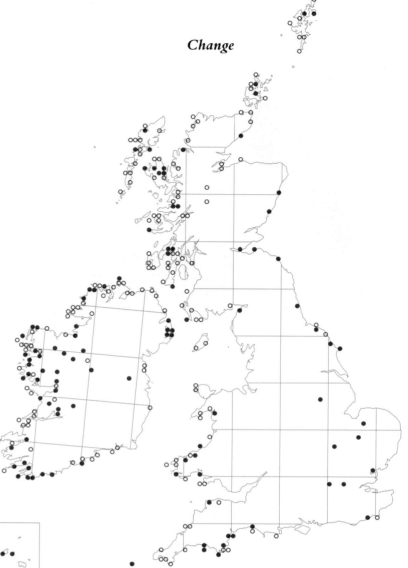

Change

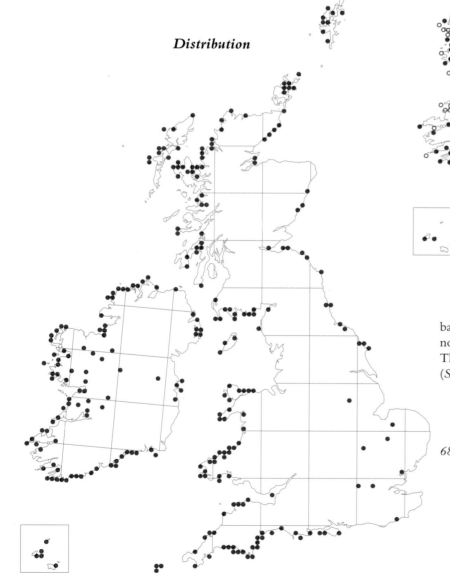

Distribution

The most recent estimate of the size of the British breeding population, based on the SCR during 1985–87, is about 7,000 pairs. This survey did not include inland colonies and an allowance for these has been made. The Irish population during the same period is estimated at 4,700 pairs (*Status of Seabirds*).

ROBIN M. SELLERS

68–72 Atlas p 50

Years	Breeding evidence		
	Br	Ir	Both (+CI)
1968–72	216	97	316
1988–91	174	93	272
% change	−20.0	−4.1	−14.3

many pairs do not attempt to breed (Aebischer 1986), all cause problems. Although there have been only trivial changes in breeding distribution since the *68–72 Atlas*, the results of Operation Seafarer (1969–70) and SCR (1985–87) indicate a 40% increase in numbers between surveys. An increase was noted in all areas except for NW Scotland and the Northern Isles, where counts were down by up to a third. Annual monitoring in Shetland suggests that the main decline occurred during the 1980s (M. Heubeck). Numbers in Ireland appear to have more than doubled, perhaps due to the introduction of legal protection (*Status of Seabirds*). Monitoring counts at a range of colonies indicate that the British population may no longer be increasing (Walsh *et al.* 1991).

The Shag has not always been common and widespread. For instance, in the early 1900s there were 10 pairs breeding between the Rivers Tay and Humber, in 1963 there were 1,900 there, in 1985–87 about 4,000 – an average increase of about 10% per annum for almost a century. The reasons for the increase are unclear but Potts (1969) put a strong case for it being due to a change in human attitudes to Shags, and other seabirds, from persecution to active protection. Certainly both the number and tameness of Shags in many parts of Britain contrast markedly with situ-

Shag

Phalacrocorax aristotelis

Any large, black long-necked bird hanging its wings out to dry or intently preening itself while standing on a rocky skerry or sea cliff is instantly recognisable as a Shag or Cormorant. The habitat often gives a clue to its identity, for the Shag is almost entirely marine and prefers rocky shores whereas the Cormorant also frequents brackish estuaries and occurs inland. A Shag on freshwater is a sorry and usually moribund bird.

The Shag is a colonial nester with a markedly northern and western distribution. The main concentrations occur in Orkney, Shetland, Inner Hebrides, Firth of Forth and NE England. Colonies in Wales, Ireland and SW England tend to be smaller and only a few pairs breed in England between the Farne Islands and Dorset. The largest colonies, with up to 2,400 nests, are on islands – notably Foula, Fair Isle, Shiants, Canna, Isle of May, Farne Islands and Lambay – where nests are built almost underground in large boulder fields or exposed on open ledges. The species is, however, not averse to breeding at low density on mainland cliffs, often low down where the nests are at risk from the sea, and are easily overlooked.

Shags obtain much of their food at or near the sea bottom in relatively shallow (<40m) water and, when breeding, probably do not usually forage more than 15km from the colony (Wanless *et al.* 1991); therefore the structure of the coast can restrict the dispersion of colonies. The paucity of pairs in SE Britain must be at least due partly to lack of nesting habitat. Birds wintering in this area appear to have trouble finding suitable roosts and numbers are occasionally blown inland.

Assessing the size of populations is not easy. The problems of counting nests on cliffs and under boulders, the protracted laying season (in which, at any time, some pairs will have yet to lay and/or will have failed), and the marked annual differences in laying date and occasional years when

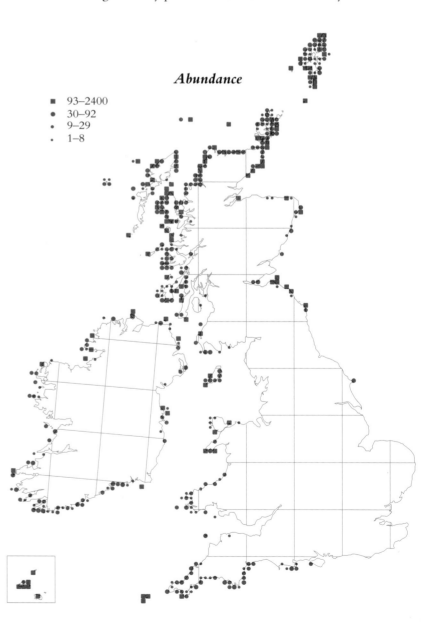

Abundance

■ 93–2400
● 30–92
• 9–29
· 1–8

ations elsewhere, where persecution or shooting 'for sport' still occur. Even so, a good food supply is essential to support the birds. During the breeding season, and probably at other times of year, British Shags rely heavily on lesser sandeels (Harris and Wanless 1992), small fish which are the main food of many seabirds, marine mammals and larger fish as well as being the target of a large fishery industry. Shags are still shot, mostly illegally, at and around fish farms, and get drowned in fishing nets and traps, or become oiled and accumulate toxic chemicals in their bodies, but generally losses are few. The greatest potential threat to the Shag is a reduction in the stocks of sandeels.

The Shag breeds only in the NE Atlantic and the Mediterranean. SCR estimated the British and Irish populations at 38,500 and 8,800 pairs, respectively: Norway may have 50,000 pairs. Together the three areas have almost three-quarters of the world population.

M. P. HARRIS

68–72 Atlas p 52

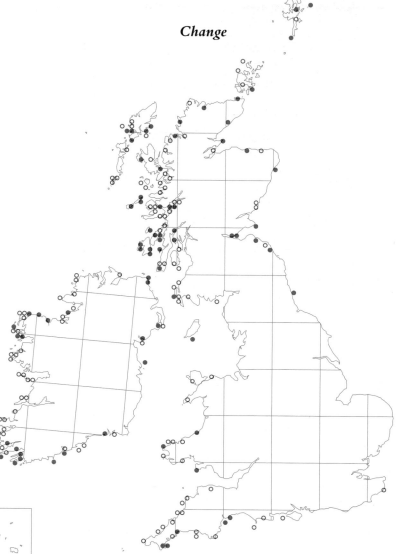

Change

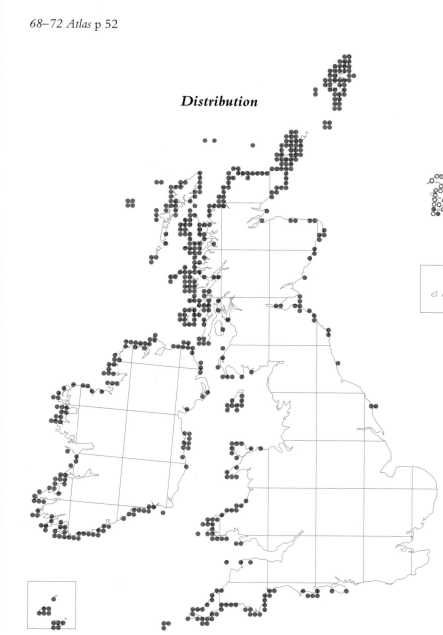

Distribution

Years	Breeding evidence		
	Br	Ir	Both (+CI)
1968–72	406	143	554
1988–91	386	123	520
% change	−5.5	−14.0	−7.6

Bittern

Botaurus stellaris

The distant booming of a Bittern from the depths of a reed bed is a potent reminder of the days when much of the low lying land in East Anglia was covered in fen and marsh.

Drainage, and particularly persecution, drove the Bittern to extinction as a breeding bird in Britain by 1900, but following recolonisation early in the century, a breeding population again slowly built up. Another decline, starting in the Norfolk Broads in the 1950s, and elsewhere during the 1960s, has persisted to the present day reducing numbers to a few pairs.

The Distribution map shows that breeding Bitterns are now confined to a handful of reed beds in Norfolk and Suffolk, and at Leighton Moss in Lancashire.

Bitterns are more widespread in winter, when Continental birds arrive, mainly in December and January, with the greatest numbers in hard weather. Most wintering birds have been recorded in S and SE England, but it is not known whether any stay on to breed the following summer (Bibby 1981a).

Numbers of breeding Bitterns peaked in the early 1950s, with about 80 booming males recorded. Thereafter a serious decline began, with numbers halving in the Norfolk Broads by 1970. Outside the Broads numbers also began falling, accompanied by a contraction in range. A national survey in 1976 revealed 45–47 boomers in the whole of Britain (Day and Wilson 1978), and by 1980 breeding had ceased outside East Anglia and Lancashire.

Bitterns are highly secretive, and the dense and inaccessible nature of their reed bed habitat makes them an extremely difficult bird to study. Male Bitterns begin to boom as early as February, and females start laying their clutch of 4–6 eggs from April onwards. Some males are polygamous. Following mating, it appears that females are left alone to incubate and feed the young, and most broods have fledged by the end of August. Breeding is probably less successful in dry summers (J. Wilson).

Bitterns feed on fish, amphibians, small mammals and large aquatic insects, with medium sized eels particularly important (Gentz 1965, *The Handbook*). Authorities differ on whether Bitterns feed at night. Recent studies by the RSPB, using radio tracking, suggest that reed bed ditches are not favoured as much as natural reed edges for feeding, and electro-fishing studies have established that fish populations in shallow, reed choked ditches are low, compared with those in regularly cleared watercourses, although the latter may offer less favourable fishing conditions for Bitterns.

The decline in numbers and range contraction in Britain could have had a number of causes. Hard winters increase mortality but available evidence suggests that Bittern populations can make good cold-winter losses relatively quickly, and that most dead birds found in this country in winter are of Continental origin (Day and Wilson 1978, Bibby 1981a).

Disturbances from reed cutting, agricultural and drainage operations and recreation may have contributed to declines at smaller, more vulnerable breeding sites, but do not explain similar declines at larger, undisturbed nature reserves.

It seems likely that more widespread environmental factors were at work, and the timing of the initial decline suggests that pesticides could have been implicated. Herons, which have a similar range of food as Bitterns (*BWP*), were found to be contaminated by organochlorines, PCBs and mercury during the 1960s and 70s (Cooke *et al.* 1982). During the same period adult mortality rates of Grey Herons increased and population recovery after hard winters was slow (Reynolds 1979).

Habitat loss and impoverishment may have affected Bitterns, particularly on the Norfolk Broads, where the population declined from 60 boomers in 1954 to six by the late 1980s. Reduced reed harvesting led to scrub encroachment, whilst failure to maintain the reed bed ditches used to transport the cut reeds resulted in vegetation choking the waterways. Recreational boating increased enormously on the Broadland rivers, causing bank erosion, loss of fringing reed, and turbidity. Water enrichment from agricultural run-off and sewage effluent caused algal blooms and created a hostile environment for aquatic invertebrates and fish. In addition, a number of reed beds were drying out. It is not known which, if any, of these factors affected Bitterns, but it seems likely that they have all contributed to the species' problems.

Where ditch management and water level control has been promoted on nature reserves, breeding Bitterns have persisted although numbers have still declined. Such sites now hold about 70% of the British population. Past estimates of breeding numbers were based on counting booming males. Recent research suggests that this may have led to overestimates, but, despite this possible bias, the seriousness of the decline and range contraction (illustrated by the Change map), cannot be doubted.

The current population of Bitterns breeding in Britain is about 16 'pairs': an insignificant proportion of the European total (excluding the former USSR), which was estimated at 2,500–2,700 pairs in 1976 (Day 1981).

JOHN UNDERHILL-DAY

Two records have been omitted from the Distribution map. To maintain the confidentiality of these two sites, open dots have been placed in the relevant 10-km squares on the Change map, else their location would have been apparent by comparison with the *68–72 Atlas* map.

68–72 Atlas p 56

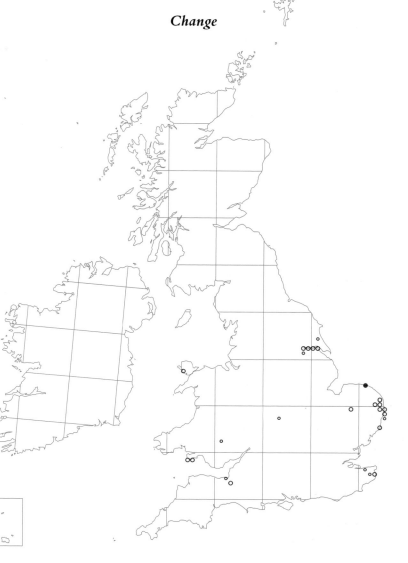

Change

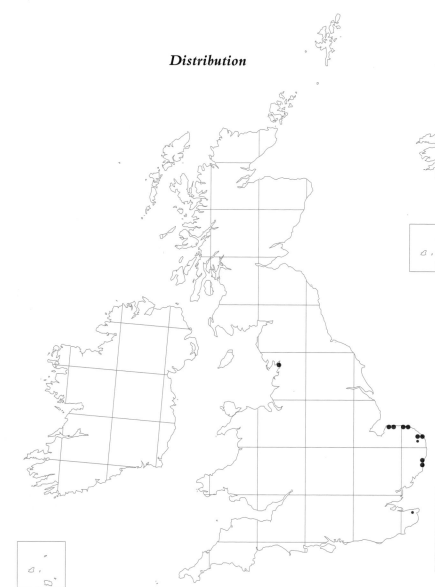

Distribution

Years	Present, no breeding evidence		Breeding evidence		All records		
	Br	Ir	Br	Ir	Br	Ir	Both (+CI)
1968–72	9	0	26	0	35	0	35
1988–91	3	0	10	0	13	0	13
			% change	−62.9	0		−62.9

Grey Heron

Ardea cinerea

Most people, even those without any interest in birds, recognise herons because of their peculiar shape. The Grey Heron's long legs, long neck and dagger-shaped bill are obviously good for stalking, striking at and seizing fish, but seem to ill-fit birds which nest in trees. This is not the case, as anyone who watches at a heronry will see: the birds are remarkably agile as they move about the trees, interacting with other herons. Nevertheless, with broad wings and long legs, some actions must be easier given a reasonable platform and room to manoeuvre, which is presumably why Grey Herons have large stick nests in the upper canopy. Such nests are often easy to see and systematic counts of occupied nests in British heronries started as long ago as 1928. A sample of heronries has been counted in every year since, with nationwide censuses in 1954, 1964 and 1985. This annual index of breeding numbers, together with ring recoveries and Atlas maps, monitors the British population.

The Distribution map shows Grey Herons to be widespread in Britain and Ireland, occurring almost everywhere except in the most mountainous of regions. The map of abundance indicating frequency of occurrence (Appendix E) shows a much more marked pattern, with many more Herons at low altitudes, along the major river systems and on the coast, consistent with the map of nest counts shown here. This is not unexpected, for a detailed analysis of the 1985 heronry census in Scotland showed that Grey Heron breeding density there varied with altitude (Marquiss 1989), such that 57% of the population was found near the coast but only 17% in the uplands. The simplest explanation is that more fish are available to Grey Herons on the lower reaches of rivers, in estuaries and in the sea. At higher elevations, in more dissected terrain, there are fewer fishing sites; waterways are narrower and fish populations less diverse and less productive.

The *68–72 Atlas* map shows that Grey Herons were less widely distributed then, although this seeming expansion may in part be because the *88–91 Atlas* included birds away from possible breeding habitat, whereas the earlier Atlas did not. In the late 1960s the Grey Heron population was still recovering from its crash following the 1962/63 severe

winter. The pace of increase slackened in the late 1970s and early 1980s, but then continued, so that the current population index is higher than ever before. There are probably almost twice as many Grey Herons breeding now than in the late 1960s. The Change map suggests that as Grey Herons have become more abundant, their breeding population has expanded into areas and habitats which supported fewer birds in the earlier period.

It may be that the recent extended series of mild winters with few prolonged cold spells has upgraded some Grey Heron foraging habitat. Feeding sites which were previously untenable in cold weather could now sustain Grey Herons over winter and in early spring, so the current breeding population has expanded into what was previously suboptimal habitat. An alternative explanation is that, for a variety of reasons, including reduced persecution, less pollution and mild winters (Mead *et al.* 1979, North 1979) Grey Herons now survive longer, and with the increased number of adults many of them are obliged to use suboptimal habitat that was previously unoccupied.

Grey Herons feed on a diversity of aquatic animals, switching between prey species according to their seasonal availability (Marquiss and Leitch 1990). In the uplands their diet is less diverse, mainly small trout and

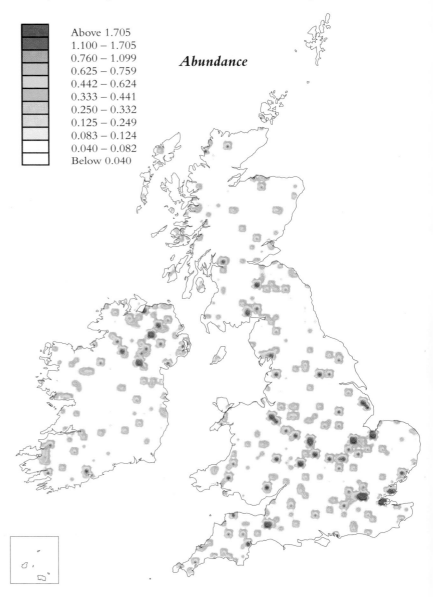

■	Above 1.705
■	1.100 – 1.705
▩	0.760 – 1.099
▨	0.625 – 0.759
▥	0.442 – 0.624
▤	0.333 – 0.441
▢	0.250 – 0.332
▢	0.125 – 0.249
▢	0.083 – 0.124
▢	0.040 – 0.082
□	Below 0.040

Abundance

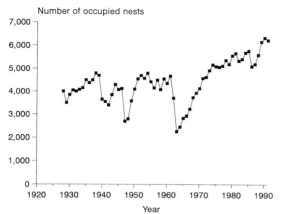

Estimated annual number of occupied nests of Grey Herons in England and Wales since 1928.

salmon, and in all regions Grey Herons exploit unprotected fish farms. Farm managers kill Grey Herons because they eat or damage many fish, though this is now considered unnecessary as most farms can be fully protected from Grey Heron predation (Carss and Marquiss 1992).

Change

Distribution

In the present century the Grey Heron population has increased quite dramatically in Europe, with range expansion northwards into Norway, Sweden and Finland, and a large population explosion in S France, bordering the Mediterranean. A complete census of heronries in 1985 revealed approximately 9,570 Grey Heron nests in Britain (*Trends Guide*). Given that the annual index increased by 8% between 1985 and 1991, it is estimated that there were 10,300 nests in Britain in 1991, and, by extrapolation, 3,650 in Ireland. However, as the 1985 figures are of unknown accuracy, these estimates must be treated with some caution.

MICK MARQUISS

68–72 Atlas p 54

Years	Present, no breeding evidence		Breeding evidence		All records		
	Br	Ir	Br	Ir	Br	Ir	Both (+CI)
1968–72	874	320	813	452	1687	772	2459
1988–91	1544	567	791	224	2335	791	3129
			% change		38.4	2.6	27.3

Mute Swan

Cygnus olor

A pair of Mute Swans on their breeding territory evoke an image of beauty combined with strength. Their elegance of form, the repose of the incubating female, the power of the male as he surges across the water to see off an intruder, and their powerful flight on throbbing wings, all give the species a special place in the public's affection.

The Distribution map shows the Mute Swan to be a widespread but predominantly lowland species, rarely occurring above 300 m. Although absent from many northerly and westerly areas, where high ground predominates, they flourish on the southern Outer Hebrides and Orkney. Within this broad-brush distribution the species is to some extent restricted by its preference in most areas for eutrophic waters for breeding. Adequate supplies of food in the form of aquatic vegetation, and sufficient space for take-off runs, are the main limiting factors. Mute Swans are highly territorial, and since territory size depends on the quality of the habitat they are thinly distributed over the suboptimal parts of their range. A national survey in 1990 (Delany *et al.* 1992) showed that 52% of occupied 10-km squares held only one or two pairs.

Breeding waters include lakes, reservoirs, canals, slow-flowing rivers and small streams. In recent years, river authorities have been zealous in cleaning and straightening rivers to improve water flow and reduce flooding, removing so much bank and aquatic vegetation as to seriously damage the nesting potential for Mute Swans and other water birds. Fortunately, this loss has been counterbalanced in many areas by the great increase of flooded gravel workings, usually in river basins and eventually producing eutrophic waters. Although many of these new wetlands are being used for recreation, a great many have been occupied by pairs of Mute Swans.

Coastal brackish lagoons have long been used for breeding by Mute Swans, offering an abundance of food and reduced tidal activity. The

Fleet, Dorset, supports a large, artificially maintained, colony (50–100 pairs) of this normally territorial species, the birds nesting as close as a few metres apart, and thriving on the vast beds of eel-grass growing in the lagoon. One other coastal site in Britain, Loch of Harray, Orkney, supports similar numbers of at least semi-colonially nesting pairs. In areas of low tidal range, as in W Scotland, pairs may breed on the sea shore, usually in sheltered bays or sea lochs.

In the Uists and Benbecula, Outer Hebrides, there is an isolated population of Mute Swans, containing well over 100 breeding pairs. They nest beside waters that may be eutrophic, mesotrophic or brackish. Productivity is significantly higher in the first named than in the other two water types (Spray 1981).

The Change map shows comparatively few, and mostly minor, differences since the *68–72 Atlas*, and given the somewhat specialised habitat requirements of the species, and its considerable conspicuousness (except when nesting at very low densities), probably most of them can be attributed to differences in coverage. Yet although the geographical distribution on a 10-km grid may have changed little in the last 20 years, this apparent stability of range conceals wide variations in population size over the same period.

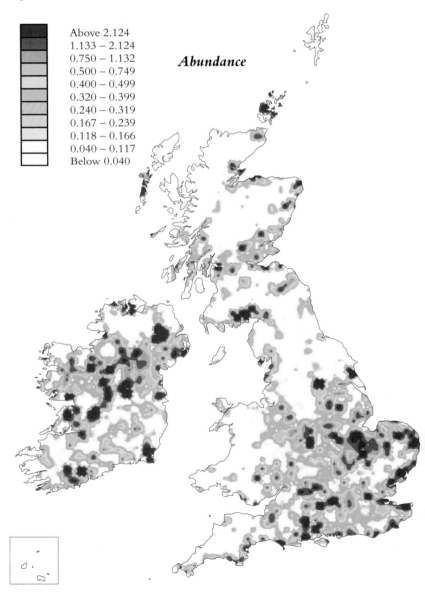

| Above 2.124 |
| 1.133 – 2.124 |
| 0.750 – 1.132 |
| 0.500 – 0.749 |
| 0.400 – 0.499 |
| 0.320 – 0.399 |
| 0.240 – 0.319 |
| 0.167 – 0.239 |
| 0.118 – 0.166 |
| 0.040 – 0.117 |
| Below 0.040 |

Abundance

During the 1950s, the population was seemingly increasing quite fast (Campbell 1960), but received a major setback from the hard winters of the early 1960s. Although there was some recovery in the following years, the numbers in certain river systems of southern and central England, notably the Thames, Warwickshire Avon and the Trent, declined rapidly and severely during the 1970s and into the 1980s (Ogilvie 1981, 1986). While habitat degradation along rivers played some part, the main reason for the declines, resulting in direct mortality as well as reduced breeding success, was found to be lead poisoning caused by the ingestion of fishing weights (Goode 1981). The very considerable increase in coarse fishing in recent decades, with up to 4,000,000 participants by the early 1970s, and also the universal use of non-degradable nylon line, greatly exacerbated the problem.

Following the introduction of suitable alternatives and the subsequent banning of lead for fishing weights in 1987, the incidence of poisoning among Mute Swans fell sharply (Sears and Hunt 1991) and numbers in the badly affected areas have begun to recover, paralleling increases in other parts of the country (Delany *et al.* 1992). Lead poisoning has also occurred in Ireland, where it is thought to have depressed populations in some areas (O'Halloran *et al.* 1991).

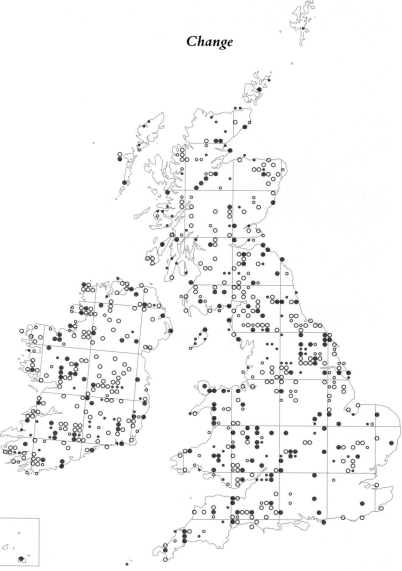

Change

Distribution

The 1990 survey estimated that there were about 25,800 birds in Britain, a 38% increase on the 1983 survey. Extrapolation from tetrad counts in 1988–91 yielded an estimate of 27,000 birds in Britain, a remarkably similar value to that from the 1990 survey. Interestingly, extrapolation of these figures to Ireland suggests that there may be a much higher population there (19,000–20,000 birds) than the previous most recent estimate of 7,000 birds (*Birds in Ireland*).

MALCOLM OGILVIE and SIMON DELANY

68–72 Atlas p 102

Years	Present, no breeding evidence		Breeding evidence		All records		
	Br	Ir	Br	Ir	Br	Ir	Both (+CI)
1968–72	155	51	1468	584	1623	635	2258
1988–91	238	146	1341	414	1579	560	2141
	% change				−2.6	−11.7	−5.1

Whooper Swan

Cygnus cygnus

The Whooper Swan is so well known as a winter visitor from the far north that it comes as a surprise to many birdwatchers to discover that a few not only summer regularly in Britain and Ireland, but also occasionally breed here. This species has a breeding distribution which stretches from Iceland across N Europe and Russia as far as E Siberia. In winter, the majority of birds breeding or summering in Iceland migrate south to Britain and Ireland, where the *Winter Atlas* shows them to be concentrated in Ireland and S Scotland. Up to 25% of birds remain in Iceland during the winter, although weather and food availability cause this proportion to vary between winters. Most of the Fenno-Scandian and Russian breeding population winters on the N coasts of Germany and Denmark, and in S Sweden (*BWP*), whilst the Siberian breeding population winters chiefly around the coastal regions of China.

The breeding grounds of this species tend to be further south than those of the Bewick's Swan. Whoopers favour the taiga and steppe south of the tundra, and usually south of the Arctic Circle. Unlike Bewick's Swans they do not breed in North America. The breeding habitat varies, ranging from large, shallow, reed-fringed lakes to small pools surrounded by forest (*BWP*). The nest, which consists of a large mound of reeds or other plants, is built close to the water's edge, often on small islands, and may be used for several years. The species tends to nest at very low densities, with just one or two pairs nesting on even the largest lakes.

Accounts of early naturalists show that there was a small but well established breeding population of Whooper Swans on Orkney, but that this died out some time between 1760 and 1780. Fea (1775) and Neill (1806) considered the extinction of this population to have been due to egg collecting and disturbance. Since that time, there have been few British breeding records, although summering birds are recorded almost every year. A pair of injured birds bred in Sutherland each year between 1910 and 1918, when they were shot by men recently returned from the war. A pair in West Highland fared better, breeding each year from 1912 until 1921, but a pair in Perthshire were poisoned by an overflow from a lead mine after breeding for three consecutive years (Baxter and Rintoul 1953). *The Handbook* makes reference to an extraordinary breeding record from Norfolk in 1928, although it is not known whether wild birds were involved. There were no breeding records between 1948 and 1977, but in Scotland in 1978 a pair reared three young, raising another two in the following year. Also in 1979, a feral pair reared one young, and since then feral birds have bred regularly in Dumbartonshire. Of the four records of pairs breeding during 1988–91, two were thought to be of wild birds, one of feral birds, and one a Whooper/Mute pairing. RBBP records suggest that at least five pairs bred in Britain in 1990, so it is clear that a number of records were not reported to the Atlas. In Ireland, although they summer regularly in the west, and there have been several records of Whoopers paired with Mute Swans, breeding was not confirmed until 1992, when a pair nested in Co. Donegal.

The Distribution map shows breeding season Whooper Swans to be concentrated, like the wintering population, in Scotland and W Ireland, with only a few summering birds in England. The proportion of breeding records was very low, but this was not necessarily due to lack of suitable breeding conditions since it is known that even in Iceland only 30% to 40% of the population breed in any one year (Rees *et al.* 1991). This low productivity is thought to apply to all breeding populations of Whooper Swans. In the Western Isles, numbers of summering non-breeders may approach 20 birds, but pairs are much rarer (Dix and Cunningham 1991). Elsewhere, summering birds are few and far between. Some are thought to be injured or suffering from poisoning, but there is no evidence to suggest that summering, and occasional breeding, in Britain and Ireland is abnormal behaviour. Indeed, Whoopers breed regularly in S Sweden at lower latitudes than in Scotland. The inability of this species to re-establish itself as a regular breeder is most likely due to its susceptibility to disturbance. Unfortunately its conspicuous white plumage also makes its nests easily found by determined egg collectors, and at least one Scottish nest was robbed during the *88–91 Atlas* period.

It is likely that 2–5 pairs of Whooper Swans breed in Britain each year, although some are probably from feral stock.

PAUL DONALD

68–72 Atlas p 447

Four records have been moved by up to two 10-km squares on the
Distribution and Change maps.

Change

Distribution

Years	Present, no breeding evidence		Breeding evidence		All records		
	Br	Ir	Br	Ir	Br	Ir	Both (+CI)
1968–72	8	0	0	0	8	0	8
1988–91	34	13	4	0	38	13	51
				% change	375.0		537.5

In Ireland, the population around Strangford Lough, now totalling about 500 birds, has spread little, but breeding in scattered sites, due to further introductions in the northwest and southwest (*Birds in Ireland*), may provide a springboard for expansion in future years.

Greylag Geese start breeding in March, the incubating female often taking virtually no food over the four week incubation period, while the male stands guard. The nests are usually hidden from view among tall vegetation, so that often the first real sign that breeding has been attempted is the sight of a party of goslings. Initially both parents remain with their brood but after a few weeks broods often amalgamate, and large crèches of up to 150 young form with many, but not necessarily all, breeding adults accompanying them.

Both the native and reintroduced birds are viewed with some concern by farmers, and the latter also by amenity landowners. Because they are non-migratory, birds are present when young plants are growing in the spring and when cereals are ripening in the autumn, both times when crops are vulnerable. Although measured yield losses have been small, usually less than a 5% yield reduction, in some circumstances, especially

Greylag Goose

Anser anser

This large grey goose has two British populations. From S Scotland, northwards, the birds are derived mainly from stock of the native wild population, now confined to the Western Isles and N Scotland. This native population has a considerable conservation value; something seldom attributed to their progeny reintroduced further south, where the indigenous populations were eliminated by the mid 1800s due to drainage of the wetlands.

Since the *68–72 Atlas*, the wild population has undoubtedly spread in Caithness, Sutherland and, to a lesser extent, in the Inner and Outer Hebrides. There they are mainly associated with coasts and lochs with islands covered in heather or grass/*Juncus*, preferring heather over 30cm high in which to nest (Paterson *et al.* 1990).

The reintroduced stock, too, has spread rapidly over much of England, although it remains very scarce in the SW Midlands, south-central and SW England. In Wales it is virtually unknown except on Angelsey. The causes of the spread are difficult to determine, but many deliberate introductions have been made by landowners and wildfowling clubs, most early ones under a WAGBI scheme which started in 1959. Natural spread has followed, aided by a large increase in the number of reservoirs and wet gravel pits over the last twenty years. The reintroduced birds choose, and breed more successfully on, the well vegetated islands (Wright and Giles 1988) which are often a feature of amenity or conservation gravel pits.

The *68–72 Atlas* showed that the major earliest successful reintroductions were in SW Scotland, the Lake District, Kent and Norfolk. Birds have clearly spread from these centres, but an even more dramatic expansion has taken place in the E Midlands from Bedfordshire, through Lincolnshire, Humberside and Yorkshire, an area where very few were present two decades ago. Details of the spread, based mainly on the wildfowl counts organised by the Wildfowl and Wetlands Trust, are given by Owen and Salmon (1988). They show that the increase by reintroduced birds has taken place between the late 1970s and 1986, at an average rate of about 13% per annum.

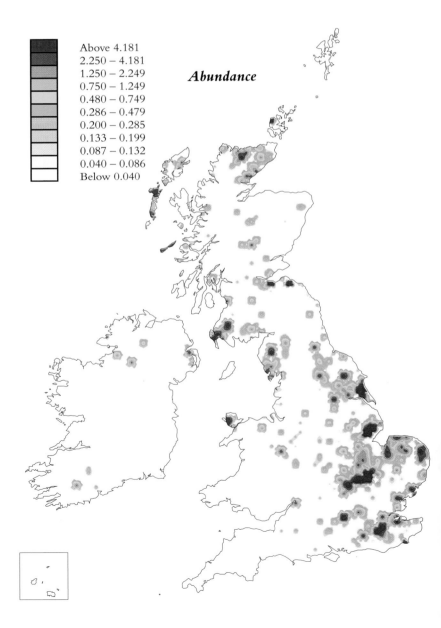

Abundance

Above 4.181
2.250 – 4.181
1.250 – 2.249
0.750 – 1.249
0.480 – 0.749
0.286 – 0.479
0.200 – 0.285
0.133 – 0.199
0.087 – 0.132
0.040 – 0.086
Below 0.040

on the low-input, low-output farming system on Hebridean machair, it can be significant. During the mid summer moulting period Greylag Geese are flightless and often with growing young. At this stage they are associated strongly with water bodies where they can take refuge if danger threatens, but they also have to forage around that site. Grazing can thus be heavy on waterside pastures, crops or emergent vegetation around the lakes. In some circumstances it can have a considerable impact on reeds, particularly where the reed beds are small or are regressing through other causes.

In some areas where there is a large amount of suitable breeding habitat, such as in E and N Norfolk, previous counts did not reveal the true numbers. Here, complete July counts from 1989 to 1991 have revealed a population of about 3,400 birds, compared with 1,250 estimated in 1986, although an average of only 30% of the July birds are juveniles. Owen and Salmon (1988) estimated the reintroduced British population to be up to 14,000 birds in 1986. Extrapolation from tetrad counts during April to July in 1988–91 yields a total British population of 22,000 Greylags, although this estimate includes both native and re-introduced birds. The

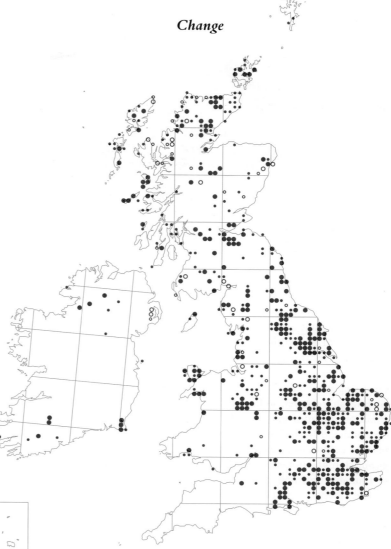

Change

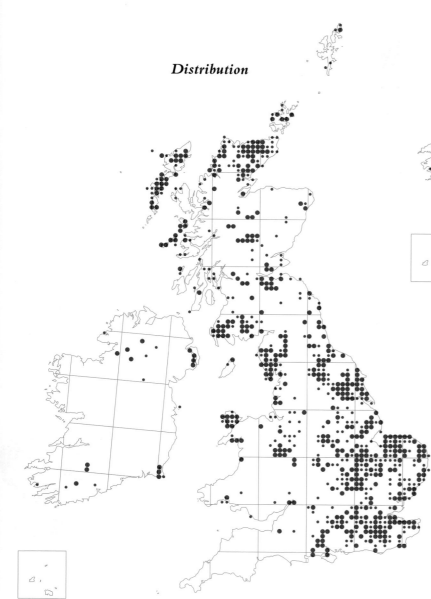

Distribution

feral population in Ireland probably numbers about 700 birds (*Birds in Ireland*).

Although the distribution of Greylags has changed little on the Outer Hebrides over the last 20 years, numbers there have more than doubled over that period. The latest (1986) census estimated a post-breeding population of 1,630 birds (Paterson 1987). *Birds in Scotland* puts the total breeding population of native Scottish birds at 500–700 pairs and the total post-breeding numbers at 2,500–3,000 individuals.

TONY PRATER

68–72 Atlas p 98

Years	Present, no breeding evidence		Breeding evidence		All records		
	Br	Ir	Br	Ir	Br	Ir	Both (+CI)
1968–72	74	1	126	7	200	8	208
1988–91	301	9	417	14	718	23	741
			% change		258.0	187.5	255.3

Canada Goose

Branta canadensis

The Canada Goose is without doubt the most abundant and familiar alien waterfowl in Britain. Whilst its honking brashness in public parks endears it to the hearts of many, to those responsible for the management of amenity grasslands, and to the agricultural community, the Canada Goose is considered as anything from nuisance to pest.

The species shows a catholic taste in breeding sites, from park lakes, reservoirs and gravel pits, to rural and suburban canal and river banks. Not surprisingly, therefore, the Distribution map shows it to be widespread in all English counties, although becoming more sparse, in terms of distribution and numbers, towards the north and in the SW peninsula. Wales, Scotland and Ireland are relatively thinly populated but in each there are marked centres of distribution, and there is no reason to believe that the population will not expand from these into large tracts of currently vacant and suitable breeding habitat.

The Change map shows the remarkable expansion which has occurred throughout Britain and Ireland since the *68–72 Atlas*, for the species is now present in approximately 530 10-km squares which were previously unoccupied. On the other hand, it is absent from nearly 80 formerly occupied squares, such losses occurring throughout the range and following no real pattern.

Where Canada Goose numbers are locally high, damage to nearby agricultural interests may be severe, especially in the spring, and the limiting of further population growth may need to be considered. One of the early methods used was to round up flightless adults and unfledged young at 'problem' sites and transport them to unoccupied waters elsewhere. However, such action, common during the 1950s and 60s, did not solve the problem but simply shifted it, providing new centres for local colonisation and thereby greatly facilitating the expansion of the species in Britain. Today translocations of this type continue, within the provisions of the Wildlife and Countryside Act (1981), under licence from the Department of the Environment. Licences will be issued, however, only if the Department is satisfied that the geese are being released into a 'secure' area from which they cannot contribute to any new, or existing, subpopulation. Clearly, whilst such conditions may be met at the time of the release, for example by wing-clipping, over subsequent seasons adults with regrown pinions, and fledged young, are likely to disperse locally. Management of translocated birds needs to be ongoing and thorough to prevent this.

Aside from shooting (a highly emotive method of control) other 'on

site' methods are available. Recent work by Wright and Phillips (1991) at Great Linford, Buckinghamshire, suggests that the most successful control strategy is to place substitute eggs in the nests of early-nesting pairs, and to remove eggs from those birds nesting later in the season. (Removing eggs from early-nesting pairs would encourage renesting elsewhere, whereas eggs removed from later nesting pairs will not be replaced. Additionally, substituting the eggs of late-nesting pairs would be wasteful of effort if those birds would not renest after clutch removal.) Since Canada Geese are long lived, reducing gosling production at a site, by using egg substitution and removal, would at best provide a brake on population growth. If the aim were to reduce the number of adult pairs attempting to breed then these methods would need to be combined with winter shooting. Restricting numbers breeding in urban environments, where shooting may be neither appropriate nor acceptable, might be more cheaply and sustainably achieved by manipulating habitats and reducing supplementary feeding. Minor changes in mowing regimes, for example, especially on reservoir banks and pond fringes, can virtually close sites to moulting adults and unfledged young, both of which avoid rank vegetation, apparently because it may harbour predators (P. Belman). Some authorities maintain that a concentrated campaign of reducing pro-

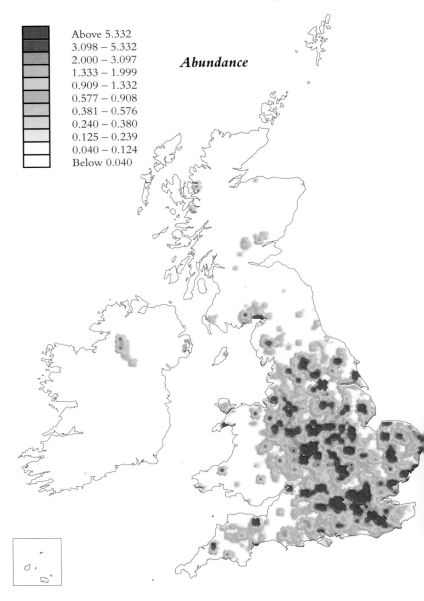

Abundance

	Above 5.332
	3.098 – 5.332
	2.000 – 3.097
	1.333 – 1.999
	0.909 – 1.332
	0.577 – 0.908
	0.381 – 0.576
	0.240 – 0.380
	0.125 – 0.239
	0.040 – 0.124
	Below 0.040

ductivity, combined with shooting, sustained over a five year period throughout the present breeding range, could eradicate the Canada Goose completely. Few would regard this as desirable, but clearly if the population continues to expand at an annual rate of 8% per annum (Owen *et al.* 1986) such measures might need to be employed.

Based on reported annual rates of increase, Owen *et al.* (1986) estimated that the British population of Canada Geese would have exceeded 50,000 birds by 1990. This prediction is borne out by extrapolation from tetrad counts during April to July in 1988–91 which yields a British population of 59,500 Canada Geese. The same calculation suggests a population of 575 birds in Ireland, although this may be too low, as *Birds in Ireland* provides an estimate of more than 700 birds.

STEVE CARTER

68–72 Atlas p 100

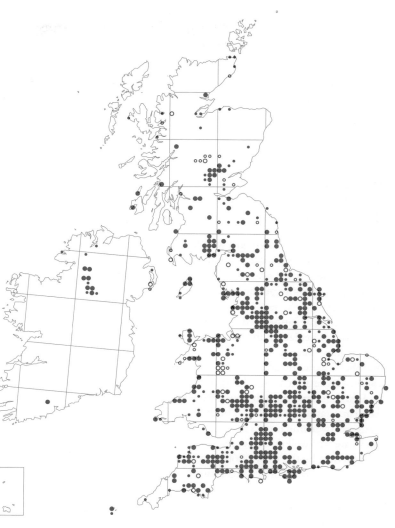

Change

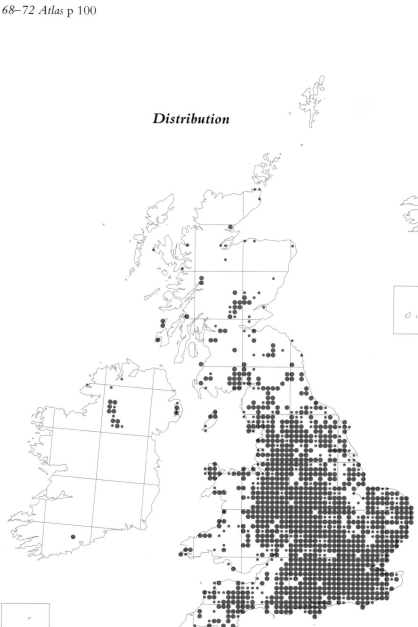

Distribution

Years	Present, no breeding evidence		Breeding evidence		All records		
	Br	Ir	Br	Ir	Br	Ir	Both (+CI)
1968–72	141	2	540	4	681	6	687
1988–91	218	8	978	11	1196	19	1215
			% change		75.6	216.7	76.9

Egyptian Goose

Alopochen aegyptiacus

A peculiar pinkish-brown bird with large, dark eye patches and huge white wing flashes; is it a goose? is it a duck? is it a bustard? Egyptian Geese are regularly taken to be one or other of these species, and as their numbers increase, so may the public become more aware of, and confused by, this exotic bird.

As with most introduced species, its distribution is tied into the history of releases. The date of its first introduction to Britain is uncertain though it may have been late in the seventeenth century. Certainly, however, 'numbers' were introduced from South Africa in the late 1700s and early 1800s. In the mid 1800s the species was well known from Devon north to East Lothian, though most were probably on waters of major East Anglian estates (Lever 1987). In this century there were pairs, or sightings, scattered over much of SE England, but the fortunes of the large estate flocks are not well documented, so gaps occur in the known history.

Certainly, by the early 1960s it was only regular in Norfolk, with a few reaching the meres of Breckland in N Suffolk. The *68–72 Atlas* reflected what was considered to be a small stable population in Norfolk with an odd record elsewhere. Since then there has been a distinct spread in Norfolk and N Suffolk, with breeding in 32 10-km squares in 1988–91 compared with just 12 in the *68–72 Atlas*. A similar result was shown by Kelly (1986) who reported confirmed breeding in 25 10-km squares in Norfolk between 1980 and 1985. Birds are also now noted outside the Norfolk/Breckland areas with first breeding records in Essex (1979), Somerset (1982), and Cambridgeshire (1988).

Quite what has fuelled this spread is uncertain. In some areas, such as the Wensum Valley in Norfolk, the increase in gravel pits appears to be coincidental with the increase there. In Broadland and Breckland, however, the number of waters has changed little over the last few decades, yet birds have spread there too. Since the Wildlife and Countryside Act (1981) came into force, a licence has been required to translocate this species, and as no licences have been issued it must be presumed that the spread in the 1980s has occurred without human assistance.

In many ways the goose is ill-adapted to British conditions. It breeds early, with nests numerous in late February and March, while the young seem to spend an inordinate time (often over one month) in the downy stage. Thus losses are high and productivity relatively low. Sutherland and Allport (1991) estimated that breeding pairs reared 1.06 young per pair in 1988. July censuses in 1989, 1990 and 1991 of the majority of the population in Norfolk, revealed 14.7%, 21.1% and 21.9% juveniles among

the 409, 412 and 521 birds respectively (A. Prater), emphasising a relatively low, but not unacceptable recruitment rate.

The three factors which are required for successful breeding are the presence of water, usually a lake or gravel pit, to provide a refuge, short grass for the goslings to feed on and, most frequently, large, old trees with holes (Sutherland and Allport *loc. cit.*). The Egyptian Goose is unusual in that it often, perhaps usually, nests in holes, although it sometimes uses thick vegetation. This indicates its closer affinity to the Shelduck than to typical geese. Indeed, in the breeding season, it is not unusual to see the male sitting on guard high up on a main bough while the female is incubating inside the tree. The birds breeding in Somerset used old nests, one of a Buzzard and one of a crow. Old trees, lakes and adjacent pasture are features of many of the large private estates in East Anglia. It is not surprising therefore, to find that most of the population is in such areas. This habitat is, however, far from being restricted to East Anglia and if the species can disperse more widely there should be little to stop its slow spread from continuing. Its capacity for movement is still uncertain, although the moulting flock of 110–150 at Holkham Park, Norfolk, clearly draws in birds from a substantial area of mid/west Norfolk.

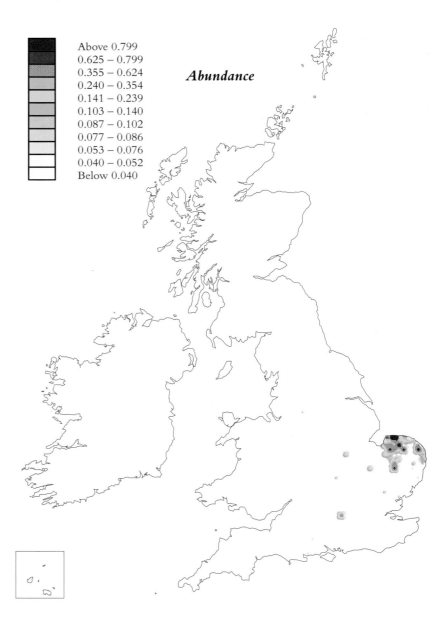

Above 0.799
0.625 – 0.799
0.355 – 0.624
0.240 – 0.354
0.141 – 0.239
0.103 – 0.140
0.087 – 0.102
0.077 – 0.086
0.053 – 0.076
0.040 – 0.052
Below 0.040

Abundance

The size of the present population in Britain is not known for certain. Atkinson-Willes (1963) estimated the winter population (restricted to Norfolk) to be 300–400 birds. Sutherland and Allport (1991) estimated it to be at the upper end of this range, despite the clear spread of records. *Winter Atlas* data suggested a total British population of 504 birds, with 429 of them in Norfolk. Extrapolation from Atlas tetrad counts during 1988–91 yields a minimum British breeding population of 660 individuals. However, as the same number of adults was counted in Norfolk in July 1991, it is surely an underestimate, and the true British population is probably about 750–800 individuals.

TONY PRATER

68–72 Atlas p 96

Distribution

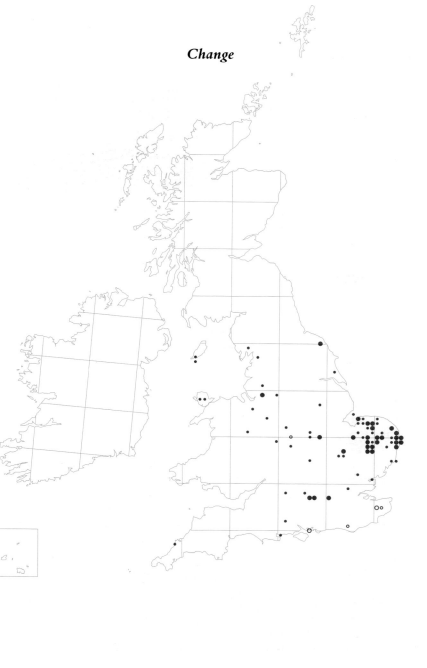

Change

Years	Present, no breeding evidence		Breeding evidence		All records		
	Br	Ir	Br	Ir	Br	Ir	Both (+CI)
1968–72	4	0	14	0	18	0	18
1988–91	48	0	39	0	87	0	87
	% change				383.3	0	383.3

Shelduck

Tadorna tadorna

The Shelduck is one of the most characteristic and colourful of estuary birds, and a familiar sight on muddy and sandy shores around Britain. It breeds wherever there is an adequate invertebrate food supply in close proximity to suitable nest sites in rabbit burrows, tree cavities or similar holes. This unusual duck is also well known for its spectacular moult migration each July, when most of the British birds join the rest of the NW European population in the Heligoland Bight off NW Germany. Other British birds stay to moult nearer home, in Bridgwater Bay and in large estuaries such as the Wash and the Forth (Bryant 1978, 1981).

On their return from the moult migration, breeding Shelducks are widely distributed throughout Britain, concentrating wherever there are estuaries and muddy shores. In Scotland, the main concentrations are found in Orkney, the eastern Firths of Moray, Tay and Forth, the Solway, SW Scotland, and between Kintyre and Islay. In Ireland most Shelducks are found around the northeast and south coasts, with lower numbers elsewhere. In England and Wales there are concentrations around the Humber, the southeast coast from the Wash to the Thames estuary, the Hampshire coast, the Severn estuary, S Wales and the Irish Sea coast.

The Abundance map is broadly similar to the Distribution one, showing that in general the densest local populations are found in the main areas of distribution. However, the inland sites, even those close to the coast, tend to have fewer pairs per 10-km square than coastal areas.

The Change map shows that since the *68–72 Atlas* the Shelduck's distribution in Scotland has remained fairly stable, apart from losses in the west and an increase in the number of inland records during the breeding season, especially in the south. In Ireland, there has been some contraction and inland breeding is still rare. The Shelduck's greatest increase has occurred in England and Wales, mainly through the colonisation of inland sites throughout England, where breeding was recorded in nearly 100 new 10-km squares. The large number of additional summering records may well indicate further scope for increase in breeding away from the coast.

The causes of changed distribution may have been an increase in total population size, leading to saturation of the original estuary and shore habitats. Shelducks defend their breeding territories and broods of ducklings very aggressively, and breeding success in crowded estuaries has been found to decrease sharply with increasing density (Patterson 1982). More dispersed populations have been shown to have significantly higher fledging success than crowded ones (Pienkowski and Evans 1982) and this

may have promoted a spread from crowded coastal areas to new inland sites. Locally, habitat changes such as the enrichment of lakes and pools by nutrient inflow may have made these sites more suitable for Shelducks.

In their spread inland, British Shelducks have become more similar to those in the eastern part of a range which extends from the west coast of Ireland to the western part of China. The eastern Shelducks breed mainly on inland freshwater or saline pools, often in fairly arid areas. But in spite of this difference in habitat, and a gap in SE Europe between the western and eastern parts of the range, there are no obvious differences in size or plumage between the birds from the two extremes (Patterson 1982).

Throughout Europe, Shelduck numbers have increased during the present century (*BWP*). This trend has been shared by the British population, which showed a steady increase from 1975 to an estimate of 42,000 birds prior to the breeding season (March) in 1986 (Owen *et al.* 1986). Extrapolation from tetrad counts during April to July in 1988–91 yields estimates of 44,200 Shelduck in Britain and 4,650 in Ireland. The two independent British estimates are remarkably similar. Assuming that rather

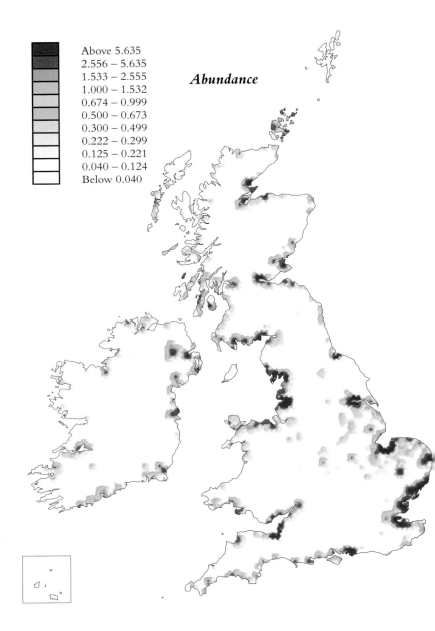

	Above 5.635
	2.556 – 5.635
	1.533 – 2.555
	1.000 – 1.532
	0.674 – 0.999
	0.500 – 0.673
	0.300 – 0.499
	0.222 – 0.299
	0.125 – 0.221
	0.040 – 0.124
	Below 0.040

Abundance

less than half of the Shelducks present in spring breed (Yarker and Atkinson-Willes 1972), there are estimated to be 10,600 breeding pairs of Shelduck in Britain and 1,100 in Ireland.

IAN J. PATTERSON

68–72 Atlas p 94

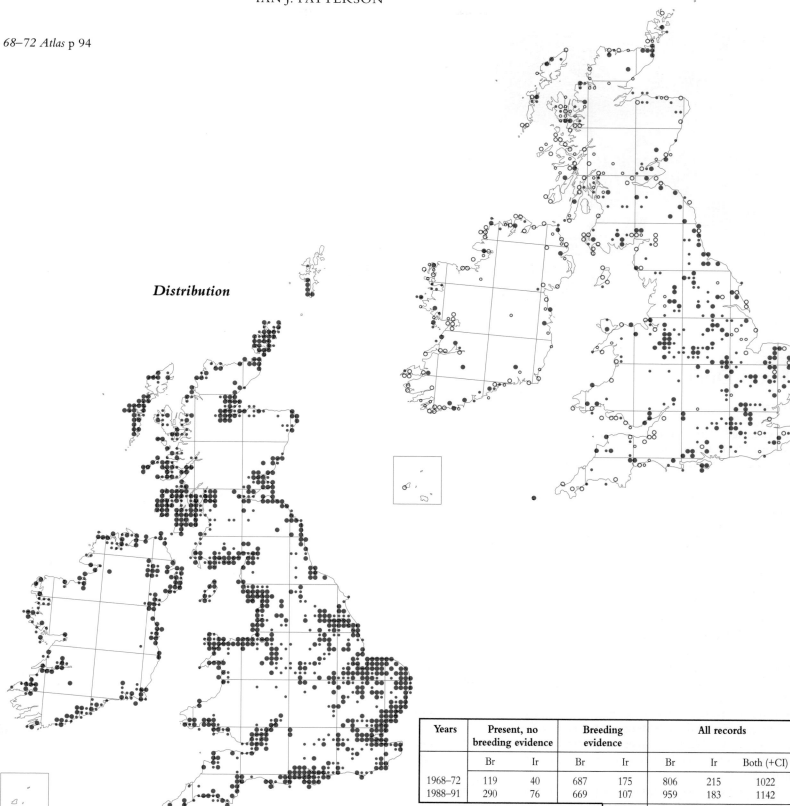

Distribution

Change

Years	Present, no breeding evidence		Breeding evidence		All records		
	Br	Ir	Br	Ir	Br	Ir	Both (+CI)
1968–72	119	40	687	175	806	215	1022
1988–91	290	76	669	107	959	183	1142
				% change	19.7	−14.9	12.3

Wood Duck

Aix sponsa

The plumage of the male Wood Duck (known to aviculturists as the Carolina Duck and in former times as the Summer Duck) is almost as spectacular as that of the only other member of the genus *Aix*, the Mandarin. This woodland, perching duck, a native of western and eastern, but not central, North America, has never been as successful a colonizer in Britain as has its relative.

Compared with that of the *68–72 Atlas*, the current Distribution map shows a slight extension of range from the species' previous main stronghold in Surrey and E Berkshire into Sussex, Kent and parts (mainly southern) of East Anglia.

The habits and habitat of the Wood Duck, its feeding, flight and nesting are all remarkably similar to those of the Mandarin. Why, then, has it failed to become as firmly established in Britain as its Oriental relation? The Wood Duck has a longer fledging period (ten weeks) than the Mandarin (eight weeks), and Savage (1952) suggested that its ducklings would thus be vulnerable for a longer period to avian and terrestrial predators, a view with which Lever (1977) and Kear (1990) concur. The *68–72 Atlas* points out that most Wood Duck breeding records have occurred in Surrey and E Berkshire, where the birds may come into competition for food and nesting sites with the well entrenched and much more numerous Mandarin; supplies of old decaying trees with suitable nesting holes have diminished in recent years due to 'improved' forestry methods.

Another divergence between the Wood Duck and the Mandarin is that, although closely related, they come from different continents and different temperate latitudes, the former occurring between 31°N and 51°N and

the latter between 36°N and 55°N. This variation in distribution, albeit small, is, as Murton and Kear (1978) point out, likely to affect laying date, clutch size, duckling weight, incubation and fledging periods, and the timing of moult and eclipse plumage. From records kept by the Wildfowl Trust (as it then was) at Slimbridge and Peakirk (respectively 52°N and 53°N), Murton and Kear found that Wood Ducks consistently lay earlier than Mandarins: over a 36 year period the earliest Wood Duck egg was on average laid on 20 March, whereas the first Mandarin egg appeared on average on 15 April. This three-week difference is confirmed by records of hatching dates kept by London Zoo during the nineteenth century. Earlier nesting, as Kear (1990) points out, evolved to suit the climate of the Wood Duck's area of origin, but in more northerly Britain means that the birds are incubating eggs and hatching ducklings when the temperature is still relatively low. Furthermore, Kear continues, Wood Duck eggs are on average 5 g lighter than those of Mandarins, as a result of which the former's ducklings hatch some 16% lighter than young Mandarins, and thus lose body heat more rapidly and require more brooding by the female. Finally, Wood Ducks that escape or are released into the wild in Britain are descended from many generations of captive-bred birds, and have thus lost much of their genetic viability.

According to Kear, the Wood Duck in Britain has three points potentially in its favour. Firstly, because it nests earlier than most other hole nesting species, it has an advantage in the competition for nest holes; but as Grey (1927) noted, British Mandarins sometimes 'dump' their own eggs in Wood Duck clutches, leaving the latter to rear the ducklings together with their own. Secondly, since both ducks cease laying around late May, the Wood Duck has a breeding season four weeks longer than the Mandarin. Thirdly, as the Wood Duck lays a slightly larger clutch (on average between 10 and 14) it can produce more, albeit lighter, ducklings. These implicit advantages, however, have so far failed to secure the species' firm establishment in Britain.

In the 19th century, the Wood Duck was almost exterminated in North America, due to loss of nesting sites as a result of widespread deforestation. Since then, however, it has more than recovered its former status, largely because of its ready adoption of artificial nesting places. Thus in England, where it was first introduced in the 1870s and where in Devon in 1895 it was described as 'breeding freely and wandering at will over the country', the population, such as it is, is of no conservation significance.

CHRISTOPHER LEVER

68–72 Atlas p 74

Change

Distribution

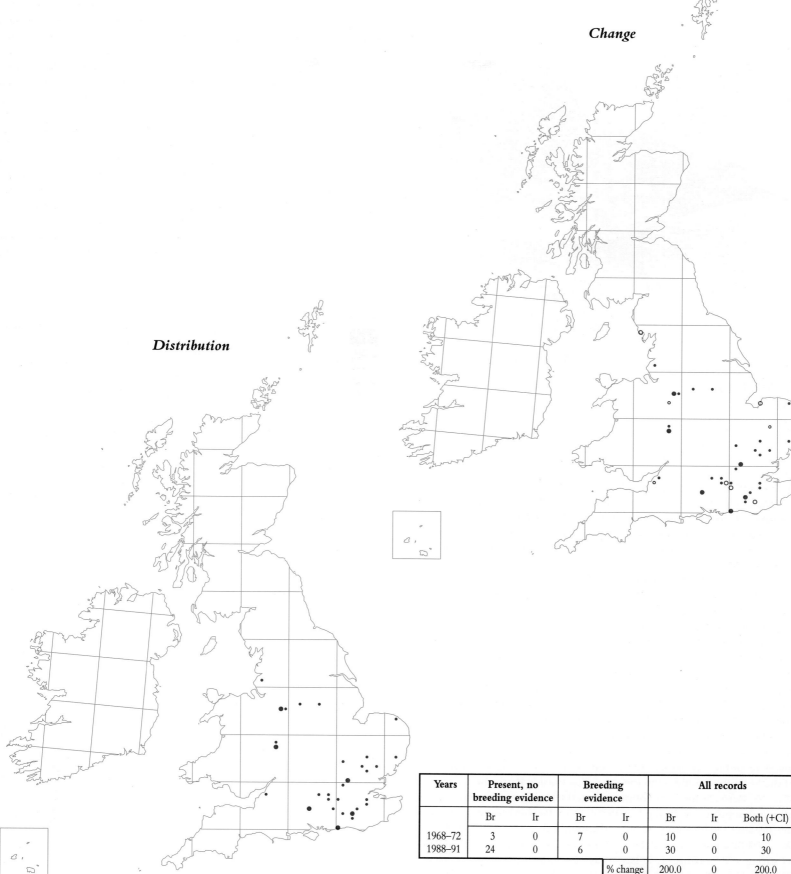

Years	Present, no breeding evidence		Breeding evidence		All records		
	Br	Ir	Br	Ir	Br	Ir	Both (+CI)
1968–72	3	0	7	0	10	0	10
1988–91	24	0	6	0	30	0	30
				% change	200.0	0	200.0

Mandarins normally nest in trees, especially oak, occupying holes up to 3m deep and to a height of 15m from the ground. Nesting boxes, e.g. in Windsor Great Park, are readily accepted, even where natural sites occur. The nest itself is a simple depression lined with down. The first eggs may be laid in late March, but the main laying period is from early April to the end of May. Incubation, by the female only, lasts 28–33 days, and the eggs usually hatch synchronously. The young leave the nest within 24 hours, by climbing the sides of the nest hole with their needle-sharp claws and then dropping to the ground. They may travel for several kilometres to find water and to feed. The survival rate of ducklings is in the range 30–40%.

The normal clutch is 9–12, but there is extreme variability due to 'dumping'. Davies and Baggott (1989a) found that the provision of nest boxes in their study area in Windsor Great Park neither affected the size of clutches laid in natural sites nor promoted 'dumping'. When both were available, the clutches in artificial sites were smaller than those in natural sites, and 25% or more of all clutches, regardless of size or site, were not incubated. According to Davies and Baggott (1989b) clutches laid in the

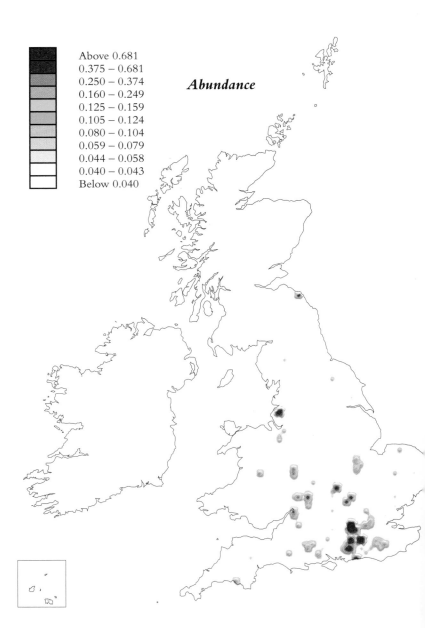

■	Above 0.681
▨	0.375 – 0.681
▨	0.250 – 0.374
▨	0.160 – 0.249
▨	0.125 – 0.159
▨	0.105 – 0.124
▨	0.080 – 0.104
▨	0.059 – 0.079
▨	0.044 – 0.058
▨	0.040 – 0.043
□	Below 0.040

Abundance

Mandarin

Aix galericulata

The breeding plumage of the Mandarin drake is by far the most spectacular of any British duck. Although this native of the Far East (principally Japan, China, Korea, and the former USSR) was first introduced to Britain before 1745, the naturalized population here is descended from escapes and deliberate releases from captivity since the early 20th century. Mandarins frequent flowing or standing fresh waters with a dense growth of marginal trees and shrubs (rhododendron is a favourite), especially where these overhang the water's edge, and where there is an abundance of reeds, sedges, or other emergent vegetation to provide shelter. The species was admitted to the British and Irish list in 1971.

The Distribution map is broadly similar to that in the *Winter Atlas*, except that in the latter a number of sub-populations were not recorded (e.g. Berwickshire). Comparison with the *68–72 Atlas* suggests that in the past two decades there has been a marked extension of the Mandarin's range. Davies (1988), however, believes there is little evidence of material change in the species' distribution during the last 60 years. Mandarins have probably always been far more numerous and widespread than previous accounts (e.g. Savage 1952 and Lever 1977) have suggested. This is almost certainly because the species' shy and secretive nature leads to under-recording, although a few new subpopulations, e.g. in Wales, have probably arisen as a result of escapes or releases from collections. However, such under-recording would have been common to all the atlas surveys. Two of the Scottish localities – on the River Tay outside Perth and on the Eye Water in Berwickshire – are of special interest, being so far removed from other subpopulations. In the late 1980s up to 100 birds were counted on the Tay after the breeding season, and around half that number on the Eye Water. Whilst the species is scattered over much of central and S England, its stronghold remains, as it has always been, in Buckinghamshire and Windsor Great Park on the Berkshire/Surrey border.

early part of the breeding season tend to be larger.

The principal reasons for the successful establishment of the Mandarin in Britain are, firstly, the existence of a vacant ecological niche for a hole nesting duck which feeds largely on aquatic invertebrates in spring and summer, and on acorns, chestnuts, and beechmast in autumn and winter; secondly, the availability of an abundance of suitable habitats; and, thirdly, because the founder stock was a particularly vigorous one which was derived directly from the wild. The Mandarin's breeding range and population will probably continue to expand and increase, but slowly, being limited by competition for nesting sites (mainly with Jackdaws and grey squirrels); by the continued availability of suitable habitats; by the species' relatively sedentary nature in Britain (in contrast with its migrations in the Far East); and by its reliance on an adequate supply of nuts (in many places supplemented by artificial feedings) to see it through hard winters. The population of wild Mandarins in Britain now almost certainly exceeds the whole of that in the Orient outside Japan (between 2,800 and 3,900 birds), and probably equals the total in that latter country.

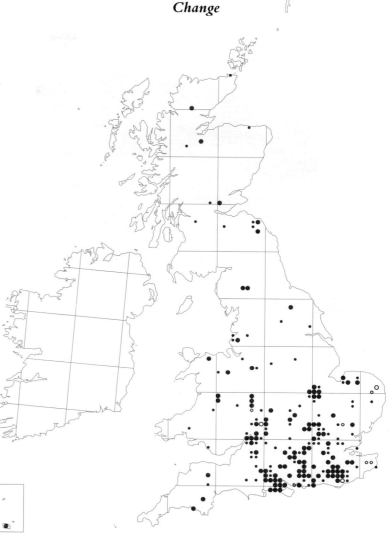

Change

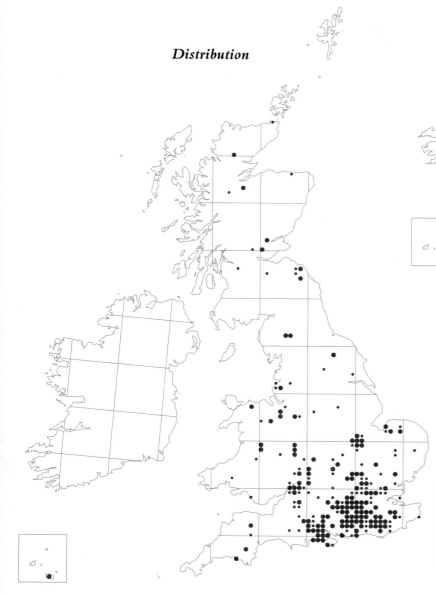

Distribution

Outside Britain, small naturalised populations of Mandarins occur in the USA (California) and possibly in Germany (Berlin) and the Netherlands. The *68–72 Atlas* estimated the total British population at 300–400 pairs. This was probably a considerable underestimate. According to A.K. Davies, the current total may exceed 7,000 individuals.

CHRISTOPHER LEVER

68–72 Atlas p 72

Years	Present, no breeding evidence		Breeding evidence		All records		
	Br	Ir	Br	Ir	Br	Ir	Both (+CI)
1968–72	11	0	28	0	39	0	39
1988–91	85	0	133	0	218	0	219
				% change	459.0	0	461.5

Wigeon

Anas penelope

An evocative whistling from the misty sea's edge indicates the presence of the Wigeon, one of our most familiar wintering ducks. Britain has a small breeding population, at the southernmost edge of the Wigeon's range. A dispersed breeder, this shy species occurs mainly along the shores of upland lakes and boglands, where it nests in marginal grass or shrub cover, or on islands. The species appears to have colonised Britain relatively recently; the first recorded nesting in Scotland was in 1834 and in England in 1897. There are two reports from Ireland, in 1933 and 1953. The present breeding range is centred around Tayside, the Grampians and the NW highlands of Scotland. The flow country of Caithness and Sutherland is also important, and there are other, smaller, populations in S Scotland and the uplands of N England. Further south, the Wigeon is present during the breeding season in many areas, particularly in the E Midlands of England and East Anglia, but rarely breeds. Some of these southerly occurrences are of injured stragglers from the winter, and some undoubtedly originate from captive-bred escapes or deliberate releases.

Although breeding south of the Humber is rare (recorded in eight 10-km squares in 1988–91, compared to eleven in 1968–72), Wigeon were present during the breeding season in a further 66 squares. The number of sightings in Ireland during the summer appears to be increasing (25 10-km squares during 1988–91) and the species may be breeding there unnoticed in some years.

The Wigeon is a grazing species in adult life, but its breeding season requirements are little understood. At Loch Leven, it nests with large numbers of other dabbling ducks and Tufted Ducks on the 42ha St Serf's Island in the loch. The Wigeon was the most dispersed of the four most numerous species nesting there between 1966 and 1973. In common with

other ducks on the loch, female Wigeons usually nested under tussocks of tufted grass *Deschampsia caespitosa* (Newton and Campbell 1975). The nest is a shallow depression or 'scrape', lined with vegetation and insulating down.

The selection of water body type for breeding was studied in a survey of the Caithness and Sutherland flows in 1988 (Fox *et al.* 1989). Water acidity was one of the most important controlling factors; Wigeon selected neutral or alkaline waters and almost totally avoided lakes where the pH was lower than 5.5. The species would, therefore, be one of the most vulnerable to the acidification of waters due to increasing afforestation and acid rain.

At Loch Leven, brood survival of Wigeon, as with other species, is low and this is attributed to the shortage in the loch of sheltered shorelines with abundant insects. Certainly, at the important breeding area around Lake Myvatn, Iceland, the abundance of Chironomid midges is a critical factor in determining duckling survival. The emerging midges are the most important food of adults and young during brood rearing (Bengtson 1972). In Caithness and Sutherland, broods of Wigeon, especially broods with well-grown young, were most commonly found along streams and at their convergence with lochs (Fox *et al.* 1989). At that stage adults and

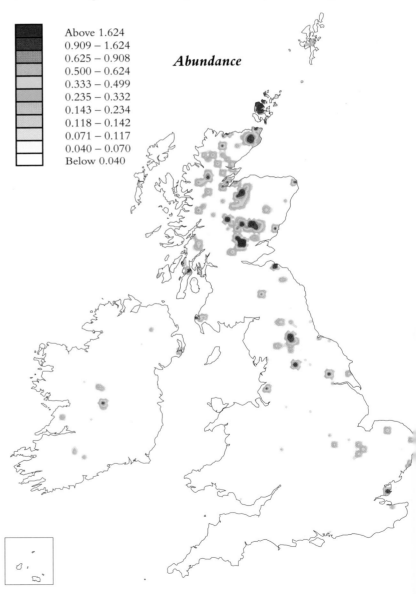

	Above 1.624
	0.909 – 1.624
	0.625 – 0.908
	0.500 – 0.624
	0.333 – 0.499
	0.235 – 0.332
	0.143 – 0.234
	0.118 – 0.142
	0.071 – 0.117
	0.040 – 0.070
	Below 0.040

Abundance

young probably graze marginal vegetation and browse underwater plants as well as feeding on insects.

Estimating the number of breeding pairs in Britain as a whole is fraught with difficulty. Using quantitative data from Wildfowl Trust surveys in the late 1960s, and some early data from the *68–72 Atlas*, Yarker and Atkinson-Willes (1971) estimated around 350 breeding pairs. The *68–72 Atlas* did not disagree with this, although it was less precise, estimating 300–500 pairs.

There are few quantitative data from more recent years, though an estimate of 30–39 pairs in 1978, and 12 broods in 1988, has been made for the Ettrick Forest in the Scottish Borders (Thomson and Dougall 1988), with 15 pairs in one area of the Grampians (R. Duncan). The 1988 survey of the Caithness and Sutherland flows yielded 18 broods. There were 107 other sightings in the area, however, and because of nest losses and the shyness of females and broods, the number of nesting pairs was thought to be very substantially greater than the number of broods seen (Fox *et al.* 1989). At Loch Leven numbers are down from the 25–30 pairs in the 1960s and 70s; between 1988 and 1991 there were, in chronological order 8, 6, 11 and 12 pairs (G. Wright).

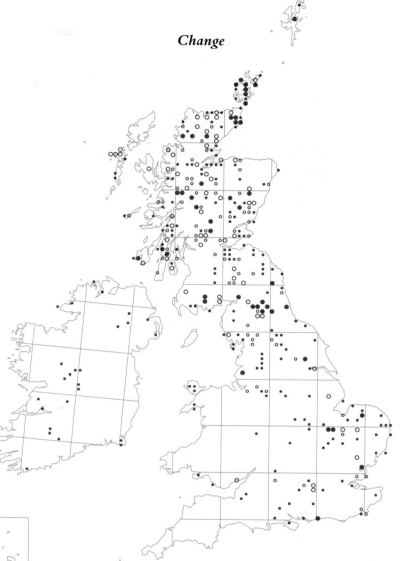

Change

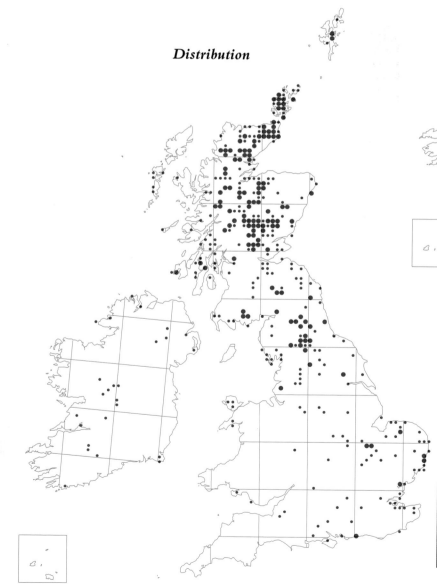

Distribution

Neither the number of 10-km squares where the species occurred, nor those where it is known to have bred during this survey, give grounds for changing the estimate of the breeding population in Britain. The previous figure of 300–500 pairs seems, therefore, still to be appropriate. This means that the number of Wigeon reared locally makes a trivial contribution to the wintering population originating from Iceland to central N Siberia, which is currently around 200,000–250,000 birds (Owen *et al.* 1986).

MYRFYN OWEN

68–72 Atlas p 66

Years	Present, no breeding evidence		Breeding evidence		All records		
	Br	Ir	Br	Ir	Br	Ir	Both (+CI)
1968–72	126	0	157	0	283	0	283
1988–91	232	25	128	0	360	25	385
			% change		27.2		36.0

to the time of the *68–72 Atlas*, had stimulated expansion to adjacent reservoirs and wetlands associated with mining subsidence.

Gadwalls are essentially vegetarian, foraging on plentiful submerged vegetation with generally lower levels of indigestible fibre than terrestrial plant material. For this reason, the species exploits the fertile nutrient-rich waters of lowland Britain where such food is in greater abundance. Even the ducklings become dependent on a diet of green material from a very early age and thus do not compete with ducklings of closely related species such as Mallard, which are largely insectivorous until the time of fledging (Sugden 1973).

The recent provision of large areas of lowland waters in the form of gravel pits and reservoirs, whose eutrophic waters support lush submergent vegetation, has created new opportunities for colonisation by the Gadwall. This increase in suitable wetlands has supported an upsurge in the breeding, passage and wintering numbers of this species (Fox 1988, Fox and Salmon 1989a) and, indeed, the present summer distribution closely resembles that of the *Winter Atlas*. The Gadwall is frequently regarded as a feral species and hence perhaps of limited conservation interest. There is no doubt that the original breeding stock, introduced to Norfolk during the

Gadwall

Anas strepera

In recent years the relatively drab plumage of the Gadwall has become better known to ornithologists as this diminutive duck has become more securely established as a breeding species.

The Gadwall's range has shown a dramatic expansion over the last 20 years, with new breeding records even in the traditional heartlands of East Anglia where the species was first introduced in the 1850s. Recent expansion into the Broads, for instance, may reflect eutrophication processes here. Numbers breeding in the Thames marshes have consolidated since the *68–72 Atlas*, but the species has spread into Sussex, Hampshire, Dorset and throughout the Midlands. Where suitable conditions exist, substantial populations have colonised Yorkshire, Lancashire and Anglesey. In Scotland, the species has shown limited expansion from its stronghold at Loch Leven to other areas in Central Region. The species also persists in areas on the Uists, Tiree and Orkney, where shallow eutrophic waters offer suitable habitat. An extensive survey of Caithness and Sutherland wetlands recently rediscovered the long established breeding population there. In Ireland, the population remains centred on Lough Neagh and in Co. Wexford, although Gadwalls now breed regularly on Belfast and Strangford Loughs (perhaps aided by liberated birds) and nesting has been reported from several other areas (*Birds in Ireland*).

Gadwalls are said to have bred on Tiree in the 19th century, while small numbers have certainly wintered there, on the Uists and on Orkney, for many years. These presumably originated from the Iceland breeding population, established in the early 19th century. The nucleus of the Scottish breeding population has, however, remained at Loch Leven, which was apparently first colonised in 1909. Between 25 and 40 pairs have bred there annually since that time. Allison *et al.* (1974) suggested that this species is very sensitive to autumn shooting pressure which, prior

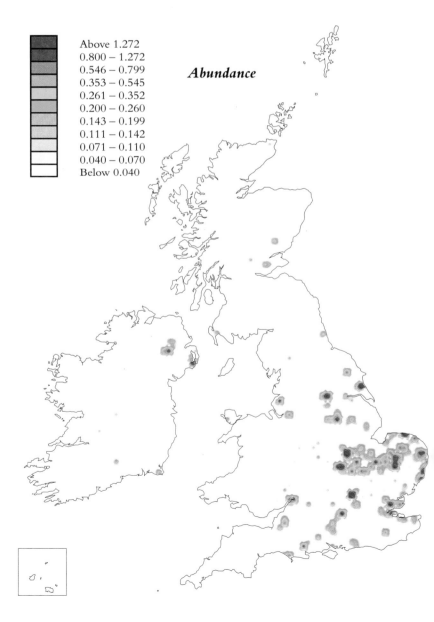

Above 1.272
0.800 – 1.272
0.546 – 0.799
0.353 – 0.545
0.261 – 0.352
0.200 – 0.260
0.143 – 0.199
0.111 – 0.142
0.071 – 0.110
0.040 – 0.070
Below 0.040

Abundance

middle of the last century, came from pinioned wild birds caught at the nearby Dersingham Decoy. Although much of the present population has descended from this source, we can never know what contribution has been made by the migrant and wintering population of genuinely wild birds induced to stay and breed here through pairing with resident birds, or settling to breed at suitable nesting habitat. We do know that a substantial proportion of birds ringed in Britain during the summer has been recovered further south, in France and Spain (Fox and Salmon 1989a).

Most artificial introductions of Gadwall in Britain took place before 1970, therefore much of the expansion over the last 20 years has been natural in the sense that colonisation has not been augmented by hand-reared birds. Since many areas of suitable habitat have yet to be colonised, there remains the opportunity for further expansion. The nature of the Gadwall's diet necessitates long periods of foraging, and consequently it is susceptible to disturbance (Mayhew 1988). This feature may affect its expansion of range as a breeding bird and may ultimately limit its future abundance as a nesting species in Britain and Ireland.

Extrapolation of a 4% per annum expansion in the breeding population of 580–590 pairs in Britain and 20 pairs in Ireland in 1983 (Fox 1988)

Change

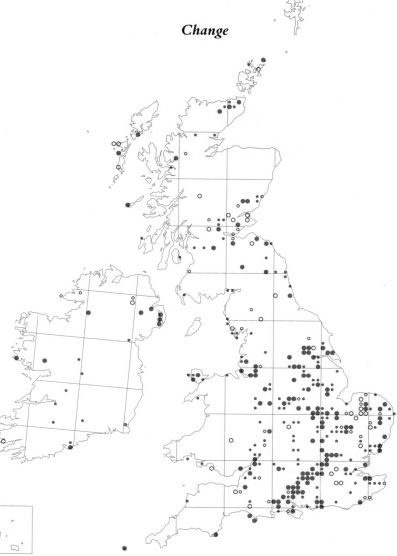

Distribution

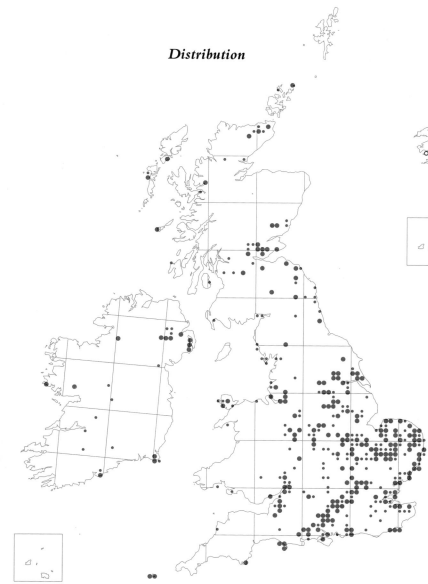

gives a population of 770 and 30 pairs respectively in 1990. Given *ca* 80 pairs on the Ouse Washes and 25–40 at Loch Leven this would seem a reasonable estimate of overall abundance. This represents 5–10% of the present population of the NW European flyway, as determined by Monval and Pirot (1989).

TONY FOX

68–72 Atlas p 64

Years	Present, no breeding evidence		Breeding evidence		All records		
	Br	Ir	Br	Ir	Br	Ir	Both (+CI)
1968–72	54	8	104	6	158	14	172
1988–91	165	12	192	13	357	25	382
			% change		125.9	78.6	122.1

Teal

Anas crecca

In Britain and Ireland, the shy and secretive Teal favours generally oligotrophic waters, in particular the moorland pools, bogs and patterned mires of northern and upland Britain (Fox *et al.* 1989). For this reason, the greatest numbers tend to nest in the north and west, although the species will also nest in lowland Britain where it adapts to a variety of well-vegetated wetland situations.

The *68–72 Atlas* described how a thinly-distributed species such as the Teal presents considerable problems in monitoring population trends. The species is not well represented on CBC or WBS plots (but see *Trends Guide*). For this reason, the results of the *88–91 Atlas* give us a first perspective on the distribution and abundance of the Teal since the *68–72 Atlas*. Although the core pattern of distribution has altered relatively little over the past 20 years, the 1988–91 fieldwork reveals an alarming contraction in range for the Teal.

Even in 1963, Atkinson-Willes was describing a decline in breeding Teal numbers in Aberdeenshire, Perthshire and Fife, while Parslow (1973) noted a reduction in parts of Galloway. The *68–72 Atlas* already charted dwindling populations in Dorset and Cornwall, counties which supported only a handful of breeding records during 1988–91. Similarly there have been substantial reductions in Hampshire, Sussex, Kent, throughout East Anglia and much of lowland Britain, while breeding declines in Scotland have been most apparent in coastal areas (Buckland *et al.* 1990). It might be expected that, in these areas, habitat loss or modification, pollution and increased human disturbance may have affected a shy species requiring densely vegetated wetlands. The picture is similar in upland Britain, however, with fewer breeding records and, in many areas, a complete absence of records from sites previously supporting breeding populations.

The distribution has been maintained in places such as Islay, in contrast to adjacent Kintyre where afforestation has had a considerable impact. In Galloway, extensive afforestation began in the 1940s, and Teal numbers have been declining since then. By contrast, Caithness and Sutherland, Inverness District and parts of the Borders have only recently started to be afforested, and to date appear to have sustained their breeding populations of Teal. Although conifer planting and consequent acid deposition could create conditions which are generally more favourable to Teal, afforestation also adversely affects sediment loads of rivers and is associated, at least in the early stages, with applications of inorganic fertiliser. The developing tree cover can also affect the abundance and diversity of predators, and hence directly affect breeding success and adult survival. Our understanding of the effects of these factors is rudimentary, and our

duck species have rarely been studied in response to such radical changes in land use.

In Ireland, the pattern is very much the same, with a dramatic contraction in range, fragmentation of breeding populations and a drop in breeding records.

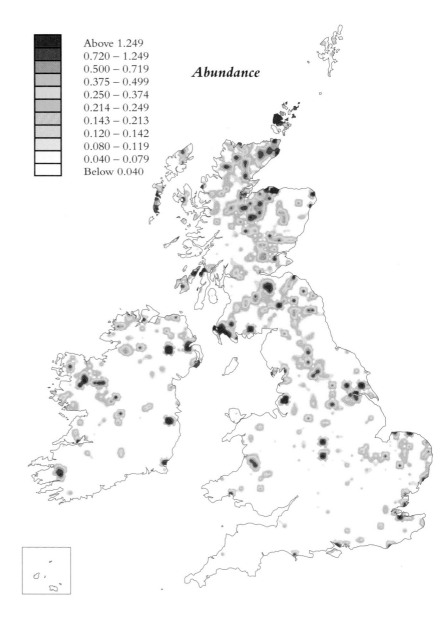

Abundance

| Above 1.249 |
| 0.720 – 1.249 |
| 0.500 – 0.719 |
| 0.375 – 0.499 |
| 0.250 – 0.374 |
| 0.214 – 0.249 |
| 0.143 – 0.213 |
| 0.120 – 0.142 |
| 0.080 – 0.119 |
| 0.040 – 0.079 |
| Below 0.040 |

Teal tend to nest closer to water than other duck species, which reduces predation risks, and, amongst studied populations, breeding success can be moderate to high (Fox 1986). The same study showed that the experimental creation of open water in an acidic peatland habitat could increase the number of Teal breeding successfully. Against this, as an important quarry species, the Teal has very low survival rates (Gitay *et al.* 1990) and is highly susceptible to winter cold weather (Ridgill and Fox 1990). Hence relatively small reductions in breeding success, caused by habitat change or disturbance, could have a profound effect on the population dynamics of this attractive little duck.

Teal breed throughout the middle latitudes of the entire W Palearctic. The NW European population originates from Iceland, Fennoscandia, the Baltic states, Russia, N Poland and Germany, and number an estimated 400,000 birds. Based on winter counts, the population has shown a very slight growth throughout the flyway (Monval and Pirot 1989).

If we (optimistically) assume that the range reduction of Teal nesting in Britain has not been accompanied by a decrease in nesting density, we can use the *68–72 Atlas* assumption of 3–5 pairs per 10-km square with breeding evidence to obtain current estimates of 1,500–2,600 and 400–675 pairs breeding in Britain and Ireland respectively, compared with the

Change

Distribution

3,500–6,000 for Britain and Ireland, combined, given in the *68–72 Atlas*. In practice, it is unlikely that such a range reduction has not been accompanied by declines in breeding density, and therefore even a conservative assessment of a reduction of the population by one third over twenty years may underestimate the real loss.

This picture gives considerable cause for concern since, assuming the estimates to be correct, Britain and Ireland have gone from holding 1.75–3% of the nesting Teal of the flyway population to 0.95–1.65% in just 20 years.

TONY FOX

68–72 Atlas p 60

Years	Present, no breeding evidence		Breeding evidence		All records		
	Br	Ir	Br	Ir	Br	Ir	Both (+CI)
1968–72	438	126	938	263	1376	389	1765
1988–91	576	114	571	74	1147	188	1335
			% change		−16.6	−51.5	−24.3

Mallard

Anas platyrhynchos

The stunning green head of the drake Mallard in spring, and the quack of the duck, are familiar to the countless thousands who take bread to the water's edge; and many Mallards, for their part, have developed an easy relationship with man. Few waters are too insignificant to provide a breeding site for Mallards, and their adaptability enables them to breed, when necessary, some distance from water, and to take rapid advantage of newly created habitats.

As a result of this adaptability the Mallard is widely distributed throughout Britain and Ireland. The Distribution map shows it to be absent only from the 10-km squares of high moorland and mountain: the highest parts of the Cairngorms, parts of central Scotland, Exmoor and Dartmoor. These absences are presumably because the waters there are too oligotrophic. The Mallard's preference for lowland farmland and river valleys is illustrated by the Abundance map. Although it is generally present throughout Wales and the southwest, it is relatively less abundant than in the lowland areas of rich arable farmland such as lie around the Wash, in East Anglia, Essex and the E Midlands. Concentrations in Lancashire and SE Scotland also indicate its preference for productive land. The pattern of distribution has altered little during the period since the *68–72 Atlas*, but some fragmentation may be detectable, particularly in Ireland. Although the number of farm ponds has been reduced during the period, there has been a significant increase in the number of man-made waterbodies such as reservoirs and gravel pits. This has probably resulted in there being little change in distribution.

Mallards nest most successfully on islands within water-bodies since they are particularly susceptible to clutch predation, and there is some evidence that greater predation occurs when nests are at high densities

(Hill 1984a). Although nests are usually well dispersed, high densities do occur on some sites such as St Serf's island at Loch Leven in Scotland (Newton and Campbell 1975). Predation has also been found to be higher on nests placed in short vegetation. A height of 1m provides the best nesting cover (Hill 1984b). Egg laying occurs from mid March (or even mid February in the south) through to the end of July, although young broods have been recorded in November.

Studies have shown that following years of high production of young, a greater proportion of birds is unable to establish 'territories' the following spring. This 'overwinter loss' operates in a density dependent manner, causing some degree of regulation of Mallard numbers on individual sites (Hill 1984c). The length of shoreline relative to the size of a water-body appears to determine breeding density. This offers scope for managing sites for wildfowl generally, and Mallards will often benefit by the creation of shallows which provide brood-rearing areas in the summer. Young Mallard ducklings feed on emerging aquatic invertebrates, and studies have shown that survival of young during the first two weeks of life is the key factor affecting autumn numbers (Hill 1984c). Where these insects are scarce, the ducklings have to range over larger areas and suffer greater mortality as a consequence (Hill *et al.* 1987). Competition with coarse

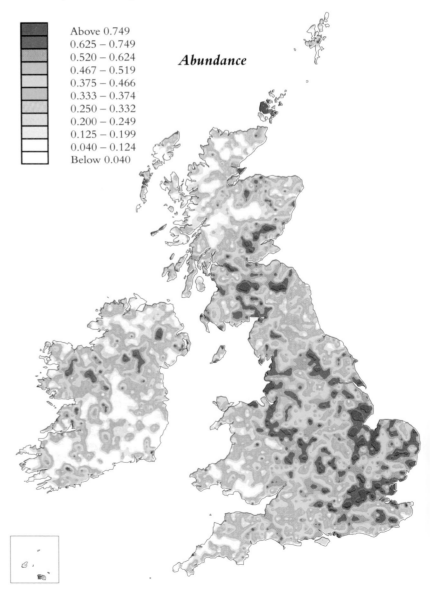

Abundance

| Above 0.749 |
| 0.625 – 0.749 |
| 0.520 – 0.624 |
| 0.467 – 0.519 |
| 0.375 – 0.466 |
| 0.333 – 0.374 |
| 0.250 – 0.332 |
| 0.200 – 0.249 |
| 0.125 – 0.199 |
| 0.040 – 0.124 |
| Below 0.040 |

fish for invertebrates has been shown experimentally to be a further factor affecting food supplies of ducklings, particularly on ecologically immature waters as exemplified by gravel pits (Hill *et al.* 1987).

Autumn numbers are augmented by the release of many birds artificially reared for shooting purposes. Those that survive to the following spring will undoubtedly add to the apparent distribution of the species. The *Trends Guide*, from farmland CBC plots, shows that the breeding population has been increasing since the early 1960s. Owen *et al.* (1986) derive a figure of about 100,000 pairs, based on a post-breeding population of 500,000 birds, with each breeding pair producing an average of 1.8 young. The Irish population is estimated at 23,000 pairs by extrapolation from the British figure.

DAVID HILL

68–72 Atlas p 58

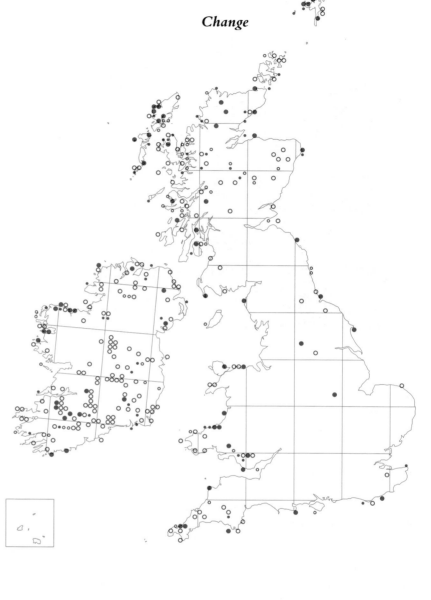

Change

Distribution

Years	Present, no breeding evidence		Breeding evidence		All records		
	Br	Ir	Br	Ir	Br	Ir	Both (+CI)
1968–72	142	57	2488	858	2630	915	3548
1988–91	199	202	2397	629	2596	831	3437
			% change		−1.2	−9.1	−3.2

Pintail

Anas acuta

The elegant Pintail is more a summer bird of the N of Europe and the Soviet Union than of the Continental oceanic western fringe. It is hardly surprising therefore that it remains a rare and local breeding species in our islands. Pintails prefer shallow unimpeded waters adjacent to grassland or other open habitat. These habitats tend to be those at greatest risk from modification, water level fluctuations (i.e. drought or floods) and the trend towards the loss of temporary habitats. In North America, habitat loss and prolonged drought have severely curtailed Pintail reproduction since 1980, contributing to a serious decline in the population there (Bartonek 1989). Shifts in distribution are therefore common and it is perhaps not surprising that the species is considered a sporadic breeder in many parts of Britain.

Breeding was first recorded in 1869 in Inverness-shire, and the species then spread to Orkney and the Shetland Isles. Subsequently, Kent was colonised in 1910 and records suggest that numbers of breeding pairs continued to increase during the early part of this century. By the 1930s breeding was reported widely throughout Scotland, although numbers were seemingly not very large. Breeding was never more than irregular in NW England, East Anglia and Kent and this was the case up to the years of the *68–72 Atlas*. By that time, the species was a rare breeder on the Scottish mainland away from Inverness-shire and Caithness.

The species summers regularly on the Suffolk coast, in Dorset, Cheshire, Angus and elsewhere, but the most frequent reports of breeding come from several sites in Orkney, where up to eleven broods have been recorded since 1985 (RBBP), and Tiree, where up to six pairs have bred over the same time period (Stroud 1989) although none during 1988–91. Up to 46 pairs of Pintails displaying on the Ouse Washes in spring 1988 dem-

onstrate the potential of this important site and the nearby Nene Washes to support a substantial number of nesting pairs (Cambridgeshire Bird Reports). Unfortunately, water levels are rarely managed in a manner suitable for the species and subsequent breeding involves relatively few pairs, although they are successful in most years. In Dumfries and Galloway, the species was a regular breeder in the late 1970s and early 1980s, and breeding was again recorded during the *88–91 Atlas* period. The species has bred in N Kent and Cumbria in the past, although proof of breeding has not been as regularly obtained in recent years.

In Ireland, the Pintail was a regular breeder at Loughs Beg/Neagh during the years 1917–38, but has been only sporadic since. There were only two records of breeding in the period between the fieldwork for the two Atlases (*Birds in Ireland*) and it is clear the species has not consolidated its breeding status there.

As well as being the source of nesting records in the Solway area (*68–72 Atlas*), feral breeding pairs have recently raised young in Lancashire (1983), Cornwall (1985), Sussex (1987) and Essex (1989), so it might be expected that this element may make an increasing but unquantifiable contribution to the population (information from local bird reports).

Parslow (1973) considered the British population to exceed 100 breeding pairs in some years, while Atkinson-Willes (1970) and the *68–72 Atlas* estimated 50 pairs. The latter estimate was based on evidence of breeding (probable or confirmed) from 46 10-km squares (including Ireland). Although the Pintail is renowned as a sporadic nesting species, records of summering and breeding birds do seem to be slightly reduced since the *68–72 Atlas*. It would appear that the total British breeding population is now unlikely to exceed 30–40 pairs, even allowing for missed and unreported instances, making this one of the rarest dabbling ducks breeding with us. There was only a single breeding record in Ireland (from Strangford Lough) during 1988–91. In terms of a flyway population of 70,000 birds, our breeding population is of little international significance. As something of an opportunist nester, the Pintail presents particularly difficult conservation problems. It probably bred more commonly earlier this century than currently, but there is little evidence for any specific factors that were instrumental in the decline. Adequate protection of the present regular breeding localities is fundamental, and enhancement of sympathetic management at summering areas may further encourage settlement.

TONY FOX

68–72 Atlas p 68

Two-year averages for Pintail nesting in Britain 1973–90

	73–74	75–76	77–78	79–80	81–82	83–84	85–86	87–88	89–90
Number of localities	5	11	13	17	17	19	16	18	18
Confirmed	7	9	9	10	8	9	8	11	10
Possible	3	12	16	24	24	17	12	15	28

Based on RBBP reports: the lower two lines relate to number of pairs breeding

Change

Distribution

Years	Present, no breeding evidence		Breeding evidence		All records		
	Br	Ir	Br	Ir	Br	Ir	Both (+CI)
1968–72	42	10	44	2	86	12	98
1988–91	68	8	17	1	85	9	94
			% change		−1.2	−25.0	−4.1

Garganey

Anas querquedula

This small, slightly built, and slender-necked dabbling duck has the distinction of being the only wildfowl species which is a summer visitor to Britain and Ireland. The sighting of a Garganey in early spring is therefore a welcome reminder of the delights of summer birdwatching to come. The drake is particularly handsome, displaying a broad white supercilium that curves from just in front of and above the eye to well down the nape, and contrasts sharply with brown breast feathers and pale grey flanks. Though relatively quiet and unobtrusive, the dry crackling or rattling call is often the first indication of the presence of a Garganey in suitable breeding habitat.

Garganeys arrive on the breeding grounds from late March onwards and breed from late April to June. They prefer rushy marshlands, wet meadows or rough grasslands intersected with ditches or dykes: a proliferation of emergent vegetation in which to feed and nest is much preferred. The species is highly secretive and breeding pairs usually remain concealed amidst vegetation. The difficulty of censusing is highlighted by the relatively small proportion of records received that involved birds known to be breeding.

In Britain and Ireland the Garganey is on the western edge of its breeding range, which extends eastwards throughout Eurasia between 40°N and 60°N, and north to 65°N in the Gulf of Bothnia (Owen *et al.* 1986). The Distribution map reveals a very scattered distribution, as is characteristic of a species which wanders widely and breeds opportunistically at sites which may be only temporarily suitable. The map certainly exaggerates the picture for any single year, as it includes records from some regularly used sites, such as the Ouse Washes (Cambridgeshire/Norfolk), and others from sites used perhaps only once.

Garganeys are much more common in England than elsewhere in Britain or Ireland. Most occur in central and southeastern areas; the majority of confirmed breeders were located in the eastern half of the country from North Yorkshire south through Lincolnshire to East Anglia and the southeast. Their principal site has long been the Ouse Washes and the Distribution map shows a marked concentration of records in this area. Numbers there are thought to have declined: there were totals of 25–35 pairs in 1952, 23–24 in 1968–69, 6–10 during the 1970s and 4–15 pairs during the 1980s (C. Carson, D. Revett). In Ireland, the species is a scarce annual visitor and breeding is very rare; between 1966 and the *88–91 Atlas* period only once was a pair thought to have attempted to breed (*Birds in Ireland*). Garganeys are also scarce but regular in Scotland and Wales, where they occasionally breed (*Birds in Scotland*); the Distribution map shows evidence of breeding at five Scottish sites in 1988–91.

The Change map reveals few obvious differences, except that there appear to be more records from Scotland now but fewer from the S Midlands and S England, including the southwest.

The habitat favoured by the Garganey has been much reduced by drainage and other agricultural improvements, probably to the detriment of this species. This has almost certainly resulted in a reduction of breeding numbers in some parts of Britain, most conspicuously in the fens of E England. The breeding population was estimated at 50–70 pairs for the *68–72 Atlas*, and records submitted to the RBBP since 1980 indicate that numbers fluctuate, generally in the range of 40–60 pairs, but rising to 100 pairs in occasional years. The only Irish breeding record during 1988–91 was of a family party with 4–5 juveniles at Tacumshin in Co. Wexford (J.P. Hills). Pirot *et al.* (1989) place the number wintering in W Africa, which will include British breeders, at around 2 million birds.

JEFF KIRBY

68–72 Atlas p 62

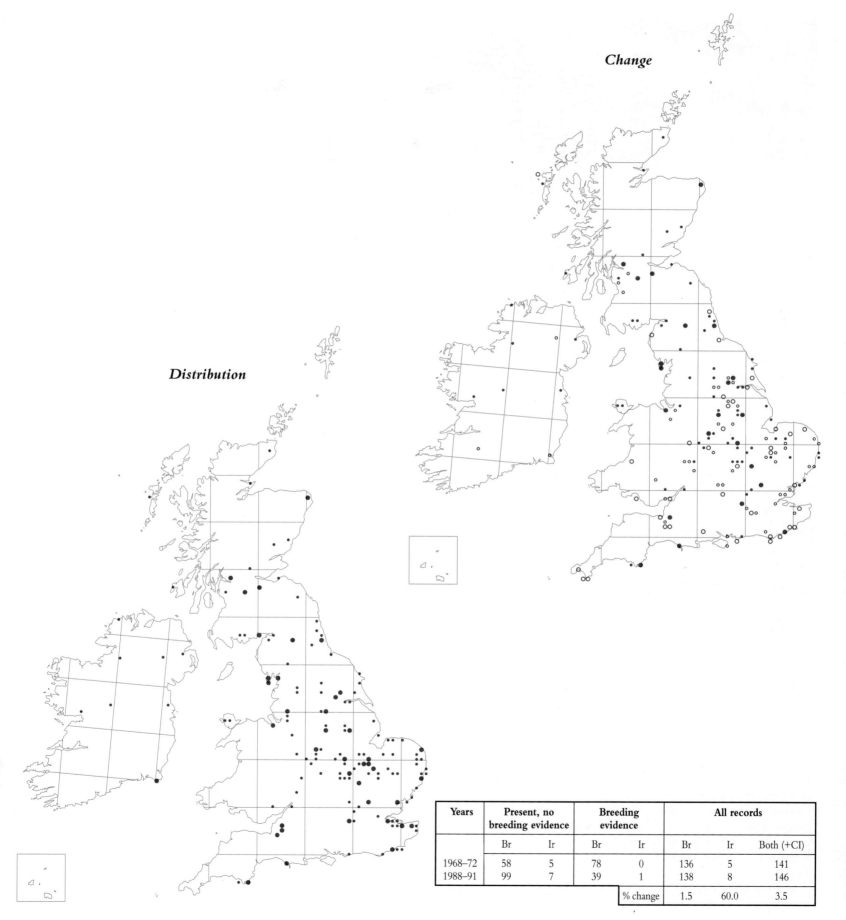

Change

Distribution

Years	Present, no breeding evidence		Breeding evidence		All records		
	Br	Ir	Br	Ir	Br	Ir	Both (+CI)
1968–72	58	5	78	0	136	5	141
1988–91	99	7	39	1	138	8	146
				% change	1.5	60.0	3.5

Shoveler

Anas clypeata

Tucked away at the back of a lowland lake or reservoir, Shovelers may go unnoticed during the late spring months. The normally distinctive chestnut, white and green of the drake will then be attaining its eclipse plumage and the female may well be sitting on eggs, concealed by tussocks of grass or sedge. Although a common duck, the habits and movements of this species have only recently become better understood.

The Distribution and Abundance maps illustrate the Shovelers' dependence on shallow, eutrophic waters. The preferred breeding habitat is usually marshland or rough pasture adjacent to shallow open water, and the species is particularly scarce in upland areas of Britain and Ireland and west of a line between Kent and Gwynedd. The eastern counties of England through to areas of the West Midlands remain strongholds. In Scotland, lowland sites between the Forth and the Grampians hold the bulk of the Scottish population, and base-rich lochs of the Uists, Orkney, Coll and Tiree provide excellent vegetation cover at the edges for nesting and a plentiful food supply for brood rearing. In Ireland, Shovelers are rare breeding birds with the records coming predominantly from Lough Neagh and the Shannon callows but with a thin scattering in the north and west.

Superficially the distribution appears to have changed little since the *68–72 Atlas*, but closer examination reveals a contraction of key areas – overall a 12% reduction in the number of 10-km squares in which the birds were recorded and a 39% reduction in the number of squares from which breeding evidence was obtained.

The provision of an effective network of protected areas and refuges and their sensitive management has led to pronounced local increases in numbers of breeding pairs. Numbers on the Ouse Washes NNR, Norfolk, for example, have increased from 20–50 pairs in the 1950s to 150–250 pairs in the 1970s (C. Carson, RSPB). Similarly, at the Lower Derwent Valley NNR, Humberside, Shovelers bred less frequently than annually in the 1960s but since the site was afforded protection in 1971 numbers have steadily increased to 43–63 pairs in the 1980s (T. Dixon, EN). It is currently estimated that about 500 pairs of Shovelers breed within the proposed Specially Protected Area network of sites. In Scotland the most important breeding waters already have reserve status – for example, Kinordy Loch NNR holds 25–35 pairs and Loch Leven NNR about 10–15 pairs.

Unfortunately, the very nature of the marshy, shallow freshwater sites upon which Shovelers depend, means that unprotected wetlands are vul-

nerable to drainage and changes in land use, particularly breeding areas near the water's edge. Whilst key areas have been afforded reserve status, it is the small marginal wetlands that have been lost as breeding sites. Changes in agricultural practices have affected lake edges and water quality. Even at Abberton Reservoir, a RAMSAR site in Essex, a decline from 10–15 pairs in the late 1960s to 1–2 pairs in the late 1980s can be attributed to the cutting of silage right up to the reservoir edge, leaving little rough pasture for nesting (R. King, WWT). Lowland sites, particularly those associated with the Washes of Cambridgeshire and Norfolk, are also prone to spring flooding: ten out of the last sixteen breeding seasons have been 'flood years' on the Ouse Washes, resulting in a fall in the number of annual successful breeding attempts, often to below 100 pairs.

Shovelers nest on the ground, making a simple hollow lined with grasses and down where typically 8–12 creamy eggs are laid from late April and are incubated by the hen for 22–23 days. The species is unusual among dabbling ducks in being territorial. In France, Shovelers benefit from the alertness and aggressiveness of breeding waders towards predators (Ibanez and Trolliet 1990) and the same may be true in this country. Shovelers occurring in Britain in winter are often not those that have bred here. Up to half of our breeding birds leave Britain and head south to winter in

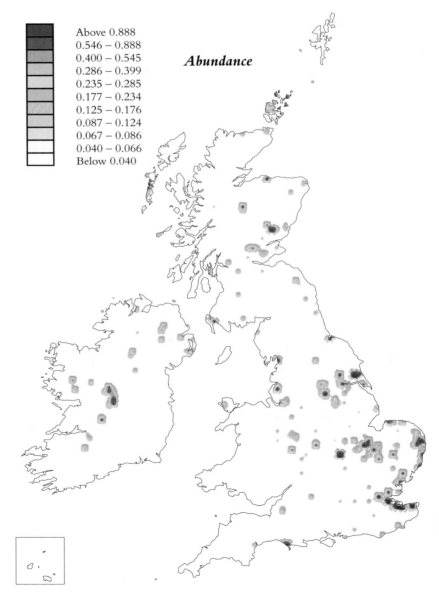

Above 0.888
0.546 – 0.888
0.400 – 0.545
0.286 – 0.399
0.235 – 0.285
0.177 – 0.234
0.125 – 0.176
0.087 – 0.124
0.067 – 0.086
0.040 – 0.066
Below 0.040

Abundance

France and the Iberian peninsula. These are mostly replaced by immigrants from NW Continental Europe and Russia (Mitchell 1990).

Throughout their NW European range, numbers of Shovelers have recently shown a period of decline after steady growth during the 1960s and 1970s (Monval and Pirot 1989; Kirby and Mitchell 1993). Breeding attempts in Britain and Ireland may now be reflecting this trend. It is evident in Britain that whilst numbers have increased on protected refuges, breeding at some other sites has declined. During the late 19th and early 20th centuries the Shoveler increased as a breeding species, not only in Britain but also over much of W Europe, and by the 1960s Britain held perhaps 500 pairs (Yarker and Atkinson-Willes 1972). The *68–72 Atlas* confirmed the distribution but estimated the breeding population at 1,000 pairs. Recently the National Wildfowl Counts show a September average of 6,000–7,000 birds (Kirby and Mitchell 1993) indicating that the breeding population estimate for Britain is still as low as 1,000–1,500 pairs. In Ireland the breeding population is probably still no more than 100 pairs.

CARL MITCHELL

68–72 Atlas p 70

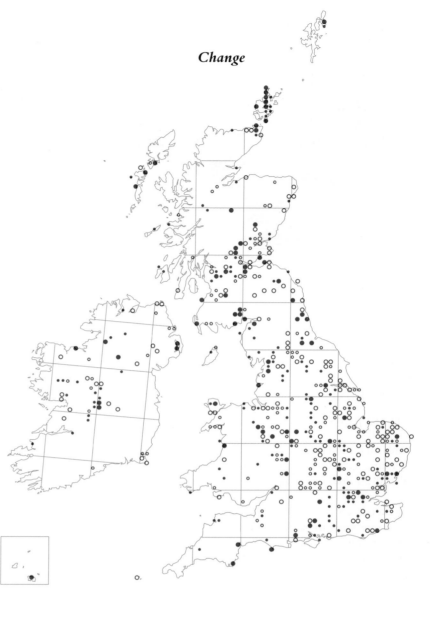

Change

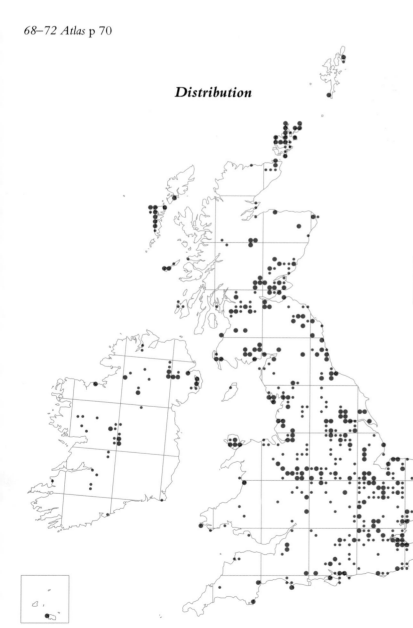

Distribution

Years	Present, no breeding evidence		Breeding evidence		All records		
	Br	Ir	Br	Ir	Br	Ir	Both (+CI)
1968–72	178	19	342	32	520	51	571
1988–91	237	32	217	13	454	45	500
			% change		−12.7	−11.8	−12.4

Itchen valleys in Hampshire, the breeding populations were undoubtedly augmented by feral birds. The Ouse and Nene Washes can hold important breeding populations when the water levels are suitable.

In the W of Britain the species may winter on relatively oligotrophic waters, but the breeding pairs are generally confined to suitably eutrophic waters such as those of Anglesey, Witchcett Pool (Dyfed) and Leighton Moss (Lancashire).

In Scotland, the pattern is similar, with the species restricted to productive lowland lochs, and the concentration of 30–40 pairs at sites in Perthshire, Angus and Fife forms the majority of the breeding population there. Other regular breeding outposts again represent 'oases' of suitable eutrophic wetland habitat in areas where generally such waters are scarce (e.g. Sand Loch in Grampian, and lowland lochs in Caithness). Most authors suggest a decline in breeding numbers during this century and this certainly seems to be the case in Dumfriesshire (D. Watson).

In Ireland, the Pochard is a scarce, but increasing, breeding bird, with regular breeding confined to Lough Neagh, Roscommon and Cork (*Birds in Ireland*).

Pochard

Aythya ferina

Although the Pochard became established as a breeding species early last century, it has not expanded rapidly within Britain and Ireland in the way that the Tufted Duck has. Indeed in some counties the discovery of more than a single brood at one site represents a significant find.

Breeding Pochards are largely restricted to the E of England and Scotland, confined to lowland waters rich in emergent and submerged vegetation. Little further study of the breeding biology of the Pochard has occurred since the work of Hori (1966), but the species does seem susceptible to disturbance and everywhere suffers low breeding success (see Fox 1991). Despite occasional single-year breeding records, the species appears relatively loyal to its more important breeding sites within Britain and Ireland.

Changes in the distribution and abundance of the species are hard to assess; observer effort for the *68–72 Atlas* located more of the breeding population in counties such as Norfolk and Essex, than in the years before or since (Fox 1991). The largest populations are thought to breed in the Breckland meres and Broads of Norfolk (Owen *et al.* 1986), although it is considered that this population may have declined in very recent years (M. Seago). The Essex coastal populations remain of considerable significance, and some of them now enjoy reserve protection. The N Kent marshes were colonised only early this century, but the population greatly expanded subsequently, there and in the Stour valley. The 'fleet' habitats adjacent to the Thames estuary and associated waters, are important. The waters are often brackish, generally eutrophic and densely vegetated, and in relatively undisturbed situations. They have, however, suffered from fertiliser, herbicide and pesticide run-off in areas of intensive agriculture, and the birds occupying them remain vulnerable to such pollution.

The species has colonised gravel pits and reservoirs throughout Britain: indeed, an important population relies entirely on such habitats in the Lea valley, north of London. In the central London parks, and on the Test and

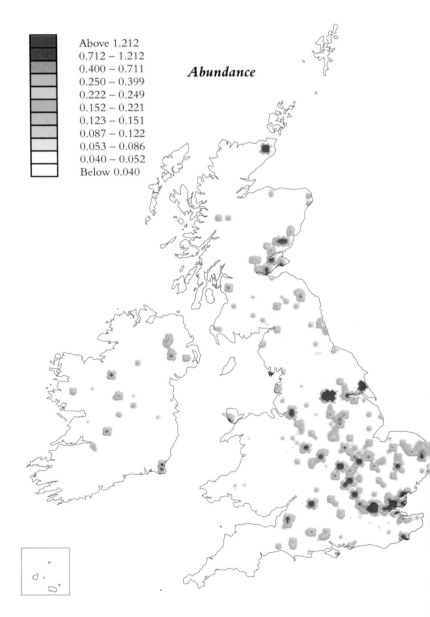

Abundance

Above 1.212
0.712 – 1.212
0.400 – 0.711
0.250 – 0.399
0.222 – 0.249
0.152 – 0.221
0.123 – 0.151
0.087 – 0.122
0.053 – 0.086
0.040 – 0.052
Below 0.040

Yarker and Atkinson-Willes (1972) estimated the British breeding population at around 200 pairs in the late 1960s, and the *68–72 Atlas* put the British and Irish population in the range of 200–400 pairs. A 1986 estimate, based on an extensive review of local bird reports, put the British population at 370–395 pairs, whilst the RBBP report for 1989 reported a maximum of 336 breeding pairs. Numbers in Norfolk may have declined a little in recent years, but in Yorkshire they continue to increase. Thus, although there were fewer reports of breeding in 1988–91 than in 1968–72, it seems likely that the British breeding population is about 400 pairs. With an estimated 20 pairs breeding on the Irish stronghold of Lough Neagh, the population in Ireland may be up to 30 pairs.

The wintering population of Pochards increased dramatically in Britain until the mid 1970s, peaking at about 50,000 or 10% of the NW European population (Fox and Salmon 1989b). Since that time, however, numbers have declined here and throughout the W Palearctic (Monval and Pirot 1989). With a flyway population estimated at 350,000, the small breeding population of Britain and Ireland is insignificant in NW Europe. Even so, its dispersed nature, susceptibility to change in water quality and dis-

Change

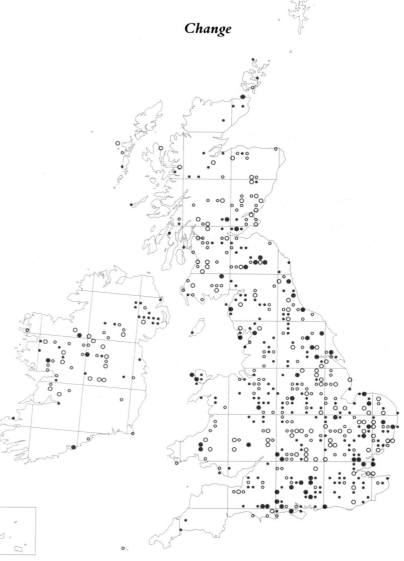

Distribution

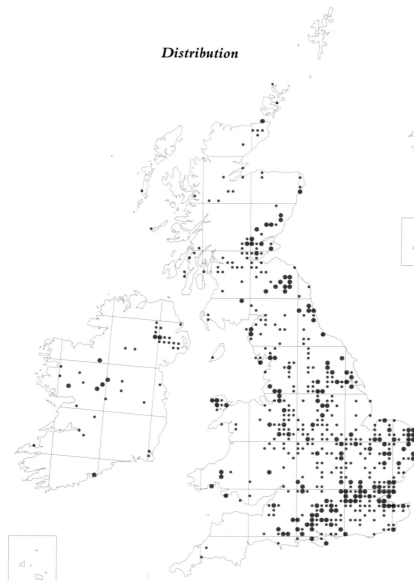

turbance, and the difficulty of effective monitoring give cause for concern, and there is no doubt that the population could be enhanced given sympathetic management.

TONY FOX

68–72 Atlas p 80

Years	Present, no breeding evidence		Breeding evidence		All records		
	Br	Ir	Br	Ir	Br	Ir	Both (+CI)
1968–72	275	31	237	20	512	51	563
1988–91	349	32	162	8	511	40	551
			% change		−0.2	−21.6	−2.1

Tufted Duck

Aythya fuligula

The male Tufted Duck has a spectacularly contrasting black and white plumage, with a black 'pigtail' or head tuft of feathers from which its name is derived. It is the commonest of the diving ducks breeding in Britain and Ireland, and has increased dramatically since the first reported breeding in 1849. It inhabits most types of inland water, from reservoirs and natural lakes to large and medium-sized sluggish rivers.

The Tufted Duck is a relatively widespread species, as shown by the Distribution map. It is essentially a lowland species, being absent from areas more than 400m above sea level, e.g. N, central and S Wales, the Pennines, and large areas of N and W Scotland. An interesting feature is its virtual absence as a breeding species in Ireland to the south of a line running from Dundalk Bay to Cape Clear. The same pattern is evident in the *Winter Atlas* and the *68–72 Atlas*, and apparently reflects the scarcity of suitable water bodies, roughly those larger than one ha, and rivers of appropriate width. Concentrations occur in N East Anglia, S Suffolk, Essex, the W Midland/Welsh border area, in SE Scotland and the NE tip of Scotland, extending into Orkney. In Ireland, the contributions of the large loughs Corrib, Mask, Conn, Ree, Derg and Neagh are of overwhelming importance.

The patterns of the Abundance map are similar to those of the Distribution map. The largest numbers tend to occur in the gravel rich areas, where Tufted Ducks successfully exploit the more mature gravel workings. This is significantly so in the Thames region, Essex, and the Severn-Trent area.

There has been much in-filling of occupied areas throughout its range in Britain in the years since the *68–72 Atlas*, from the S coast of England to Orkney (where it is now widespread). However, the Change map shows a worrying loss of occupied squares in Ireland, the reasons for which remain unclear. Although this apparent decline may be related to poorer

coverage of Ireland during the 1988–91 survey, breeding numbers have declined markedly at Lough Neagh, from 1000 pairs in the 1960s to 300 in 1987 (Davidson 1987). This has been attributed to increased competition for food by the expanding roach population.

With the provision of more gravel pits in lowland England, the species has rapidly become more numerous, but part of that growth has been due to a parallel increase in what is now its major food item – the freshwater zebra mussel, which was first discovered in the London docks in 1824 (Olney 1963). It is conceivable that the mussel has greatly aided the spread of the species.

Tufted Ducks are carnivorous, with animal material forming over 80% of their diet. Gravel pits and their associated islands together provide a good food supply and cover for nesting. Studies have shown that the height of vegetation at the nest site affects breeding success (Hill 1984a and b), for predation by Crows and Magpies often accounts for over 50% of all clutches. Studies have also shown that nests closer together are more susceptible to predation than those farther apart (Hill 1984a). Under high density breeding conditions the species nests semi-colonially as, for

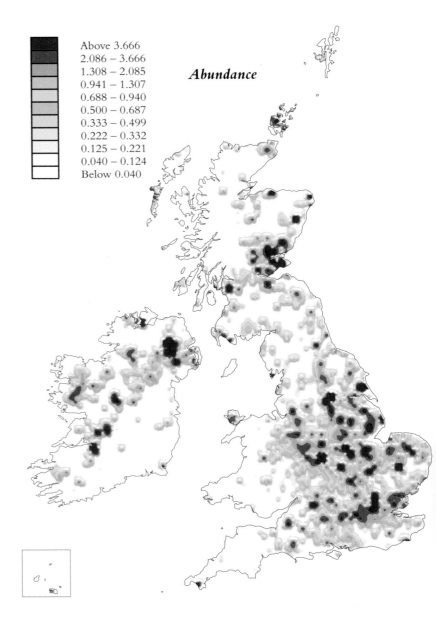

Above 3.666
2.086 – 3.666
1.308 – 2.085
0.941 – 1.307
0.688 – 0.940
0.500 – 0.687
0.333 – 0.499
0.222 – 0.332
0.125 – 0.221
0.040 – 0.124
Below 0.040

Abundance

example, on St Serf's island on Loch Leven in Scotland. At high densities, desertion as a result of increased nest parasitism can also cause significant losses.

Tufted Ducklings feed themselves from the moment of hatching, initially by catching emerging Chironomid midge larvae, and, from four or five days of age, by diving. The amount of time which the ducklings spend diving increases until by four or five weeks of age most food is taken by this method (Hill and Ellis 1984). The foraging success of ducklings increases with the density of Chironomid midge larvae (Giles 1990). Water less than 5m in depth is preferred, which explains the Tufted Duck's greater abundance on lowland, eutrophic waters, than on oligotrophic ones at higher altitudes.

Owen *et al.* (1986) provide the latest estimate of 7,000–8,000 pairs of Tufted Ducks breeding in Britain. By extrapolation, the population in Ireland would be 1,750–2,000 breeding pairs. *Birds in Scotland* suggests that the Scottish breeding population may have stabilised, but the *Trends Guide* indicates that numbers continue to increase in England.

DAVID HILL

68–72 Atlas p 78

Change

Distribution

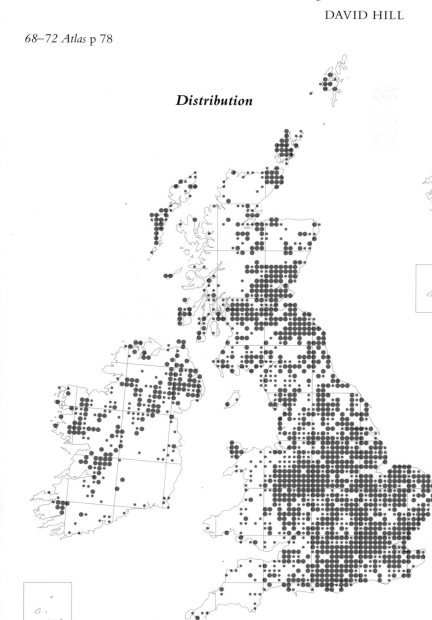

Years	Present, no breeding evidence		Breeding evidence		All records		
	Br	Ir	Br	Ir	Br	Ir	Both (+CI)
1968–72	321	65	969	267	1290	332	1622
1988–91	465	113	1019	139	1484	252	1739
				% change	15.1	−24.1	7.2

produce more offspring over their lifetimes if they avoid breeding in years when conditions are poor.

Variations in the breeding success of Eiders at the Sands of Forvie are determined mainly by duckling survival (Mendenhall and Milne 1985). Less than 10% of ducklings survive to fledging in most years, but up to 55% in occasional 'good' years. Herring Gull predation causes most of the duckling losses, and is most severe during wet and windy weather. Eider ducklings which are short of food are particularly vulnerable to gull predation, for poorly nourished ducklings do not respond properly to the alarm signals of accompanying females (Swennen 1989).

Males defend their mates during the pre-breeding and laying periods but take no part in incubation or in the care of their young. In areas where groups of Eiders nest together, broods often amalgamate to form crèches attended by several females. Some females remain with crèches for only a few days but others attend for several weeks, establishing stable bonds with other adults and young. The main advantage of crèching appears to be that individual ducklings are less likely to be predated than if they were members of a single brood.

Most British Eiders move only short distances between their breeding and wintering areas, although many of the birds that breed at Forvie and

Eider

Somateria mollissima

As spring approaches, activity intensifies within Eider flocks. Males throw their heads back and coo in the displays which are such an attractive feature of this, our largest duck. Meanwhile the females are busy feeding, as they must soon have enough reserves to lay their eggs and to survive 26 days of incubation without food.

Britain is at the southern limit of the range of this Arctic-breeding species. Eiders have an exclusively coastal distribution in Britain and Ireland, extending south to Coquet Island in the east and Walney Island in the west. Concentrations occur in SW Scotland, Orkney, Shetland and E Scotland, mainly around the firths of Forth and Tay. There are particularly large breeding colonies at the Sands of Forvie, Grampian, and Walney Island, Cumbria. The distribution pattern is probably determined by the availability of the marine invertebrates on which Eiders feed. The concentrations on the E coast of Scotland are associated with the extensive estuarine mussel beds which are favoured feeding grounds.

The distribution of Eiders is little changed since the *68–72 Atlas*. They have spread a little further down the E coast of Ireland, and a few more 10-km squares are occupied around Walney Island. The small numbers of records further south are of non-breeding birds – a category not recorded in the *68–72 Atlas*. Numbers have increased over the last 20 years, at an average rate of 2–3% per annum (S.R. Baillie, J.C. Coulson 1984), and if these rates are typical, they suggest a 60% increase in Eider populations between the two Atlas periods.

Eiders have high adult survival rates and low reproductive rates. Females start to breed at 2–4 years of age (Baillie and Milne 1982). Young females lay later in the year, have smaller clutches and start incubation at lighter weights than established breeders. In some years a substantial proportion of those females which have nested previously do not attempt to breed (Coulson 1984). In such long-lived species, individuals are likely to

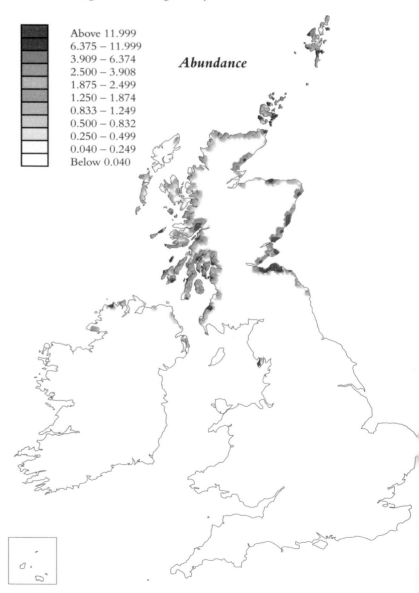

	Abundance
	Above 11.999
	6.375 – 11.999
	3.909 – 6.374
	2.500 – 3.908
	1.875 – 2.499
	1.250 – 1.874
	0.833 – 1.249
	0.500 – 0.832
	0.250 – 0.499
	0.040 – 0.249
	Below 0.040

in Northumberland converge on the Firths of Tay and Forth for the winter. Females are faithful to the areas where they were reared but males frequently change colonies both before reaching breeding age and subsequently (Baillie and Milne 1989). A small number of males ringed at Forvie has even been recovered from Scandinavia. This dispersal pattern reflects the social organisation of Eiders, the pair formation taking place in winter and males returning to their mate's natal colony.

The *68–72 Atlas* suggested a population size of about 20,000 pairs. Increasing this by 2.5% per annum over 19 years gives an estimate of 32,000 breeding females in 1989. An independent estimate can be derived from the *Winter Atlas* counts. The mid winter total of 72,400 birds can be converted to an estimate of 29,300 breeding females in 1983, using demographic data from Forvie to correct for the average proportion of first-year birds and the average sex ratio of 1.23 males per female. Adding population growth at 2.5% per annum gives an estimate of 32,800 breeding females in 1989. If the distribution of Eiders between Britain and Ireland is similar to that shown by the *Winter Atlas* counts, then there were about 31,000 breeding females in Britain and 1,000 in Ireland. These estimates

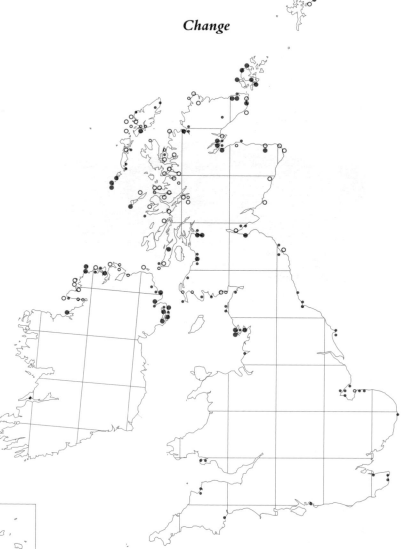

Change

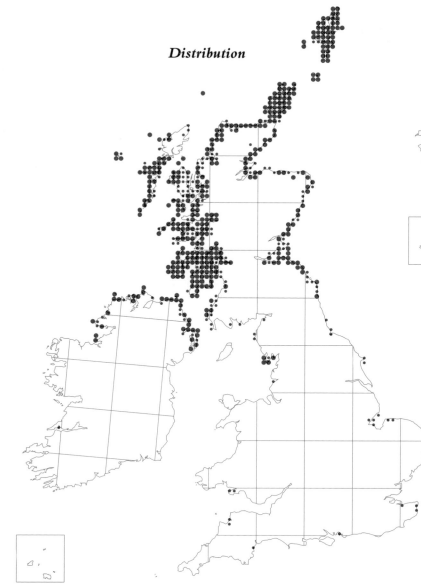

Distribution

differ only slightly from those of 77,000 Eiders in Britain and 1,500 in Ireland calculated from tetrad counts during April to July in 1988–91, suggesting British and Irish populations of 31,200 and 600 pairs respectively.

The number of Eiders in the W Palearctic was estimated to be about three million in 1989, the British population being only about 3% of this total. Nevertheless, the Eider remains our second commonest duck, its population being exceeded only by that of the Mallard.

STEPHEN R. BAILLIE

68–72 Atlas p 86

Years	Present, no breeding evidence		Breeding evidence		All records		
	Br	Ir	Br	Ir	Br	Ir	Both (+CI)
1968–72	43	8	417	30	460	38	498
1988–91	129	14	359	31	488	45	533
				% change	5.7	18.4	6.7

Common Scoter

Melanitta nigra

Common Scoters are known to many observers only as lines of rather dumpy black shapes flying low over a wintry sea. In summer they are to be found in strongly contrasting breeding habitats, ranging from large limestone lakes in W Ireland to remote lochans in the Scottish flow country. Across N Fennoscandia and the former USSR, Common Scoters breed at low densities on the open tundra, wherever sufficient vegetation cover exists (*BWP*). The flow country breeding habitat resembles the tundra; the limestone lakes habitat in Ireland, where nesting densities can reach six nests per ha, is atypical.

The Distribution map reflects the mainly NW breeding distribution, with a scattering of summering non-breeding birds. The Irish population, estimated at 65–70 pairs (Partridge 1989), is concentrated on four lowland lakes – Lower Lough Erne, Loughs Conn and Cullin, Lough Corrib and Lough Ree. In Ireland, Common Scoters nest in dense vegetation, mainly on scrub-covered islands or uninhabited stretches of shoreline. The Scottish breeding population extends from Loch Lomond and Islay, by the lochs and lochans in the Inverness glens to the flow country of Sutherland and Caithness. On moorland sites, nests are in long heather at least 10m from the water's edge (*Birds in Scotland*).

Whilst the overall breeding distribution of the Common Scoter has altered little since the *68–72 Atlas*, there have been some significant local changes. The population on Lower Lough Erne, formerly the species' main British and Irish stronghold, has suffered a severe decline (J. Magee, RSPB). In contrast, the Conn/Cullin population (38 pairs) seems to have remained stable over the same period. Comparative research at two sites indicates that a deterioration in water quality at Lough Erne has reduced

food availability, thereby affecting breeding success (Partridge 1989). Feral mink are also thought to have had an impact on breeding females, as have roach, by taking bottom-dwelling invertebrates.

Lough Ree was apparently colonised during the 1970s, and by 1984/85 the population was reported to have reached at least 40 pairs (Ruttledge 1987). This may have been an overestimate, for the population in 1987 was 10–20 pairs and is likely to be in decline due, probably, to water pollution. At Lough Corrib, the other main Irish site – also colonised during the 1970s – numbers seem to have stabilised at 10 pairs.

Due to the remoteness of many of the Scottish breeding sites, less is known about this segment of the population, which numbers some 100 pairs (Partridge *op. cit.*). No records were received in 1988–91 for Loch Lomond – the best-documented Scottish site – which previously held up to nine pairs (Mitchell 1977) though birds still occasionally occur (J. Mitchell). On Islay, one site remains occupied with up to seven pairs present during the mid 1980s. No records were received for mainland Argyll, though breeding has been recorded intermittently in the recent past.

The flow country holds the most important concentration of Scoters in Britain, with an estimated 55 pairs. Birds are found on all sizes of water, from small pools of only a few ha through to large lakes of 200ha or more, but an analysis of RSPB survey data for 133 lochs found no useful predictors for Common Scoter presence (M. Avery).

The importance of the hills and glens north of Fortwilliam is confirmed by the Distribution map. Here Scoters breed both on large lochs in the glens and on smaller hill lochs. The population in this area is estimated at about 35 pairs and is thought to have increased slightly since 1968–72 (R. H. Dennis).

Common Scoters spend most of the year on the coast, feeding largely on marine molluscs. The fact that the males spend less than two months inland during the breeding season may indicate that the species is poorly adapted to finding food on freshwater (Gardarsson 1979). The availability of a specific food may determine whether or not a particular site is selected for breeding. At Lough Conn, birds fed to a large extent on the common bithynia snail.

Common Scoters appear to be vulnerable to changes in water quality which affect food supply, and this must be the major threat to the British and Irish populations.

The spread of feral mink westwards in Ireland and northwards into the Highlands also poses a threat to this and other nesting duck. Mink damage to waterfowl is a controversial subject, but studies of Common Scoters in Iceland (Gardarsson 1979), of Eiders in Sweden (Gerell 1985), and unpublished observations on Loch Lomond (J. Mitchell) have shown that the impact can be substantial.

J. K. PARTRIDGE

68–72 Atlas p 84

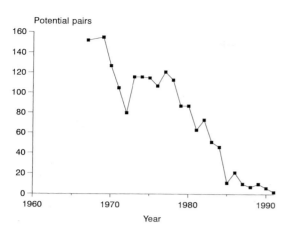

The number of potential pairs of Common Scoter on Lower Lough Erne, 1967–91. No count was performed in 1968, and those in 1972 and 1985 were adversely affected by bad weather.

Change

Distribution

Years	Present, no breeding evidence		Breeding evidence		All records		
	Br	Ir	Br	Ir	Br	Ir	Both (+CI)
1968–72	24	6	18	9	42	15	57
1988–91	35	8	16	8	51	16	67
				% change	21.4	6.7	17.5

Goldeneye

Bucephala clangula

This attractive duck is a typical breeding bird of boreal forests, where it nests in holes in trees. It is a common winter visitor to Britain and Ireland, mainly in coastal areas, where it is often classed as a seaduck. In spring, Goldeneyes return to freshwater where the spectacular displays of the drakes take place prior to nesting. In Scotland, the species has increased rapidly since it was first proved to breed in 1970 (MacMillan 1970).

It is a feature of the species that some winter visitors from Scandinavia may linger on in this country, well into May, and give the impression of potential local breeders, when in fact a few days later they have departed for their northern breeding grounds. This will explain the wide distribution of records south to the Channel coast. Nevertheless, some of the sightings in N Scotland will relate to Scottish-reared pre-breeders, as well as others which have dispersed from the nesting areas in June. The moulting areas for Scottish males are not properly documented but appear to involve movements to Argyllshire and the Moray Firth.

During the *68–72 Atlas* there were only two 10-km squares with confirmed breeding, and for security reasons these were displaced on the published map. Those records related to Strathspey, and the Distribution map shows a continued presence there. It does not, however, reveal how faithful the birds have been to the lochs and rivers of Badenoch and Strathspey, where the majority of the birds still nest. Some colonisation of areas to the north of the main breeding population has occurred and is likely to continue. The breeding record shown for Lancashire relates to feral birds from the WWT's Martin Mere reserve.

Although the exact nest site of the initial pair was unknown, nearly all birds since have used nest boxes. The provision of boxes has been carried

out for over two centuries in Sweden, where there may now be over 75,000 of them available. In Scotland, about 20 boxes were available in the early 1970s, increasing to over 80 in Strathspey by 1980, and approaching 700 in Britain ten years later. Birds readily find nest boxes, even those several km from the water, and once an individual starts to nest it tends to return to the same box, or nearby, in later years. The provision of nest boxes certainly helped the birds to colonise Scotland, for there are no suitable nest holes excavated by Black Woodpeckers as on the Continent (Dennis and Dow 1974).

The nest box scheme has been organised by the RSPB with assistance in latter years from the Forestry Commission. All boxes likely to be used are checked by RSPB staff and other ornithologists, and an annual newsletter detailing the year's activities has been circulated from the Society's Highland office. Eggs are laid from early April, the average clutch being 9–11 eggs. Extra eggs are frequently 'dumped' in nests by other females, probably young birds, leading to clutches of from 22 to 28 eggs. Between 22% and 33% of clutches are not incubated and are abandoned; of the remainder about 80% are successful, but not all eggs hatch. The mean brood size leaving boxes in the last ten years has ranged from 7.85 to 9.86 young per box. Failures include the destruction of eggs by humans as well as by Jackdaws and pine martens; nests being taken over by other nesting birds such as Goosanders, Tawny Owls and Jackdaws; and females being killed or disturbed. Soon after hatching, most young are taken to the River Spey, where they grow up; surveys later in the summer suggest that about a third survive to fledge (Dennis 1987a and Newsletters 1–12).

Following successful breeding in 1970, a pair nested at the same locality near Aviemore in 1971 and 1972. Since then the population has continued to grow (Dennis 1987a). Although it is likely that an increasing number of breeding attempts in natural sites is missed, the known totals for occupied nests during the earlier years of the Atlas were 87 in 1988, 85 in 1989 and 95 in 1990. In 1984, two of these pairs nested to the north of the main site and up to four pairs have continued to nest (in three localities) away from the main range.

The British population is expanding rapidly, and in some places in Strathspey the Goldeneye is the commonest breeding duck; but these numbers are tiny when compared to the very large breeding populations of Sweden, Finland and Russia. At the southern limit, smaller numbers breed in Germany, Poland and Czechoslovakia.

ROY DENNIS

68–72 Atlas p 82

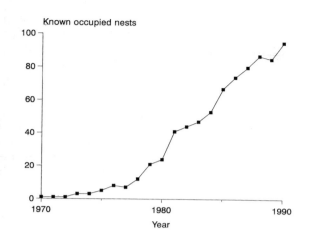

Annual number of known occupied nests of Goldeneyes in Britain.

Change

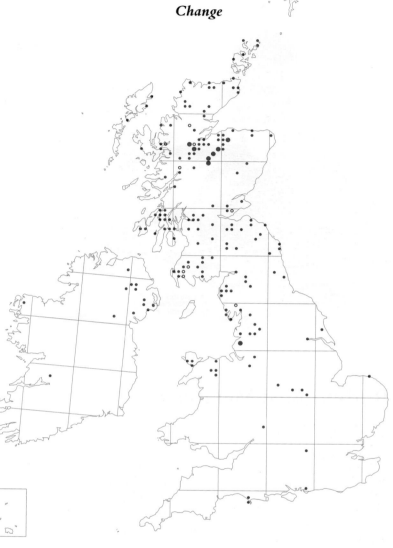

Distribution

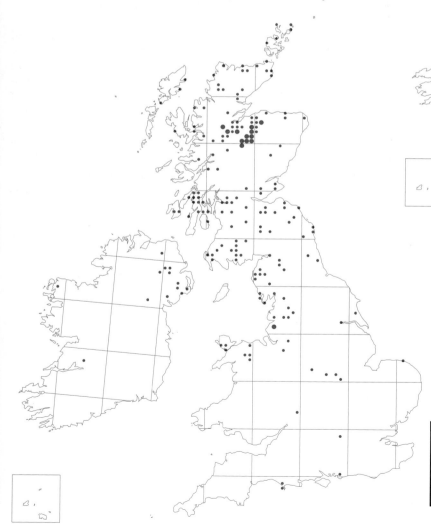

Years	Present, no breeding evidence		Breeding evidence		All records		
	Br	Ir	Br	Ir	Br	Ir	Both (+CI)
1968–72	14	0	5	0	19	0	19
1988–91	160	13	13	0	173	13	186
				% change	810.5		878.9

Red-breasted Merganser

Mergus serrator

The Red-breasted Merganser occurs as a breeding species in a variety of freshwater and tidal habitats, from the tranquillity of the English Lake District to the tempestuous Atlantic coast of Ireland. The species is widely dispersed during the breeding season but there are records from a single site on the River Tay, near Perth, of colonial nesting Mergansers with up to seven pairs breeding on an island measuring about 150m × 150m (Martin 1988).

The species has a wide but predominantly northern and western maritime distribution in Britain and Ireland, with the population at its most abundant at sites around the Irish Sea, in W central Ireland and along the NW seaboard of Scotland extending up through Orkney and Shetland. Although breeding records inland are widespread, particularly north of the Central Lowlands of Scotland, numbers involved are typically low.

Comparing this distribution with that in the *68–72 Atlas* reveals marked regional differences. In Scotland there is a suggestion of a reduction in the breeding population from scattered areas of the NW coast, Shetland, parts of the Central Lowlands, and the Southern Uplands. The overall impression in Scotland is one of contraction and the same holds true for Ireland, with the species much less widespread inland and absent from many previously recorded sites. In England the population remains concentrated in the northwest with only scattered breeding records outside this area. In Wales there appears to have been a southern extension of breeding range since the *68–72 Atlas*, but for inland sites Griffin (1990) reported a marked decrease in numbers between 1981 and 1990 which, it was suggested, may have been due to competition from the increasing Goosander population over this period. Whilst this may or may not be true, the fact that inland breeding has also dramatically declined in Ireland

(where competition with Goosanders could not occur), indicates the existence of other controlling factors.

Red-breasted Mergansers suffer heavily from both illegal persecution and licensed control in the name of fisheries protection, with reported kills in Scotland of 247 in 1989, 418 in 1990 and 235 in 1991 over the three month season (February to April inclusive). Birds breeding on estuarine and coastal sites are perhaps less affected than those inland, although with the recent enormous increase in the number of coastal fish farm installations in Scotland the potential for conflict is high. A study in Argyll, however, found that although Mergansers often passed close to holding facilities they did not attack stock (Carss 1989).

As with the Goosander, recent research has shown game fish to be important constituents of the diet of Mergansers. Carter and Evans (1986) estimated that 87% of fish eaten by Mergansers on the River North Esk, Tayside, over the period of the smolt run (April to June) in 1986 were salmonids, whereas Feltham (1990) reported a figure of approximately 70% derived from a larger sample of birds examined from several rivers in NE Scotland over the smolt runs of 1987 and 1988. Despite the magnitude of these values, no effect of the control of these predators (or of Goosanders) on fish populations has been demonstrated.

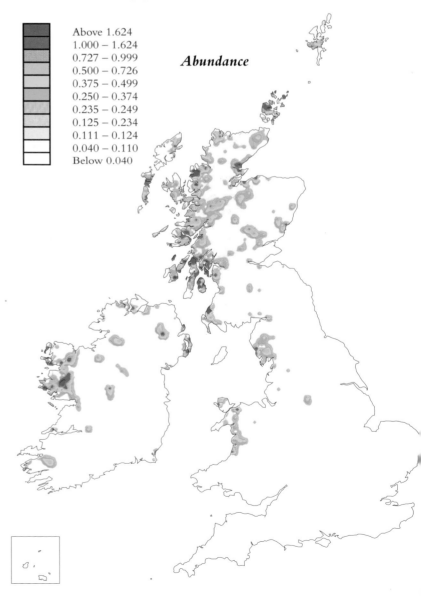

	Above 1.624
	1.000 – 1.624
	0.727 – 0.999
	0.500 – 0.726
	0.375 – 0.499
	0.250 – 0.374
	0.235 – 0.249
	0.125 – 0.234
	0.111 – 0.124
	0.040 – 0.110
	Below 0.040

Abundance

The 1987 Sawbill Survey, with its emphasis on river coverage, provided little information on the density of breeding birds outwith freshwater sites. A Shetland Bird Club survey suggested a maximum of 100 pairs on the islands (with little evidence of recent change), whilst on the west coast, on Colonsay and Oronsay, only two and three broods respectively were recorded during complete coastal counts. Densities were also low on rivers; spring fieldwork in 1987 recorded the density of adult male Mergansers as 0.05, 0.02, and 0.06 per km respectively on 2,592km of 104 Scottish rivers, 1,456km of 45 English rivers and 1,043km of 31 Welsh rivers. Assuming each male to correspond to a potential breeding pair, then a single pair of birds occurs per 20km of Scottish river, per 50km of English river and per 16.7km of Welsh river. However, as Mergansers are common in coastal districts, it is not possible to calculate national population estimates from these figures.

Extrapolation from tetrad counts during April to July in 1988–91 yields estimates of 4,300 Mergansers in Britain and 1,400 in Ireland, suggesting that the minimum British and Irish populations are of 2,150 and 700 pairs respectively.

                                                            STEVE CARTER

68–72 Atlas p 90

Change

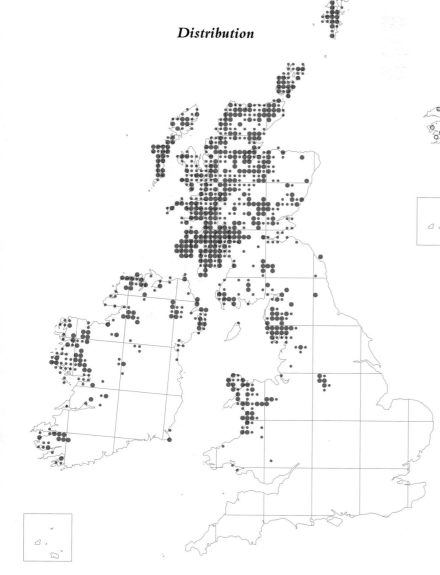

Distribution

Years	Present, no breeding evidence		Breeding evidence		All records		
	Br	Ir	Br	Ir	Br	Ir	Both (+CI)
1968–72	199	73	467	175	666	248	914
1988–91	297	97	377	70	674	167	841
			% change	1.5	−32.7	−7.8	

Goosander

Mergus merganser

The Goosander, although a typical breeding bird of upland rivers, may be found nesting in a variety of other habitats ranging from meandering lowland reaches to moorland tarns.

This is reflected in its current distribution. In Scotland north of the Central Lowlands breeding records are widely scattered, but concentrations are evident along the Great Glen, in the N Grampians, and along both the Dee and the Tay. It is in the Southern Uplands, however, extending into the border counties of Cumbria and Northumberland and further south into Lancashire and parts of Yorkshire, that breeding populations are most concentrated. Elsewhere in England breeding is largely confined to parts of the S Pennines and to a handful of records from the SW peninsula. In Wales the Goosander is widespread during the breeding season, but in Ireland the map shows only two records. This overall pattern repeats that found by the BTO's national Sawbill Survey in 1987.

Comparison of the present Distribution map with the *68–72 Atlas* shows two important features: firstly, a scattering of losses and gains throughout Scotland and, secondly, a major southward expansion of range which has been most pronounced in N England and in Wales. The single confirmed Welsh breeding record of the *68–72 Atlas* had increased to at least 10 pairs by 1977 and to an estimated 100 pairs by 1985 (Tyler 1985). Although recent work has shown continuing colonisation of new rivers within Wales, no further increases in numbers appear to have taken place (Griffin 1990). This may be attributable to human persecution (see below).

Additional notable range expansion into the rivers of the SW peninsula and S Pennines has also occurred, breeding being first recorded in Devon in 1980, and in Derbyshire in 1981. More recently Goosanders have been confirmed breeding in Shropshire (1987) and in Herefordshire (1988).

The reasons for such swift and dramatic changes in breeding distribution are unknown. The Goosander is, however, a relatively recent addition to the British and Irish avifauna having been first recorded nesting as recently as 1871. It therefore seems likely that the essentially southward expansion noted over the last 20 years is a continuation of that initial colonisation.

The success, not only of initial colonisation but also of subsequent range expansion, is even more remarkable in view of the intensely persistent and often illegal persecution suffered by this species in the name of fisheries protection. Under the provisions of the Wildlife and Countryside Act (1981) evidence of 'serious damage' to fisheries interests is required before a licence for the control of Goosanders is issued. To date such licences

have been issued only in Scotland, with recent reported kills of Goosanders during the shooting season (February to April inclusive) totalling 383 in 1989, 673 in 1990 and 504 in 1991. Although recent research has shown that young salmon and trout may comprise up to 80% of the Goosander's diet on some river systems (Carter 1990), no beneficial effect of predator control in terms of increases in the number of adult fish returning has been demonstrated or justified.

In Ireland the *68–72 Atlas* showed a single confidential breeding record. This was the first confirmed breeding of the species there and although a single pair bred annually over the next decade, the species was absent throughout the 1980s. Atlas fieldwork in 1988–91, however, resulted in another breeding record.

On totals of 2,592 km of 104 Scottish rivers, 1,456 km of 45 English rivers and 1,043 km of 31 Welsh rivers surveyed as part of the Sawbill Survey in spring 1987, the density of adult male Goosanders was 0.18, 0.21 and 0.08 per km respectively. If each of these corresponds to a potential breeding pair, then on the rivers surveyed, one pair occurs per 5.6km of Scottish river, per 4.8km of English river and per 12.5km of Welsh river.

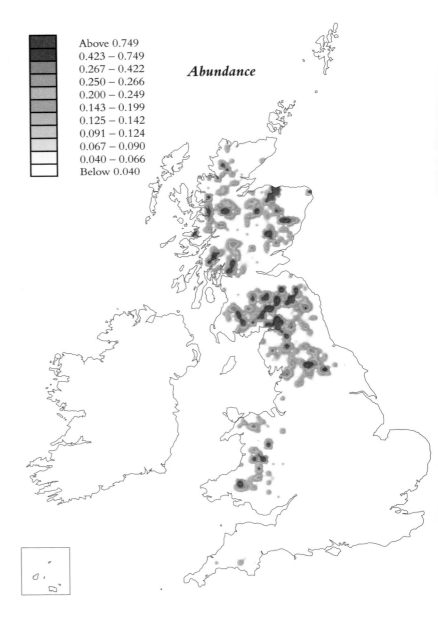

| Above 0.749 |
| 0.423 – 0.749 |
| 0.267 – 0.422 |
| 0.250 – 0.266 |
| 0.200 – 0.249 |
| 0.143 – 0.199 |
| 0.125 – 0.142 |
| 0.091 – 0.124 |
| 0.067 – 0.090 |
| 0.040 – 0.066 |
| Below 0.040 |

Abundance

From the total lengths of river and taking into account regional variation in density, these values can be extrapolated to provide an estimate of 2,700 adult male Goosanders in Britain. This estimate, however, will be prone to a degree of error, because pairs occurring away from rivers are not included, some birds may have been winter migrants from Scandinavia, and some local breeding birds may not have returned from their wintering sites elsewhere by the time of the spring survey period. Nevertheless, this estimate, which may be taken as equivalent to the number of breeding pairs (e.g. Haapanen and Nilsson 1979), compares favourably with the *68–72 Atlas* estimate of 1,000–2,000 pairs and that of 1,250 pairs proposed by Meek and Little (1977), in view of the considerable range expansion since then.

STEVE CARTER

68–72 Atlas p 92

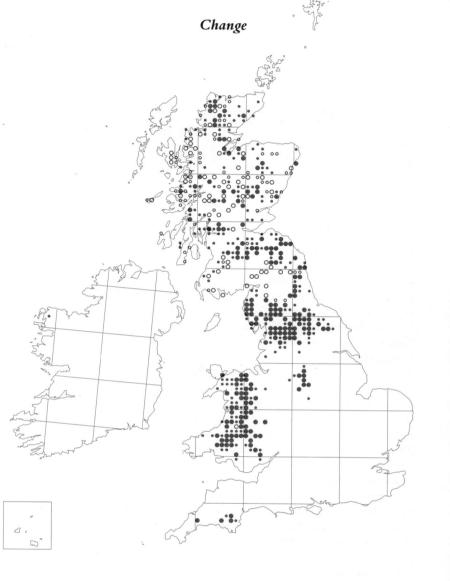

Change

Distribution

Although it is known that the 1988–91 Goosander breeding record in Ireland was in the same 100-km square as that in 1968–72, the full 10-km grid reference was not supplied by the observer. In the production of the Change map it has been assumed that both records were from the same 10-km square, and the record thus does not feature on the Change map.

Years	Present, no breeding evidence		Breeding evidence		All records		
	Br	Ir	Br	Ir	Br	Ir	Both (+CI)
1968–72	152	0	259	1	411	1	412
1988–91	276	1	398	1	674	2	676
			% change		64.0	100.0	64.1

Ruddy Duck

Oxyura jamaicensis

Male Ruddy Ducks, unmistakable with their bright chestnut-red breeding plumage and cobalt-blue bills, are especially conspicuous while performing their characteristic 'bubbling' display. Only 30 years ago they were rare in Britain, restricted to a small number of breeding sites in the Midlands and southwest. Today, in summer months, they are a common sight on reed-fringed pools and gravel pits throughout the midland and N counties of England.

The Distribution map shows that the stronghold of the British breeding population of Ruddy Ducks remains in the Midlands, where the birds first bred 30 years earlier, but there are now notable concentrations breeding in Cheshire, Greater Manchester, Yorkshire and on Anglesey. During the 1980s their breeding range continued to expand, mainly northwards. Although the Abundance map shows that it is more usual to find Ruddy Ducks breeding at low densities, some sites have held exceptional numbers, such as Woolston Eyes in Cheshire, where 20 broods were reared in 1991 (although this site has since been drained). Ruddy Ducks now breed regularly in Fife and Tayside in Scotland, and in small numbers in Ireland. These differences in distribution are well illustrated in the Change map.

Of all of the species covered in this Atlas, Ruddy Ducks have shown one of the greatest changes in distribution since the *68–72 Atlas*. The colonisation of unpopulated areas will no doubt continue for the fore-seeable future as there is a great abundance of suitable breeding habitat. Ruddy Ducks prefer small lakes surrounded by emergent vegetation such as common reed, rushes or reedmace, in which they build their nests above water. Fluctuations in water level, causing flooding or making the nest inaccessible to the incubating female, are the main factors limiting breeding success. Eggs have a high hatching rate, and chick survival is also high (Joyner 1975), even though ducklings are usually abandoned at only three weeks of age (B. Hughes). Other factors contributing to the amazing increase in numbers of Ruddy Ducks in this country include brood parasitism by some females, which may then lay full clutches in their own nests, and the ability to rear two broods per season (Palmer 1976, C. Wilson).

Ruddy Ducks appear to be exploiting a previously unoccupied niche in the British wetland ecosystem, and there seems to be no serious competitive interaction between Ruddy Ducks and other species, given the present population size (B. Hughes). Brood females are notoriously aggressive, however (Hughes 1990), and there is obvious potential for competition in the future as the expected population increase continues.

Including Britain, Ruddy Ducks have been reported from at least 13 European countries, from Iceland to Italy, and the colonisation of other European countries will doubtless occur over the next decade. Indeed, there have already been five breeding records from the Continent. This has serious implications for the small Spanish population of the White-headed Duck. There have now been about 80 sightings of Ruddy Ducks from the Andalucian breeding grounds of the White-headed Duck where male Ruddy Ducks are dominant over male White-headed Ducks (Pintos Martin and Rodriguez de los Santos 1992). In consequence, hybrids between the two species have already been observed. Similar hybrids in captivity at the WWT are fertile. It is likely that there will also be a high degree of niche overlap between the Ruddy Duck and the White-headed Duck, causing direct competition.

Ruddy Ducks tend to breed in small groups and are consequently difficult to census at that time of year. A more accurate assessment of the population size can be obtained from winter counts. From 1960 to 1980 the number of Ruddy Ducks wintering in Britain increased at an exponential rate, slowing to an annual increase of approximately 10% during the 1980s. The WWT's National Wildfowl Count for January 1991 gave a projected population level of over 3,400 Ruddy Ducks in

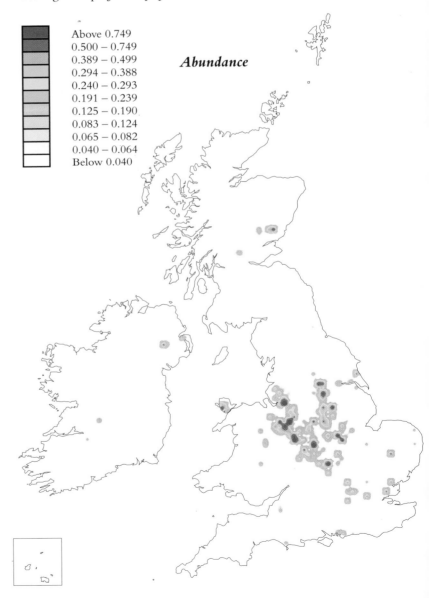

Above 0.749
0.500 – 0.749
0.389 – 0.499
0.294 – 0.388
0.240 – 0.293
0.191 – 0.239
0.125 – 0.190
0.083 – 0.124
0.065 – 0.082
0.040 – 0.064
Below 0.040

Abundance

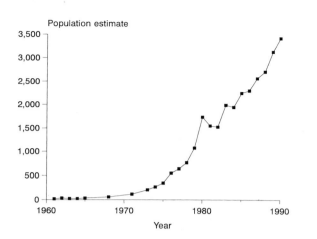

Population estimate

Estimated annual total of Ruddy Ducks in Britain. Figures up to 1982 are from Owen *et al.* 1986, and later figures are from National Wildfowl Counts. Note that 1960 refers to the winter of 60/61, 1970 to 70/71 etc.

Change

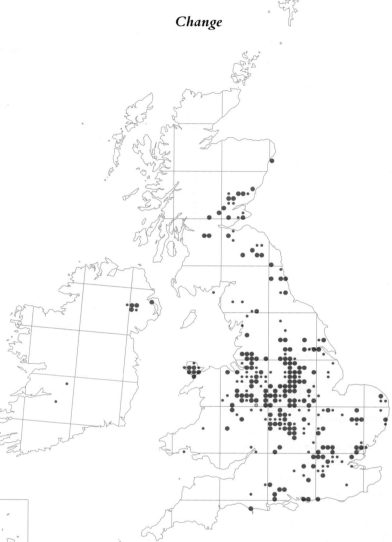

Distribution

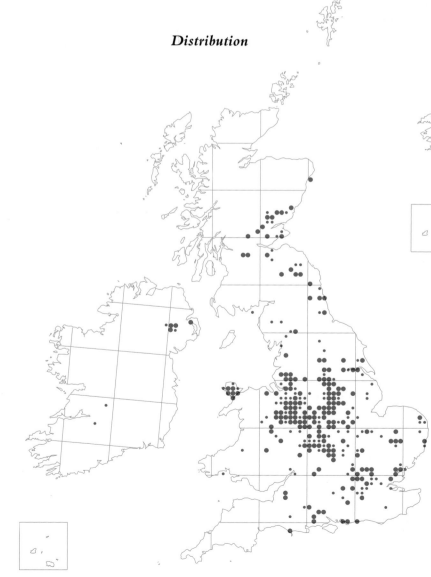

Britain, assuming a 90% count efficiency (after Owen *et al.* 1986). This represents a breeding population of about 570 pairs, if the post-breeding population is similar to that in 1975, when the formula 'population = pairs × 6' (Owen *et al.* 1986) was developed. In Ireland, by extrapolation, there are estimated to be 15–20 pairs of breeding Ruddy Ducks. The *Trends Guide* estimated a breeding population of 350–400 pairs in 1988. The figure indicates that the population continues to increase, contrary to the *Trends Guide* claim that numbers have stabilised.

BARRY HUGHES

68–72 Atlas p 88

Years	Present, no breeding evidence		Breeding evidence		All records		
	Br	Ir	Br	Ir	Br	Ir	Both (+CI)
1968–72	6	0	13	0	19	0	19
1988–91	108	4	184	4	292	8	300
				% change	1436.8		1478.9

Honey Buzzard

Pernis apivorus

The Honey Buzzard is unique among raptors in its heavy dependence in the breeding season on the larvae and adults of wasps and bees. The young are usually fed almost exclusively on wasp larvae unless wasp populations are very low, and the birds' breeding season is timed so that they are feeding young when this principal food is most abundant. Honey Buzzards often have very large home ranges (up to 40km²) which are probably related to wasp density. Thus, sightings, even of displaying birds, do not necessarily indicate a nest nearby. The conspicuous display appears mainly to serve as an advertisement of home range occupancy (*BWP*).

The Honey Buzzard is among the most numerous raptors in the world, but it excites a disproportionate interest among British birdwatchers and egg collectors which has led to the suppression of much information about its breeding distribution in Britain. Interpretation of the Distribution map is confounded by the fragmentary nature of the picture it presents. Some records were submitted to the RBBP in confidence, whilst some breeding groups are known to exist which remain unreported because of fears, however unfounded, that information will find its way to egg collectors, or to twitchers and nest-seeking birdwatchers. These fears are supported by comparisons of the number of raptors, including Honey Buzzards, in disturbed and undisturbed areas in the Netherlands and Germany (Voous 1977). For example, in two areas of similar size and biotype near Nijmegen, one of which was heavily used for recreation and with housing and roads, held one pair of nesting Honey Buzzards; the other, which had few houses and roads and where access was restricted, held six pairs.

No comparisons between 1968–72 and 1988–91 are possible because, so far as can be determined, the information in the earlier Atlas was as incomplete as that in the present one. The most that can be deduced is that the species has a widespread but very fragmented distribution and occurs in at least six British counties, mostly as isolated pairs or small groups. Their fortunes over time may be affected by their isolation, or by local changes in wasp density, and by persecution, disturbance and land use changes. According to *Red Data Birds*, the population documented by the RBBP is fairly stable at 8–15 pairs, but the total population may be as high as 30 pairs, not all of which may breed each year.

Honey Buzzards have long been known to occur in the New Forest, Hampshire, but hitherto no data have been released by the group which has monitored this population. It is now believed that this degree of confidentiality is attracting more potential disturbance than it prevents, and it is thus possible to report on the position. Between 1961 and 1980 6–9 pairs of Honey Buzzards occupied home ranges annually, not all of which attempted to breed. One or more unpaired individuals were also present in most years. Between 1982–85 the population declined to two pairs, and no more than that have attempted to breed since. Analysis of a range of ecological parameters is in progress to determine the cause of the decline.

There have been many studies of the Honey Buzzard's population and breeding biology in Europe, from which it appears that density is highly variable, probably reflecting that of its main prey. The species is evidently influenced by low summer temperatures and high summer rainfall, both of which reduce wasp numbers and availability, and hence reduce Honey Buzzard density and productivity (e.g. Kostrzewa 1989).

The only published study of Honey Buzzards in Britain is of 1–2 pairs in Nottinghamshire during 1971–79 (Irons 1980). There, low productivity was attributed mainly to cold or wet summers. In nine years only four young were known to have fledged. Honey Buzzards breed throughout the temperate and boreal regions of Europe and eastward into the boreal zone of Asia. They migrate to winter in equatorial Africa and the numbers recorded at sea crossings, and elsewhere where they are funnelled into a narrow front, suggests a total Palearctic population of some hundreds of thousands of birds (Porter and Beaman 1985).

COLIN TUBBS

68–72 Atlas p 114

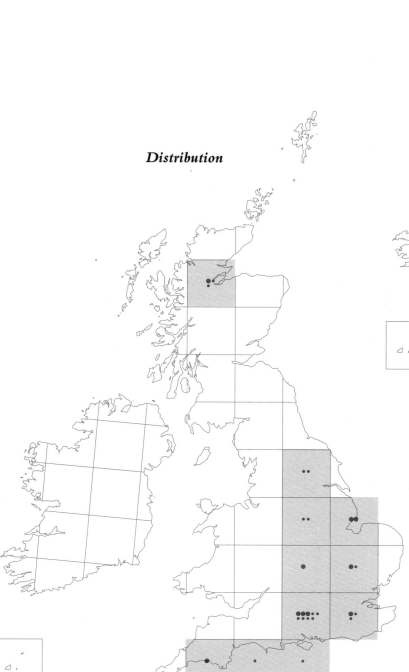

Change

Distribution

Years	Present, no breeding evidence		Breeding evidence		All records		
	Br	Ir	Br	Ir	Br	Ir	Both (+ CI)
1968–72	4	0	7	0	11	0	11
1988–91	17	0	10	0	27	0	27
				% change	145.5	0	145.5

Red Kite

Milvus milvus

By about 1900 the Red Kite in Britain was reduced by human persecution to a tiny relict population, and since that time it has been typically associated with the high open sheepwalks and hanging oakwoods of central Wales. This picture was modified in recent years by a gradual reoccupation of more productive lower ground, where Red Kites now breed successfully in pastoral farmland, out of range of the open hills. Much of Britain still offers a similar environment, and a progressive expansion of Red Kites from the Welsh heartland may be predicted, unless they are actively discouraged by man. The number of 10-km squares found to contain territorial pairs in 1988–91 is more than double that in 1968–72, from 20 to 45, and those with non-territorial records increased from 14 to 40. However, two current reintroduction schemes, in Scotland and England, may change the breeding distribution far more radically before the next Atlas survey.

Welsh Red Kites breed chiefly in oakwoods, often adapting the disused nests of corvids and Buzzards. Defended territories are small, and adjacent nests may be as little as 300m apart, though spacing of 3-5km is more usual. Wide hunting ranges, extending up to 10km or more from the nest, are exploited in common with other kites. Laying may start as early as 25 March and most pairs have eggs by mid April, but stragglers lay as late as the second week of May. Most young fly in early or mid July. The majority of nest failures occur about hatching time, from unviable eggs or the loss of small young, often during wet weather; but egg collecting remains a serious problem, accounting for about 10% of nests in recent years (Walters Davies and Davis 1973; Davis and Newton 1981; unpublished data).

The Red Kite's food spectrum is broad; almost any accessible animal material may be taken. Live prey, especially medium sized birds and small mammals, are important in late spring, and rodents continue to be taken through the autumn. Invertebrates (mainly earthworms and beetles) are exploited, particularly outside the breeding season. Waste from refuse dumps is taken mainly in late autumn and winter, and sheep carrion during its peak availability between February and April, bridging the 'hungry gap' (Walters Davies and Davis 1973; Davis and Davis 1981).

Concern about the possible impacts of large-scale afforestation of the upland hunting area provoked an investigation in the 1970s, which looked at Kite breeding in relation to a wide selection of environmental parameters. No firm evidence was found that the afforestation of sheepwalks had influenced the overall distribution or the breeding performance of Red Kites in Wales. The birds' wide hunting range and the fragmented nature of most new forests ensured the continued accessibility of suitable feeding areas. Nests at higher altitudes were later and produced fewer young. Young were fledged more often from clutches on the better agricultural land than on poor, but fewer pairs reached the laying stage on good land, perhaps due to human disturbance (Newton *et al.* 1981). It is planned to repeat this programme, with much larger samples, in 1992.

Adult Welsh Kites are sedentary, though some leave the higher nesting areas and shift to lower altitudes in autumn. Juveniles disperse more widely, and while most remain in central Wales, marked individuals have been found as far away as Kirkcudbright, Kent, and Cornwall. Tagging has demonstrated a marked westward shift within Wales by juveniles in their first autumn, and a compensatory eastward movement in spring. Most survivors breed within about 15km of their birthplace.

Some 40% of young Kites survive to enter the breeding population, at ages of two to seven years, the average being about three and a half years. Annual survival of adult Kites may be over 90%, so they are potentially long lived (Newton *et al.* 1989; unpublished data). From post-mortems, however, it seems that half of all Welsh Kites die prematurely by ingesting illegal poison baits intended for Crows and foxes. Excessive adult mortality, coupled with low production of young, for many years inhibited increase; but in the post-war period adult survival and, more recently, the proportion of successful nests, have gradually improved. The annual rate of population increase is currently over 10%.

The world status of the species has lately been reviewed (Evans and Pienkowski 1991). The very fragmented breeding distribution covers central and S Europe and NW Africa, the chief populations being in Germany, France, and Spain. The breeding stock is estimated at between 11,000 and 13,000 pairs.

In Wales the spring population is counted annually by the Kite Committee. In 1991 some 92 territorial pairs were located, and 77 proved to breed, and there were in addition over 70 unmated (mainly immature) individuals. In 1992, reintroduced Red Kites bred for the first time in England (four pairs produced nine young) and Scotland (one pair produced a single youngster).

PETER DAVIS

68–72 Atlas p 112

One record has been moved by up to two 10-km squares on the Distribution and Change maps.

Change

Distribution

Years	Present, no breeding evidence		Breeding evidence		All records		
	Br	Ir	Br	Ir	Br	Ir	Both (+CI)
1968–72	14	0	20	0	34	0	34
1988–91	40	0	45	0	85	0	85
				% change	150.0	0	150.0

KHEF

White-tailed Eagle

Haliaeetus albicilla

In many parts of N and W Britain last century, the White-tailed Eagle was commoner than the Golden Eagle. Over 100 eyries were known in Britain and at least 50 more in Ireland (Love 1983). It was then a familiar and conspicuous bird of the coast, but by 1916 its raucous cry was finally silenced by a prolonged campaign of persecution by man. All birds of prey suffered but, being our largest raptor, the White-tailed Eagle was regarded as a potential threat to lambs. As a carrion-feeder it was particularly susceptible to poisoning.

The first attempts at reintroducing the species took place in 1959 in Argyll (Sandeman 1965), and on Fair Isle in 1968 (Dennis 1968, 1969), but they involved only seven birds and proved unsuccessful.

In 1975 the then NCC began a long-term reintroduction programme, based on the Isle of Rum, a National Nature Reserve in the Inner Hebrides (Love 1989). Eighty-two White-tailed Eagles (39 males and 43 females) from N Norway were released, eight of which have since been recovered dead. At least one was the victim of illegal persecution.

The first – unsuccessful – breeding attempt in the wild took place in 1983, but no chick was fledged until 1985, in which year the release programme ended. The RSPB has since assumed a major role in nest protection and survey.

In 1986 the original pair produced two chicks, as they did in 1987 when a second pair fledged a single young. A new pair reared the only two young for 1988, and all three raised a total of five young in 1989. Only two chicks fledged from two nests the following year but in 1991 all three pairs, plus a new one, reared a record total of seven young. At least three of the 22 wild-bred young have been reported dead, causes unknown.

For much of the duration of the Atlas project 11 pairs have maintained territories, their localities of necessity being kept confidential. Eight pairs have laid clutches at some time or another, but four of them have repeatedly failed to produce any young and, in 1991, three of these pairs broke up and some of the adults dispersed.

Successful pairs have largely remained faithful to their nesting territory, and do not range far in the winter months. Immature White-tailed Eagles are, however, prone to wander in the first two or three years of their life. Sightings of reintroduced birds have ranged from Shetland to Northern Ireland. Recent records from the S of England have probably all involved Continental young; one bird shot in 1984 had been ringed in a German eyrie. It is assumed that all nesting sites have been located, although the nature of the terrain in the W Highlands and Islands, where all known eyries are situated, renders it possible that breeding attempts may have gone unrecorded.

In Scotland, White-tailed Eagles feed mainly on hares, rabbits and seabirds, and take carrion in the winter (Watson *et al*. 1992). Fish are probably under-recorded, since the smaller bones can be digested and rarely appear in pellets. Some of the Scottish pairs kill many Fulmars, and these may be the source of the DDE and PCBs recorded from three analysed eggs: in only one of these might the residues have been sufficient to cause infertility (Love 1989). The shell thicknesses of 11 unhatched eggs are within the normal range, if at the lower end of the scale (M. Marquiss).

Three-quarters of the known eyries were on cliff ledges, the remainder in trees. A few eyries have been improved artificially by members of the Protection Team. Egg laying has begun as early as the first week of March, but the less experienced pairs sometimes do not lay until early April. Two eggs are the normal clutch but one and three eggs have also been recorded. Incubation is about 40 days and the young take to the wing after about 12 weeks. Weather and inexperience appear to account for some of the nest failures, but research is continuing (Elliott *et al*. 1991). The option remains to import fresh young from Norway if it is deemed necessary for a more viable population. In 1992 a reintroduction programme was initiated in the W of Ireland.

JOHN A. LOVE

Change

Distribution

Years	Present, no breeding evidence		Breeding evidence		All records		
	Br	Ir	Br	Ir	Br	Ir	Both (+CI)
1968–72	0	0	0	0	0	0	0
1988–91	0	0	9	0	9	0	9
			% change			0	

Marsh Harrier

Circus aeruginosus

No birdwatching visit to the fens and marshes of East Anglia would today be complete without catching a glimpse of the v-shaped silhouette of a distant Marsh Harrier quartering the reed beds with slow, untiring flight. It seems hard to believe that at the time of the *68–72 Atlas* this was Britain's rarest regularly breeding raptor.

The Distribution and Abundance maps show that coastal Norfolk, Suffolk and the Broads hold most of the breeding and summering population, with a few records elsewhere, mostly in the east between Kent and the Humber. Outside this area, breeding has been recorded from as far apart as the southwest, Lancashire and Scotland, which must give hope for future spread and colonisation away from the strongholds in East Anglia. Surprisingly, there were no records from coastal Dorset, where breeding was regular between 1949–62.

This is a far more encouraging picture than that shown by the *68–72 Atlas*, as the Change map reveals. Marsh Harrier numbers had been declining over much of NW Europe during the 1950s and 1960s, and in Britain, by 1971, were down to a single breeding pair, at Minsmere in Suffolk. In Britain the evidence suggested that this was not due to loss of reed bed nesting habitat or to persecution, but was caused by the effects of organochlorine pesticides and (particularly in Scandinavia) other environmental contaminants. Since the withdrawal of these compounds there has been a remarkable recovery in breeding numbers, with a mean annual increase during the 20 years to 1991 of 19.6% (Underhill-Day 1984, Jorgenson 1985).

As the British population grew, breeding spread beyond the large reed beds and, in the last 10 years, 40% of all nests have been in small reed beds (< 25ha), and 10% in arable crops. During this period nearly 15% of nests have failed, two-thirds of them before hatching. The only natural predators of Marsh Harrier chicks in Britain are foxes and possibly stoats. Marsh Harriers generally start laying at the end of April, and the main natural causes of failure during incubation are predation and flooding. Desertion, and the disappearance of adults, have frequently been recorded, however, partly perhaps because this species is particularly sensitive to disturbance.

Although overall failure rates have not differed significantly between nests in large, or small reed beds and in crops, half of failures in the latter, but only 16% in the former, have been caused directly or indirectly by humans. In the ten years to 1988, 22 incidents of persecution of Marsh Harriers were recorded. Thus, the evidence suggests that pesticides are no longer a threat, but that persecution and disturbance, while presently at low levels, might prejudice any further rise and spread of the population if they increase.

In the intensively farmed arable areas where many Marsh Harriers now nest, their main foods were found to be small birds, ducklings, gamebird chicks and mammals, particularly rabbits. The mean live weight of prey was about 140g (Underhill-Day 1985). The fact that they take young gamebirds might lead to conflict with game interests, but one study showed no significant difference in wild gamebird productivity between areas where Marsh Harriers bred and where they were absent (Underhill-Day

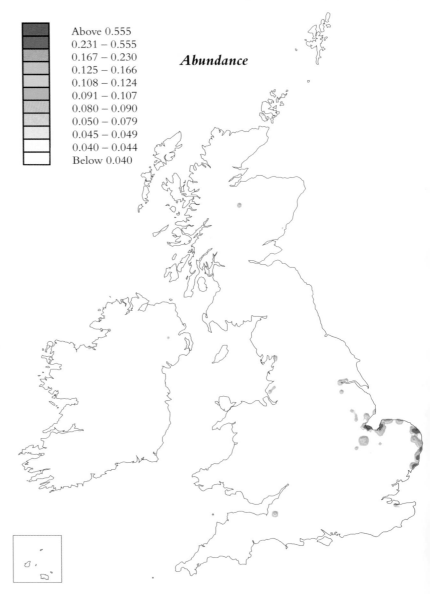

Above 0.555
0.231 – 0.555
0.167 – 0.230
0.125 – 0.166
0.108 – 0.124
0.091 – 0.107
0.080 – 0.090
0.050 – 0.079
0.045 – 0.049
0.040 – 0.044
Below 0.040

Abundance

1989). Reared pheasant poults are mostly too heavy at release to form part of the normal prey.

Marsh Harriers lay large clutches (mean 4.6 eggs) which hatch at the beginning of June and the young fledge about mid July. In 1990 there were at least 87 Marsh Harrier nests in Britain, fledging 213 young. About 16% of summering birds in recent years were non-breeders, suggesting a total British summering population of around 190 birds. Like some other harrier species, Marsh Harriers can be polygamous and the number of polygamous males has slowly increased over the last 20 years, with about 15% of all breeding males having two or more mates. This has possibly allowed more females to breed (Underhill-Day 1984).

A further factor which might have helped survival rates has been the increasing number of birds, mainly females, overwintering here, and thus avoiding the risks and stresses of migration to the Mediterranean and N Africa where most birds winter (*Winter Atlas*).

The nominate race of the Marsh Harrier breeds throughout the Palearctic, from Britain east to Mongolia, with six further races in Asia, Australasia and Africa. There have been increases in most NW European countries during the last 20 years and the most recent estimates suggest a

Change

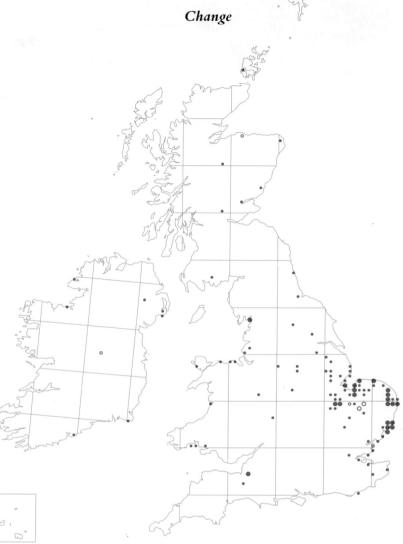

Distribution

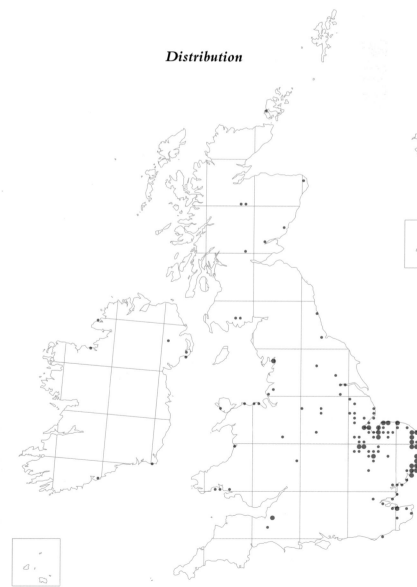

total breeding population of 4,100 pairs (Gensbøl 1984).

There seems no reason why breeding Marsh Harriers should not spread outside East Anglia in larger numbers during the next few years. The Distribution map shows that although birds have been widely recorded during the summer, most records were in the eastern parts of England, and it is there, in the coastal counties, that further expansion might be expected initially.

JOHN UNDERHILL-DAY

To maintain confidentiality, four extralimital breeding records have been downgraded to 'seen' records (= small dots) on the Distribution and Change maps.
68–72 Atlas p 116

Years	Present, no breeding evidence		Breeding evidence		All records		
	Br	Ir	Br	Ir	Br	Ir	Both (+CI)
1968–72	20	1	6	0	26	1	27
1988–91	82	7	32	0	114	7	121
				% change	338.5	600.0	348.1

Hen Harrier

Circus cyaneus

There are few sights more exhilarating than a displaying Hen Harrier sky diving, roller coasting and cork screwing over a heather hillside on a bright spring day. Equally impressive is the aerial food-pass, performed with precision over the ground nesting site.

Hen Harriers show a strong preference for undulating moorland, usually below 500m, with a lush covering of heather, but they will also breed in young forestry plantations. Thus, the proliferation of upland afforestation schemes in the last 40 years has favoured this species by providing, at least in the early years of tree growth, enhanced hunting and nesting grounds free from human disturbance. New areas can be colonised if this forestry occurs on grassland or heavily grazed moorland; habitats which are normally avoided.

In contrast to their extensive winter distribution (*Winter Atlas*) Hen Harriers are confined in the breeding season to N and W Britain, the Isle of Man and Ireland. They are particularly widespread in Scotland, with breeding strongholds in Orkney, E Highland, and Strathclyde north and west of the Clyde. The population in Ireland occurs mainly in the southwest, in counties Kerry, Limerick, Cork, Clare, Tipperary and Laois, and in the north in Fermanagh, Tyrone, and Antrim. Gwynedd and Clwyd are the centres for the small Welsh population. In England breeding was confined to nine 10-km squares in the north and northwest.

Lack of preferred habitat will account for many of the obvious gaps in breeding distribution, such as on Lewis and Harris, on Rannoch Moor and over much of the NW Highlands, the Cairngorm plateau, the bogs of W Ireland, S Wales and the English Lakes. This is because Hen Harriers avoid deer forest, extensive mires, continuous high ground, sheep walk and acid grasslands which dominate many of these regions.

New areas have been colonised since the *68–72 Atlas*, noticeably around the Great Glen area in E Highland, on Skye and the whole of the Isle of Man. The species has also consolidated its position in the Southern Uplands and N Strathclyde but any such increases are partly offset by population declines and range contractions in E Grampian and Tayside, Central, N England and much of S Ireland.

There are several reasons for these declines. O'Flynn (1983) attributed the decrease in the Irish population, which occurred from the late 1970s, to a combination of maturing plantations and the use of marginal land for agriculture following Ireland's entry into the EC. Bibby and Etheridge (1993) discussed the problems associated with the loss of habitat caused by maturing forests and the species' reluctance to recolonise areas of clear-

felled timber. Their report into the status of Hen Harriers in Scotland highlighted the impact of persecution on managed grouse moors. During 1988–89 only 14% of Harrier nests on grouse moors successfully raised young as opposed to 40% of nests on other heather moors and 66% of nests in young plantations. Moreover, they presented evidence not only of frequent nesting failure on keepered ground, but of the deliberate shooting of at least 80 breeding adults each summer in Scotland alone. The scale of this widespread destruction constitutes a considerable annual drain on the population. It can cause local extinctions and is probably preventing harriers from colonising many sporting moors, particularly in E Scotland and the Borders and throughout the English northern counties.

The Hen Harrier breeds across a wide area of Europe, Asia and North America. Recent estimates in W Europe, excluding Britain and Ireland, quoted in *Red Data Birds*, gave 4,160–6,610 pairs. During 1988–89 Bibby and Etheridge (1993) found an average of two pairs per 10-km square in Britain, where breeding was probable or confirmed, from which they derived a population of about 570 pairs in Scotland with an additional 60 pairs in England, Wales and the Isle of Man. By extrapolation, the population in Ireland would be about 180 pairs. The *68–72 Atlas* estimate, also based on two nests per 10-km square, gave 500–600 pairs in Britain, and

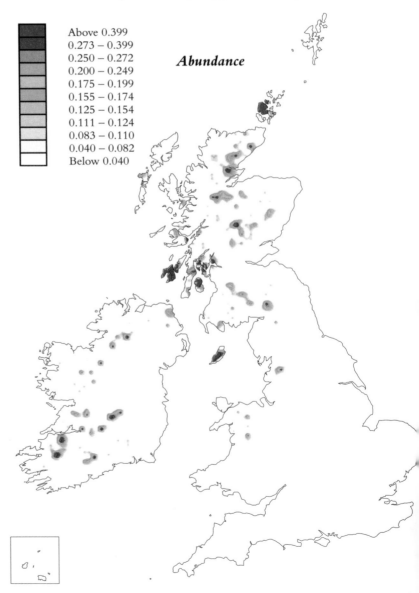

Above 0.399
0.273 – 0.399
0.250 – 0.272
0.200 – 0.249
0.175 – 0.199
0.155 – 0.174
0.125 – 0.154
0.111 – 0.124
0.083 – 0.110
0.040 – 0.082
Below 0.040

Abundance

200–300 pairs in Ireland. Therefore, over the last 20 years the total population size has probably remained the same in Scotland, but has declined in Ireland and England.

Hen Harriers are capable of breeding in their first year and have a mean brood size of 3.4 young per successful nest (Bibby and Etheridge 1993); given favourable conditions they have the potential to increase their numbers rapidly. At the time of the *68–72 Atlas* there was much optimism for the future of this species (Brown 1976, Watson 1977), which was still in the process of a population increase and range expansion that began during the 1939–1945 war. Unfortunately, at the current level of persecution and rate of habitat loss, there must be genuine grounds for concern for the long term future of this much vilified but spectacular raptor.

BRIAN ETHERIDGE

68–72 Atlas p 118

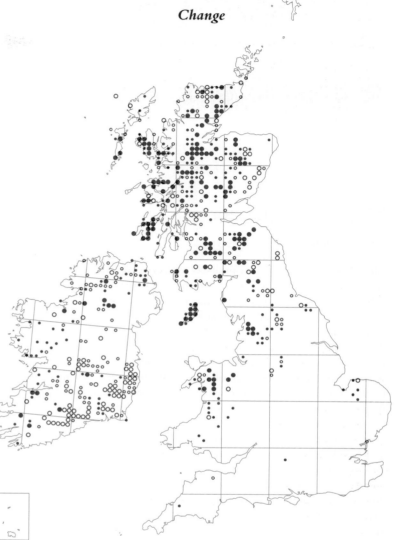

Change

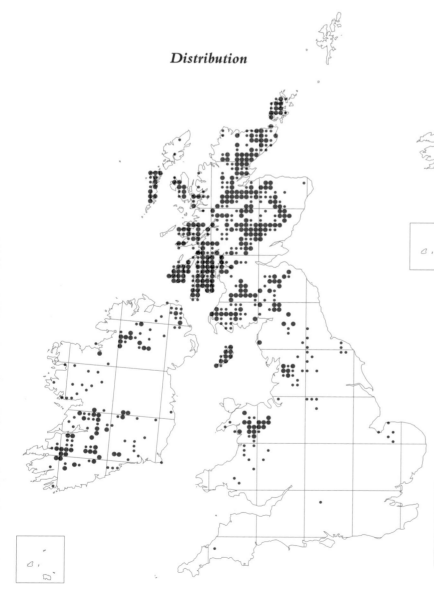

Distribution

Years	Present, no breeding evidence		Breeding evidence		All records		
	Br	Ir	Br	Ir	Br	Ir	Both (+CI)
1968–72	180	88	193	104	373	192	565
1988–91	212	77	286	46	498	123	621
			% change		33.5	−35.9	9.9

Montagu's Harrier

Circus pygargus

The Montagu's Harrier has probably always been a rare bird in Britain, and although nesting has been reported from many different areas and habitats over the years, few birdwatchers are familiar with this, the smallest of the four European species of harrier.

Sadly, the numbers are still very low, and the Distribution map confirms that breeding is confined to a few areas of S and SE England. The Change map shows a marked contraction, with no summering in South Wales where there were a number of records in 1968–72, and no breeding records for Devon and Cornwall, for many years the stronghold of this species in Britain. Breeding has also ceased in Ireland where the last recorded nest was in 1971.

In the mid 1950s, up to 30 pairs of Montagu's Harriers bred in a season, and breeding had been recorded in 22 counties in England and Wales, as well as in Scotland and Ireland. But numbers declined from the late 1960s, and despite a partial revival during 1963–68, continued to fall until there were no breeding records in 1974, and only one in 1975. There were 12 nests in 1990 and at least six in 1991 (Elliott 1988, Underhill-Day 1990, Image 1991).

The reasons for these fluctuations are hard to unravel, but similar declines in much of NW Europe during the 20 years to 1979 (Gammell 1979), suggest a widespread environmental cause. It is known that Montagu's Harriers in Britain started to lay thin-shelled eggs after the introduction of DDT in 1947, and it seems possible that the initial decline was caused by pesticide ingestion. The partial recovery from 1963 (which coincided with a recovery by other raptor populations as organo-chlorine pesticide use declined), did not however continue after 1968. It is believed that British Harriers, as well as those from neighbouring European countries, winter in the Sahel region of Africa, where drought began in 1968, and this could have affected winter survival and thus breeding numbers. In Sweden and Denmark, where breeding Montagu's Harriers, in common with other northerly populations, may winter further south, and in Russia where the population winters in SE Africa, numbers have been stable or increasing (*BWP*).

Montagu's Harriers have not been recorded wintering in Britain. Both sexes arrive during the first week of May, and start nest building within a couple of weeks. Three to five eggs is the usual clutch size, laid in the last week of May, and hatching occurs about 30 days later. Only the female incubates the eggs, and in Britain the mean fledged brood size is 2.3 per nest. About 4% of males have two or more mates. Some 20% of nests fail from natural causes, mostly predation by foxes, with 95% of failures occurring during incubation (Underhill-Day 1990).

The smaller size and manoeuvrability of Montagu's Harriers compared with other harrier species equip them for hunting small, agile prey, and studies have shown that small passerines form a large part of their diet in N Europe and Britain. They also catch small mammals, including young rabbits up to 300g, but in England mean live prey weight is only 35–50g (Schipper 1973, Underhill-Day 1990).

In the past, Montagu's Harriers have been heavily persecuted, suffering particularly from egg collectors. Nearly half of all known failures during 1900–83 were due to collecting. In recent years, however, nests have been protected, and since 1980 no failures from persecution or disturbance have been recorded.

Until 1968, no Montagu's Harrier nests had been found in arable crops, but since then, most nests have been in winter wheat, barley or oilseed rape, and during the last 10 years nest failures in these habitats have been only about 15%. It seems probable that without active steps to protect nests and young from farming operations, nest failures would have been much higher (Elliott 1988). In SW France, one study found that up to 90% of known nests in crops were destroyed by agricultural machinery (Berthemy *et al.* 1983).

Another cause of concern is the decline in the number of open farmland birds (*Trends Guide*), some of which, such as Skylarks, Meadow Pipits and Turtle Doves, form an important part of the diet of breeding Montagu's Harriers.

From 1983, nearly all young Montagu's Harriers have been ringed and colour ringed, and some of these birds have been identified as breeding adults in the same areas in subsequent years, in company with unringed birds. This suggests that the present small breeding population derives from British bred stock and from immigrants, and lends support to the efforts that have been made to protect the small number of nests in recent years.

The future of this attractive species as a breeding bird in Britain is still precarious, with an overall decline and range contraction since the *68–72 Atlas*. Numbers are still low enough for the species to be seriously vulnerable to natural fluctuations, and protection from disturbance and persecution will be necessary for some years.

JOHN UNDERHILL-DAY

68–72 Atlas p 120

Five records have been moved by up to two 10-km squares on the Distribution map. Montagu's Harriers bred in two of these during 1988–91 only, and these records have been moved on the Change map in the same manner as on the Distribution map. In the remaining three 10-km squares, however, Montagu's Harriers occurred during both Atlas periods. To maintain the confidentiality of these locations, which might otherwise be apparent by comparison of the Change and 1968–72 maps, six false records (three filled and three open dots) have been placed on the Change map. Although the 100-km square grid reference of the 1968–72 Irish probable breeding record was supplied to the BTO, the 10-km square reference was not.

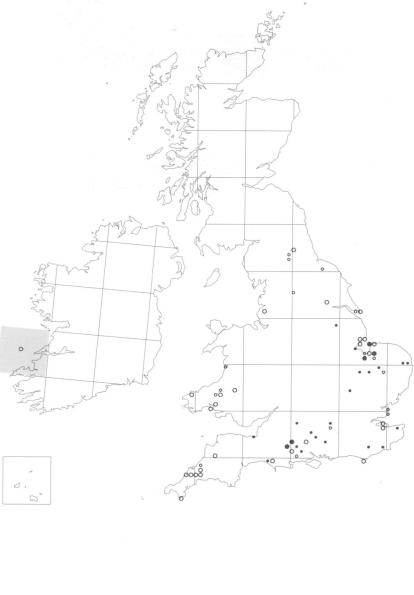

Change

Distribution

Years	Present, no breeding evidence		Breeding evidence		All records		
	Br	Ir	Br	Ir	Br	Ir	Both (+CI)
1968–72	20	0	30	1	50	1	51
1988–91	27	0	5	0	32	0	32
				% change	−36.0		−37.3

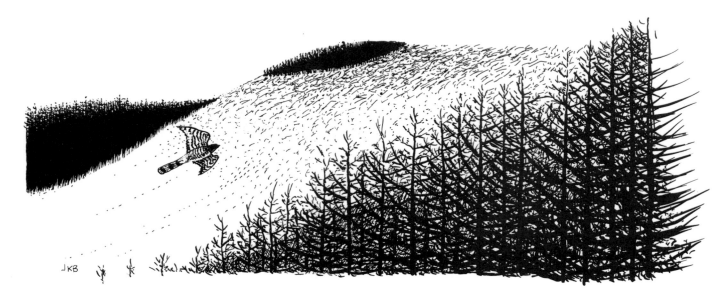

Goshawk

Accipiter gentilis

For such large birds, Goshawks are remarkably elusive because they use cover skilfully when hunting, and exclusively use woodland to roost and nest. Most Goshawk sightings are fleeting glimpses of low-flying birds, or more prolonged views of soaring birds, particularly in late winter and early spring when they display over their nesting territories. By far the commonest of Goshawk 'experiences', however, is a pile of blood-stained Woodpigeon feathers and a ravaged sternum (Petty 1989, Plate 5) decorating an upturned tree stump in woodland.

As a result of deforestation and persecution, the Goshawk was exterminated in Britain by the late 19th century, with only sporadic breeding thereafter. From the mid 1960s breeding became regular and by 1980 there were records of about 60 pairs, including 39 active nests (Marquiss and Newton 1982). At that time Goshawks were established in at least 13 widely separated areas. Breeding has apparently ceased in two of these areas but the true Distribution map shows that it still occurs in ten of them and in many other localities. The occupied squares are tightly clustered around some of the long established areas. This is to be expected because Goshawks are resident and young birds (particularly males) do not disperse far (e.g. Anon 1990) so breeding populations tend to become locally 'saturated' before spreading out into new areas. The reoccupation of the Netherlands proceeded in this fashion, the species taking over 15 years to spread from a few nests on the German border, 160km westwards to the forests of the coastal dunes (Thissen *et al.* 1981).

The increase and spread of the British population have been hampered by nest robbing and persecution. The theft of eggs and young by egg collectors and hawk keepers reduced production substantially in the 1970s (Marquiss 1981). However the breeding population is now much larger and many nests are in extensive woodland where they can be difficult to find, so that nest robberies, though still frequent, no longer dominate production, which is near the expected natural level of about 2.3 young per breeding pair (Petty 1989, Anon 1990). In contrast, the illegal killing of Goshawks by game managers continues apace (Cadbury 1991) and has probably increased. It appears to have exterminated one small population, prevented the proper establishment of another and halted the expansion of a third. Thus spread is now largely confined to state-owned forest. We

can only speculate on the overall impact of persecution because the total number of Goshawks killed remains unknown.

An important aspect of both academic and applied interest concerns the provenance of present day British Goshawks. They were largely, if not wholly, derived from imported birds which escaped from hawk keepers or were deliberately released. The Goshawks which started breeding in the mid 1960s were mainly small birds from Central Europe, whereas those established in the early 1970s were much larger and paler birds, mainly from Finland (Marquiss 1981). In places these populations are now close enough to mix and the outcome, large, small, or intermediate, should be of general interest both for taxonomy, and the application of animal reintroduction schemes.

Goshawks breed throughout the boreal and temperate zones of the northern hemisphere and are abundant. They do not have stringent habitat requirements, nesting in woodland and hunting a variety of medium-sized mammals and birds in woodland and open country. Breeding density varies; in Britain active nests have been spaced 6–8km apart in some areas but as close as 2km on richer ground, and in one study four pairs averaged 1.5km apart (Anon 1989). There is thus the potential for a substantial population of several thousand pairs. Goshawks are not easily counted and estimates of the numbers breeding in Britain have been heavily influenced by the efforts of a few observers who search out nesting places. The problems of disturbance, egg robbing and persecution have engendered secrecy so that until recently, conventional monitoring has been ineffective; the *68–72 Atlas* map was inaccurate and the reports of the RBBP show only rough trends. S.J. Petty (Forestry Commission Wildlife and Conservation Research Branch) has access to more information than any other authority, so his figure of about 200 pairs in Britain during 1988, is probably the most accurate recent estimate.

This is only a small proportion of the European population, which now probably exceeds 40,000 pairs (Bijlsma 1991). The British population is as yet of little international significance except as an interesting case history of an unplanned reintroduction using birds of various provenances.

MICK MARQUISS

68–72 Atlas p 110

Thirty-five records on the Distribution map, and slightly fewer (33) on the Change map have been moved by up to two 10-km squares. Fewer were moved on the Change map because Goshawks bred in some 10-km squares during both Atlas periods and thus do not feature on this map.

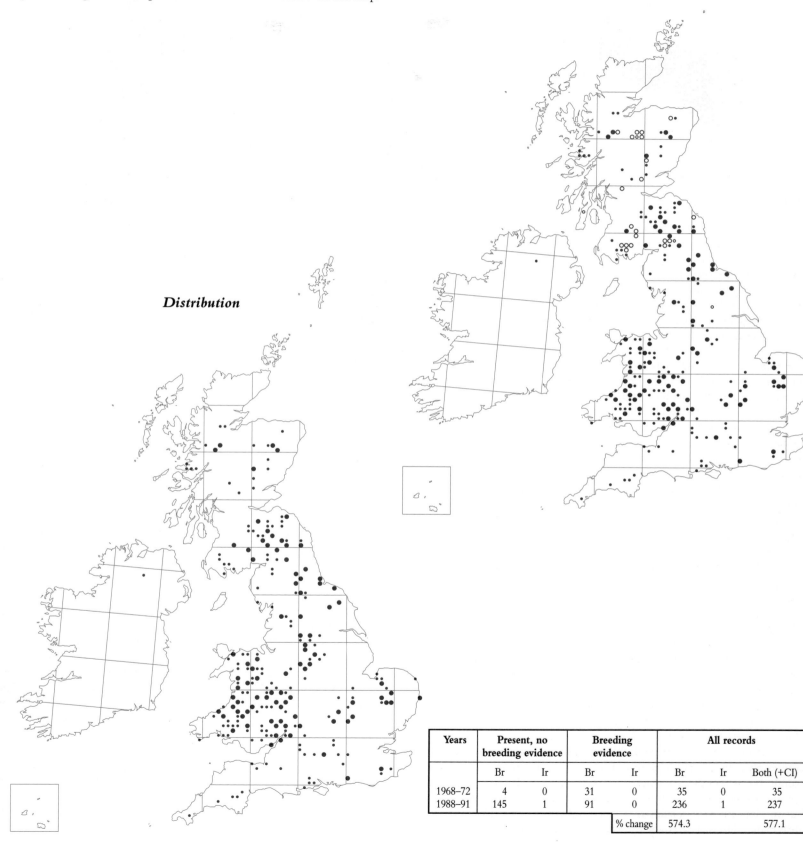

Change

Distribution

Years	Present, no breeding evidence		Breeding evidence		All records		
	Br	Ir	Br	Ir	Br	Ir	Both (+CI)
1968–72	4	0	31	0	35	0	35
1988–91	145	1	91	0	236	1	237
			% change		574.3		577.1

Sparrowhawk

Accipiter nisus

The secretive Sparrowhawk is one of the most widespread and abundant raptors in Britain, second in numbers only to the Kestrel. It breeds over most of the country, wherever there is suitable woodland and scrub, hunting whatever small birds are available locally. Its absence from parts of the Scottish Highlands and from some of the Northern and Western Isles can be attributed to lack of nesting habitat rather than lack of suitable prey.

The distribution and abundance of the species have increased considerably since the *68–72 Atlas* survey, as the population has continued to recover from the pesticide impacts of the late 1950s (Newton 1986). At that time, the Sparrowhawk population of Britain declined markedly, following the introduction of organochlorine pesticides in agriculture. These are extremely persistent, highly fat-soluble, and so readily accumulate in the bodies of birds. The Sparrowhawk, preying upon a wide range of other birds, accumulates concentrations of these chemicals large enough to depress reproduction and survival. The insecticide DDT came into wide use from 1947, and led to shell thinning and egg breakage, with a consequent reduction in the breeding rate. The more toxic cyclodiene compounds, such as aldrin and dieldrin, were used from 1956, and poisoned many adult Sparrowhawks. By 1960 population decline was evident over most of the country, and the species had almost disappeared from some eastern arable districts, where cyclodiene use was greatest (Newton and Haas 1984). Since then, following progressive reduction in the uses of the offending chemicals, the Sparrowhawk has gradually recovered. Compared with the *68–72 Atlas* survey, the species is much more strongly represented in the arable districts of E England. Although there is still some residual contamination and measurable shell thinning, the species can now be considered to have fully recovered in numbers. Indeed, it may now be more abundant than at any time this century, as the widespread afforestation of upland districts has provided new nesting habitat where previously there was none. In addition, since the *68–72 Atlas* survey the species has spread into many towns and cities, nesting in such places as parks and cemeteries. It now breeds in central London, and numerously in such cities as Bristol and Edinburgh. The population of Edinburgh alone numbers 30–35 pairs.

For nesting, the Sparrowhawk favours conifer woods, but it also nests commonly in broadleaved woods and tall scrub. It prefers forests and larger woods, but where these are lacking it will nest in small patches less than 1ha in extent, in riparian trees, or even in thick overgrown hedgerows, but not, apparently, in isolated trees. Pairs tend to nest in the same restricted areas in different years, in some cases forming traditional nesting territories

used over long periods. Such territories are recognisable to the human observer by the groups of nests of different ages, as the birds normally build a new structure each year.

In continuous forest, nest groups tend to be regularly spaced, reflecting the territorial behaviour of breeding pairs, and the mean distance between nests varies from 0.5km to several km, depending on soil fertility and

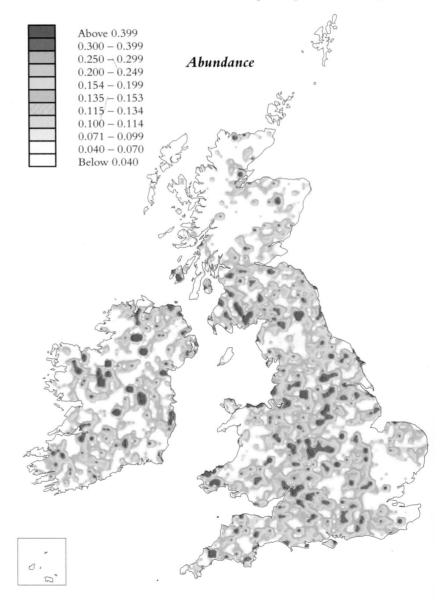

	Above 0.399
	0.300 – 0.399
	0.250 – 0.299
	0.200 – 0.249
	0.154 – 0.199
	0.135 – 0.153
	0.115 – 0.134
	0.100 – 0.114
	0.071 – 0.099
	0.040 – 0.070
	Below 0.040

Abundance

elevation. Close nest spacing occurs in lowland woods on fertile soil, with high songbird densities, while wide spacing occurs in upland forest on infertile soil with low songbird densities (Newton *et al.* 1986). Although dependent on woodland, Sparrowhawks also hunt in other habitats, from farmland and gardens to open treeless terrain and sea coasts, wherever small birds occur.

Sparrowhawks normally breed for the first time in their first, second or third year of life, depending on the availability of nesting territories (Newton 1986). They lay 3–6 eggs in a clutch, and usually at least half the nests in a given area produce young each year. Food shortage – which leads to non-laying or egg desertion – is the commonest cause of nest failure. Now that the pesticide problem has receded, the main threats to the species come from game preserving interests.

Suitably-aged forest and woodland covers about 9% of Britain's land surface, and is distributed mostly in small patches extending from sea level to more than 610m (2,000 feet). From knowledge of the relationship between nest spacing and altitude, and estimates of the area of appropriate-age woodland present in different altitude zones in Britain, Newton (1986) estimated the maximum number of Sparrowhawk pairs likely to breed in

Change

Distribution

Britain each year at around 32,000 pairs. The importance of low ground was shown by the fact that about 30,000 of these pairs would have bred below 300m (1,000 feet) elevation, and only 2,000 above this level. The extensive conifer plantations on high ground support Sparrowhawks at only low densities. By extrapolation, the population in Ireland is estimated at 11,000 pairs, bringing the total population for Britain and Ireland to around 43,000 pairs. To these should be added the non-breeders, mainly younger birds. Their numbers are hard to estimate but in some areas they could be as numerous as the breeders.

IAN NEWTON

68–72 Atlas p 108

Years	Present, no breeding evidence		Breeding evidence		All records		
	Br	Ir	Br	Ir	Br	Ir	Both (+CI)
1968–72	547	184	1276	616	1823	800	2623
1988–91	667	336	1511	327	2178	663	2845
			% change		19.6	−17.0	8.5

JB

Buzzard

Buteo buteo

Buzzards soaring over their nesting territories are a familiar and charac-teristic spectacle of spring in much of W and N Britain. Why then are they absent from much of central and E England and Scotland and most of Ireland?

The literature suggests that Buzzards once bred throughout most of Britain and perhaps of Ireland, but were reduced by persecution in the 18th, 19th and early 20th centuries to a relict distribution in the W of England, Wales and Scotland. After 1915, reduced persecution by gamekeepers permitted a gradual eastward recolonisation in Britain and, in the late 1950s, of N Ireland. However, early in the 1950s the rabbit (then a main prey of the Buzzard) went into catastrophic decline through myxomatosis, and in the 1950s and 1960s organochlorine pesticide con-tamination of other prey species reduced productivity and hence the possibility of continued range expansion (Moore 1957). Even now, breed-ing Buzzards are few and scattered in the eastern and central counties of England, much of the Central Lowlands of Scotland, and parts of E Scotland. They remain absent from most of Ireland.

There was little change in distribution in Britain between 1968–72 and 1988–91. The BTO Survey of 1983 showed that there had been some expansion of range since 1968–72, but that E and NE Scotland were exceptions to this trend, due, apparently, to persecution (Taylor *et al.* 1988). Fewer outlying pairs were recorded in the far S and E of the Buzzard's range in England in 1988–91 than in 1968–72, and the popu-lations in West Sussex and the Brecks had all but disappeared. On the other hand, Buzzards had considerably extended their distribution in Ireland where, in 1968–72, confirmed and probable breeding were reported from 19 10-km squares, and possible breeding from a further six. In 1988–91 the figures were respectively 44 and 49 with most of the spread in the north. Furthermore, the Change map documents a slight eastward expansion along the edge of its main range in England, Wales and S Scotland, with new populations in N Islay and NE Caithness.

The Abundance map suggests that the densest populations occur in the SW peninsula of England, in Wales and in parts of W Scotland and the Great Glen.

Numbers are greater where farm sizes are relatively small, with mainly pasture, abundant hedgerows and small woods (often on steep escarpments) and a high frequency of small ungrazed steep slopes of grassland and scrub. Such areas tend to sustain numerous small rodents and rabbits, the principal prey of most Buzzard populations. A constraint on Buzzard densities in upland areas is the limited availability of crag and tree nest sites.

The Buzzard is a versatile predator and, even where its principal prey species are scarce, it is able to persist at moderate or low densities by turning to other prey such as birds, squirrels, reptiles and moles. Thus, provided that established populations produce surplus young for col-onisation, there seems no reason why the species should not reoccupy much more of its former breeding range. In fact, much of central and E England contains apparently suitable habitat but it is possible that per-secution by gamekeepers is still a limiting factor. Picozzi and Weir (1976) reported 223 Buzzards killed on four estates in Scotland in 1968. Of 52 young Buzzards found dead within 20km of their Speyside study area, 28

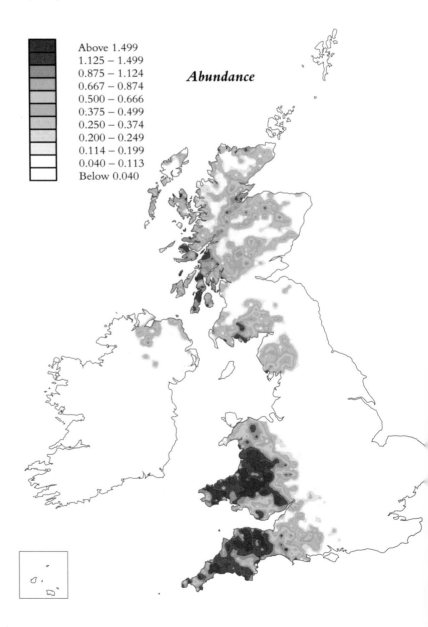

Above 1.499
1.125 – 1.499
0.875 – 1.124
0.667 – 0.874
0.500 – 0.666
0.375 – 0.499
0.250 – 0.374
0.200 – 0.249
0.114 – 0.199
0.040 – 0.113
Below 0.040

Abundance

were poisoned and eight shot or trapped. The RSPB recorded 302 incidents of Buzzard persecution during 1971–87: 98 shot, 16 trapped and 188 poisoned. In an analysis of 238 deaths reported during 1975–89, Elliott and Avery (1991) confirmed that most were due to alpha-chlorolose, mevinthos and strychnine. They found that Buzzards were especially subject to persecution at the edge of their range, and that this continues to limit the numbers of birds available to colonise new areas. It may also have been responsible for the loss of isolated colonising pairs at the eastern fringe of the range, where conspicuous gaps in Buzzard breeding distribution coincide with well-keepered estates.

The Buzzard breeds almost throughout the Palearctic, north of the Mediterranean to the limit for the boreal forest. Its range also extends eastward across the boreal forest zone of Asia. In Europe, estimates since the 1960s suggest a population in excess of 100,000 breeding pairs. In Britain, Moore (1957) estimated about 12,000 breeding pairs after the decimation of the rabbit by myxomatosis; Tubbs (1974) suggested 8,000–10,000 breeding pairs in the early 1970s; and Taylor *et al.* (1988) estimated 12,000–17,000 territorial pairs in 1983. There is a rough consistency between these estimates which accords with the relatively static breeding

Change

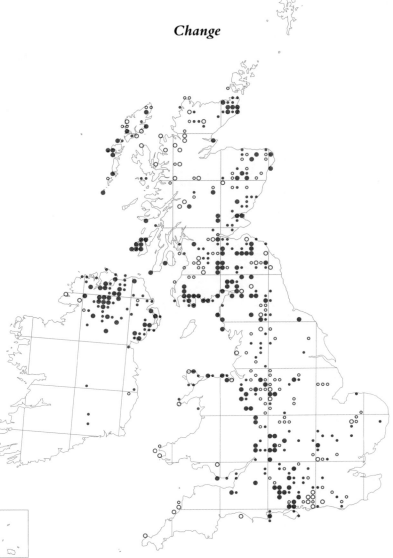

Distribution

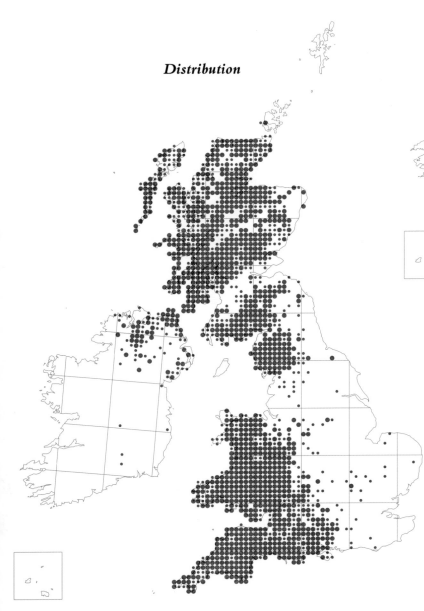

distribution since the early 1950s. The Northern Ireland population was estimated at more than 100 pairs in 1986 (*Birds in Ireland*) and that of the Republic at 26 pairs in 1989–91 (Norriss 1991). Given the continuing expansion in Ireland it is likely that the population there is at least 150 pairs.

COLIN TUBBS

Three records in Ireland and one in Britain have been moved by up to two 10-km squares on the Distribution and Change maps.

68–72 Atlas p 106

Years	Present, no breeding evidence		Breeding evidence		All records		
	Br	Ir	Br	Ir	Br	Ir	Both (+CI)
1968–72	312	6	1114	19	1426	25	1451
1988–91	370	49	1174	44	1544	93	1637
			% change		8.4	272.0	13.0

Golden Eagle

Aquila chrysaetos

Confined to the more remote hills and glens of NW Britain, the Golden Eagle is one of the few year-round inhabitants of this comparatively harsh upland landscape. A secure nesting site in a crag or ancient Scots pine, a sufficiency of food in the form of Red Grouse, mountain hare or similar sized open-country prey, and freedom from persecution or excessive human disturbance are the Golden Eagle's minimum requirements. The availability of deer or sheep carrion, particularly in winter, may also be important in some areas.

It occurs throughout the Scottish Highlands and most of the Hebrides and a handful of pairs maintains a precarious existence in SW Scotland, whilst a single pair continues to breed in NW England; there has been no breeding recorded in Ireland for over 30 years. Sightings in peripheral localities in SE Scotland and N England are mainly of immature or sub-adult birds which can travel several hundred kilometres from their natal site. That these birds do not settle and breed is probably a reflection of the amount of deliberate persecution and incidental disturbance by people in these areas, although shortage of nest sites may be a factor in some localities.

At first sight there are few differences in range between the two Atlas periods, although there are fewer breeding records in the *88–91 Atlas*. A comprehensive survey of the species in 1982–83 (Dennis *et al.* 1984), found confirmed or probable breeding in 287 squares, a figure very close to the 1968–72 result. The lower equivalent figure for 1988–91 probably comes about because less emphasis was put on proving breeding during the 1988–91 survey, rather than as a result of any reduction in the number of pairs breeding, or change in range.

There is no single explanation for the slight distributional changes detected between 1968–72 and 1988–91. Small increases in Ross and

Cromarty and E Sutherland may reflect improved coverage, whilst additional records in the S Grampians may indicate a modest reduction in the persecution of birds of prey on some estates in that area (A. G. Payne). Slight reductions in the number of squares holding breeding eagles in SW Scotland (Argyll and Galloway) are consistent with the putative effects of afforestation on Golden Eagle hunting ranges (Marquiss *et al.* 1985). Losses in the NE Grampians may reflect increased persecution or changes in land use such as increased deer culls which, through a reduction in winter carrion, were linked to an earlier decline in the E Grampians (Watson *et al.* 1989).

Watson *et al.* (1989) measured population parameters in nine areas of the Highlands and Islands, and related differences to land use and ultimately to food. They showed that density was related to the amount of carrion available in late winter, and thus highest densities occurred in W Scotland where sheep carrion was particularly plentiful. The Abundance map supports this finding. By contrast, breeding success was highest in the E Highlands and on the biologically richer lands of the Inner Hebrides, areas with greater amounts of live prey such as grouse, hares and rabbits.

A recent review of the Golden Eagle population in Europe west of the Urals (Watson 1992) put the total at about 5,600 pairs; within this area

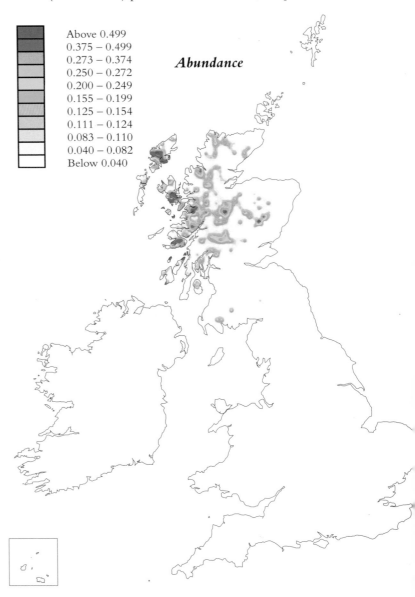

	Above 0.499
	0.375 – 0.499
	0.273 – 0.374
	0.250 – 0.272
	0.200 – 0.249
	0.155 – 0.199
	0.125 – 0.154
	0.111 – 0.124
	0.083 – 0.110
	0.040 – 0.082
	Below 0.040

Abundance

there are at least five distinct ecological groupings. The British population is part of a NW European montane group, which includes the populations in Norway and much of Sweden, and comprises some 1,880 pairs. Other groupings are centred on the E Baltic lowlands (810 pairs), the Alpine mountains (840 pairs), the W Mediterranean mountains (1,470 pairs) and the Balkans (600 pairs). The best estimate for the British population is some 420 breeding pairs in 1982 (Dennis *et al.* 1984), and although a repeat survey was undertaken in 1992 the data are not yet available to modify this estimate. Using 1982 data the British population, is around 7.5% of the European population or 22.3% of the NW European ecological grouping.

JEFF WATSON

Three records have been moved by up to two 10-km squares on the Distribution map. One of these was obtained during a timed visit, and this record has been moved in the same manner on the Abundance map.

68–72 Atlas p 104

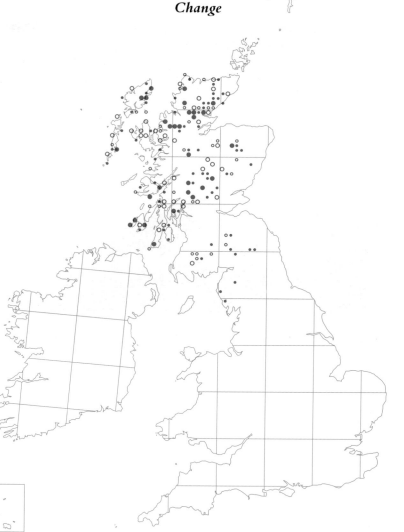

Change

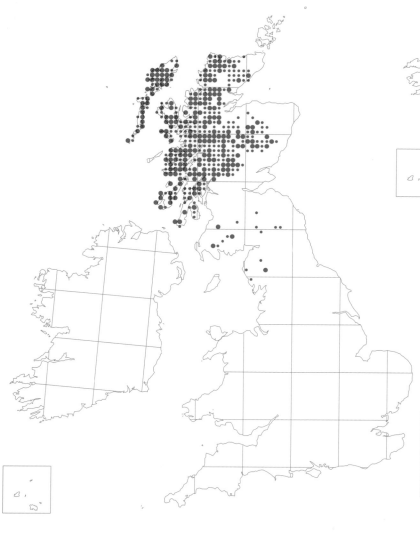

Distribution

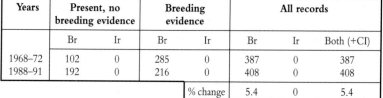

Years	Present, no breeding evidence		Breeding evidence		All records		
	Br	Ir	Br	Ir	Br	Ir	Both (+CI)
1968–72	102	0	285	0	387	0	387
1988–91	192	0	216	0	408	0	408
				% change	5.4	0	5.4

Osprey

Pandion haliaetus

The spectacular sight of an Osprey diving into water to catch a fish is once again a normal part of the Scottish scene, and people elsewhere in Britain and Ireland can see this large raptor on passage with increasing frequency. The famous RSPB site at Loch Garten has attracted one and a half million visitors to view the Ospreys at their eyrie.

When, in the 1950s, Ospreys returned to breed again in Scotland, they chose to nest in the Scandinavian-like pine forests, but now they are to be found from inland glens to coastal estuaries throughout a large part of Scotland. They occur mainly in the east and central areas, with no reoccupation, as yet, of their ancient haunts in the NW Highlands. During the survey period of this Atlas, breeding was recorded in 39 10-km squares, covering an area of 220km from north to south and 150km from east to west.

Immature birds return to the nesting areas in the summers prior to breeding and the spread of sightings in Scotland relates to these birds. Many of them make visits to occupied nests, and such sightings of 'intruder' Ospreys at the eyries are common. Observations in England, Wales and Ireland may involve passage birds, but evidence of summering in new areas is ever increasing, and the habit is often the forerunner of breeding attempts.

Comparison with the *68–72 Atlas* illustrates how successful the recolonisation has been. Breeding in 11 10-km squares by 1972 had increased to over three times this number by 1988–91.

The breeding population in Scotland is regularly monitored by RSPB staff and other ornithologists. Despite nest protection schemes, the robbing of eggs has been a serious problem during the Atlas survey years. Eleven out of 49 nests were robbed in 1988, and nine from a similar number of pairs in 1989. Overall, between 1954 and 1990, 9.3% of clutches were illegally taken (Dennis 1991). Other failures are due to storm damage, nest competition between adults, human disturbance and the death of chicks, especially soon after hatching, due to bad weather and shortage of fish. Overall breeding success in the last ten years has ranged from 1.12 young per nesting pair in 1987, to 1.68 young in 1981.

The pattern of expansion has generally involved a pair building an eyrie in a new area, to which they return the following season and lay eggs. As long as this pair or eyrie remains successful, new pairs are attracted to the area in subsequent years to form a loose colony. Not all 'trial' nests result in successful colonisation of new areas, and from observations of intruders at established nests it is clear that most young birds prefer to gain access to established sites.

To date all successful nests have been built in trees, the majority (70%) in Scots pine, with smaller numbers in seven other species of tree (Dennis 1987b). Trial nests have been built on electricity pylons (twice) and a high stone monument, but no birds have used natural rock pinnacles as occurred in the last century in the NW Highlands. Since the early 1970s, storm damage to natural eyries has encouraged pairs to become established in new areas. Of 70 nests used in recent seasons, 41 were natural, 14 were originally built by Ospreys but had been restored by man, and 15 were artificial nests in new areas.

The world population has been estimated to be 25,000–30,000 pairs (Poole 1989), with about 3,000 pairs in Scandinavia and the Baltic countries. The Mediterranean population is 60–70 pairs, with only the Corsican population increasing in recent years (J.-C. Thibault). One pair recolonised mainland France in 1984 and there are now four pairs there (J.-F. Terrasse). The Scottish population is increasing annually, with 62 pairs in 1990 and 72 pairs found at nests in 1991, although not all pairs lay eggs each year. This recolonisation is well documented (Dennis 1983a). As the population grows it is more difficult to locate new sites and to be certain of exact numbers, but there is every likelihood that numbers will continue to increase, with the colonisation of new areas. Eventually the species should become re-established throughout suitable areas of Britain.

ROY DENNIS

68–72 Atlas p 122

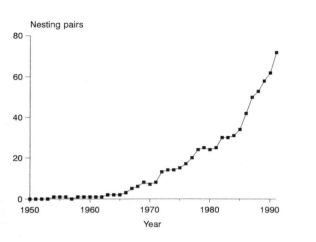

Annual number of nesting pairs of Ospreys in Scotland.

Distribution

Change

With the exception of NH91 (the well known Loch Garten site) all records on both the Distribution and Change maps are placed centrally within the 100-km squares in which breeding occurred during 1988–91. Ospreys bred in NH91 during both Atlas periods, thus this 10-km square does not feature on the Change map.

Years	Present, no breeding evidence		Breeding evidence		All records		
	Br	Ir	Br	Ir	Br	Ir	Both (+CI)
1968–72	14	0	11	0	25	0	25
1988–91	129	2	39	0	168	2	170
				% change	572.0		580.0

Kestrel

Falco tinnunculus

The Kestrel is familiar to many people, partly because it is so widespread and common, and partly because its hovering flight distinguishes it from other British raptors. Although Kestrels feed mainly on voles, they are adaptable predators and can find food and nest sites almost anywhere, including the centres of our largest cities.

This adaptability doubtless explains their extensive distribution in Britain and Ireland. They are found in most habitats, though numbers are highest during good vole years in the rough grassland of upland areas such as S Scotland and the Pennines (Village 1990). Kestrels have long been scarce in NW Scotland, and their hold on some of the islands is tenuous. The last reliable breeding record for Shetland was 1905 (*Birds in Scotland*), though a bird was seen there during the present survey. These areas may have a low food supply, linked to poor quality of the soil and a depleted small mammal fauna (Shetland has no voles).

The Abundance map suggests that the scarcity in NW Scotland extends in some degree to Ireland, SW Wales and parts of SW England. The present strongholds seem to include parts of SE Britain, particularly East Anglia. This is in marked contrast to the *68–72 Atlas*, when Kestrels were still scarce in East Anglia and the Cambridgeshire fens because of the lingering effects of pesticides (*68–72 Atlas*, Village 1990). Populations in these areas recovered in the 1970s, but there was no long-term increase in a fenland population monitored during the next decade (Village 1990).

The increase in SE England contrasts with a decline in more western areas. This is also apparent in the CBC index, where the national decrease since 1972 was largely due to plots in western England and Wales (*Trends Guide*). There is no obvious, single factor that might explain this decline, or any reliable data indicating how long it has been happening. Kestrel breeding density is difficult to measure accurately (Village 1990), and even large changes in abundance might go unnoticed if they occurred over

many years. The decline in the Scottish Isles seems to have lasted most of this century (*Birds in Scotland*), but no change in status has been reported previously for Ireland (*Birds in Ireland*). Perhaps it is only in marginal areas that a general decline becomes evident through extinctions.

The factors controlling breeding density may vary from place to place. Kestrels do not build nests and rarely breed unless they find a suitable site such as a rock ledge, disused stick nest, tree hole or somewhere on a building. Breeding density may be limited by lack of nest sites in extensive open areas with plenty of voles, but this is unusual in Britain.

In most habitats, density is likely to be regulated by the supply of food. Kestrels are largely migratory in upland areas of the north and west (*Winter Atlas*). Breeding density in one area of young conifer plantations in S Scotland was determined by the abundance of voles in spring (Village 1990). Kestrels in this study bred in disused Crow nests, which were unevenly distributed across the landscape. In good vole years, more Kestrels settled in the area in spring and more were able to acquire nest sites because territories were generally smaller than in poor vole years. In peak vole years, some pairs bred only a few metres apart in the small shelterbelts.

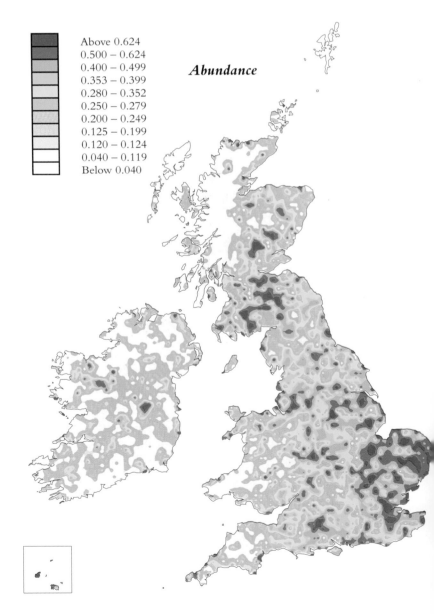

	Abundance
	Above 0.624
	0.500 – 0.624
	0.400 – 0.499
	0.353 – 0.399
	0.280 – 0.352
	0.250 – 0.279
	0.200 – 0.249
	0.125 – 0.199
	0.120 – 0.124
	0.040 – 0.119
	Below 0.040

In SE England, however, most Kestrels are sedentary, so breeding density depends on both spring and winter food supply. A study in arable farmland in the E Midlands found that mortality was high in harsh winters, particularly among juveniles (Village 1990). Individuals that did survive until spring were more likely to breed if voles were plentiful than if they were scarce. Most pairs nested in holes in trees but even in arable fenland, where these were scarce, there was no evidence that shortage of nest sites limited density. Instead, the fragmented vole habitat caused a low food supply in most springs, and this directly prevented some birds from breeding.

Where breeding density in Britain has been measured accurately, it seems to vary from about 36 pairs per 100km² in good vole years in grassland, to 10 pairs per 100km² in intensive arable farmland (Village 1990). The higher of these densities would be comparatively unusual, so a mean density for the whole of Britain is probably nearer 20 pairs per 100km². Applying these figures to the 2,481 occupied 10-km squares gives lower and upper limits to the total British population of 25,000 and 89,000, with a likely average of about 50,000 pairs. By extrapolation from this latter figure, the Irish population is about 10,000 pairs. The apparent

Change

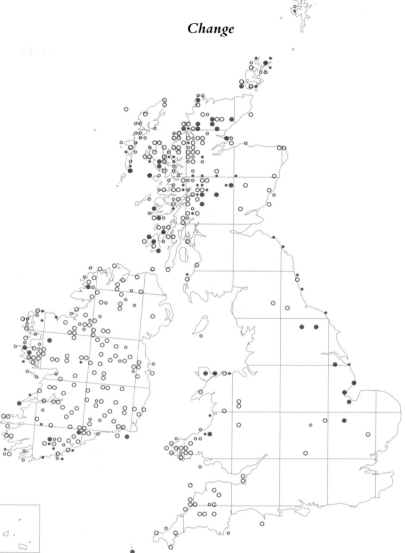

Distribution

decline in the west may not be threatening to a species that occurs across Europe, Asia and Africa, but it needs to be verified and explained. If it is genuine, it might be either an interesting natural shift in distribution or, more worryingly, a detrimental side effect of habitat change.

ANDREW VILLAGE

68–72 Atlas p 130

Years	Present, no breeding evidence		Breeding evidence		All records		
	Br	Ir	Br	Ir	Br	Ir	Both (+CI)
1968–72	283	164	2307	783	2590	947	3542
1988–91	489	416	1992	388	2481	804	3298
			% change		−4.1	−15.0	−7.0

JB

Merlin

Falco columbarius

The Merlin is a small dashing falcon which breeds on moorland and winters on lower ground or coasts. At the time of the *68–72 Atlas* little was known about the bird in Britain and Ireland but recent studies have shed more light on its status and ecology.

The Distribution map shows Merlins breeding thinly across the major areas of heather moorland. Proving breeding can be time consuming, and some of the small dots undoubtedly reflect breeding places where the observer missed proof of breeding. Non-breeding and hunting birds can be seen more widely and some small dots are away from known or likely breeding areas. The distribution pattern picks out clearly the Welsh mountains, the Pennines, and the Southern Uplands. In the Scottish Highlands, Merlins are more abundant on the drier more easterly hills, and seem surprisingly thinly spread in the Western Highlands. This might be because of poorer land productivity in the wetter and colder west, but this is also an area thinly endowed with ornithologists. The Distribution map of the Red Grouse suggests that heather moorland is more extensively distributed than Merlins. The Abundance map, however, which shows the same eastern bias in the Highlands is corrected for such variation in observer numbers, thus the effect is likely to be real.

Merlins nest either on the ground in the cover of heather or in old Crow nests (Bibby 1986, Newton *et al.* 1986). Ground nesting is a remarkable feature of Merlins in Britain relative to a world perspective. It seems likely that ground nesting is possible only because of the scarcity of mammalian nest predators on moors managed for grouse shooting. Merlins require extensive open ground for hunting, and take a wide range of small birds (Bibby 1987). The reasons why they prefer heather moor are obscure, because their main prey species, Meadow Pipits and Skylarks, are as common on grassy areas as on heather moors. Some pairs have been displaced by extensive afforestation especially in Northumbria and SW Scotland (Newton *et al.* 1986, Bibby and Nattrass 1986), but forestry has

not generally been on a scale to threaten Merlin populations greatly. Indeed, in recent years, pairs have started nesting on the edges of conifer plantations in Wales, Northumbria and Ireland. It is not known when this habit developed because forest nests are not easy to find. The change seems to have started in the late 1970s and is probably still both developing and being discovered. These pairs still feed primarily on open country birds.

The Change map shows some real reductions, especially in SW England, Anglesey, SW Scotland and the Borders. In some more marginal areas, this is probably due to loss of habitat as heather moorland has been converted to grass, by grazing or reclamation. The distribution in the Highlands looks to have expanded, but this may reflect the much increased level of interest in Merlins by raptor study groups. There may have been some declines in Ireland, but differences in coverage for the two Atlases is a possible explanation. The apparently improved status in Northern Ireland reflects an increased local interest which the Republic has not shared. Forest nesting seems to be widespread in Ireland and birds may be overlooked where there are few raptor specialists.

Merlins seem to have declined in numbers from about the 1950s, with pesticide contamination implicated. Organochlorine contamination has

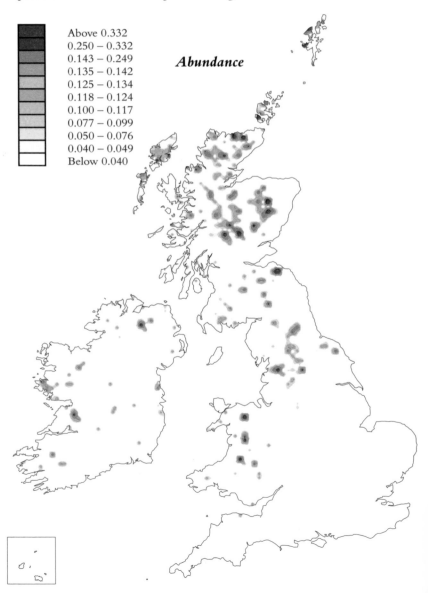

Above 0.332
0.250 – 0.332
0.143 – 0.249
0.135 – 0.142
0.125 – 0.134
0.118 – 0.124
0.100 – 0.117
0.077 – 0.099
0.050 – 0.076
0.040 – 0.049
Below 0.040

Abundance

recently declined somewhat, but remains unsatisfactory (Newton and Haas 1988). Northern populations in Orkney, Shetland and, possibly, the Hebrides carry inexplicably heavy mercury burdens. *Red Data Birds* and *Birds in Scotland* record concern over recent Merlin losses when other raptors have increased as a result of reduced pesticide contamination. Loss of moorland habitat is seen as a continuing threat. Trends in Wales and Northumbria suggest that Merlins may now be in a phase of recovery, partly associated with their recent use of forest margins. In most parts of its world range the Merlin is a bird of forest margins interspersed with open ground, and British and Irish birds may now be becoming more typical in their nesting habitats.

An estimate for British numbers in 1983–84 was 550–650 pairs (Bibby and Nattrass 1986). By extrapolation, the size of the Irish population in 1988–91 (assuming no change in the British population) was 110–130 pairs. Greater knowledge may well increase these numbers and reveal whether declines reported in the last twenty years are now slowing or being reversed. The increase in number of 10-km squares with breeding

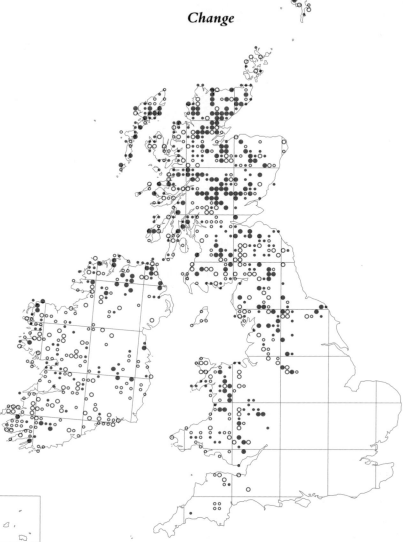

Change

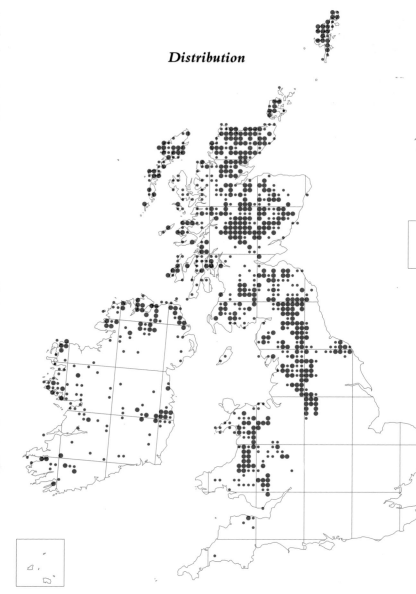

Distribution

evidence in Britain is certainly encouraging. The species has a circumpolar distribution and is much more numerous in Scandinavia than in Britain and Ireland. As a result, whilst our numbers form the entire EC population, they are trivial in European and World terms.

COLIN BIBBY

Two or three pairs of Merlins bred on the Isle of Man in 1991. However the 10-km square grid reference(s) were not supplied by the observer so these records are not shown on the maps.

68–72 Atlas p 128

Years	Present, no breeding evidence		Breeding evidence		All records		
	Br	Ir	Br	Ir	Br	Ir	Both (+CI)
1968–72	259	143	339	102	598	245	843
1988–91	307	100	386	58	693	158	851
				% change	15.9	−35.5	0.9

Hobby

Falco subbuteo

Mention of this slender-winged falcon conjures up an image of a Swift-like shape hawking insects on a warm May evening over a patch of pines on some southern heath or downland. But these are habitats under threat, so what of the Hobby's future here?

Heaths apart, Hobbies frequent farmland with pine clumps or mature broadleaved hedgerow trees (e.g. Campbell and Ferguson-Lees 1972), and the maps here are most reassuring. That for Change demonstrates a marked northward spread. The fact that some 60 mainly southwestern 10-km squares have seemingly lost Hobbies since 1968–72 might suggest a population shift but, in all, there are 429 'new' squares on what remains an evenly patterned Distribution map.

The traditional stronghold – Dorset to Berkshire and West Sussex – had begun to spill over the Severn-Wash line by 1968–72, but only into 27 10-km squares; by 1988–91 this figure had grown over sevenfold to 194. All counties west and north to Gwent, Powys, Shropshire, Derbyshire and Humberside were involved, along with more isolated breeding records for North Yorkshire and Tyne and Wear.

A comparable sixfold increase, from 17 to 104, east of a line from the Wash south to Beachy Head included a spread into the ideal habitats of East Anglia's Breckland. But the colonisation of Scotland – a possibility in 1968–72 – did not develop: the four records there, and the three in Ireland, probably related to off-course migrants or yearling wanderers. As in SW peninsular England, higher summer rainfall may be a limiting factor.

Hobbies were recorded infrequently during timed tetrad visits, but this is hardly surprising for a species that is often unobtrusive except in courtship or with young. Indeed, Fuller *et al.* (1985) believed that Hobbies had been seriously under-recorded in the *68–72 Atlas*. Even so, the evidence of increasing numbers and range cannot be doubted.

Hobbies breed in disused nests, especially of Crows. In Britain, 92% of clutches, usually three eggs, are completed during 6–25 June (Fiuczynski and Nethersole-Thompson 1980). Incubation, fledging and independence each take around 30 days so the families hatch in mid July, fledge in mid August and leave about mid September.

Although largely insectivorous for much of the year, Hobbies switch to bird prey when breeding – like Eleonora's Falcon, whose very late broods coincide with autumn migration through the Mediterranean. Summer nesting by Hobbies similarly allows the chicks to be raised when young birds are plentiful, notably inexperienced juvenile hirundines and Swifts (e.g. *BWP*, Parr 1985) along with Skylarks and roosting sparrows.

Pesticides seem to have had little effect on this species (Fiuczynski and

Nethersole-Thompson 1980), perhaps because of its high proportion of aerial prey. Shooting (especially on migration), egg collecting and, not least, nest predation by Crows and squirrels after human disturbance appear the main threats.

The Hobby breeds from Iberia to Japan, and winters mainly in southern Africa and N India. Although we think of it as being on the northern fringe of its range in Britain, over much of Eurasia it reaches the Arctic Circle.

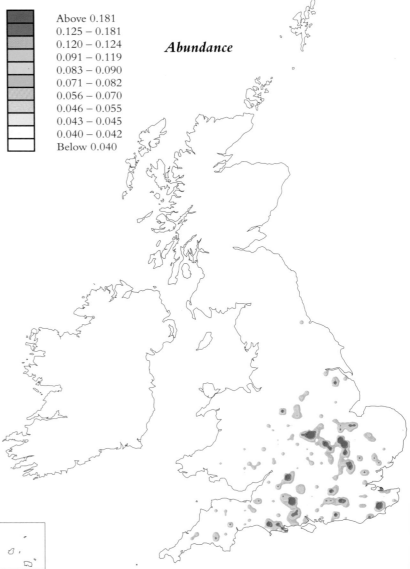

Above 0.181
0.125 – 0.181
0.120 – 0.124
0.091 – 0.119
0.083 – 0.090
0.071 – 0.082
0.056 – 0.070
0.046 – 0.055
0.043 – 0.045
0.040 – 0.042
Below 0.040

Abundance

In the late 1970s the West Palearctic held perhaps 20,000 pairs (based on *BWP*, Gensbøl 1984): Britain, Netherlands and European USSR were the only countries reporting increases, though subsequently Portugal rose from 'few' to 'a minimum of 300 pairs' (Rufino *et al.* 1985). Cade (1982) calculated the world population at 30,000–60,000 pairs, but the upper figure must be nearer the mark.

In the 1950s the British population was put at 60–90 pairs. The *68–72 Atlas* concluded: 'close to, or may exceed, 100 pairs'. But, finding 3–4 pairs per 100 km² in the S Midlands, Fuller *et al.* (1985) proposed 'a conservative average of two pairs' for all recorded squares. That would suggest over 500 pairs in 1968–72, and a remarkable 1,250 in 1988–91, but many peripheral Atlas observations will relate to yearling wanderers.

Compare these figures with the RBBP data in the table; annual reported maxima increased nearly sevenfold, from 65 pairs in 1973 to 435 in 1990. The *Trends Guide* found that in 1968, 1978 and 1988 Hobbies occupied 1%, 2% and 5% of CBC plots, with a clear increase from 1976; thus, 100 pairs in 1968 might represent 500 by 1988. All in all, a 1988–91 breeding population of 500–900 pairs seems a fair estimate.

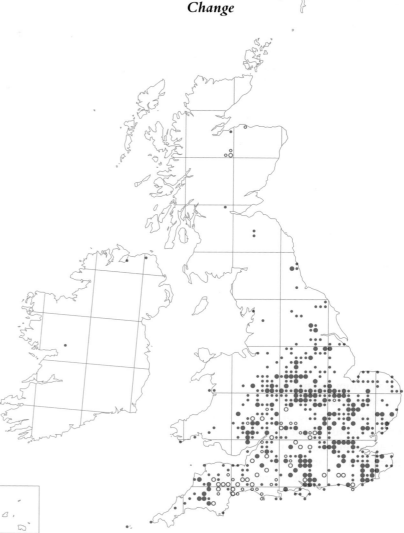

Change

Distribution

The important monograph by Fiuczynski (1987) includes a 17-page bibliography.

JAMES FERGUSON-LEES

Two-year averages for Hobbies in Britain 1973–90

	73–74	75–76	77–78	79–80	81–82	83–84	85–86	87–88	89–90
Confirmed	36	49	61	72	74	87	95	106	147
Possible	45	89	77	72	107	149	175	195	269
Max total	81	138	138	144	181	236	270	301	415
Min Young	38	56	87	79	76	98	122	147	222

Based on RBBP reports: the first three lines relate to pairs breeding, the last to young reared.
68–72 Atlas p 124

Years	Present, no breeding evidence		Breeding evidence		All records		
	Br	Ir	Br	Ir	Br	Ir	Both (+CI)
1968–72	122	0	137	0	259	0	259
1988–91	395	3	230	0	625	3	628
			% change		141.3		142.5

Peregrine

Falco peregrinus

The almost cosmopolitan Peregrine has acquired a modern symbolism, to add to the glamour it inherited down the ages, from the earliest days of falconry, some 4,000 years ago. The collapse of its population in many parts of the world through the effects of organochlorine insecticides became a *cause célèbre*, at the forefront of the new age of concern for the problems of global pollution, and revealing the potential hazards for humankind itself.

The *68–72 Atlas* charted initial recovery in the British and Irish populations, from the trough of 1963–64, when the Peregrine maintained its previous numbers and breeding success only in parts of the Scottish Highlands. Decline in numbers and/or breeding performance in most other regions ranged from total to significant (Ratcliffe 1980). The Distribution and Change maps amount to a celebration of its fuller recovery. Numerically, the overall British population is now at its highest known level, but the species remains below pre-1940 numbers in Sussex and Kent, E Yorkshire, the N and W Highlands, the Hebrides, Orkney and Shetland. Overall, the Irish population appears also to be at its highest known level.

Remarkably, numbers in parts of the E Highlands, S Scotland, N England and Wales are now at least double those ever known before. This super-recovery has involved both increase in breeding density and occupation of new or long-deserted breeding haunts. Increase in breeding density has often involved doubling or even trebling of pairs in old territories. The highest breeding density is now eight pairs per 10-km square (one pair per 12.5 km^2), and many 10-km squares have four or five pairs. Expansion into new breeding areas has involved much increased use of 'walk-in' nest sites in tiny crags, and genuine ground nests in at least eight different areas. A tenth of the population is nesting in quarries, many still worked. Nesting on buildings has recently occurred in at least 12 localities, and in old Raven nests in Scots pines in two widely separated areas (two in one, both failed; one in the other, successful). Many small dots indicate prospecting birds or pairs in possible new nesting areas.

Recovery was fuelled first by the phased withdrawals or reductions in use of the organochlorine insecticides – in particular, dieldrin, aldrin and heptachlor – regarded as the main cause of the post-1955 population crash, by inducing much increased adult mortality. Sub-lethal effects, notably eggshell thinning caused by DDT, also contributed to the decline by reducing breeding performance. Contamination of the Peregrine popu-

lation by dieldrin residues is now very low, but levels of DDE (metabolite of DDT) still remain sufficient to cause slight eggshell thinning in some districts (Newton *et al.* 1989). All these organochlorines are now banned under EC regulations, but their highly persistent residues will remain in environmental circulation for some time.

The failure of some depleted population segments to recover is mysterious. In SE and E coast areas of England lingering pesticide effects seem possible, but in far northern coastal parts of Scotland, where Peregrines feed much on seabirds, the more probable cause is a cocktail of pollutants acquired through the marine food chain. Analyses of falcons and their eggs from this last region show higher levels of PCBs and mercury than for inland birds, but these are only two of many marine pollutants which could affect Peregrines.

Super-recovery has depended on other factors besides the removal of pesticide problems. The publicity given to the Peregrine's plight, combined with an active campaign of protection, has helped to reduce adult mortality and increase output of young. During 1977–82, the established breeding population in S Scotland showed the astonishingly low annual mortality of 10% (Mearns and Newton 1984). The resulting pressure of new birds seeking nesting places could explain occupation of previously

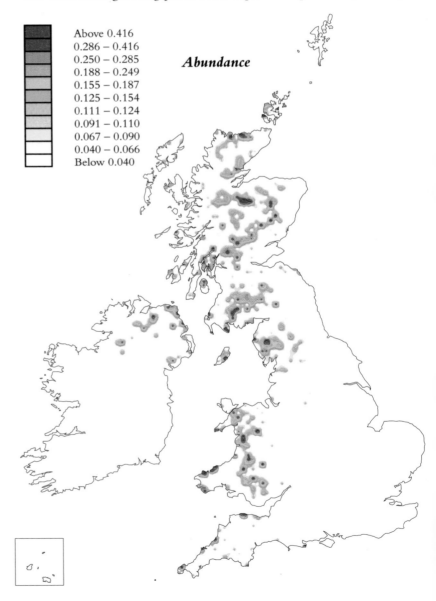

Abundance

| Above 0.416 |
| 0.286 – 0.416 |
| 0.250 – 0.285 |
| 0.188 – 0.249 |
| 0.155 – 0.187 |
| 0.125 – 0.154 |
| 0.111 – 0.124 |
| 0.091 – 0.110 |
| 0.067 – 0.090 |
| 0.040 – 0.066 |
| Below 0.040 |

vacant areas, but increase in density presumes a slackening in territorial demands, and this seems unlikely without an increase in food supply. The main possibility here is the undoubted increase in availability of domestic pigeons, which in many districts are the major food item for Peregrines during the breeding season.

Ringing recoveries are mostly within 100km of birthplace and show that, although the British and Irish populations mingle, there is hardly any emigration to continental Europe. The population recoveries in France, Germany and Switzerland have thus depended on their own performances. The former winter influx of Continental Peregrines to these islands was evidently mainly from Scandinavia, where the species remains much reduced and rare.

The countrywide survey of the Peregrine breeding population in Britain in 1991, organised by the BTO, RSPB, NCC and regional raptor groups, showed at least 1,100 pairs. Northern Ireland had another 99 pairs and the Republic of Ireland an estimated 265 pairs, giving an overall total of at least 1,465 pairs. Mean fledged brood size was 2.22 young and productivity 1.28 young per territorial pair.

DEREK RATCLIFFE

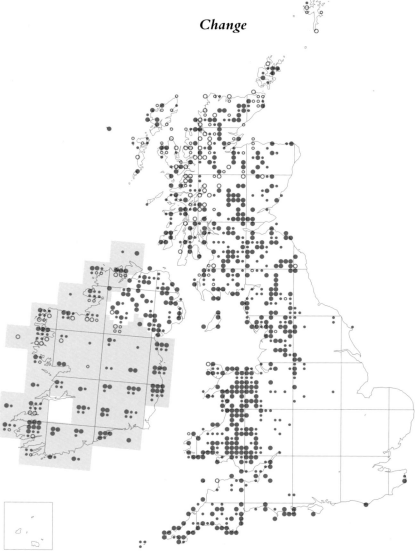

Change

Distribution

All records from the Republic of Ireland are centralised within 50-km squares on the Distribution and Change maps. To maintain this confidentiality the Abundance map only includes data from the UK.

68–72 Atlas p 126

Years	Present, no breeding evidence		Breeding evidence		All records		
	Br	Ir	Br	Ir	Br	Ir	Both (+CI)
1968–72	157	38	358	71	515	109	624
1988–91	328	116	720	174	1048	290	1338
			% change		103.9	166.1	114.4

Red Grouse

Lagopus lagopus scoticus

The Red Grouse is a subspecies of the Willow Ptarmigan, a bird with a circumpolar distribution that inhabits sub-Arctic habitats, but it differs from the Willow Ptarmigan in a number of interesting ways. Firstly, it feeds almost exclusively on a diet of heather, and consequently its distribution throughout Britain is restricted to heather-dominant moorland. Secondly, it does not moult into a white plumage, making it obvious and vulnerable to predators during snowy conditions. Thirdly, given the correct conditions, Red Grouse can reach remarkably high densities: post-breeding densities of 650 birds per km² have been recorded in North Yorkshire (Hudson 1986). This characteristic, together with the facts that it is a gamebird, good to eat and a challenging target, has encouraged owners and managers of upland estates to manage their grouse populations to produce a harvestable surplus.

Densities of grouse, estimated from bag records, show large scale variations, both throughout their distribution and through time. The greatest numbers shot per km² are centred in North Yorkshire and decline the further north and west the population is located. Three factors account for most of this variation: numbers of grouse shot per km² increase with an index of heather productivity (based on summer temperature and rainfall); July temperature; and the density of keepers employed on the estate, taken as a measure of the degree of fox and Crow control rather than a measure of habitat management (Hudson 1992). The Abundance map shows lower densities in Wales and Ireland than elsewhere.

Grouse populations have shown a long-term decline in numbers, and this is reflected in the Change map. In some parts of their distribution such as Wales, the Pennines, the Lake District and the Scottish Borders, severe overgrazing by sheep has caused the loss of heather-dominant moorland, and consequently of suitable habitat. Additional loss has occurred through the sale of moorland for commercial afforestation. Even after excluding estates that are no longer managed as grouse moors, the numbers shot have fallen, overall by 50% over the past 60 years. This decrease has occurred principally during two periods, the first between 1930 and 1950, and the second between 1975 and 1985. The rate of decline during both these periods was greater in populations in the north and west of the country. In S and E Scotland the decline in grouse density has been associated with an increasing population of foxes, following the recovery of rabbits after myxomatosis, and a decrease in the number of keepers employed on grouse moors (Hudson 1992).

Grouse populations throughout most of their range exhibit cyclic fluctuations in numbers, with the period of these fluctuations increasing with latitude, from 4–5 years in N England to 6–8 years in Scotland. Small isolated populations, and those on moors with low rainfall, tend not to show such cyclic changes in numbers. Two hypotheses have been proposed for these cyclic fluctuations and have been summarised and presented clearly by Lawton (1990). The first hypothesis concerns changes in the spacing behaviour of grouse, which causes cyclic changes in density through the differential behaviour between kin and non-kin. Given certain

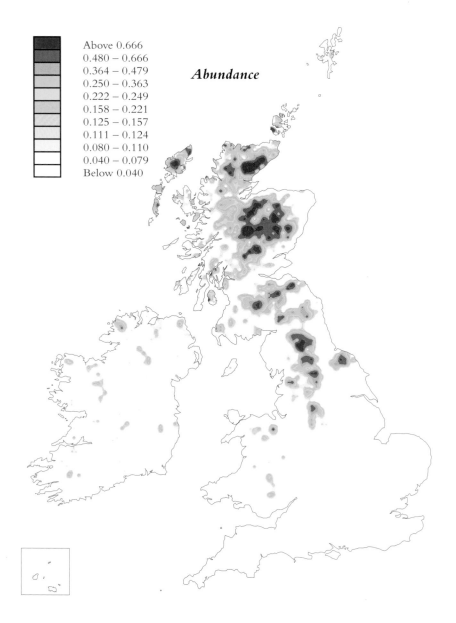

Abundance

	Above 0.666
	0.480 – 0.666
	0.364 – 0.479
	0.250 – 0.363
	0.222 – 0.249
	0.158 – 0.221
	0.125 – 0.157
	0.111 – 0.124
	0.080 – 0.110
	0.040 – 0.079
	Below 0.040

parameters, computer models can create cycles which match the amplitude and period of the wild Grouse cycles (Moss and Watson 1991). While there is evidence that spacing behaviour determines density in NE Scotland, the evidence for differential aggression and changes in genetic relatedness with population density requires further study.

The second hypothesis is that population cycles are caused by the delayed density-dependent effects of a nematode parasite (*Trichostrongylus tenuis*) on the breeding population of Red Grouse. Extensive monitoring and intensive experiments have demonstrated the effects of the parasite on breeding production and survival (Hudson *et al.* 1992) and, when combined into a formal mathematical model, demonstrate that such effects are sufficient to cause the population cycles observed (Dobson and Hudson 1992). Grouse populations which do not exhibit cycles have lower levels of infection, and variations in the period of the cycle are influenced by differences in the growth rate of the grouse population and the climatic effects on the life cycle of the parasite. These hypotheses are, however, not necessarily mutually exclusive and it would be naive to expect parasites to be the sole cause of all population fluctuations. It seems likely that parasites interact with other factors to influence patterns of population change.

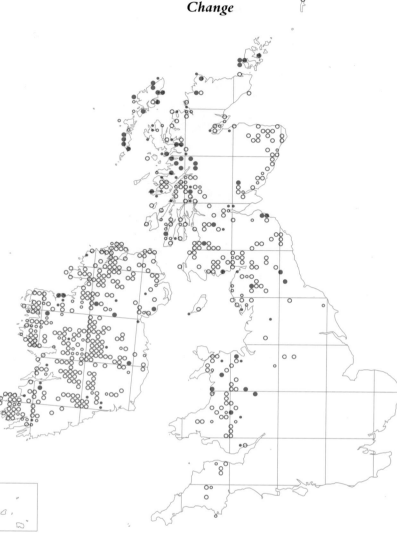

Change

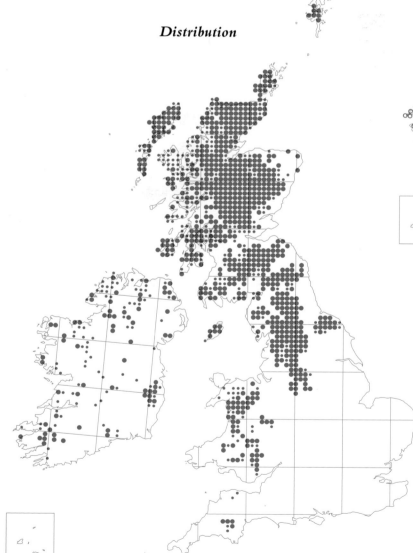

Distribution

The average grouse bag for Britain is in the region of 450,000 per annum (Hudson 1992). Given average breeding and harvesting conditions this would give an estimated breeding population of 250,000 pairs. In Ireland grouse populations have declined dramatically since 1920 and densities are low; an estimated figure would be between 1,000 and 5,000 pairs.

PETER J. HUDSON

68–72 Atlas p 132

Years	Present, no breeding evidence		Breeding evidence		All records		
	Br	Ir	Br	Ir	Br	Ir	Both (+CI)
1968–72	110	51	972	370	1082	421	1503
1988–91	196	77	749	64	945	141	1086
				% change	−12.7	−66.4	−27.7

Ptarmigan

Lagopus mutus

The Ptarmigan is beautifully camouflaged in summer and autumn, blending into the lichen-covered rocks and short mountain vegetation. One often hears it before seeing it, for the cock crows loudly, a belch-like and almost mechanical snore. Otherwise the Ptarmigan is secretive, relying on its camouflage for protection. Often one can approach within a few metres before a bird moves. Ptarmigan are most conspicuous when the cocks display frequently in March–May, and again in August–September when birds flock.

Ptarmigans are today restricted to the Scottish Highlands, where they breed on Arctic-alpine ground, the largest area of near-natural land in Britain. They became extinct in the Lake District and Southern Uplands at the end of the 1700s and early 1800s respectively. They occur on all large areas of Arctic-alpine ground except in the Hebrides, where distribution is more patchy, and in the Southern Uplands, where birds have been present in recent decades following an introduction. Some isolated smaller hills have birds only in years of generally high numbers.

The distribution has not changed materially since 1968–72. Abundance on many areas varies greatly from year to year. On the infertile granite of the Cairngorms, the fluctuations show a cyclic pattern with an average period of about ten years. On the richer rocks and soils of a nearby hill-range south of Braemar, where density and breeding success are higher (Watson 1979), fluctuations are more frequent and irregular. Locally, densities are higher than recorded in several other countries in the world distribution. In one of these rich areas, in 1989, Stuart Rae started an intensive study of habitat at all seasons, using marked birds. In summer 1990 he recorded a density of 50 adults per km², and two young per adult were reared.

Most territorial cocks pair with one hen, but a few with two and some are unmated. Hens mostly start to lay in mid-late May, exceptionally at the end of April or early May, the typical clutch being 6–9 eggs. Only one brood per year is reared, but birds often relay if the first clutch is destroyed. Chicks are almost adult-sized by mid August.

Breeding success is very poor in years with heavy summer snow or rain storms. The incubating hen can usually sit through snow, except prolonged deep fresh snow with heavy drifting, but chicks less than 10–14 days old need some invertebrate food and more than half the chicks on one study area in 1990 died during a period of prolonged rainfall. Predation also accounts for some egg and chick loss, and can be severe. At Cairn Gorm, Crows were attracted to high ground by food scraps dropped by tourists, and have greatly reduced the breeding success of Ptarmigan by robbing many eggs (Watson 1982).

Hens usually weigh less than cocks, but put on much fat in March–May before breeding and by April–May weigh more than cocks (Watson 1987a). There is some evidence of a maternal effect on chick survival, probably through the hen's condition in spring affecting the quality of her eggs (Moss and Watson 1984).

The adults' main food consists of blaeberry, crowberry and other heaths. They also eat various other plants, especially in summer, and take many

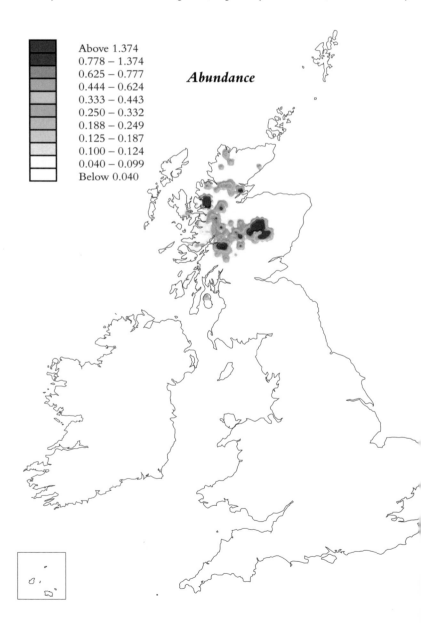

| Above 1.374 |
| 0.778 – 1.374 |
| 0.625 – 0.777 |
| 0.444 – 0.624 |
| 0.333 – 0.443 |
| 0.250 – 0.332 |
| 0.188 – 0.249 |
| 0.125 – 0.187 |
| 0.100 – 0.124 |
| 0.040 – 0.099 |
| Below 0.040 |

Abundance

berries in autumn. As a result of over-grazing by sheep, heaths have declined in some areas and been replaced by grass. This has led to poorer conditions for Ptarmigan, and possibly accounts for their disappearance from the Southern Uplands and Cumbria as annual breeders (Ratcliffe 1990). Predation by foxes and Crows may also be involved: foxes have greatly increased in association with more carrion from dead sheep, following higher sheep stocking levels.

Scottish Ptarmigan have parasites similar to Red Grouse, but, unlike them, seldom have very large numbers of the damaging caecal threadworm *Trichostrongylus tenuis*. Ptarmigan live well above the altitude limit for ticks, and so do not suffer from the tick-borne virus disease of louping ill, which can cause severe mortality in Red Grouse.

Breeding Red Grouse and Ptarmigan often overlap in altitude, but seldom in habitat. Red Grouse need the cover of heaths more than 15cm high, whereas Ptarmigan prefer shorter vegetation and obtain cover from rocks and topographic variations in the ground.

Scotland is an oceanic outpost of the birds' large circumpolar distribution on rocky tundra down to sea level in the Arctic and on more southerly rocky mountain ranges. In the E Highlands, Scottish Ptarmigan

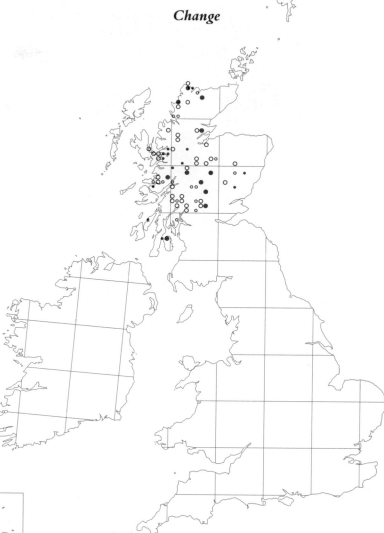

Change

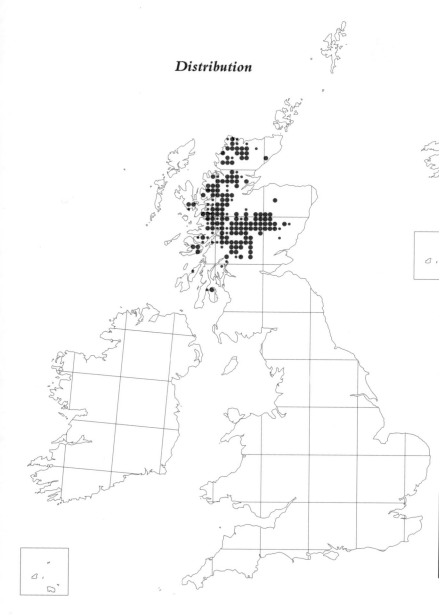

Distribution

breed down to 800m altitude, in the W Highlands down to 600m, and in the NW down to 200m, matching the lower limits of Arctic-alpine vegetation towards the more exposed NW seaboard. The most recent population estimate is a minimum of 10,000 pairs (Ratcliffe 1990), but an accurate figure cannot be given because so few counts have been done.

ADAM WATSON and STUART RAE

68–72 Atlas p 134

Years	Present, no breeding evidence		Breeding evidence		All records		
	Br	Ir	Br	Ir	Br	Ir	Both (+CI)
1968–72	33	0	162	0	195	0	195
1988–91	40	0	133	0	173	0	173
				% change	−11.3	0	−11.3

birds, together with populations in N Scotland, have shown the greatest rate of decline. The remaining British stronghold of the species is now in Tayside (Hudson 1989). Overall, numbers shot in N Britain have decreased by 95% since 1900 and it seems reasonable to suppose that this is a true reflection of the rate of decline in Black Grouse density. While demographic data on them are sparse, there is evidence that the decline in Black Grouse has been fastest in those areas with poor breeding success.

Overall, the decline in Black Grouse has probably been caused by loss of habitat through agricultural intensification. In the lowlands, the loss of woodland edge, agricultural improvement of heathlands and the specialisation of farming practices have reduced the diversity of habitats so favoured by Black Grouse. In the uplands there has also been a loss of habitat diversity on the edge of moorland areas, where grazing pressure and burning practices have replaced scrub woodlands and associated grass and heather stands with bracken and heather. Overgrazing also has a secondary effect, removing the vegetation which supports the invertebrates

Black Grouse

Tetrao tetrix

The lekking behaviour of birds is an interesting phenomenon, and a collection of some 20 or 30 Blackcocks displaying on a traditional lek site, with their evocative bubbling calls, is an exciting experience that few forget. Cocks gather at lekking areas in the early morning to compete for prime areas of the lek and to attract females, and cocks with central locations obtain most of the copulations (Krujit and Hogan 1967).

An examination of historical records shows that the distribution of Black Grouse has shrunk dramatically this century. Previously they were common throughout the southern counties of England from Lincolnshire and Norfolk south to Hampshire and Cornwall, whereas the Distribution map shows that they are now totally absent. Furthermore, since the production of the *68–72 Atlas* the Black Grouse populations on Dartmoor, Exmoor and the Quantocks are now believed to have become extinct, while the populations in the Peak District and Wales have declined further. In Wales a recent survey recorded fewer than 300 cocks, while in Staffordshire only 14 cocks were recorded, and this in a county where in the 1880s a total of 252 birds were shot in one day. In contrast, lek counts indicate that local populations in parts of the Scottish Borders, Tayside and the N Pennines have increased within the past few years.

The continued loss of Black Grouse from the southern edge of their distribution is not restricted to Britain. In Holland, Belgium and much of central Europe the Black Grouse population has declined to a low level or has become totally extinct. In S Europe, only the Alps now carry a significant population, and this is decreasing on the marginal areas through the effects of disturbance, agriculture and forestry. Only in the boreal forests of Scandinavia and Russia are Black Grouse populations still relatively healthy, and increasing in some areas through silvicultural practices.

In N Britain, bag records indicate that at the turn of the century the greatest densities of Black Grouse were found in SW Scotland, but these

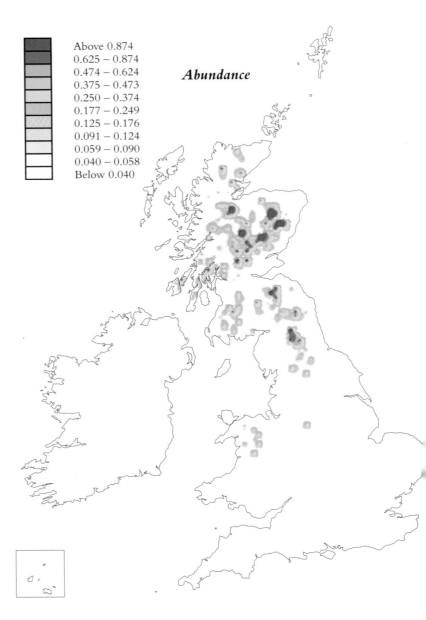

Above 0.874
0.625 – 0.874
0.474 – 0.624
0.375 – 0.473
0.250 – 0.374
0.177 – 0.249
0.125 – 0.176
0.091 – 0.124
0.059 – 0.090
0.040 – 0.058
Below 0.040

Abundance

taken by chicks during the first two weeks of life. In particular, the loss of bilberries, which can carry high densities of lepidopteran larvae – a rich source of protein – could have a significant effect on chick growth and survival. Grazing animals in the uplands may also compete with Black Grouse for the nutritionally rich flowers of cotton grass frequently taken by females prior to laying and incubation.

Numbers of Black Grouse increase when ground is planted for forestry and they will remain in such areas until the forest trees form a canopy. The reason for this is not clear, but exclusion of grazing animals may be important in increasing vegetation height and diversity, which in turn leads to more invertebrates and better cover for nests and chicks. In Scandinavia, higher breeding success in young plantations is associated with lower levels of nest predation (Brittas and Willebrand 1991). In some years predation may reduce the number of chicks reared, with three times as many juveniles per hen in areas where foxes and Crows were controlled by gamekeepers than in areas without predator control (Baines 1991).

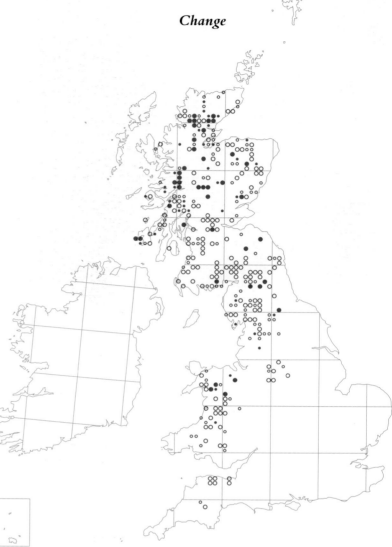

Change

Distribution

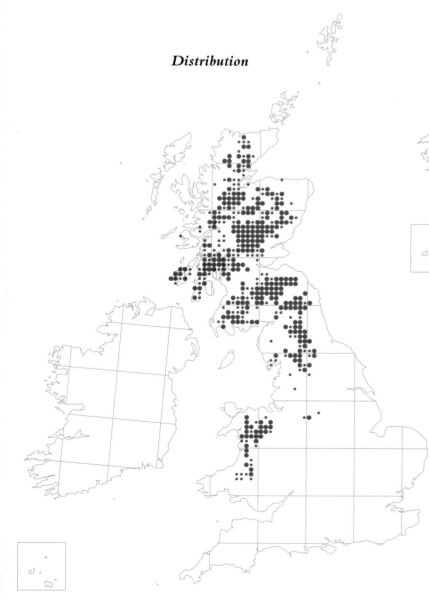

Black Grouse are absent from Ireland despite a number of attempted introductions. The British population has not been estimated accurately but the number of breeding females is probably in the region of 10,000–15,000 individuals, with perhaps a slightly higher population of cocks, given the average sex ratio of 1.3 cocks per hen.

PETER J. HUDSON and DAVID BAINES

68–72 Atlas p 136

Years	Present, no breeding evidence		Breeding evidence		All records		
	Br	Ir	Br	Ir	Br	Ir	Both (+CI)
1968–72	140	0	463	0	603	0	603
1988–91	154	0	278	0	432	0	432
				% change	−28.4	0	−28.4

The Capercaillie is a lekking species and cocks take no part in rearing the young. Hens are mated in late April, and typically lay 5–12 eggs in a scant nest on the ground, beginning incubation in early May. Hatching occurs in early June and the chicks' survival is related to the number of rainy days they experience. This may not be a simple process of wetting the chicks, however, but could involve reducing the time when they can forage for insect food (Moss 1985).

If the chicks are to survive the summer they must grow quickly – about as fast as broiler chickens. Since hens are smaller than cocks, they grow more slowly and their food requirements are less. This may be why, in wet summers when breeding success is poor, more hens than cocks survive to independence (Moss and Oswald 1985).

The Scottish population, probably a few thousand birds, forms a minute fraction of the world population, whose distribution more or less coincides with that of the Scots pine. Over much of its range, Capercaillie numbers are declining in parallel with the disappearance of mature Scots pine and oak forests, as these are felled by man.

R. MOSS

68–72 Atlas p 138

Capercaillie

Tetrao urogallus

The Capercaillie is the biggest grouse in the world, adult cocks weighing over 4kg and hens almost 2kg. It is essentially a bird of mature Scots pine forests, which have a relatively open canopy and allow enough light to reach the ground to permit the growth of blaeberry, the chicks' main food. The spelling 'Capercailzie' is sometimes seen, but the 'z' is not a 'z' at all, it is an old Scottish consonant pronounced almost like a 'y'.

The Capercaillie became extinct in Britain in the 18th century and was reintroduced into Scotland in the 19th. The easterly bias in its distribution reflects the distribution of Scots pine, but is also related to rainfall (Moss 1985). At its peak, early in the 20th century, breeding was recorded from Golspie in Sutherland in the north, to Stirling in the southeast, and from Cowal, Argyllshire, in the west to Buchan, Aberdeenshire, in the east.

The current Distribution map shows a marked contraction since the *68–72 Atlas*. A decline in numbers and distribution has certainly occurred since the early 1970s, and this perception is the cause of the voluntary ban on shooting recently suggested by the Scottish Office. Even so, the Distribution map probably underestimates the birds' present range. A new estimate of distribution and abundance is now (1992) being prepared through a joint initiative by the Forestry Commission, the Game Conservancy, the ITE, the RSPB and other organizations. The causes of the decline are uncertain but possibilities include habitat destruction (Moss *et al.* 1979), changed silvicultural practices leading to a reduction in blaeberries, wetter June weather (Moss and Oswald 1985), increased numbers of foxes and crows, over shooting and mortality due to hitting deer fences.

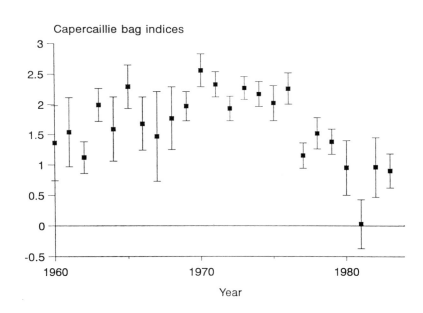

Index of Capercaillie numbers shot annually. Means ($\pm$ SE) from six large estates: each annual value for an estate was $\log_e$ ($10 \times$ annual bag/mean annual bag for 1964–75), excluding zero returns. This pattern is also reflected in other, less complete, bag records as well as population counts (Moss and Weir 1987). After 1983, few Capercaillies were shot and bag records are too sparse to interpret. P. Hudson provided some of the data.

Change

Distribution

Years	Present, no breeding evidence		Breeding evidence		All records		
	Br	Ir	Br	Ir	Br	Ir	Both (+CI)
1968–72	48	0	134	0	182	0	182
1988–91	32	0	34	0	66	0	66
				% change	−63.7	0	−63.7

GBB

Red-legged Partridge

Alectoris rufa

In many parts of E England this gaudy partridge is now far better known than its distant relative the Grey Partridge. This is not only because it is more numerous, but also because it is less crepuscular and draws more attention to itself by calling frequently. Although the species appears more assertive than the Grey, it is in fact usually sub-dominant to it, for example in territorial disputes.

The Red-legged Partridge has been scarce for most of the period since its introduction in the 18th century, only gradually reaching the present distribution by the 1930s. Some expansion has taken place since, for example northwards through Yorkshire, but the change has been one of increase within its 1930s range, rather than of expansion out of it. The expansion since 1968–72 into nearly 375 new 10-km squares will have been mostly due to released birds. Breeding was recorded in only 42% of these squares, compared to an overall occurrence of breeding of 87% for the 1968–72 distribution. However, a few self-sustaining populations may have been established.

In comparison with the Grey this species is much more tolerant of wooded landscapes, yet more restricted to areas of arable crops, being especially common where sugar beet is grown. On the very open grassland areas of Salisbury Plain it has always been uncommon, and in the more open parts of the South Downs it is now almost absent.

Red-legs are less affected by the reduction in cereal crop insects than the Grey and this has helped them. On the other hand, however, Red-legs are more affected by predation of clutches during the breeding season, due to the reduction in predator control consequent upon the decline of the Grey. This susceptibility to nest predation is somewhat counteracted by the fact that a pair of Red-legs can each incubate separate clutches (Green 1984). Without this ability, presumably, the high clutch losses would have prevented establishment in Britain.

Even allowing for changes in predator control, the interpretation of recent variations in numbers and of changes in distribution has been difficult, due to the release of captive-reared birds, which began in 1963, increasing rapidly over the period to 1984. The aim of this releasing was to maintain numbers of partridges shot, despite the declining numbers of Greys, but this has not been achieved.

In September 1970, more than 2,000 Chukars (*Alectoris chukar*) and *A. chukar × rufa* hybrids were released on the South Downs. In the following year hybrids were again released, there and in two other localities. By 1972 the releases were much more widespread, ranging from Hampshire to Angus. Some birds seen at this time were reported as Rock Partridges (*A. graeca*) (for example, *Brit. Birds* **65**, p 404), but that partridge has never been released in the UK.

The advantage of the hybrids is that they lay twice as many eggs and are generally more suited to captivity. The disadvantage is that they breed much less well than Red-legs in the wild. The reason for the poor performance is not known, but the original stocks came from Cyprus via a game farm in Italy, where they were bred for docility and egg production.

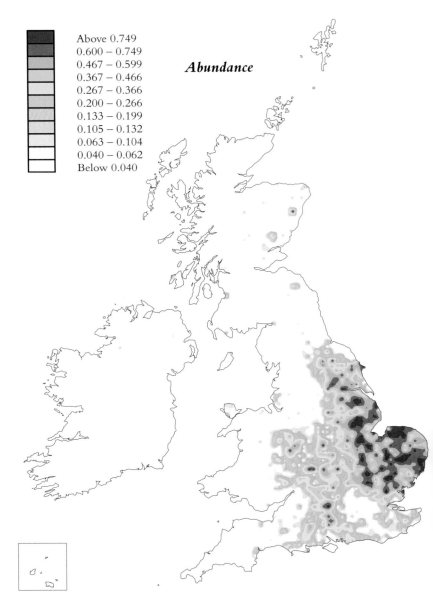

■	Above 0.749
▨	0.600 − 0.749
▨	0.467 − 0.599
▨	0.367 − 0.466
▨	0.267 − 0.366
▨	0.200 − 0.266
▨	0.133 − 0.199
▨	0.105 − 0.132
▨	0.063 − 0.104
▨	0.040 − 0.062
□	Below 0.040

Abundance

In the field, pairs of hybrids, or the relatively infrequent mixed pairs of hybrids and Red-legs, produce only one seventh as many young as pure Red-leg pairs. Moreover, it has been shown that the shooting of a mixed population of wild Red-legs and captive bred and released Chukar hybrids will eventually drive the Red-legs to local extinction. This would not happen if the released birds were Red-legs (Potts 1989). The system of licences under which the hybrid releases were allowed ceased in 1992.

When the effects of predator control, releasing and shooting are discounted, Game Conservancy data suggest that there has been a marked decline in stocks of wild Red-legs since 1985 and recent declines have been reported from several areas of France, NW Italy, Spain and Portugal; the four other countries where it is found.

The British population was estimated at 90,000 territories during 1988–91 by extrapolation from mean CBC densities. However, this estimate is surely too low, as the Game Conservancy estimates that prior to the start of the shooting season there may be up to 1,500,000 individuals, including released birds. If this latter figure is correct, it is interesting that wild stocks of both partridges are now approximately level at 750,000 each, post-

Change

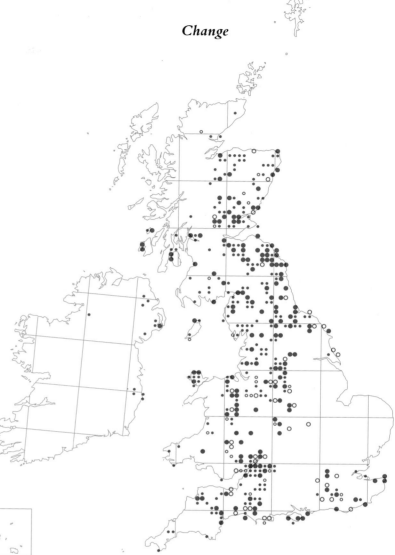

Distribution

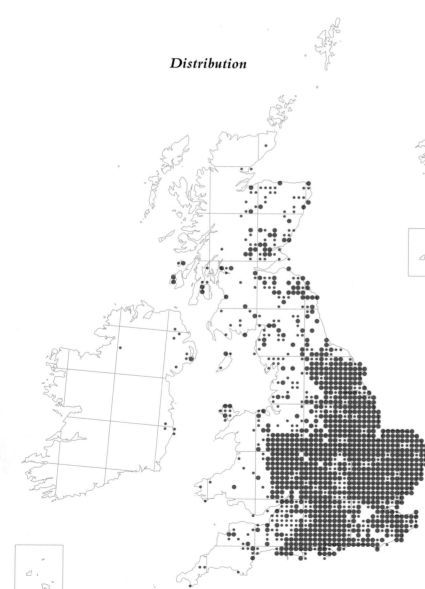

breeding, whereas up to the 1950s Greys outnumbered Red-legs by about 20:1. This changed ratio is considered to be all due to the decline of the Grey, half directly and half indirectly through reduced shooting pressure consequent upon the decline of the Grey (Potts 1980).

There are no self-sustaining populations in Ireland.

G. R. POTTS

68–72 Atlas p 140

Years	Present, no breeding evidence		Breeding evidence		All records		
	Br	Ir	Br	Ir	Br	Ir	Both (+CI)
1968–72	124	0	795	0	919	0	919
1988–91	320	11	894	1	1214	12	1226
			% change	32.1			33.4

Grey Partridge

Perdix perdix

It is astonishing to realise that this species was at one time so common that, had it not declined, it would today be the tenth most numerous species of bird in Britain and Ireland. Formerly, on many farms in the eastern counties, it was the most abundant species. Although still numerous in a few specially favoured localities where great efforts are made to conserve them, numbers have generally declined for at least 40 years. It could be argued that the decline will eventually drive numbers down in a way comparable to those of the Corncrake.

The Change map shows a clear pattern of retreat – but in the opposite direction to the Corncrake. Most notable has been the near disappearance from Ireland, large parts of SW Scotland, NW England, Wales, Cornwall, Devon and parts of Dorset. Other notable recent absences are from the central Fens and from the W Weald.

On average the range has retreated 50km eastwards since the *68–72 Atlas*, mainly from the areas where even in its heyday it was least abundant. The CBC shows the decline of breeding stocks between 1968–72 and 1988–91 as approximately 75%. The highest numbers, up to 60 birds per km² in August, are now in a very limited area of N Norfolk and, to a lesser extent, on some traditional ley farms on Salisbury Plain and from Nottinghamshire northwards to parts of E Scotland. Exceptionally, on the margins of some N Pennine grouse moors with abundant rushes, numbers have not declined at all.

In Ireland at the end of the 1991 breeding season a survey was organised which identified two very small, but just possibly viable, populations on areas harvested for bog peat in Co. Offaly (Kavanagh 1992). The total area concerned was 120km², and counts in 1991 suggested a spring density approaching one pair per km². Outside this area, and excluding localities where birds are known to have been released, the species has been encountered in six other counties, all of which records are included in the map.

The Grey Partridge has been the subject of much research here and overseas. One finding has been that a rate of survival of just under 30% for the first six weeks of life is generally the minimum necessary to maintain numbers.

Between 1952 and 1962, chick survival rates declined from 45% to less than 30%, but then not significantly during the next three decades. The example from the South Downs study area of the Game Conservancy is given in the Figure. In a computer modelling study it was calculated that over 40% of the decline in the UK has been due to decreased chick survival rate (Potts 1980).

The absence of a marked decline in chick survival rate during most of the population decrease has led some recent authors (Watson 1987b, Green and Hirons 1991) to query the effect of chick survival on population status. These authors have failed to grasp that a non-trending long-term average chick survival rate below 30%, the minimum necessary to maintain numbers, can result in a steadily progressive decline of around 7% per annum (Potts 1986).

The cause of the decline in chick survival is generally considered to be the use of herbicides in cereals which has reduced the amount of insect food available for chicks (Potts 1986). Extensive Game Conservancy work has shown that chick survival rates can be restored towards the mean levels obtained prior to the use of herbicides, and above 30%, either by the use of selective unsprayed 'conservation headlands', or by the traditional system of undersowing cereals with ley pasture. This system is of great benefit to sawflies (Aebischer 1990), a favourite chick food (Potts 1970).

The status of the Grey Partridge, which occurs widely throughout North America, Europe and Asia, varies considerably from country to country. In general, it has declined most at the periphery of its range, least around latitude 49°N. Following the very poor season of 1991, numbers in Britain in spring 1992 were estimated at 150,000 pairs by the Game

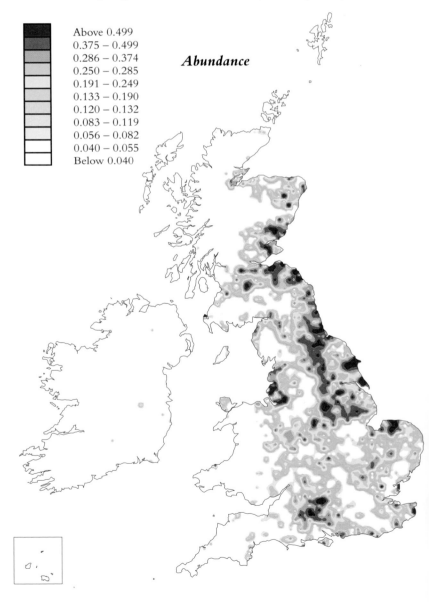

Above 0.499
0.375 – 0.499
0.286 – 0.374
0.250 – 0.285
0.191 – 0.249
0.133 – 0.190
0.120 – 0.132
0.083 – 0.119
0.056 – 0.082
0.040 – 0.055
Below 0.040

Abundance

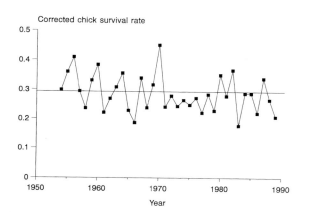

Corrected chick survival rate

Corrected chick survival rates for Grey Partridges at North Farm from 1954–89, after removal of the effects of weather, predation and rearing. The solid horizontal line gives the threshold below which population change is negative. The correction allows for the annual effects of post-hatching weather.

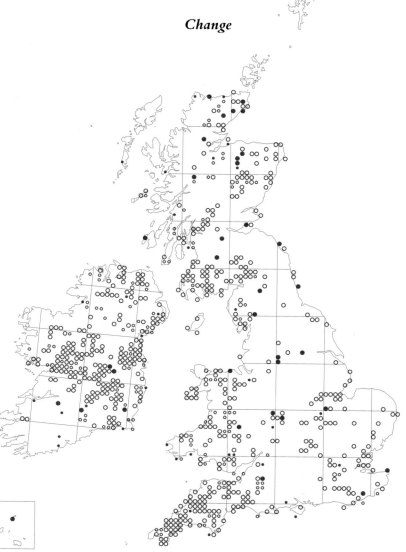

Change

Distribution

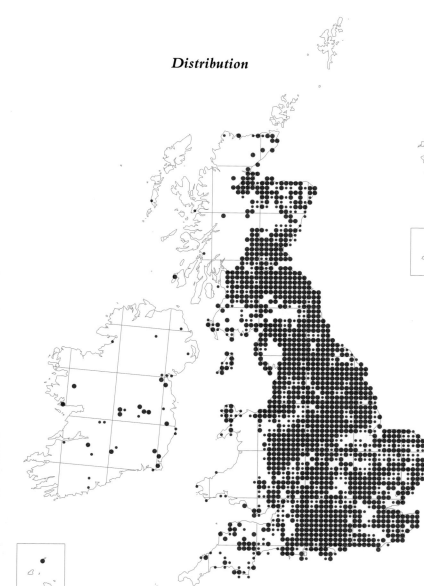

Conservancy. This estimate was slightly greater than that of 140,000 territories extrapolated from mean CBC densities for the period 1988–91. However, the two estimates are sufficiently similar for confidence to be placed in them. Such a population would represent about 5% of the world total. The Irish population is now small, and may number less than 500 pairs.

G. R. POTTS

68–72 Atlas p 142

Years	Present, no breeding evidence		Breeding evidence		All records		
	Br	Ir	Br	Ir	Br	Ir	Both (+CI)
1968–72	120	48	1885	207	2005	255	2260
1988–91	351	20	1278	15	1629	35	1665
			% change		−18.7	−86.3	−26.3

Quail

Coturnix coturnix

Occasionally heard, but rarely seen, the enigmatic Quail remains one of our most poorly known birds. As Europe's only migratory game bird, and one which traditionally fetched a good price at market, it was mercilessly trapped along Mediterranean coasts until the regulation of spring hunting in 1937. Its preference for breeding in dense, tall herbage, and its diminutive size, ensure that 'the field for modern investigation offered by the Quail is as considerable as the practical difficulties attending it' (Moreau 1951).

The Quail shows great variation in numbers between years, apparently brought about by invasions from the Continent. These have been well recorded and peak years have occurred in 1870, 1893, 1947, 1952–53, 1964, 1970 and 1983. By good fortune the *88–91 Atlas* period included a peak year, 1989. Prior to this, the greatest influx was of 600 in 1964 (Parslow 1973) although the *Trends Guide* suggests that 1970 was equally good. During peak years earlier in the century numbers were lower, in the range of 100–300 birds (Moreau 1951). Because the Distribution map includes data from three 'normal' as well as one peak year, only by presenting the distribution by year (Appendix E) can the true patterns be seen. Not only is the increase in number of records in 1989 apparent, but Quails were distributed further north and west than in 1988 and 1990. In these years there were few records from the Welsh Marches, Wales, Scotland and Ireland, but in 1989 there were many. More Quails were recorded in Shropshire in 1989 than in all previous years combined, since records began in 1957 (J. Sankey). Astonishingly, of five Quail nests recorded this century in the Faeroe Islands, three were in 1989 (J. Jens-Kjeld) so the species clearly reached the northernmost tip of its range in reasonable numbers.

The traditional strongholds of the species remain, as in the *68–72 Atlas*, on the calcareous soils of Wiltshire and Dorset. Elsewhere, however, the link with soil type seems less clear than in the earlier survey (particularly around Cambridgeshire, Bedfordshire and Hertfordshire), suggesting, as first proposed by Moreau (1951), that land use is an equally important predictor of distribution. Certainly, the large population in the Welsh Marches, and smaller ones in N England and Scotland, were not apparent in the *68–72 Atlas* even though it, too, included a Quail year.

Because the Quail shows a preference for level landform, clear of trees and shrubs, it is at home in cereal fields, winter wheat being a favourite (*BWP*). The species' apparent increase in numbers since the Second World War was no doubt assisted as much by the extension of arable farming as by the cessation of hunting (Moreau 1951).

The mating system of the Quail is variable, from monogamy to promiscuity (*BWP*). Males arrive on the breeding grounds before females and set up territories, from which they call regularly. Often this is the only indication that they are present and most censuses (including those for both Atlases) have relied largely on vocal registration. Clutches of 8–13 eggs, laid in a shallow scrape hidden well within the maturing crop, are incubated by the female alone. Very late clutches (in August and September) are probably second clutches, although they may be laid by late-arriving females that have attempted to breed elsewhere in Europe. Although unsubstantiated, it is thought that females arriving in Italy from N Africa in mid to late summer have already bred and may do so again.

The cause of the invasions is unknown. Zuckerbot *et al.* (1980) suggested that, early this century, low Quail numbers were associated with poor rainfall in the Sahel, although the evidence is weak. The large numbers recorded during 1989 and the increasing numbers recorded during peak years this century, despite ever worsening conditions in the Sahel, do not support this idea. Moreau (1951) suggested that influxes into Britain and Ireland are related to spring drought in France and Spain (particularly with southeasterly winds to help the birds on their way), although this remains to be tested.

In 1989 a survey of Quails, conducted by requests to county recorders, received counts of calling males from a total of 719 sites. The mean count at each site was 1.78 Quails or 3.88 per occupied 10-km square. The Atlas data for 1989, which incorporate these data and those of a similar survey in Scotland (Murray 1991a), show that Quails were recorded in 670 10-km squares in Britain and 24 in Ireland, yielding estimates of 2,600 calling males in Britain and 90 in Ireland. The Scottish survey revealed 750 males. The RBBP recorded a maximum of 1,655 pairs in Britain but suggested this could be an underestimate. Either way, it is clear that the influx of

1989 was the highest on record. The British population during non-invasion years is about 100–300 pairs (RBBP) and that of Ireland less than 20 males, with the majority of the population found in Co. Kildare (*Irish Birds*).

DAVID GIBBONS and STEVE DUDLEY

68–72 Atlas p 144

Distribution

Change

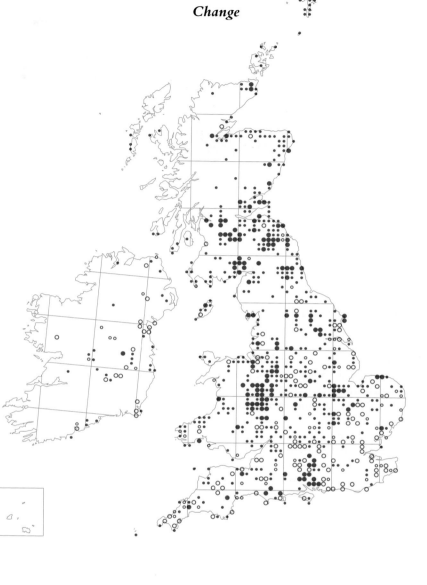

Years	Present, no breeding evidence		Breeding evidence		All records		
	Br	Ir	Br	Ir	Br	Ir	Both (+CI)
1968–72	100	10	305	23	405	33	438
1988–91	571	30	233	4	804	34	838
				% change	98.5	3.0	91.3

Pheasant

Phasianus colchicus

The crowing of a cock Pheasant on a March or April morning is one of the most distinctive sounds of the lowland countryside. Despite being such a common and widely distributed species, the Pheasant is not a native. It was introduced to Britain by the Normans, or possibly by the Romans, and into Ireland in the 16th century, and now breeds in almost every county.

Pheasants are sedentary birds and there are few differences in distribution between this survey and the two previous Atlas studies. In February the cocks begin to establish exclusive territories along the boundaries between permanent cover and open ground. Most are found along woodland edges or, in treeless country, reed-filled ditches. In many areas there are more males than suitable territories and the extra birds remain non-territorial. Early in March the hens visit territories of a number of males and eventually settle to breed with one (Hill and Robertson 1988).

Pheasants are polygynous, and males may attract numerous females to their territories, where they feed in a group while being guarded by the male against disturbance. The number of hens choosing to breed with each territorial male varies enormously, from none to over 20, but the average harem size is two or three. The cock provides the hens with security while they are building up their body reserves for the rigours of egg production and incubation. The hens feed at dawn and dusk, usually on growing shoots of arable crops, and then retire to the safety of woodland or reeds to digest their meal.

Although there are large populations of wild Pheasants, especially in the southern and eastern portions of Britain and Ireland, in autumn numbers are greatly increased by the release of birds hand-reared for shooting. These released birds have three effects on the breeding population. Firstly, they may cause the Pheasant's range to spread into areas in the north and west where they would not normally be self-sustaining. Secondly, they can increase the spring population density, although this affects only the numbers of females and non-territorial males. The density of territorial males appears to be limited by habitat quality (Woodburn and Robertson 1990). Lastly, hand-reared birds have poor breeding success (Hill and Robertson 1988) and, as the proportion of these birds in the breeding population increases, so productivity declines (Robertson and Dowell 1990).

The density of male territories is highest along the edges of woodlands rich in shrubby cover 30–200cm in height. Edges bordering arable fields provide three times as many territories as those bordering grassland

(Woodburn and Robertson 1990). A model of Pheasant breeding density in Britain, based on the regional availability of woodland edges, arable land and reed-filled ditches, provides a distribution similar to that of the Abundance map (Robertson *et al.* 1989).

Females begin to nest at the end of April. Early nests are located in woods and hedgerows, although replacement nests are usually in growing crops. Pheasants lay large clutches, with an average of 11 eggs. Clutches of 30 or more are not uncommon but are thought to be the work of more than one female (Hill and Robertson 1988). The chicks are led from the nest soon after hatching, and feed on invertebrates, usually in arable crops.

Excluding moorland and other mountainous regions, the average breeding density of Pheasants is 3.9 territorial males and 7.3 females per km^2. This may be compared with 5.9 territorial males per km^2 recorded in the species' native range in China (Zhang 1991). On areas managed for Pheasant shooting this average figure rises to 10.3 territorial males, while the number of females and non-territorial males is determined largely by the extent of hand-rearing. This increase in territory density appears to be due to increased woodland planting and preservation on areas managed for shooting, leading to twice the length of woodland edge compared with random samples of countryside (Robertson *et al.* 1989).

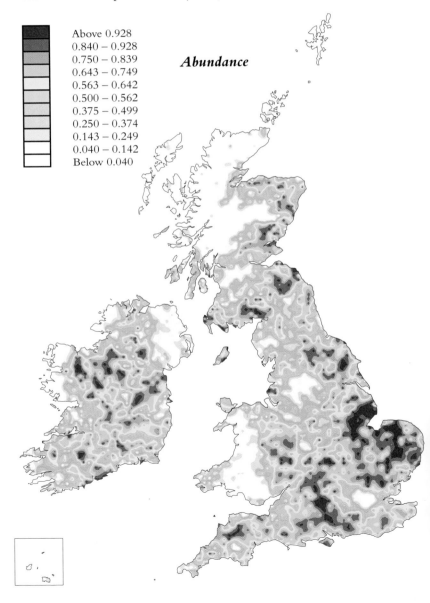

Above 0.928
0.840 – 0.928
0.750 – 0.839
0.643 – 0.749
0.563 – 0.642
0.500 – 0.562
0.375 – 0.499
0.250 – 0.374
0.143 – 0.249
0.040 – 0.142
Below 0.040

Abundance

The model of Pheasant breeding density in relation to regional variations in habitat quality suggests a British breeding population of 850,000 territorial males, 650,000 non-territorial males, and 1,600,000 females (Robertson *et al.* 1989). By extrapolation there would have been 530,000 male and 570,000 female Pheasants in Ireland during 1988–91.

PETE ROBERTSON

68–72 Atlas p 146

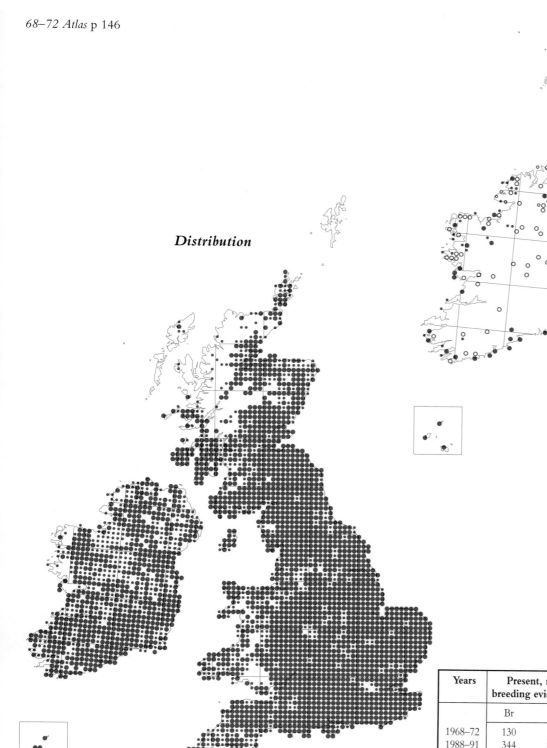

Change

Distribution

Years	Present, no breeding evidence		Breeding evidence		All records		
	Br	Ir	Br	Ir	Br	Ir	Both (+CI)
1968–72	130	43	2120	808	2250	851	3103
1988–91	344	251	1925	593	2269	844	3123
			% change		0.9	−0.7	0.6

Golden Pheasant

Chrysolophus pictus

Golden Pheasants are secretive and elusive birds from the uplands of central China. Often the only indications of their presence are the harsh 'ker-cheek' calls of the males in spring and their exotic feathers found moulted in the summer

They became established as a result of a series of introductions during the 19th and 20th centuries. The present distribution is centred around these sites, particularly in the Brecklands and Thetford Forest in Norfolk and Suffolk, the South Downs in Hampshire and West Sussex, Anglesey and parts of Galloway in SW Scotland. They are sedentary birds with quite specific habitat requirements and their distribution is similar to that observed in previous Atlas surveys, although there has been an increase in isolated records. This may reflect new releases, for the species is widely kept in captivity.

Their main habitat appears to be thicket-stage coniferous woodland over 10 years old. At two sites near Petersfield in Hampshire they occur in mature yew woodland and in areas of unmanaged hazel coppice under oak, but these appear to be exceptions. Nevertheless, in all cases they seem to select dense dark woodland with a relatively bare floor.

Very little is known of the breeding habits of this species. Despite their exotic plumage and extravagant displays, in which the male erects his neck feathers to form a ruff around his eye, they are extremely difficult to observe. It appears that in February the males form small flocks on suitable territories and display to each other to claim ownership. By March the territories are established and the males are solitary and dispersed throughout the suitable habitat. They call from their territories at dawn and dusk. The calling area appears to be extremely small, possibly just one tree. In dense woodland, calling takes place on the ground although in more open oak/hazel they have been observed parading and calling on low branches.

Nothing is known about their breeding behaviour in the wild and the females are rarely seen, but males of the closely related Lady Amherst's Pheasant are thought to be polygamous and the same is probably true of this species.

At the Petersfield sites, territories have been mapped for many years. These sites appear to hold self-sustaining populations, the last release occurring in 1975 at the latest. Territory density is surprisingly high and is stable with between 20 and 25 calling males per km² (R. Williamson).

The Golden Pheasant's native food consists of leaves and tender shoots of shrubs, low-growing bamboos, rhododendron flowers and invertebrates. The chicks need a high protein diet in captivity and are almost certainly insectivorous as are Pheasant chicks (Johnsgard 1976). In Britain they are known to eat ants and to come to feed sites provided by gamekeepers. They nest on the woodland floor and, in Britain, eggs are laid in April and May (*BWP*). One particularly interesting feature of this species is the reluctance of the female to leave the nest. It has been recorded that the females sometimes become covered in spiders webs during incubation (Goodwin 1948).

Golden Pheasants are rare in their native China. Although no detailed population figures are available they are still snared for their meat and feathers. Unsuccessful introductions have taken place in France, New Zealand, Colombia, USA, Canada, Hawaii and Tahiti (Long 1981). At present the only viable populations appear to be those in Britain and China. Although common in captivity the scarce wild populations make it an international rarity. There is a real need for research on this bird to ensure the British population remains healthy, and to suggest methods of conserving the remaining Chinese stocks.

This species continues to hold its own and the *Winter Atlas* estimate of 1,000–2,000 birds remains our best guide. It is unlikely ever to become a common bird due to its sedentary habits and specific habitat requirements, but sporadic releases may lead to the establishment of further population centres in the future.

PETE ROBERTSON and DAVID HILL

68–72 Atlas p 148

Change

Distribution

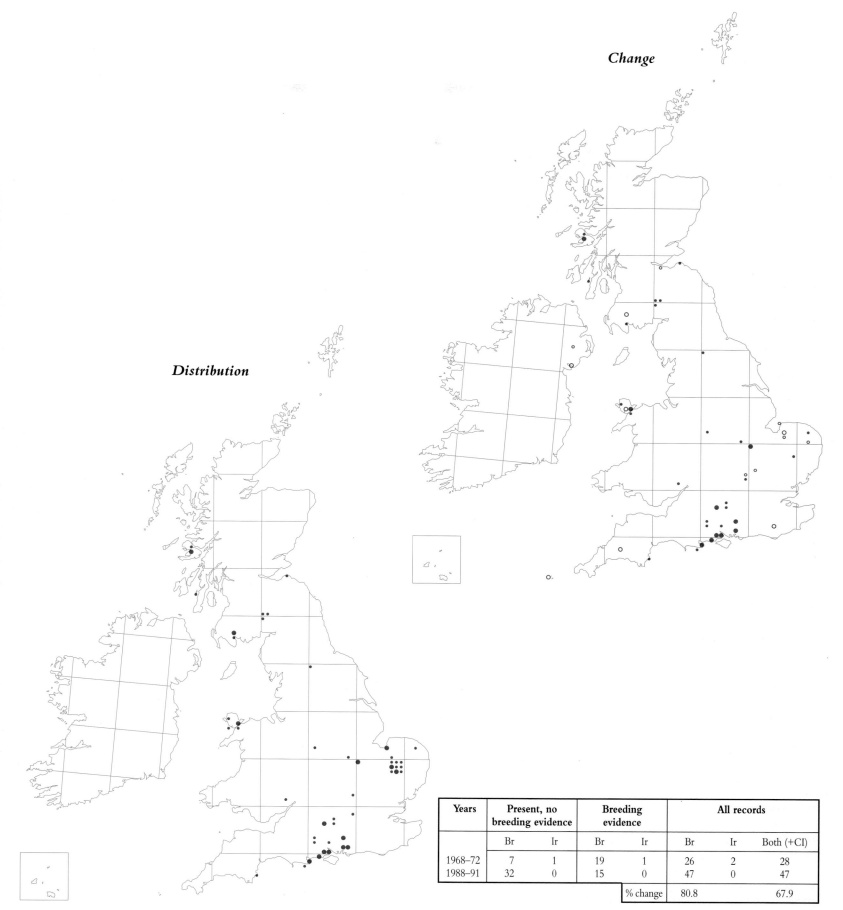

Years	Present, no breeding evidence		Breeding evidence		All records		
	Br	Ir	Br	Ir	Br	Ir	Both (+CI)
1968–72	7	1	19	1	26	2	28
1988–91	32	0	15	0	47	0	47
				% change	80.8		67.9

Lady Amherst's Pheasant

Chrysolophus amherstiae

Lady Amherst's Pheasant is a native of SW China (Szechwan, Yunnan and W Kweichow) and adjacent parts of Tibet and upper Burma, and was first brought into Britain in the early 19th century. It is held to be the most beautiful of all breeding birds in Britain, and was introduced for ornamental reasons.

As detailed in the *68–72 Atlas*, the first known attempts to liberate Lady Amherst's Pheasants were made around the turn of the century at Mount Stuart (Bute) and Woburn (Bedfordshire). The former failed due to interbreeding with Golden Pheasants, but the Woburn stock continued and spread beyond the confines of the estate. Further releases took place at Beaulieu Manor Woods, Hampshire, in 1925, Richmond Park, Surrey, in 1928–29 and 1931–32 (24 released), Whipsnade Park, Bedfordshire, in the 1930s, and at Elveden, Suffolk, in 1950 (Lever 1977). The Woburn and Whipsnade groups formed the feral population now living in Bedfordshire, adjacent parts of Buckinghamshire and Hertfordshire.

In its native range in China, Lady Amherst's Pheasant occurs on wooded slopes, in bamboos or other thickets and dense bushes. It lives at greater altitudes and in colder regions than the Golden Pheasant (Cheng 1963, Johnsgard 1986). In Britain it favours relatively young conifer plantations, prior to thinning, in which the closed, dense canopy leads to open ground below, with a diminished plant cover. Such sites, as well as more mature ones, usually contain some tall shrubs, including brambles and rhododendron (*BWP*).

The Distribution map shows just how locally the species occurs. It is now mainly confined to Bedfordshire where the population was estimated at more than 250 birds in 1979 (Trodd and Kramer 1991). The main sites there were along the Greensand Ridge, including Heath and Reach, Battlesden, West Woburn, Aspley Heath, Woburn Park, Eversholt, Steppingley, Millbrook, Clophill, Haynes and Old Warden. These records document a decline in both occurrence and density in outlying sites with relatively 'stable' population centres based at Woburn, Millbrook, Whipsnade and Maulden Woods. The Change map shows a distinct retraction in range in support of the local records and identifies a probable decline in the species.

Lady Amherst's Pheasant is extremely sedentary in Britain. This quality, together with the distribution of suitably-sized blocks of habitat, has probably limited its ability to colonise other parts of the country. It hybridises with the Golden Pheasant, which has been suggested to be a reason for the decline in its population, although other causes include habitat change, predation pressure and, perhaps, cold winters and wet springs. The climate is probably an unlikely cause for decline as the species lives at up to about 3,000m in China, experiencing harsh winters, wet springs and wet summers.

Attempts to introduce the species to Hawaii, Colombia and New Zealand have all failed. At present China and Britain are the only countries with breeding populations (Long 1981).

As in the case of the Golden Pheasant, its ecology, including its diet in the wild, is virtually unknown. However, recent Chinese studies (Han *et al.* 1990) found an average of 6.2 territorial males per km^2 in secondary deciduous forest and bush, with slightly lower densities in secondary pine. During the breeding season males display in small open glades within the forest. Males are probably polygamous in the wild, even in Britain. In China, Han *et al.* (1990) reported two or three females associating with each male. Incubation is carried out solely by the female, which nests on the ground in thicket stage woodland. After hatching, the chicks feed on insects for the first few days, taking berries and fruit later on (Han *et al.* 1990).

The current Bedfordshire population is estimated at 100–200 birds (Trodd and Kramer 1991) and represents the majority of the British population.

DAVID HILL and PETE ROBERTSON

68–72 Atlas p 150

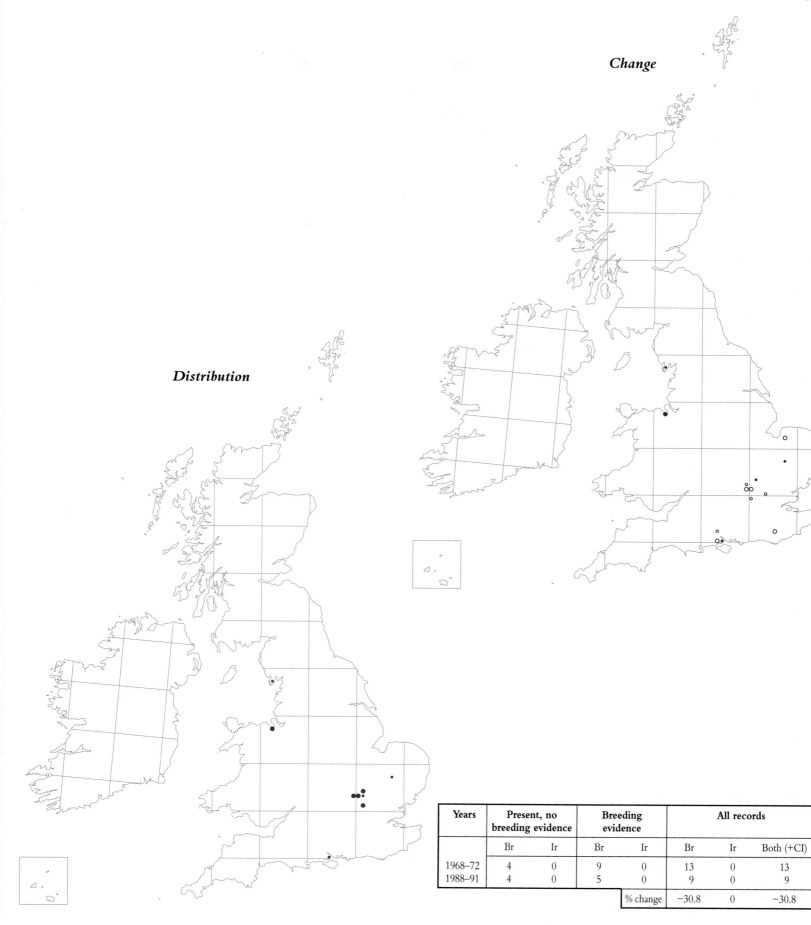

Change

Distribution

Years	Present, no breeding evidence		Breeding evidence		All records		
	Br	Ir	Br	Ir	Br	Ir	Both (+CI)
1968–72	4	0	9	0	13	0	13
1988–91	4	0	5	0	9	0	9
				% change	−30.8	0	−30.8

Water Rail

Rallus aquaticus

Atlas recorders consistently place the Water Rail among the 'Top Ten' most elusive of birds, more often to be heard than seen during the breeding season. At best, the cautious observer may glimpse a tall slender-bodied crake, walking with a high stepping gait and head raised across soft mud; or bolting head down into dense marshy vegetation. More frequently observers hear the diagnostic pig-like squeals and grunts ('sharming') or repeated 'kek' calls, delivered most regularly at dawn, soon after dusk, or when nesting birds are disturbed by day.

BTO nest records indicate that Water Rail breeding territories usually embrace a mosaic of static or slow moving freshwater, normally with open mud, and always expanses of tall emergent vegetation. The latter usually include blocks of sedge, rush, reedmace, common reed or reed-grass, though 1½–2ha of swampy riverine vegetation, and even small damp pockets in field corners, may hold a pair (Flegg and Glue 1973, Bayliss 1985).

These conditions occur widely, and the Distribution map shows that Water Rails currently breed extensively, if thinly, over much of Britain and Ireland. In England and Wales they breed most regularly around the coastal fringes of the southeast, the central southern counties, East Anglia, marshes and remnant mosses of Humberside, Vale of York and Wales. In Scotland, most Water Rails are in the Border region, and waterways from the Clyde, north to Speyside and the Great Glen. The Highlands and flow country are generally avoided, as are the outer isles save Canna, Eigg, Mull, Colonsay, Islay and Orkney. In Ireland, there are breeding concentrations in the northeast, notably around Lough Neagh, and, especially, along water courses of the River Shannon complex.

The Abundance map emphasises the relative importance of Water Rail populations in East Anglia and particularly in Ireland.

Today the true status of this elusive bird is uncertain in many counties for it is reluctant to fly, and no aquatic vegetation is too dense for it. Water Rails are overlooked, maybe considerably so, and although the general pattern of distribution remains as widespread as during the *68–72 Atlas*, there are signs of a substantial thinning of the population throughout much of its range in Britain and Ireland.

Water Rails are omnivorous and eat chiefly insects and their larvae, but also snails, small vertebrates such as frogs and fish, carrion and vegetable matter. Low-lying eutrophic waters on rich soils hold more of such suitable foods, but some Water Rails breed in acidic, low-lying bogs and upland mires (Bayliss 1985).

Loss of Water Rail habitat during inter-Atlas years was locally attributed to canalization of waterways, the partial drainage of bogs, urban development of coastal grazing marsh and loss of vegetation along water courses. Today an increasingly high proportion of pairs is to be found in nature reserves and, especially in central and E Britain, in mineral extraction sites, all places which may be managed for the species' needs.

The Water Rail is especially prone to changes of water level. Nests, holding clutches of 5–13 eggs, are generally constructed in very shallow sedge-choked water. Pairs are normally double brooded giving a protracted breeding season from mid March to July, even August (Flegg and Glue 1973). Birds are especially vociferous when establishing territories between late March and the end of April, and confusion may arise from migrants which call from their winter territories and on passage (Bayliss 1985). Britain and Ireland receive birds from N and E Europe in winter (Flegg and Glue 1973) but there is no evidence that home-bred stock migrate. The generally warm and dry summers of 1988–90, that produced drought conditions especially in southern parts, will have reduced the amount of suitable nesting ground for Water Rails, and possibly their breeding numbers.

The *Trends Guide* repeats the population estimate of 2,000–4,000 pairs

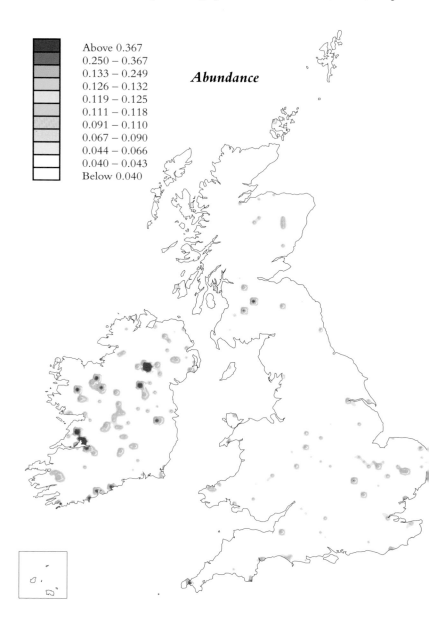

	Above 0.367
	0.250 – 0.367
	0.133 – 0.249
	0.126 – 0.132
	0.119 – 0.125
	0.111 – 0.118
	0.091 – 0.110
	0.067 – 0.090
	0.044 – 0.066
	0.040 – 0.043
	Below 0.040

Abundance

in Britain and Ireland based upon the *68–72 Atlas*. It mentions the Water Rail's vulnerability to severe cold winters (Dobinson and Richards 1964), implying that several cold winters during the early 1980s may have reduced numbers. More recent exceptionally mild winters in the late 1980s could have allowed the replacement of such losses, but quantitative information is not available. Tape lures used to stimulate 'song' in Scottish localities during the 1980s have revealed concentrations of 20–30 pairs (*Birds in Scotland*). Following the assumptions of slightly more than 2–4 pairs per occupied 10-km square used in the *68–72 Atlas*, there are estimated to be 450–900 pairs of Water Rails in Britain, with nearly twice as many (850–1700) in Ireland. Careful, more widespread use of tape lures would probably confirm the belief that these estimates are minimal. A more accurate assessment of the population size of the Water Rail is called for, especially in view of current indications of decline.

DAVID E. GLUE

68–72 Atlas p 152

Change

Distribution

Years	Present, no breeding evidence		Breeding evidence		All records		
	Br	Ir	Br	Ir	Br	Ir	Both (+CI)
1968–72	282	63	351	214	633	277	911
1988–91	205	87	215	89	420	176	597
			% change		−33.6	−36.5	−34.5

Spotted Crake

Porzana porzana

Spotted Crakes are most often found in extensive wetlands as these provide the conditions they prefer. They are amongst the most difficult of species to detect and, unless the observer is very lucky, hearing the characteristic song is the only hope of success. This is a sharp ascending whistle, likened to a whip crack, uttered as often as 80 or 90 times per minute and over long periods, especially on fine nights. Although birds have been recorded throughout the year it is likely that the breeding population winters in sub-Saharan Africa.

Over the period of the *88–91 Atlas* birds were recorded in various wetlands throughout Britain and at one in Ireland. Most records were in the northern half of mainland Scotland but there was a regular, well-recorded, population each year in Cambridgeshire. Because their call is so penetrating, and their habitat requirements are so precise, it is most unlikely that any regular breeding sites have been missed. On the other hand, Spotted Crakes have a history of sporadic territory-holding (and probably breeding) on quite small wetlands, and it is very likely that a number of these were not detected.

The centres of population of this species have changed since the *68–72 Atlas*, with only two 10-km square records common to both surveys. Speyside is the sole area represented by more than a single 10-km square in both 1968–72 and 1988–91, while the SW Scotland population, recorded in five 10-km squares in 1968–72, failed to provide a single record in 1988–91. There were fewer records in the Northern and Western Isles, and on the W coast of Scotland, than in 1968–72. The Irish report is probably only the fifth summer record there since breeding was proved, for the first and only time, in about 1851 (*Birds in Ireland*).

It is believed that in times past this was a much more numerous breeding species in Britain, but it decreased during the 18th and 19th centuries, as drainage of wetlands progressed. Yarrell (1882–84) considered it a much reduced but still not uncommon bird in the remaining suitable habitats.

He believed it most often to be found 'in the maritime counties of the east coast'. *Red Data Birds* describes it as a sporadic breeding bird during much of the 20th century, and what data on distribution and numbers there are for this period are shown on the accompanying Table.

Most major wetlands in Britain which Spotted Crakes might colonise are already protected as reserves. The species is nowhere common in W Europe and has probably declined everywhere (except possibly in Sweden) over the last few decades (*BWP*). Careful management of reserve areas to provide the preferred conditions might establish small, but regular, breeding populations in several parts of Britain. Unfortunately there is no real indication of a sustained increase at present, although the most recent years reported by the RBBP (1989 and 1990) have some of the highest counts of territorial birds in the last decade.

	Distribution	
Period	No. counties with probable nesting	
1926–37	*ca* 10	
1963–81	25	

	Numbers	
Years	No. singing males	No. sites
1985	3	2
1986	4	3
1987	18	7
1988	10	6
1989	21	14
1990	17	10

CHRIS MEAD

68–72 Atlas p 154

Two records have been moved by up to two 10-km squares on the
Distribution and Change maps.

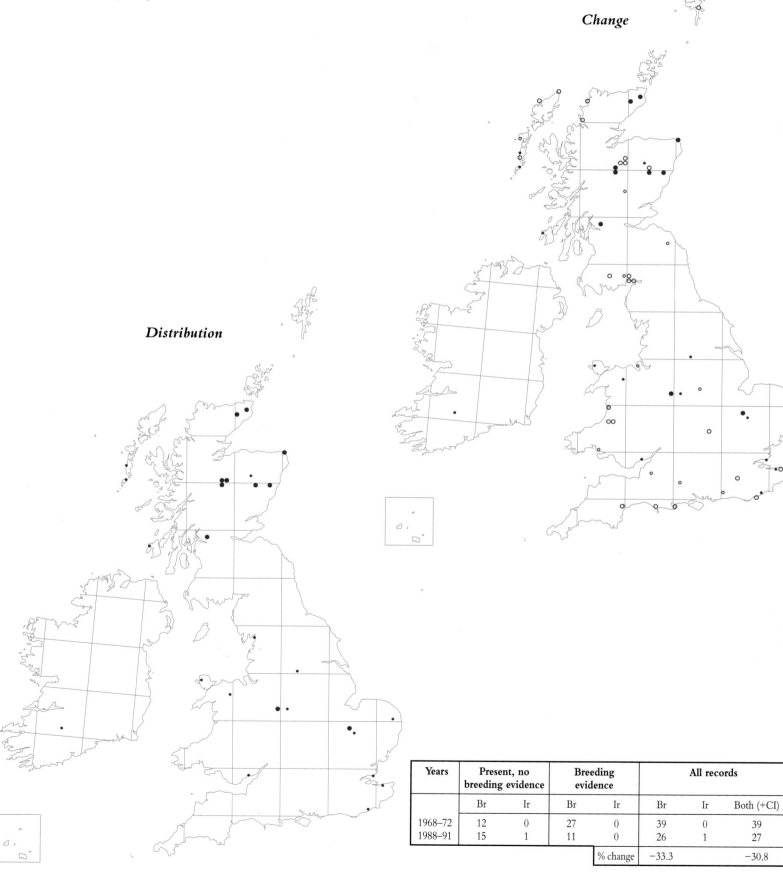

Change

Distribution

Years	Present, no breeding evidence		Breeding evidence		All records		
	Br	Ir	Br	Ir	Br	Ir	Both (+CI)
1968–72	12	0	27	0	39	0	39
1988–91	15	1	11	0	26	1	27
				% change	−33.3		−30.8

Corncrake

Crex crex

The secretive Corncrake, with its rasping, monotonous 'crek crek', conjures up images of flower-rich meadows of the Hebridean machair. Frustratingly difficult to see, and always farther from you than you think, this remarkable bird has been declining in Britain for 150 years. Is its disappearance as a breeding species only a matter of time?

The Distribution and Change maps are little comfort. In Britain Corncrakes are now confined almost entirely to NW Scotland, and particularly the Hebridean islands. In Ireland the main areas are the Shannon callows, and the north and west. Corncrakes are difficult to count accurately, as their main vocal period is between midnight and 3am (Hudson *et al.* 1990), and even during this period not all birds 'crake' each night. Not surprisingly, therefore, Atlas fieldwork did not give a good indication of variation in numbers, so the Abundance map shown here is taken from the RSPB and IWC 1988 census (Hudson *et al.* 1990, Mayes and Stowe 1989). The Distribution map reflects the Corncrake's main habitat: grass late-cut for hay or silage.

The Change map emphasises the severe extent of contraction of range. Declines were recorded during the fieldwork for the *68–72 Atlas* and these have continued. The number of occupied squares has fallen alarmingly by 1988–91, with only 21 remaining outside Scotland and Ireland. In all parts of mainland Scotland and in Shetland, Corncrakes have become much rarer or have disappeared altogether, and in Ireland, too, the contraction in range has been severe and widespread, particularly in the southwest, south and east.

The cause of the Corncrake's long decline has been linked to the mechanisation and earlier cutting of the hay harvest. The former destroyed more nests than the traditional scything of hay, and the introduced grass species could be cut earlier, so more nests, young and adults were destroyed. More recently, the introduction of silage, which encourages even earlier

cutting, has worsened the problem. In many areas cutting is now so early that suitable nesting habitat no longer exists in the breeding season. Even in the remaining strongholds the trend towards agricultural 'improvement' threatens to bring cutting dates further forward. During the period of the Atlas fieldwork the Corncrake population in Northern Ireland declined dramatically, from 122–134 males in 1988 to 25 in 1991, as more farmers were encouraged to produce silage and, accordingly, the cutting dates were brought forward.

The conversion of meadows to permanent sheep pasture is another threat, and in N Scotland many meadows have been abandoned in favour of sheep ranching. Recent research by the RSPB indicates that habitat change continues on the breeding grounds, as does the earlier timing of mowing, and both can be linked to the continuing decline in numbers of Corncrakes.

Despite being reluctant fliers on the breeding grounds, Corncrakes migrate to winter in SE Africa: there is no evidence to suggest a wintering area in W Africa. It has been thought that problems in their winter quarters may have led to the Corncrakes' decline, but information from Africa provides no evidence that the bird and its habitat are threatened in winter,

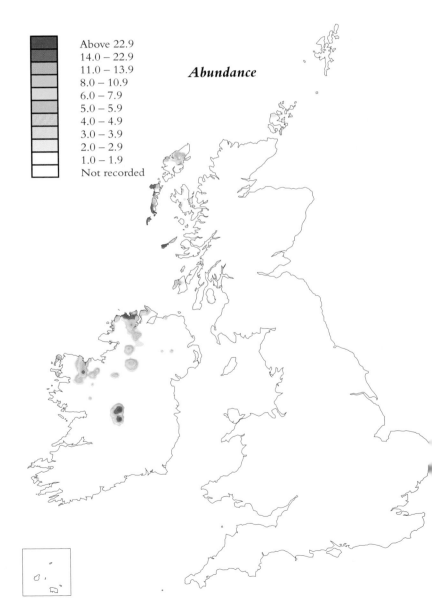

Above 22.9
14.0 – 22.9
11.0 – 13.9
8.0 – 10.9
6.0 – 7.9
5.0 – 5.9
4.0 – 4.9
3.0 – 3.9
2.0 – 2.9
1.0 – 1.9
Not recorded

Abundance

nor of processes or changes there that could account for a decline spanning 150 years (Stowe and Becker). On migration, there may be a problem associated with the centuries-old tradition of Quail netting (which incidentally kills Corncrakes) in autumn on the N African coast, particularly Egypt. It is possible that some W European birds travel south into Africa via Egypt.

The Corncrake's mating system is still poorly understood. Females can be successively polyandrous, leaving their first brood after about three weeks. Males appear to take no part in rearing young, and move around during the season in the hope of attracting more females (Stowe and Hudson 1991).

The decline in Corncrake numbers and the contraction of range have affected most breeding areas in Europe. Hudson *et al.* (1990) estimated a W European population of 5,000–7,500 calling males, with several thousands in E Europe. The situation in the former Soviet Union is less clear: it appears that declines have occurred in the Baltic states but in the east the Corncrake is still widespread and common.

The British population was estimated at 2,640 pairs during 1968–72, falling to 730–750 calling birds in 1978–9 (Cadbury 1980) and to 550–

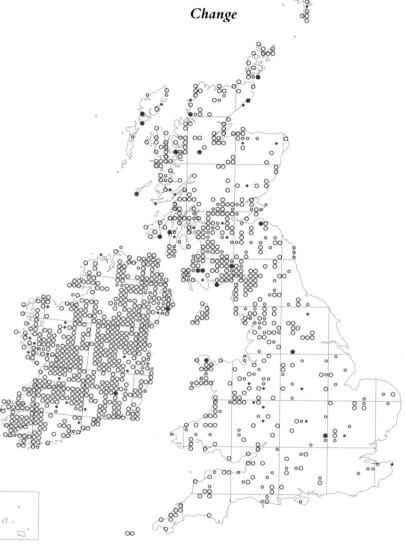

Change

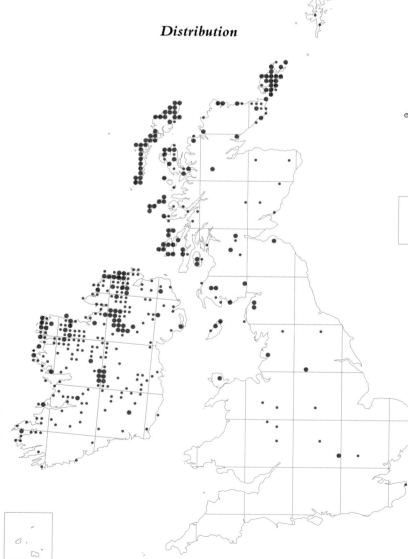

Distribution

600 in 1988 (Hudson *et al.* 1990). In Ireland, 1,500 were recorded in 1978 (O'Meara 1979), falling to 903–930 in 1988 (Mayes and Stowe 1989). Surveys by the RSPB of the main areas in Britain suggest a decline to 470 males in 1991. The future for the Corncrake looks bleak.

TIM J. STOWE

68–72 Atlas p 156

Years	Present, no breeding evidence		Breeding evidence		All records		
	Br	Ir	Br	Ir	Br	Ir	Both (+CI)
1968–72	133	36	527	792	660	828	1488
1988–91	56	188	105	58	161	246	407
				% change	−75.6	−70.3	−72.6

Moorhen

Gallinula chloropus

The adaptable Moorhen breeds on any lowland freshwater body, from the smallest farm pond to the fringes of the largest lake, wherever there is sufficient bankside or emergent vegetation.

As a result, it is widespread throughout lowland Britain but absent from the higher ground of England and Wales, parts of S and SW Scotland, and most of the Scottish Highlands. It is most common in SE England, East Anglia, and the W Midlands, with pockets of high abundance in Lincolnshire, Humberside and Yorkshire. Elsewhere in Britain, densities are low save in the S Tayside region of Scotland. In Ireland, where it is also absent from high ground, its pattern of abundance is more patchy. Native Moorhens are largely sedentary, so it is no surprise that the Abundance map is similar to that in the *Winter Atlas*.

The breeding density of Moorhens declines as river gradients increase (Marchant and Hyde 1980), and more generally with reduced land fertility (Sitters 1988). The reason for the latter decline is probably because only waters on fertile soil produce the necessary eutrophic conditions, with sufficient food for survival and emergent vegetation for nesting, the species being almost entirely absent from unproductive acidic upland waters.

Although the pattern of distribution has altered little since the *68–72 Atlas*, there have been changes at the extremities. The population in the Outer Hebrides has recently declined, and the species may now be extinct as a breeding bird on Lewis and Harris (*Birds in Scotland*): there were no breeding records from these islands between 1988 and 1991. Similarly, there were none from Skye or Mull, even though Moorhens bred there during 1968–72. The populations on Islay and Orkney, however, remain healthy. The decline of the NE Scotland population is thought to be due to increased land drainage (Buckland *et al.* 1990), whereas the causes of the small declines in the rest of Scotland, central Wales and the SW peninsula remain unclear. In Ireland there is evidence for a contraction of range in counties Donegal, Sligo, Kerry and Cork. This may be linked to the great increase in mechanisation of agriculture since the 1960s (*Birds in Ireland*) and its inevitable impact on land drainage, but such apparent

changes have to be treated with caution, due to differences in the level of coverage between the two Atlases.

Land drainage is not the only threat to Moorhen populations, as the species forms an important part of the diet of feral mink. Following their appearance in the wild in the 1950s, mink are now widespread in Britain and expanding throughout Ireland. The decline in Moorhen numbers at some sites has been attributed to mink predation, and WBS data show that low densities of Moorhen are associated with the presence of mink (Birks 1990). On Harris, mink fur farms were established in the late 1950s, and the first feral mink reported in 1969. By 1988 there were estimated to be up to 7,500 females on Lewis and Harris. Prior to the arrival of fur farms, 60 pairs of Moorhens bred on these islands, but by 1982 the species was extinct (S. Tyler). Despite these local extinctions, there is no evidence of a detrimental effect on national population levels (*Trends Guide*), and although it is tempting to blame the losses in parts of the Moorhen's range on mink, the case has yet to be proven.

Sympathetic management of waterways can increase Moorhen numbers. Dredging, and the removal of potential nesting vegetation from banks, reduces the density of breeding pairs and the proportion that lay second clutches (Taylor 1984). Moorhens may be encouraged to breed by ensuring

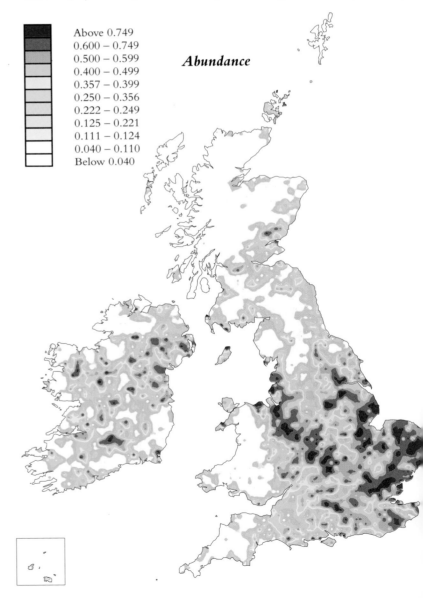

Abundance

	Above 0.749
	0.600 – 0.749
	0.500 – 0.599
	0.400 – 0.499
	0.357 – 0.399
	0.250 – 0.356
	0.222 – 0.249
	0.125 – 0.221
	0.111 – 0.124
	0.040 – 0.110
	Below 0.040

that dredging disturbance is restricted to one bank at a time, thus leaving vegetation suitable for nesting on the opposite bank.

The egg laying period is from mid March to the end of August, and about one-third of pairs attempt a second brood. Third and fourth broods do occur but are rare. Clutch size varies from 1–21 eggs, but the commonest number is seven. Establishing clutch size is complicated because more than one female may lay in the same nest: in one study, 75% of clutches were laid by single females, 16% by two, 7% by three and 2% by four females (Gibbons 1986). Breeding success varies greatly, but about two-thirds hatch at least one egg. Many clutches are lost to predation, mainly by Crows and stoats, but Moorhens will continue to lay after failures, and up to five repeat layings have been recorded. Thus, numbers can rapidly build up after a population decline such as occurred in the severe winter of 1962–63.

During 1988–91 there were an estimated 240,000 Moorhen territories in Britain and 75,000 in Ireland.

DAVID W. GIBBONS

68–72 *Atlas* p 158

Change

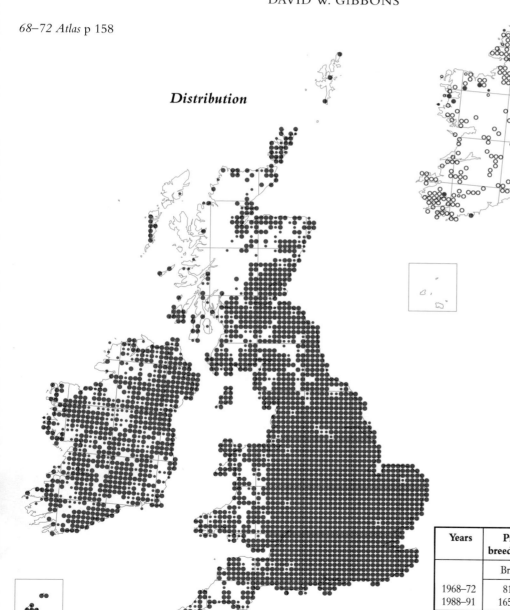

Distribution

Years	Present, no breeding evidence		Breeding evidence		All records		
	Br	Ir	Br	Ir	Br	Ir	Both (+CI)
1968–72	81	35	2159	870	2240	905	3149
1988–91	165	150	1867	564	2032	714	2758
			% change		−9.2	−21.0	−12.6

Coot

Fulica atra

The chunky, charcoal-plumaged Coot is a familiar breeding bird on many lowland freshwaters, including large ponds, reservoirs, flooded gravel pits and slow-moving rivers. Highly gregarious in winter, it changes character in the breeding season to become aggressively territorial, contesting ownership of waterside nesting areas with puffed-up posturing and lunging attacks on rivals.

Coots are commonest and most widespread in lowland areas of England, especially the Midlands, southern and SE England. In Wales, densities are lower and the breeding distribution more patchy. The bulk of the Scottish population is in the fertile lowlands south and east of the Highland Boundary Fault, particularly the coastal farmland areas of Fife, Tayside and Grampian. Coots are absent from most of the Highlands and Islands, except for the nutrient-rich lochs of the southern Outer Hebrides and Orkney. In Ireland, as in Britain, they are absent from many upland areas, including parts of Kerry, Cork, SW Clare, Galway, Mayo, Sligo, Donegal and Londonderry. The distribution of Coots is very similar in winter and summer, although winter abundance, boosted by an influx of migratory birds from NW Europe joining the sedentary native population, is often greater on favoured waters. The Coot's selection of relatively large, shallow, nutrient-rich waters with abundant bottom vegetation for food and some emergent plants for nest anchorage and concealment explains its absence from most upland areas, where breeding habitat of this type is rare. Upland waterways are also generally unsuitable for Coots. WBS data show that most riparian territories are on waters below an altitude of 50m, mainly in southern and eastern areas of England (Marchant and Hyde 1980) and regional surveys in the Hertfordshire and London areas indicate that even in the core of the Coot's British range only a small percentage of the population breeds on rivers and canals.

After an expansion in Scotland early this century, numbers fell and distribution there contracted (*Birds in Scotland*), with Coots no longer breeding regularly on Shetland and numbers declining on Orkney. This Scottish range reduction has continued in the last 20 years, most noticeably in the coastal areas along the southern fringe of the Moray Firth and in NE Scotland, although an apparent lack of suitable habitat in some of these now abandoned areas of NE Scotland makes it seem surprising that Coots ever bred there (Buckland *et al.* 1990). In Ireland, there has been a contraction of the breeding range in several counties, and locally the species has become more sparsely distributed. As with changes in Moorhen distribution in Ireland since the 1960s, these reductions in Coot range could be a result of land drainage associated with agricultural intensification, although conclusive evidence for this is not available.

In the Coot's British heartlands of lowland England, the breeding distribution has remained largely unchanged since the 1960s. Coots here benefited greatly from the availability of new freshwater breeding habitats through the construction of reservoirs from the last century onwards and from the flooding of sand and gravel workings since the 1930s (Parslow 1973). When unmolested, Coots are very tolerant of human presence

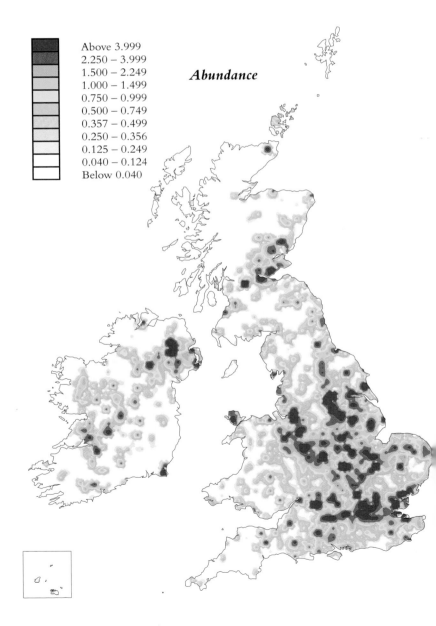

Above 3.999
2.250 – 3.999
1.500 – 2.249
1.000 – 1.499
0.750 – 0.999
0.500 – 0.749
0.357 – 0.499
0.250 – 0.356
0.125 – 0.249
0.040 – 0.124
Below 0.040

Abundance

(*BWP*) and so in lowland areas can further benefit from the creation of large ponds and urban lakes, provided suitable emergent vegetation or islands are available for nesting.

Winter Coot flocks break up in February and March, when migrants depart for Continental breeding grounds and residents assert claims to the multi-purpose breeding territories where they mate, nest and feed. The egg laying period is from the second week in March to mid July, with six eggs in a typical clutch, although this varies from three to eleven (*BWP*). In favourable summers some pairs rear a second brood but one is more usual. For such a large and easily observed lowland bird, it is surprising that very little is known about Coot densities, and there is still considerable uncertainty about the size of the population (*Trends Guide*). Precise data are also very limited for other parts of the range in the W Palearctic. Extrapolation from tetrad counts during April to July in 1988–91 yields minimum British and Irish populations of 46,000 and 8,600 birds respectively. However, as an unknown proportion of these may be non-breeding birds, it is difficult to translate these figures into breeding pairs.

KENNY TAYLOR

68–72 Atlas p 160

Change

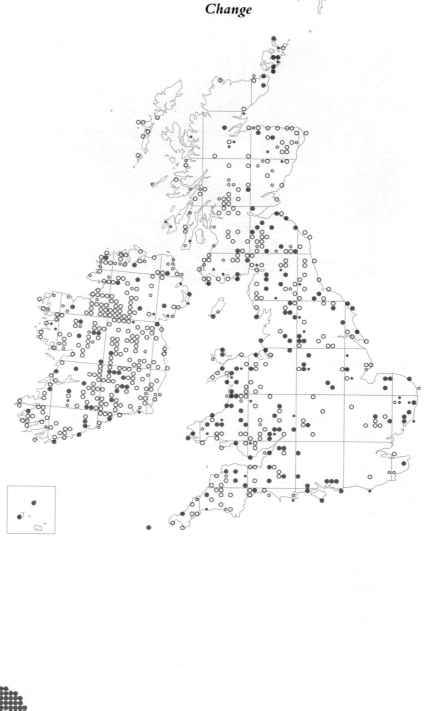

Distribution

Years	Present, no breeding evidence		Breeding evidence		All records		
	Br	Ir	Br	Ir	Br	Ir	Both (+CI)
1968–72	108	42	1582	523	1690	565	2256
1988–91	133	60	1470	294	1603	354	1963
				% change	−5.1	−37.2	−13.0

RI

Oystercatcher

Haematopus ostralegus

The clamorous piping parties and shrill display flights of this conspicuous and strongly territorial wader are familiar features of much of our coastline and along many river systems of N Britain. Indeed, breeding Oyster-catchers draw attention to themselves by their mobbing and distraction behaviour. Their nests are relatively easy to find. This is a bold and adaptable species that nowadays occupies diverse inland, as well as ancestral, coastal, nesting habitats.

The Distribution map shows clearly the two patterns of breeding, though some of the records with no breeding evidence in England and Wales may refer to wandering birds rather than to potential breeders. Due to increasing disturbance and development in the south, coastal Oystercatchers there are becoming ever more confined to nature reserves. Elsewhere, particularly large concentrations nest in reserves in Norfolk, NW England and on Welsh islands. In Ireland, the smaller proportion of breeding records compared to the *68–72 Atlas* probably reflects lower emphasis on proof of breeding in the 1988–91 survey. Inland nesters are almost ubiquitous in NW England north of the Ribble Valley, and throughout most of Scotland. In S Britain, inland nesting is especially evident in East Anglia, with scattered pairs across the N Midlands from the Wash and Humber to the Mersey and NE Wales. In Ireland inland breeding is clearly rare, though the scatter of dots in the north and west might indicate prospecting birds.

Coastal Oystercatchers nest on shingle beaches, dunes, salt marshes, and rocky shores including cliffs, skerries and vegetated tops of islands. Breeding, however, is limited to relatively undisturbed coasts where stocks of mollusca and other marine prey are uncovered daily by a sufficiently wide tidal range.

Inland, Oystercatchers nest mainly on riverine shingle beds, lake shores and in fields (especially arable) along the river valleys, with some pairs

even on heathland, or in open birch or conifer woodland, within reach of rivers.

The breeding range has spread inland since the late 19th century, or even earlier in E Scotland, by colonisation up the main rivers in NW England, and SW and E Scotland (Buxton 1962, Dare 1966). By the 1960s, Oystercatchers were moving away from water and into Lapwing and Curlew habitats in some areas. This radical expansion was probably caused by behavioural change, rather than by human pressure on shore nest sites in the north. It also necessitated advancing the nesting season in order to utilise the earlier peak availability of terrestrial foods – chiefly earthworms and tipulid larvae (Heppleston 1972). A parallel spread on to farmland has occurred in all countries around the North Sea, but most dramatically in the Netherlands (*BWP*, Piersma 1986).

The Abundance map must be interpreted with caution for whilst Oystercatchers have a basically linear distribution along coasts and rivers, they breed locally at high concentrations in loose 'colonies', especially on islands and in reserves. The map understates the true position in, for example, Norfolk and Pembrokeshire, but probably reflects well the relative abundance of inland breeders in Scotland and N England.

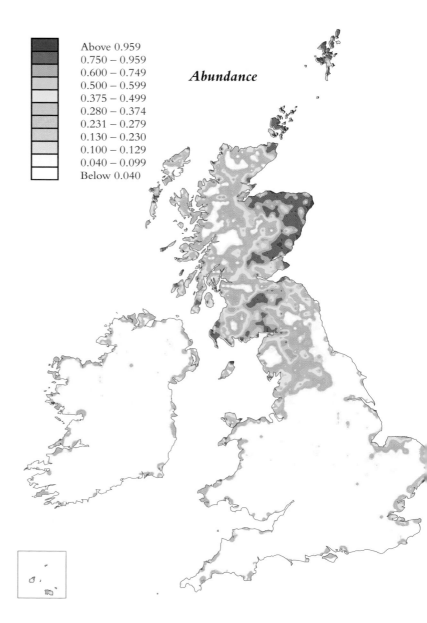

	Above 0.959
	0.750 – 0.959
	0.600 – 0.749
	0.500 – 0.599
	0.375 – 0.499
	0.280 – 0.374
	0.231 – 0.279
	0.130 – 0.230
	0.100 – 0.129
	0.040 – 0.099
	Below 0.040

Abundance

The Change map confirms the continuing range expansion inland in England since the *68–72 Atlas* as well as in mid and NE Wales. This has been most striking in E English counties in the belts of intensive arable farming from East Anglia northwards to the Humber and Vale of York; and to the south of the Firth of Forth. For coastal breeders, the only prominent change is the spread up the banks of the Severn estuary. Losses and gains elsewhere may simply be related to differences in coverage between the two Atlases.

Population estimates indicate the magnitude of the Oystercatcher's increase in Britain in the past 30 years; from 19,000–30,000 pairs in the early 1960s (with about 70% in Scotland (Dare 1966)), to 33,000–43,000 pairs by the mid 1980s (Piersma 1986). The Irish population, by contrast, seems to have remained stable at 3,000–4,000 pairs (Dare 1966, Hutchinson 1979, Piersma 1986), perhaps because Oystercatchers there have yet to make the 'breakthrough' inland. The combined population of Britain and Ireland represents about 20% of the European Oystercatchers breeding outside the former USSR (Piersma 1986). Our population is comparable with that of Norway, and is exceeded only by the rapidly increasing population (90,000 pairs) in the Netherlands.

Change

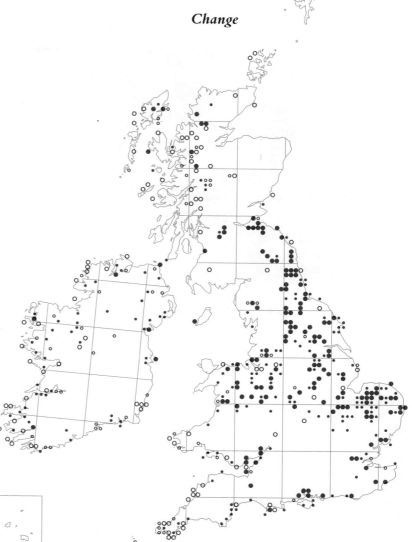

Distribution

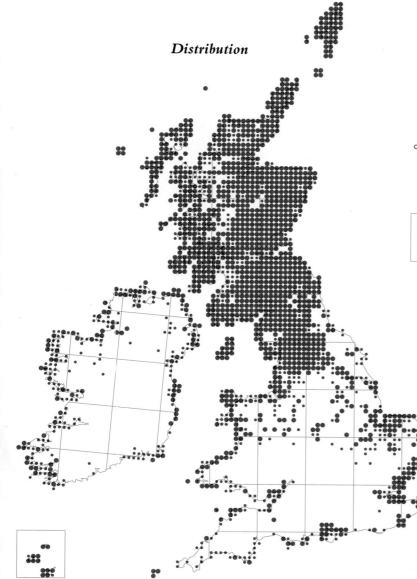

British Oystercatchers winter predominantly around our own shores, inland haunts then being deserted. Many juveniles, however, emigrate from all parts of Britain to winter in W France and Iberia. Adults begin returning to their inland territories before the end of winter, even in Scotland.

The recent intensification of agriculture in Britain has clearly not impeded the Oystercatcher's colonisation of inland habitats. Given the Netherlands' experience, it is conceivable that this highly successful wader will continue pushing further south into the central England farmlands and the river valleys of Wales.

PETER DARE

68–72 Atlas p 162

Years	Present, no breeding evidence		Breeding evidence		All records		
	Br	Ir	Br	Ir	Br	Ir	Both (+CI)
1968–72	135	84	1397	179	1532	263	1800
1988–91	327	169	1375	94	1702	263	1979
			% change	11.4	0.8		9.8

Avocet

Recurvirostra avosetta

Thanks to its being adopted as a logo by the RSPB, the Avocet's striking black and white plumage and upward curved bill are familiar to thousands who have never seen one in the flesh.

In England, breeding Avocets favour shallow, brackish coastal lagoons, with bare or sparsely vegetated low islands, but in the central Asian steppe and in East Africa they breed far inland.

Until the early 18th century Avocets bred regularly on the E coast of England from the Humber to Kent. In 1818, they still bred at Orfordness, Suffolk, close to a current stronghold of the species on Havergate. Breeding ceased in Norfolk in the mid 1820s; at the mouth of the Trent, N Lincolnshire (now Humberside) about 1840; and from Romney Marsh, Kent in the mid 1840s. About that time, improvements to the sea defences not only facilitated the drainage of marshes but reduced the frequency with which the sea walls were breached. As a consequence, the brackish pools where Avocets (and malaria-carrying mosquitoes) bred were lost. People living near the marshes may have taken the birds and their eggs for food, but as the Avocets became rare, collectors of eggs and specimens may have been responsible for the temporary extinction of the species in Britain.

Avocets recolonised England 100 years later, in the early 1940s, perhaps because sea walls tended to fall into disrepair during the Second World War, and also because areas were closed to the public for security reasons. Following sporadic nesting in Essex and Norfolk, Avocets became re-established in Suffolk in 1947. Until the period of the *68–72 Atlas*, they bred regularly only in that county, with most at two RSPB reserves, Havergate Island and Minsmere (Cadbury and Olney 1978, Cadbury *et al.* 1989).

Since the mid 1970s there has been a considerable range expansion and population increase, and the species now breeds regularly in Norfolk, Suffolk, Essex and Kent. The main colonies are all on reserves, where there is protection from human disturbance. In N Norfolk, Avocets have bred at Cley (Norfolk Naturalists' Trust) from 1977 and at Titchwell (RSPB) from 1984. There were 60 and 49 pairs respectively at these two sites in 1990. In Suffolk, up to 132 pairs have bred on Havergate Island and 69 pairs at Minsmere; in 1990 their respective populations were 100 and 47 pairs. In N Kent, Avocets have nested annually at Elmley marshes (RSPB) since 1986, with 24 pairs in 1990. In total, at least 18 sites were occupied by breeding Avocets in 1990. Recent studies, analysing data from Havergate and Minsmere, showed that population processes are similar at both sites. As the number of pairs of Avocets increased, so the number of young reared per pair declined (Hill 1988). This has since been shown to occur at two Norfolk localities as well. Consequently birds at newly colonised sites generally produce more young per pair than those on older established sites. Annual fluctuations in the number of breeding birds were related to the level of chick mortality the previous year, and the main factor regulating population size was found to be competition for breeding space. Colour ringing has shown that Avocets do not exhibit

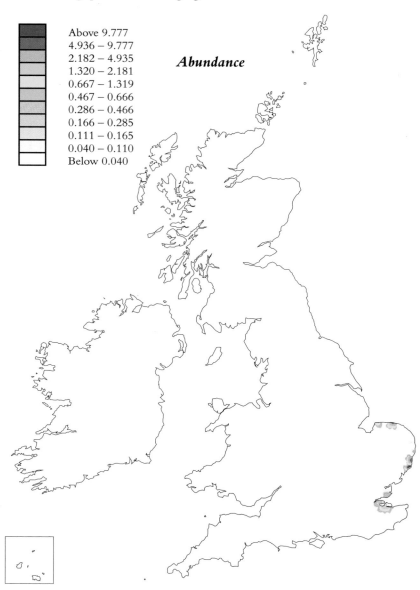

Above 9.777
4.936 – 9.777
2.182 – 4.935
1.320 – 2.181
0.667 – 1.319
0.467 – 0.666
0.286 – 0.466
0.166 – 0.285
0.111 – 0.165
0.040 – 0.110
Below 0.040

Abundance

a strong attachment to their natal site when breeding for the first time, but once they have bred they tend to return to the same area in subsequent years. They usually breed in their second or third year. Immigration from the Continent may have contributed to the increase in the English population.

On a number of sites, management has been aimed at encouraging Avocet colonisation and then increasing breeding numbers, largely by creating nesting habitat, reducing water salinity, and controlling competition from breeding Black-headed Gulls (Cadbury and Richards 1978, Hill 1988). Hydrological management at Havergate has been aimed at curbing the hypersaline condition of the lagoons, thereby benefiting the Avocet's main food prey – the common ragworm, the shrimp *Corophium volutator*, the prawn *Palaemonetes varians*, and the larvae and pupae of the chironomid *Chironomus salinarius* (Mason 1986, Cadbury and Richards 1978). This type of management can improve chick survival, but a computer model shows that the population cannot increase once the breeding habitat is at 'pair carrying capacity' (Hill and Carter 1991). Such management, however, has probably helped the colonisation of other sites in England.

Change

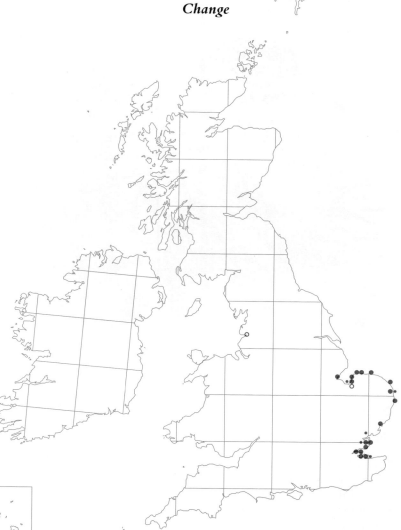

Distribution

RBBP figures estimate the size of the breeding population in England at about 400–500 pairs, which represents about 2% of 19,600 pairs in W Europe (Piersma 1986). The Avocet is included in *Red Data Birds* but it is not a threatened species in world or European terms. Nevertheless, it will probably remain very local in Britain and in Ireland (where it has bred) because it depends on a scarce, specialised habitat.

DAVID HILL and JAMES CADBURY

68–72 Atlas p 198

Years	Present, no breeding evidence		Breeding evidence		All records		
	Br	Ir	Br	Ir	Br	Ir	Both (+CI)
1968–72	1	0	7	0	8	0	8
1988–91	6	0	22	0	28	0	28
			% change		250.0	0	250.0

RI

Stone Curlew

Burhinus oedicnemus

The wild cries of Stone Curlews still break the silence of spring and summer nights in open country in a few parts of S and E England, but the range of this summer visitor has contracted substantially so that it is now one of Britain's rarest and most threatened birds (*Red Data Birds*).

Stone Curlews are most active at night and are difficult to find by day. They seek out stony, sandy soils and sit immobile, almost invisible, the streaks and stripes on their light-brown plumage disrupting their outline. As a result, the Distribution map owes much to recent systematic survey work by RSPB fieldworkers. There are two main population centres, the Breckland of Norfolk and Suffolk, and Wessex, centred on Salisbury Plain. There are also precariously small sub-populations in Berkshire, the South Downs, the Cambridgeshire/Essex border, E Suffolk and N Norfolk. The breeding record in Lincolnshire refers to a ringed female, hatched in Breckland, which was found incubating a clutch of eggs, but was not known to have had a mate. A map of abundance based on intensive fieldwork will be published elsewhere.

The species' distribution is restricted to free-draining, sandy soils with a high proportion of chalk rubble or flints. This association is almost certainly attributable to selection for a background upon which the bird and its eggs and chicks are well camouflaged. Frequent records of nesting on concrete rubble on airfields are the exception which proves this rule. Another factor may be the tendency for free-draining soils to warm up early in the spring and hence to have enhanced availability of the invertebrates upon which Stone Curlews feed.

Stone Curlews nest on open ground with short, sparse vegetation or bare, stony soil. Breckland grass heaths and chalk downland grazed by

rabbits and livestock hold the highest population densities; up to seven pairs per km². Only fragments of these formerly extensive habitats remain, most having been converted to arable farmland or forestry plantations. Some of the largest surviving patches on Salisbury Plain and in Breckland persist because of their use as military training areas. As a result of a long-term reduction in grazing pressure, the vegetation on many of the surviving areas of formerly short grassland has become too tall and dense for Stone Curlews. This change began with the myxomatosis epidemic amongst rabbits in the 1950s and has been exacerbated by a decline in the demand for low quality grazing for sheep and cattle. More than two-thirds of the population now nest on arable farmland among spring-sown crops such as sugar beet, barley and carrots. Population densities on arable farmland rarely exceed one-tenth of those on the most suitable fragments of short grassland (Green 1988a).

There is evidence of a long-term decline in Stone Curlew population and distribution, the species having disappeared from the northern part of its range in Yorkshire and Lincolnshire in the first half of the 20th century (Parslow 1973). The Change map shows that the range contraction has continued. The reasons for the decline appear to include reductions in the area of semi-natural short grassland and changes in farming practices, which affect the suitability of arable land for Stone Curlews. In former strongholds a reduction has taken place in mixed farming which provided the birds with ley pastures, rich in invertebrate prey, close to tilled land suitable for nesting (Green 1988a).

Recent research by the RSPB on the ecology of Stone Curlews has shown that they forage mainly by night and hunt for earthworms and arthropods on the soil surface in areas of short vegetation. Two eggs are laid in a scrape on stony ground and the chicks are usually reared near the nest site on invertebrates brought to them one by one by the parents. Eggs and chicks on spring tillage are endangered by farming operations, particularly the rolling of cereals and inter-row hoeing of root crops. Protection of eggs and young from these hazards by RSPB fieldworkers has recently produced substantial increases in the breeding success of pairs nesting on farmland (Green 1988a).

A programme of colour ringing has shown that exchanges of individuals between sub-populations of Stone Curlews are comparatively rare: only two cases are known of young hatched in Wessex moving to breed in Breckland, and no dispersal has yet been seen in the opposite direction.

Surveys of breeding pairs, and estimates using data on individually marked birds, put the current population at 150–160 pairs (Green 1988a). The estimate of 300–500 pairs in the *68–72 Atlas* was based on less detailed information, but the evidence of range contraction and changes in numbers in some sample areas support the suggestion that the population has continued to decline.

R. E. GREEN

68–72 Atlas p 202

Two records have been moved by up to two 10-km squares on the Distribution map. Stone Curlews bred in these 10-km squares during both Atlas periods. To maintain the confidentiality of these two locations, which would otherwise be apparent by comparison of the Change and 1968–72 maps, four false records (two filled and two unfilled dots) have been placed on the Change map.

Change

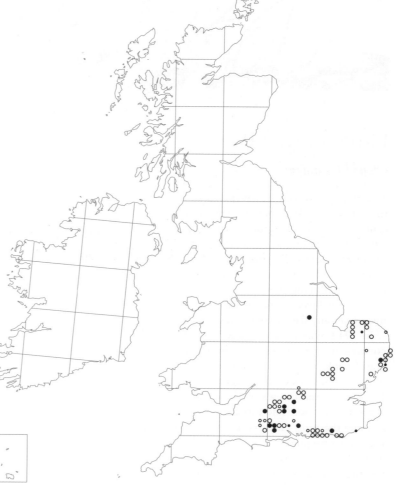

Distribution

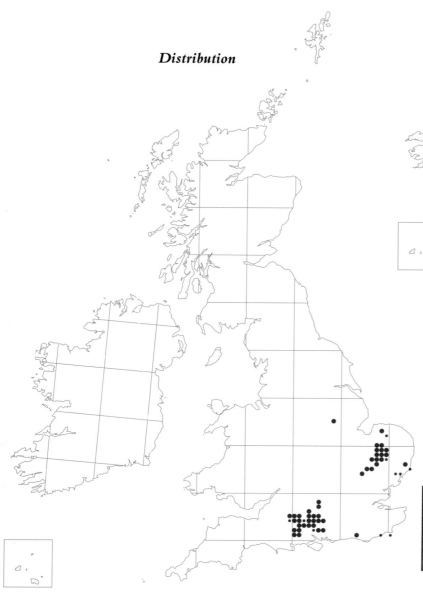

Years	Present, no breeding evidence		Breeding evidence		All records		
	Br	Ir	Br	Ir	Br	Ir	Both (+CI)
1968–72	12	0	81	0	93	0	93
1988–91	10	0	44	0	54	0	54
				% change	−41.9	0	−41.9

Little Ringed Plover

Charadrius dubius

The arrival and subsequent spread of breeding Little Ringed Plovers in England and Wales is one of the ornithological success stories of the 20th century. The first recorded breeding pair was located at Tring in 1938, when three young reached the flying stage (Leslie and Pedler 1938). Other pairs arrived in subsequent years and, despite the attentions of egg collectors in the late 1940s, the species flourished. Today, during migration, it is worth looking at all areas of shallow fresh water to see if there are any of these small waders feeding round the edge, or running like clockwork toys across any flat, sparsely vegetated ground nearby.

In 1940, when Little Ringed Plovers were relatively unknown in this country, they were described in *The Handbook* as "frequenting sand and gravel banks on rivers or borders of lakes and *sometimes* (the italics are mine) in disused sand or gravel-pits". Although this still applies in parts of Europe, in countries where these natural habitats are scarce, the birds have come to rely more and more on man-made nesting areas, and these have become the most favoured breeding habitat in this country, with natural sites the exception. Gravel and other pits and quarries, coal mining complexes, waste and industrial dumps, reservoirs and sewage farms together accounted for 90.8% of all summer records in 1984 while only 3% were in natural habitats (Parrinder 1989). This preference for man-made sites, especially those where work is in progress, has undoubtedly had an influence on breeding success, for the human presence keeps daytime predators from the nest (which is always on the ground) and from the non-flying young.

The Little Ringed Plover, by reason of its specialised breeding sites, is relatively easy to monitor. Censuses were made in 1973 and 1984 when observers visited all sites thought to be suitable, and during the intervening 11 years the population increased by just over 30% to 608 summering pairs in England and Wales (Parrinder 1989). Quite notable during the period was the expansion into NW England, shown clearly on the Change map. In Scotland, however, there has been very little change, apart from occasional sightings; one pair bred in 1968 and breeding was again proved (in Fife) during the period of the *88–91 Atlas*. Ireland has not been colonised and there have been very few sightings there, although a single bird was present in suitable habitat in Co. Tipperary during April and May 1991. Possibly the most notable change in range since the *68–72 Atlas*, however, has been the dramatic colonisation of Wales. In 1989, as a result of two very dry summers, extensive river shingle banks became exposed on the Tywi river and 14 pairs of Little Ringed Plovers were located. In 1991 it was estimated that the population

in Wales was some 60 pairs, 50 of which were on river shingle (S.J. Tyler).

Although the present situation is satisfactory, there is some concern for the long-term future of the species. Gravel and mineral deposits become exhausted and the man-made sites are put to other uses, so that areas

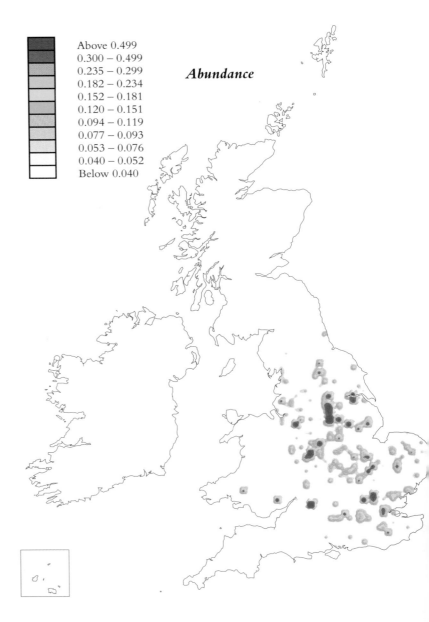

Abundance	
Above 0.499	
0.300 – 0.499	
0.235 – 0.299	
0.182 – 0.234	
0.152 – 0.181	
0.120 – 0.151	
0.094 – 0.119	
0.077 – 0.093	
0.053 – 0.076	
0.040 – 0.052	
Below 0.040	

where Little Ringed Plovers bred through a succession of years may be lost for ever. The demand for building materials has, however, been such that new pits are constantly being excavated to replace the old, while some organisations and some landowners make special after-care provision by maintaining the sparsely vegetated areas and shallow pools which the birds favour. One such former gravel pit in Suffolk may rightly be regarded as a Little Ringed Plover haven. It is in the middle of farmland and consists of 3ha of water with vegetated islands, but is bordered on one side by a sand and shingle 'beach' which looks ideal for nesting. Each year since its creation in 1984 four or five pairs have spent the summer there, but they frequently select nest sites in adjacent fields of late-germinating crops, such as sugar beet and linseed (John Wilson).

The 1984 survey revealed a mean of 2.55 summering pairs per occupied 10-km square. Assuming, conservatively, that densities within occupied squares have not changed since 1984, the British population of Little Ringed Plovers is estimated at 825–1,070 pairs. The lower estimate considers only 10-km squares with breeding evidence, the higher reflects all occupied squares.

EILEEN PARRINDER

Change

Distribution

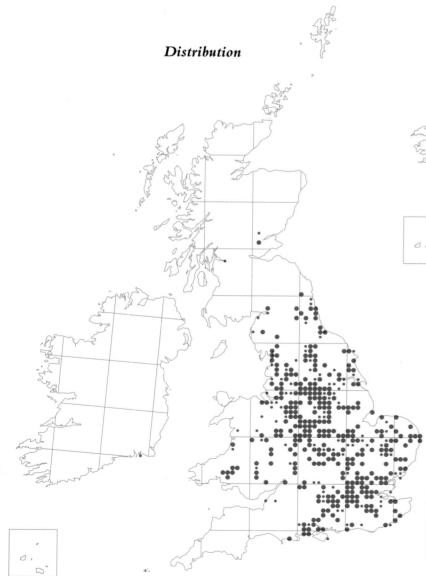

One record has been moved by up to two 10-km squares on both the Distribution and Change maps. The Co. Tipperary record (see text) is not shown, as the grid reference was withheld by the observer.

68–72 Atlas p 168

Years	Present, no breeding evidence		Breeding evidence		All records		
	Br	Ir	Br	Ir	Br	Ir	Both (+CI)
1968–72	28	0	260	0	288	0	288
1988–91	97	1	324	0	421	1	422
				% change	46.2		46.5

are taken by surprise in the open, they crouch motionless, relying on their cryptic colours and patterns for protection.

The *Winter Atlas* showed that Ringed Plovers were found all around the coast, including areas where they do not breed in the milder SW of England, Ireland and Wales. Many breeding birds from Norfolk leave in the late autumn but perhaps in most areas, including the Outer Hebrides and Orkney, breeders are resident. A return to breeding sites may start as early as mid February in mild winters but the characteristic 'butterfly' display is infrequent until late March.

The coastal distribution of the Ringed Plover has changed little since the *68–72 Atlas*, no doubt because the favoured habitats are limited in extent, essentially stable and have been little affected by agricultural changes. Indeed, in a survey of breeding birds in SE England, about 70% of pairs were on nature reserves or other well-protected sites (Prater 1989). Although recreational pressure is severe on many sites, birds are still present, but Pienkowski (1984) demonstrated that nests of pairs within a kilometre of an access point to the shore at Lindisfarne had a very low survival. Here 50+ visitors a day were noted and fewer than 2% of nests survived

Ringed Plover

Charadrius hiaticula

Although under substantial pressure from human disturbance, this small plover has managed to hold on to most of its favoured sandy and shingle beaches. It has at the same time taken advantage of gravel pits and other wetland sites to expand inland in England.

Ringed Plovers can be found breeding around all low coasts in Britain and Ireland. However, in S Wales, SW England and southernmost Ireland they are present in only a few most favoured sites and are absent from many small beaches where in similar rocky areas in Scotland they might be expected to occur. The southernmost and very small breeding populations in Europe are in NW France, so perhaps this virtual absence is related to edge-of-the-range effects, combined with intensive disturbance by holidaymakers. In England the extensive beaches around the Greater Thames and up to Humberside hold the majority of the population, but the species' stronghold is on the islands off Scotland. Here the birds in the Uists and Benbecula in the Outer Hebrides, Orkney and Shetland make up over 40% of the population of Britain and Northern Ireland (Prater 1989).

Over most of Britain and Ireland the species is attracted to sandy or shingle beaches, where they nest in exposed positions, relying on camouflage to protect sitting birds and unguarded eggs. In the Hebrides most Ringed Plovers – almost one-fifth of the British population – breed on the calcareous shell sand of the machair. This thin strip of agricultural land, farmed traditionally, is of outstanding importance for breeding waders, combining, as it does, extensive agriculture with limited chemical inputs and much fallow land (Fuller *et al.* 1986).

Within a day of hatching the young are taken to the edge of any nearby water-body to feed, but mostly they remain within a short distance of cover, where they can hide should danger approach. Often the chicks crouch under leaves of sea kale or other sand and shingle plants. If they

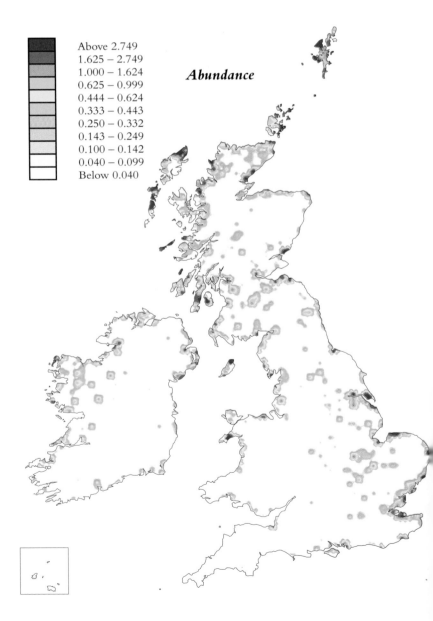

Abundance

Above 2.749
1.625 – 2.749
1.000 – 1.624
0.625 – 0.999
0.444 – 0.624
0.333 – 0.443
0.250 – 0.332
0.143 – 0.249
0.100 – 0.142
0.040 – 0.099
Below 0.040

compared with 43–58% where the number of visitors was fewer than five per day.

The main change since the *68–72 Atlas* has been for more pairs to set up territories inland. Over much of eastern and central England there has been a noticeable colonisation, documented by Prater (1989). Yet even though the inland population has doubled, it still contributes only 17% of the English total. No doubt the increase has been associated with an increase in the number of gravel pits, lakes and reservoirs: 40% of the inland birds choose these sites, but river beds (19.5%) and industrial sites (20%) are also important. The spring of 1984 was dry, and the resulting low water levels may have aided the spread inland. Likewise in the very dry summers of 1989 and 1990 there were many inland breeding records.

Prater (1989) estimated that in 1984 some 8,480 pairs bred in Britain. Numbers in England and Wales had increased by 21% in the 11 years since the previous survey, but in Scotland the trend was unclear, but it was not thought to have increased much. Although it is not possible to update the British estimate, a figure of 1,250 pairs in Ireland can be calculated by extrapolation: somewhat less than the estimate of more than 2,000 pairs

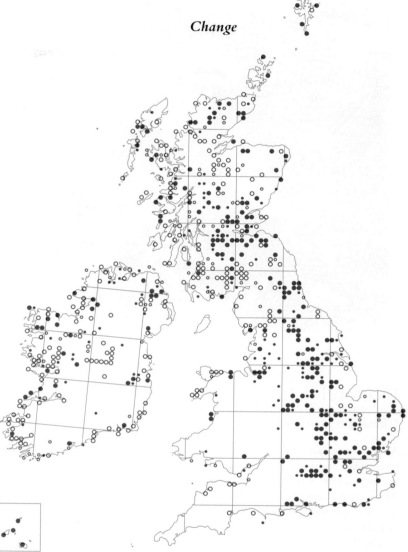

Change

Distribution

given in *Birds in Ireland* for 1979. The British and Irish total is thus probably slightly less than 10,000 pairs. The importance of this total should be recognised for it is almost 80% of the temperate breeding population of the nominate race of Ringed Plovers.

TONY PRATER

68–72 Atlas p 166

Years	Present, no breeding evidence		Breeding evidence		All records		
	Br	Ir	Br	Ir	Br	Ir	Both (+CI)
1968–72	119	42	797	271	916	313	1229
1988–91	213	71	812	173	1025	244	1274
			% change		12.2	−21.5	3.9

Dotterel

Charadrius morinellus

After a long haul to the high ground you reach what seems like an unchanged and eerily silent place; then you glimpse this most beautiful wader running in fits and starts on the skyline. The Dotterel is one of our Arctic-alpine species, virtually confined to the Scottish Highlands, but found much more widely on the dry tundras and fellfields of Asia and Scandinavia.

The Distribution map shows that the stronghold of this species in Britain is the Scottish Highlands, although a few pairs (about 1–4 each year) nest on the montane heaths in the Lake District, N Pennines and S Scotland. The central E Highlands have most of the suitable habitat and about 65% of the British population; 15% is in the SW Highlands and 18% in the N and W Highlands (Galbraith *et al.* 1993). Dotterels may have bred on Islay in 1990 (Petty 1991). Breeding was not recorded in Wales during 1988–91 and there has only ever been a single breeding record in Ireland (1975).

Nesting habitat consists mainly of fairly level summit plateaux of moss and lichen heaths, or open *Juncus tritidus* fellfield, though south of the Highlands these can be quite grassy. Some nest on more sloping, mixed prostrate dwarf-shrub heaths. Dotterels tend to avoid bogs and snowbed hollows. Nesting occurs at above 700m in the central Highlands, descending to 640–700m in Ross-shire and to 460m in Sutherland.

There seem to have been marked changes since the *68–72 Atlas*, when the population was estimated at 100 pairs. This was increased to 150 pairs (Nethersole-Thompson and Nethersole-Thompson 1986, *Birds in Scotland*), and then to at least 600 pairs in the mid 1980s (Watson and Rae 1987). Ratcliffe (1990) and Galbraith *et al.* (1993) suggest that Dotterels are now more extensive and numerous than in the period 1930–70. Breeding densities have at least doubled on some hills (range 0.2–9.3 nests

per km²) and on others breeding has been proved for the first time. Three factors may point to a higher tendency now for Dotterels to settle and breed in Britain: numbers on passage through Britain have increased since the late 1950s; the British climate cooled in the 1960s and 1970s; band there has been an extension of late snowlie in Norway over the past 30 years (P.S. Thompson, D.B.A. Thompson and J.A. Kalas). Watson and Rae (1987), however, argue that now more birdwatchers detect more Dotterels.

There were certainly more breeding Dotterels in N England during the mid 19th century (Ratcliffe in Nethersole-Thompson 1973). Their demise there despite the much more buoyant situation in the Scottish Highlands may point to habitat deterioration inflicted by grazing sheep, more disturbance and more acidic deposition (Ratcliffe 1977, Thompson *et al.* 1987, Thompson and Baddeley 1991, Thompson and Brown 1992). Although nesting Dotterels seem insensitive to current levels of disturbance (e.g. Watson 1988) the combination in N England and N Wales of much disturbance, more sheep (which seem to be on the tops earlier further south) and grassier, less natural heaths may deter many birds from settling when they arrive in May from Morocco en route to Arctic and sub-Arctic regions.

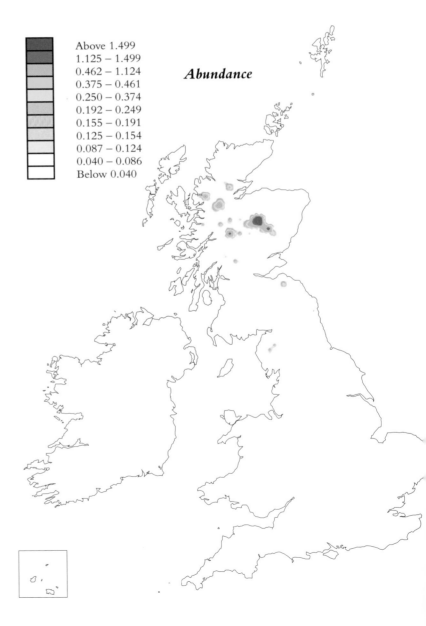

Abundance

Above 1.499
1.125 – 1.499
0.462 – 1.124
0.375 – 0.461
0.250 – 0.374
0.192 – 0.249
0.155 – 0.191
0.125 – 0.154
0.087 – 0.124
0.040 – 0.086
Below 0.040

Since 1987, the NCC (now SNH) has undertaken a major study of Dotterels and montane habitat. This has revealed marked differences in breeding density and productivity not related merely to geology as suggested by Nethersole-Thompson (1973) and Nethersole-Thompson and Watson (1981). The more productive tops export young birds to sustain other populations. Males are more site faithful on productive sites, and their young also tend to be more philopatric. The ceiling on breeding density remains to be discovered; so too does any role of females in limiting numbers and influencing dispersion. On some sites female display groups court males (Nethersole-Thompson 1973), with the brighter coloured females being the more dominant and mating first with the brighter (and in better condition) males (Owens 1992, SNH). Productivity on each site has ranged from zero to 1.19 young per adult. It is affected by severe weather in June and July, by predation pressure and nest losses to trampling by sheep and deer.

The British population was estimated to be 840 pairs (based on 330 pairs found during the 1987–88 NCC survey, but corrected for birds overlooked and suitable habitat covered), although it was probably nearer 950 pairs. Britain has virtually the entire EC population, but only 1–1.5%

Change

Distribution

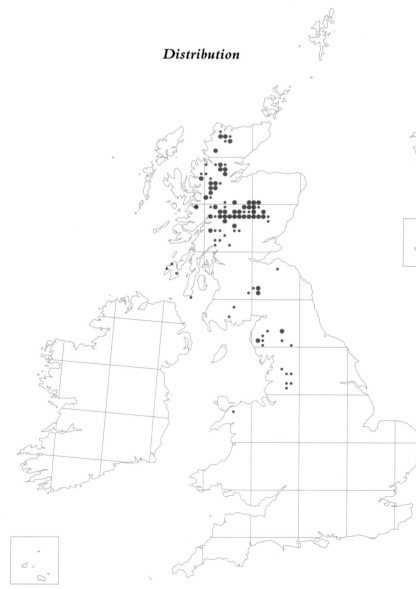

of the European population. Precise European estimates are complicated by some individuals breeding both in Norway and Scotland in the same year. The EC Directive on the Conservation of Wild Birds (79/409/EEC) obliges the British government to protect habitat needed to sustain the EC population. Three montane Special Protection Areas have been proposed and others are under consideration. The Arctic affinities of the British Dotterel, its beauty, its rarity and its likely sensitivity to habitat and climate change secure its place as one of our most fascinating breeding birds.

D.B.A. THOMPSON and D.P. WHITFIELD

68–72 Atlas p 172

Years	Present, no breeding evidence		Breeding evidence		All records		
	Br	Ir	Br	Ir	Br	Ir	Both (+CI)
1968–72	14	0	32	0	46	0	46
1988–91	51	0	48	0	99	0	99
				% change	115.2	0	115.2

Golden Plover

Pluvialis apricaria

This is the bird that so evocatively captures the character of the uplands. In early spring, the male advertises his territory with a complex looping song flight accompanied by plaintive but penetrating songs. Golden Plovers are readily located during establishment and defence of territories from March through to early May and later, during the chick-rearing period, when they utter their monotonous 'too' cautions.

The *68–72 Atlas* indicated a breeding distribution in Britain matching the full extent of heather moor, blanket bog, acidic grasslands and montane tops, but biased towards the western and northern uplands in Ireland. The 1988–91 Distribution map reveals marked changes. The distribution of Devon's small population (9–17 pairs during 1979–86) seems to have changed only slightly. Minor range extensions have occurred in the Western Isles, Islay, Mull, Skye, localised parts of W Scotland and in Co. Donegal. The Pennines have the most southerly major populations in Britain. Despite concerns about increasing public use of the Peak District National Park (Yalden and Yalden 1988) there is no evidence of any long-term declines in the Pennines, though the eastern moors may be more widely used as the extent of those in the northwest and southwest contracts.

Elsewhere in Britain there have been marked range reductions. The greatest losses appear to have occurred in Borders, Dumfries and Galloway, Tayside, Grampian, southern and eastern parts of Highland, Central and Strathclyde. There have also been range reductions in Wales (notably south of the Plynlimon range) and parts of Ireland.

Golden Plovers space their nests with remarkable regularity, the favoured moors having nests 400m apart (5–8 pairs per km²). Territoriality seems to involve a balance between habitat quality, predation risk and breeding numbers, with surplus non-breeders seen in some areas (Parr 1980). The highest recorded density is 16.4 pairs per km² on fertile, hummocky limestone turf in the Pennines (Ratcliffe 1976, 1990). Densities are gen-

erally high (> 10 pairs per km²) on heather-dominated short-rotation burned moors, especially those juxtaposed with pastureland (containing more of the favoured prey, earthworms and leatherjackets). The more barren moors and bogs of the W Highlands and Ireland have lowest densities (< 1 pair per km²). However, many grouse moors have few Golden Plovers, and the high breeding densities in parts of the acidic Pennines is a mystery.

Afforestation is the most obvious cause of the decline, with over a million ha of the uplands planted since the early 1920s, and post-war semi-natural habitat loss from this cause running at 1–1.8% per annum (Thompson *et al.* 1988). Although in NE Norway Golden Plovers nest within forests having a short-sward understorey, in Britain tree planting is inimical to them because they prefer to breed on the less steep and better drained moors (Ratcliffe 1976, Haworth and Thompson 1990) which are also valued for tree planting. Burning ceases on moorland adjoining plantations (which generally harbour more predators, especially Crows and foxes), and it is possible that birds avoid sites where visibility is restricted by ranker vegetation. Data on the amount of land under timber (Forestry Commission 1990a) indicate that Tayside, Borders and Dumfries

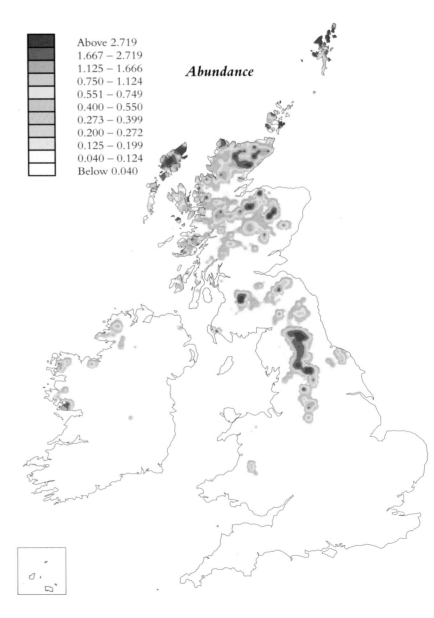

Abundance

Above 2.719
1.667 – 2.719
1.125 – 1.666
0.750 – 1.124
0.551 – 0.749
0.400 – 0.550
0.273 – 0.399
0.200 – 0.272
0.125 – 0.199
0.040 – 0.124
Below 0.040

and Galloway all have extensive plantations, and in these regions there is a sustained contraction of Golden Plover distribution. In Caithness and Sutherland, afforestation in the last ten years may have displaced 19% of the population nesting there, representing a loss of over 2% of the British population (e.g. Thompson *et al.* 1988).

But there are complications. There are fewer keepers now than in the late 1960s. More disturbance by walkers, especially with dogs off the lead (and locally from hang gliders), may deter some birds from settling. Heather is giving way to grassland under intense grazing pressures from sheep (half the remaining heather moorland in England and Wales is likely to deteriorate much further). Hill pastures are not fertilised as much as in the early 1970s. Acidic deposition (notably of nitrates) is increasing so that some western moors may have partly lost their mossy cover, though the Penines population seems unaffected (Ratcliffe 1990, Thompson *et al.* 1993).

One population in NE Scotland declined from over 100 birds in the mid 1970s to zero in 1990 (Parr 1992). This study revealed that whilst breeding success was sufficient to maintain the population early on, losses (emigration and mortality) associated with severe winters matched the decline. Birds were more susceptible to harsh weather because the preferred

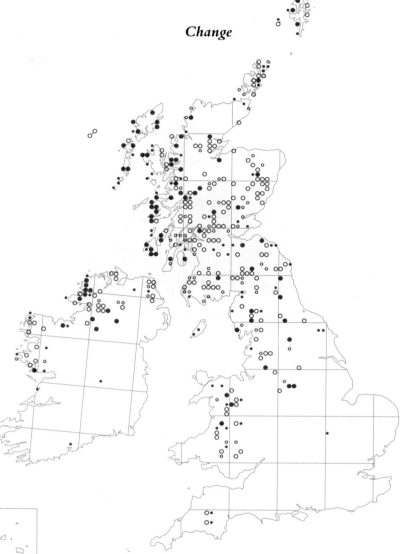

Change

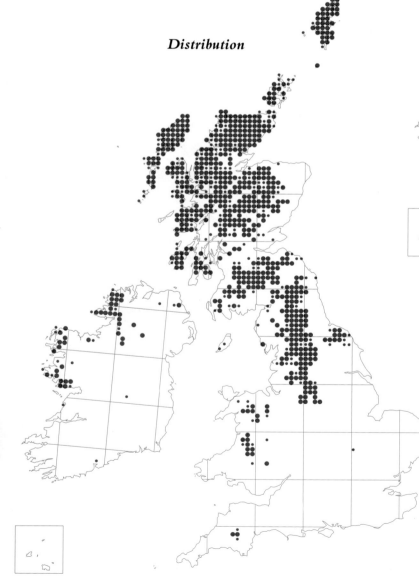

Distribution

permanent pastures had been ploughed up (Barnard and Thompson 1985). Predation of eggs and young, associated with afforestation and lack of keepering, further reduces the population: habitat loss means that the more productive populations have to subsist in smaller areas, and so export fewer birds.

Currently, the British population is estimated to be 22,600 pairs compared with 29,400 at the time of the *68–72 Atlas*. In Ireland a former population of 600 pairs may now be as low as 400. Britain and Ireland hold over 99% of the EC population, about 3% of the European, and 2.5% of the world populations.

D. B. A. THOMPSON and G. BOOBYER

68–72 Atlas p 170

Years	Present, no breeding evidence		Breeding evidence		All records		
	Br	Ir	Br	Ir	Br	Ir	Both (+CI)
1968–72	100	10	749	56	849	66	915
1988–91	154	17	630	40	784	57	814
				% change	−7.7	−13.6	−8.1

Lapwing

Vanellus vanellus

There are few more attractive birds of open countryside than the Lapwing, our most numerous breeding wader, and displaying males are amongst the most evocative sights and sounds of spring.

Despite a substantial decline since 1962 (*Trends Guide*), the species remains widely distributed. The Distribution map shows that 83% of British and 48% of Irish 10-km squares held Lapwings, the comparable figures in the *68–72 Atlas* being 91% and 68%. The Change map shows that the decline has affected distribution most in the Republic of Ireland, W Wales, NW Scotland, Devon and Cornwall. The Abundance map, however, reveals how relatively small are the numbers in S and E England, where a regional CBC index shows a consistent long-term decline (Shrubb and Lack 1991). South of the Mersey–Humber line, densities were low and the population is now concentrated in N England, and S and E Scotland. Lancashire and Yorkshire, the Lothians, Tayside and the Northern Isles seem to be the most important areas.

Over 90% of pairs in England and Wales breed on agricultural land (Shrubb and Lack 1991) and there is no doubt that changing agriculture has had major impacts on Lapwings. Nevertheless overall population changes may not correlate with cropping changes because Lapwings are affected more by the declining availability of true mixed farms rather than any changes in acreage of individual crops. They prefer to nest in fields cultivated in the spring and in particular select those which are close to, preferably adjacent to, grass. Such sites combine bare land, the most productive nest sites, with grass, the best habitat for rearing chicks (Galbraith 1988, Shrubb 1990, Shrubb and Lack 1991). Such mixed-crop combinations were widespread in Britain until the mid 1960s. Today they are scarce in lowland England, where tillage is now dominated by autumn sown cereals and oil seed rape, and the amount of intermixed grassland

has declined. Shrubb (1990) showed that in cereal growing counties Lapwing numbers declined steeply as the amount of autumn cereals grown increased.

The areas with higher numbers of Lapwings shown by the Abundance map are broadly those with spring tillage. This practice remains important in East Anglia, on parts of the Wessex chalk and in Shropshire and Cheshire, all of which have local concentrations of Lapwings. In N England and on lower ground in Scotland nearly half of the tillage is cultivated in spring, mainly for cereals. In these cereal areas a relatively high ratio of grass to tillage, of 60:40, suggests that favourable crop combinations remain common.

The decline of mixed farming has been as characteristic of the far west as the southeast. Grassland now covers over 90% of farmland in Ireland, Wales and W Scotland, and 80% in Devon and Cornwall. In addition, in England and Wales sharp increases in stocking rates have led to declining nest success by grassland-nesting Lapwings because of rising losses from trampling, desertion and increased farm activity; in many upland areas grassland Lapwing populations may now rear too few young to be self supporting (Shrubb 1990). Similar effects have been observed in Holland

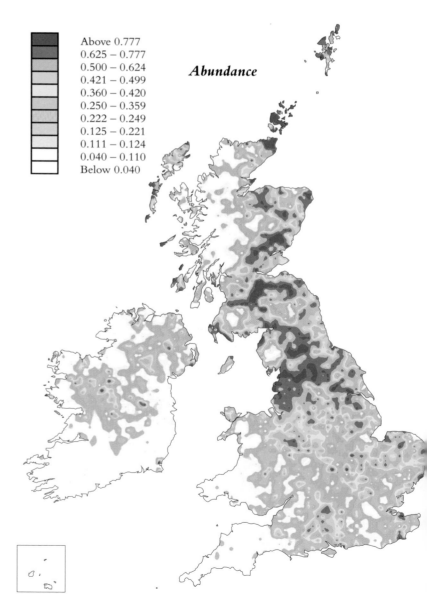

Above 0.777	
0.625 – 0.777	
0.500 – 0.624	*Abundance*
0.421 – 0.499	
0.360 – 0.420	
0.250 – 0.359	
0.222 – 0.249	
0.125 – 0.221	
0.111 – 0.124	
0.040 – 0.110	
Below 0.040	

(Beintema and Muskens 1987) and presumably apply elsewhere in Britain and Ireland. In Ireland such factors have probably been particularly marked since the country joined the EC. The spread of ensilage, now occupying 26% of improved grass in England and Wales and which, unlike hay fields, Lapwings avoid, has increased the dependence on stocked fields, and therefore exacerbated these effects. The most heavily stocked areas are now being deserted entirely (Shrubb 1990, Shrubb and Lack 1991).

Possibly the large Lapwing breeding populations that persist, for example, in N England, are now regularly replenished from successful sources in nearby areas of mixed farmland. Spring tillage in N England held an average density of seven pairs per km² in 1987 (Shrubb and Lack 1991, Table 4), twice that of the highest regional grassland density. Where such replenishment sources are disappearing, as in Ireland, Wales, W Scotland and SE England, Lapwings perhaps cannot maintain their numbers as farming of grassland intensifies.

Shrubb and Lack (1991) estimated 123,100 pairs in England and Wales in 1987 (95% confidence limits about 110,000–138,000), Thom, *Birds in Scotland*, estimated 75,000–100,000 pairs in Scotland, and Partridge, in *Birds in Ireland*, gave 5,250 pairs for Ulster. There is no recent estimate for

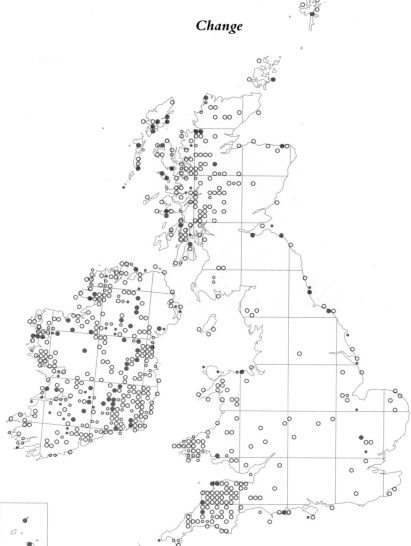

Change

Distribution

the Irish Republic but, as the abundance over the occupied parts of Ireland is very uniform, the average of 44 pairs per occupied 10-km square, calculated from Partridge's figure, is likely to be a good approximation for the whole island. This leads to an estimate of 21,500 pairs for Ireland. Thus the present population of Britain and Ireland is 205,000 to 260,000 pairs.

MICHAEL SHRUBB and PETER LACK

68–72 Atlas p 164

Years	Present, no breeding evidence		Breeding evidence		All records		
	Br	Ir	Br	Ir	Br	Ir	Both (+CI)
1968–72	75	96	2498	595	2573	691	3264
1988–91	249	147	2091	344	2340	491	2833
			% change		−9.0	−28.8	−13.1

Dunlin

Calidris alpina

Whether trilling close by, almost confidingly, or skipping off the nest in escape, every encounter with this dapper little wader emphasises the joy of upland birdwatching. While Dunlins also breed in low-lying saltmarshes and machairs, it is wet moorland and montane habitats which typify the species' range in the breeding season.

The main clusters of occurrence and breeding density are on the flows of Caithness and Sutherland, the high peaty bogs and mosses of the Grampians and Pennines, the peatlands of the Northern and Western Isles and also the machair of the Western Isles. The peatlands of the Flow Country, with their complex mosaic of dubh lochans, hold 39% of the British breeding population, usually at densities of 1–3 pairs per km², but in places up to 11 pairs per km² (Stroud *et al.* 1987). In other parts of the Sutherland peatlands, densities reaching the equivalent of 25 pairs per km² have been recorded, pairs nesting less than 200m apart (Nethersole-Thompson and Nethersole-Thompson 1986). In the Uists the base-rich machair, especially the damper areas, is much preferred to the more acid, less productive peaty moorland. About a third of the British population breeds here (Fuller *et al.* 1986) at densities equivalent to more than 300 pairs per km² in some places, some nests being less than 10m apart (Etheridge 1982).

Small breeding pockets are mainly restricted to the Southern Uplands, the estuarine marshes and bogs of NW Ireland and the hills of central Wales. Dartmoor retains the distinction of having the most southerly breeding Dunlins in the world.

There has been a downturn in the Dunlin's fortunes in the last 20 years in Britain, and in Ireland, too, where 44% fewer 10-km squares held breeding pairs during 1988–91 compared with 1968–72. Perhaps the most notable feature of the Change map is the disappearance of breeding

Dunlins from many parts of SW Scotland. It is likely also that breeding densities in many 10-km squares are smaller now than in the *68–72 Atlas*. However, the losses in SW Scotland are, at least in part, made good by gains in NW Scotland and the Western and Northern Isles. Summering, non-breeding Dunlins occur at several localities on the E and S coasts of Britain.

Blanket afforestation with alien conifers has been the main threat to Dunlins in the uplands. For example, the extent of breeding habitat loss in Caithness and Sutherland has been such that a reduction of 17% in the Dunlin population has been attributed directly to afforestation (Stroud *et al.* 1987). Over roughly the same period of time Avery and Haines-Young (1990), using data on pool-complex habitat loss derived from satellite photographs, found a similar decrease in Flow Country Dunlin numbers. SW Scotland has had by far the greatest area of upland habitat afforested in recent times (Avery and Leslie 1990) and it is likely that this has been responsible for the decline there (Ratcliffe 1990).

Research shows that afforestation does indeed affect the behaviour of Dunlins. In the breeding season most are recorded at distances greater than 800m from newly afforested areas, with fewer birds seen between

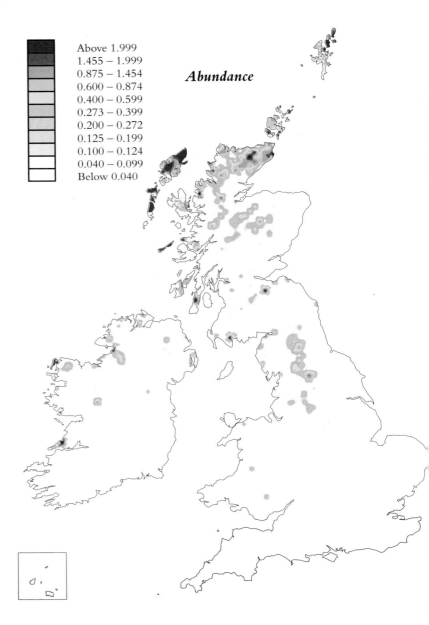

Above 1.999
1.455 – 1.999
0.875 – 1.454
0.600 – 0.874
0.400 – 0.599
0.273 – 0.399
0.200 – 0.272
0.125 – 0.199
0.100 – 0.124
0.040 – 0.099
Below 0.040

Abundance

400m and 800m from the trees. Lowest numbers are recorded close to (i.e. within 400m of) the plantation edges (Stroud *et al.* 1990). The reasons for this apparent avoidance of plantations are poorly understood. It may be due to changes in upland vegetation consequent upon tree planting, or because the forests act as refuges for predators (see Thompson *et al.* 1988, Stroud *et al.* 1990).

The Dunlin normally lays four eggs in a nest often situated in a small tussock. On the central Grampian tops, for example, the site is usually in *Nardus* grassland. There, throughout the breeding season, most birds are seen feeding in bogs although individuals off duty during the incubation period are more likely to be seen on *Rhacomitrium lanuginosum* heaths. There they may feed, but they also keep a watch for predators, often paging a Golden Plover (K. Duncan). The (mainly insectivorous) diet of Dunlins breeding in our uplands has not been studied in detail.

The Dunlins breeding in Britain comprise more than 90% of the EC population (thereby meriting special protection under EC legislation). In a wider European and W Palearctic context the size of the British and Irish population is less noteworthy. Within these islands, however, the numbers breeding in the Flow Country and Western Isles are of crucial

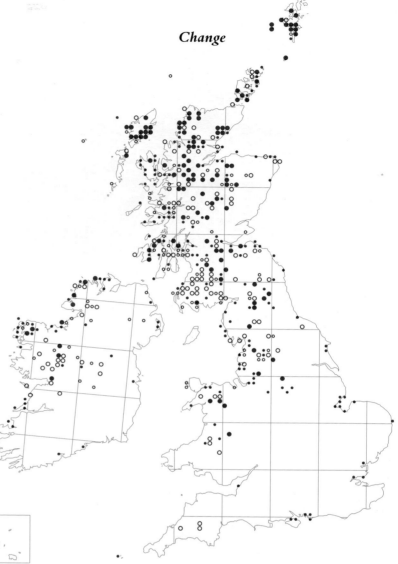

Change

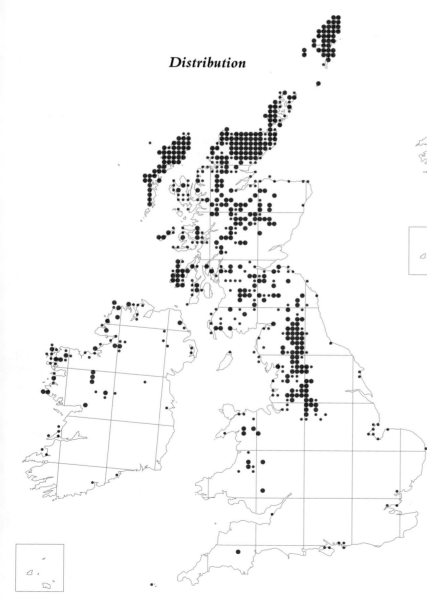

Distribution

significance in indicating the value of our blanket bog and machair as internationally important ecosystems.

Stroud *et al.* (1987) estimated the British breeding population to be 9,900 pairs and *Red Data Birds* estimates 9,150 pairs. The Irish population, surveyed in the mid 1980s, numbers about 175 pairs (*Birds in Ireland*).

J. B. REID

68–72 Atlas p 194

Years	Present, no breeding evidence		Breeding evidence		All records		
	Br	Ir	Br	Ir	Br	Ir	Both (+CI)
1968–72	112	25	360	39	472	64	536
1988–91	216	47	353	22	569	69	638
				% change	20.6	7.8	19.0

Ruff

Philomachus pugnax

A visit during spring to one of the few remaining traditional arenas used by courting Ruffs (males) and Reeves (females) is perhaps one of the most spectacular birdwatching events that Britain has to offer. The appearance of a breeding male, with its huge, variously coloured ruff and long ear-tufts, is satisfying enough, but to see several of these handsome waders, each varying in plumage pattern and displaying energetically to entice and court females, is simply magnificent. Such display, termed lekking, is crucial to the polygamous mating system of Ruffs, and involves much wing fluttering, jumping, hovering and the raising and lowering of ruff and ear tufts. The position of bill and tail are also important, and together these actions are used by the very promiscuous males to attract females into territorial compartments of the arena, known as courts. Much aggression between males also occurs during lekking, involving the use of head jerks, strutting, crouching, bill thrusting, charges and attacks – behaviour contributing to the considerable excitement of the event.

Males occupy leks as soon as they arrive from their winter quarters in Africa, and virtually all are present by mid-April. Reeves arrive mostly from the second week of April. The main breeding habitat is low lying, wet, grassy meadows, which were once fenland and marshes but are now usually grazed in summer and flooded in winter. Males may spend most of the daylight hours on or close to their arenas, which are usually placed on slight elevations, mounds or dykes, but sometimes on flat, muddy or marshy areas, and lekking activity reaches a peak shortly after sunrise. Once fertilized, females lay and incubate in almost complete isolation from the males and are very inconspicuous. They nest solitarily, though several may nest in the neighbourhood of the lek. Once the young have hatched, the Reeve is much more visible and will protest loudly in response to intrusions by predators.

Britain and Ireland are on the western edge of the breeding range of the Ruff, which extends through the Low Countries and N Scandinavia and eastwards across Eurasia (*BWP*). The Distribution map shows no breeding records of Ruff at all in Ireland, where males in breeding plumage are very rare and have never been suspected of breeding (*Birds in Ireland*). There was only one record from Wales. The species displays regularly in Scotland in areas as widely scattered as Dumfriesshire, Argyll, Caithness,

Shetland and the Hebrides, but only one nest has been found to date, in the Inner Hebrides in 1980 (*Birds in Scotland*). The majority of records came from England, with summering birds at a number of localities, mainly in northwestern and eastern areas, and breeders at just ten sites. Of these, the Ouse Washes in Cambridgeshire/Norfolk remain the stronghold, though numbers have declined considerably; 5–21 females during the 1970s but only 1–6 females during the 1980s (C. Carson, D. Revett). Breeding has also recently occurred in Essex, North Yorkshire, Northumberland and Lancashire. The species used to breed in many more areas of England but declined during the 18th and 19th centuries due to land drainage and human predation.

Breeding Ruffs can be classified on the basis of differences in territoriality and behaviour into a number of discrete types (see *BWP*). Certain males, termed 'residents', remain on the same territory throughout the breeding season. Others, named 'marginals', remain on the edge of the arena, and try to obtain residence. Residents and marginals normally have dark coloured plumages. By contrast, other males, known as 'satellites', rarely fight or threaten, are mainly white or light coloured, and are often tolerated on residences, where they behave opportunistically. They may be as successful as other males in mating (Van Rhijn 1983) and the different rôles are probably genetically based. Van Rhijn (1985) speculates that evolutionary disruption of a double clutch system, where the first clutch is tended by the male and the second by the female, is perhaps the reason for the mating system adopted by the ancestral Ruff. This might have given rise to the behavioural dimorphism observed in Ruffs today. Satellite males would have evolved from males selected to care for offspring (and hence the first clutch); the other males from those selected to compete for additional copulations, where attributes not including their caring qualities would be selected.

In any one year, probably fewer than five females currently breed in Britain (RBBP), though females may easily be overlooked. Around 93% of the 250,000 birds breeding in Europe are considered to breed in N Sweden and Finland (Piersma 1986).

JEFF KIRBY

68–72 Atlas p 196

Change

Distribution

Years	Present, no breeding evidence		Breeding evidence		All records		
	Br	Ir	Br	Ir	Br	Ir	Both (+CI)
1968–72	6	0	8	0	14	0	14
1988–91	32	0	10	0	42	0	42
				% change	200.0	0	200.0

upland areas, in Britain and Ireland. Declines in population have also been documented in detail for certain lowland wet grassland areas, which formerly held concentrations of Snipe (Robins 1987), and on lowland farmland in general (*Trends Guide*). In some areas, lowland wet grassland has been converted to arable land with virtually complete loss of breeding Snipe. Even where grassland remains, improved drainage has been installed, which reduces the duration of the season during which soil invertebrates are available to Snipe. Another consequence of drainage and intensification of management of wet grassland is that grazing is begun earlier in the spring and at higher stocking rates. This leads to increased rates of destruction of nests and broods, caused by trampling livestock (Beintema and Muskens 1987, Green 1988b). In the lowlands, Snipe are becoming increasingly restricted to areas that are being managed, at least in part for nature conservation; either as reserves or where statutory bodies make payments available to farmers to maintain high water tables. Similar effects of grassland management are likely to have occurred on the margins of upland areas where grassland has been improved. There is no information on population trends in the uplands.

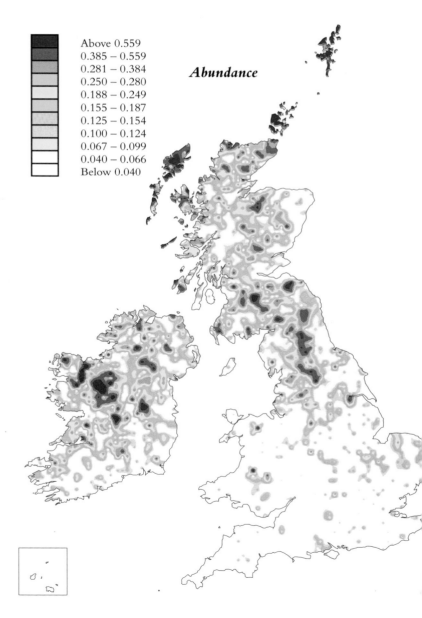

	Abundance
	Above 0.559
	0.385 – 0.559
	0.281 – 0.384
	0.250 – 0.280
	0.188 – 0.249
	0.155 – 0.187
	0.125 – 0.154
	0.100 – 0.124
	0.067 – 0.099
	0.040 – 0.066
	Below 0.040

Snipe

Gallinago gallinago

Snipe are usually secretive birds, spending much of their time in the cover of vegetation, but in the breeding season they become conspicuous as they display in the air over bogs and marshy pastures. They circle their territories, and the drumming sound produced by their tail feathers during power dives is unmistakable.

The Distribution map shows Snipe to be widespread on moorland bogs and marshy rough pastures in N England, Wales, Scotland and Ireland, and patchily distributed in the lowlands where they are found in fens, wet pastures in river valleys and on coastal grazing marshes. The Abundance map suggests that Snipe are more numerous in N Britain and Ireland. However, they tend to be very local in the south, but live at relatively high densities where they do occur, whereas they are thinly spread at low densities in the northern uplands (Smith 1983). A survey of Snipe on lowland wet grassland in England and Wales, in 1982, showed that more than 70% of the birds in this habitat were to be found in just four areas (Smith 1983). The distribution of Snipe is strongly correlated with the presence of marshy ground, because breeding success is influenced by the duration of the period, during spring and summer, when the soil remains moist enough for the birds to probe for earthworms and cranefly (tipulid) larvae (Green 1988b). Bogs and poorly drained pastures subject to winter flooding are favoured by breeding Snipe, because until midsummer they are able to find food close to the nest, without long flights to pools. Where soils dry out in May or early June there is little opportunity for successful breeding, and re-nesting after losing eggs or chicks is scarcely possible.

The Change map shows a contraction in the distribution of Snipe which is particularly apparent in the lowlands, and on the margins of

Estimates of total population are subject to large margins of error, because of difficulty in censusing this species and the lack of comprehensive survey results for upland areas. Smith (1983) estimated that there were about 2,000 pairs on lowland wet grassland in England and Wales. Piersma (1986) estimated a total of 30,000 pairs for Britain and 10,000 pairs for Ireland. However, on the basis of detailed surveys, Partridge and Smith (1992) put the population of Snipe, in 1985–87, in Northern Ireland alone at 5,700 pairs, so Piersma's estimate for the whole of Ireland must be too low.

R. E. GREEN

68–72 Atlas p 174

Change

Distribution

Years	Present, no breeding evidence		Breeding evidence		All records		
	Br	Ir	Br	Ir	Br	Ir	Both (+CI)
1968–72	316	43	1915	853	2231	896	3127
1988–91	495	190	1311	451	1806	641	2447
			% change		−19.1	−28.4	−21.8

Woodcock

Scolopax rusticola

In Britain and Ireland the Woodcock is an unique wader. Its association with woodland and its crepuscular habits give it a very special fascination for birdwatchers, most of whom will be familiar only with its roding flight. Some will have seen it silently weaving through the trees when disturbed, but only a fortunate few will have known it as a breeding bird.

It is primarily a lowland species but it can also be found at higher altitudes if suitable habitat – more or less extensive woodland – exists. The woodland structure is more important than its species composition, and dense cover is avoided. The most favoured woods are moist, and waterlogged and dry woods are less densely populated (Kalchreuter 1982). As one would therefore expect, the Distribution map shows it to be widely spread throughout Britain, with notable gaps only in the high ground and low-lying areas such as the fens. More surprising is its scarcity in the southwestern parts of England and Wales (major wintering areas) and the Cotswolds. The Irish distribution is much patchier.

The Change map indicates a marked reduction in the number of 10-km squares occupied, but no overall change of range. The species is probably under-recorded in general, and this may have been particularly so during 1988–91 when most fieldwork was done during diurnal, timed tetrad visits.

There were few breeding records up to the 18th century, but by the end of the 19th almost every British and Irish county had a breeding population. Climatic changes were probably responsible for this increase, although it has also been attributed to the cessation of shooting in the breeding season (Kalchreuter 1982). Substantial expansion continued into the 20th century, due to the increasing availability of suitable habitat in the form of young conifer plantations (Avery and Leslie 1990). Numbers in Ireland increased (*Birds in Ireland*), and Woodcocks became abundant in some of the extensive conifer plantations of SW Scotland (*Birds in*

Scotland). The *Trends Guide* reports a recent progressive decline, becoming steeper after 1980. CBC monitoring is, however, biased towards S England, taking little account of the extensive afforested areas of the north, and may not accurately reflect the population level.

Young age classes of trees are preferred, birds being generally found only near the edges of older stands until overmaturity renders the plantation once more attractive. Maturing conifer woodlands have also been used to explain population fluctuations by Kalchreuter. The period 1950 to the mid 1970s saw a major expansion of afforestation in Britain and Ireland, and the 1968–72 fieldwork coincided with the period of optimum suitability. During the 1980s many of the forests reached the thicket stage, and the 1988–91 fieldwork would therefore have taken place at a time of falling numbers. The Change map indicates that the greatest reductions have occurred in Ireland and N Britain where the bulk of the afforestation took place. It is not yet established that restocked areas, following felling, are as suitable for Woodcocks as the comparable stages following initial planting.

Radio tracking indicates that successive polygamy is practised, with the pair bond lasting only whilst the male accompanies the female during

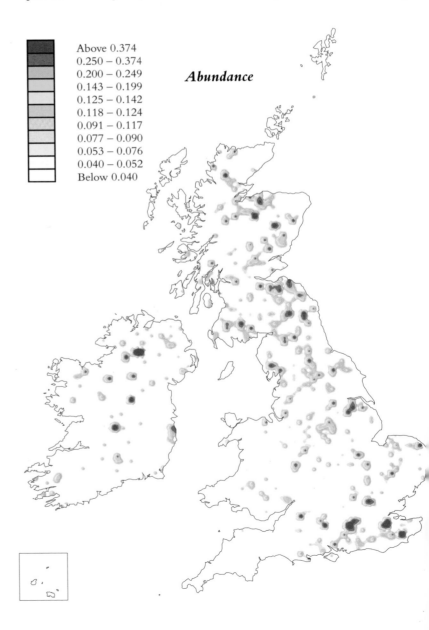

Above 0.374
0.250 – 0.374
0.200 – 0.249
0.143 – 0.199
0.125 – 0.142
0.118 – 0.124
0.091 – 0.117
0.077 – 0.090
0.053 – 0.076
0.040 – 0.052
Below 0.040

Abundance

laying. The general locality from which two birds are flushed is a useful area for the prospective nest finder to search (Campbell and Ferguson-Lees 1972). During incubation the female relies on her cryptic colouration to avoid detection, and as she emits very little scent is not easily found, even with trained dogs. Once hatched, the brood is escorted to favoured feeding areas but does not appear to wander as widely as do many other waders (Nethersole-Thompson and Nethersole-Thompson 1986).

Nesting success is correlated with temperatures at hatching time (Clarke 1974), and in general this species has low levels of infertility, high hatching rates and low fledging mortality (Kalchreuter 1982). The recent series of dry springs is likely to have reduced breeding success as well as reducing the overall numbers breeding.

Woodcocks breed widely across Europe between approximately 50°N and 65°N. This is a particularly difficult species to survey: nests are hard to find and the relationship between the numbers of roding males and breeding females is not clear. Estimates of breeding density are unreliable, but *BWP* gives a European population (excluding Russia) of some 160,000–190,000 'pairs'. The *68–72 Atlas* guessed at 10–25 pairs per 10-km square with probable or confirmed breeding, giving a British and Irish

Change

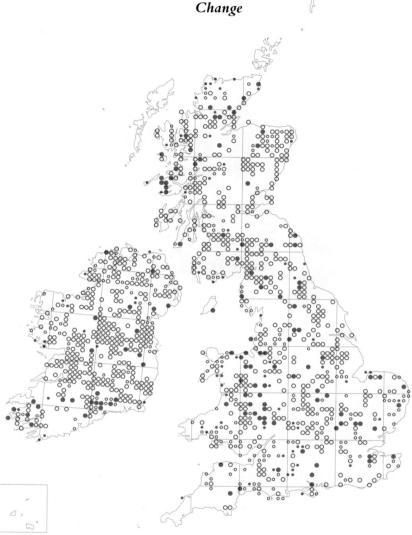

Distribution

population of the order of 19,000–47,000 'pairs'. The Scottish population probably represented half of this estimate and the species was more numerous overall in the N than in the S of Britain. Using the same basis for estimation the present population would be some 8,500–21,500 'pairs' for Britain and 1,750–4,500 'pairs' for Ireland, although it is likely that these are only minimal estimates.

JERRY LEWIS and STEPHEN J. ROBERTS

68–72 Atlas p 176

Years	Present, no breeding evidence		Breeding evidence		All records		
	Br	Ir	Br	Ir	Br	Ir	Both (+CI)
1968–72	254	70	1439	430	1693	500	2193
1988–91	287	67	917	112	1204	179	1383
			% change		−28.9	−64.1	−36.9

JB

Black-tailed Godwit

Limosa limosa

For anyone familiar with the Black-tailed Godwit only as a somewhat drab bird on a winter shore, to see it in its richest russet plumage, and displaying, is a memorable experience.

The species is particularly associated with lowland wet grasslands, and formerly bred widely in Britain. Until the early 19th century, it was especially common across much of East Anglia and parts of Yorkshire. Drainage, egg collecting and shooting were blamed for its extinction as a regular breeder by the late 1820s or 1830s. Recolonisation began with sporadic nesting in the fens of Cambridgeshire and Norfolk in the 1930s, coinciding with a marked increase in numbers nesting in the Netherlands. The species has bred regularly on the principal site, the Ouse Washes, since at least 1952 (Cottier and Lea 1969). Indeed, these washlands in Cambridgeshire and Norfolk have been the main breeding site in Britain, holding over 90% of the population until the mid 1970s, when there were up to 64 pairs. A succession of deep and extensive spring floods in the 1980s seriously disrupted breeding, and few young were reared (Green *et al.* 1987). This resulted in the Ouse Washes population declining to less than 20 pairs (about half the national population), but there has been a slight recovery since 1987.

In the late 1970s, a satellite colony became established at the nearby Nene Washes where breeding had been sporadic in the previous decade. It is probable that some of the birds displaced from the Ouse Washes by flooding settled there. Numbers at the Nene Washes have increased to 13 or 14 pairs, most of which are now concentrated on the RSPB reserve, where management has much improved conditions for Godwits and other breeding waders.

Black-tailed Godwits nested on the Somerset Levels in 1963, with up to six pairs annually since the early 1970s, until pump drainage lowered the water table; since 1985 only a single pair has attempted to breed. In the Lower Derwent Valley in Yorkshire, one or two pairs bred almost annually in the 1970s but only sporadically in the 1980s. Small numbers have bred on coastal grazing marshes, and even salt marshes, in E England from the Humber to Kent. Since 1972, breeding has been almost annual in Suffolk (up to five pairs, at more than one site) and Kent (up to five pairs, but usually two or three).

Godwits breeding in England and S Scotland have been assumed to belong to the nominate race *limosa*, but the one or two pairs that have bred in Caithness (1946), Orkney (irregularly since 1956) and Shetland (most years since 1949) are apparently of the Iceland race *islandica*. These northern birds frequent small marshes at the edge of moorland. Two or three pairs bred in Ireland most years in the period 1975–87, and displaying birds were reported from the Shannon callows during 1988–91 although neither nest nor young were located. These were also considered to be of the *islandica* race (*Birds in Ireland*).

At the Ouse Washes, Godwits begin returning to their breeding grounds in March, and the first nests are usually in the last two weeks in April, but a week or so earlier on the Nene Washes. When breeding, Black-tailed Godwits are usually loosely colonial (4–5 pairs per 100ha) and such groups provide effective communal defence of eggs and young against Crows, Kestrels and other predators (Green 1986, Green *et al.* 1990). Godwits nest in damp, tussocky pastures which have been grazed fairly heavily by cattle the previous summer.

On their breeding grounds, Black-tailed Godwits feed to a considerable extent on earthworms, which they find by probing. For this reason they favour sites with a fairly high water table and a soft, peaty substrate. They also feed on aquatic insects in shallow pools and ditches. Since the chicks feed largely on insects which they pick off vegetation, they are less dependent on moist soil than the adults (Green *et al.* 1987).

In the late 1970s, about 60 pairs bred in Britain, but in the 1980s, largely as a result of the decline at the Ouse Washes, the average was nearer 40 pairs. During 1988, 1989 and 1990, 33–36 pairs were confirmed to have bred in Britain (RBBP), and it is likely that 2–3 pairs breed in Ireland each year (*Birds in Ireland*). These populations are small indeed in comparison with the 10,000–30,000 pairs of *islandica* and 102,000–123,000 pairs of *limosa* nesting elsewhere in Europe (Piersma 1986).

JAMES CADBURY and JEFF KIRBY

68–72 Atlas p 182

Two records have been moved by up to two 10-km squares on the Distribution map. Black-tailed Godwits were recorded in six 10-km squares in Ireland during 1968–72 although the records were not published in the *68–72 Atlas*. Five of these were adjoining 10-km squares and to preserve the confidentiality of the location they have been omitted from the Change map. Godwits were recorded in the sixth 10-km square during both atlas periods, and thus this square does not feature on the Change map.

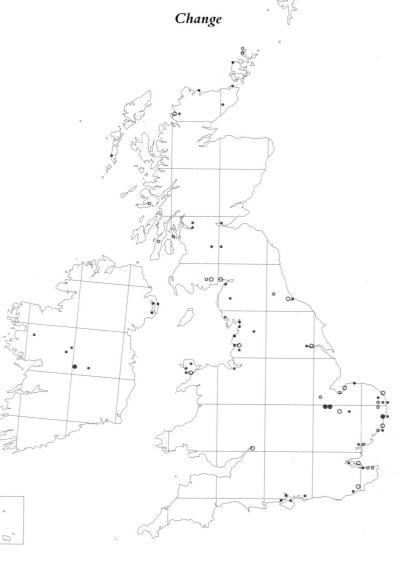

Change

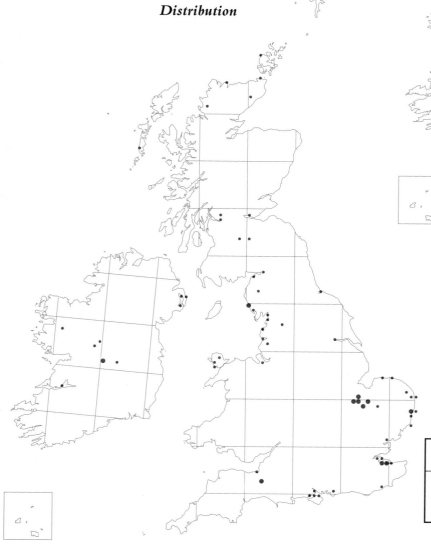

Distribution

Years	Present, no breeding evidence		Breeding evidence		All records		
	Br	Ir	Br	Ir	Br	Ir	Both (+CI)
1968–72	18	3	25	3	43	6	49
1988–91	47	8	12	1	59	9	68
				% change	37.2	50.0	38.8

Whimbrel

Numenius phaeopus

One of the most evocative sounds of the Northern Isles of Scotland is the bubbling bell-like call of the Whimbrel on its breeding grounds.

The general distribution of the Whimbrel is boreal: the nominate sub-species *N. p. phaeopus* reaches its southern limit in Norway and in Britain, where the entire British population is confined to Scotland.

About 95% are found on the exposed heathlands and moorlands of Shetland. Here, latest estimates put the population at 410–470 pairs (Richardson 1990). Elsewhere breeding Whimbrels are extremely rare, with fewer than 25 pairs spread between the Western Isles, the N Scottish mainland, and Orkney. The Whimbrel has been recently confirmed as a breeding species in Sutherland and Ross and Cromarty, although not during 1988–91.

The pattern of breeding distribution remains similar to that reported by the *68–72 Atlas*, though Whimbrels have increased in number within that range.

The *68–72 Atlas* reported the total British population to be around 200 pairs, of which 75% were present in Shetland. It is clear that the numbers in Shetland have increased considerably over the past 20 years, at an average rate of 7.5% per annum. The species has also extended its distribution within the islands. This is a continuation of a trend from at least the mid 1950s. Then, for example, only 10–12 pairs were present on the island of Fetlar, while the only two known locations on Mainland Shetland held seven pairs between them (Venables and Venables 1955).

Now, over 70 pairs nest on Fetlar, at densities up to 21 pairs per km², and 180–220 pairs are present at around 40 locations on Mainland. Breeding was reported from Foula in 1989. These increases may be due, in part, to previous under-recording. Certainly geographical coverage of Shetland increased considerably during the 1980s through large-scale ornithological surveys (NCC and RSPB). Even so, data from areas which have been counted systematically indicate that the increase has indeed been real, with numbers growing by between 9% per annum on Orkney and 25% per annum on Fetlar. This rate of increase has been most marked since the mid 1970s. Continuing, cooler, wetter summers may well be the main factor responsible for the increase (Richardson 1990), and Grant (1991) has shown that, with an adult survival rate of at least 89%, chick production is greater than that needed simply to offset adult mortality, and could account for the population growth. The increase in breeding numbers in Orkney and Shetland has not been mirrored elsewhere in Britain, where the species remains rare.

Whimbrels tend to breed in loose aggregations on higher, drier open hill land. Mostly there is little overlap with Curlews, which select lower-lying, damp, semi-improved grasslands. Whimbrels nest in a range of habitats including deeply hagged wet blanket-bog. Their favoured habitat, however, is the dry short sward *Calluna/Carex* basic heathlands overlying serpentine rock. Re-seeded areas and former improved land are avoided. Preferred ground is slightly sloping, with less than a 5° incline, generally open, to provide visibility against predators, yet with localised shelter in the form of small-scale hummocks and runnels.

The grass-lined nest is usually built on a low *Calluna* hummock. The mean clutch size is four, and egg laying, which begins in May, continues into June. Incubation lasts around 26 days. Hatching begins at the end of the first week in June with the peak occurring between 13–20 June. The chicks fledge 28–30 days after hatching. Adult Whimbrels attend their young constantly up to fledging, and most broods remain within 400m of the nest site. By mid July groups of Whimbrels, adults and fledglings, move off the hill lands to the richer feeding areas of agricultural re-seeded land. From mid July onwards, birds begin their migration southwards.

Above 2.999
2.143 – 2.999
1.000 – 2.142
0.800 – 0.999
0.375 – 0.799
0.300 – 0.374
0.200 – 0.299
0.125 – 0.199
0.083 – 0.124
0.040 – 0.082
Below 0.040

Abundance

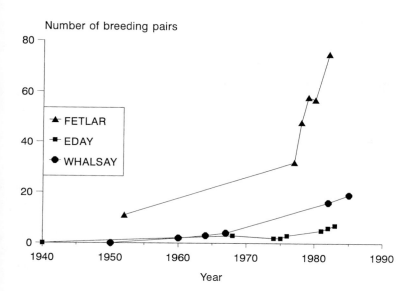

Number of pairs of breeding Whimbrels on the islands of Whalsay and Fetlar (Shetland) and Eday (Orkney).

Change

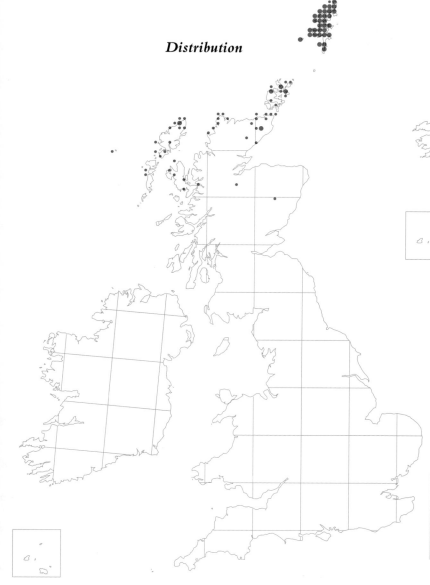

Distribution

Associated breeding species most commonly include Arctic Tern, Common Gull and Arctic Skua. The last mentioned provides a valuable aerial defence against marauding larger gulls and Great Skuas. Grant (1991) found that predation accounted for 45% of known egg losses, and chicks less than ten days old are particularly vulnerable to gulls and skuas.

M. G. RICHARDSON

68–72 Atlas p 180

Years	Present, no breeding evidence		Breeding evidence		All records		
	Br	Ir	Br	Ir	Br	Ir	Both (+CI)
1968–72	28	0	31	0	59	0	59
1988–91	47	0	36	0	83	0	83
				% change	40.7	0	40.7

Curlew

Numenius arquata

With its bubbling song and undulating display flight in spring and early
summer the Curlew is one of the most characteristic birds of our moorlands. Frequently nesting at altitudes of up to 600m (*Birds in Scotland*),
damp upland and northern moorlands are its traditional haunts in the
breeding season. However, as described in the *68–72 Atlas*, during this
century it has colonised many lowland regions and occupied agricultural
habitats such as pastures and cereals, although its rate of expansion into
these habitats was considered to have slowed down by the 1950s.

As a breeding species it is widespread throughout much of Britain but
is absent from most of SE England, and is sporadic in SW England, NW
Scotland and parts of Ireland. The Abundance map shows that it is most
common in the N Pennines, the Southern Uplands of Scotland, parts of
the E Highlands, Caithness, Orkney and Shetland. Smaller areas of high
abundance also occur in northern and central parts of Ireland, N and
central Wales, and on the W coast of Britain, between Anglesey and Islay.

Thus, despite its expansion into lowland agricultural habitats, this species
is still most abundant in upland and northern regions where there are
extensive areas of moorland and rough grazing. Breeding densities show
that unimproved habitats are preferred by nesting Curlews. For example,
on Scottish arable and dry pasture, nesting densities were less than one
pair per km^2 compared with 2.4–2.9 pairs per km^2 on rough grazing
(Galbraith *et al.* 1984). On eight marginal farmland sites in N England
densities declined from 10 pairs per km^2 to one pair per km^2 as the
proportion of improved grassland on sites increased (Baines 1988). Densities of 14–18.5 pairs per km^2 have been recorded on calcareous marshes
and raised bogs in the Pennines (Williamson 1968), but the highest
recorded nesting densities appear to be in Orkney where they average 12

pairs per km^2 on extensive areas of moorland above 100m, and approximately 55 pairs per km^2 on fragments of wetland and moorland below
100m (M.C. Grant). Despite the exceptionally high densities on these
fragments of semi-natural habitat in Orkney, relatively few pairs nest on
the surrounding improved grasslands.

The paucity of Curlews in NW Scotland indicates that suitable breeding
conditions do not necessarily occur in all upland regions, but it is presumably important that suitable nesting habitat should lie within reasonable
proximity of pastures, or other cultivated habitats, where the adults can
feed.

It is apparent from the Change map that there has been no further
expansion of the breeding range in the last 20 years, and over much of the
country the distribution is similar to that presented in the *68–72 Atlas*.
There would, however, appear to have been a considerable decline in S
and E Ireland and also localised declines in W Scotland, the Midlands,
Pembrokeshire, Devon and Dorset. Given the habitat preferences of
nesting Curlews, recent agricultural improvements, such as land drainage
and re-seeding of moorlands, are possible causes of declines. Further losses
of nesting habitat are likely to have resulted from increased afforestation.

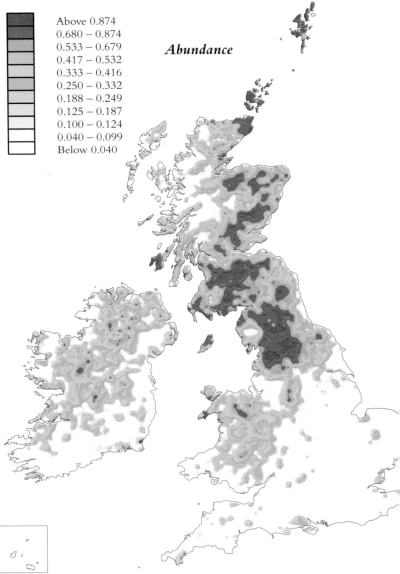

Above 0.874
0.680 – 0.874
0.533 – 0.679
0.417 – 0.532
0.333 – 0.416
0.250 – 0.332
0.188 – 0.249
0.125 – 0.187
0.100 – 0.124
0.040 – 0.099
Below 0.040

Abundance

This is considered to have caused population declines in certain parts of Ireland (*Birds in Ireland*).

Movements on to the breeding ground may begin as early as February but laying usually starts in the second half of April and continues into late May. As with most wader species, the usual clutch size is four and incubation, which usually starts after laying of the third or fourth egg, lasts for a period of 26–30 days (*BWP*, M.C. Grant). Nests appear to be vulnerable to predators, with relatively low rates of nest survival reported from several localities. For example, in N England only 59% of clutches on unimproved grassland, and 23% on improved, hatched chicks (Baines 1989), and in Orkney approximately 50% of the clutches studied were successful (M.C. Grant). Hooded Crows and Ravens were suspected of being the main egg predators on Orkney, apparently being quite capable of locating nests in the deepest heather. The Orkney broods ranged over relatively short distances, most remaining within 300m of their nests until fledging (approximately five weeks after hatching). They made extensive use of improved grasslands but showed strongest selection for semi-natural habitats with tall, rank vegetation. Areas of heather-dominated moorland were avoided by the broods. Preliminary findings in Orkney indicate that

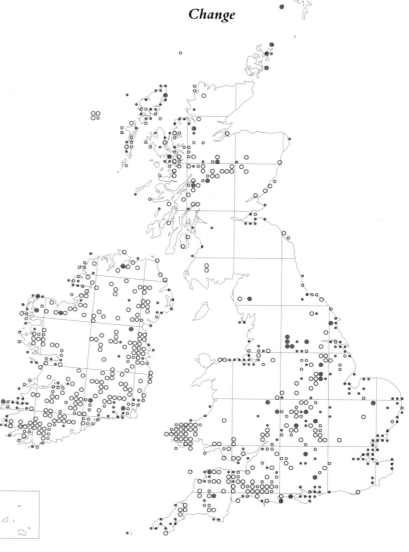

Change

Distribution

breeding adults of both sexes show high levels of site fidelity.

The most recent population estimates are 33,000–38,000 pairs in Britain and 12,000 pairs in Ireland (Reed 1985). Combined, these figures represent at least 35% of the estimated European breeding population (Piersma 1986).

MURRAY GRANT

68–72 Atlas p 178

Years	Present, no breeding evidence		Breeding evidence		All records		
	Br	Ir	Br	Ir	Br	Ir	Both (+CI)
1968–72	233	139	1714	697	1947	836	2783
1988–91	539	406	1354	265	1893	671	2564
				% change	−2.8	−19.6	−7.9

Redshank

Tringa totanus

During the spring and summer the sight and sound of a scolding Redshank, perched on a post or wall, are familiar to most birdwatchers. Its habit of persistently alarm-calling close to an intruder has earned it the titles of 'yelper', 'watchdog' and 'warden of the marsh'.

Breeding Redshanks are found on wet grasslands inland and on coastal saltmarshes, where they occasionally reach densities of over 100 pairs per km^2 (Cadbury *et al.* 1987, P.S. Thompson). In the Western Isles of Scotland, breeding densities on the unique machair are also high, with up to 50 pairs per km^2 (*Birds in Scotland*). Except where breeding occurs at very low density, the noisy Redshank will not go unnoticed.

Redshanks are back on their breeding grounds from mid February onwards. The neat nest often has grass pulled over in a distinctive canopy, and the four eggs are laid from early April. Both sexes share in the incubation and, on hatching after 22 to 25 days, the chicks are led to the nearest wet areas, where they remain with the parents until fledging some 4–5 weeks later.

From the Distribution and Abundance maps, it is clear that the greatest concentrations occur in the Northern and Western Isles, NE and S Scotland and NW England.

Notable concentrations also occur on the Inner Hebrides, particularly Islay, and around the coastal marshes of S and SE England. Redshanks are virtually absent from the extreme SW of England and are sparse in Wales. In Ireland, the distribution is more local, with marked gaps in the south and east counties. In the N of Ireland, breeding Redshanks are mainly to be found in the northeast in Antrim and Co. Down and in the northwest in Co. Fermanagh. In NW Scotland, Ireland and Wales, Redshanks are largely absent from upland areas, which offer little in the way of wet meadows, the favoured inland breeding habitat. In contrast, upland areas in N England (but not the Lake District) with their rich mosaic of

agriculturally improved and unimproved grasslands, support high numbers of breeding Redshanks.

The biggest reductions have occurred in NE and central Scotland, inland S England and in central Ireland. Implicated in this decline is the loss of wet grasslands to drainage and agricultural intensification. A survey in the early 1980s detected this decline and linked it with agricultural improvement resulting in fragmentation of breeding habitat (Smith 1983). Since then, the situation has further deteriorated, with most concentrations in the southern half of England now confined to the Ouse and Nene Washes, Essex, the N Kent Marshes, the Solent and the Somerset levels. Although still present throughout much of the Midlands and inland SE England, the Redshank now occurs only in small numbers (see Abundance map), with pockets of birds presumably hanging on in remnants of suitable habitat.

Despite the recent decline in inland wet grassland sites, Redshanks continue to do well in coastal areas. In 1985, an RSPB/NCC survey of the breeding waders on saltmarsh revealed that the Redshank was the most widespread of waders breeding there. In over 24 sites surveyed, densities exceeded 50 pairs per km^2, with saltmarshes in NW England and East

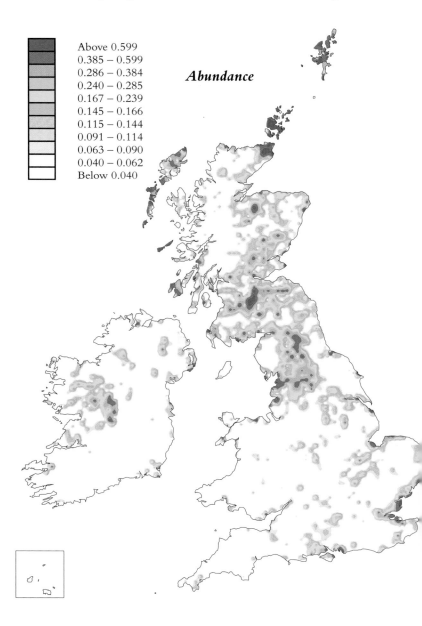

| Above 0.599 |
| 0.385 – 0.599 |
| 0.286 – 0.384 |
| 0.240 – 0.285 |
| 0.167 – 0.239 |
| 0.145 – 0.166 |
| 0.115 – 0.144 |
| 0.091 – 0.114 |
| 0.063 – 0.090 |
| 0.040 – 0.062 |
| Below 0.040 |

Abundance

Anglia holding particularly large numbers. Redshank densities were highest on sites on which grazing pressure was medium, and were lowest on heavily grazed sites (Cadbury *et al.* 1987).

Breeding Redshanks are site faithful, tending to return to the same area each year to breed. In a study carried out on the Ribble saltmarsh, NW England, breeding site fidelity was found to be influenced by sex, years of breeding experience and the previous year's breeding success. Females generally dispersed farther than males; and young, inexperienced birds farther than older birds. Breeding success influenced divorce rates and dispersal patterns, with females dispersing much farther if they had bred unsuccessfully in the previous year (Thompson and Hale 1989).

Like most monogamous waders, young Redshanks are philopatric, and tend to return to their natal area to breed (Thompson and Hale 1989). More experienced Redshanks (i.e. previous breeders) breed earlier and lay larger eggs (which are more likely to hatch) compared with birds breeding for the first time (Thompson and Hale 1991).

An extremely useful monograph on the Redshank incorporating many references is available in the German monograph series (Stiefel and Scheufler 1984).

Change

Distribution

The British Redshank population was estimated at 30,600–33,600 pairs in the early-mid 1980s (Reed 1985). Extrapolating from this figure, the Irish population is estimated to be 4,400–5,000 pairs.

PATRICK S. THOMPSON

68–72 Atlas p 188

Years	Present, no breeding evidence		Breeding evidence		All records		
	Br	Ir	Br	Ir	Br	Ir	Both (+CI)
1968–72	246	79	1426	172	1672	251	1923
1988–91	427	117	1046	96	1473	213	1686
			% change		−11.8	−15.1	−12.2

Greenshank

Tringa nebularia

The Greenshank, one of Britain's rarest and most elegant birds, inhabits some of the remotest parts of Britain. In the early spring, birdwatching visits to the breeding areas may be hampered by persistent rain, wintry showers and gusting winds. Now and again, the skies clear to reveal an awesome landscape, with breathtaking views of rugged mountains and sparkling rivers and lochs. Sightings of Greenshanks may be brief.

Greenshanks are most common in Sutherland, W Caithness, Wester Ross, W Inverness and the Western Isles, the distribution being closely associated with areas of high rainfall and poorly drained acidic peat soils (Nethersole-Thompson and Nethersole-Thompson 1979).

During the breeding season (April to July), Greenshanks mainly frequent the pool-dominated and boulder-strewn blanket bogs of the NW and central Highlands of Scotland ('deer forest'), where they breed at up to 600m above sea level. In Britain, a few still breed in Scots pinewood, a habitat more familiar to Greenshank on the Continent. A number of birds breed in recently afforested moorland, such areas typically being deserted within a few years of planting.

Greenshanks are back on territory from late March, the male announcing his return with a series of glorious switchback song flights. Periods of song continue until a territory has been staked out and a mate won. Generally, clutches are completed in the first week of May, but may be later in cold, wet springs (Thompson *et al.* 1986). Throughout most of the Greenshank's world range, the nest is usually sited by a piece of dead timber or tree stump. In Scotland, however, most nests are beside a rock or slab. Only in a few areas in Strathspey and Deeside are nests still to be found in forest clearings. Greenshanks are highly site faithful, returning to the same territory each year. Both birds incubate, in a nest considered one of the most difficult to find. Once incubation has begun, the birds are

secretive and easily overlooked. Despite the difficulties in nest finding, successful breeding is readily apparent following hatching, when the pair become noisy and demonstrative. Females typically desert the brood, which may move considerable distances, within two weeks of the hatch, leaving the male to fend for the young which fledge in 4–5 weeks.

Since 1968–72 there has been a slight change in distribution, with fewer Greenshanks recorded, particularly in the central Highlands. In the 1950s many of the Greenshank's traditional forest marsh haunts in Strathspey and Deeside became overgrown and dried out, rendering them unsuitable (Nethersole-Thompson and Nethersole-Thompson 1986). More recently, the much publicised afforestation of the vast blanket peatlands (flows) of Caithness and E and central Sutherland has resulted in the loss of much breeding habitat, with an estimated 130 breeding pairs considered lost (Stroud *et al.* 1987). Interestingly, no such change in distribution is apparent from the 1988–91 data. To the north, breeding is sporadic in Shetland (*Birds in Scotland*) while the southern limit for regular breeding is Rannoch Moor. One Irish site was occupied in the early 1970s.

Breeding densities seem to be highest where feeding habitats (pool complexes, rivers) are closely packed. On grouse moors and drier bogs,

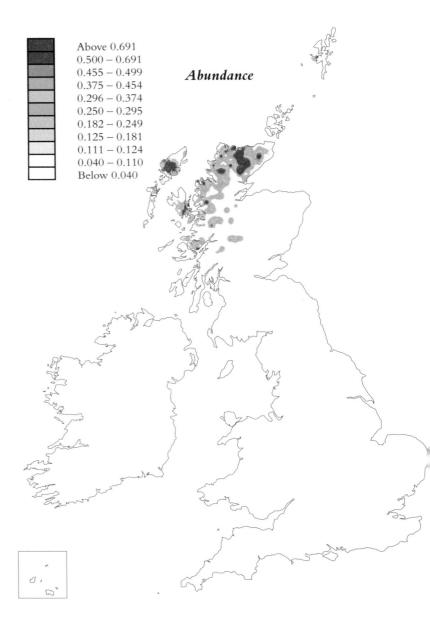

Above 0.691
0.500 – 0.691
0.455 – 0.499
0.375 – 0.454
0.296 – 0.374
0.250 – 0.295
0.182 – 0.249
0.125 – 0.181
0.111 – 0.124
0.040 – 0.110
Below 0.040

Abundance

densities are lower. In one study area in NW Sutherland breeding densities varied from 0.2 breeding pairs per km^2 in 1966 to 0.7 breeding pairs per km^2 in 1976 (Nethersole-Thompson and Nethersole-Thompson 1986).

In NW Sutherland, numbers were highest following years when June was relatively dry and when estimates of breeding success were high (Thompson *et al.* 1986). More recently, there has been a decline in the Sutherland study population, despite favourable conditions during the chick rearing period in June. This decline is considered to be associated with the increased use of all-terrain vehicles through part of the study area, with some important chick-rearing pool complexes drained and denuded of their rich vegetation cover. The population decline was particularly marked because river-based territories which had been regularly occupied since 1964, were most damaged. Territories experiencing the greatest disturbance are now vacant; others least damaged ten years ago and now recovering are being reoccupied (Thompson and Thompson 1991).

A monograph on the Greenshank based on over 50 years of study gives details of its breeding biology (Nethersole-Thompson and Nethersole-

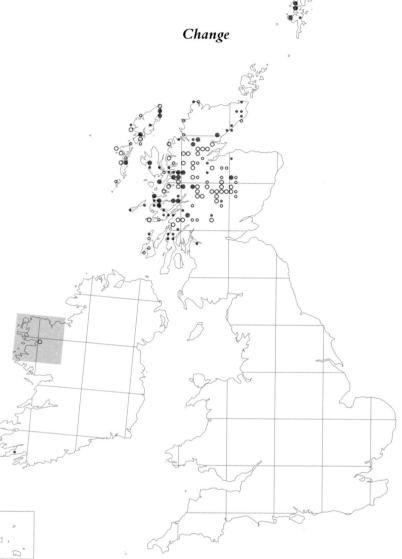

Change

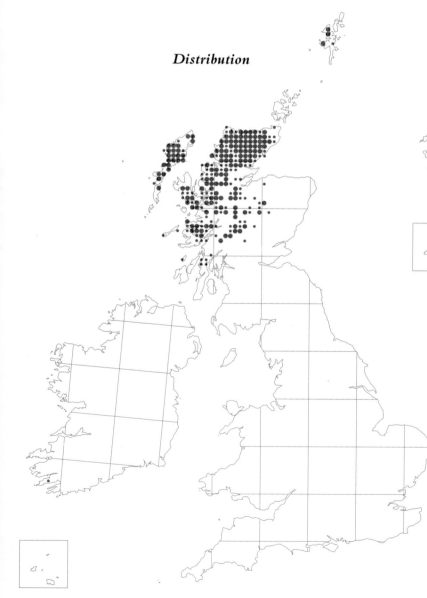

Distribution

Thompson 1979). An estimated 1,100 to 1,600 pairs of Greenshanks breed in Britain, less than 1% of the world population but the only major concentration of Greenshanks in the EC.

PATRICK S. THOMPSON

Only the 100-km (and not the 10-km) grid reference of the 1968–72 breeding record of Greenshank in Ireland was supplied to the BTO.

68–72 Atlas p 190

Years	Present, no breeding evidence		Breeding evidence		All records		
	Br	Ir	Br	Ir	Br	Ir	Both (+CI)
1968–72	66	0	187	1	253	1	254
1988–91	94	1	149	0	243	1	244
			% change		−4.0	0	−3.9

CR.

Wood Sandpiper

Tringa glareola

In the nesting season, this delightful wader is to be found in the tree-fringed mires of the boreal and tundra lands of Scandinavia and Russia. In Scotland, very small numbers breed in similar boggy places, and add a special thrill to summer evenings on the peat mosses.

The Distribution map shows that this species still has a very tenuous foothold as a breeding bird in Scotland. Twenty years ago, Wood Sandpipers bred (probable and confirmed) in 14 10-km squares and were present in another four squares. In 1988–91 breeding occurred in four squares and the species was present in four others.

This is a difficult species to locate unless one is familiar with its display and song flight, or passes near adults with young, when they can be very noisy and demonstrative. Fortunately, all the main sites are checked annually, and the results of these observations have been included on the maps.

In Scotland, nests are often up to 1km from the area where the young are reared, and may be near flushes in lightly wooded country. Some pairs have nested in more open bog, with few or no trees. Sometimes three birds rather than a pair may be present, but no detailed study has been made of the breeding behaviour of this species in Scotland.

The RSPB reserve at Insh Marshes, in Badenoch and Strathspey, has been the most successful site for the species, with breeding recorded in 13 of the 20 years when birds have been present: the best years were 1968 (4–6 pairs) and 1980 (5–7 pairs). In the last few years, no successful breeding has occurred on the reserve and only sporadic sightings have been made. Forestry in the nesting areas above the marshes has probably affected the birds, and this may now have been exacerbated by house building. At another site, in W Inverness-shire, where breeding was proved, the glen has been ploughed and planted with conifers. A third regular site has also been subject to afforestation, and yet another narrowly escaped being damaged by a mining development. On the credit side, at one nature reserve an area of marsh has been re-flooded and the species has nested regularly since 1983. Other sites, which have been used for breeding on single occasions, have been deserted or not used again, without any obvious changes of habitat taking place.

This is a common breeding species in N Europe, especially Sweden,

Finland and Norway, where there is an estimated population of nearly half a million pairs. About 100 pairs nest in Denmark, and a further 50 nest in W Germany, although both countries report a decline in numbers (*BWP*). Clearly, the Scottish population is insignificant in European terms, but it is another example of the Scandinavian influence on the avifauna of N Scotland.

The first singing bird was located in the Highlands in 1947, but breeding was not proved until 1959 when a pair nested in N Sutherland. This was not the first breeding record for Britain and Ireland, as a pair nested in Northumberland in 1853 and, possibly, in 1857. The species remained extremely scarce in the 1960s but by the end of the decade a small population had built up. About 20 different locations, from Perthshire north to Caithness, have been involved, and in the best years, 1968 and 1980, numbers just reached double figures. Since then they have declined and it is likely that the breeding population was about 5–6 pairs between 1988 and 1991, rarely as high as ten pairs. A decade earlier the actual population was probably 10–20 pairs.

The following data compiled from RBBP reports shows the fortunes in recent years.

	1985	1986	1987	1988	1989	1990
No. of localities	3	2	3	4	6	2
Confirmed (pairs)	2	2	3	3	2	1
Possible (pairs)	1	1	1	1	4	1
Maximum (pairs)	3	3	4	4	6	2

ROY DENNIS

68–72 Atlas p 184

Four records have been moved by up to two 10-km squares on the
Distribution map. Wood Sandpipers were recorded in three of these
squares during both Atlas periods, and thus only a single record has been
moved on the Change map, the others not featuring on the map at all.

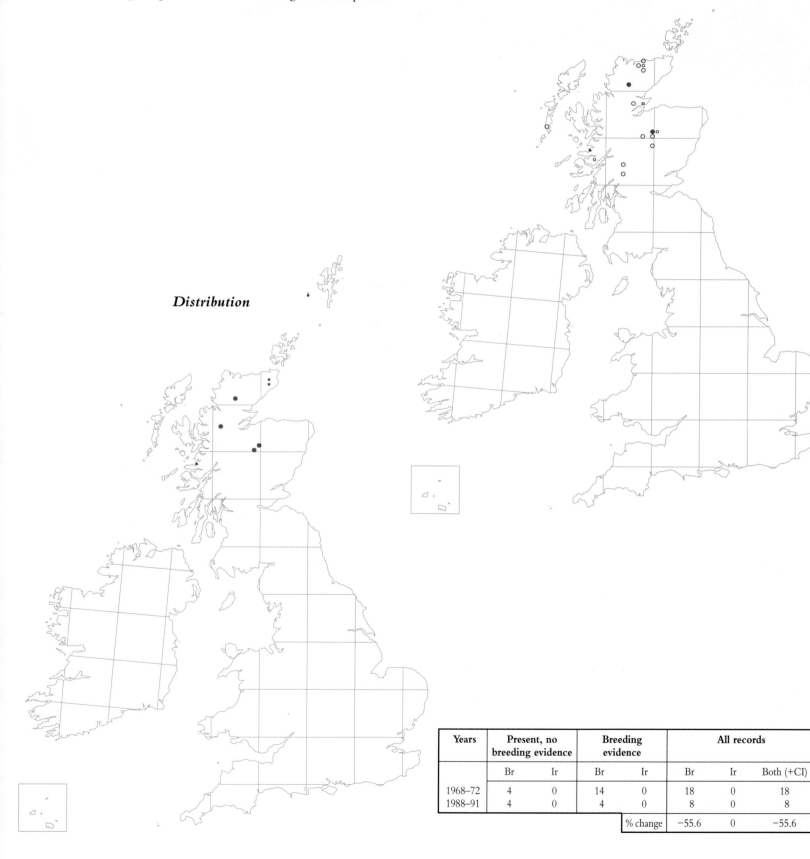

Change

Distribution

Years	Present, no breeding evidence		Breeding evidence		All records		
	Br	Ir	Br	Ir	Br	Ir	Both (+CI)
1968–72	4	0	14	0	18	0	18
1988–91	4	0	4	0	8	0	8
				% change	−55.6	0	−55.6

Common Sandpiper

Actitis hypoleucos

Skimming low over the water on stiff wings, the Common Sandpiper is a characteristic sight on hill streams, reservoirs and lochs, but only for its short, three-month breeding period; Victorian texts call it the Summer Snipe. It seems surprising that its breeding habitat is so specific, and its occupation of that habitat so brief, for on migration it appears at a much wider range of sites, muddy as well as stony, including sewage farms, lowland ponds, reservoirs, and estuaries.

The ecological requirements of the chicks seem to be more exact than those of the adults, and stony shorelines provide them with both food and cover. Small flies and beetles are abundant in the damp riparian zone, where chicks spend nearly all their feeding time (Yalden 1986). When threatened, they hide under stones and undercut banks. Upland lochs and reservoirs provide appropriate habitat because of falling water levels during the drier summer months, while upland streams in their middle reaches benefit from occasional flash floods which scour out steep banks and deposit bare shingle banks. Thus the species is widely distributed in the upland N and W of Britain and Ireland, but breeds only very sporadically in lowland sites. Scattered summering individuals recorded in England were well outside the usual breeding range although breeding has occurred in a number of SE English counties. The WBS suggested that peak densities (0.6 pairs per km) occurred at the relatively shallow gradient of 5m per km (Marchant and Hyde 1980), but detailed studies in the Peak District found the highest density (4.7 pairs per km) at a gradient of 13m per km (Holland *et al.* 1982); WBS plots seem to be relatively poor Common Sandpiper habitats.

The WBS, *Trends Guide* and detailed studies all suggest that the population is fairly stable from year to year. This is in accord with the high (79%) adult annual survival in most years, although severe late April weather may cause exceptional mortality (Holland and Yalden 1991). Many local studies have, however, drawn attention to contractions on the fringes of the species' range, a retreat which is highlighted by the Change map; note the absences along the Welsh borders, the North Yorkshire Moors, central and NE Scotland. Most marked is the trend in Ireland, first detected by the *68–72 Atlas*. Now, the species apparently breeds in less than half its former range, although poorer coverage in Ireland during 1988–91 may account for the lower figures. The reduction is most marked in the southwestern counties of Clare and Kerry, and in the midland counties of Roscommon, Longford and Leitrim.

As a species favouring linear habitats which may also be popular with anglers, hikers and others, it is certainly vulnerable to recreational disturbance and this is mentioned as the major cause of decline in virtually every local study (Yalden 1992). It can, however, breed in territories only 100–150m long, and quite small sanctuaries will serve to protect it. Small

exclosures, artificial gravel islands and zoning, as well as close seasons for angling, could all help this species.

At world, European and British levels, the species is abundant enough to cause no great conservation concern. Densities are difficult to judge for a species of such linear habitat, and those over most favoured areas (up to 4.7 pairs per km) cannot be simply extrapolated to lengths of rivers and shorelines which include unsuitable stretches. In the Peak District, densities on longer stretches of river and reservoir shore ranged from 0.7 to 2.4 pairs per km, and there were, overall, 11 pairs per 10-km square (in 18 squares). The Abundance map suggests that the species is most abundant in Scotland, and an overall figure of 15 pairs per 10-km square with breeding evidence seems reasonable. On this basis, the population may now be 15,800 pairs in Britain and only 2,500 pairs in Ireland. These values are a tiny proportion (1.8% and 0.3% respectively) of the European population which is estimated at 882,000 pairs, the vast majority of which (96%) is in Fennoscandia (Piersma 1986).

D. W. YALDEN

68–72 Atlas p 186

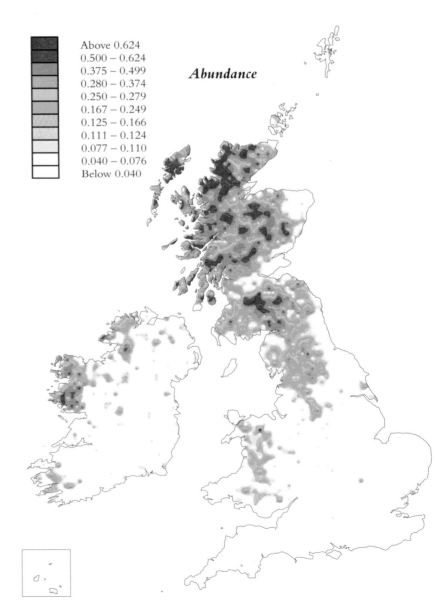

	Above 0.624
	0.500 – 0.624
	0.375 – 0.499
	0.280 – 0.374
	0.250 – 0.279
	0.167 – 0.249
	0.125 – 0.166
	0.111 – 0.124
	0.077 – 0.110
	0.040 – 0.076
	Below 0.040

Abundance

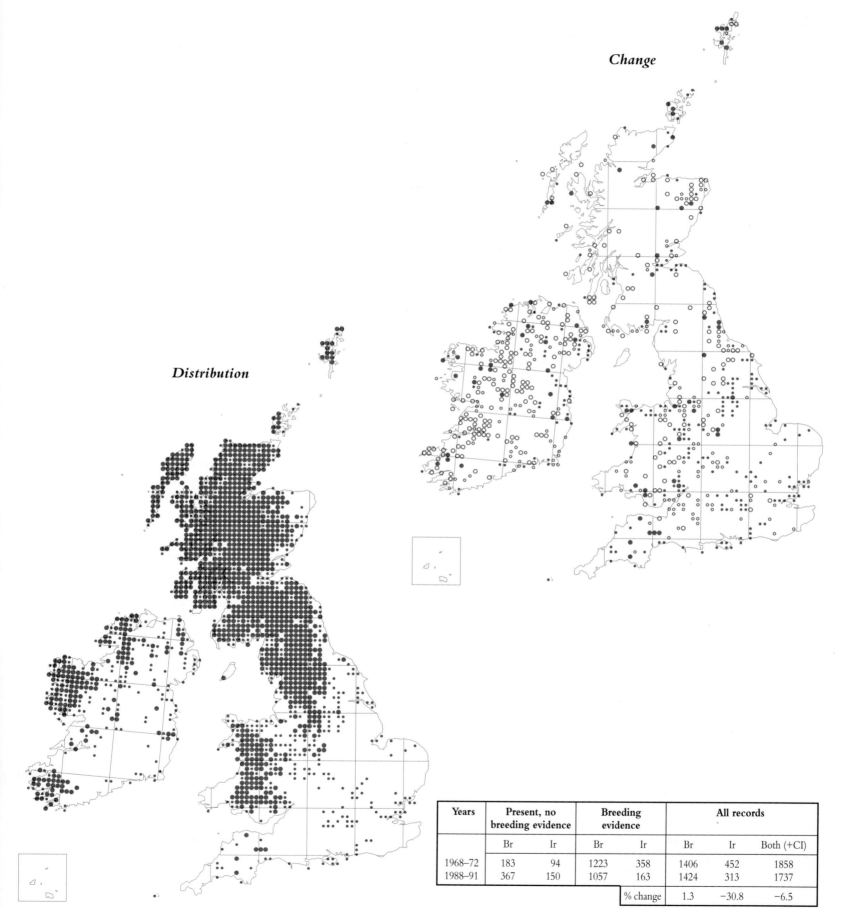

Change

Distribution

Years	Present, no breeding evidence		Breeding evidence		All records		
	Br	Ir	Br	Ir	Br	Ir	Both (+CI)
1968–72	183	94	1223	358	1406	452	1858
1988–91	367	150	1057	163	1424	313	1737
	% change				1.3	−30.8	−6.5

Red-necked Phalarope

Phalaropus lobatus

This delightful wader is one of the latest migrants to reach Britain and Ireland in the spring, rarely doing so before late May – much as it does in its Arctic breeding areas.

The Red-necked Phalarope, like the Dotterel, shows a reversal of the normal sex roles at breeding time. The brighter coloured females, who precede the males by a few days on the nesting grounds, are usually monogamous but sometimes serially polyandrous, each pair bond lasting only a few days (Hilden and Vuolanto 1972).

The present and historical distributions of the species are similar, although numbers have varied. The main populations have always been in the far north and west, with the island groups of Shetland, Orkney, and the Hebrides, and Co. Mayo, being the most important. Some sites as far south as counties Kerry and Wexford in Ireland have been occupied for short periods, as during the *68–72 Atlas* period. Most birds, however, are faithful to traditional sites, though there is occasional breeding elsewhere. Some sites, too, have lost their Red-necked Phalaropes for several seasons, as happened in Co. Mayo, where no birds were seen for several years in the 1970s.

The *88–91 Atlas* shows a population divided between W Ireland, the Western Isles, the N Scottish mainland and Shetland much as in the *68–72 Atlas*. One pair bred in Argyll in 1989 although the record was not formally submitted to the Atlas. The most recent estimates, of up to 20 pairs in Britain (*Red Data Birds*, RBBP), and 0–1 pairs in Ireland, during 1988–91, may be compared with 'about 45 pairs' in 1970 (Everett 1971). The long-term trend has been of decline, probably due to habitat loss combined with climate change.

Attention from egg collectors over the years and the associated disturbance may have contributed to the decline in Scotland during the 19th century, when they were regarded as fairly numerous on Shetland, Orkney and the Outer Hebrides (Everett 1971). Numbers increased to at least 100

pairs in the 1920s, including about 40 pairs at the Co. Mayo site. More recently some loss of habitat has occurred due to agricultural improvements (Everett 1971), and disturbance by birdwatchers and photographers is now a problem at some sites.

The stronghold of the species today is in Shetland, where nesting is usually amongst flooded peat cuttings or other shallow pools. In the Hebrides and Co. Mayo the sites are more often in hay fields or rough pasture near machair pools. Research by the RSPB in Shetland and the Hebrides (Yates *et al.* 1983) established that breeding sites had a number of common features: all included a mix of open water, emergent vegetation and marsh, and were usually neutral to alkaline (pH 7–9). Open water need not be extensive but pools most used for courtship and mating were usually at least 30m² in area. The most important areas were emergent vegetation and marsh which were used for feeding.

In common with many wetland species, the Red-necked Phalarope has lost suitable habitat in recent years, but with several of the main breeding sites now being managed as reserves. This has been done successfully at sites in Shetland and in Ireland, with areas that had become unsuitable now being used again.

The Red-necked Phalarope is a species particularly suited to conservation by reserve acquisition and management, and it is perhaps encouraging that against a general climate of decline, populations using managed sites are holding their own (M. Ausden and C. Evans).

The species is a common breeding bird between 60° and 70° north, with a Holarctic distribution. There is a large population of 50,000–100,000 pairs in Iceland, 50,000 pairs in Sweden and populations also in Finland and Norway (Piersma 1986). The numbers in Britain and Ireland are thus not significant.

ROBERT A. CHAPMAN

68–72 Atlas p 200

One breeding record, from the Western Isles, has been omitted from the Distribution and Change maps, and three further records have been moved by up to two 10-km squares. Phalaropes occurred in only two of these three 10-km squares during 1988–91, and these records have been moved on the Change map in the same manner as on the Distribution map. In the remaining 10-km square, however, breeding occurred during both Atlas periods. To maintain the confidentiality of this location, which would otherwise be apparent by comparison of the Change and 1968–72 maps, two false records (one filled and one open dot) have been placed on the Change map.

Change

Distribution

Years	Present, no breeding evidence		Breeding evidence		All records		
	Br	Ir	Br	Ir	Br	Ir	Both (+CI)
1968–72	2	0	16	4	18	4	22
1988–91	4	0	5	1	9	1	10
% change					−50.0	−75.0	−54.5

in Britain has hardly changed over the last 150 years. Most of the small colonies in the W of Scotland have long histories. There were colonies on Jura in 1772, on Coll in 1898, in the Western Isles in 1888, and on the coast of Sutherland in 1887. Colonies in Caithness have been documented since 1868 (Furness 1987). All of these colonies have held from only a few to a couple of dozen pairs. Between 1969 and 1987 several of them have grown in numbers and a few new ones have arisen. The most pronounced increase has been on Handa, where one pair nested in 1970 and 30 territories were occupied by 1986.

The main concentrations of breeding Arctic Skuas occur in Orkney and Shetland, and these populations also increased between 1969 and 1985. The 'Operation Seafarer' census in 1969 undoubtedly underestimated skua numbers because it excluded inland nesting seabirds. A careful analysis of numbers in Shetland from 1969 to 1986 suggested that Arctic Skuas had increased by less than 1% a year overall (Ewins *et al.* 1988). Numbers in Orkney seem to have increased faster, from perhaps 360 pairs in 1969 to 716 pairs in 1974–75, 1,034 pairs in 1982 and about 1,190 pairs in 1985–87 (*Status of Seabirds*). Increases in Arctic Skua numbers in the 1970s and early 1980s may simply reflect the increases in populations of terns, Kittiwakes and Puffins, from which the Arctic Skua obtains most of its food while breeding.

Arctic Skua

Stercorarius parasiticus

The mewing cries and aerial displays of this falcon-like seabird are an evocative feature of coastal moorlands of the extreme N and W of Scotland. Although one of the scarcest seabirds breeding in Britain, the Arctic Skua is well known for its dark and light colour phases, the functions of which are still not fully understood (Caldow and Furness 1991), and its habit of kleptoparasitism. In Scotland, Arctic Skuas gain most of their food during the breeding season by chasing and robbing terns, auks and Kittiwakes of the fish they are carrying. Although the species they rob breed in large numbers around much of the coastline of Scotland, Ireland and Wales, the Arctic Skua is confined as a breeding species to the Northern Isles, Caithness and a few sites on the islands west of Scotland (Handa, Western Isles, Coll, Jura and Colonsay have held breeding colonies in recent years). It is difficult to see why there are no Arctic Skuas breeding in association with seabird colonies further south, but Scotland is at their southern breeding limit and the species may be unable to colonise warmer climates, regardless of food availability. Arctic Skuas can breed in areas of the Arctic where the temperature rarely exceeds freezing. They have thick insulative plumage and a high metabolism and may simply find summer conditions too warm in England, Wales or Ireland. The southern limit of their breeding range appears to coincide with the 14°C mean July isotherm in Britain and elsewhere (Furness 1990).

Arctic Skuas are occasionally seen from the coast during the breeding season but outside the established breeding range. Such records are likely to be of immature birds, not necessarily from the Scottish population, since Arctic Skuas do not start breeding until two to seven years old (O'Donald 1983, Furness 1987). The breeding distribution of Arctic Skuas

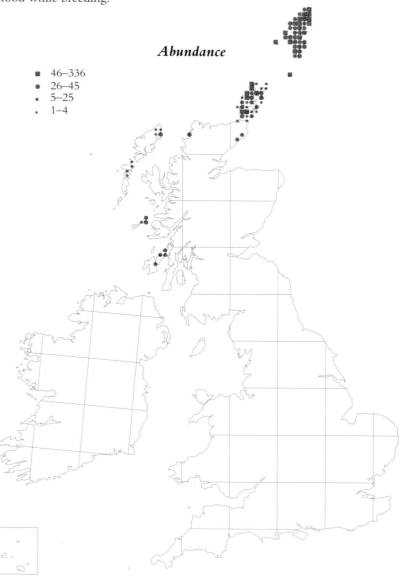

Abundance

■   46–336
●   26–45
•   5–25
·   1–4

In 1985–87, when there were estimated to be about 3,350 pairs of Arctic Skuas breeding in Scotland, about 57% nested in Shetland, 36% in Orkney, 2% in the Western Isles, 2% in Caithness and 3% in Sutherland, Argyll and Bute (*Status of Seabirds*). Since that survey conditions for skuas have altered dramatically in Shetland. Arctic Skuas have been severely affected by a shortage of sandeels, the staple prey of Arctic Terns, Kittiwakes and Puffins. As a result of the food shortage, few of these host seabirds have reared young and so few carry fish to the colony. Breeding success of Arctic Skuas was close to zero from 1986 to 1990 and numbers in the main Shetland colonies have fallen. On Foula, for example, there were 185 pairs in 1986 but only 130 in 1988 and 98 in 1990. Thus the 1985–87 census data may already be out of date and, in the Arctic Skua's Scottish stronghold in Shetland, breeding failures in recent years might lead to further declines in numbers. Although many other possible threats to Arctic Skua populations in Britain have been considered (Furness 1987, *Status of Seabirds*), the current scarcity of sandeels for seabirds in Shetland seems to be by far the most serious conservation issue (though the Foula colony unexpectedly bounced back to 141 pairs in 1991 when sandeels were once again abundant around Shetland).

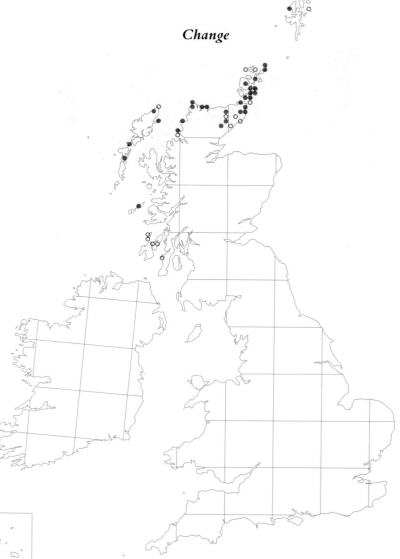

Change

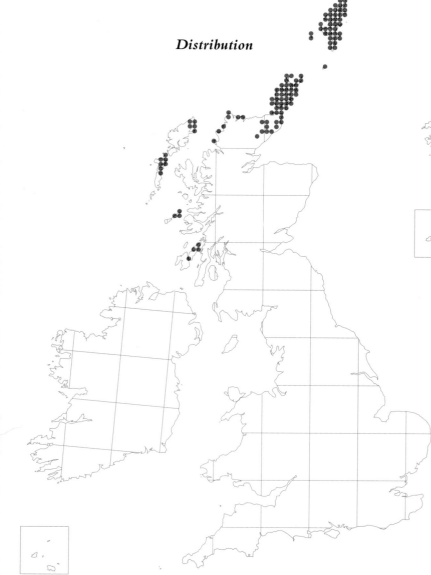

Distribution

The Scottish population of Arctic Skuas represents only 1–3% of the estimated world population (*Status of Seabirds*). However, as one of the less numerous breeding seabirds in Britain and Ireland, and with major breeding failure and population decline affecting the main area of its British breeding range, the species undoubtedly requires urgent attention from conservation bodies.

ROBERT W. FURNESS

68–72 Atlas p 206

Years	Breeding evidence		
	Br	Ir	Both (+CI)
1968–72	99	0	99
1988–91	113	0	113
% change	14.1	0	14.1

Great Skua

Catharacta skua

The Great Skua, or Bonxie, is famous for its habits of chasing and robbing seabirds of their food, killing and eating birds from the size of Storm Petrel to Whooper Swan, killing lambs and ewes in distress and clouting on the head anyone injudicious enough to wander near to its nest.

The SCR recorded 7,900 pairs of Great Skuas in 1985–87, of which 5,647 were in Shetland, 2,000 in Orkney, 28 in the Western Isles, 82 in Sutherland and three in Caithness, Ross and Cromarty (*Status of Seabirds*). As with the Arctic Skua, the Great Skua breeds predominantly in areas much colder than N Scotland. The closely related South Polar (or McCormick) Skua has been recorded near the South Pole, and will incubate when buried by several cm of snow (Furness 1987). Thus the southern extent of the Great Skua's breeding range may be set by mid-summer temperature; the breeding range lies north of the 13°C mean July isotherm and that of the S hemisphere populations lies south of the same midsummer isotherm. Certainly, it is difficult to see any other reason why Great Skuas do not breed further south in Scotland, because their diet during the breeding season is quite broad and includes sandeels caught at the surface, fish stolen from auks, gulls and Gannets, discards scavenged from behind fishing boats, and seabirds and mammals killed or scavenged at or near colonies. All these foods appear to be readily available in areas such as the Firth of Forth, the Clyde and the Irish Sea.

By the 1890s, the only Great Skua colonies in Britain, at Foula, Hermaness and Ronas Hill, were close to extinction, but protection measures saved the population and an increase began, the breeding numbers doubling about every ten years up to 1970. Whilst this increase started because the birds were protected from human persecution, it is difficult to imagine that the populations would have risen so much if changes in fish stocks and fishing practices had not also provided new food supplies in the form of discards and increased stocks of sandeels. With the increase in numbers, new areas were colonised, and chicks colour ringed at Foula have been found breeding in Orkney, the Hebrides, Sutherland, Norway, Spitsbergen, Bear Island (Furness 1987) and Russia. As numbers increased in Shetland the rate of increase fell, but emigration appears to have increased, for numbers grew at a faster rate in the new colonies. It is striking that despite the spread of the species, colonisation from Shetland and Orkney has tended to be northwards into the Arctic rather than to further south in Scotland. In 1989, however, breeding did occur on Coll, although the record was not submitted to the Atlas and so is not shown here.

The reduced availability of sandeels around Shetland in recent years has had a considerable effect on Great Skuas there. Numbers of occupied territories on Foula have fallen, by 8% from 1975–89, but numbers of nonbreeders on 'clubs' fell by 57% over the same period (Klomp and Furness 1990). The drop in number of nonbreeders precedes a further influence, of very low breeding success since 1987, that will soon be evident. The change seen so far appears to be due to an increased adult mortality and hence more rapid recruitment (Hamer *et al.* 1991). This suggests that a faster population decline will occur at Shetland colonies over the next few years, for the depleted pool of nonbreeders will receive few new members from the weaker cohorts fledged since 1987.

To add to the Great Skua's problems with food in Shetland, changes are occurring in the whitefish fishery that will have adverse effects. Increase in minimum net mesh size will result in less discarding and fewer small discards – the only ones Great Skuas can handle efficiently in competition with Gannets and Great Black-backed Gulls (Hudson and Furness 1989). Other threats to Great Skua populations (Furness 1987, *Status of Seabirds*) are much less serious than these changes in food availability in Shetland, at least in the medium term. Great Skuas share with Gannets the dubious distinction of carrying the highest loads of mercury of all British seabirds.

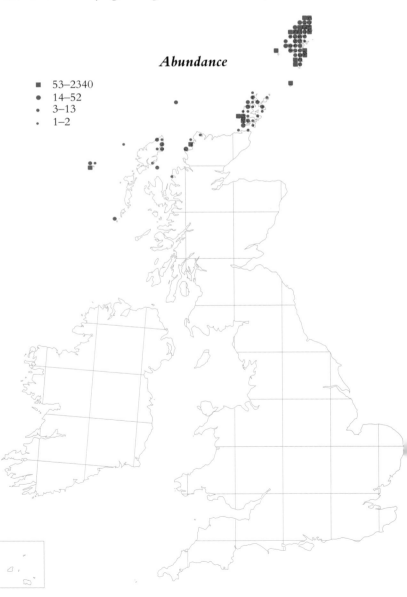

Abundance

- ■ 53–2340
- ● 14–52
- • 3–13
- · 1–2

Although mercury levels are increasing in Great Skuas and some other British seabirds, a recent study found no relationship between mercury burden and breeding success and survival of individual adults (Thompson *et al.* 1991).

The Scottish population of Great Skuas (7,900 pairs) represents 58% of the N hemisphere total, with 5,400 pairs in Iceland, 250 pairs in the Faeroe Islands, 62 pairs in Svalbard, four pairs in Norway and one in Jan Mayen (*Status of Seabirds*). Of these, 5,647 pairs nest in Shetland, and so the current breeding and survival problems there affect the principal concentration of the species. This is clearly a situation of considerable concern.

ROBERT W. FURNESS

68–72 Atlas p 204

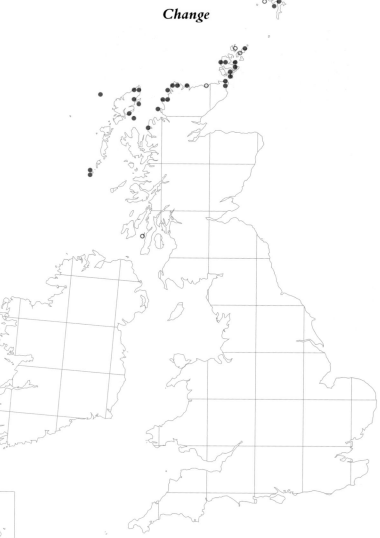

Change

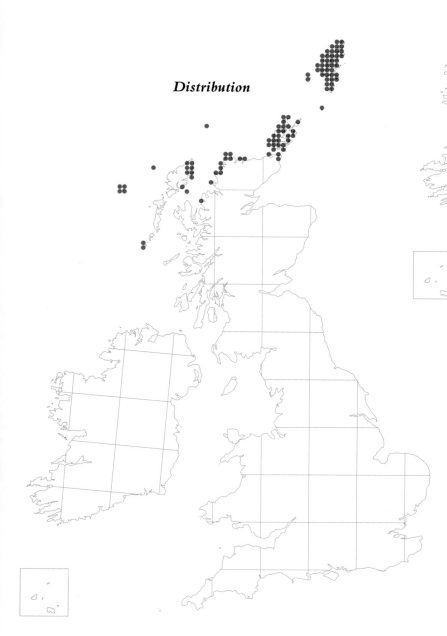

Distribution

Years	Breeding evidence		
	Br	Ir	Both (+CI)
1968–72	70	0	70
1988–91	97	0	97
% change	38.6	0	38.6

Mediterranean Gull

Larus melanocephalus

An exciting find at any time of year, the Mediterranean Gull presents a particular challenge to observers in Britain and Ireland during the breeding season, when a single bird or pair of birds may sit hidden amongst vast numbers of breeding Black-headed Gulls. The black, rather than brown, head, the heavy red bill and a distinctive nasal call are the features which most usually attract the attention of patient observers.

At the time of the publication of *The Handbook* (1940), the Mediterranean Gull was considered to be an extreme rarity in Britain, with only four accepted records, and it was unknown in Ireland. Although it still remains very uncommon, and whilst the increased ability of observers to identify it should not be underestimated, it appears that since the 1950s there has been a real increase in the numbers of this beautiful gull visiting our shores. This increase has followed a general increase in the world population of the species, the bulk of which breed in a few mainly island colonies around the Black Sea (*BWP*). Even as recently as 1960, it was thought by at least one authority that the species showed a relict range distribution and was likely to be heading for natural extinction (Voous 1960). Since then, the Black Sea colonies have increased from an estimated 93,500 pairs in 1961 to at least 170,000 pairs in 1977 (*BWP*), while in 1982 numbers there were thought to exceed 330,000 pairs (Siokhin *et al.* 1988). The bulk of this breeding population, which represents over 95% of the total world population, winters in the Mediterranean (*BWP*).

During this period of rapid increase in the heart of the breeding range, the species has been more widely recorded throughout W Europe, together with a number of breeding records. A small central European breeding population was established in Hungary in the 1950s, since when breeding has been recorded in most W European countries. The first breeding record in Britain was in 1968, when a pair nested in a colony of over 10,000 pairs of Black-headed Gulls in Hampshire (Taverner 1970). It was eight years before the species was recorded nesting successfully again in Britain, although there were regular observations of birds pairing with Black-headed Gulls to produce hybrid young. Since 1979, breeding has been annual, with a maximum of 11 confirmed pairs in 1990. This increase in breeding records reflects the trend shown in other countries in central and W Europe. In Holland, the Mediterranean Gull has been a regular breeder since 1970, with small colonies of up to ten pairs forming inside larger colonies of Black-headed Gulls (Meininger and Bekhuis 1990). At least 125 pairs bred in the Netherlands in 1990 and the colonies of the Dutch/Belgian delta now contain almost half of the central and W European breeding population of this species.

Given the concentration of breeding birds in the delta area of the Netherlands, it is perhaps not surprising that results of the *88–91 Atlas* fieldwork show breeding records in Britain to be concentrated in SE England. Six of the seven breeding records fall east of a line drawn from the Wash to the Solent, and RBBP records show that there were up to three further records of breeding in S England that were not reported to the Atlas organiser. Although breeding was not proved in Ireland, there were several records of single birds visiting gull colonies during the breeding season, with one bird returning for two consecutive seasons (10-km square G20), and future colonisation has been predicted (Madden 1987). In all areas, breeding or summering records are usually correlated with the presence of Black-headed Gull colonies, although in Scotland there have been several records of birds summering in colonies of Common Gulls (RBBP).

The breeding distribution is markedly different from that found in winter, which the *Winter Atlas* showed to be concentrated along the SW coast of England. This difference seems likely to be related to temperature, with the winter distribution corresponding well to the areas of mildest weather in winter, and the breeding distribution to the warmest areas in summer. This suggests that the Mediterranean Gull is unlikely to become a common bird in Britain or Ireland.

Records submitted to the RBBP over the years indicate that the breeding success of this species in Britain has been very low, as might be expected for a recent colonist on the extreme northwest limit of its breeding range. It may well improve if the species is able to establish breeding colonies. During 1988, 1989 and 1990, 5–16 pairs of Mediterranean Gulls bred in Britain annually (RBBP).

PAUL DONALD and JOHAN BEKHUIS

Two-year averages for Mediterranean Gulls breeding in Britain, 1973–90.

	73–74	75–76	77–78	79–80	81–82	83–84	85–86	87–88	89–90
Confirmed	0	1	1	2	3	3	2	3	9
Max total	0	1	2	4	5	7	7	9	13

Based on RBBP reports: the first line refers to records of confirmed breeding, the second to possible maximum breeding pairs. Note: figures are rounded up to the nearest whole number.

68–72 Atlas p 448

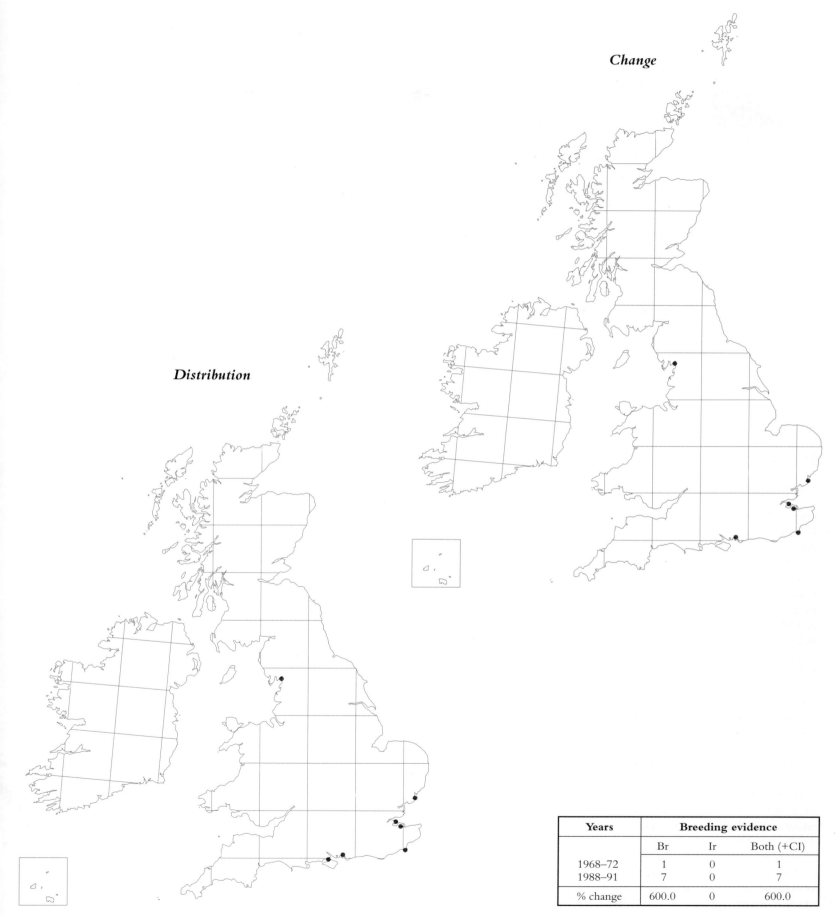

Change

Distribution

Years	Breeding evidence		
	Br	Ir	Both (+CI)
1968–72	1	0	1
1988–91	7	0	7
% change	600.0	0	600.0

GBB

Black-headed Gull

Larus ridibundus

One of the most 'inland' of British and Irish seabirds, the Black-headed Gull breeds colonially at sites ranging from coastal saltmarshes and sand dunes to freshwater lakes, marshes, gravel pits and upland tarns. Even those breeding coastally may feed inland to a great extent.

The species' coastal distribution is not as wide as that of the Herring Gull, reflecting the unsuitability of cliff-bound coasts, and is more comparable with that of some of the terns. Coastal colonies are largest in S and SE England and around the Irish Sea, including Bank's Marsh in the Ribble Estuary, Lancashire (20,000 pairs in 1985) (*Status of Seabirds*) and the Pennington-Keyhaven and Needs Ore Point colonies in the Solent (14,700 pairs in total in 1991) (Aspinall and Tasker 1992). Colonies on the coasts of W Scotland and W Ireland, and in Orkney and Shetland, are generally much smaller.

Inland colonies outnumber coastal ones, particularly in NW England, Wales, NW Ireland and much of Scotland, and include concentrations of 33,000 pairs at Lough Neagh in Northern Ireland (*Status of Seabirds*) and 25,000 individuals at Sunbiggin Tarn in Cumbria (J. C. Coulson). There are, however, few inland colonies in S England (where distribution is predominantly coastal), in S and SE Ireland and in NW Scotland. Except in the latter case, this may reflect absence of suitable marshy habitat inland; many artificial sites such as reservoirs and clay pits are available in S England, but the extensive coastal marshes may be preferred. The map in Appendix E shows that summering, non-breeding birds are widespread.

Throughout Britain and Ireland, there has evidently been some turnover of colonies (and occupied squares) since the *68–72 Atlas*. The most notable changes are the apparent loss of colonies in Scotland and NW Ireland.

The latter loss may be in part due to changes in coverage, but that in Scotland is less likely to be. Further work on inland populations of gulls in these regions is warranted.

Recent population changes have been documented for the coastal population only, with a slight increase (of about 7%) between 1969–70 and 1985–87 for Britain and Ireland as a whole (*Status of Seabirds*). Most of the increase occurred on English coasts (30 + %), although there may have been a slight decrease since a 1973 survey (Gribble 1976). A 55% decrease was noted for Scottish coasts, especially SE Scotland, and the small coastal population in Wales also decreased. In Ireland, numbers on the NE coast increased by 70%, but changes elsewhere were unclear.

On a county by county basis, some major changes in coastal numbers have been noted since 1969; for instance a tenfold increase in East and West Sussex from 260 to 2,900 pairs. However, such changes often reflect the fortunes of one or two large colonies and may sometimes indicate movements between colonies. Inter-colony movements by breeding birds are known to occur inland (Gribble 1976), at least, in response to such factors as disturbance and changes in water levels. Establishing whether or not there is a balance between increases and decreases at nearby colonies is difficult, however, as regular, accurate counts are usually not available (especially for inland colonies).

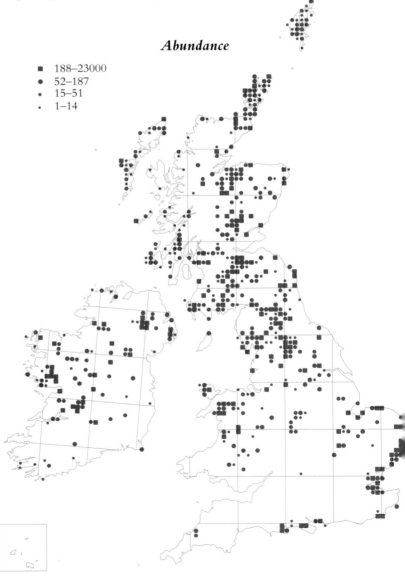

Abundance

- ■ 188–23000
- ● 52–187
- • 15–51
- · 1–14

The breeding success of this species is typically very variable, with recent figures for coastal colonies generally ranging from 0.05 to 1.0 chicks fledged per pair (Walsh *et al.* 1991). As with terns, colonies are very susceptible to mammalian predators, flooding by high tides and disturbance. At some colonies, licensed egg collecting takes place on a large scale, but first clutches may in any case be lost to high tides. Uncontrolled egg collecting elsewhere may have a more serious influence, especially on small colonies.

About 84,200 pairs breed on British and Irish coasts (77,400 Britain, 6,800 Ireland) (*Status of Seabirds*). Available figures for inland colonies (SCR and this Atlas) include at least another 47,000 pairs in Ireland, 30,000 pairs in Scotland and 40,000 pairs in England and Wales. The minimum population is thus 201,200 pairs (147,400 Britain, 53,800 Ireland), or at least 16% of the European and 12% of the world population.

PAUL M. WALSH and MARK L. TASKER

68–72 Atlas p 216

Change

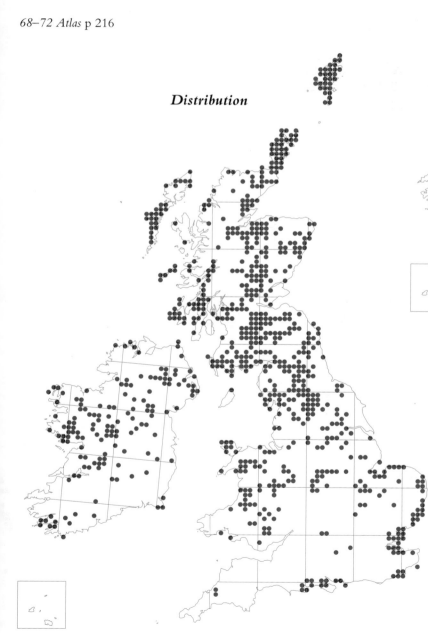

Distribution

Years	Breeding evidence		
	Br	Ir	Both (+CI)
1968–72	827	281	1108
1988–91	671	145	816
% change	−18.8	−48.4	−26.3

Common Gull

Larus canus

The high-pitched mewing calls of this gull give away the location of its colonies or nest sites: most often located inland on moorland or bogs, but frequently on small coastal islands. The species nests regularly with other gull species or with terns, but there are few large colonies.

This is a breeding bird of the N and W of Britain and Ireland, and few breed in England or Wales. In Scotland its absence from the Central Lowlands probably indicates a lack of suitable breeding sites, or of feeding areas. As with other ground-nesting seabirds, Common Gulls require sites relatively free of mammalian predators. Such sites include small islands in both freshwater and sea lochs, the middle of mires – even small ones, the tops of moorland hills and, more recently, artificial sites such as the roofs of buildings, or fenced enclosures around industrial installations.

Inland in Scotland, the breeding gulls feed on pastures and cultivated fields (Douse 1981). Nearer the coast the inter-tidal areas are used. It would be interesting to examine whether there is a relative lack of pasture and other feeding areas, or of potential breeding habitat, in the Central Lowlands.

Although there appears to have been little real change in overall breeding distribution since the *68–72 Atlas*, there has been a worrying decline in the number of 10-km squares in which breeding was recorded in both Britain and Ireland. Unfortunately, little is known of the change in numbers over recent years as the inland population received relatively little attention in the past. The coastal population, however, increased from 13,000 to 15,700 pairs between 1969–70 and 1985–87 (*Status of Seabirds*), but since the coastal population comprises only about a fifth of the overall British and Irish population, the significance of this change is not known. The birds breeding in SE England are believed to derive from the Continental population that migrates to Britain in the winter (*68–72 Atlas*). The numbers of breeding birds and colonies in SE England have increased in recent years, but the total population is still very low.

In NE Scotland, the number of colonies appears to have decreased, and the population to have become concentrated on fewer sites. This concentration has perhaps been caused by draining of lowland mires for agricultural purposes, or more frequently by afforestation. Two colonies hold perhaps 50% of the British and Irish population (Tasker *et al.* 1991): one on the Correen Hills, the other on the Mortlach Hills. Both these groups of relatively low, heather-covered hills are managed for grouse shooting, and it seems likely that predator control has made them attractive for the gulls also. Both colonies appear to have grown in size in recent

years, but assessment of the degree of growth is hampered by a lack of good censuses in the past. Afforestation or a relaxation in management for grouse would be likely to reduce the Common Gull population on these hills.

Little has been published on the breeding ecology of Common Gulls in these islands. A small colony on Lough Corrib in Co. Galway produced averages of between 0.32 and 1.12 fledged chicks per pair over a seven-year period, with an overall average of 0.66 (Whilde 1984). Most egg losses were caused by nest flooding during heavy rain or large changes in water level. Predation of eggs from completed clutches was not a problem at this island colony. Mortality of chicks was highest immediately after hatching, again mostly caused by flooding of nests (Whilde 1984).

During 1985–87, the British population was estimated to be about 68,000 pairs, with 53,000 of them inland, while the equivalent figures for Ireland were 3,600, with 2,700 inland (*Status of Seabirds*). Together, the British and Irish populations comprise some 15% of the European population of about 488,000 pairs, and about 12% of the world population.

MARK L. TASKER and PAUL M. WALSH

68–72 Atlas p 214

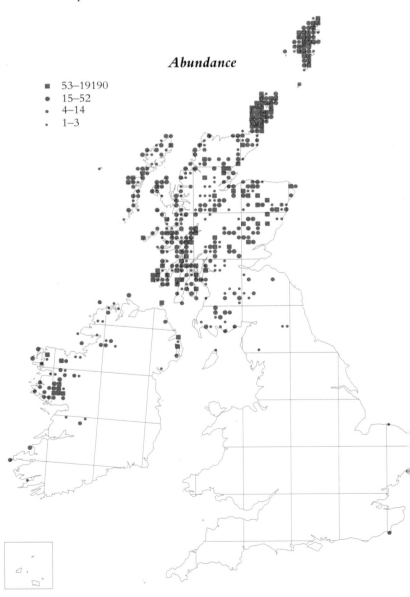

Abundance

- ■ 53–19190
- ● 15–52
- • 4–14
- · 1–3

Change

Distribution

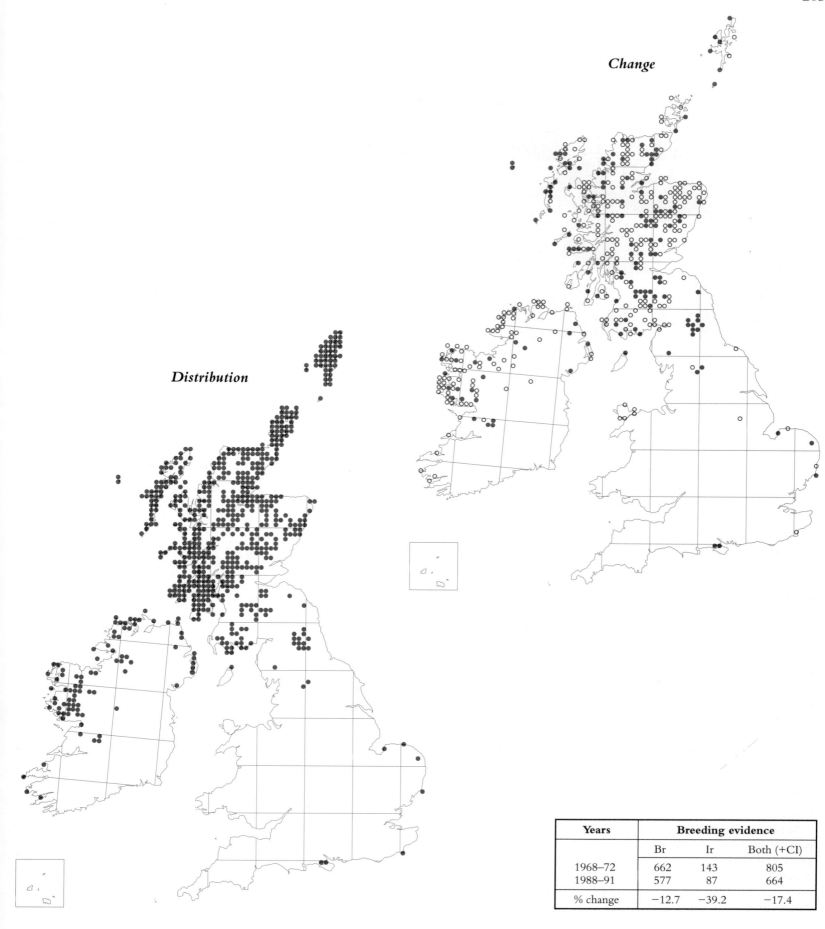

Years	Breeding evidence		
	Br	Ir	Both (+CI)
1968–72	662	143	805
1988–91	577	87	664
% change	−12.7	−39.2	−17.4

which, even allowing for the inland nesters, is a much smaller number. Recent censuses have shown that in marked contrast to the Herring Gull, however, the Lesser Black-backed population has continued to increase; in 1985–87, an estimated 64,400 pairs bred on British and Irish coasts, an increase of 29% since Operation Seafarer. Skomer, Walney, Little Cumbrae and Orfordness are the largest coastal colonies, while the largest inland colony is at Abbeystead in the Pennines.

The rise of the colony at Orfordness is particularly interesting. Since this colony was formed in the mid 1960s numbers there have increased to over 5,000 pairs; predation and disturbance are low, and it remains one of the few large gull colonies which is not culled (*Status of Seabirds*). The increase seems to be largely due to immigration of birds from other areas. Like the Herring Gull, the Lesser Black-backed has also continued to increase in urban colonies since the last census in 1976 (Monaghan and Coulson 1977), especially in central Scotland, where new rooftop colonies are being formed (A. Wood).

There have also been marked declines in some areas. The disappearance of an inland colony of around 8,000 pairs at Flanders Moss, near Stirling,

Lesser Black-backed Gull

Larus fuscus

The Lesser Black-backed Gull is something of a puzzle. Much less well studied than the closely related Herring Gull, it is just as much of a generalist. Although coastal breeding birds normally have a more marine diet than Herring Gulls in the same area, inland colonies of the Lesser Black-backed are nevertheless the more common. And why does the majority of the population undertake an extensive migration to wintering grounds on the Atlantic coasts of Iberia and North Africa, while the Herring Gull successfully overwinters in Britain and Ireland?

Like the Herring Gull, the Lesser Black-backed is a colonially nesting seabird. It is, however, smaller, more colourful, noisier and certainly has a sharper bill. It is widely distributed throughout Britain and Ireland and its overall distribution has changed little since the *68–72 Atlas*. A few new inland colonies have been established, and summering, non-breeding birds are frequently encountered inland (see map in Appendix E). Compared with the Herring Gull, the Lesser Black-backed has a more southerly distribution, with a marked concentration in central Scotland, and on the W coast of England and Wales, where 58% of the coastal breeding population occurs (*Status of Seabirds*). Though sharing many breeding areas with the Herring Gull, the Lesser Black-backed tends to prefer nesting in the more vegetated areas rather than on rocky outcrops, which may in part account for its increased occurrence as an inland breeder.

In common with that of many other seabirds, the Lesser-backed Gull population in Britain and Ireland increased with the introduction of general bird protection legislation, but has remained considerably smaller than that of the Herring Gull. Just over 50,000 pairs were estimated nesting on the coasts of Britain and Ireland during Operation Seafarer in 1969 (Cramp *et al.* 1974) compared with over 300,000 pairs of Herring Gulls,

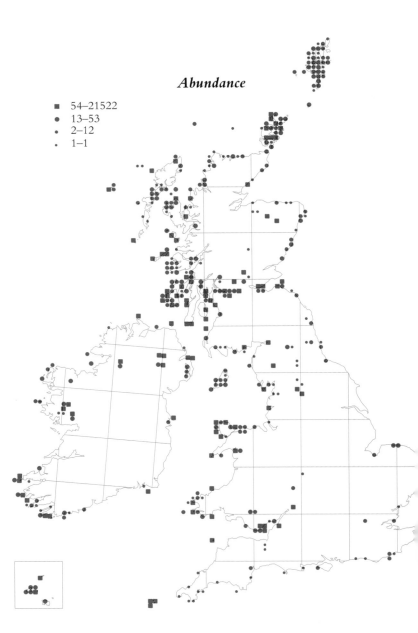

Abundance

- ■ 54–21522
- ● 13–53
- • 2–12
- · 1–1

appears to have been due to a combination of disturbance and predation (A. Whitelaw). Culling as a reserve management practice has been extensive, and, as with the Herring Gull, this has undoubtedly disrupted population processes such as recruitment (Coulson 1991). The large scale cull in the moorland colony at Abbeystead (initially claimed to be to improve water quality but clearly having more to do with sporting interests) may be partly responsible for the decline in numbers at Walney, which have fallen from 17,500 pairs in 1969 to 10,000 in 1985 (*Status of Seabirds*). Recent breeding failures of Lesser Black-backeds on Skomer have been attributed to a decline in waste from Irish Sea trawlers. That food is likely to be important has been shown by experimental studies in the Bristol Channel, where the provision of supplementary food during egg laying, in a colony where egg production was poor, increased the average clutch size to the normal level (Hiom *et al.* 1991).

There are clearly substantial changes taking place in the Lesser Black-backed Gull population in Britain, such as the differences in population trends between colonies, and also the increasing trend for birds to over-winter in Britain (see *Winter Atlas*). 83,500 coastal and inland pairs of

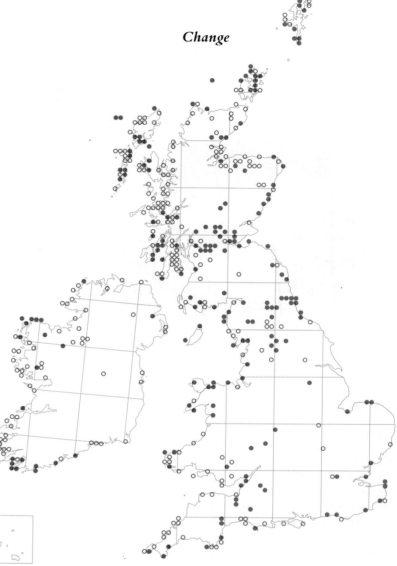

Change

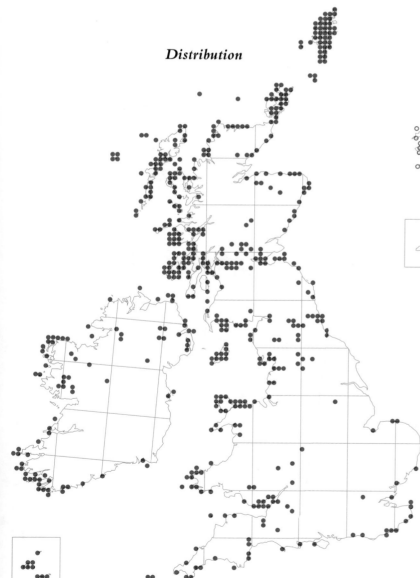

Distribution

Lesser Black-backed Gulls breed in Britain, with a further 5,200 in Ireland (*Status of Seabirds*). Given that in N Norway the Lesser Black-backed Gull is fast disappearing as a breeding species, we should pay this bird more attention.

PAT MONAGHAN

68–72 Atlas p 210

Years	Breeding evidence		
	Br	Ir	Both (+CI)
1968–72	431	118	554
1988–91	434	81	525
% change	0.5	−31.4	−6.4

tunistic ability to exploit new food sources such as refuse tips. The population explosion led to many environmental problems, including the expansion of colonies into areas formerly occupied by terns, and colonisation of towns and moorland near reservoirs, with the potential spread of disease to man and domestic livestock (Monaghan 1983). In an attempt to manage the population, 19th century persecution was replaced by control measures, initially involving egg and nest destruction, supplemented from the early 1970s onwards by extensive culling of adults. Many tens of thousands of breeding birds have been removed from the population, mainly by poisoning with narcotic bait. The extent to which these drastic measures were justified is discussed in some detail by Lloyd *et al.* (*Status of Seabirds*). The Herring Gull is one of the few seabirds not protected by law, and hence, in addition to organised management operations, eggs can be collected and birds shot by private individuals. The effects of culling operations on the population dynamics of the Herring Gull are not well understood at present, but may have been considerable. In addition to the obvious reduction of the breeding population, they disrupted recruitment outside the areas where culling actually took place (Coulson 1991).

Herring Gull

Larus argentatus

Although a very common bird the Herring Gull is full of surprises. For an essentially coastal seabird the apparent ease with which it has invaded inland areas is remarkable. The extent of the population increase throughout much of its range during the first two-thirds of this century was not foreseen, nor was the extent of the decline apparently taking place now in Britain and elsewhere in N Europe.

Like most gulls, the Herring Gull is a very adaptable, unspecialised bird. It breeds in a wide range of habitats, forming noisy colonies on steep cliffs, rocky outcrops, screes, beaches, moorland and even on inhabited buildings. It can feed, for example, on intertidal organisms, but also on seeds, roots, fruits, invertebrates and small vertebrates in agricultural habitats. It can catch, scavenge and steal fish, and can make good use of the waste products of human populations. We should not, then, be surprised that its distribution is so widespread in Britain and Ireland. Less than 100 years ago, Herring Gulls occurred rarely inland, but the Distribution map shows that there are now several inland breeding colonies, particularly in the Central Lowlands of Scotland, and non-breeding birds are seen regularly inland (see map in Appendix E), especially in winter (*Winter Atlas*). The majority of Herring Gulls still breed on the coast, and here the overall range of the Herring Gull has changed little since 1968–72.

On the other hand, the abundance of the Herring Gull has changed considerably. During much of the present century, it increased dramatically, but the extent of the increase was not really appreciated until Operation Seafarer in 1969 (Cramp *et al.* 1974). It has been estimated that, between the 1940s and early 1970s, the Herring Gull population increased by about 13% per annum, and annual adult survival was over 95% (Chabrzyk and Coulson 1976). Much of the cause of this increase probably lies with the enforcement of bird protection legislation, and with the bird's oppor-

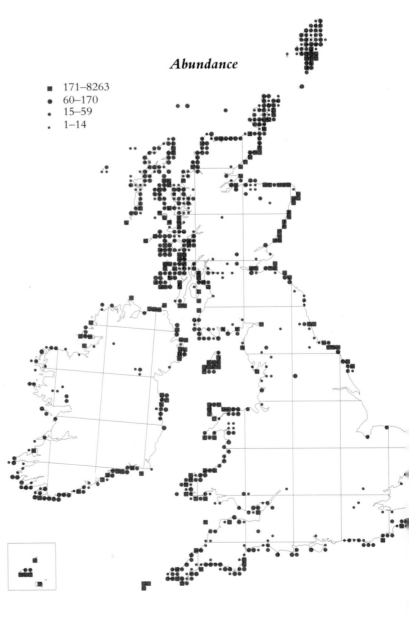

Abundance

- ■ 171–8263
- ● 60–170
- • 15–59
- · 1–14

Recent census information shows that the Herring Gull population breeding on the coasts of Britain almost halved between 1969–87, from an estimated 283,900 pairs to 146,700 pairs (*Status of Seabirds*). The decline in Ireland was somewhat less, from 59,700 to 44,200. There was considerable regional variation in this change in status, the decline being most marked in SW England and the adjacent areas of S Ireland and N Scotland. The Change map also suggests that a number of, generally small, colonies have been lost from the W coast of Ireland. The population has, however, continued to increase in the Firth of Clyde area and in nearby areas of Northern Ireland. The notable increase at Orfordness in Suffolk may have been fuelled to some extent by immigrants from the Netherlands and elsewhere, though relevant ringing data are limited (*Status of Seabirds*). Urban nesting, which increased greatly in the early 1970s (Monaghan and Coulson 1977), has apparently continued to increase, though a full repeat census has not been carried out. Nevertheless, an estimated total of 161,000 coastal and inland pairs of Herring Gulls breed in Britain, with a further 44,700 in Ireland.

The cause of the unexpected downturn in Herring Gull numbers is currently being investigated but may involve, in addition to the effects of

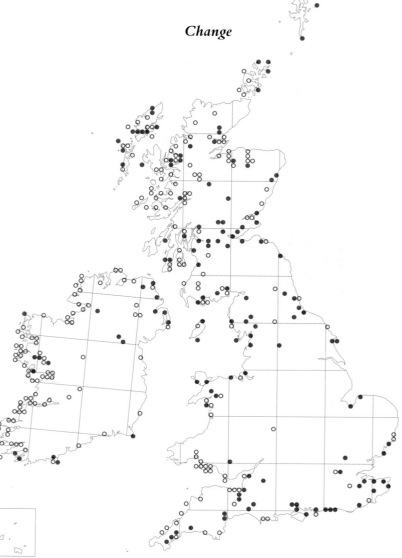

Change

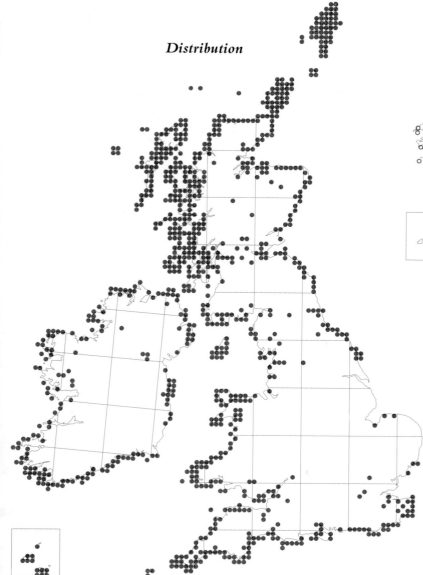

Distribution

culling, an increased incidence of diseases such as botulism, increased predation by foxes, and changes in food availability. Current analyses of ringing recoveries, excluding culled areas, suggest large regional variations in adult mortality rates, which are being examined in relation to environmental factors. No doubt the Herring Gull has more surprises in store for us.

PAT MONAGHAN

68–72 Atlas p 212

Years	Breeding evidence		
	Br	Ir	Both (+CI)
1968–72	738	225	968
1988–91	729	163	904
% change	−1.1	−27.6	−7.3

Great Black-backed Gull

Larus marinus

The Great Black-backed Gull is the largest, most maritime and least common of the five *Larus* gulls which breed regularly in Britain and Ireland. An omnivorous opportunist, it is equally skilled at snatching offal from fishing boats far out at sea or catching other seabirds as prey in high-speed aerial pursuits close to the shore.

Most Great Black-backed Gulls are coastal breeders, nesting either in colonies or in groups of a few pairs, typically on islands, offshore stacks and rocky headlands. Very small numbers nest inland on moorland and beside lochs. The most striking features of the breeding distribution are the absence of breeding birds on the E and S coasts of Britain between the Firth of Forth and Chichester Harbour and the concentration of breeders along N and W coasts of Britain and Ireland. This is likely to reflect both the Great Black-backed Gull's selection of nest sites which are relatively safe from human disturbance, on rocky rather than sandy coasts, and its use of the food-rich northern and western waters for scavenging and fishing.

In 1985–87 about 70% of the total coastal population bred in Scotland (*Status of Seabirds*), the majority of them in Orkney, Shetland and the Western Isles. Only one Scottish island colony outside Orkney, North Rona (733 pairs), held more than 200 pairs. The rest of the population was divided between Ireland (19%), England – mainly SW – (7%), Wales (2%), Isle of Man (2%) and the Channel Islands (1%). Two Irish colonies – the Duvillaun Islands off Mayo and Inishtooshkert in the Blasket Islands off Kerry – held more than 200 pairs in 1985–87, with most breeders elsewhere being in counties Donegal, Dublin, Cork, Galway and Wexford. Away from the Northern Isles, only a few pairs nest inland in Britain or Ireland.

From a low ebb near to extinction last century, there was a strong and widespread increase in the breeding population, beginning around 1880 and continuing throughout much of this century in Britain and Ireland (Parslow 1973). This was part of a more general spread and increase over much of the range on both sides of the Atlantic (*BWP*), of unknown cause, but conceivably fuelled by the greater availability of discarded fish and offal at sea and, later, by access to rubbish at landfill sites.

The general pattern of breeding distribution after this expansion has changed very little since 1970. Slight expansion of the range of the numerically stable Scottish population, including occupation of inland sites but involving very few pairs, has occurred in Caithness and Sutherland and on the island of Jura. Little change is apparent in the distribution of

the fairly stable English population since 1970, or in the Irish population which increased in numbers in western and northern coastal areas in the period 1969–87 (*Status of Seabirds*) despite there being apparent losses of many, probably small, colonies on the W coast. In Wales, there was a 54% reduction in estimated breeding pairs in the period 1969–87, and fewer squares are now occupied in Gwynedd and Dyfed, where most of the small Welsh population breed.

Laying of the typical three-egg clutch is from mid April to mid July, with main fledging from late June to mid August (*BWP*). The Great Black-backed Gull has gained a certain notoriety in the past as a breeding-season predator of other seabirds. At low breeding numbers, other seabirds may suffer relatively heavy adult mortality from predation by specialist seabird-killing gulls, as is the case in the low density part of the puffinry on Dun, St Kilda (Harris 1984). There is evidence that gulls which are specialist seabird-killers tend to be solitary nesters, exploiting the most abundant seabird chicks available locally (Hudson 1982, Taylor 1985). By contrast, the more numerous, colonially nesting Great Black-backed Gulls on Great Saltee, Wexford, feed mainly fish and crabs to their chicks (Hudson 1982).

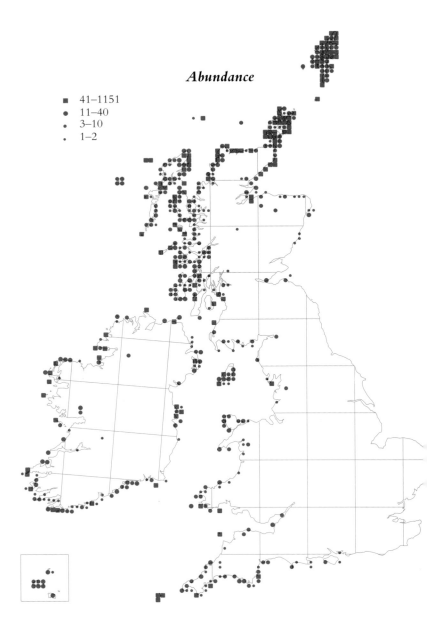

Abundance

■ 41–1151
● 11–40
• 3–10
· 1–2

The species breeds only in the North Atlantic, with the world population probably less than a quarter of a million pairs (*Status of Seabirds*). The British population in 1985–87 was estimated at almost 19,000 pairs – virtually unchanged overall from 1968–72 – and the Irish population was estimated at 4,500 pairs. The suggested British and Irish total of over 23,000 pairs represents about 20% of the estimated W European population and perhaps nearly 10% of the world population.

KENNY TAYLOR

68–72 Atlas p 208

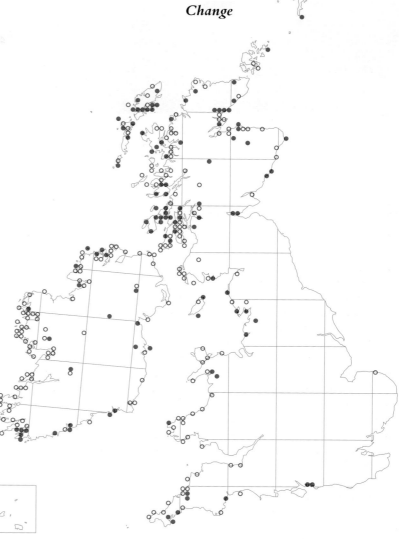

Change

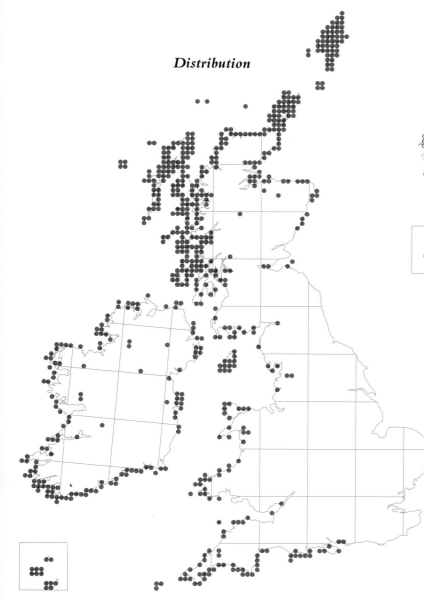

Distribution

Years	Breeding evidence		
	Br	Ir	Both (+CI)
1968–72	505	192	702
1988–91	486	137	636
% change	−4.4	−28.3	−10.9

Kittiwake

Rissa tridactyla

Although Kittiwakes are the most oceanic of our gulls in winter, during the breeding season they are the most abundant gull in British and Irish coastal waters. The steep cliff faces typically used for breeding are often shared with Guillemots, and include some of the largest and most impress-ive seabird colonies in Britain. Smaller colonies may also occur, sometimes on cliff faces too featureless for any other seabird.

A high proportion of the population breeds in N Scotland and along North Sea coasts south to Humberside, where the Bempton-Flamborough cliffs (83,700 occupied nests in 1986) hold possibly the largest colony in the North Atlantic (*Status of Seabirds*). Other major northern colonies with over 10,000 pairs each, include St Abb's Head (Borders), Fowlsheugh (Grampian), Fair Isle (Shetland), West Westray (Orkney) and Handa (Sutherland).

Colonies are quite widespread on other coasts, but generally much smaller and separated by longer stretches of unoccupied coastline. In particular, English coasts from Lincolnshire clockwise to Dorset have few colonies. To a large extent, this reflects a shortage of suitable cliffs. The only East Anglian colony (at Lowestoft in Suffolk) includes nests on derelict buildings and a pier. Buildings are also occupied at several other colonies along the North Sea coast, including a flour mill at Gateshead (Tyne and Wear), 16km up the River Tyne.

A major population expansion has taken place during this century, with many new colonies revealed by surveys in 1959 and 1969 (J.C. Coulson in Cramp *et al.* 1974, Coulson 1983). There have been fewer changes in breeding distribution since the *68–72 Atlas*, mainly the formation of new colonies in Britain, including the colonisation of East Sussex. Over the same period, breeding numbers on coasts from Suffolk to Sussex have increased dramatically, from only 80 pairs in 1969 to 3,760 pairs in 1985–87 (*Status of Seabirds*). Populations in other regions (apart from Orkney, NW England and NW Ireland) have also shown net increases since 1969, mainly by less than 30% (but by 110% in NE Ireland and 170% in NE England). The total British and Irish population was 22% higher in 1985–87 than in 1969.

There have been both increases and decreases since 1969 as shown by the results of long-term monitoring at sample colonies, and a 1979 survey which included good coverage of England, Wales and the Isle of Man (Coulson 1983). The 1979 survey showed numbers had declined at many colonies on the W coast of Britain and in SE Ireland since 1969. Later results, from 1985–87, suggest that some recovery has occurred in most

western regions. In contrast, numbers in NE England more than doubled between 1969 and 1979, largely reflecting growth of the Bempton colony, but showed little further change by 1987 (*Status of Seabirds*). Long-term trends for sample colonies in E Scotland also suggest some slowing down, or temporary reversal, of growth in the mid 1980s, while the large Shetland and Orkney populations have generally been declining since the mid 1970s (*Status of Seabirds*, Walsh *et al.* 1991).

Coulson (1983) suggested that food availability during the breeding (or immediate pre-breeding) season was the most likely factor behind regional variations in population status. The implication was that food availability was highest at North Sea colonies. Even there, long-term changes in fish stocks may have been affecting Kittiwakes adversely. Thus at the North Shields warehouse colony, studied since the 1950s, Coulson and Thomas (1985) recorded a gradual decline in Kittiwake numbers, breeding success and adult survival, paralleling a decline in North Sea herring stocks.

Available information on Kittiwake diet, mainly at North Sea colonies, indicates that while Clupeids (sprat and herring) are important in spring, sandeels are the most important item during chick rearing (e.g. Coulson and Thomas 1985, Harris and Wanless 1990, *BWP*). In Shetland, there is strong evidence that recent declines in breeding success of Kittiwakes, and

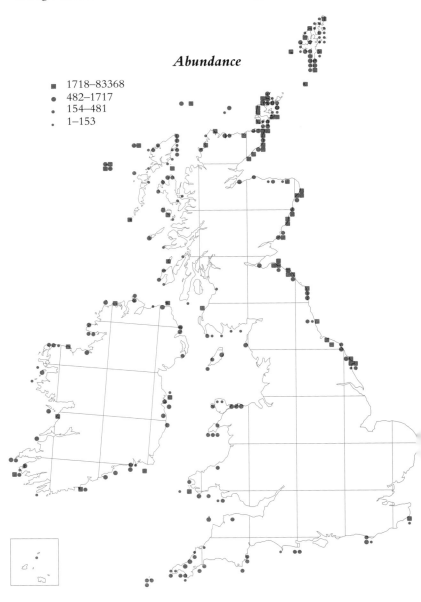

Abundance

- ■ 1718–83368
- ● 482–1717
- • 154–481
- · 1–153

other seabird species feeding from near the surface, are linked to a reduction in sandeel availability (Heubeck 1989, Harris and Wanless 1990). For Kittiwakes, the decline in success was first noted in 1985–86, and by 1988 the majority of colonies failed totally (mainly at the chick stage). Productivity was less than 0.1 fledgling per pair that year (compared to averages of about 1.0 per pair for other North Sea regions in most years). Poor success continued in 1989 and 1990 (Walsh et al. 1991). In comparison, other North Sea colonies showed a less marked reduction in success in 1988 (Harris and Wanless 1990), but a more obvious reduction in 1990. Productivity at west coast colonies has fluctuated more irregularly, but colonies in SW Britain and SE Ireland are currently the least successful outside of Shetland (averaging less than 0.5 chicks per pair in 1989–90).

A total of 543,600 pairs breed in Britain and Ireland (50,200 Ireland, 493,400 Britain; *Status of Seabirds*). This represents about 8% of the world population, and about 30% of the European population (including Iceland).

PAUL M. WALSH and MARK L. TASKER

68–72 Atlas p 218

Change

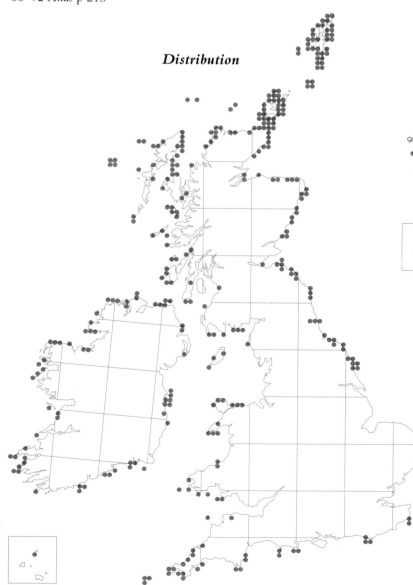

Distribution

Years	Breeding evidence		
	Br	Ir	Both (+CI)
1968–72	239	78	318
1988–91	252	62	315
% change	4.2	−20.5	−1.9

Sandwich Tern

Sterna sandvicensis

Their large size, shaggy crests and relatively short tails distinguish Sandwich Terns from other tern species breeding in Britain and Ireland. They are among the first migrants to return each spring, and by mid April they are forming densely-packed colonies at traditional locations around the coast.

The distribution of Sandwich Terns has changed little since the *68–72 Atlas*. A few new colonies have been formed (e.g. Loch of Strathbeg, Dungeness and Langstone Harbour), helped, probably, by the creation of nature reserves, whereas a few small colonies (e.g. Minsmere, Tentsmuir) and some large ones (Ravenglass) have ceased to exist, perhaps temporarily. In Wexford, Sandwich Terns have recently recolonised Lady's Island Lake, possibly encouraged by programmes of predator control. As with other terns, their distribution follows no very obvious pattern and this highlights how little we understand seabird ecology. We do not know why there is an Orkney population which moves from island to island and yet Shetland has never produced a breeding record. Why, too, are the Inner Hebrides and Western Isles usually shunned by Sandwich Terns despite their colonies of Little, Common and Arctic Terns?

Most major colonies are on nature reserves and are monitored annually. Breeding success is estimated at many colonies, but there appears to be no current research directed at the ecology or behaviour of this species.

A study in The Netherlands (Veen 1977) investigated the nesting association between Sandwich Terns and Black-headed Gulls. Why do Sandwich Terns frequently nest with Black-headed Gulls when Black-headed Gulls often eat Sandwich Tern eggs and chicks? Veen found that most of the eggs taken by Black-headed Gulls were from deserted clutches and thus had little effect on the terns' productivity. The protection that the terns derive from the aggressive anti-predator responses of Black-headed Gulls, and from the fact that predators might sometimes prefer Black-headed Gulls' eggs to Sandwich Terns' eggs, explains why this nesting association persists.

Sandwich Terns winter in West Africa where they are trapped and killed. Trapping intensity varies from year to year, and Green *et al.* (1990) found, from an analysis of ringing recoveries, that cohorts of first-year Sandwich Terns which experienced much trapping produced fewer breeding adults than cohorts which escaped intensive trapping. This is the strongest evidence to date that trapping of terns in winter influences the subsequent level of breeding populations. It is, however, possible that years

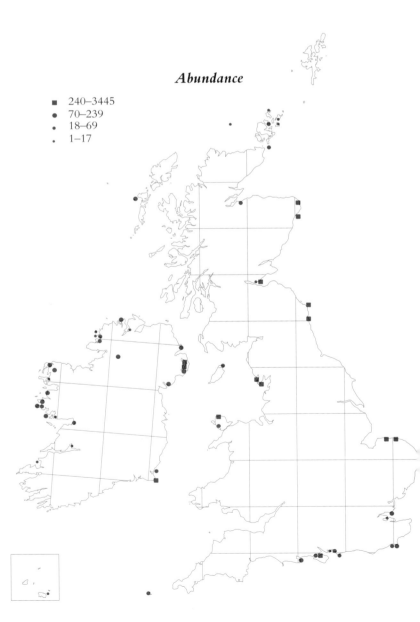

Abundance

■ 240–3445
● 70–239
• 18–69
· 1–17

when many terns are caught are also years when environmental conditions are adverse too (e.g. in years when food stocks are low or weather is bad). Terns may then be forced onshore and become susceptible to trapping, so the causal link between trapping and population levels remains unproved.

In 1985–87 the Irish population was *ca* 4,400 pairs and the British population was *ca* 14,000 pairs (*Status of Seabirds*). This represented a 50% increase since 1969–70. Numbers have fallen back slightly since 1985–87 but the five-year trend is for little change in numbers (Walsh *et al.* 1991).

MARK AVERY

68–72 Atlas p 228

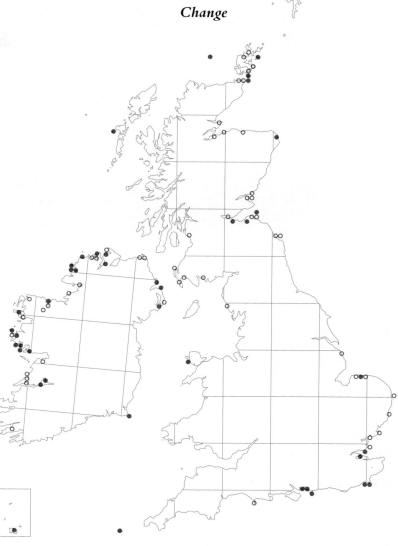

Change

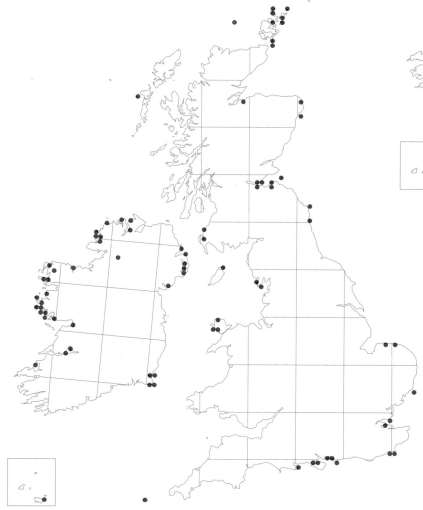

Distribution

Years	Breeding evidence		
	Br	Ir	Both (+CI)
1968–72	58	34	92
1988–91	43	37	81
% change	−24.6	8.8	−11.0

Roseate Tern

Sterna dougallii

In a genus in which all its members are attractive, the Roseate Tern is a singularly elegant and beautiful bird, but, despite its name, its recent past has been far from rosy and its future is uncertain.

Roseate Terns nest at scattered locations around the British and Irish coasts, and always with other terns. Almost all colonies are on small islands, which may limit this species' potential distribution. The occasional colonies which form on mainland sites tend neither to grow nor to endure. The answers to questions such as why are there no colonies in W Scotland, why are there only small numbers in W Ireland and why have Roseate Terns not become established in Orkney, must be left to future research.

Since the *68–72 Atlas*, there have been a few minor shifts. In the Firth of Forth, Roseate Terns no longer breed on Fidra and until recently have favoured Inchmickery. They returned to The Skerries, Anglesey, in 1987 to nest for the first time since 1952. The major change, however, has been the loss of the very large colony on a sand bar in Wexford Harbour which was washed away by gales in the mid 1970s. These birds apparently moved to another Wexford site, Lady's Island Lake, where they fared badly before deserting a few years later. Lady's Island Lake has recently been recolonised by up to 76 pairs.

The Roseate Tern Distribution maps for both Atlases are remarkably similar considering that the population has crashed by about 80% since 1969 (*Status of Seabirds*). Now the British and Irish population numbers about 490 pairs (Walsh *et al.* 1991), which is close to a third of the size of the Wexford Harbour colony in 1969. Monitoring of breeding numbers has occurred at most colonies in most years since 1969 and the *88–91 Atlas* map has not thrown up any surprises.

Since 1987 the main British and Irish colony has been Rockabill, Co. Dublin, where up to 332 pairs have nested. Rockabill is wardened by the Irish Wildlife Service and the Irish Wildbird Conservancy. Recent studies there have shown that breeding success has been high, more than one chick being fledged per pair (Walsh *et al.* 1991), and the continued success of this colony will have a major influence on the prospects of the Roseate Tern remaining as an Irish and British breeding species (Avery 1987).

Outside Britain and Ireland, numbers of Roseate Terns have also fallen in Brittany, France, where about 100 pairs bred in 1990, compared with 600 pairs in the late 1960s (Henry and Monnat 1981). Recent surveys

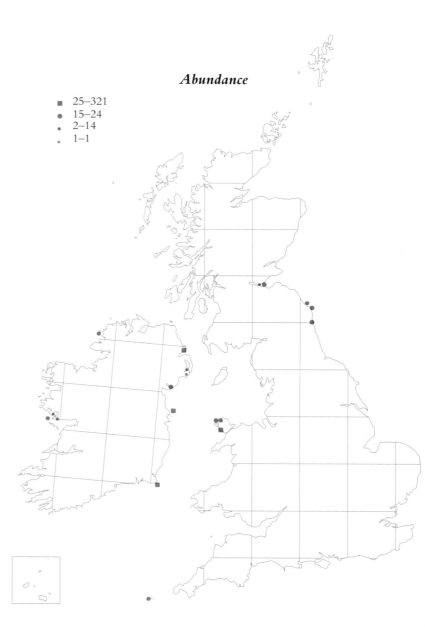

Abundance

- ■ 25–321
- ● 15–24
- ● 2–14
- · 1–1

(E.K. Dunn and A. del Nevo) have shown that the 1,000 pairs in the Azores, Portugal, now comprise about two-thirds of the European population.

Because the decline in numbers has affected most colonies in Britain, Ireland and France, it seems very likely that these apparently separate declines have a common cause. One possible threat to the whole breeding population is human persecution on the wintering grounds (Avery and del Nevo 1991).

The Roseate Tern was the only British breeding species to qualify as a Red Data bird on all four criteria; rare breeder, localised breeder, severe decline and international importance (*Red Data Birds*). The current British population of fewer than 90 pairs would not now qualify on the grounds of international importance. The Irish population is currently just over 400 pairs (Walsh *et al.* 1991).

MARK AVERY

68–72 Atlas p 224

Change

Distribution

Years	Breeding evidence		
	Br	Ir	Both (+CI)
1968–72	24	22	46
1988–91	19	11	30
% change	−17.4	−50.0	−33.3

Common Tern

Sterna hirundo

The sight of Common Terns noisily defending their colony is a welcome addition to any gravel pit or reservoir. Because they breed at inland sites they are probably the most familiar of the five terns regularly breeding in Britain and Ireland, but, despite their name, they are not the most numerous of the five, being outnumbered by Arctic and Sandwich Terns (*Status of Seabirds*).

Most of the population breeds in widely dispersed colonies along the coast of Britain and Ireland, usually on bare shingle expanses but also on rocky islets. There are many colonies in the Inner Hebrides which frequently change location, possibly in response to predation. In the period 1986–90 nine colonies have each held more than 300 pairs. These have been widely distributed between Northern Ireland, W Scotland, NE and S England. From 1969 to 1982 over 1,000 pairs bred at Coquet Island in Northumberland but numbers decreased to 650 pairs in 1990. In 1987 and 1989 the largest colony, of over 700 pairs, was in SW Scotland. Over 300 pairs bred at Leith Docks in Edinburgh in 1987, illustrating their tolerance of disturbance. As with the Little Tern, the coasts of SW England and mid S Wales are sparsely populated, possibly due to a shortage of undisturbed beaches. We are accustomed to seeing Common Terns nesting on flat ground, yet in the Azores they nest on steep cliffs and ledges (E.K. Dunn). If only they were to adopt this habit in Britain and Ireland their range could change substantially.

In common with other terns, there have been relatively few changes in the distribution of coastal colonies, although the size of individual colonies has varied, and there have been notable losses from some areas (e.g. SW and NE Scotland). In Ireland, several coastal colonies have been lost from around Wexford and Co. Kerry since the *68–72 Atlas*, and Whilde (1985) suggested that there were signs of a general shift in distribution towards the N of Ireland. Perhaps the same is happening in Britain, for between 1969–70 and 1985–87 numbers in England decreased while the Scottish population increased (*Status of Seabirds*).

Most inland colonies are in central and E England, E Scotland and central Ireland. Formerly, in England, Common Terns nested on gravel bars in river channels as they still do in parts of Scotland. Recent river management has largely destroyed this habitat in England and instead they nest on islands and banks in gravel pits. In central Ireland there are several small colonies amongst the scattered loughs.

The largest changes have been inland. In Ireland there were approximately 60 inland 10-km squares where breeding was probable or confirmed in 1968–72; by 1984–91 this had halved. Several colonies were flooded at

Lough Derg (Shannon valley) due to engineering works at the Ardnacrusha Power Plant but a new colony of 35 pairs has been established on a man-made island (Reynolds 1990). The number of inland colonies has increased in SE and N W England and around the Tees. Common Terns have colonised many new gravel pits along riverine floodplains and have even started nesting in the London area. Numbers there have increased steadily since the first pair bred in 1963 at Cheshunt in the Lee valley (Hertfordshire), from over 10 pairs by 1970 to around 200 pairs in 1990 (Roberts 1991).

Although numbers of Common Terns in Britain and Ireland have declined this century (*Status of Seabirds*), their ability to occupy man-made sites has probably prevented a greater decline due to habitat loss at inland sites. Purpose-built rafts and islands have been colonised at many sites, both inland and on the coast. On the rafts and platforms at Breydon Water (Norfolk) numbers have increased from 17 pairs in the first year, 1977, to 129 pairs in 1990. Numbers built up during the first seven years after rafts were installed at Rye Meads (Hertfordshire) but have remained stable at 36–42 pairs for the past 12 years (Roberts 1991). Although Common Terns will rapidly colonise new sites, they may still be limited by food, and to be successful require easy access to gravel pits and rivers.

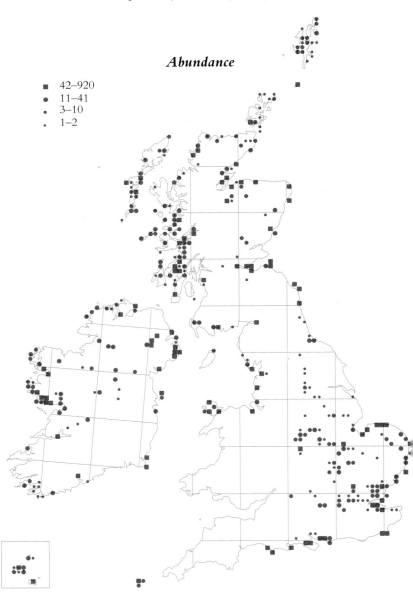

Abundance

- ■ 42–920
- ● 11–41
- • 3–10
- · 1–2

Between 1984 and 1987 the populations of Common Terns were estimated at 12,900 pairs in Britain, and 3,100 in Ireland. The Scottish population had increased by 1,900 pairs since 1969–70 but numbers had decreased elsewhere in Britain, as in Ireland. Britain and Ireland hold 3–6% of the world population of Common Terns (*Status of Seabirds*).

JANE SEARS

68–72 Atlas p 220

Change

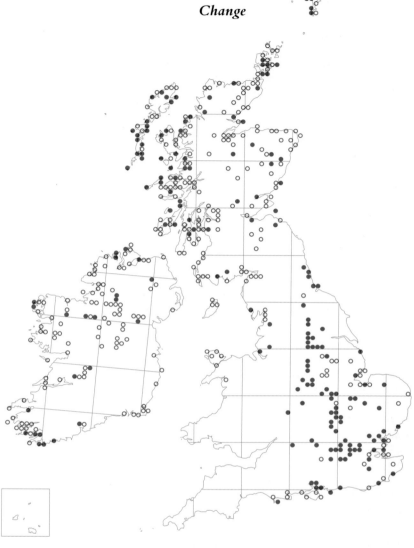

Distribution

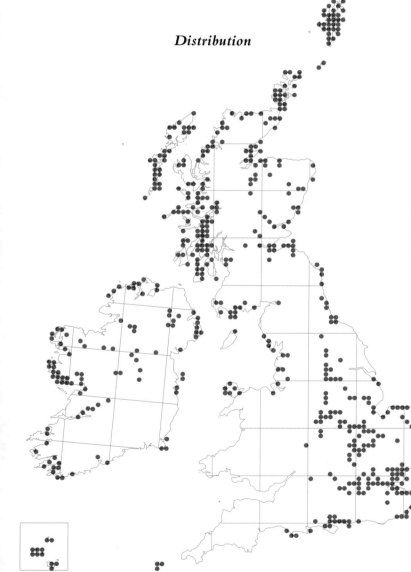

Years	Breeding evidence		
	Br	Ir	Both (+CI)
1968–72	458	173	636
1988–91	426	98	535
% change	−7.0	−43.4	−16.9

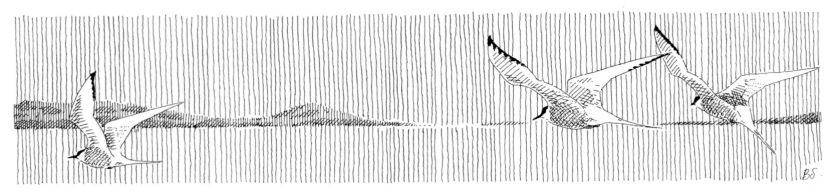

Arctic Tern

Sterna paradisaea

Arctic Terns nest in colonies which they defend ferociously by diving on intruders. Anyone who has been rapped on the head by an Arctic Tern knows that this is an unforgettable experience.

The Arctic Tern's distribution has remained fairly constant since the *68–72 Atlas*. The species is widely distributed throughout Shetland, Orkney, the Western Isles, Inner Hebrides and along the W coast of Scotland. Anglesey is normally the southernmost breeding locality on the W coast of Britain. On the British E coast regular colonies occur as far south as Coquet Island, Northumberland, while occasional pairs nest in colonies of other terns in Norfolk and along the S coast of England. In Ireland, the species is found patchily all around the coast.

Unlike Common Terns, few Arctic Terns nest far inland and their distribution is much more northerly. The Arctic Tern's virtual absence from East Anglia is difficult to explain. The existence of many large tern colonies must surely be indicative of the availability of suitable nesting habitat and of a plentiful food supply including sandeels. Might it be that this northerly species finds S England's climate too warm and dry? If that is so the next Atlas may show the adverse effects of global warming on Arctic Terns.

The stable nature of the range hides large changes in the status of the Arctic Tern since the *68–72 Atlas* (*Status of Seabirds*). Most British Arctic Terns nest in Orkney and Shetland, where over 800 colonies have been used in the past 20 years. In 1980, a complete survey of Orkney and Shetland (Bullock and Gomersall 1981) discovered that there were many more pairs of Arctic Terns than had previously been thought; about 64,900 compared to about 40,000 in 1969–70. Unfortunately, in 1990 it was discovered that numbers had fallen by 55% on Shetland and by 45% on Orkney (Walsh *et al.* 1991). Sampled Shetland colonies showed a further decline in 1991. The reason behind this drop in numbers was that since 1984 Arctic Terns on Shetland had failed to breed successfully because of a lack of sandeels on which they depended (Monaghan *et al.* 1989). The recent pattern on Shetland is for Arctic Terns to return in smaller numbers each year, lay their clutches but then desert during incubation or in the first few days of their chicks' lives. In 1990 there was considerable debate among Shetland's resident and visiting ornithologists about whether even one Arctic Tern fledged from Shetland's 500 + colonies. Breeding success on Orkney has not been closely monitored and so it is not known whether declines in productivity have occurred there as well as Shetland, but unsystematic observations have not uncovered widespread chick starvation. Perhaps numbers have declined on Orkney because of a lack of recruits from Shetland colonies.

There has been considerable concern as to the role of the Shetland sandeel fishery in the crash of sandeel stocks (Avery 1990). It is not really an argument about whether overfishing has occurred, because the data are not good enough to allow any proper decision, but rather a debate as to

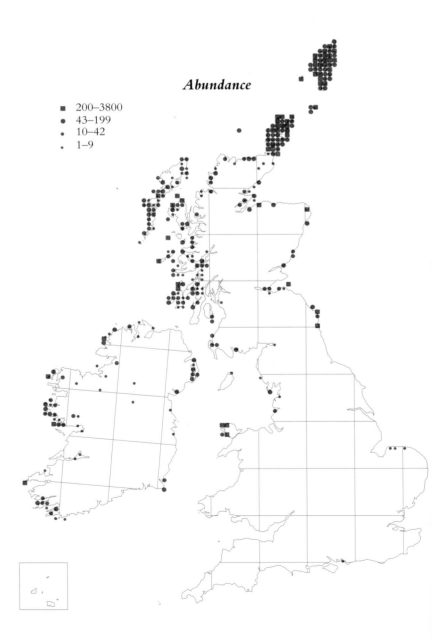

Abundance

- ■ 200–3800
- ● 43–199
- ● 10–42
- · 1–9

whether the crash in stocks of sandeels, and the associated poor breeding success of many seabirds, merits a precautionary approach to further fishing (Avery and Green 1989). No consensus has been reached on this subject but, perhaps belatedly, the Scottish Office Agriculture and Fisheries Department announced the closure of the Shetland sandeel fishery on 6 March 1991. This measure was introduced to protect the stock rather than to limit any possible environmental side-effects on Arctic Terns and other seabirds. Nevertheless, if overfishing has contributed to both the crash in sandeel numbers and the Arctic Tern breeding failures it seems that this threat has now been removed, at least temporarily. The whole Shetland sandeel debate has highlighted the fact that little is known about many of the fish on which seabirds feed; we still await the sandeel breeding atlas.

Because of its localised distribution, and the international importance of its population in Britain (*Red Data Birds*), the Arctic Tern is a species of high conservation significance. The current decline in numbers means that this species is likely to qualify for the next *Red Data Birds* as a declining species and not as a species for which Britain has populations of international importance. The Irish population numbers about 2,500 pairs,

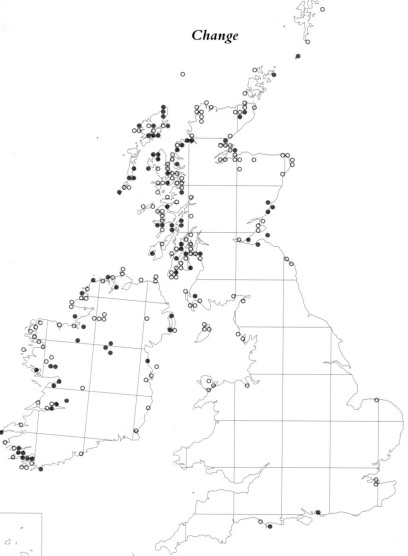

Change

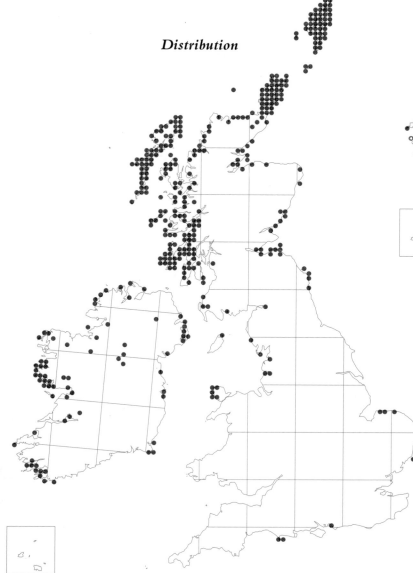

Distribution

although some of these figures date from 1983–84 (*Status of Seabirds*). The 1989 Orkney and Shetland Arctic Tern survey suggested a total population for the Northern Isles of about 30,000 pairs, which in turn suggests a British population of about 44,000 pairs. The effects of all of the past years' breeding failures have still to work their way through to the breeding population, since Arctic Terns first breed at 3–4 years of age.

MARK AVERY

68–72 Atlas p 222

Years	Breeding evidence		
	Br	Ir	Both (+CI)
1968–72	358	89	447
1988–91	303	72	375
% change	−15.7	−19.1	−16.4

Little Tern

Sterna albifrons

April marks the return of these dainty summer visitors, which breed in small but conspicuous colonies on sand and shingle beaches around much of our coastline. In Britain and Ireland the Little Tern is almost entirely a coastal breeder but elsewhere in its range it also breeds inland. Although colonies are widely distributed, Little Terns are absent or scarce in SW England, S Wales, SW and NE Ireland, Orkney and Shetland. The coast south from Lincolnshire to Hampshire holds over half the British and Irish breeding population and about half of all colonies. For a species which is very susceptible to disturbance, it is surprising that the highest concentration of colonies is in the most densely populated part of Britain.

Although Little Tern colonies tend to be small, averaging 30 pairs (J. Sears and M. Avery), several colonies have held over 100 pairs each. In recent years these have been at Blakeney Point, Holkham and Great Yarmouth in Norfolk, and at Langstone Harbour in Hampshire. Foulness in Essex was the largest colony from the late 1970s to mid 1980s, numbering over 300 pairs in three separate years.

While there have been large changes in the size of individual colonies, Little Tern distribution has remained relatively constant over the last 20 years. The most notable changes have been a decrease in the number of Welsh and Irish colonies, with further losses from Kent and Sussex. The *68–72 Atlas* recorded probable or confirmed breeding at nine colonies in Wales; in 1988 this was reduced to three colonies and by 1990 only one of these, at Gronant, was occupied. Despite this, the Welsh population has remained fairly constant over the last ten years. This concentration into fewer sites is of concern because such high densities sometimes attract increased predation (O'Briain and Farrelly 1990, *Status of Seabirds*). In 1990 the Gronant colony failed completely due to fox predation.

In 1968–72, the coast of SE Ireland between Dublin and Waterford Harbour held 16 colonies where breeding was probable or confirmed. The SCR recorded only eight colonies along the same coast in 1985–87. Most of the losses were in Co. Wexford and Co. Wicklow, where colony desertions due to human disturbance were recorded by Whilde (1985). There have also been colony losses on the W and S coasts of Ireland.

Elsewhere there have been relatively few changes in distribution. In England, the Ravenglass colony disappeared when the site became overgrown, but two small new ones were formed. As a result of protection through fencing and wardening by the RSPB, a new colony on the beach at Great Yarmouth has grown from five pairs in 1977 to 201 pairs in 1990 (Walsh *et al.* 1991).

The Little Tern is of high conservation priority in Britain because it breeds in internationally important numbers and is concentrated into relatively few sites (*Red Data Birds*). Its preference for sand and shingle beaches makes it particularly vulnerable to human disturbance, a factor which restricts the number of suitable sites. Predation and tidal flooding are other factors limiting Little Tern productivity.

During the last 20 years much effort has been expended on Little Tern protection schemes (Haddon and Knight 1983). Most of the important colonies are currently wardened and several are fenced to deter humans and ground predators. At colonies prone to tidal flooding, nests have been moved up the beach. At the largest Irish colony, in Co. Wicklow, wire exclosures were constructed around individual nests in 1989, which reduced predation by rats and foxes. Despite such intensive protection schemes, colonies can still fail. At the Wicklow colony in 1990 a Kestrel learnt to combat the exclosures, and a freak tide washed away several nests (O'Briain and Farrelly 1990).

Such protection schemes are likely to have contributed to the increase in the British and Irish population during the mid 1970s. Average productivity, however, does not appear to have increased in response to greater protection. At 87 British colonies monitored over 21 years, annual

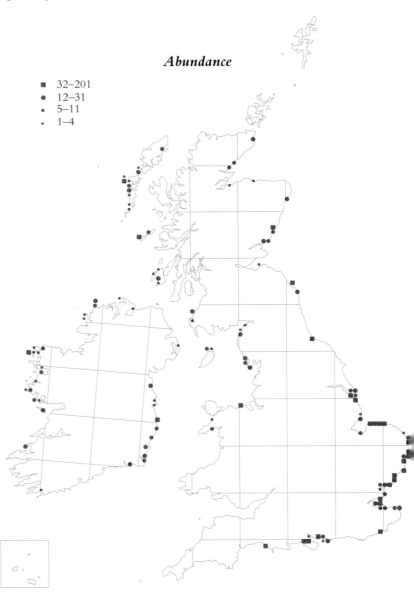

Abundance

- ■ 32–201
- ● 12–31
- • 5–11
- · 1–4

productivity was highly variable, fluctuating between 0.23 and 0.90 young per pair with an average of 0.56 (J. Sears and M. Avery). Although site protection can reduce disturbance, complete failures due to flooding and predation are not uncommon.

The SCR recorded 2,430 pairs of Little Terns in Britain, and 390 pairs in Ireland. This represents around 3% of the world population of 70,000–100,000 pairs and 15% of the European total (*Status of Seabirds*).

JANE SEARS

One record has been moved by up to two 10-km squares on all three maps.

68–72 Atlas p 226

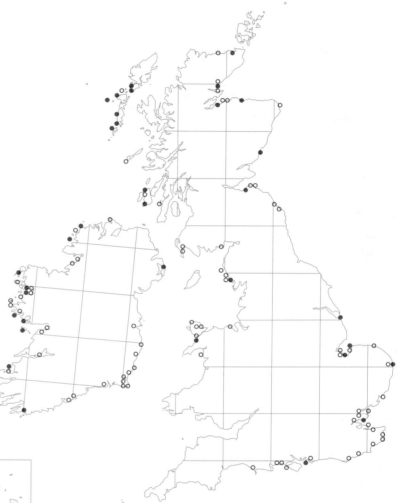

Change

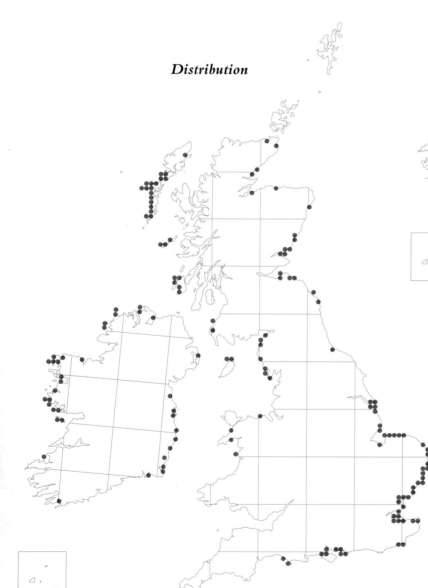

Distribution

Years	Breeding evidence		
	Br	Ir	Both (+CI)
1968–72	135	52	187
1988–91	110	36	146
% change	−18.5	−30.8	−21.9

Guillemot

Uria aalge

The constant streams of adults arriving at and departing from sheer cliffs or from cascades of boulders at their feet, and the loud 'murrings' of these noisy birds, make a visit to a Guillemot colony one of the more memorable of ornithological experiences. This is a common seabird and although the largest colonies tend to be in the more isolated and wilder parts of the country, where the habitat is suitable Guillemots breed relatively close to centres of human population.

The Guillemot is a marine species, eschewing even brackish water. Its preferred nesting habitats are ledges, sheer cliffs or boulders at their bases, and the tops of isolated stacks. Britain and Ireland lack the large colonies on flat, low islands found elsewhere. *Status of Seabirds* gives median colony size in England, Wales, and Ireland as 100 birds whereas in Scotland the figure is 400–500. Most of the birds, however, breed in much larger assemblies.

Colonies are sited where suitable nesting habitats occur close (say within 50km) to good feeding areas. Guillemots feed their young on single fish, which entails much toing and froing, so birds nest as close as possible to their food. Hence, the largest colonies tend to be where there is a good feeding area close to a limited land area, as at the ends or corners of islands or island chains, e.g. Hermaness, Foula, Noss and Sumburgh Head in Shetland and islands off otherwise unsuitable coasts, as at Britain's largest colony on Handa (98,700 birds, SCR). Mainland sites are quite acceptable (Fowlsheugh in Grampian, Bempton in Humberside) as long as the cliffs give protection from ground predators.

Scotland has most of the Guillemots in Britain and Ireland and SCR recorded 28 individual colonies with more than 10,000 birds (the unit used in Guillemot surveys). In N and E Scotland Guillemots breed on most suitable cliffs, but in England the only sizeable concentrations are in the north and east, and none breed between Flamborough Head and the Isle of Wight, presumably due (partly, at least) to lack of suitable habitat.

Colonies in SW England and Wales are smaller, but Ireland has four well dispersed large colonies.

Guillemot colonies are traditional and many have been known for centuries, having been used as sources of eggs, flesh and feathers. Despite the apparent gain or loss of a few minor colonies, there has been little change in distribution in recent years.

The *68–72 Atlas*, reporting on the total population, noted 'heartening signs of stability, or even of an increase despite occasional pollution calamities'. This optimism was well founded. Counting Guillemots is difficult, and single counts give inaccurate estimates of population. The SCR surveys in 1985–87 found that the total British and Irish population had approximately doubled since the Operation Seafarer estimate in 1969–70. The population of SW Scotland increased by 380% and even those of Wales and SW England, areas about which the *68–72 Atlas* had been pessimistic, had increased by 130% and 65% respectively.

Regular counts show considerable regional differences in patterns of change, and that some populations are now declining. Numbers in the south and west increased throughout the 1970s and 1980s whereas in the north and east numbers peaked in the late 1970s or early 1980s. In the

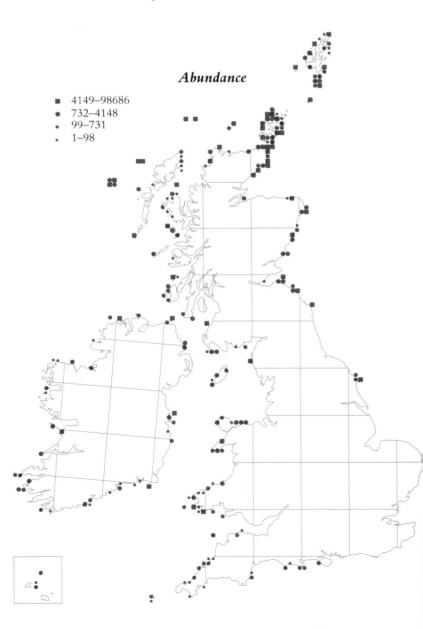

Abundance

- ■ 4149–98686
- ● 732–4148
- • 99–731
- · 1–98

North Sea, decreases started in the far north and spread towards the south, and there is a significant relationship between latitude and the rate of decrease, the numbers at northern colonies declining more rapidly (Harris 1991). Breeding success and adult survival have remained high, but there has been a marked reduction in the number of immatures returning to colonies (Harris and Wanless 1988).

Oiling, shooting, and drowning in fishing nets, are obvious causes of death: chemical poisoning, disease and starvation are less so, but could well be critical. There are large regional and temporal differences in the causes of mortality (Mead 1989) making it difficult to decide which factor(s) might have caused the widespread increases and some subsequent declines in the last 20 years. Change in food availability is a strong possibility, and there is a negative link between the numbers of sprats (an important prey) in the North Sea and the first winter mortality of Guillemots (Harris and Bailey 1992). Many more auks have been seen in the southern North Sea in recent years, suggesting a change in winter distribution which matches a southern shift of sprats (Camphuysen 1989). Only detailed population studies can hope to discover the factors influencing the numbers of these seabirds.

Change

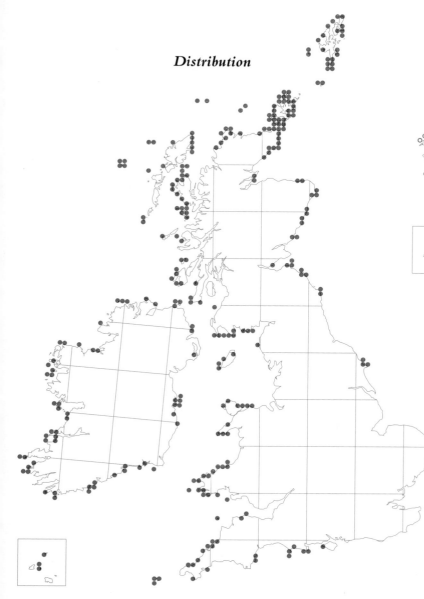

Distribution

Guillemots breed widely in the North Atlantic and North Pacific. Apart from a few hundred birds in France and Iberia, Britain is the southern limit of the species' range in the NE Atlantic. *Status of Seabirds* estimated 5–6 million birds in the Pacific and 3–5 million in the Atlantic, including 950,000 in Scotland, 63,000 in England, Isle of Man and Channel Islands, 34,000 in Wales and 153,000 in Ireland.

M. P. HARRIS

68–72 Atlas p 232

Years	Breeding evidence		
	Br	Ir	Both (+CI)
1968–72	225	74	302
1988–91	212	59	274
% change	−5.8	−20.3	−9.3

Counts in one unit are converted to the other, using correction factors (Birkhead and Nettleship 1980) but this adds yet another source of error and all population estimates must be treated with caution. *Status of Seabirds* gives a good indication of the problem of interpreting counts but there is no doubt that the total population of Britain and NE Ireland (but not the rest of Ireland) increased substantially between 1969–70 and 1985–87. There have been too few counts since to assess the current situation although numbers in Shetland may be declining (Walsh *et al.* 1991).

The Razorbill and the Guillemot have similar, and unusual, breeding strategies. Both adults incubate the single egg, brood and feed the chick which, when three weeks old and only a third grown, leaves the colony at dusk, often by jumping from a sheer cliff. The male then takes the chick to sea, maintaining contact by call, and feeds it for 2–3 months while it continues to grow and learns to feed itself.

During this time the male has its main moult and is flightless. The female remains at the colony for a few days or weeks, visits the nest site and may consort with other males. She then leaves and moults. In some

Razorbill

Alca torda

The black-and-white Razorbill cuts a dashing, if somewhat malevolent, figure strutting above its cliff colony or growling and aggressively shaking its head at any bird, however big, rash enough to approach its nest site.

This is a truly marine bird which rarely enters even sea lochs or estuaries. It breeds on cliffs and under boulders on coasts. It is less dependent on sheer cliffs than is the Guillemot and sometimes nests in rabbits' burrows, but in the entrance rather than in the dark recesses so liked by the Puffin. The bulk of the British and Irish population is in Scotland with the larger colonies, which contain over 10,000 birds (the usual census unit) being on Handa – which also has Scotland's largest Guillemot colony – Berneray and the Shiants (*Status of Seabirds*). There are three other colonies with more than 5,000 birds and Razorbills are also common in E Scotland.

In extreme NE England there is a small colony on the Farne Islands, but about three-quarters of England's Razorbills occur at Bempton/Flamborough and there are none at all between there and the Isle of Wight. This is probably due in part to lack of suitable nesting habitat. Ireland has large colonies at Horn Head and Rathlin Island in the north, and concentrations at Great Saltee and at Bull and Cow Rocks in the south, although the latter two colonies were not surveyed during 1985–91 and are thus not included on the maps. The largest Welsh colony is on Skomer. Relatively little change in distribution occurred between the 1969–70 (Operation Seafarer) and 1985–87 (SCR).

Despite its out-going appearance and behaviour the Razorbill is a difficult species to survey as many nest out of sight and pairs breeding in isolation or among thousands of Guillemots are easily overlooked. Nesting is usually at a fairly low density, except in a few boulder screes where thousands may concentrate. Estimates of colony size are based on counts of birds or occupied sites, but neither method is particularly satisfactory.

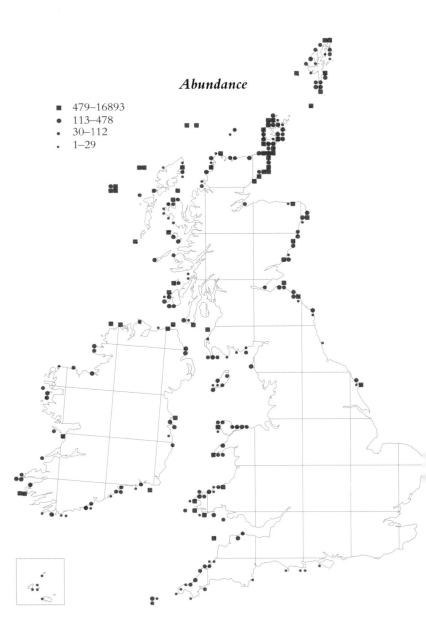

Abundance

- ■ 479–16893
- ● 113–478
- ● 30–112
- · 1–29

NE colonies many Guillemots and a few Razorbills return to the colonies in October and visit infrequently in the early morning during the winter.

We are ignorant of the factors influencing Razorbill distribution and numbers. They and their eggs are good to eat, and both were taken in quantity until the start of this century, but there are too few old accounts of Razorbill numbers to allow a detailed assessment of earlier populations. Protection has presumably allowed numbers to increase.

Little is known of the causes of Razorbill mortality except that some are oiled, shot (far fewer in the 1980s following protection in Norway in 1979) or caught in nets. Losses to the last cause may have increased in recent years (Mead 1989), but the seriousness of this threat first came to light in the 1970s when large numbers of salmon drift nets started to be set off W and S Ireland (Whilde 1979). It is unlikely to be coincidence that the only colonies in Ireland where counts were consistently higher in 1985–88 than in 1969–70 were in the east and northeast, where such nets were not commonly used (*Status of Seabirds*).

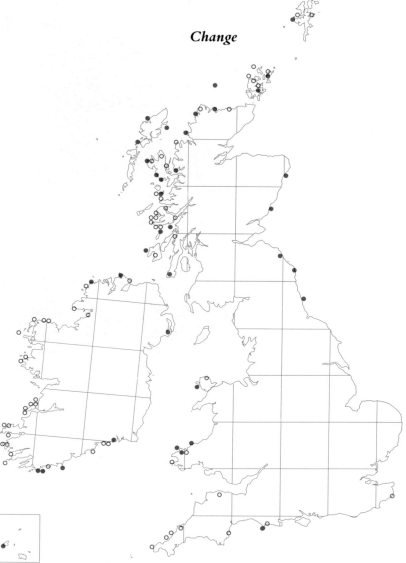

Change

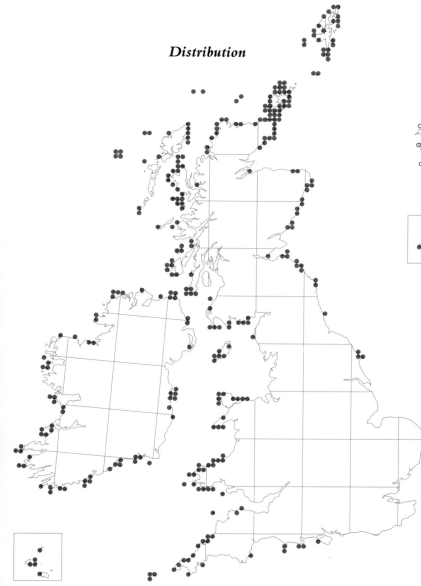

Distribution

SCR estimated 148,000 and 34,000 birds in Britain and Ireland, respectively. The world population is extremely difficult to assess due to the almost insurmountable problems of censusing the large and remote Iceland colonies. *Status of Seabirds* 'allowed' Iceland 450,000 pairs and suggested Britain and Ireland had 20% of a world population of 500,000–700,000 pairs.

M. P. HARRIS

68–72 Atlas p 230

Years	Breeding evidence		
	Br	Ir	Both (+CI)
1968–72	241	82	327
1988–91	233	63	301
% change	−3.3	−23.2	−8.0

Black Guillemot

Cepphus grylle

Away from their breeding range, Black Guillemots are among the more elusive of our seabirds. In districts where they do breed, these dainty auks are a pleasant part of the coastal scene throughout the year, and tend to frequent inshore areas away from large numbers of other seabirds.

Black Guillemots breed in largest numbers in Shetland and Orkney, with many in Caithness, on the W coast of Scotland, and on the Western Isles. Few breed in England or Wales, but the species is present all around Ireland and on the Isle of Man. Black Guillemot distribution extends to the Arctic, and the birds living in Britain and Ireland are at the southern edge of the species' range. Why its present range should end here is not known, but is presumably related to lack of safe nest sites and oceanic influences on its food supply.

Ewins and Tasker (1985) showed that Black Guillemots nested either at high density on islands free from mammalian predators (principally rats), or at lower density at cliff sites inaccessible to mammals. There are few such sites on the English and Welsh coasts of the Irish Sea, possibly explaining the absence of Black Guillemots from most of this area. In historical times, they nested at St Abbs Head in Berwickshire, but now the most southerly breeding site in the North Sea is at Muchalls in Kincardineshire.

It is best to census Black Guillemots during the early morning in April, when they gather to display on the water off their colonies (Ewins 1985). The distribution mapped here, and the numbers of individuals reported for Britain and Northern Ireland, are based largely on such surveys made in the past ten years (*Status of Seabirds*). Recent surveys were the first to be carried out in April, and thus it is difficult to determine if there have been any changes in numbers and in detailed distribution. Surveys in Shetland have indicated that numbers of Black Guillemots there have increased recently (M. Heubeck). There have been no major changes of range since the *68–72 Atlas*.

American mink that have escaped from fur farms have seriously reduced the Black Guillemot population in parts of Scandinavia (Barrett and Vader 1984). Black Guillemots are also vulnerable to oil pollution, and a spill at the Sullom Voe oil terminal in 1978 almost wiped out the breeding colonies in that part of Shetland. Some 12 years later, the numbers in the area do not seem to have fully recovered.

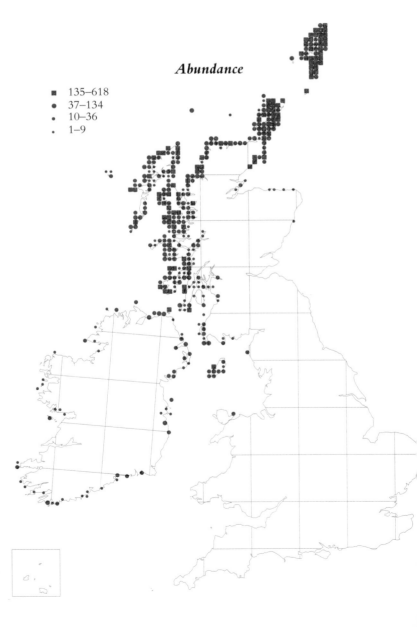

Abundance

- ■ 135–618
- ● 37–134
- • 10–36
- · 1–9

Ewins (1989) studied the breeding biology of Black Guillemots in Shetland in the early 1980s. Nest sites susceptible to predation by Crows tended to have only single-egg clutches. Feeding conditions were good in these years and adult survival, clutch size and breeding output were very high. In more recent years, sandeel stocks and food availability have been low, and breeding productivity appears to have dropped at some colonies (Walsh *et al.* 1991). Both food availability and nest site quality are likely to be important in determining individual breeding success.

All suitable parts of the Scottish coast have now been surveyed in April for Black Guillemots. The British breeding population is currently estimated to be about 37,000 individuals (update from *Status of Seabirds*). Only parts of the Irish coast have been surveyed in April, with 630 birds being found at this time. A combination of this figure with counts at other times of year suggests that the Irish population is about 3,000 birds (*Status of Seabirds*). The British and Irish total represents about 20% of the European population, but about a half of the biogeographic population (*Status of Seabirds*).

MARK L. TASKER and PAUL M. WALSH

68–72 Atlas p 234

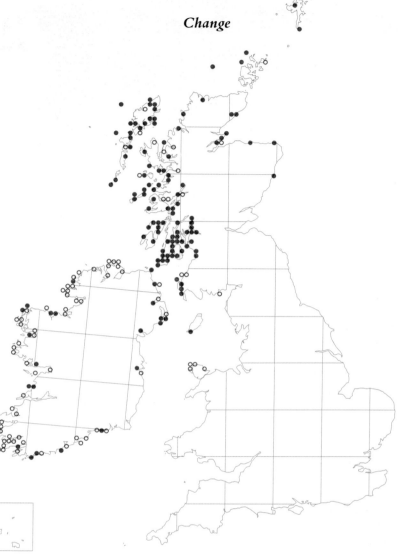

Change

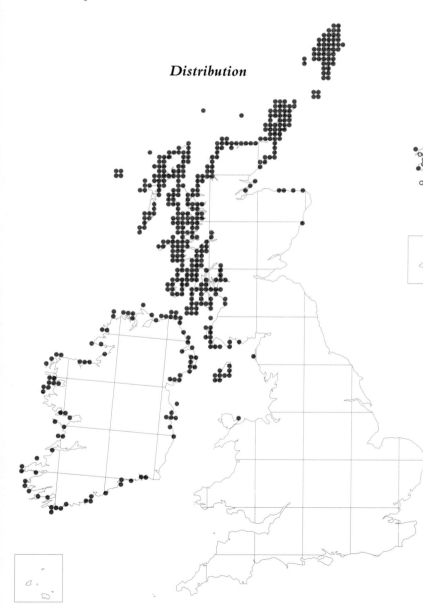

Distribution

Years	Breeding evidence		
	Br	Ir	Both (+CI)
1968–72	295	126	421
1988–91	383	90	473
% change	27.8	−28.6	10.9

Puffin

Fratercula arctica

Everyone can recognise a Puffin but it takes a visit to one of our more spectacular seabird breeding islands to see one. The Puffin is, however, not a rare bird – Britain and Ireland have more breeding Puffins than Herring Gulls, or more than the total of Cormorants and Shags combined. An aura of doom and depression surrounds the Puffin and, to general eyes, extinction threatens at least once a decade! In the early 1980s I considered that the general state of Puffindom was far better than at any time this century (Harris 1984). Can we still be that optimistic?

The Puffin is a classic seabird, spending most of the year in the open ocean and visiting coastal areas only to breed. It is very sociable and typically nests in large colonies. The nest is always underground. Even very small colonies are easy to find as birds spend a considerable time bobbing around on the sea beneath. Determining the size of a colony is not easy and estimates of populations are best based on counts of burrows. Difficulties of getting into colonies or finding burrows mean that some-times birds have to be counted but numbers at a colony vary dramatically even within a day and counts give only a rough idea of colony size. Combining counts to give total population estimates or comparing counts made by different people in different years are both difficult and prob-lematical.

This is a coldwater species and Britain and Ireland are on the southern fringe of its range. It is also a bird of the open sea and most Puffins breed in N and W Scotland: the largest colonies are on St Kilda (total of 155,000 burrows), Shiants (last counted in 1970 when there were 76,000 burrows), Foula (48,000) and Sule Skerry (47,000; figures from SCR). Slightly smaller, but still substantial, colonies occur at Hermaness, Fair Isle and the Isle of May. There are relatively few colonies on the numerous islands in the Inner and Outer Hebrides, but in Shetland, Orkney and the mainland of NE Scotland Puffins breed at many places with steep cliffs where the birds are safe from foxes, dogs, cats and inquisitive humans.

In England, the Farne Islands, Coquet Island and Bempton together have about 28,000 pairs, but elsewhere the Puffin is almost unknown as a breeding bird. Wales has concentrations on Skomer and Skokholm. The strongholds in Ireland lie in the southwest – the Blaskets, the Skelligs and Puffin Island. Small numbers are found on the Channel Islands and the Isle of Man.

The changes of distribution occurring between the surveys in 1969–70 (Operation Seafarer) and 1985–87 (SCR) are trivial. Both the historical record and recent documentation make confusing reading but it is clear

that many populations declined at various times between the end of the last century and the middle of this (Harris 1984). The report of Operation Seafarer and the *68–72 Atlas* were both pessimistic about the future, but *Status of Seabirds* detected relatively little change during the subsequent 15–20 years.

Two studies monitor changes in numbers in a statistically acceptable way. On Dun, St Kilda, the population in the main colony was estimated at 19,000 burrows in 1990 (Walsh *et al.* 1991). This compares with 29,600 in 1987 but all such estimates have wide 95% confidence intervals (e.g. ± 4,400 in 1990) and the 1990 count was still higher than the 18,000 estimated in 1978 (Harris and Rothery 1988). It is not possible to say what is happening. The situation is much clearer on the Isle of May in SE Scotland, where annual counts gave increases of almost 20% per annum during 1973–81. The rate of increase then slowed quite abruptly and there has been no increase at all during 1985–91 (Harris and Wanless 1991).

Many factors have been implicated in declines at specific colonies earlier this century – human exploitations, pollution, rats, gulls, erosion – but at many colonies there was no obvious cause. Harris (1984) argued for some unifying, underlying cause and suggested oceanic change acting through their food. Work on the Isle of May during the last 20 years detected an

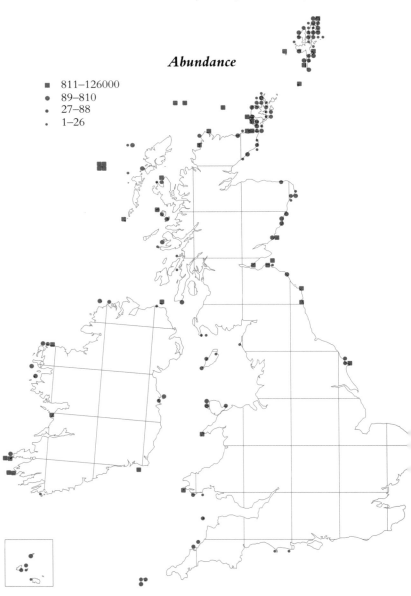

Abundance

- ■ 811–126000
- ● 89–810
- • 27–88
- · 1–26

approximate doubling of the annual mortality rate of breeding adults between the 1970s and the 1980s, which coincided with the cessation of the colony increase and, also, with a reduction in the numbers of sprats in the North Sea (Harris and Wanless 1991, Harris and Bailey 1992). Conditions away from the colony during the winter appear to be critical for the Puffin.

The Puffin breeds only in the northern North Atlantic with the largest population (guessed at 8–10 million birds) being in Iceland. *Status of Seabirds* suggests that Britain's 900,000 and Ireland's 41,000 individual Puffins might be 8% of the total world population.

M. P. HARRIS

68–72 *Atlas* p 236

Change

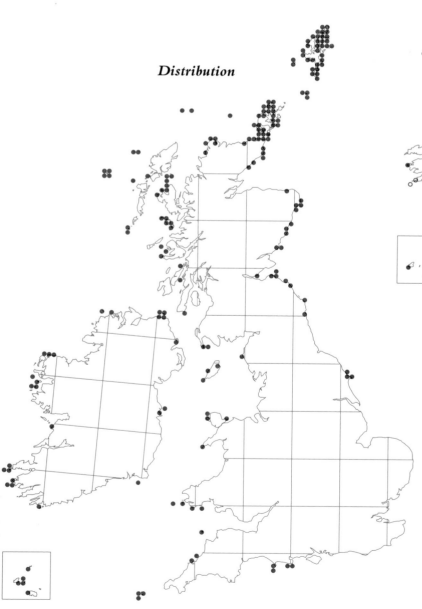

Distribution

Years	Breeding evidence		
	Br	Ir	Both (+CI)
1968–72	181	31	216
1988–91	151	25	181
% change	−17.1	−19.4	−6.7

A further change in the Rock Dove, not apparent from the Distribution maps, is that the populations both of W Scotland and Ireland continue to be diluted by birds showing Feral Pigeon characteristics (*Birds in Scotland*, *Birds in Ireland*). It is unclear whether this is due to range extension by Feral Pigeons or to homing or racing pigeons that have failed to return to their lofts.

Feral Pigeon numbers are closely attuned to the food supply (Murton *et al.* 1972, Haag 1987), and the distributional changes over the last 20 years presumably reflect changes in food availability. In towns, the expansion of fast food outlets and their associated waste is thought to have increased the birds' food supply, but the evidence is only circumstantial. In rural areas, increasing emphasis on cereal husbandry and on-farm storage of the harvested crop may have provided extra food, as may the rapid expansion in the area of oilseed rape (Inglis *et al.* 1990).

Rock Doves have lengthy breeding and moult seasons (*BWP*) and Feral Pigeons have built on this background to the extent that, in the favourable climates and year-round food supplies provided by city centres and industrial premises, at least a proportion of them breed throughout the year, raising 5–6 broods. Young birds can breed at 6–7 months old, so that

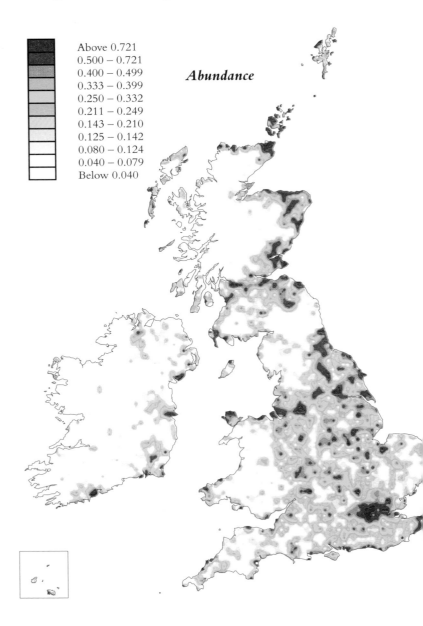

Rock Dove and Feral Pigeon

Columba livia

Through their co-existence with man in towns and cities, Feral Pigeons, though cherished and fed by many, are often regarded as a cause for complaint, and much effort and expense are devoted to moving them or to attempting to reduce their numbers. Although it is one of our most familiar birds, ornithologically the species is surprisingly poorly known. Its ancestor, the Rock Dove, is primarily a bird of coastal cliffs of the N and W of Britain, but hybridization between the two forms casts doubt on the genetic integrity of Rock Doves in all but the most isolated colonies.

Presumed Rock Doves inhabit coastal areas of N and W Scotland, and the offshore isles, and the coast of N, W and S Ireland (*68–72 Atlas*), whereas Feral Pigeons inhabit much of lowland Britain, but are absent from higher ground and from most of central Ireland. Particular concentrations occur in the major conurbations of SE England, W Midlands and the industrial N and NE of England.

Comparison of the early 1970s distribution with that of 1988–91 shows striking changes. The species has disappeared from many parts of Ireland. In the coastal N and W of Ireland this probably involves Rock Doves, while Feral Pigeons have similarly vacated many parts of the centre and north. In England, on the other hand, Feral Pigeons have occupied many areas, especially in the centre, south and east of the lowlands, from which they were absent in the early 1970s, and have clearly spread into rural areas. Because Rock Doves and Feral Pigeons are largely sedentary, and the winter distribution in the early 1980s (*Winter Atlas*) was similar to that revealed by the *88–91 Atlas*, it seems that most of this change occurred in the 1970s.

potential productivity is high (Johnson and Johnson 1990). Shortage of nest sites, however, is often a limiting factor (Haag 1987) and birds thus excluded from breeding form a reservoir that militates against efforts to prevent damage by seeking to reduce numbers directly.

In attempts to reduce damage, due largely to fouling, it has been found that large numbers of Feral Pigeons have to be culled, at great expense, to achieve even a small reduction in overall numbers. Unless the culling effort is sustained, numbers quickly recover. Locally, more success in damage alleviation is achieved by excluding birds from specific areas with netting, or from building ledges by using appropriately spaced sprung wires or other devices. However, this simply moves the birds to other buildings. Although reproductive inhibitors are often advocated as humane alternatives to culling in attempts to reduce numbers, many problems have to be overcome, including selection of a suitable chemical and identification of a safe and selective baiting technique. Even then, there is considerable doubt whether chemosterilisation would produce the desired result, since immigration from untreated areas could maintain high populations. The only way to achieve a permanent reduction in numbers is to remove the birds' food supply (Feare 1990), necessitating improvements in urban hygiene and possibly changes in public attitudes to city pigeons.

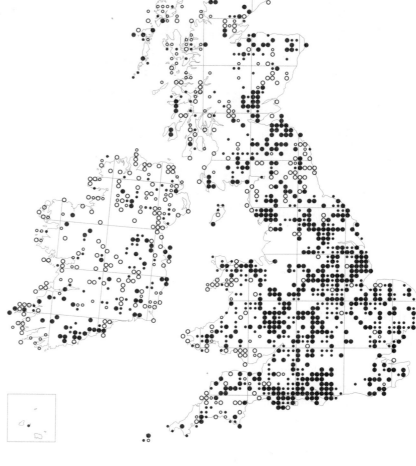

Change

Distribution

There have been no attempts to census Feral Pigeons/Rock Doves. Sharrock (*68–72 Atlas*) guessed that the British and Irish population probably exceeded 100,000 pairs. Whatever the population then or now, the present distribution suggests that there are now fewer Rock Doves but more Feral Pigeons, with a 39% increase in the number of occupied 10-km squares in Britain since 1968–72.

CHRISTOPHER J. FEARE

68–72 Atlas p 240

Years	Present, no breeding evidence		Breeding evidence		All records		
	Br	Ir	Br	Ir	Br	Ir	Both (+CI)
1968–72	322	93	1177	311	1499	404	1906
1988–91	659	135	1427	210	2086	345	2443
% change					39.2	−14.6	27.8

can be correlated with the expansion of arable farming (O'Connor and Mead 1984, O'Connor and Shrubb 1986a). Stock Doves prefer to find their food on lightly vegetated ground or in open conditions. A few decades ago, much arable land offered ideal feeding conditions, with fallows available in summer and large areas of stubbles in winter – even the crops themselves did not present the dense stands of plants now familiar to us.

Then organochlorine seed dressings were introduced and a steep decline in numbers ensued, with the centre of gravity of the British population shifting significantly to the west, away from the areas of most intensive arable farming (O'Connor and Mead 1984). The ban on these chemicals in the early 1960s led to a steady recovery over 15–20 years. The earlier peak was not quite matched, for herbicide use, annual production without fallows, and autumn ploughing, all denied the Stock Dove its year-round food supply (O'Connor and Shrubb 1986a). There is, however, evidence that the breeding season is now more spread through the year, rather than showing a June peak (O'Connor and Mead 1984).

In some areas the clearance of hedges and the loss of large trees to Dutch elm disease, may have severely reduced supplies of potential nest

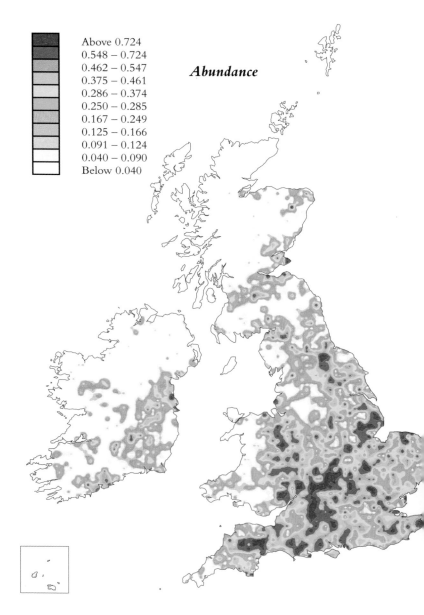

Abundance

Above 0.724
0.548 – 0.724
0.462 – 0.547
0.375 – 0.461
0.286 – 0.374
0.250 – 0.285
0.167 – 0.249
0.125 – 0.166
0.091 – 0.124
0.040 – 0.090
Below 0.040

Stock Dove

Columba oenas

This grey and black, medium-sized pigeon is found over most of Britain, save the far north, and in SE Ireland. It is primarily a bird of parkland, the forest edge and wooded farmland, where it nests mostly in holes in trees, although there are areas with healthy breeding populations and no trees. Here, buildings, though not usually inhabited ones, cliffs and even rabbit holes may harbour nests.

It is not a bird which most birdwatchers give much thought to, but its current distribution, though short of the maximum probably achieved some 40 years ago, is very much wider than it was in the mid 19th century. At first the species spread north and west from SE England. Breeding in Scotland was first recorded as late as 1866 (*Birds in Scotland*), and in Ireland in 1877 (*Birds in Ireland*). Now there are few areas in England, Wales or lowland Scotland which do not, at the very least, have a scattering of breeding pairs. In Ireland the far west still lacks birds, as does the central area surrounding Co. Cavan.

Comparison with the *68–72 Atlas* shows that, whilst the general pattern of distribution has remained the same, the range has contracted slightly in the last 20 years. The species is now absent from areas in W Cornwall, W Wales and parts of Scotland, as well as from much of Ireland north and west of a line from Belfast to Cork. This decline in range in Ireland was noticeable by the time of the *Winter Atlas*. These changes, and the history of the species in Britain and Ireland, are intimately related to the progressive changes in agricultural practice, for it is basically in the farming environment that our Stock Doves find their food.

The initial increase, starting during the 19th century and continuing until the introduction of cyclodiene-based seed dressings in about 1950, sites. Nevertheless the immediate prospects for this species, as agricultural practices change and farmers are encouraged to use less intensive techniques, are good. The species readily uses nest boxes and may take advantage of nest box schemes aimed initially at such birds as Barn and Tawny Owls and Kestrel. It is theoretically quite possible for one of these predators to raise a brood early in the year, and for a pair of Stock Doves then to raise a couple of broods in the same box.

Stock Doves occur over most of temperate Europe, and the expansion in Britain is mirrored in the Netherlands and France (*BWP*). More recently the populations in Scandinavia, Poland, Germany, Switzerland and other countries farther east have been declining. In Britain the current population levels measured by the CBC are about double those at the time of the *68–72 Atlas*, and during 1988–91 there were an estimated 240,000 Stock Dove territories in Britain and 30,000 in Ireland.

CHRIS MEAD

68–72 Atlas p 238

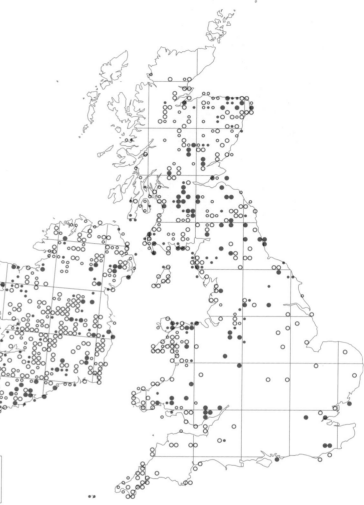

Change

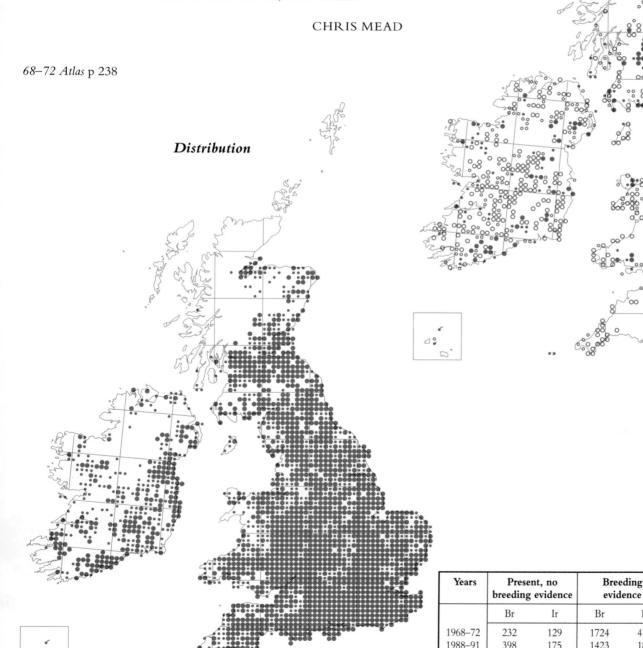

Distribution

Years	Present, no breeding evidence		Breeding evidence		All records		
	Br	Ir	Br	Ir	Br	Ir	Both (+CI)
1968–72	232	129	1724	411	1956	540	2499
1988–91	398	175	1423	188	1821	363	2190
				% change	−6.8	−32.7	−12.3

A.H

Woodpigeon

Columba palumbus

Throughout spring, male Woodpigeons may be seen in an upward-soaring flight which ends in a stall and one or more wingclaps. This is a territorial display serving to warn other males that the area is occupied. In April and May, woods and copses echo to the 'coo-coo-coo co co' territorial song of the species and this sound is often followed by a thrashing amongst the leaves, as territory holder and intruder come to blows with their wings.

Woodpigeons breed almost everywhere in Britain and Ireland except on the higher hills and mountains; and are increasingly found in the centres of our cities. Nevertheless, the highest concentrations occur in the major arable farming areas, namely central, S and E England and the central lowlands of Scotland. The Woodpigeon is economically the most import-ant bird pest in Britain for although it feeds on a wide range of natural foods (e.g. weed seeds, acorns, beech mast and a variety of berries), these are insufficient to maintain the large population throughout the winter months. Today the acreage of certain crops, in particular oilseed rape, is the major factor limiting population size (Inglis *et al.* 1990). The intro-duction of new oilseed varieties has allowed this crop to be grown further north and as a result Woodpigeon numbers in N England and Scotland have risen.

The Woodpigeon has a long breeding season and there are records of eggs in every month of the year (Murton 1965), but data from a 22-year study of the population within a wood near Carlton, in Cambridgeshire (Inglis *et al.* 1990), show that most egg and nestling production is during June to September. Analysis of the BTO's Nest Record Cards of the Woodpigeon shows that the clutch size is remarkably constant. Two eggs were laid in 94% of recorded nests, with 5% of nests having only one egg, and 1% with three. There were no significant differences in average clutch size between nests in rural, suburban or urban areas.

The main cause of breeding season mortality is predation on the eggs, and in the Carlton study area this is greater now than in the 1960s. Over the period of the research each nesting pair produced an average of 2.1 fledged young per summer. There were, however, wide fluctuations in breeding success. For example in 1966, 3.2 young per pair were produced, whilst in 1980 it was only 1.2 young per pair. Fledging success varied

little; on average 87% of the eggs which hatched resulted in fledged young. The percentage of eggs hatching was more varied, ranging from 68% in 1966 to only 36% in 1981, and with an overall average of 53%.

Corvids are the main egg predators and the increase in predation rates mirrors an increase in the Magpie population. Murton and Isaacson (1964) found that even after removing large numbers of Woodpigeon eggs there were more than enough young to replace adult losses. Later, as the graph shows, in the late 1970s and early 1980s the number of young produced per pair barely compensated for adult mortality.

In the 1950s and 1960s Woodpigeons survived the winter by feeding mainly on clover and weeds present on pastures and leys. In the late 1960s and early 1970s the areas of grass and clover declined as more land was put to cereal production. In addition, the increased use of herbicides and a switch to winter sown cereal varieties meant less fallow land. The Woodpigeon population fell markedly (Inglis *et al.* 1990) until oilseed rape was introduced in the mid 1970s. This crop filled the gap left by the clover and the population began to recover.

Extrapolation from mean CBC densities suggests a British population of some 2,100,000 territories in Britain. Murton (1965) estimated that, in Britain, in the early 1960s there were approximately 5,800,000

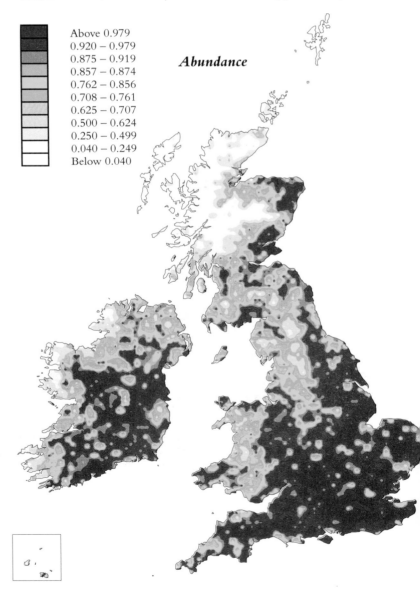

Above 0.979
0.920 – 0.979
0.875 – 0.919
0.857 – 0.874
0.762 – 0.856
0.708 – 0.761
0.625 – 0.707
0.500 – 0.624
0.250 – 0.499
0.040 – 0.249
Below 0.040

Abundance

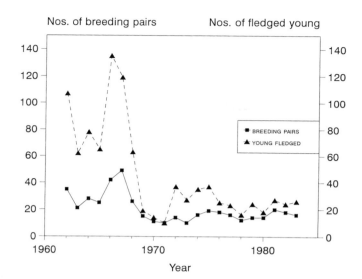

Nos. of breeding pairs   Nos. of fledged young

- ■ BREEDING PAIRS
- ▲ YOUNG FLEDGED

Numbers of nesting pairs and fledged young of Woodpigeon at the Carlton study site.

Change

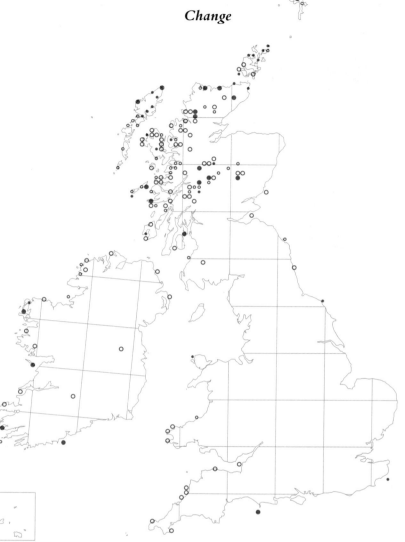

Distribution

Woodpigeons during the main breeding period. He based this estimate on data derived from regular population scans of a 1,050ha study site in Cambridgeshire which he considered typical of areas of intensive arable farming. Monitoring on this site is continuing (A.J. Isaacson and I.R. Inglis). Assuming that the changes in numbers on this site are representative, then a rough estimate for Britain would be 5,100,000 breeding Woodpigeons (or 2,550,000 pairs) in 1990. This and the CBC figure are very similar, and can be used to calculate an Irish population of 800,000 territories – 970,000 pairs.

IAN INGLIS

68–72 Atlas p 242

Years	Present, no breeding evidence		Breeding evidence		All records		
	Br	Ir	Br	Ir	Br	Ir	Both (+CI)
1968–72	91	14	2482	943	2573	957	3535
1988–91	202	131	2308	814	2510	945	3469
			% change		−2.3	−1.2	−2.0

Collared Dove

Streptopelia decaocto

The Collared Dove is a remarkable bird with an astonishing history. Watching a pair displaying in a suburban garden, it is all too easy to take these birds for granted and to forget that when *The Handbook* was published (in 1941), the nearest Collared Doves were probably to be found in Austria.

Today the Collared Dove is widespread across most of Britain but is less common in Ireland. It is noticeably absent from higher ground and moorland, for instance in the Scottish Highlands, the Lake District and Exmoor. The diet of Collared Doves in Britain consists largely of cereal grain, although they will take weed seeds, invertebrates and seed from bird tables (*BWP*). Much of the grain taken is spillage from stores such as barns and silos, and thus Collared Doves are often associated with human habitation, sometimes so closely as to become pests. They tend to avoid open expanses of countryside unbroken by farm buildings. In these respects they differ from the congeneric Turtle Dove, which feeds mainly on weed seeds and is sensitive to human disturbance. Why Ireland has not been colonised as effectively as Britain by Collared Doves is unclear; the largest numbers there are associated with the cereal growing regions of the east (*Birds in Ireland*). The Abundance map shows a bias in numbers towards the S and E of England, presumably associated with the arable production of that region.

The 1930s saw the start of an unparalleled range expansion by the Collared Dove. From Turkey and Yugoslavia they spread like wildfire across NW Europe, colonising an estimated 2.5 million km² in just 40 years (*BWP*). This meteoric spread of Collared Doves has been described in detail elsewhere (Fisher 1953, Hudson 1965, 1972, *68–72 Atlas*). Breeding was confirmed for the first time in Britain in 1955, and over the next 10 years the population increased exponentially (Hudson 1965). By 1970, although the annual rate of increase had slowed from 100% to 25%, Collared Doves bred in every county in England, Wales and most counties in Scotland and Ireland (Hudson 1972). Thus the *68–72 Atlas* was produced at about the time when the period of rapid colonisation was ending. There has been comparatively little change in distribution since that time, although slight increases are apparent in Wales, parts of Ireland, N England, S Scotland and the Great Glen. In contrast, the farmland CBC index increased tenfold between 1970 and 1982 (*Trends Guide*). Since then there has been a slight decline, possibly reflecting a growth beyond the optimal population size. CBC plots are largely confined to open countryside and do not include towns and suburban areas, where Collared Doves reach their highest densities. Thus the increasing index

might indicate an overspill from built-up areas (*Trends Guide*).

What triggered the sudden range expansion is not well understood, but the potential for high reproductive output was surely a contributory factor to the remarkable colonisation. Throughout its range the Collared Dove has a long breeding season: mid February to early October in Britain (Coombs *et al.* 1981, Robertson 1990). In continental Europe, eggs have been recorded in all months although the main breeding season is of similar duration to that in Britain, but starting and finishing about one month earlier (Robertson 1990). Clutch size is almost invariably two (means of 1.86 in an urban site (Coombs *et al.* 1981); 1.89 in a rural site (Robertson 1990)). Overall urban breeding success was found to be 26%, and the rural equivalent 41%. In the rural location, losses of eggs and chicks were attributed mainly to predation, especially early in the breeding season. Despite relatively high losses, Collared Doves can achieve high output by multiple brooding. They frequently start a new clutch while still attending to dependent fledged young, and sometimes when they still have young in the nest (Robertson 1990). One urban pair was observed to lay seven clutches in one season, although most of these failed (Coombs *et al.* 1981), whilst in the rural study one pair laid nine clutches. Robertson (1990) recorded a mean of 3.8 clutches per pair per year, with some pairs

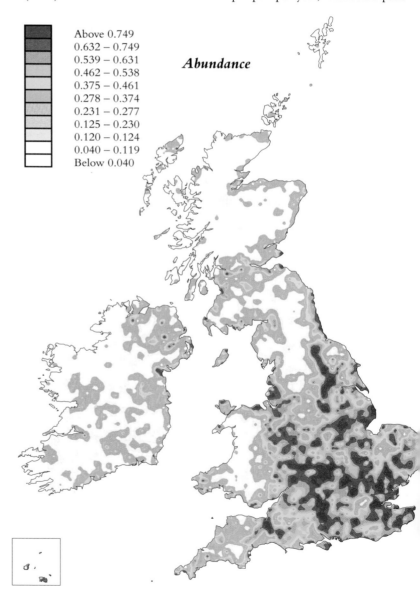

Abundance

Above 0.749
0.632 – 0.749
0.539 – 0.631
0.462 – 0.538
0.375 – 0.461
0.278 – 0.374
0.231 – 0.277
0.125 – 0.230
0.120 – 0.124
0.040 – 0.119
Below 0.040

rearing up to five broods. Overall production was 2.5 young fledged per pair per year in the urban study and 3.1 in the rural area. Annual mortality was calculated as 39% for adults and 69% for juveniles, the main cause being shooting (Coombs *et al.* 1981). When the species was still protected, prior to 1977, mortality rates were probably considerably lower and far exceeded by the number of young.

During 1988–91 there were an estimated 200,000 Collared Dove territories in Britain. By extrapolation there would have been 30,000 territories in Ireland.

ANDY EVANS

68–72 Atlas p 246

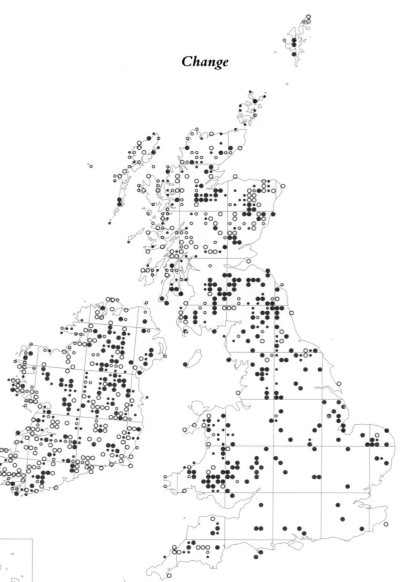

Change

Distribution

Years	Present, no breeding evidence		Breeding evidence		All records		
	Br	Ir	Br	Ir	Br	Ir	Both (+CI)
1968–72	310	110	1762	466	2072	576	2653
1988–91	384	218	1826	341	2210	559	2783
			% change		6.7	−2.8	4.6

Turtle Dove

Streptopelia turtur

In late April or early May a gentle purring call signals the return of Europe's only migratory dove from its winter quarters in Africa.

Turtle Doves are associated with agriculture and prefer warm, dry conditions. They are lowland birds, rarely breeding above 350m in continental Europe (*BWP*), and require open ground for feeding. These factors may to some extent explain why Turtle Doves are confined largely to the fertile S and E of Britain, thus avoiding higher ground and the increased rainfall of the west and north. Although there is a regular spring passage 'overshoot' to Ireland, only a handful of pairs have bred there (*Birds in Ireland*). Breeding was first recorded in SE Scotland in 1946, but there have been few records since (*Birds in Scotland*). The Abundance map emphasizes the southeast bias in the distribution, and presumably the area occupied provides more of the open, arable countryside required by the species.

It is often stated that the distribution of Turtle Doves is related to that of fumitory but this theory should be treated with caution. Whilst, at some sites, fumitory seeds undoubtedly form a large percentage of the diet (Murton *et al.* 1964) the bird is by no means dependent on this food source (*BWP*). Moreover the relationship between the distributions of the plant and the bird is not always good. In Devon, for example, fumitory was recorded in 263 tetrads and breeding Turtle Doves in 200, yet only 56 tetrads contained both species (Sitters 1988). Thus there was no relationship at the local level, but taking the county as a whole there was, however, a broad relationship. This is probably a coincidence resulting from the similar ecological requirements (light, fertile soils) of both species.

At the time of the *68–72 Atlas*, the CBC index had been steadily increasing for ten years and the population was probably at a long-term high. Since then the index has fallen by about 60% (*Trends Guide*) and a

substantial reduction in range is apparent from the Change map. The areas most affected by the decline have been Wales, W England and the SW peninsula. Reductions in range are also apparent in NE and central England and the Midlands. Why Turtle Doves should have declined so greatly is unclear. They spend at least part of the winter in the Sahel region of Africa and it is possible that they have been affected by drought, although there is no direct evidence for this (*BWP*). They are heavily persecuted on migration; over 100,000 are shot annually on Malta alone (*BWP*). Whether this level of hunting has affected the population is difficult to discover. During the breeding season the diet of Turtle Doves consists almost entirely of weed seeds, cereal grain rarely being taken (Murton *et al.* 1964). It is possible that the increasing use of herbicides over the past 20 years has reduced the availability of these seeds. Certainly the pattern of range contraction bears some similarity to that of the Grey Partridge, which has declined largely due to herbicides killing the food plants of insects (Potts 1984). In one study, Turtle Doves fed mainly on hay fields and clover leys in early summer, switching to stooked wheat and pea fields later in the year (Murton *et al.* 1964). The recent change from hay to silage and the associated earlier mowing may have reduced seed availability and thus affected the bird. Similarly, increased use of fertilizer

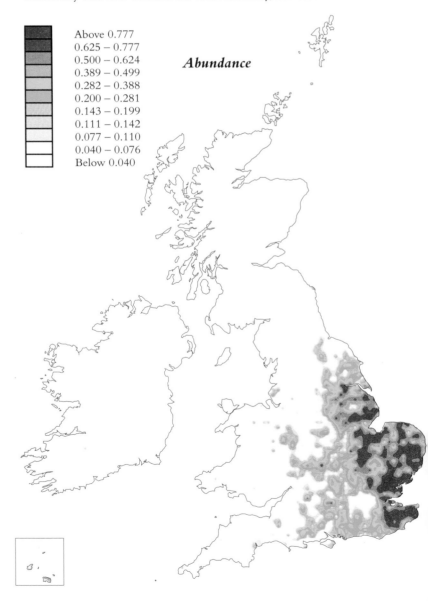

Abundance

Above 0.777
0.625 – 0.777
0.500 – 0.624
0.389 – 0.499
0.282 – 0.388
0.200 – 0.281
0.143 – 0.199
0.111 – 0.142
0.077 – 0.110
0.040 – 0.076
Below 0.040

on grassland will have promoted crop growth at the expense of weed diversity (*Trends Guide*). Further applied research is needed to understand the reasons for the decline of the species.

First clutches are laid at the beginning of May, and breeding continues to the end of August. Murton (1968) analysed the nest record cards held by the BTO. Most nests were in farmland, scrub or deciduous woodland and the majority were sited in hawthorn or elder, usually 2–2.5m above the ground. The mean clutch size was 1.9 eggs; incubation lasted 13–14 days and the nestling period was about 20 days. Breeding success (the percentage of eggs giving rise to fledglings) varied between 30% and 50% through the season. Most losses were due to predation of eggs. Pairs are normally double brooded (*The Handbook*).

Declines in population have also been recorded in Portugal, Belgium, the Netherlands, Germany and Greece (*BWP*). Recent increases have been reported from Latvia, Lithuania and Estonia.

During 1988–91 there were an estimated 75,000 Turtle Dove territories in Britain. Because of the marked decline shown by its CBC index, the Turtle Dove is a candidate species for *Red Data Birds*.

ANDY EVANS

68–72 Atlas p 244

Change

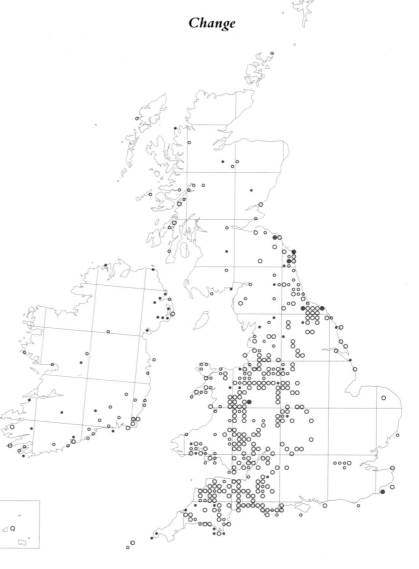

Distribution

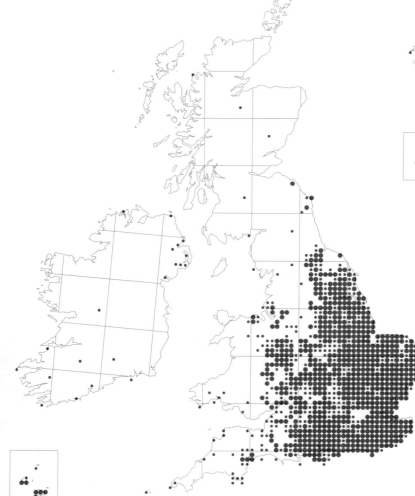

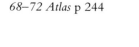

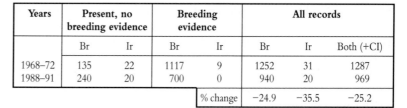

Years	Present, no breeding evidence		Breeding evidence		All records		
	Br	Ir	Br	Ir	Br	Ir	Both (+CI)
1968–72	135	22	1117	9	1252	31	1287
1988–91	240	20	700	0	940	20	969
				% change	−24.9	−35.5	−25.2

Ring-necked Parakeet

Psittacula krameri

The Ring-necked Parakeet, a native of much of the Afrotropical and Oriental regions (Forshaw 1980), seems the most incongruous of birds to find at large in the British countryside. Yet it is well established in several localities, principally in SE England.

Almost uniformly yellowish green, with a long tail and a pinkish neck ring on the male (an alternative name is the Rose-ringed Parakeet), it is a fast-flying bird which screeches when at rest and in flight, and usually occurs singly, in pairs or in small parties. For nesting, it uses tree holes in gardens, parks and orchards, where it competes with several native species, notably owls, Jackdaws and woodpeckers; and the same site may be used in subsequent years. The nest is lined with a layer of wood detritus, and sometime between January and June a clutch of 2–4 eggs is laid. Incubation, which seldom begins until the clutch is nearly complete, occupies 22–24 days, and is by the female only. The young hatch synchronously and are cared for by both parents. In the early stages, at least, they are brooded almost continuously. Fledging takes place in 40–50 days, and the young become independent of their parents a few weeks later. Hatching and fledging success in England is reported by B. Hawkes (*BWP*) to be very high.

The Distribution map suggests that since the time of the *68–72 Atlas* this species has considerably increased in numbers and range. Against this trend, sub-populations in NW England, mainly on Merseyside and in the Greater Manchester area, present as recently as the *Winter Atlas* years, have not been recorded, or have disappeared. No records were submitted to the Atlas during 1988–91, and none appeared in the Greater Manchester Bird Report during the same period.

Ring-necked Parakeets first appeared in the wild in Britain in 1969 (Hudson 1974), and were admitted to Category C of the British and Irish List in 1983. By then the total population of this generally under-recorded species was estimated to number between 500 and 1,000 birds, most of them in Kent and the Thames valley (Ferguson-Lees *et al.* 1984). With the exception of the NW region of England already discussed, records submitted subsequently to county bird reports do not show any reduction in numbers, even after cold winters. Indeed, according to the *Trends Guide* a slow increase seems to have occurred, at least around London. J.J.M. Flegg (BBC Radio 4, 9 May 1991) estimated the total British population as numbering several thousands.

Three separate sources have been suggested as the origin of the naturalised population in England; it may have come from free-flying 'homing' birds that failed to return to their aviaries; it may derive from birds which escaped from pet shops or exotic bird farms (since London is one of the principal centres of commercial aviculture), or it may come from birds released by returning sailors when they realised the expense involved in a lengthy period of quarantine. In all probability, each of these sources has contributed to the present population, the success of which is due at least partially to the species' ready acceptance of food provided in winter by suburban households (Lever 1977). B. Hawkes in the *Winter Atlas* recorded no deaths, even in the harshest weather.

In parts of its native range the Ring-necked Parakeet is a serious pest of various crops (especially fruit), so if it were ever to become widely established in, for example, Kent, it could cause serious economic damage. In Britain it appears to have no predators other than man. Elsewhere in Europe small populations are established in Belgium, Germany, and the Netherlands. The species is also established in the wild in many other countries throughout the world outside its natural range (Lever 1987).

CHRISTOPHER LEVER

68–72 Atlas p 452

Change

Distribution

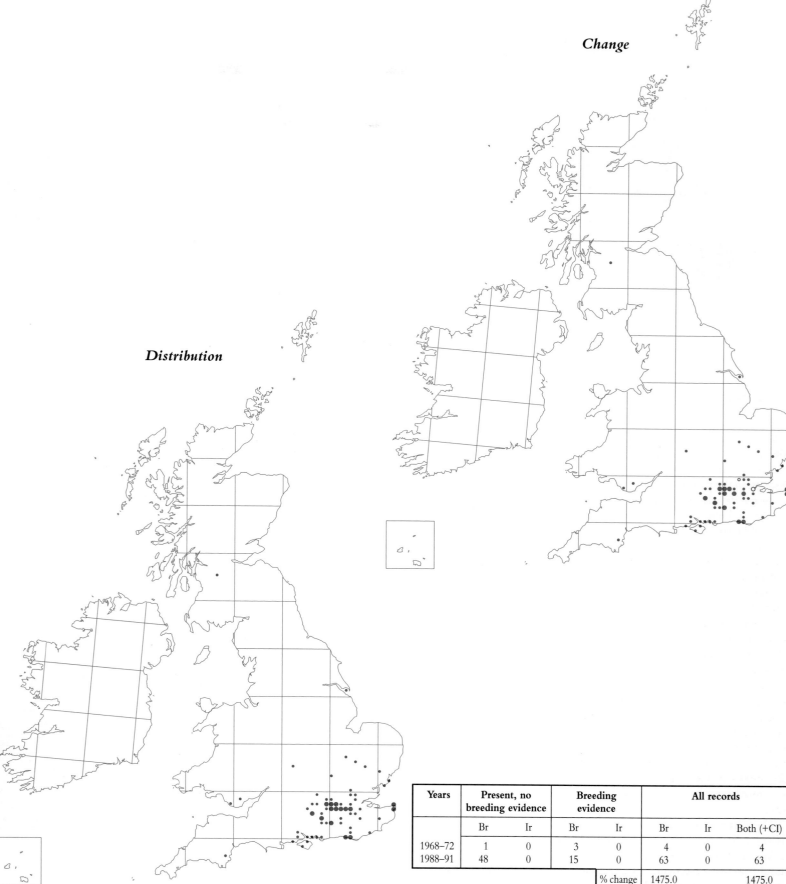

Years	Present, no breeding evidence		Breeding evidence		All records		
	Br	Ir	Br	Ir	Br	Ir	Both (+CI)
1968–72	1	0	3	0	4	0	4
1988–91	48	0	15	0	63	0	63
% change					1475.0		1475.0

Cuckoo

Cuculus canorus

Few people are unfamiliar with the Cuckoo, which announces its return from Africa in April with characteristic song. On arrival at its breeding destination the male proclaims its presence from any suitable vantage point, and some days later the unmistakable bubbling call of the more secretive female can be heard during courtship; the prelude to the Cuckoo's most unusual life-style. The male's distinctive song, heard from late April until the end of June, makes it unlikely that its presence would be missed, so the Distribution map of this species is probably as accurate as any.

Although in the extreme northwest of its European breeding range, the Cuckoo occurs commonly throughout Britain and Ireland, occupying most habitats and breeding wherever there are suitable hosts and supplies of its favourite food – hairy caterpillars. It is most common in S and central England, East Anglia and western districts of Scotland and Ireland. The Distribution map reflects the availability of its preferred food and hosts, while the Abundance map reflects their relative abundance. Thus in S and central England the main host is the very abundant Dunnock, while in East Anglia the colonial Reed Warbler can attract the attention of several Cuckoos in a small area. In northern and western districts of Britain and Ireland the main host is the Meadow Pipit, a relatively sparsely distributed species, especially on higher ground. Over 50 species of host have been recorded in Britain, and throughout Europe the list extends to over 100 (Wyllie 1981). In Britain, however, the three main hosts accounted for 80% of breeding records in 1939–71 (Glue and Morgan 1972).

The Change map shows some decline in the last 20 years, especially in Ireland where the Meadow Pipit is the principal host (which appears also to have declined). This suggests a direct link between Cuckoo distribution and that of one of its main hosts, although a similar effect could have been caused by variations in the level of coverage.

Egg laying begins in May, the resident hosts receiving Cuckoo eggs before the summer migrants. Each female is thought to belong to a host-specific gens, laying about 10–25 eggs in different nests of her preferred species. Studies of individual females have shown that they lay most of their eggs in the nests of one host species, but that occasional deviations occur (Chance 1940). Females which parasitise Meadow Pipits sometimes lay in nests of Tree Pipits, and those which use Reed Warblers occasionally use Sedge Warblers. We do not know whether a female Cuckoo selects a particular host through genetic inheritance or by a form of imprinting on the host while a chick. Experiments to examine host selection have so far failed.

The female Cuckoo spends many hours observing host nests to ensure that each egg is laid during the host's egg laying period and before incubation has began. An host egg is removed by the Cuckoo when laying its own egg, which happens on alternate days, on each occasion in a different nest, in the afternoons. Eggs are laid quickly, in as little as 10 seconds, and this reduces the host's chance of spotting the deception. Discrimination by hosts has resulted in the evolution, in some cases, of perfect egg mimicry by Cuckoos. In Britain most of their eggs are a

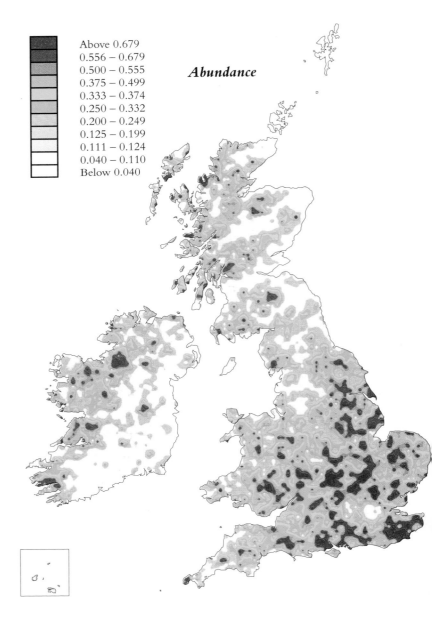

Above 0.679
0.556 – 0.679
0.500 – 0.555
0.375 – 0.499
0.333 – 0.374
0.250 – 0.332
0.200 – 0.249
0.125 – 0.199
0.111 – 0.124
0.040 – 0.110
Below 0.040

Abundance

reasonable colour match to those of Reed Warblers and Meadow Pipits (which are themselves variable), but contrast markedly with the bright blue eggs of Dunnocks. A poor degree of discrimination by the Dunnock has allowed successful exploitation by Cuckoos (Davies and Brooke 1991). There is, however, no evidence that frequent parasitism has adversely affected a population of any host species. Nest Record Cards show that the frequency of parasitism on a national level is generally below 3% of nests for any host (Glue and Morgan 1972) but may reach up to 50% at a local level, especially, for instance, with a small colony of Reed Warblers.

Elsewhere in Europe only about a dozen species are regularly victimized and in some of these the degree of egg mimicry is very high. In E Europe, for example, the Great Reed Warbler hosts Cuckoo eggs which can be identical in size, colour and pattern, whereas in Scandinavia unmarked blue eggs are found in nests of Redstarts, and in Germany eggs perfectly matching those of Robins and Garden Warblers (Baker 1942).

The decline in breeding numbers reported since the 1950s in Britain may have continued during the last two decades. Apart from a possible decrease in Meadow Pipit numbers, factors such as habitat loss with associated loss of food and host nests, and an annually fluctuating abun-

Change

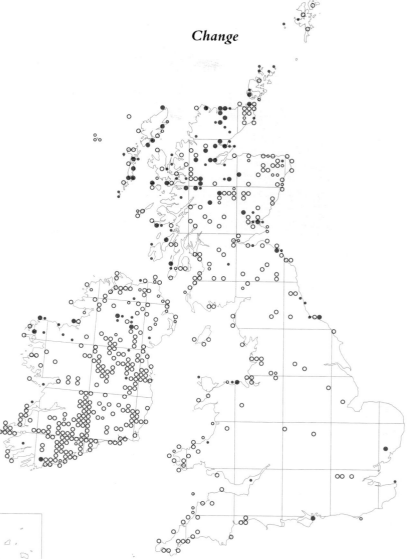

Distribution

dance of caterpillars caused by extreme weather conditions, may have contributed to the decline. Despite the reduction in number of occupied 10-km squares, there seems no reason to change the *68–72 Atlas* estimate of 5–10 pairs per occupied square, giving 1988–91 populations of 13,000–26,000 pairs in Britain and 3,000–6,000 in Ireland.

IAN WYLLIE

68–72 Atlas p 248

Years	Present, no breeding evidence		Breeding evidence		All records		
	Br	Ir	Br	Ir	Br	Ir	Both (+CI)
1968–72	108	60	2440	883	2548	943	3496
1988–91	679	413	1739	293	2418	706	3136
			% change		−4.9	−25.1	−10.3

Barn Owl

Tyto alba

Barn Owls were once a common feature of farmland and other areas of rough grazing in Britain, their ghostly white forms often seen flitting along hedgerows and ditches as twilight approached. Numbers dwindled seriously as agriculture intensified and suitable habitat was reduced, and now they are all too rare a sight. There is no doubt that they have suffered a major decline since the early part of this century. The *68–72 Atlas* drew attention to the decline that had taken place since the 1930s survey by Blaker (1934), whilst a more recent survey during 1982–85 by the Hawk Trust confirmed that this decline had continued (Shawyer 1987).

The Barn Owl is one of the more difficult of birds to survey accurately, for it is largely nocturnal, much less vocal than other owls, and its numbers fluctuate greatly between years (closely following the three year cycle of their main food supply, voles) (Taylor *et al.* 1988). This makes it difficult to compare the results from different surveys, although some general patterns are clear.

The Distribution map shows that the population is now restricted to a number of strongholds and is absent from much of the intervening land. The Hawk Trust survey found a similar result, with the bulk of the population occupying relatively few pockets of suitable habitat. There appears to be no clear pattern in the distribution of these higher density areas, with several along the W coast of Britain but also some southwards along the E coast from Yorkshire. The apparent absence of Barn Owls from much of the central part of the country may in part reflect the influence of climate: mortality rates can be high in periods of severe cold and snow (Shawyer 1987, Percival 1991). This is perhaps not unexpected for a species which has reached the northern edge of its global range in Britain.

The Change map suggests that there has been a further decline in Barn Owl distribution over the whole country since the *68–72 Atlas*. Differences in survey techniques make comparison of results questionable, and it is likely that part of this change could be due to the change in methods between the two Atlases rather than a real change in Barn Owl numbers. Furthermore, it is possible that some of the patchiness shown on the Distribution map may reflect observer knowledge, with con-

centrations in areas in which Barn Owl enthusiasts are particularly active.

Recent analyses of the BTO's Barn Owl Nest Record Cards and ringing recoveries suggest that the future for Barn Owls in Britain is less bleak than had been supposed. Both breeding success and survival rates have increased on a national basis since the mid 1970s (Percival 1991) and there is good reason to hope that this may signal the beginning of a slow process of recovery. A recent study by ITE (Newton *et al.* 1991) provides some additional information which may explain the main cause for the changes in Barn Owl numbers: levels of organochlorine residues (particularly those of dieldrin and aldrin) in Barn Owl corpses have declined while their breeding success and survival rates have increased, paralleling changes shown among other birds of prey, such as the Sparrow-hawk. This finding strongly indicates that these pesticides may have played an important role in reducing Barn Owl populations in the past, and now that their use has been restricted the population is beginning to recover. Britain is by no means alone in having suffered a decline in its Barn Owl population. Most other European countries have reported similar decreases, including France, the Netherlands and Sweden (*BWP*), and numbers have been reduced in many parts of the United States.

The first estimate (from Blaker's survey of 1932) of the total number of Barn Owls in England and Wales was 12,000 pairs. Subsequently the *68–72 Atlas* suggested 4,500 – 9,000 pairs in the whole of Britain and Ireland and the Hawk Trust survey located 4,400 pairs in Britain, 600–900 pairs in Ireland and 40 pairs in the Channel Islands during 1982–85. Though all these surveys give an approximate idea of numbers, none, including the *88–91 Atlas*, has been able to tackle the problems of annual fluctuations associated with cycles in vole populations, which makes it very difficult to produce a more accurate estimate. There is a clear and urgent need to establish a thorough and integrated monitoring scheme for this most attractive and evocative species.

STEVE PERCIVAL

68–72 Atlas p 250

Change

Distribution

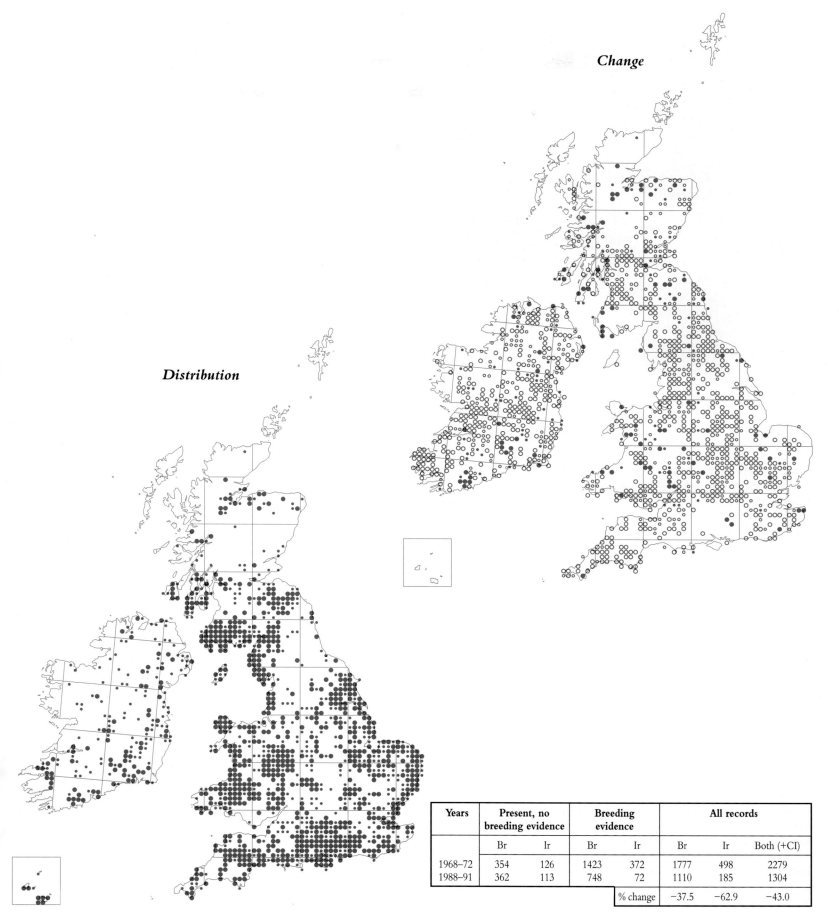

Years	Present, no breeding evidence		Breeding evidence		All records		
	Br	Ir	Br	Ir	Br	Ir	Both (+CI)
1968–72	354	126	1423	372	1777	498	2279
1988–91	362	113	748	72	1110	185	1304
% change					−37.5	−62.9	−43.0

Little Owl

Athene noctua

The long legged, plump form of the Little Owl is a regular sight over much of lowland Britain today. Single birds, less often pairs, perch daily in a prominent position in open country, frequently betraying the presence of a nest site. Others may be detected, when courtship reaches a peak in March and April, as the male delivers a loud territory-proclaiming 'hoo-, hoo-, hoo' and the female yelps and shrieks in reply.

The Little Owl is not native to Britain. Birds were released in England as early as 1842, in Yorkshire, and extensively during the 1870s, and by 1900 regular breeding was established in Bedfordshire, Northamptonshire, Rutland and Kent. From these, and later centres, there followed a population explosion during 1910–30 (detailed in *The Handbook*). Thereafter, the population spread more slowly, with local decreases. Some of these declines, especially in southeast counties, were attributed to pesticide poisoning from 1955 into the early 1960s, but by then the Little Owl had bred in all the historic counties of England and Wales (*Trends Guide*).

The Distribution map reflects the Little Owl's affinity for open country, well endowed with hedges, copses and orchards (Glue and Scott 1980). Breeding was confirmed widely, if unevenly, over most of lowland England and thinly into Scotland.

Fewer Little Owls breed on higher ground, where winter survival may be a problem (Glue 1973): hence their current absence from much of Cornwall and, for example, the Cambrian Mountains and Lake District. Of the BTO's nest records, 85% of nests lie below 400 feet (122m), just a few pairs nesting on upland moor fringes and treeless hill pasture at 1,000 feet (305m) in Lancashire and Yorkshire.

In Scotland, where breeding was first proved in 1958 at Edrom, Berwickshire, breeding is confined to southern counties. There have been just four records of Little Owls in Ireland, from 1903 to the latest in Wicklow in December 1981, but breeding has never been confirmed.

The Change map shows there are now fewer Little Owls over much of the intensively cultivated parts of East Anglia, Lincolnshire, the E Midlands and SW England. These apparent losses are partly offset by a strengthening of the population in much of the Welsh borders and N England.

The *68–72 Atlas* map depicted a population having recovered from low levels after losses during severe winters of 1961/62 and 1962/63. During the inter-Atlas years the CBC shows the population apparently fluctuating in a 3–5 year cycle. This may well reflect the relative abundance of small rodent prey, shown to be a key component of the Little Owl's diet when they are abundant (Hibbert-Ware 1937–38). Birds are an important food source only in the breeding season, and then most commonly Starlings, House Sparrows and thrushes: rarely gamebird chicks or poultry. This helped to clear the Little Owl's tainted name and reduce persecution, so that today it is actively encouraged by most farmers.

Little Owls most commonly nest in hollow deciduous trees, showing a strong local association with orchards and pollarded timber (Glue and Scott 1980). They will nest harmoniously alongside Barn Owls and Kestrels, but compete directly with, and may be killed by, Tawny Owls (Mikkola 1976). The Little Owl can, however, exploit many more lower

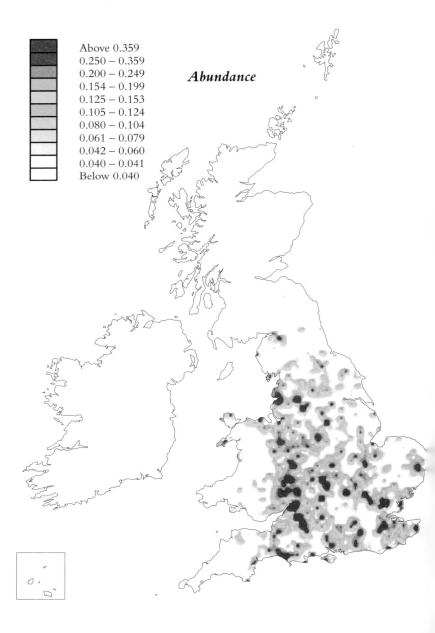

Above 0.359
0.250 – 0.359
0.200 – 0.249
0.154 – 0.199
0.125 – 0.153
0.105 – 0.124
0.080 – 0.104
0.061 – 0.079
0.042 – 0.060
0.040 – 0.041
Below 0.040

Abundance

cracks and cavities in smaller trees, ruins and farm outbuildings, and even the burrows of rabbits. The continued loss of suitable hedgerow and streamside trees, and of orchards, plus the decay and demolition of long-frequented farm buildings, may have combined to reduce nest site availability, especially in arable eastern parts.

Territorial behaviour may start as early as January and is in earnest by March. The usual clutch of 3–4 eggs (average 3.6) is laid almost exclusively in April and May. Incubation takes four weeks, and young leave the nest after 30–35 days. Although they may disperse within 4–5 weeks, broods of 1–3 (average 2.4) may be seen with parents into September (Haverschmidt 1946, Glue and Scott 1980). Replacement clutches are infrequent and second broods rare. Pair bond and site tenacity appear strong. Breeding is not attempted in some years, instances attributed to cold late springs (e.g. 1991), poor food supplies, and nest site disturbance. British ringed birds show a random dispersal of immatures during their first winter, with 93% moving less than 50km, and most settling within 10km of their birthplace.

The CBC population index for the two Atlas periods shows broadly comparable levels. The index increased in the late 1970s, after a succession of mild winters, and fell in the colder ones of 1981/82 and 1984/85. The

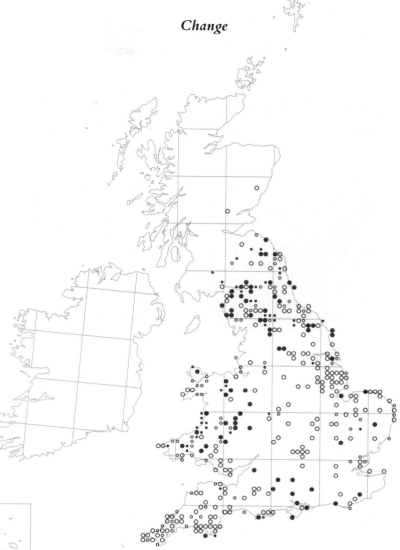

Change

Distribution

CBC regional density rankings (calculated for 1988) confirm highest levels in western regions (*Trends Guide*), as shown by the Abundance map.

A conservative figure of 5–10 pairs per occupied 10-km square suggests a British population of 6,000–12,000 pairs. Nearby, on the Continent, declines have been recorded in Denmark, Netherlands, Belgium, France, West Germany and East Germany, chiefly attributed to habitat destruction and pesticide poisoning (*BWP*).

DAVID E. GLUE

68–72 Atlas p 254

Years	Present, no breeding evidence		Breeding evidence		All records		
	Br	Ir	Br	Ir	Br	Ir	Both (+CI)
1968–72	202	0	1179	0	1381	0	1381
1988–91	286	0	942	0	1228	0	1228
				% change	−11.0	0	−11.0

Tawny Owl

Strix aluco

The Tawny Owl, generally the most familiar of our owls, is primarily a woodland bird but is also found commonly in farmland and gardens where there are enough trees to provide secure nest and roost sites. Despite this familiarity, remarkably little is known about its national population. There have been several detailed local studies which have shown the close dependence that the species has on its food supply, mainly small mammals, for breeding success and survival, and how its populations are regulated (Southern 1970, Petty 1987). Our only source of knowledge of its overall population trends, however, is the CBC, which suggests that unlike those of many of our other raptor species such as the Sparrowhawk and the Kestrel, numbers have remained relatively stable over the last 30 years (*Trends Guide*).

Comparison of the results of the *88–91 Atlas* with those of the *68–72 Atlas* gives further support to the idea that the population has not changed greatly in recent times, although there is evidence of a decline in NE Scotland which supports the findings of Buckland *et al.* (1990). During both Atlases Tawny Owls were generally widespread throughout Britain but only sparsely distributed in the Highlands and the fenlands of E England, and absent from the Northern and Western Isles, reflecting the lack of mature woodland in these areas. The species remained absent from the Isle of Wight and the Isle of Man, probably resulting from its poor dispersive ability, which is presumably the reason for its continued absence from Ireland. As the *68–72 Atlas* was not quantitative it cannot be used to detect changes in Tawny Owl numbers, only their overall distribution. This probably explains why the increases in Tawny Owl abundance that must have taken place recently in, for example, new areas of forestry plantation (Avery and Leslie 1990), are not apparent from the results. Tawny Owls have probably simply increased in areas where there already were small numbers present during the 1968–72 survey. The new data on abundance from the timed counts will allow future changes to be monitored more effectively.

In agreement with the findings of the recent BTO Tawny Owl survey (Percival 1990), which used methods more tailored to the specific requirements of this species, the Abundance map does not suggest any clear regional pattern in Tawny Owl numbers. The Tawny Owl survey found that the amount of woodland in a 10-km square was more important in determining the number of Tawny Owls it supported than its geographical location, thus emphasizing the importance of habitat availability.

Recent research at the BTO has provided more evidence that the British Tawny Owl population has been relatively stable in recent years (Percival 1990). By analysing the BTO's Nest Record Cards and data from the Ringing Scheme, it has been shown that Tawny Owl breeding success and survival rates have been fairly constant, at least over the last 15 years, with slight increases in breeding success in the north of the country (perhaps resulting from improved feeding habitat after afforestation of new areas) and in survival rates in the southeast (perhaps caused by a reduction in pesticide residues in the environment, though the Tawny Owl seems to have been much less affected by these than other raptors such as the Sparrowhawk and the Barn Owl).

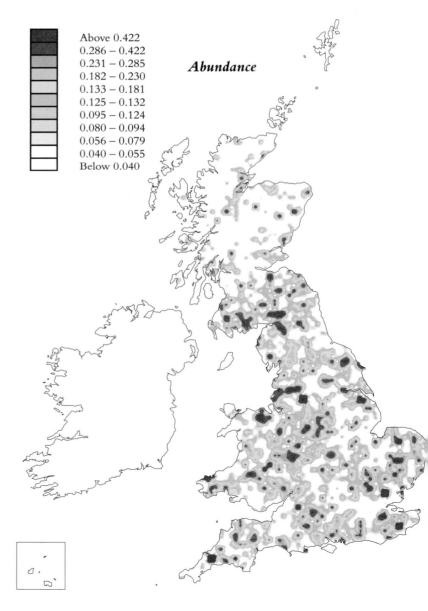

Above 0.422
0.286 – 0.422
0.231 – 0.285
0.182 – 0.230
0.133 – 0.181
0.125 – 0.132
0.095 – 0.124
0.080 – 0.094
0.056 – 0.079
0.040 – 0.055
Below 0.040

Abundance

This pattern of stability in Tawny Owl populations is common to most European countries. Some increases have been reported, particularly in Finland and the Netherlands, where range expansion has been attributed to increasing afforestation (*BWP*). Only Sweden reported a decline in the 1960s (thought to be caused by pesticides), though there has been a subsequent recovery in numbers.

Several estimates of the Tawny Owl population in Britain have been made in the past. Parslow (1973) thought there were 10,000–100,000 pairs, and the *68–72 Atlas* put the estimate at the upper end of this bracket, 50,000–100,000 pairs. From the information available from the *88–91 Atlas* and the Tawny Owl survey, it is possible to refine this estimate further. An average density of 10 pairs in each occupied 10-km square (a figure derived from the Tawny Owl survey) would give a total of more than 20,000 pairs in Britain. This is very much lower than previous estimates but the density figure was obtained from only a single year, in which small mammal populations, and hence owl numbers, were generally low. It is still, however, the best estimate available.

STEVE PERCIVAL

68–72 Atlas p 256

Change

Distribution

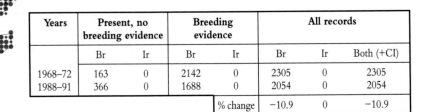

Years	Present, no breeding evidence		Breeding evidence		All records		
	Br	Ir	Br	Ir	Br	Ir	Both (+CI)
1968–72	163	0	2142	0	2305	0	2305
1988–91	366	0	1688	0	2054	0	2054
				% change	−10.9	0	−10.9

Long-eared Owl

Asio otus

Perched motionless by day close to a tree trunk in a copse or wood the Long-eared Owl is easily overlooked, but close inspection shows the bright orange-yellow eyes and oblique 'eyebrows', which often give a strangely feline appearance. Many owls are encountered by chance; some as they hunt with slow wavering flight over open country; still more through hearing the long, drawn out cooing moan 'oo-oo-oo' of the male in early spring.

The primary needs of nesting Long-eared Owls are a sturdy nest platform in the cover of scrub or woodland, and open ground nearby supporting populations of potential small mammal prey. In Britain, field voles are the chief food item and in Ireland, wood mice and common rats; everywhere fewer amphibians, beetles (Coleoptera) and roosting small birds are taken during the breeding season (Glue and Hammond 1974). These requirements are met in a wide range of habitats from coastal thickets to scrub exceeding 500m altitude. Most owls occupy isolated plantations, copses or overgrown hedges in mid altitudes, especially where surrounded by moor, rough grassland or farmland (Glue 1977a).

This ability to exploit an array of habitats is reflected in the widespread if patchy range, though an unknown proportion of pairs will have gone undetected. Breeding was recorded in most Scottish mainland counties, with concentrations in the coastal parts of northern counties. Elsewhere in Scotland, they were surprisingly scarce over much of the SW and W coast counties, with breeding recorded on few islands: Arran, Colonsay, Mull and Eigg, were occupied but it was absent from former haunts on Rum, the Outer Hebrides and Shetland. The species was scarce over much of SW Britain.

The factors restricting the Long-eared Owl population are unclear.

Shortage of preferred habitat and competition with the larger Tawny Owl probably contribute, though both owls have been shown to co-exist in Scotland where differing ages of afforestation occur (Petty 1985) and where the Tawny Owl has extended its range northwards in Britain this century. Thus there is some overlap of food and nest site requirements with the Long-eared Owl, which the Tawny Owl may physically dominate (Mikkola 1976). The scarcity of Tawny Owls on the Isle of Wight and their absence from the Isle of Man may help explain their healthy Long-eared Owl populations. In Ireland, where the Tawny Owl is absent, the Long-eared Owl is widely distributed, nesting in virtually all counties.

The Change map suggests a thinning and loss of numbers in parts of central Wales, SE England, the E Midlands and Humberside, N Solway and Grampian, and absence now from many outlying Scottish islands. There, Long-eared Owls are irregular nesters, usually on open hill ground in deep heather (*Birds in Scotland*). Similarly, the Irish population now appears to be more sparse, and notably absent from much of its former haunts in counties Kerry, Limerick, Clare and Londonderry. All maps should be interpreted with great caution, for observers spent varying times recording in darkness, especially in prime conifer woodland habitat, and differed in their knowledge of this owl's elusive nesting habits.

The afforestation programme during inter-Atlas years is likely to have benefited Long-eared Owls and some of the newly occupied squares since the *68–72 Atlas* will probably be in such areas. They are largely sedentary, with few ringed owls dispersing more than 100km from their birthplace. Site faithful pairs often occupy the same copse or wood in successive years, though invariably use a fresh nest platform. The provision of artificial stick nests and purpose built platforms has met with some success from Cambridgeshire, north to SW Scotland. Males mark territories from late January to the end of March. Incubating birds normally sit very tightly, and the roosting mate may often be found within 50m. The clutch size is usually 3–5 eggs (mean 3.9) and brood size usually 2–3 young (mean 2.4), with replacement clutches regular but second broods rare (Glue 1977a).

A study in recently afforested parts of Eskdalemuir during 1975–79 showed that the density, and proportion of pairs attempting to breed, was highest in years of vole abundance, though the percentage of hatched young which fledged was highest when vole numbers were scarce, with the implication that only 'better quality' pairs attempted to breed in poor conditions (Village 1981).

Long-eared Owls breed widely across much of Eurasia, south to NW Africa and northwards far into the boreal conifer forests. The *68–72 Atlas* provided a population estimate for Britain and Ireland of 3,000–10,000 pairs. Assuming that the contraction in range has not been accompanied by a reduction in density within occupied squares, the same calculation now would yield a population of 2,200–7,200 pairs. Timed count data show that this estimate is divided equally between Britain and Ireland with 1,100–3,600 pairs in each. On the Continent numbers are known to fluctuate with rodent populations (e.g. Sweden, USSR), while increases (e.g. Belgium, Netherlands) and probably decreases (e.g. France, Czechoslovakia, Italy) have been reported (*BWP*).

DAVID E. GLUE

68–72 Atlas p 258

Change

Distribution

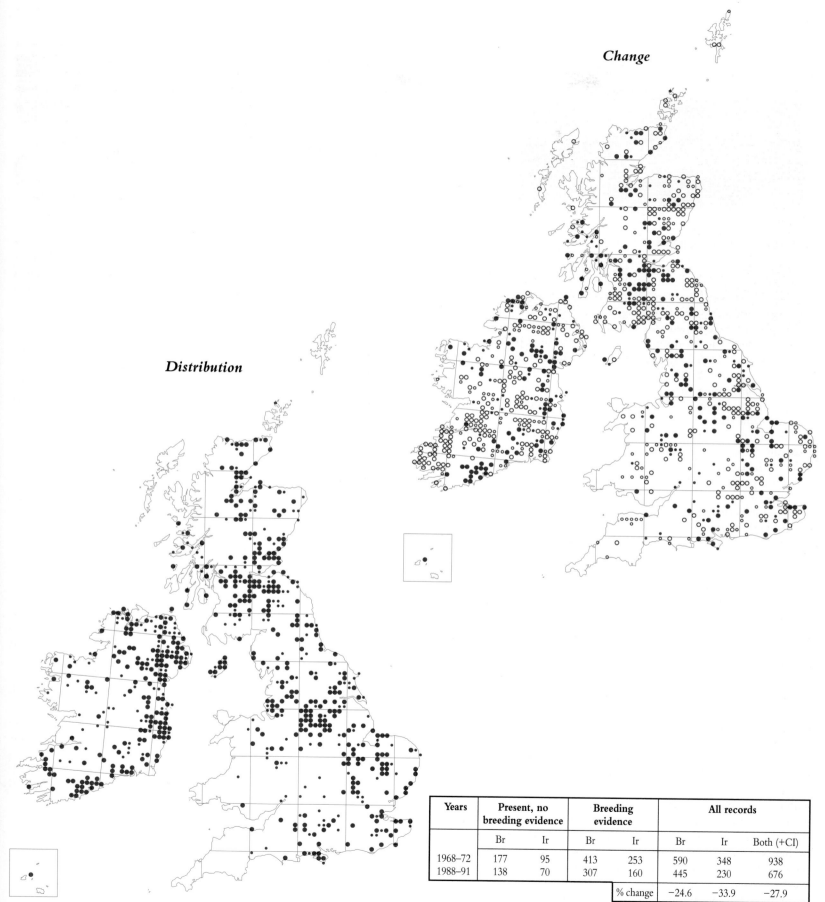

Years	Present, no breeding evidence		Breeding evidence		All records		
	Br	Ir	Br	Ir	Br	Ir	Both (+CI)
1968–72	177	95	413	253	590	348	938
1988–91	138	70	307	160	445	230	676
				% change	−24.6	−33.9	−27.9

Short-eared Owl

Asio flammeus

The sight of a graceful, buoyant Short-eared Owl hunting over a moor or heath has made many a birdwatching outing.

The primary requirements for successful nesting by Short-eared Owls are an extensive tract of open ground, a substantial population of small mammal prey, and freedom from persistent disturbance by ground predators including man. BTO Nest Records show that these needs are most often met on heather moorland, with many nest sites being found in young forestry, although a few owls nest in coastal rough grazing, marsh and sand-dune slacks.

The Distribution map should be interpreted with caution, for nests, especially those deep in heather and coarse grasses, are very difficult to locate at the egg stage when most hens sit tightly. More pairs are located when the young have hatched and males may then hover and 'bark' at intruders. When the young are a week and more old, food may be relayed by parents in daylight but, at earlier stages, first visits occur around nightfall and even during the fledging period most foraging occurs after 1700 and before 0800 GMT (Lawton Roberts and Bowman 1986, BTO Nest Records).

The Distribution map shows, for England, pairs or pockets of population in some SE counties and upland areas from the N Staffordshire moors, north to the Scottish border. The Welsh population is concentrated on moorland and afforested tracts flanking central and N Cambrian Mountains. Similar habitats are occupied in the Isle of Man, first colonised about 25 years ago.

In Scotland breeding was recorded in most mainland counties and in greatest numbers in the Southern Uplands and the foothills along the S and E fringes of the Cairngorms and Grampians. Despite recent habitat deterioration there is a healthy population on Orkney, dependent on the Orkney vole, and pairs or pockets of owls on islands of the Inner Hebrides, but an absence on Shetland, Lewis and Harris.

In Ireland there was a number of summer sightings of Short-eared Owls, but no confirmation of nesting. Breeding was first attempted, in Co. Mayo, in 1923, and was first successful in Co. Galway, in 1959. There followed an increase in sightings before pairs bred in counties Kerry and Limerick in 1977, and Co. Kerry in 1985, with the possibility of undetected pairs in other years (*Birds in Ireland*). Interestingly, the bulk of the diet of these successful pairs comprised bank voles, itself a recent colonist first noted in 1964. The absence of field voles in Ireland has been suggested as a factor contributing towards the lack of nesting; Short-eared Owls wintering there rely chiefly on common rat and wood mouse (Glue 1977b).

Although field voles almost everywhere are the main item of diet, it is not unusual for owls to specialise on alternative prey: for Britain, Glue (1977b) describes the Skomer population preying largely on Skomer vole, birds on Rum (Inner Hebrides) taking pygmy shrews, brown rat and wood mouse, others in Norfolk concentrating on brown rat.

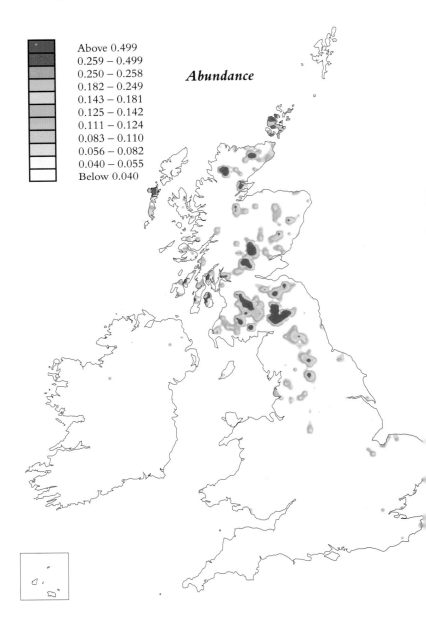

Above 0.499
0.259 – 0.499
0.250 – 0.258
0.182 – 0.249
0.143 – 0.181
0.125 – 0.142
0.111 – 0.124
0.083 – 0.110
0.056 – 0.082
0.040 – 0.055
Below 0.040

Abundance

This is a remarkably mobile hunter. Young owls disperse randomly, and often travel far, British-bred ringed birds frequently undertaking long-distance movements, including to Ireland, Spain and Malta.

Throughout Britain afforestation during the inter-Atlas years, notably the early 1970s, will have caused a redistribution of owls, with some areas now providing suitable newly afforested habitat, while forests elsewhere have matured to the pole stage and are no longer suitable. Reclamation and drainage of coastal and inland marsh have contributed to local declines.

The current distribution of nesting Short-eared Owls closely matches that of heather moorland. During a study of heathland-breeding owls in Clwyd, Lawton Roberts and Bowman (1986) found that the numbers and distribution of those feeding chiefly on small mammals with fairly stable populations (shrews and wood mice), were themselves more stable, with relatively well defined nesting places, territories, hunting ranges and fairly constant clutch sizes. In the absence of a marked peak in the cyclical populations of field voles during the 88–91 Atlas (with local exceptions in 1991), the map is likely to have included fewer specialist grassland owls. Field vole specialists exhibit distinctly unstable breeding patterns, reflecting the greater variation in their food supply. Typically, breeding numbers

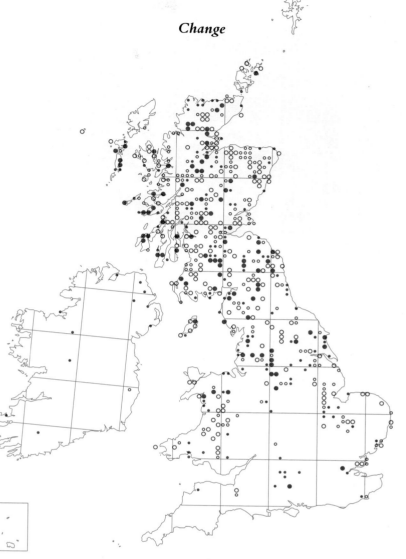

Change

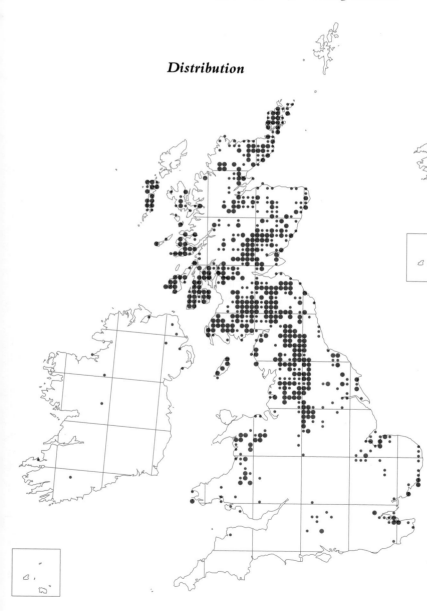

Distribution

vary greatly, site occupation is erratic, territory and hunting ranges are small but variable in relation to prey numbers, whereas clutch sizes are large for an owl of this size (Lockie 1955a, Mikkola 1983).

The 68–72 Atlas provided estimates of as few as 1,000 pairs breeding in Britain in years of poor vole abundance. With an unknown number of elusive undetected pairs, especially on moorland, but the ability to achieve at least five pairs per 10-km square of suitable habitat when food is plentiful, the British population is likely to be in the order of 1,000–3,500 pairs. Regular nesting has yet to be established in Ireland.

DAVID E. GLUE

68–72 Atlas p 260

Years	Present, no breeding evidence		Breeding evidence		All records		
	Br	Ir	Br	Ir	Br	Ir	Both (+CI)
1968–72	259	1	542	0	801	1	802
1988–91	298	11	381	0	679	11	690
	% change				−15.2	1000.0	−14.0

Nightjar

Caprimulgus europaeus

The rise and fall of the Nightjar's churring song is one of the most evocative sounds of the summer heathland night, but this strange and fascinating bird has declined dramatically over the past century (Stafford 1962, *68–72 Atlas*) – a decline that, alas, has continued between 1972 and 1991.

Although the Nightjar breeds from S England to the W coast of Scotland, it is only in S England that extensive areas are occupied, with major concentrations in the New Forest and Dorset, the Surrey/Hampshire borders, Sussex and Kent and the East Anglian Brecks and Sandlings.

The Change map emphasises this decline, with many fewer squares occupied, but (with the exception of Ireland, where Nightjars have declined from 93 10-km squares to only 11) the reduction in occupied squares has not been matched by a decrease in range. Although there are now no Nightjars in the Inner Hebrides or around the Moray Firth, the distribution of northern populations in Galloway and Arran, Northumberland and the North Yorks Moors is comparable to that in 1968–1972. In central N Wales there is a small increase. It is through the centre of the range that the decline has been most serious. The formerly widespread population of the N Midlands has been reduced to scattered pockets. That in Middlesex, Bedfordshire and Cambridgeshire has virtually disappeared and there has been a significant decline around the main population cores along the S coast of England and East Anglia.

Most of this decline had taken place before the 1981 national survey reported by Gribble (1983). Along with several special studies of the species during the 1980s, Gribble's survey gave an insight into possible reasons for the decline. Traditionally in Britain they nest on heathland and other agriculturally unimproved ground, but by 1981 over half the churring males found were on woodland sites, mainly recently felled 20th century conifer plantations.

Although most Nightjars are found at low altitudes, on light, sandy soils, the North Yorks Moors population occupied both drier podzolic soils and wetter clays (Leslie 1985). Nor was its distribution there related to altitude, but occurred over a range from 75–300m, and in N Wales the forest clearfells used by Nightjars are all above the 300m contour

(Westlake). Neither the relative stability of more northerly Nightjar populations in Britain, nor their insensitivity to altitude, suggest that climate change is the main cause of their decline.

In England significant areas of rough ground, along with their Nightjars (Gribble 1983), were lost between 1968 and 1991, but there is also evidence that surviving semi-natural heathlands may have become less suitable. Nightjars need bare ground for their nest site. They nest amongst the dead debris that follows forest harvesting but will not nest on growing vegetation. As the control of accidental fires has improved and grazing declined, heaths have changed. Experimental management at Minsmere in Suffolk involved the clearing of small glades and bare patches on *Calluna* heathland, and the removal of invading birch and Scots pine tree cover. Nightjars there increased from five pairs in 1982 to 40 in 1989 (Burgess *et al*. 1990).

The increased felling of conifer plantations has had the greatest positive effect on Nightjar populations since 1972 (Gribble 1983, Leslie 1985). In Thetford Forest, numbers have increased synchronously with felling area, from 90 pairs in 1974 to 123 in 1978, 168 in 1981 and approximately 300 in 1989. Newly felled plantations, and plantations up to five years after replanting, hold the highest densities, declining progressively up to 17 years after replanting.

Nightjars could also be affected by factors outside the nesting territory, such as availability of insect prey. In Dorset they were found to range an average of 3.1km and up to 6km from the nesting territory to feed, preferring broadleaved woodland and avoiding *Calluna* heath (Alexander and Cresswell 1990). In contrast, in Thetford forest they stayed closer to the nesting territory, which was usually felled conifer forest, but showed a preference for grass heath when they did leave their territory.

Sharrock, in the *68–72 Atlas*, was cautious about this species, qualifying the estimate of 3,000–6,000 pairs. Gribble (1983) suggested that the 1968–72 population was nearer the bottom end of this range. The 1981 survey, which had as good a coverage as could be expected, found 2,100 churring males (excluding Ireland).

The *88–91 Atlas* found 33 more squares occupied than in 1981 and some populations have increased. The 1992 RSPB/BTO Nightjar survey located over 3,000 churring males, an overall increase since 1981, and suggested that the contraction in range had slowed. The survey further highlighted the importance of forestry plantations, with over 50% of the population in this habitat. The Irish population is very small indeed and may be less than 30 pairs. Nightjars winter in Africa south of the Sahara. Decreases have been identified in most W European countries, possibly linked to habitat loss, disturbance, and a decrease in large insects due to the use of pesticides (*BWP*).

RODERICK LESLIE

68–72 Atlas p 262

Distribution

Change

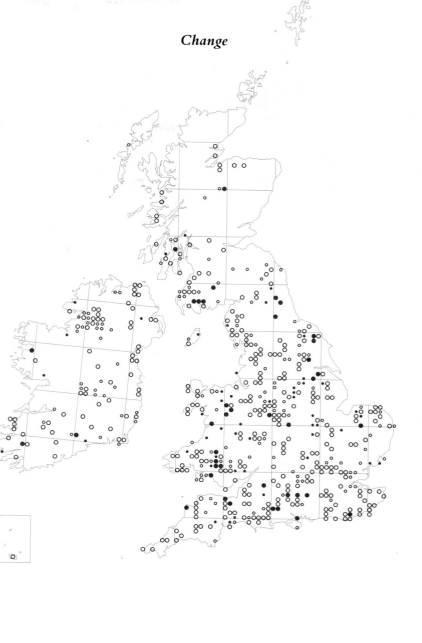

Years	Present, no breeding evidence		Breeding evidence		All records		
	Br	Ir	Br	Ir	Br	Ir	Both (+CI)
1968–72	104	31	458	62	562	93	656
1988–91	84	8	190	3	274	11	285
			% change		−51.2	−88.2	−56.6

Swift

Apus apus

The distinctive screaming of Swifts at the end of April or in early May is a far more appropriate harbinger of summer than the sight of the first Swallow. The bird's scimitar shape, scything through the sky, is very distinctive. Yet, despite our daily familiarity with the bird, it remains one of our least-known species, especially in terms of the numbers nesting in Britain and populations trends.

Swifts arrive in Britain at the end of April. Their diet is made up wholly of aerial insects and small spiders, the latter dispersing in the sky attached to their long gossamer threads. Such food is much more plentiful in warm than in wet and windy weather, and so the Swift's breeding ecology is probably more sensitive to weather conditions than that of any other British or Irish bird. They start to breed as soon as the weather allows them to catch enough food to get into breeding condition. Exceptionally, this is as early as 11 May, but usually they do not start to lay until around the 20th of the month. Early breeders tend to lay three eggs and later ones two. Eggs are usually laid at two-day intervals and incubation starts before the last egg is laid, the eggs hatching asynchronously after an incubation period of about 19 days. In a normal year the first eggs start to hatch from about 10 June. The growth rate of the young varies according to the weather. In fine weather they may grow steadily, but if poor conditions develop and the parents are unable to find food for them, they may survive several days with almost no food, becoming torpid for long periods. As a result, the fledging period varies with the weather: the young may fledge as early as five weeks after hatching in a favourable summer or take as long as seven weeks in a cool one. The fledglings are entirely independent of their parents and receive no food once they have left the nest. They leave the country as soon as they have fledged and the parents follow within a few days. Hence one can truly say of this species that the timing of its autumn migration is fixed by the weather during the breeding season.

The Distribution map shows what one would expect. This, the most aerial of all our birds, is dependent on a rich supply of insects which it can catch on the wing. It is therefore commonest in the warmer and drier S and E of Britain, and becomes progressively scarcer in the north and on the western, cooler and damper, wind-blown coasts, where its food is less abundant. The Abundance map shows this trend even more clearly, with concentrations in the east, especially in East Anglia and the London area, and declining abundance as one goes to the west and north.

Although there are old records of Swifts nesting in holes in large trees, and a few still nest in cracks in cliffs, the Swift now nests almost exclusively in buildings, preferring nest sites that are 5m or more above the ground, so giving it a good opportunity to drop from the nest and get up speed. Preferred nesting sites are under the eaves of taller, old houses, or even larger buildings. As a result, nesting records are scarce in the areas of low human habitation. In small towns comprising mainly two-storey, modern houses, the local church may provide the only suitable nesting sites.

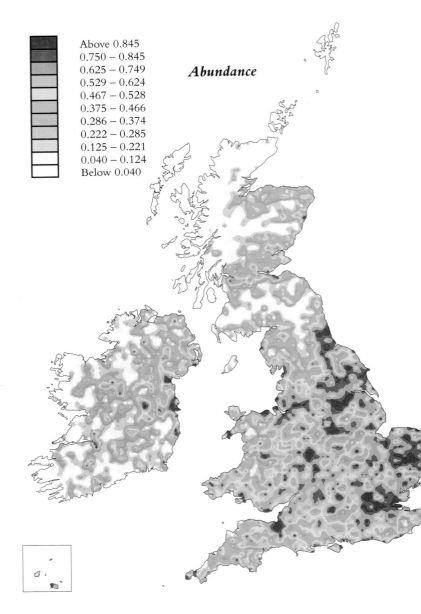

	Above 0.845
	0.750 – 0.845
	0.625 – 0.749
	0.529 – 0.624
	0.467 – 0.528
	0.375 – 0.466
	0.286 – 0.374
	0.222 – 0.285
	0.125 – 0.221
	0.040 – 0.124
	Below 0.040

Abundance

Again, the Abundance map gives some indication of this, with the largest concentrations, outside the southeast, being in urban areas.

In spite of our familiarity with the bird, we know remarkably little about its abundance, and the Abundance map, apart from confirming the low numbers in areas of low human populations, tells us little more than the Distribution map.

No one has been able to make an estimate of its numbers in any reliable, quantitative way. In some southern, lowland cities, with many suitable nesting sites in the eaves of tall houses, there may be many hundreds nesting within a 10-km square but, again, accurate estimates have not been made. In the *68–72 Atlas* it was guessed that there might be 100,000 breeding pairs in Britain and Ireland. There is no reason to believe or disbelieve this figure, nor is there any good evidence that numbers have changed significantly since then. If this figure is correct, then there will be approximately 80,000 pairs in Britain, and 20,000 in Ireland. As Swifts breed most successfully in warm, dry summers the good Atlas summers of 1989 and 1990 should have enabled them to breed well.

CHRISTOPHER PERRINS

68–72 Atlas p 264

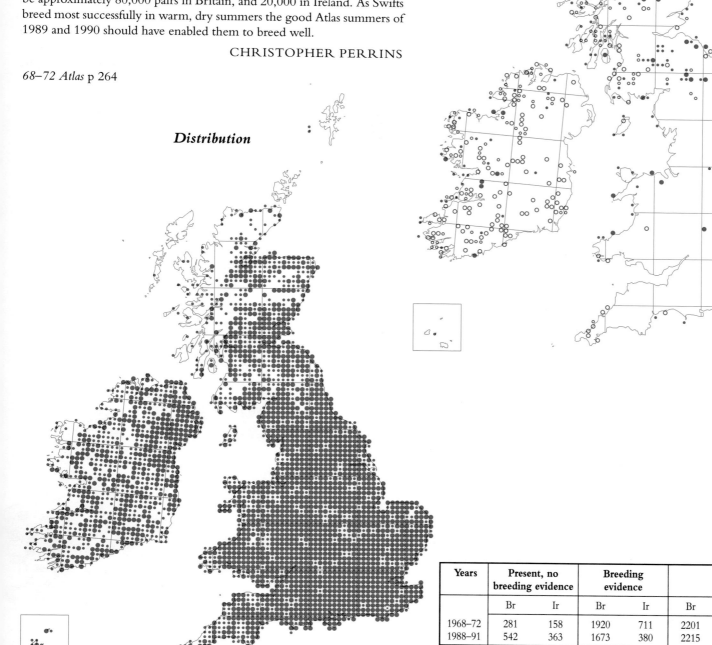

Change

Distribution

Years	Present, no breeding evidence		Breeding evidence		All records		
	Br	Ir	Br	Ir	Br	Ir	Both (+CI)
1968–72	281	158	1920	711	2201	869	3073
1988–91	542	363	1673	380	2215	743	2971
	% change				0.7	−14.4	−3.5

Kingfisher

Alcedo atthis

The electric blue flash of a passing Kingfisher brightens any birdwatcher's day. Many lowland rivers, streams, and waters hold resident birds throughout the year – but some only after a series of mild winters. The British and Irish populations of this species are the most important in the whole of N Europe. Further east, the severity of the Continental winters forces breeding birds to migrate and population densities are also generally much lower than here (*BWP*).

Kingfishers require relatively shallow and slow-moving freshwater, with thriving populations of small fish on which to feed, and vertical banks of fairly soft material where they can excavate their nesting burrows. Over much of lowland England and Wales such conditions are available and Kingfishers are present. Their prey species, however, are very susceptible to water pollution and there are, unfortunately, many individual stretches of otherwise suitable water where fish stocks have been reduced or eliminated. Industrial pollution and contamination by agricultural run-off, particularly of animal waste slurries and liquor from silage clamps, may be implicated in the lack of Kingfishers in various areas on the maps (for example in S Yorkshire).

Other blank areas, as in parts of Hampshire and Lincolnshire, are clearly caused by lack of suitable waters. In some circumstances recent management of watercourses, with meanders straightened and the streams' banks graded, has also caused the loss of Kingfishers as a breeding species. In Scotland most Kingfishers are to be found in the southern third of the country – particularly in the lowlands bordering the Solway. In Ireland there are well-scattered records, with rather fewer in the west than the east.

Although the Kingfisher is a spectacular and colourful species, with a characteristic and far-carrying call, it is surprisingly easy to overlook. It may therefore have been under-recorded in areas where the bare minimum of fieldwork was undertaken and few supplementary records submitted. Comparing the Distribution map with that from the *68–72 Atlas*, it would appear that Kingfishers are now sparser in many parts of England and Wales. In Scotland, however, the recorded distribution has consolidated and crept north. There are many more records along the Clyde, where *Birds in Scotland* reports that action by the Clyde River Purification Board in the 1970s was very effective. On the River Tay there is now a maximum of four 10-km squares with records of breeding (none previously), while in the Great Glen there are six compared with one during the *68–72 Atlas* years. Kingfishers were recorded in many fewer squares in Ireland than in

the *68–72 Atlas*, although the overall range there is relatively unchanged. The Kingfisher is a species whose numbers crash badly following severe winters. For both summer Atlas surveys this did not prove a problem as there were no exceptionally severe winters shortly before either. Furthermore, it is well adapted to make up its numbers quickly by being able to produce large broods (up to six nestlings are regularly reared) three or four times a year (*BWP*). The species has suffered direct human persecution, because of the damage individuals have been alleged to cause to fish fry. It has also been taken for its plumage, but such barbarity should be well in the past now. Direct poisoning from pesticides has been recorded but many more will surely have been lost to effects of pollution on the food supply itself.

As public awareness of the importance of water quality, and legislation to enforce good standards, are now higher on the political agenda, the prospects for this species might seem to be good. Recent dry summers have, however, led to declining water quality in even the largest water bodies and reduced or completely dried up the flow in many streams and rivers. This threatens the fishes on which Kingfishers feed and also often increases the risk of predation at nesting holes normally protected by the water beneath their vertical faces. In addition a run of cold winters could

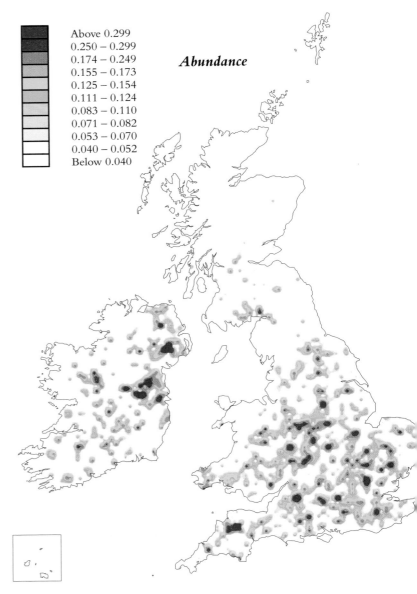

Above 0.299	
0.250 – 0.299	
0.174 – 0.249	*Abundance*
0.155 – 0.173	
0.125 – 0.154	
0.111 – 0.124	
0.083 – 0.110	
0.071 – 0.082	
0.053 – 0.070	
0.040 – 0.052	
Below 0.040	

always cause considerable mortality. Thus its future well being is by no means assured.

The overall population trend on the WBS (*Trends Guide*) has declined since indexing began in 1974, punctuated by cold-winter effects in 1979 and, particularly, 1982. Nonetheless, in many areas this is still a species frequently encountered. Following the assumption of 3–5 pairs per occupied 10-km square used in the *68–72 Atlas*, but taking into account the different relative densities in Britain and Ireland shown by the timed counts, the Kingfisher population in Britain is estimated at 3,300–5,500 pairs and that in Ireland at 1,300–2,100.

CHRIS MEAD

68–72 Atlas p 266

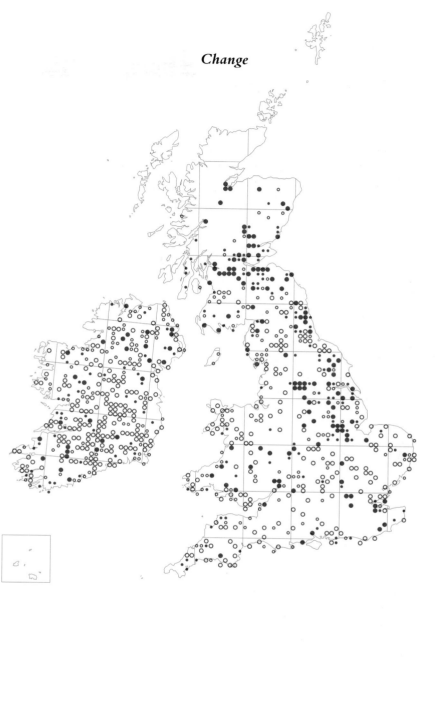

Change

Distribution

Years	Present, no breeding evidence		Breeding evidence		All records		
	Br	Ir	Br	Ir	Br	Ir	Both (+CI)
1968–72	289	128	1016	385	1305	513	1819
1988–91	408	156	816	150	1224	306	1531
				% change	−6.2	−40.4	−15.8

Wryneck

Jynx torquilla

With its plumage a subtle patchwork of fawns and browns, and its engaging habit of rotating its head and neck in a snake-like action, this little woodpecker holds a special place in the affections of those who know it. It was never a common bird this century, but it was quite widely distributed in England and Wales (Monk 1955, 1963, Peal 1968). Today it is probable that ringers manning a small number of coastal observatories see more of the species than do the army of birdwatchers active in every county. In short, it is now little more than a scarce passage migrant, while in Ireland it has always been a rare vagrant.

At the time of the *68–72 Atlas* there were high hopes that the recent proven breeding at three sites in the Scottish Highlands in 1969 (Burton *et al.* 1970) was the start of recolonisation. Through the early 1970s pairs were indeed reported breeding each year, with at least four in 1978 – their best year – but between 1981 and 1984 there was no proven breeding anywhere in Britain.

The Change map shows the enormous decline in the distribution of the species, most marked in SE England and especially in Kent (Taylor 1981a).

The reasons for the Wryneck's decline are still not fully understood. In the English strongholds large gardens, parkland and orchard-type habitat still exist and in the Scottish Highlands large areas of open pine and birch woodland remain relatively unchanged since the birds first bred there in the late 1960s. Both these habitat types still offer a reasonable supply of natural holes or holes created by other species, especially woodpeckers. Where there is a scarcity, nest boxes have been used. Pesticide use has certainly increased and in some areas may have damaged the populations

of ants, the birds' major food source. However, this is unlikely to have affected the Scottish sites. Furthermore, the decline really started from the mid 1800s, well before worries of habitat loss and pesticide use. Wetter, cooler summers do not suit ants, on which Wrynecks rely, but there is little evidence to show that such factors have operated. The Scottish population has been at its highest when good numbers of birds have been recorded in spring at Scottish bird observatories, and good breeding may be directly linked to the presence of a 'blocking high' over Scandinavia in spring, with the associated easterly winds deflecting Wrynecks to N Britain. At one site, though, a pair returned to the same nest for three seasons, indicating that their presence was not due to the vagaries of the weather.

In sympathetically managed woodland, the leaving of large dead trees *in situ* at felling times could help provide the right nesting habitat. In important breeding areas, longer rotations before major fellings, aimed at creating the right conditions for natural regeneration rather than clearfell and replanting, would help maintain insect and ant populations.

In spring Wrynecks reach Britain any time between early April in England and late May in Scotland. In the early phases after arrival they are quite noisy. They may evict other nesting species, and between bouts of mating, when the male can be seen enveloping the female with wings drooped, nest excavation continues as the hole is deepened or enlarged. Once egg laying is complete, with a clutch of from seven to ten eggs, the birds become almost silent, and very difficult to locate. If they have been located in the area about seven to ten day's earlier, this silence is usually a good sign of progress.

Outside Britain, Wrynecks breed quite widely in Europe, from Scandinavia south to Spain and Portugal, but in the western populations, especially in Denmark, the Netherlands, Belgium and France, major declines have taken place similar to those in Britain.

It seems probable that factors affecting the food chain, rather than direct loss of habitat, have played a key role in the decline. Unless they can be identified and corrected, it seems likely that by the time of the next atlas there may be little progress to report. Meanwhile, some benefit to the birds arise from not publicising the exact location of birds thought or known to be nesting.

The British population of Wrynecks is clearly very small, with only single confirmed breeding attempts in 1988 and 1989 and none in 1990, although in most years there are records of other birds in suitable breeding habitat.

British population estimates from RBBP reports and elsewhere

Yr	1850	1954	1958	1966	1979	1980	1981	1982
Est.	high	150–400	100–200	20–30	1–19	1–14	0–2	0–10
Yr	1983	1984	1985	1986	1987	1988	1989	1990
Est.	0–15	0–10	1–9	1–9	1–10	1–10	1–8	0–6

From 1979 onwards the ranges refer to confirmed pairs and maximum total pairs.

STEWART TAYLOR

68–72 Atlas p 276

Change

Distribution

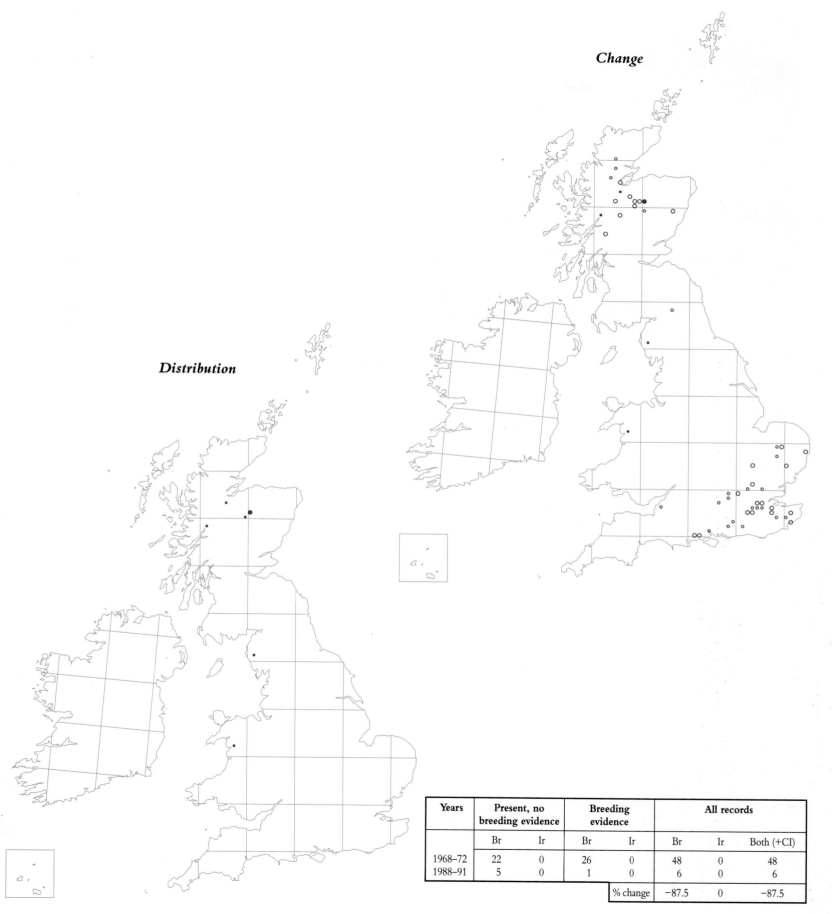

Years	Present, no breeding evidence		Breeding evidence		All records		
	Br	Ir	Br	Ir	Br	Ir	Both (+CI)
1968–72	22	0	26	0	48	0	48
1988–91	5	0	1	0	6	0	6
			% change		−87.5	0	−87.5

Green Woodpecker

Picus viridis

The largely green body plumage with yellow rump and crimson crown, combined with a clear, liquid, laughing 'yaffle' cry, make the Green Woodpecker one of our most conspicuous woodland birds. It is the largest of the three British woodpeckers and although regularly found in deciduous woodland is equally a bird of well-timbered farmland, parkland, common, and even large gardens.

The Distribution map shows that the Green Woodpecker is scarce or absent from the elevated treeless parts of Dartmoor, the Cambrian Mountains and Pennine spine, though it is better equipped to occupy upper stretches of wooded valleys than the spotted species. Few occupy a broad band of E England, including the Fens, much of which is intensively managed arable farmland with limited mature woodland.

Northern England has seen substantial fluctuations in distribution since the early 19th century, when Northumberland was temporarily the northern breeding limit before a southwards retraction of range. Northumberland was recolonised in about 1925, and the Lake District first occupied in about 1945 (Temperley and Blezzard 1951).

Scotland saw scattered records from the 1920s, then regular occurrences during the 1940s, before breeding was finally confirmed in Selkirk in 1951. Today the species nests north to Ross-shire and west to Wigtown and Dumbarton. Since the *68–72 Atlas*, breeding has been confirmed in Kinross and Kincardine (both 1973), Wigtown (1975), Aberdeen and Inverness (both 1981) and Banff (*ca* 1982) (*Birds in Scotland*). Individuals have reached certain islands – Bute (1970), Islay (1978), Mull (1978) and Coll (1982), but none was recorded there during the 1988–91 fieldwork. The Abundance map emphasises the strength of Green Woodpecker numbers in the well-wooded English southeast and Home counties extending westwards to the New Forest, in timbered land flanking the River Severn complex, the Welsh Marches and S Wales, plus scattered

concentrations in the Brecklands of East Anglia, Devon, the Lake District and Tayside.

The Change map suggests that there are now fewer Green Woodpeckers in SW England, W Wales, parts of Lincolnshire, Humberside and SW Scotland, and this may reflect changes in farmland management. But the map also demonstrates the continuing spread over parts of central and E Scotland, where policy estates provide ideal habitat. In this northwards expansion the Green Woodpecker traversed inhospitable ground, and it seems likely that Caithness, Sutherland and even the Outer Hebrides may eventually be reached.

The nest chamber is usually sited 2–6m high in the main trunk of a mature tree. Oak, ash, birch – and prior to the 1970s, elms – were preferred. Trees chosen for nesting are invariably sound externally but may have a rotten centre although this species is less reliant upon rotten wood than the spotted woodpeckers (Hågvar *et al.* 1990).

The Green Woodpecker has the most protracted breeding season of the three British woodpeckers, and may lay eggs any time between mid March and the end of June. Clutches of 5–6 eggs are normal, with generally 2–5 young reared (Glue and Boswell). Pairs are site-faithful, and immatures dis-

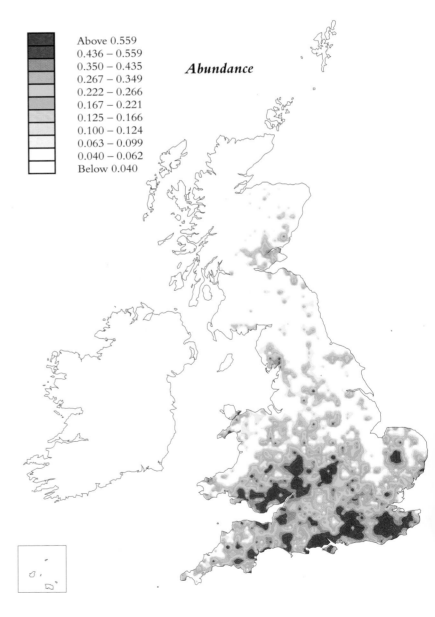

	Above 0.559
	0.436 – 0.559
	0.350 – 0.435
	0.267 – 0.349
	0.222 – 0.266
	0.167 – 0.221
	0.125 – 0.166
	0.100 – 0.124
	0.063 – 0.099
	0.040 – 0.062
	Below 0.040

Abundance

perse only short distances, few travelling more than 20km. Pairs are generally easily detected while proclaiming territory in spring, and sometimes when young give food-begging cries at the nest hole. On the other hand, drumming is rare, and later in the nesting season pairs are easily missed.

Since 1968 the Green Woodpecker CBC indices for woodland and farmland have fluctuated broadly in parallel. A general increase from 1968–75 may well have included late years in the recovery phase from the severe winter of 1962/63. This is known to have inflicted substantial local losses on Green Woodpecker populations (Dobinson and Richards 1964), for the species is very sedentary, largely ground feeding, and especially vulnerable to protracted frosts and deep snow which bar access to its soil invertebrate prey.

Following a period of stability from 1976–81, an exceptionally cold spell during the severe winter 1981/82 caused a decline in Green Woodpecker numbers, lasting four years. The loss was greatest on farmland, and might have been linked with the decline in sheep husbandry and the local demise of the rabbit, due to fresh outbreaks of myxomatosis. These herbivores help to create the short, thick turf which supports the dense ant colonies so attractive to the Green Woodpecker. Extensive ploughing, whether for

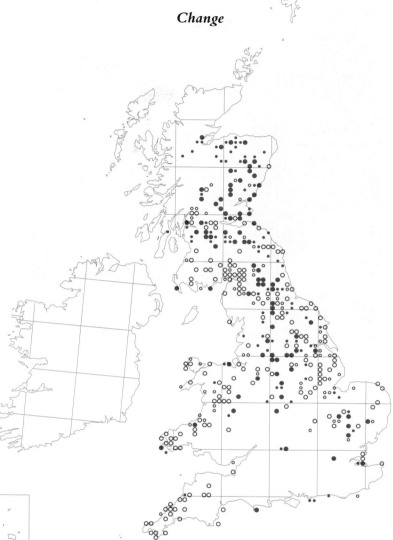

Change

Distribution

conversion to ley pasture or cereal production, would likewise effect the food supply.

A succession of unusually mild winters in the late 1980s is likely to have benefited the species, as suggested by the CBC indices. By 1988–91 the index values were broadly comparable with those of 1968–72. The *68–72 Atlas* reported a British population of 15,000–30,000 pairs, based upon an estimated 10 pairs in each occupied 10-km square, and 20 in the more favourable ones. There are, however, few such, and a subsequent revision of the CBC average densities on plots countrywide (Hudson and Marchant 1984) showed slightly less than 10 pairs per 10-km square to be a more realistic figure, giving a British population of 15,000 pairs during 1988–91.

DAVID E. GLUE

68–72 Atlas p 270

Years	Present, no breeding evidence		Breeding evidence		All records		
	Br	Ir	Br	Ir	Br	Ir	Both (+CI)
1968–72	219	0	1404	0	1623	0	1623
1988–91	425	0	1130	0	1555	0	1555
			% change	-4.1	0		-4.1

Great Spotted Woodpecker

Dendrocopos major

The Great Spotted Woodpecker is our most widespread and numerous woodpecker and is likely to be encountered anywhere where there are trees. It's sharp 'tchick..tchick..' call is often heard long before the bird is seen. Birds advertise by drumming vigorously on any suitable branch. The sound carries far and can be heard on warm days from January onwards, although most often from March to May. At the height of the breeding season the birds often perform noisy chases through the woods, terminated by displays and copulation (Blume 1977, Glutz von Blotzheim and Bauer 1980, *BWP*).

The Distribution map shows the birds to be present throughout much of England and Wales, with gaps in the uplands and the fens, where there are few trees. In Scotland, at the edge of their range, they are absent from the higher ground, substantially so from Caithness and Sutherland, and from large areas of the Central Lowlands, the northwest and the offshore islands. Although some of these areas are treeless, and therefore unlikely to support Great Spotted Woodpeckers, others would appear to have suitable stands of trees yet still lack them. No Great Spotted Woodpeckers breed in the Isle of Man or Ireland.

The species has undergone large changes of range in the past. In the 17th and 18th centuries it was said to breed as far north as Sutherland, but by the early 19th century had largely retreated from N England and totally from Scotland. It then spread north again during the late 19th and early 20th century to something like its current range (*BWP*).

The Great Spotted Woodpecker is most abundant in the well-wooded southern counties of England (Kent, Sussex and Hampshire) with other concentrations in Gloucestershire, Gwent and through the Welsh Marches. In N England and Scotland it is generally less abundant although, inexplicably, it appears to be numerous at the very edge of its range in the Great Glen.

Although the overall numbers of occupied 10-km squares were similar in this and the *68–72 Atlas*, the Change map shows apparent losses in Scotland, parts of Wales and SW England. Some of these declines may simply reflect poorer coverage, but in NE Scotland there was a significant reduction of occupied 10-km squares between 1968–72 and 1981–84 (Buckland *et al.* 1990). Over the same period the CBC indices for farmland and woodland doubled, although they have subsequently remained stable (*Trends Guide*). The CBC indices are, however, heavily biased towards sites in S and E England and are therefore unlikely to reflect accurately trends elsewhere. The increase through the 1970s was thought to be the result of Dutch elm disease (*Trends Guide*), but if this was the only factor it is difficult to explain the sustained high numbers after the dead and dying elm trees were gone.

Although the species feeds mainly on invertebrates extracted from dead or dying timber they also take a wide variety of seeds (Glutz von Blotzheim and Bauer 1980, *BWP*). During the breeding season, the young are fed on moth larvae (Smith 1987). They can be very noisy in the nest for a few days before fledging, and it is then easier to locate the nest and prove breeding; even so nests are often overlooked.

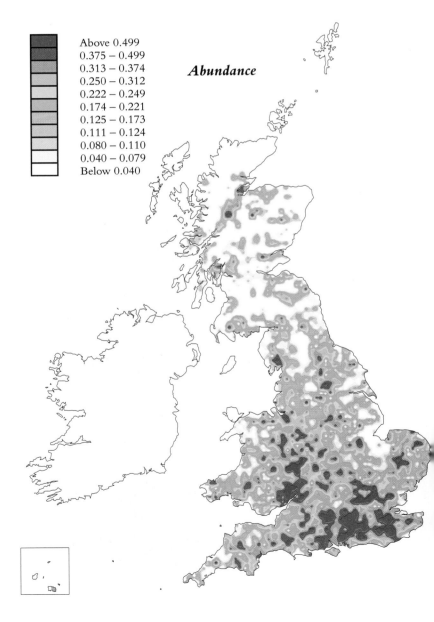

Above 0.499
0.375 – 0.499
0.313 – 0.374
0.250 – 0.312
0.222 – 0.249
0.174 – 0.221
0.125 – 0.173
0.111 – 0.124
0.080 – 0.110
0.040 – 0.079
Below 0.040

Abundance

Nesting density is at its highest in mature broadleaved woodlands. A review of 37 studies throughout NW Europe gave a mean of 0.1 pairs per ha in mature broadleaved woodland, and only 0.06 pairs per ha in mature coniferous woodland (K. W. Smith). Younger stands are much less favoured: in managed pine plantations only stands over 30 years old supported breeding Great Spotted Woodpeckers (Muller 1988), while in oak plantations the minimum age was 40 years (Ferry and Frochot 1970).

The minimum area of woodland used for breeding appears to be 2–3ha (Moore and Hooper 1975), and isolated small woods are less likely to be occupied (Opdam et al. 1985).

In the *68–72 Atlas* the population was thought to be 30,000–40,000 pairs, based on an estimate of 15–20 pairs per occupied 10-km square. A simple extrapolation from mean CBC densities suggests that there were 100,000 Great Spotted Woodpecker territories in Britain in 1988–91. However this could be a serious overestimate as the CBC probably samples optimal habitat for this species, and takes no account of lower densities in less favoured types of woodland. Breeding Great Spotted Woodpeckers are so reliant on substantial woodland blocks that is possible to use the figures for woodland area published by the Forestry Commission (Forestry

Change

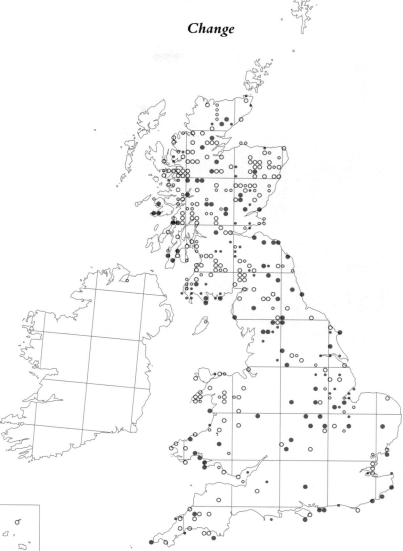

Distribution

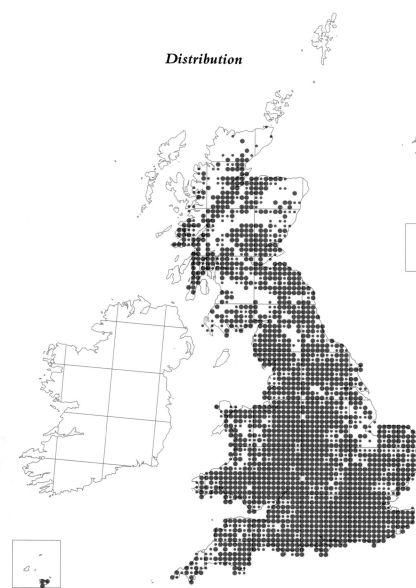

Commission 1984) and density estimates from different woodland types and ages, to arrive at a more realistic estimate of 25,000–30,000 pairs. This must, however, be treated with caution until it is confirmed by systematic survey data.

KEN SMITH

68–72 Atlas p 272

Years	Present, no breeding evidence		Breeding evidence		All records		
	Br	Ir	Br	Ir	Br	Ir	Both (+CI)
1968–72	273	1	1777	0	2050	1	2053
1988–91	361	0	1598	0	1959	0	1962
	% change		−4.4				−4.5

Lesser Spotted Woodpecker

Dendrocopos minor

The Lesser Spotted is our smallest and most sparsely distributed wood-pecker. It is almost exclusively associated with broadleaved woodlands, where it usually forages high in the trees, exploring the smallest branches with rapid sewing machine like pecks. Only the size of a House Sparrow, it calls and drums sparingly and is very easily overlooked. The first hint of its presence is often the soft 'pew..pew..pew' call, but unless the bird then moves, it is still difficult to locate. In spring, Lesser Spotted Woodpeckers drum less frequently than Great Spotted, and the drum is usually of longer duration and much softer than that of their larger relative.

Although Lesser Spotted Woodpeckers breed in extensive broadleaved woodlands they can also be found in the breeding season in parkland and orchards, and appear to favour river valley alders. Sitters (1988) found that, in Devon, they were restricted to the more fertile land at low altitudes. In Britain they are at the very NW edge of their European range and the Distribution map shows them to be largely restricted to Wales, S and W England and the Midlands, with relatively few records in the north. They just reach Lancashire and Cumbria in the west, and Northumberland in the east. There is a gap in this distribution with very few occupied squares in a narrow band from the Wash to Land's End.

The Abundance map does not suggest any major trends, overall, with the possible exception of higher numbers in Kent, Essex and the coast of Suffolk.

In the *68–72 Atlas*, Lesser Spotted Woodpeckers were recorded in 889 10-km squares, whereas in 1988–91 they were found in only 792. The northern limits of the range are apparently unchanged between the two Atlases, with the possible exceptions of a retreat from Lancashire and Cumbria, and a small expansion into NE England. The major change is an overall decrease in the number of occupied 10-km squares throughout

the range, which is particularly marked in the band from the Wash to Land's End. It is probable that this represents a real reduction in numbers rather than changes in methodology between the two surveys. In the *68–72 Atlas* there were two records for Scotland, but the authenticity of these has subsequently been questioned (*Birds in Scotland*).

Over the period between the two Atlases the CBC index for the Lesser Spotted Woodpecker increased by about a factor of two but, since 1980, has fallen significantly until it is now back, roughly, to its initial level (*Trends Guide*). Of our three woodpecker species the Lesser Spotted has been the most affected by the spread of Dutch elm disease (Osborne 1982). At first they appeared to benefit from the abundant food supply of beetles and their larvae beneath the bark of dying and dead trees. Once the elms had disappeared from the countryside it is perhaps not surprising that, with the loss of this abundant food supply, numbers returned to their former level. BTO Nest Record data show that, before the epidemic, elms were one of the most important nesting trees for Lesser Spotted Woodpecker (D. Glue) and their loss would have been a further factor in the decline.

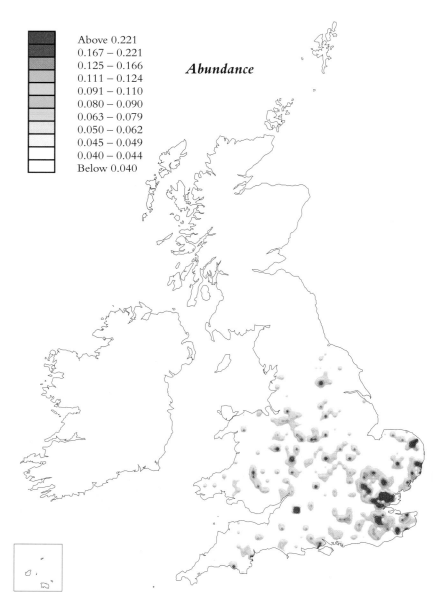

Above 0.221	
0.167 – 0.221	
0.125 – 0.166	
0.111 – 0.124	
0.091 – 0.110	
0.080 – 0.090	
0.063 – 0.079	
0.050 – 0.062	
0.045 – 0.049	
0.040 – 0.044	
Below 0.040	

Abundance

One of the most quoted examples of changes of breeding numbers during the Dutch elm disease episode was that of a 52 ha Kentish woodland, which held 1–3 pairs annually before the epidemic, increasing to 15 pairs whilst the elms where dying (Flegg and Bennett 1974). Numbers have now fallen and, in 1990, Lesser Spotted Woodpeckers were recorded only as occasional visitors to the site (A. Parker).

A new nest cavity is usually excavated each year and can be placed in any decaying timber (*BWP*). The height can range from 1–2m, to high in the crown of mature trees (D. Glue), and the entrance hole is often placed on the underside of a limb.

Although it is difficult to be certain, the gap in the distribution, from The Wash to Land's End, which has developed since the *68–72 Atlas* may be related to Dutch elm disease. The counties in a band from Somerset through to Northamptonshire were ones where hedgerow elms formed an important part of the woodland cover, and their loss may have been particularly significant for Lesser Spotted Woodpeckers.

In suitable woodlands, breeding density is typically 0.01–0.03 pairs per ha but, because the species can nest outside the extensive woodlands, it is

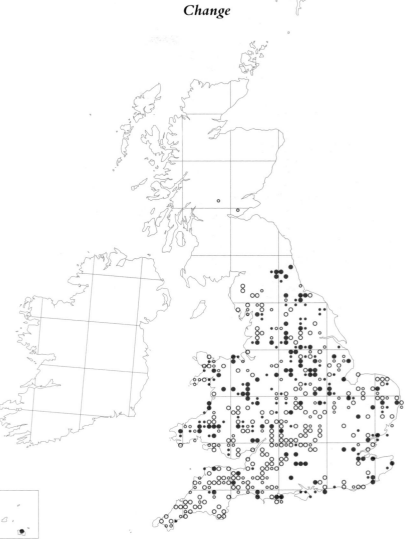

Change

Distribution

not possible to use this density to estimate the overall population. In *BWP* the population was estimated to be 3,000–6,000 pairs. Although this is little more than a guess, it is probably a reasonable one and could be improved only by extensive fieldwork.

KEN SMITH

68–72 Atlas p 274

Years	Present, no breeding evidence		Breeding evidence		All records		
	Br	Ir	Br	Ir	Br	Ir	Both (+CI)
1968–72	247	0	642	0	889	0	889
1988–91	274	0	516	0	790	0	792
			% change	−11.1	0		−11.0

Woodlark

Lullula arborea

As the Woodlark requires only a few scattered trees to act as song posts, its English name is really rather inappropriate. The German 'Heidelerche', which translates as 'Heath Lark', is more apt, as is the French 'Lulu' which describes the beautiful fluting song.

Recent research indicates that lack of suitable habitat is the principal factor restricting the British population at present, and that the most limiting element of habitat is absence of bare ground or short vegetation for foraging (Bowden 1990, H.P. Sitters and R.J. Fuller).

The Distribution map shows that the population is confined mainly to five discrete areas: the southwest (principally Devon), the New Forest, the Hampshire/Surrey border, Breckland and part of the Suffolk coast. This is much the same distribution that was revealed by a BTO survey in 1986, when each sub-population was found to occupy somewhat different habitats. In that year, a maximum of 219 pairs was found in S England. There were then 20 pairs in the southwest in mainly agricultural habitats, 32 pairs in the New Forest on grazed heathland, 79 pairs on the Hampshire/Surrey border mainly on heathland where vegetation had been cleared by fires or military exercises, 48 pairs in Breckland in recently cleared or restocked forestry plantations and 18 pairs in similar habitat on the Suffolk coast with seven pairs on nearby heathland (H.P. Sitters and R.J. Fuller).

These differences in habitat selection are a feature for which the Wood-lark is probably unique among British breeding birds. The various sub-populations also exhibit differences in migratory behaviour. In East Anglia, breeding sites are deserted in winter. On the Hampshire/Surrey border, many birds remain close to the breeding sites but others move away. Southwestern populations are even more sedentary. Therefore the tend-ency to migrate decreases from east to west, probably the result of lower

winter temperatures in E England compared with the more maritime climate of the southwest (Sitters 1986).

The Change map shows a marked contraction in range since 1968–72, with a 62% reduction in the number of 10-km squares occupied. The species has ceased to breed in Wales and in several counties in S England where it was once regular. The fact that it has recently done well in areas of good habitat indicates that the principal reason for loss is habitat change or destruction. In some areas loss of foraging opportunity can be blamed. On the Breckland heaths, for example, a reduction in the rabbit population through myxomatosis led to much feeding habitat becoming overgrown. Elsewhere, particularly in Hampshire and Dorset, Woodlark habitat has been lost through the development of heathland for housing, agriculture and forestry (H.P. Sitters and R.J. Fuller).

Despite the contraction in its range, the size of the British Woodlark population has not greatly diminished. During 1968–72, the population was estimated at 200–450 pairs. Subsequently numbers dropped to 160–180 pairs in 1975 but later increased, particularly on the Hampshire/Surrey border and in Breckland. In both areas the species benefited from the creation of new habitat. On the Hampshire/Surrey border, heath fires in 1974 and 1976 cleared large areas of scrub. Similarly, habitat was created in the Breckland pine forests when extensive areas were felled and replanted, giving rise to large areas of bare ground. Principally for these reasons, the British population increased to 400–430 pairs in 1981 (Sitters 1986).

It is likely that severe weather during the 1981/82 winter was the cause of high mortality among the more sedentary Woodlark populations. On the Hampshire/Surrey border, numbers fell from 163 pairs in 1981 to 61 in 1982. In contrast, the migratory populations of East Anglia were little affected with 56 pairs in 1981 and 46 in 1982. The British population was estimated at 210–230 pairs in 1983.

There has been no countrywide census of British Woodlarks since 1986. There has, however, been a series of intensive surveys covering the five main subpopulations in different years, the results of which indicate a total British population during 1988–91 of about 350 pairs (see Table).

Research in the Breckland forests by Bowden and Hoblyn (1990) shows that Woodlarks, there, breed only in places where trees have been recently clearfelled or restocked. A study of the replanting scheme for Breckland shows that there will be a period towards the end of the century when little suitable habitat will be available. Bowden and Hoblyn therefore propose that at that time young plantations be managed by the removal of ground vegetation so that they remain suitable for longer than would otherwise be the case. Similarly, heathland populations are probably best conserved by clearing scrub and long grass, wherever this does not occur through the effects of fire or grazing.

Year	Area	No. Pairs	Surveyed by
1988	Hants/Surrey border	100–111 pairs	RSPB
1988	Breckland	43 pairs	RSPB and Forestry Commission
1989	Breckland	63 pairs	ditto
1991	Breckland	93 pairs	Forestry Commission
1990	Suffolk coast	78 pairs	Suffolk Wildlife Trust
1990	New Forest	51–54 pairs	RSPB
1991	Devon	17 pairs	Devon Birdwatching and Preservation Society

HUMPHREY SITTERS

68–72 Atlas p 278

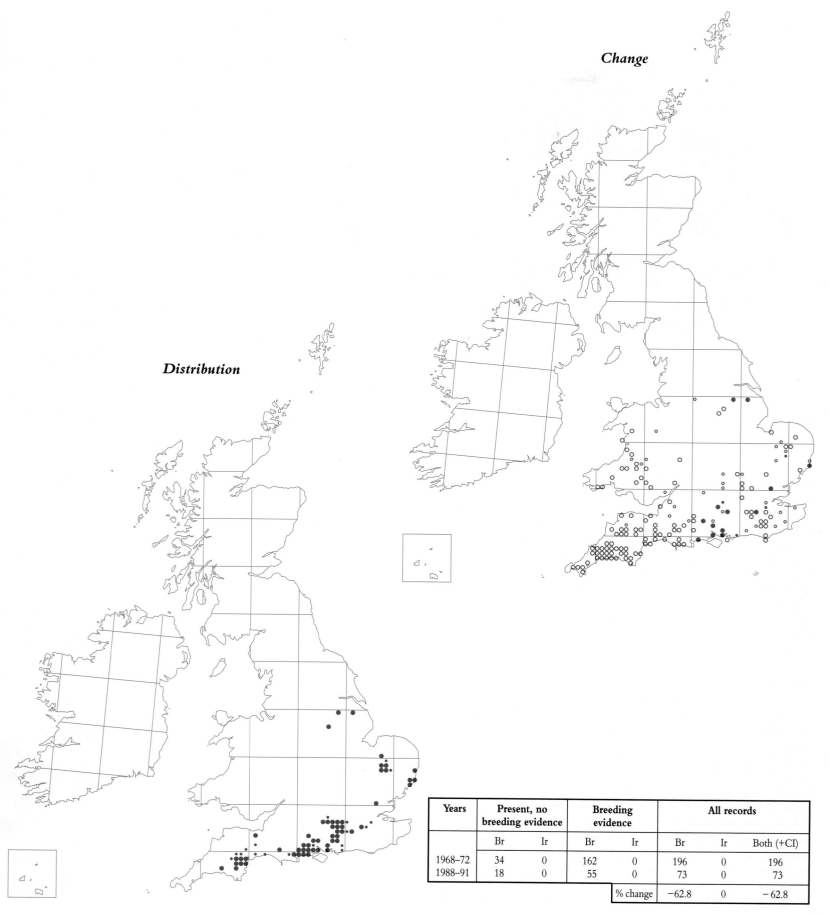

Change

Distribution

Years	Present, no breeding evidence		Breeding evidence		All records		
	Br	Ir	Br	Ir	Br	Ir	Both (+CI)
1968–72	34	0	162	0	196	0	196
1988–91	18	0	55	0	73	0	73
			% change		−62.8	0	−62.8

Skylark

Alauda arvensis

The song of the Skylark, delivered in flight, is an almost constant accompaniment to a springtime visit to open country, virtually anywhere in Britain and Ireland. The present distribution of the Skylark illustrates how ubiquitous open habitats are in farmland. The Abundance map indicates that the species is most common in lowlands of E Britain, where mixed or arable farming predominates, and least common in the highest parts of the Scottish Highlands. On farmland CBC plots, Williamson (1967) found that population densities of Skylarks were higher on land mainly given over to cereals and root crops, than in areas where pastures and leys predominated. Cereal grain and leaves are important components of the diet of Skylarks (Green 1978), so it is possible that the geographical variation in the species' abundance is mainly caused by variation in the availability of cereals as food. The habitat preferences of Skylarks are nevertheless probably more complex. For example, Schlapfer (1988) found that crop diversity was important, Skylarks in Switzerland preferring areas with a mixture of crops.

After a long period of stability of Skylark numbers on CBC plots in the 1960s and 1970s, there has been a decline since about 1980 to perhaps half of the former level (*Trends Guide*). This decline in Britain is not strongly reflected in the Change map, suggesting that it has not been accompanied by a marked range contraction. In Ireland, however, there is evidence of a reduction in the number of occupied squares in the east. O'Connor and Shrubb (1986a) showed that Skylark population fluctuations, in the period 1975–83, were correlated with the percentage of grassland in England and Wales which was ley grass under five years old. Since the density of Skylarks tends to be higher on young leys than on older grass, it may be that the Skylark's decline is related to the decline in the area of ley grassland. Another factor is likely to have been the increase in autumn sown rather than spring sown cereals. Autumn sowing reduces the area of cereal stubble left over winter, with consequent loss in winter food supplies of grain and weed seeds. Spring sown grain is itself an important source of food in the early spring when other food is scarce (Green 1978). Furthermore, from early in the breeding season, autumn

Above 0.959
0.875 – 0.959
0.778 – 0.874
0.714 – 0.777
0.625 – 0.713
0.524 – 0.624
0.455 – 0.523
0.375 – 0.454
0.250 – 0.374
0.040 – 0.249
Below 0.040

Abundance

sown cereals become too tall and dense to be much used by Skylarks (Schlapfer 1988).

During 1988–91 there were an estimated 2,000,000 Skylark breeding territories in Britain and 570,000 in Ireland.

R. E. GREEN

68–72 Atlas p 280

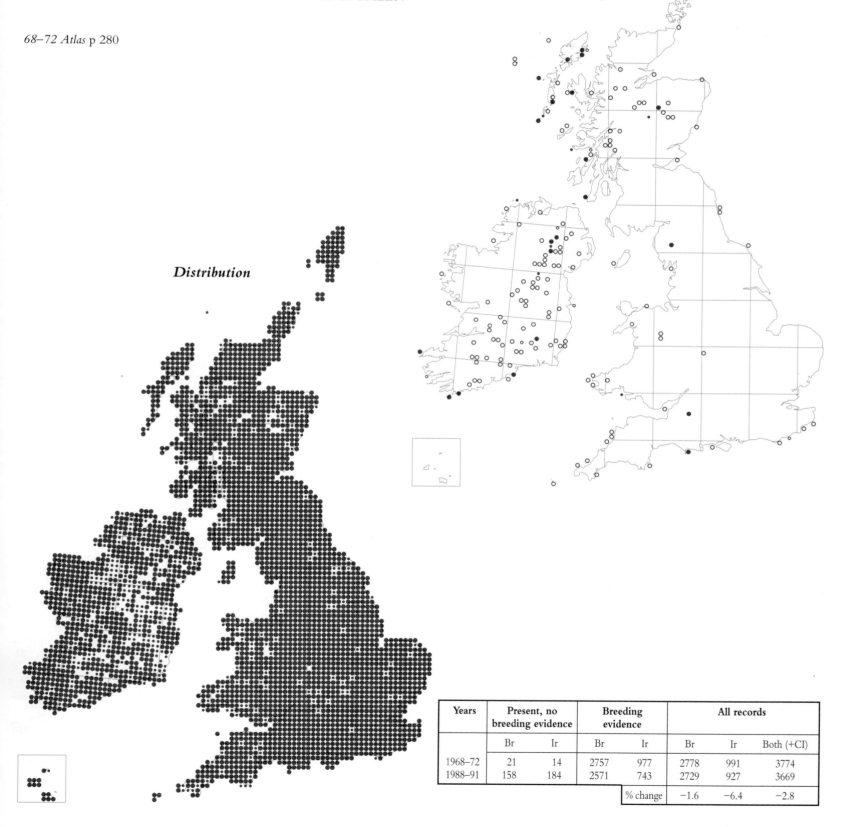

Change

Distribution

Years	Present, no breeding evidence		Breeding evidence		All records		
	Br	Ir	Br	Ir	Br	Ir	Both (+CI)
1968–72	21	14	2757	977	2778	991	3774
1988–91	158	184	2571	743	2729	927	3669
				% change	−1.6	−6.4	−2.8

Sand Martin

Riparia riparia

The Sand Martin is normally one of the first summer migrants to reach British shores, but numbers are largely dependent upon conditions in the wintering grounds bordering the Sahel region of Africa. Recent population crashes attributed to severe Sahel droughts occurred in 1968–69 (Cowley 1979) and 1983–4 (Jones 1987).

The Distribution map shows Sand Martins to have a widespread but patchy distribution over Britain and Ireland. The most obvious gap occurs in central S England, but Sand Martins are scarce also in NW Scotland, and mostly absent from the smaller islands, the Outer Hebrides and the Northern Isles (*Birds in Scotland*). In general, they are considerably less widely distributed than was recorded by the *68–72 Atlas*. They remain absent from large areas of S and E England where chalk and limestone formations predominate, but further gaps in distribution have occurred since 1968–72, and this reduction of range is probably associated with a real decrease in numbers. The Distribution map shows breeding in about half the number of 10-km squares recorded during 1968–72.

Sand Martin colonies may each comprise hundreds (exceptionally thousands) of pairs. As they are largely dependent on sandy river banks or on gravel pits for nesting sites, their distribution is in part constrained by geology. The expansion of sand and gravel quarrying in Britain during post-1945 years has made many new nesting sites available, and in central Scotland in 1982 (Jones 1986), 84% of nests were in such quarries. It is not clear whether this increase in nesting habitat led to an increase in numbers of breeding pairs between 1945 and 1968. Data from continental Europe (1930–73) suggest that high population densities (> 0.5 pairs per km^2) have been recorded only since about 1965 (Kuhnen 1975).

Studies in Britain and Europe over the past 25 years suggest that Sand Martins reached their lowest numbers in 1985 (Persson 1987, *Trends Guide*). The two periods of dramatic population decline were in 1968–69 and 1983–84. The 1968–69 decline was also noted from counts of passage birds off Ireland (*Birds in Ireland*). Following each crash the decline has continued for at least one further year. This is possibly because young birds are virtually eliminated during the crash. Cohorts of first-year breeders may then take several years to become re-established in quantity, during which time older birds die. Natural selection favoured small birds during the 1983 crash (Jones 1987), but since 1983 the central Scotland population has increased to 79–94% of 1982 numbers, and birds are now significantly larger than during the 1984 population trough (D.M. Bryant and G. Jones).

Sand Martin densities are also influenced by weather in the breeding season. During favourable years, many pairs may rear two broods. If spring weather is cold and wet, however, return to the breeding areas and nesting may be delayed, so that there is not time to rear a second brood. Recent advances in studies of Sand Martin biology are reviewed by Turner and Rose (1989).

Densities of breeding Sand Martins have been reported from several

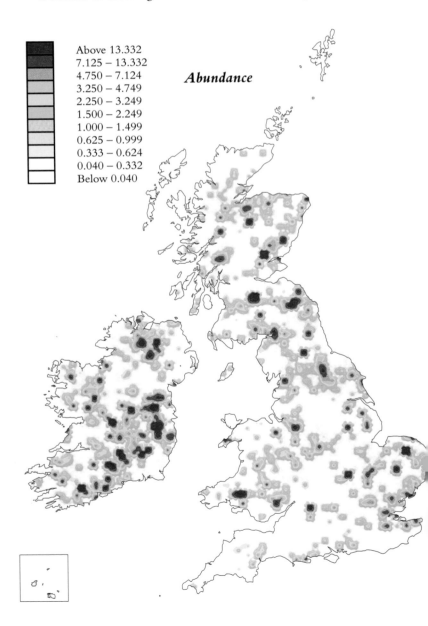

Abundance

	Above 13.332
	7.125 – 13.332
	4.750 – 7.124
	3.250 – 4.749
	2.250 – 3.249
	1.500 – 2.249
	1.000 – 1.499
	0.625 – 0.999
	0.333 – 0.624
	0.040 – 0.332
	Below 0.040

European studies. Persson (1987) quotes an average of 2.54 pairs per km² (range 0.7–4.0 pairs per km²) in Scania, S Sweden, between 1964 and 1985; Kuhnen (1975) gives 0.1–1.2 pairs per km² from studies in 30 areas of continental Europe. A breeding density of 1.97–2.11 pairs per km² was estimated in central Scotland during 1982 (Jones 1986), when numbers were highest during the last decade. Cowley (1979) recorded an average of 0.8 pairs per km² in N Nottinghamshire between 1965 and 1978.

The *68–72 Atlas* suggested well over 100 pairs per 10-km square before the 1968 population crash in those areas where breeding was proved and, wrongly, gave a breeding population for Britain and Ireland of close to 1,000,000 pairs. This calculation should have given an estimate of 250,000–500,000 pairs. By the early 1970s, average numbers slumped to 16% of those in 1968 (assuming the data from two British studies are representative of Britain as a whole), or to 32% of 1968 numbers if two Continental census results are included. Thus by the early 1970s only 40,000–160,000 pairs may have bred in Britain and Ireland, and the population in 1985 could have been even smaller. Numbers in 1968 were exceptionally high, and current levels may be more typical of early 1960s values, though there is no doubt that a decline is apparent since 1968.

Change

Distribution

Extrapolation from tetrad counts during 1988–91 yields minimum estimates of 77,500 Sand Martin nests, annually, in Britain, and 49,500 in Ireland. However, many colonies will have been missed or were inaccessible. Realistic upper limits to the British and Irish populations would be, perhaps, 250,000 and 150,000 nests respectively. The contraction in distribution since 1968–72 gives some cause for concern. If Persson's (1987) theory of periodicity in Sand Martin population size is correct, we should expect another population crash in the mid 1990s.

GARETH JONES

68–72 Atlas p 286

Years	Present, no breeding evidence		Breeding evidence		All records		
	Br	Ir	Br	Ir	Br	Ir	Both (+CI)
1968–72	288	104	1755	739	2043	843	2889
1988–91	568	242	991	353	1559	595	2160
% change					−23.7	−29.3	−25.3

Swallow

Hirundo rustica

Summer and good fortune come with the Swallow, so not surprisingly any ups and downs in its numbers are readily noted, especially by those people favoured with pairs breeding on their property. Local declines and increases are not unusual for this species, but there also seems to have been a more widespread recent decline.

The Swallow is well distributed throughout Britain and Ireland. It is absent from, or scarce in, only a few areas, particularly in the upland areas of NW Scotland, the Outer Hebrides and Shetland. Its main habitat is lowland farmland. Although the Swallow breeds at up to 1,800m in the Alps and to 3,000m in the Caucasus, in Britain it rarely breeds in the uplands. Villages often provide suitable nest sites but the Swallow is less abundant in the centres of towns.

The present distribution is almost identical with that found in the *68–72 Atlas*, the main change being an increase in records from NW Scotland and in the Western and Northern Isles where they have bred more regularly since the early 1970s (*Birds in Scotland*).

Swallows are more common in Ireland and parts of E England and less common in N England and Scotland. The CBC rankings for farmland in 1988 showed that density was greatest in Northern Ireland and W England and least in S and E England.

Swallow populations often fluctuate, especially locally, from year to year. With an average clutch size of four or five (up to seven) eggs and two or sometimes three broods a year, they can theoretically soon make up for a bad year. Numbers dropped in 1974, as shown by the CBC, but then recovered in the late 1970s. However, during the 1980s there has been a steady decline, especially in S and E England, with some recovery in 1988. The populations in N and central Europe have also declined recently (*BWP*).

Most of the mortality probably occurs on migration and in winter quarters, particularly in bad weather. Cold, wet autumn weather in Europe, for example, particularly at high altitude, can cause serious losses. The Sahara is another hazard, although the drought in the Sahel seems not to have affected numbers. In the last decade, South Africa, where the Swallows from Britain winter, has suffered a drought which will have reduced their winter food supply of insects such as swarming ants. Møller (1989) found that mortality among Swallows in his Danish study areas was associated with rainfall in the winter quarters, being highest in dry conditions. The effect seemed to be carried over into the breeding season, with drought in South Africa being followed by smaller clutches in the

next breeding season, perhaps because birds were arriving back in poorer condition. Our British population, as well as others in Europe, has probably suffered similarly. A previous drought in South Africa in the 1960s was marked by a westward extension of the British Swallows' winter quarters. In the breeding quarters, the weather, and its attendant influence on insect numbers, can also have dramatic effects: bad weather can delay breeding and reduce the number of birds attempting second or third broods.

In Britain, the local density of breeding birds at the start of the season is probably determined by the availability of insects and of nest sites. Swallows breed in close association with humans in buildings such as barns, outhouses and garages, under bridges and in culverts, and there are now very few natural nest sites such as caves. They breed in groups, but with their nests usually a few metres apart. The average size of the group depends on the local population density, but increases when there are more birds looking for nest sites (Møller 1991); or when large insects are plentiful (Møller 1987). The diet consists of aerial insects, particularly large flies such as hoverflies, horseflies, robberflies and bluebottles. Changes in farming practices that reduce the supply of nest sites and large insects can thus harm Swallow populations and have probably been partly responsible for declines in Britain, especially in the east and south. Swallows

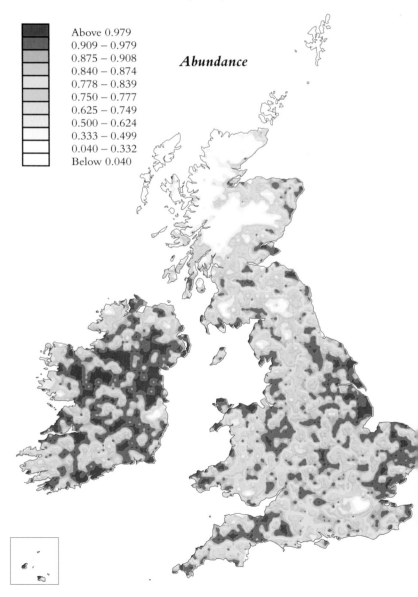

Abundance

	Above 0.979
	0.909 – 0.979
	0.875 – 0.908
	0.840 – 0.874
	0.778 – 0.839
	0.750 – 0.777
	0.625 – 0.749
	0.500 – 0.624
	0.333 – 0.499
	0.040 – 0.332
	Below 0.040

particularly like farms with plenty of traditional buildings and livestock, where nest sites are almost always available, where flies are likely to be abundant and where grazing animals stir up insects from the vegetation for flying Swallows to catch. Improved farm hygiene, intensive livestock rearing, modern buildings and the use of pesticides will make a farm less suitable for Swallows.

During 1988–91 there were an estimated 570,000 Swallow territories in Britain, and 250,000 in Ireland.

ANGELA TURNER

68–72 Atlas p 282

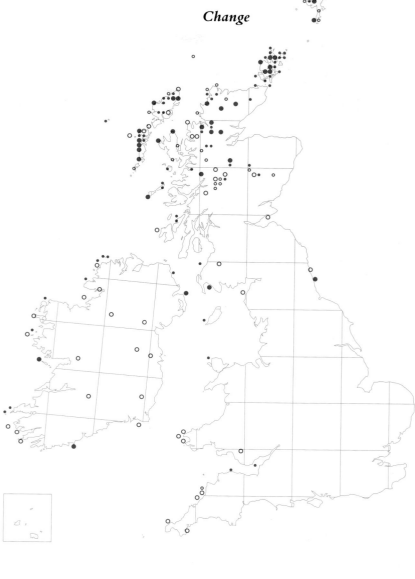

Change

Distribution

Years	Present, no breeding evidence		Breeding evidence		All records		
	Br	Ir	Br	Ir	Br	Ir	Both (+CI)
1968–72	85	6	2515	981	2600	987	3592
1988–91	169	65	2457	917	2626	982	3622
				% change	1.1	−0.4	0.7

House Martin

Delichon urbica

While its calls lack the insistence of screaming Swifts, an emphatic 'prrt' commonly betrays the presence of this aerial feeder. At close range its blue-black and white plumage is striking and, along with a tendency to nest near man, makes it an easy species to track down. Signs of House Martins can be found year round: their conspicuous mud nests cling to buildings and cliffs, and often stay in place over winter.

The House Martin is common throughout most of Britain and Ireland, but is very local on exposed and coastal areas in the west and north. Hence, over much of the Scottish Highlands and Islands, S W Scotland, W Ireland and the uplands of N England and Wales, breeding birds are thinly distributed or absent. Three factors seem to govern distribution. Suitable nest sites must be present, usually on buildings nowadays, but sometimes on cliffs (Clark and McNeil 1980). The sparsely populated areas of high ground provide rather few suitable sites, whereas elsewhere they are usually plentiful. Secondly, warmth, sunshine, and very moderate rain and winds, are best for easy availability and exploitation of aerial insects (Bryant 1975). Again, the north and west fare least well: poor weather is commoner there, increasing the chance of nesting failure. A third factor is the capacity of the locality to supply insects in the quantity and variety required for successful nesting. Poor upland soils with sparse vegetation rate badly as insect providers, whereas alluvial lowlands are productive of insects – but carry a risk of impoverishment by intensive agriculture. When all three factors are favourable, populations can flourish. The highest densities occur in East Anglia, parts of the S and W of England, the West Midlands, North Yorkshire and Humberside, with lesser concentrations in coastal Wales, parts of E Ireland, and E central Scotland. An eastern bias amongst the best areas is broadly consistent with the density rankings in the *Trends Guide*. All are areas characterised by mixed agriculture. The relatively barren, often inclement uplands of the north and west are able to support

fewer birds, although some areas support healthy populations, probably because local conditions are particularly suitable. In Scotland, for example, this is typically a sheltered glen, with a mix of woodland, loch and farm, favoured equally by human communities and House Martins for its microclimate and fertility. More generally, proximity of tree cover to houses with eaves appears to be attractive (Turner 1982).

Differences in ecology, particularly feeding habits and nest site availability, allow the relatively small differences between the distributions of aerial feeding species to be explained. Swifts feed highest on small, weather dependent species, whereas Swallows feed lowest on larger insects. Between these levels, House Martins are closer to Swifts, and Sand Martins resemble Swallows (Waugh 1979). The ranges of Swallow and Sand Martin are therefore similar, allowing for the latter's more limited nest site availability. Of the hirundines, the House Martin penetrates least into the exposed conditions of the north and west, but remains less restricted than the Swift, which fares particularly badly during spells of bad weather. Similarities in ecology also mean that House Martins and Swifts share an ability to nest in towns (Tatner 1978, *Trends Guide*). Nesting is usually in colonies, which in Britain range from loose gatherings of a few pairs to

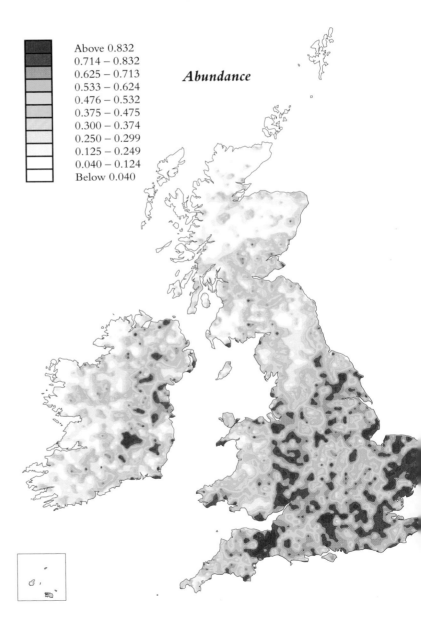

Abundance

	Above 0.832
	0.714 – 0.832
	0.625 – 0.713
	0.533 – 0.624
	0.476 – 0.532
	0.375 – 0.475
	0.300 – 0.374
	0.250 – 0.299
	0.125 – 0.249
	0.040 – 0.124
	Below 0.040

several hundred closely packed nests. Most reviewers of the status of House Martins in Britain have concluded that the population is stable or showing a slow long-term decline (Parslow 1973, *Trends Guide*). Good evidence of decline is lacking, however. A presumption of stability often contrasts with local experience, where losses are noted but gains usually pass unrecorded. The conflict can be reconciled, however, if in fact nesting has tended to become more dispersed. Some of the larger colonies have been replaced by several smaller nesting groups, perhaps spread amongst new houses, where their presence is less recorded.

Since almost all squares in England, Wales, S Scotland and E Ireland were occupied in both the first and second Atlas periods, it is necessary to examine the fringes of distribution to detect any changes. Western Ireland and parts of E and S Scotland have fewer occupied squares than in 1968–72 whereas, in contrast, there are now marginally more records from Orkney and the far N of the Scottish mainland. In no area, however, are major changes evident, so an assumption of a static distribution, showing some peripheral changes probably related to wider population fluctuations, is all that present evidence allows.

Change

Distribution

Assuming the density of 100–200 pairs per occupied 10-km square used in the *68–72 Atlas* still holds, there would now be 250,000–500,000 pairs of House Martins in Britain, and 70,000–140,000 in Ireland.

DAVID BRYANT

68–72 Atlas p 284

Years	Present, no breeding evidence		Breeding evidence		All records		
	Br	Ir	Br	Ir	Br	Ir	Both (+CI)
1968–72	95	56	2330	839	2425	895	3325
1988–91	166	115	2227	695	2393	810	3217
				% change	−1.3	−9.4	−3.5

density. The species was entirely absent from Ireland in the *68–72 Atlas* but was recorded in a small number of 10-km squares in the *88–91 Atlas*. Hutchinson (*Birds in Ireland*) considered that it may be a rare but overlooked species.

In S and E Britain, Tree Pipits mainly use open habitats such as young conifer plantations and coppice, woodland rides, heaths, commons and scrubby downland. Rarely does the species occur in closed canopy lowland woods. In N and W Britain, however, it is characteristic of mature sessile oak and birch woods. Heavy grazing causes many of these upland woods to have extremely sparse field and shrub layers. The tree canopy can also be very open, especially in old native pine woods which can also support high densities (Buckland *et al.* 1990). This type of woodland structure is seldom found in lowland woods.

In a Swiss study, however, Meury (1989) found two-thirds of nests in marsh, a habitat which is virtually never used in Britain. Interestingly, dry conifer forests of N Finland generally carry lower densities in their young stages than in the mature stands (Helle 1985). With the possible exception of native pine woods, this contrasts with Britain, where the youngest

Tree Pipit

Anthus trivialis

Tree Pipits are amongst the earliest summer visitors to arrive and their conspicuous songflights can be heard from early April until late June. Habitats used by Tree Pipits differ strikingly from one part of the country to another, though it seems that two essential features are tall song posts and suitable feeding sites. The birds feed mainly on the ground, in sparsely vegetated sites largely devoid of dense shrubs and trees.

Tree Pipits are widely distributed, both in the extreme S of England and north and west of a line running approximately from Bristol to York, though this line was less marked in the *68–72 Atlas*. The Abundance map suggests that the bulk of the British population is centred on the upland regions – from Devon, through Wales and N England into the central Highlands. The species is absent from the extreme coastal fringe of many counties, and the fact that they are scarcest in the most intensively arable parts of England corresponds with the finding of Sitters (1988) that they prefer low grade agricultural land. Presumably this is because their preferred habitats are found in such areas. The relative sparsity of Tree Pipits in central and E England can also be linked to an apparent avoidance of heavy clay and the peaty soils of East Anglia.

Comparison of the two summer Atlas distributions suggests that a striking decline has occurred, particularly in central and S England. This could be a continuation of the gradual decline in S England that was thought to be underway before the *68–72 Atlas* (Parslow 1973). A possible recent cause is that many lowland conifer plantations are no longer suitable for Tree Pipits. On the other hand, the storms of October 1987 and January 1990, locally created new habitat for the species. In the uplands, large areas of suitable habitat continue to be created as a result of afforestation and restocking with conifers. With the possible exception of Kintyre, however, there is little evidence of any range expansion that could be linked with forestry, though there may have been local increases in

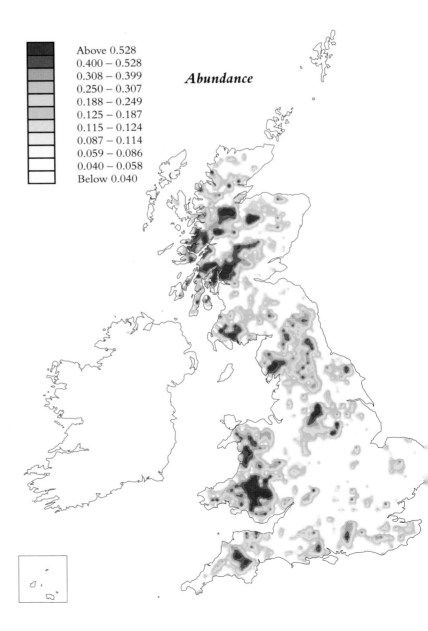

Above 0.528
0.400 – 0.528
0.308 – 0.399
0.250 – 0.307
0.188 – 0.249
0.125 – 0.187
0.115 – 0.124
0.087 – 0.114
0.059 – 0.086
0.040 – 0.058
Below 0.040

Abundance

conifer plantations can become unsuitable for the species after as little as four years of growth, but in new upland afforestation strong colonisation may not occur until the ninth year (Sykes *et al.* 1989). Within conifer forests, Tree Pipits prefer edges where a recent clearfell abuts an old stand (Sykes *et al.* 1989). Another ephemeral habitat is provided by coppice, especially of sweet chestnut. The species is closely associated with the first four years of chestnut growth (Fuller and Moreton 1987).

Like the Wood Warbler, the Tree Pipit can be polyterritorial. Meury (1989) found that one-third of the males in his study area simultaneously defended two territories. The average size of the defended areas was 1.5ha and the average distance between them was 1.1km. Polyterritoriality has not been proved in Britain. During 1988–91 there were an estimated 120,000 Tree Pipit territories in Britain.

R. J. FULLER

68–72 Atlas p 396

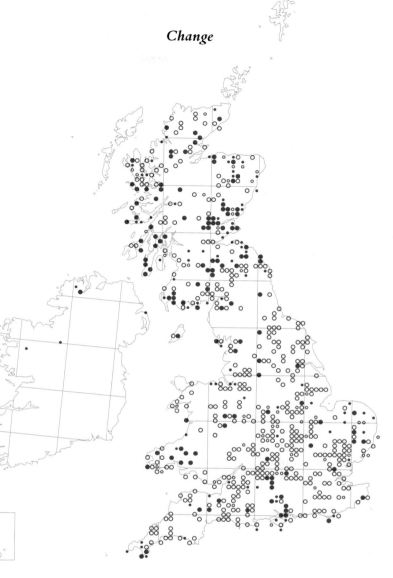

Change

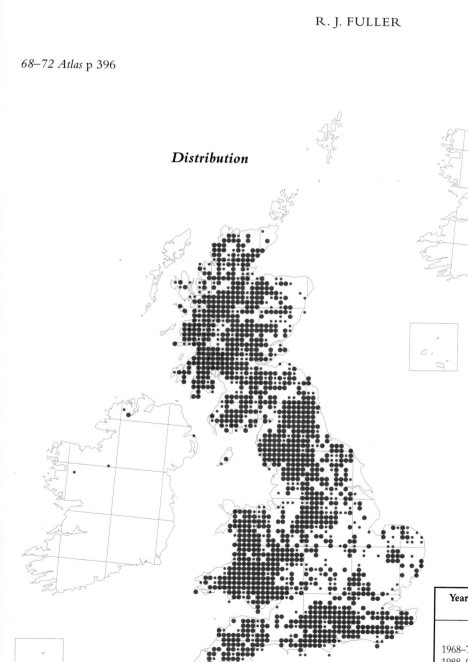

Distribution

Years	Present, no breeding evidence		Breeding evidence		All records		
	Br	Ir	Br	Ir	Br	Ir	Both (+CI)
1968–72	108	0	1685	0	1793	0	1793
1988–91	309	4	1215	1	1524	5	1529
				% change	−15.0		−14.7

Meadow Pipit

Anthus pratensis

The Meadow Pipit nests from sea level to altitudes of over 1,000m, occupying various habitats, including saltmarshes, flood meadows, chalk grassland, lowland heaths, grazed fens and bogs, and uplands. At altitudes above 500m it is the commonest nesting passerine.

As a result of this adaptability, it is widely distributed, breeding in every county. It does, however, show major regional variations in density, being thinly distributed in the English lowlands, especially on the heavier clay soils of central and S England. It is very common in N England and mid Wales, the majority of birds occurring on hill farms and moorland, and there are localised pockets of abundance in SW England and East Anglia. In Scotland and Ireland the Meadow Pipit is a widespread and abundant breeding bird, but it occurs less frequently in the south and east of both countries.

Breeding densities of Meadow Pipits vary with habitat, as shown by the CBC: 2.3 pairs per km^2 on farmland compared with 25–50 pairs per km^2 on moorland, upland sheepwalks, young conifer plantations, chalk downland and saltmarsh (*Trends Guide*). Densities on chalk grassland and lowland heath are affected by grazing pressure. Meadow Pipits prefer thicker grass to short-cropped turf, and vegetation changes due to grazing lead to Skylarks replacing them. The pipits do tolerate the presence of more scrub than Skylarks, but numbers decrease in dense open canopy scrub (Fuller 1982).

Over the past 20 years there have been distributional changes in each country. In England the declines have been most evident in the southwest (through Cornwall to Wiltshire), parts of East Anglia, the southeast and the Midlands. Scotland shows little change, although breeding densities increase in the early stages of afforestation (*Birds in Scotland*). Deep ploughing of moorland prior to planting trees leads to a doubling of numbers (Williamson 1975), which then remain high until the trees are about five years old. Densities of less than one pair per km^2 have been found in hayfields in the Lothians (*Birds in Scotland*).

In Ireland there is evidence of a decline in the central and southeastern regions, which may be linked to land-use changes such as increased ploughing up of old grasslands and the burning of moors and bogs (*Birds in Ireland*).

A steep decline in Meadow Pipit numbers was evident in 1985 and 1986, from which there has been only a slow improvement. Local population declines in the English lowlands could be due to the conversion of grassland to arable, and the loss of marginal land to cultivation and afforestation, but on a national scale the 1980s decline was too sudden to be attributable to land use change. As most British Meadow Pipits winter in SW Iberia (*Winter Atlas*) the low populations cannot be blamed on cold British winters. However, with a severe 1984/85 winter in S Europe and the cool, wet, British summers of 1986–88, climate may well be influencing British Meadow Pipit population levels. Breeding numbers increased in 1989, with fewer birds having succumbed in the relatively mild winter of 1988/89 (*Trends Guide*).

The laying period is from April to June, with most pairs attempting a second brood. Clutch size varies from three to five eggs. At high altitudes birds breed later and lay smaller clutches, but experience less nest predation (Coulson 1956). Meadow Pipits are a major host species for Cuckoos, although recently the proportion of nests parasitised has declined (Brooke and Davies 1987).

During 1988–91 there were an estimated 1,900,000 Meadow Pipit territories in Britain, and 900,000 in Ireland.

STEVE WOOLFALL

68–72 Atlas p 394

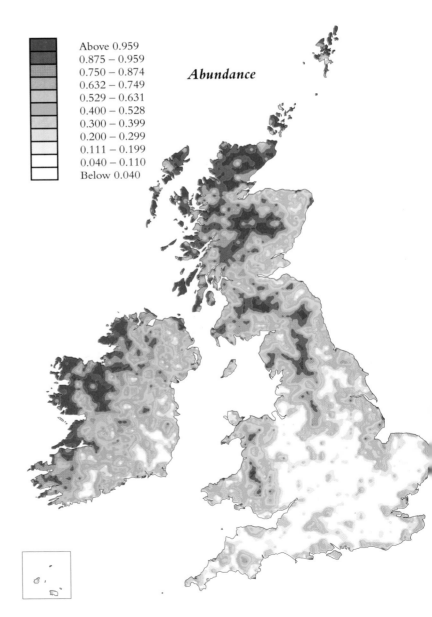

Above 0.959
0.875 – 0.959
0.750 – 0.874
0.632 – 0.749
0.529 – 0.631
0.400 – 0.528
0.300 – 0.399
0.200 – 0.299
0.111 – 0.199
0.040 – 0.110
Below 0.040

Abundance

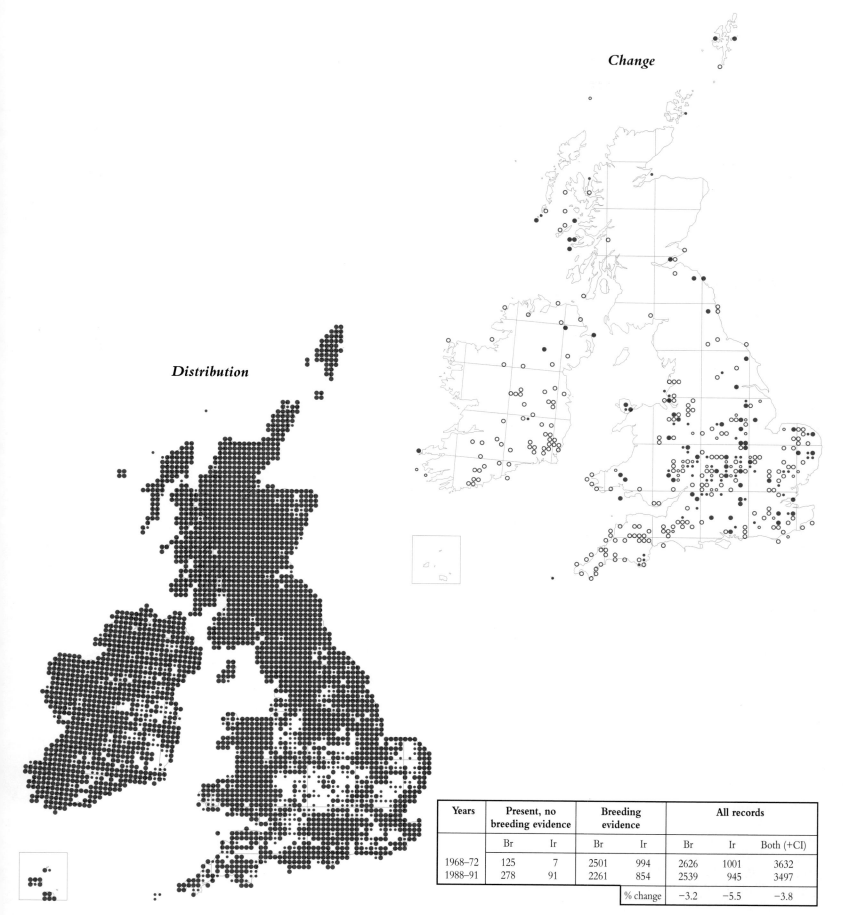

Change

Distribution

Years	Present, no breeding evidence		Breeding evidence		All records		
	Br	Ir	Br	Ir	Br	Ir	Both (+CI)
1968–72	125	7	2501	994	2626	1001	3632
1988–91	278	91	2261	854	2539	945	3497
				% change	−3.2	−5.5	−3.8

Rock Pipit

Anthus petrosus

A coastal encounter in any season with a robust, olive-shaded pipit with smoky grey outer tail feathers and long dark legs indicates a Rock Pipit. Males perform an eye-catching display flight from late March to early July, which is accompanied by a song rather fuller, louder and more metallic than that of the Meadow Pipit. Its persistent 'tsip' alarm note is similarly stronger.

All forms of the Rock Pipit have occurred in Britain. Rock Pipits breeding in Britain and Ireland belong to the subspecies *Anthus p. petrosus*, which also inhabits NW France. Elsewhere in Europe *Anthus p. littoralis* breeds along the coastline and on the islands of Denmark, W Norway and the Baltic countries.

In Britain and Ireland *petrosus* pipits are normally territorial throughout the year (Gibb 1956). Peak numbers, from late September, are likely to be post-breeding aggregations, though a few may travel further: e.g. a bird ringed on Lundy (Devon) was recovered on the Wicklow coast of Ireland in the following breeding season.

The Distribution map of the Rock Pipit reflects its liking for rocky shorelines. The gap in distribution from N Humberside around to the Isle of Wight is broken only by isolated pockets on the chalk cliffs of Kent and Sussex. In the west, the Rock Pipit is absent only where cliffs give way to sandy beaches and built-up areas, and in the flat, estuarine expanse from the Wirral north to mid Cumbria. The species breeds around virtually the entire W and N Scottish coastlines, and on most islands. On Skye and Mull, recent inland records of nesting at 300m and 500m respectively have been noted (*Birds in Scotland*). Rock Pipits breed widely on Lewis and Harris, where low hummocky ground and small islets suffice. They are scarce or absent from much of the coast south from Wick to Tayside. Ireland enjoys a virtually unbroken coastal distribution, the small gaps in

the east coinciding with extensive sandy bays and coastal resorts.

For nesting, Rock Pipits normally require a cavity close to the intertidal zone, typically set in a steep rocky bank. Some nests are in grass or heather banks, typically in a cavity amongst the roots of littoral plants or loose stones and sites, may be used over a period of many years (Campbell and Ferguson-Lees 1972). Clutches, normally of 4–5 eggs, are laid from early April through to the end of July. Pairs are usually double brooded, with some young in the nest until mid August (BTO Nest Records). Rock Pipit pairs are widely duped by the Cuckoo, and act as successful foster parents (Glue and Morgan 1972).

Because their nests are generally in inaccessible habitats, safe from man's activities, most are successful. On the other hand the recent intensification of farming, with ploughing almost flush to the cliff edge thereby reducing coastal grassland and dwarf scrub, and increased recreational activities, are both potential threats to the Rock Pipit's habitat.

In the nesting season Rock Pipits rely heavily on small marine molluscs, amphipods, adult and larval flies, and small worms. The impact of increased marine pollution in recent decades is unknown but mortality through oil and its derivatives is unlikely to have caused significant losses.

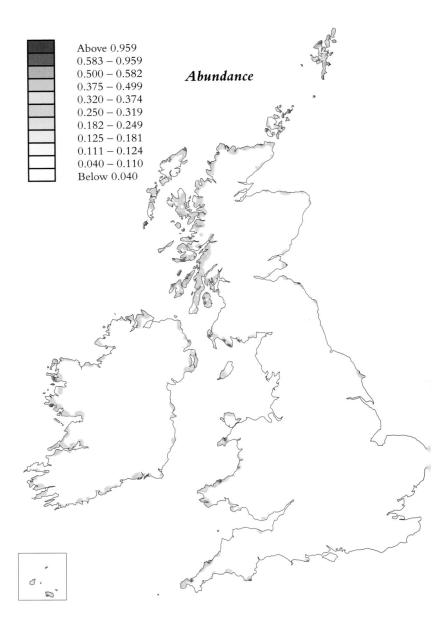

Above 0.959
0.583 – 0.959
0.500 – 0.582
0.375 – 0.499
0.320 – 0.374
0.250 – 0.319
0.182 – 0.249
0.125 – 0.181
0.111 – 0.124
0.040 – 0.110
Below 0.040

Abundance

Any assessment of Rock Pipits numbers is strongly influenced by the length and quality of the coastline. In Britain, territory length in favourable habitat is of the order of 200–250m, perhaps longer where densities are low (Gibb 1956). On St Kilda, Williamson (1964) estimated 150 breeding pairs in an area of some 6km², with about one half along the cliffline and the remainder inland. Information from 11 scattered sites throughout Britain and Ireland gives an average density of 3.5 territories per km², although the samples may be biased towards more accessible, well stocked habitats. The densities varied from 0.9–1.5 territories per km on the outlying island of Soay, Western Isles, to as high as 4.8–6.0 on Skokholm (Dyfed) and 3.6–6.0 territories per km in Cornwall (Gibb 1956).

The combined Rock Pipit population in Britain and Ireland, based on fieldwork undertaken during the *68–72 Atlas* was thought to exceed 50,000 pairs. This figure relied on an estimate of 50 or more pairs per occupied 10-km square: a credible yardstick, bearing in mind the convoluted nature of the typical coast occupied. With no population density studies during the inter-Atlas years, a similar calculation now indicates British and Irish populations of 34,000 and 12,500 pairs respectively. These estimates lend support to the general impressions of declines

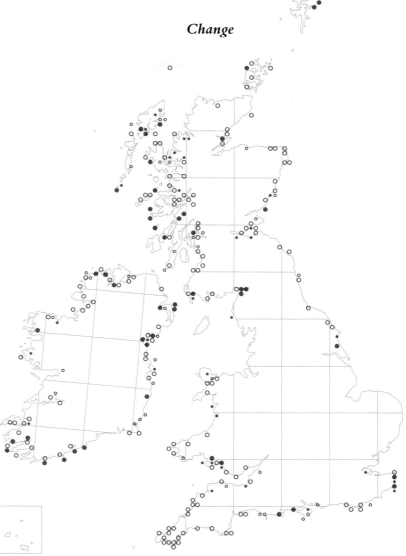

Change

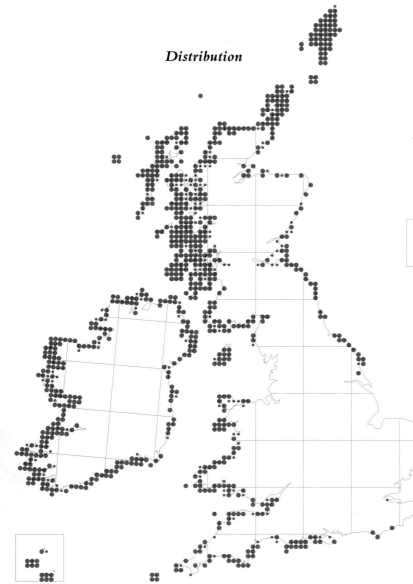

Distribution

in Rock Pipit populations on parts of the E coasts of Scotland and Ireland, and the S and SW coast of England since 1968–72, endorsed by the Change map.

Elsewhere in NW Europe the Scandinavian Rock Pipit *littoralis* appears to be thriving, with extensions of range, or consolidation, in parts of Denmark, Germany, Norway and, notably, Finland (*BWP*).

DAVID E. GLUE

68–72 Atlas p 398

Years	Present, no breeding evidence		Breeding evidence		All records		
	Br	Ir	Br	Ir	Br	Ir	Both (+CI)
1968–72	80	22	655	266	735	288	1028
1988–91	100	32	554	227	654	259	927
			% change		−11.0	−9.8	−10.7

Yellow Wagtail

Motacilla flava

The canary-yellow male singing from a fence post or hedgetop is still a frequent sight in the spring in some lowland damp pastures. This migrant, wintering in West Africa, is the most pipit-like of the three species of wagtail breeding in Britain. It prefers broad valleys along the lower reaches of rivers, usually nesting in water meadows, damp cattle-grazed pastures and marshes, or at the edges of lakes and on sewage farms.

The Yellow Wagtail is widely but patchily distributed throughout England. It breeds sparsely in S Scotland, where *Birds in Scotland* documented a decrease in the first half of this century, and the isolated population in the south of the Clyde valley has contracted markedly since 1968–72. It is virtually absent from W Wales, SW England and Ireland, although prior to the 1930s it bred more widely in these areas (*68–72 Atlas, Birds in Ireland*). It occurs more frequently in SE England, in coastal Essex and Kent, and in those counties around and southwest of the Wash.

The Yellow Wagtail appears to have contracted its British range still further in the last two decades, with birds either no longer breeding, or occurring much less frequently in many of their previous haunts. Examples of loss in S Wales include the low-lying pastures of the Gwent Levels, and S and W Glamorgan. The species has also disappeared from parts of coastal S England, from the farmlands of Dorset, Hampshire and Wiltshire, and from a scattering of other areas throughout its range. Drainage of wet fields, the intensification of agriculture and the replacement of grasslands with cereals may explain these losses. Some pairs do, however, breed in cereal crops, as for example alongside the R Severn in Montgomeryshire, Powys.

Mason and Lyczynski (1980) found that 52% of nests recorded on BTO nest record cards were close to water. Some two-thirds were associated with grasslands, and 96% were on the ground, 79% of these being sheltered by a tuft or tussock of grass. Although some clutches are laid in mid or late April, the peak of laying occurs from 11–20 May. Late or replacement clutches may be laid until late July but true second broods are rare (Mason and Lyczynski 1980). The usual clutch size is five or six, with a mean of 5.18 eggs. Nests are particularly vulnerable to wet weather, mowing and to trampling by livestock. The overall success rate of eggs was calculated as 51.5%; 67.6% of eggs hatching and 75.8% of hatched eggs producing fledged young (Mason and Lyczynski 1980).

A study of the diet and foraging behaviour of a large population of Yellow Wagtails, at Elmley Marshes in Kent, showed that flies (Diptera and Nematocera) and spiders accounted for 80% of the prey items (C.

Bell). Although smaller numbers of damselflies and beetles were caught, these comprised over half of the biomass of nestlings' diet. It was concluded, however, that birds exploited a range of foraging opportunities in a variety of habitats, showing great behavioural flexibility. The high density of breeding pairs at Elmley was probably due to the great productivity and diversity of insects in riparian habitats there.

Davies (1977a) found that small flies, mainly midges and fruitflies, were the favoured prey of flocks of Yellow Wagtails on Oxford pastures in early spring, although birds sometimes fed solitarily, catching dungflies.

Adult birds at Elmley commonly foraged by hovering over long grass and dropping to catch prey, a very different strategy from those described by Davies (1977a) and from that observed in Africa. There, the birds usually walk in short, grazed pastures or wade in shallow water, picking up items from the ground or from the water surface; sometimes they may rush forward to peck at prey (run-picking) or make a short sally from the ground and catch prey in mid air (flycatching).

The British race *flavissima* now also breeds in SW Norway, in N France and in the Netherlands (Glutz von Blotzheim 1985), where it nests alongside the blue-headed race *flava*, the two behaving as separate species (van Dijk 1975). In Ireland the scattered breeding records since 1956

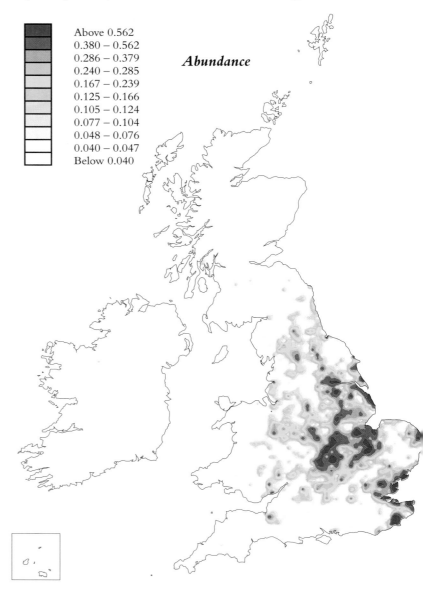

Abundance

	Above 0.562
	0.380 – 0.562
	0.286 – 0.379
	0.240 – 0.285
	0.167 – 0.239
	0.125 – 0.166
	0.105 – 0.124
	0.077 – 0.104
	0.048 – 0.076
	0.040 – 0.047
	Below 0.040

refer mainly to *flavissima*, but some birds have been recorded with the characteristics of both *flava* and the ashy-headed race *cinereocapilla*. Records of the blue-headed race collected during 1988–91 are shown separately on a map in Appendix E.

During 1988–91 there were an estimated 50,000 Yellow Wagtail territories in Britain. Breeding was not recorded in Ireland.

STEPHANIE J. TYLER

68–72 Atlas p 404

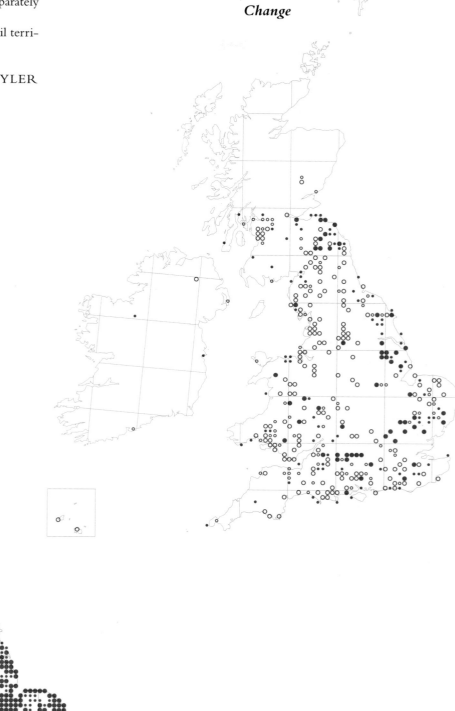

Change

Distribution

Years	Present, no breeding evidence		Breeding evidence		All records		
	Br	Ir	Br	Ir	Br	Ir	Both (+CI)
1968–72	141	2	1014	2	1155	4	1161
1988–91	288	3	759	0	1047	3	1050
				% change	−9.4	−25.0	−9.6

Grey Wagtail

Motacilla cinerea

The graceful, aerobatic flight of the long-tailed Grey Wagtail, as it fly-catches over water, is a characteristic sight on upland streams.

The Grey Wagtail is widely distributed in mainland Britain and Ireland, although it is scarce or absent from much of the lowland eastern and central counties of England. It is scarce, too, in the Outer Hebrides, Orkney and Shetland. It has a preference for feeding by fast-flowing watercourses, bordered by broadleaved trees, where there are rocks, riffles and areas of shingle (Ormerod and Tyler 1987a, 1990). Hence, it is most numerous in the upland areas in the N and W of Britain and in Ireland, although where there are suitable watercourses it breeds down to sea-level. In lowland areas, millstreams, weirs, artificial waterfalls as at the outflows of reservoirs, and canal locks are sought out (Tyler 1972). During the winter months migrants from Scotland and the N of England, as well as from the Continent, occur more widely in lowland Britain (*Winter Atlas*, Tyler 1979).

Since the *68–72 Atlas*, the Grey Wagtail has spread as a breeding bird farther into eastern counties of England, such as Lincolnshire and Norfolk. Two mild winters during fieldwork for the *88–91 Atlas* probably contributed to this expansion in the breeding population.

Unlike that of Dippers, the breeding abundance of Grey Wagtails is little affected by the acidification of watercourses, because the Wagtails take a greater range of insect prey, in terms of families and size, and more of non-aquatic origin (Ormerod and Tyler 1987a, 1991). Numerically, the most important prey are adult flies, although spiders and caterpillars can be valuable in the diet of nestlings. This is particularly so where streams are lined with broadleaved trees. Indeed, Grey Wagtails breed there in greater numbers than along streams in open moorland, while on streams among conifers they are particularly scarce. Ormerod and Tyler (1991) have attributed this to the greater biomass of insects available as prey along streams fringed by broadleaved trees.

The breeding performance of Grey Wagtails appears little affected by water quality or bankside habitat, with clutch and brood sizes being similar at all Welsh sites (Tyler and Ormerod 1991). Birds do, however, lay later in conifer and moorland sites than in broadleaved woodland or in pastures lined by broadleaved trees. Nestlings are heavier at broadleaved sites than elsewhere, probably because of earlier laying at the former, whilst those at acidic sites show reduced tarsal length (Tyler and Ormerod 1991).

If, in early spring, temperatures are high, birds start laying in early or mid March, but the peak in laying is during late April, with a second peak in late May (Tyler 1972, Ormerod and Tyler 1987b). The laying of repeat, second and occasional third clutches continues into August. Tyler (1972) gave a mean clutch size of 4.93 for a UK sample of nests. In SE Wales, the mean size of 147 clutches was 5.07, the clutches laid in May being larger than those laid in April or June (Ormerod and Tyler 1987b). The overall success rate of nests was low, with only 42% of eggs giving rise to fledged young (Tyler 1972), but nests were increasingly successful as the season advanced.

The breeding population in Britain and Ireland was estimated as 25,000–50,000 pairs in the *68–72 Atlas*. The *Trends Guide* showed that the population increased until about 1976 but then fell markedly following the severe winters of 1978/79, 1981/82 and 1984/85. The scale of decline may, however, be exaggerated by the location of WBS and CBC plots in sub-optimal lowland habitats. There was a marked increase in Grey Wagtails in Wales from 1987 to 1989, and a particularly large breeding population in Wales in 1990, which Ormerod and Tyler attributed to mild winters since 1987/88. Tyler *et al.* (1987) estimated 2–4 pairs per occupied tetrad in Gwent, although Sitters (1988) thought 1–2 pairs a reasonable figure in Devon. Assuming an average of two pairs per occupied tetrad throughout Britain and Ireland, but an average of only 10 occupied tetrads

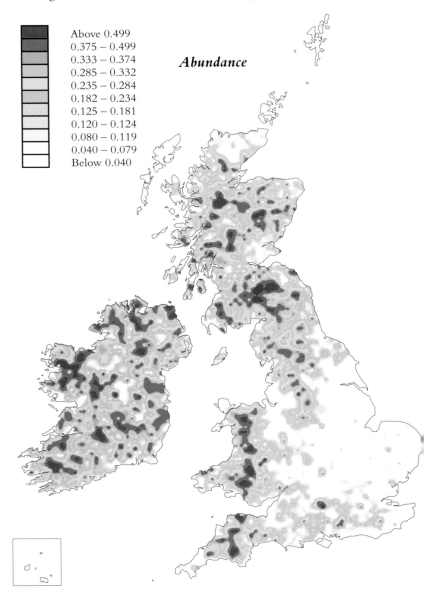

Above 0.499
0.375 – 0.499
0.333 – 0.374
0.285 – 0.332
0.235 – 0.284
0.182 – 0.234
0.125 – 0.181
0.120 – 0.124
0.080 – 0.119
0.040 – 0.079
Below 0.040

Abundance

per 10-km square, this would suggest 20 pairs per occupied 10-km square
(10–20 pairs was quoted in the *68–72 Atlas*), thus giving British and Irish
populations of 34,000 and 22,000 pairs respectively during 1988–91.

STEPHANIE J. TYLER and STEVE J. ORMEROD

68–72 Atlas p 402

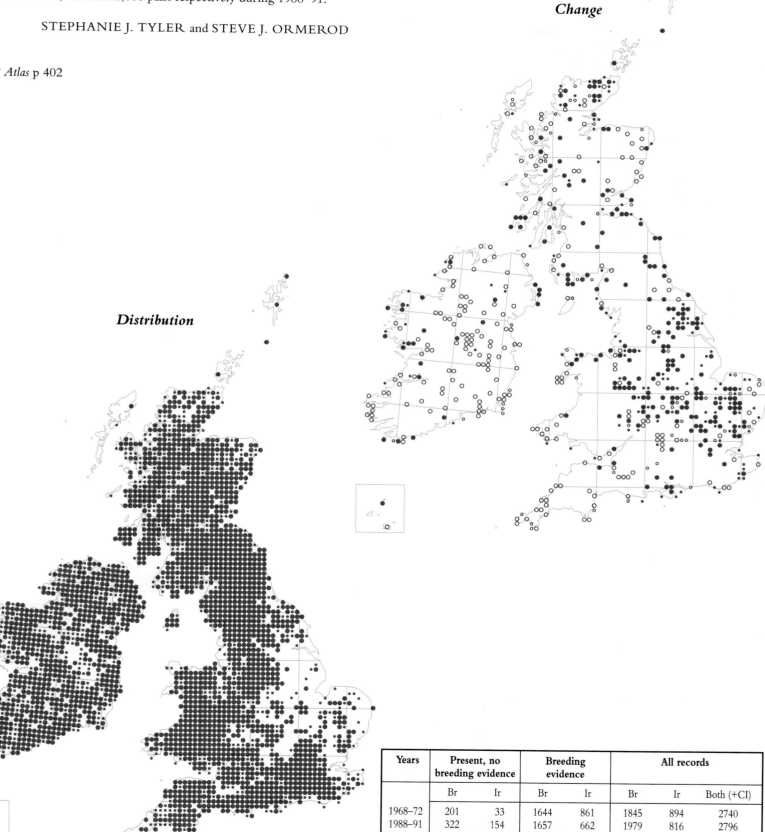

Change

Distribution

Years	Present, no breeding evidence		Breeding evidence		All records		
	Br	Ir	Br	Ir	Br	Ir	Both (+CI)
1968–72	201	33	1644	861	1845	894	2740
1988–91	322	154	1657	662	1979	816	2796
			% change		7.3	−8.6	2.1

Pied/White Wagtail

Motacilla alba

The slim pied form, and the constantly wagging tail, make this a most familiar species, known to many as 'Willy Wagtail' or 'Water Wagtail'. It is ubiquitous and occurs in a wide range of urban and rural habitats. With its confiding nature, it often breeds in proximity to man, and sometimes in rather unusual nest sites, such as in vehicles or working machinery, or in the nests of other birds, especially Blackbirds.

Pied Wagtails occur commonly throughout the mainlands of Britain and Ireland, on the Isle of Man, on the Scottish islands of the Inner and Outer Hebrides, and on Orkney. Their adaptability as to habitat and nest sites readily explains this wide distribution.

Little change in distribution has occurred over the last 20 years, although the Pied Wagtail has apparently expanded its range in Orkney, Shetland and the Outer Hebrides. It is, however, now more abundant in the N and W of Britain and in much of Ireland, than in lowland England. This is in agreement with the *Trends Guide*, which documents a decrease in E England and an increase in Scotland and W Ireland. These changes were attributed to the species' preference for mixed farmland, which in E England has largely been replaced by arable farming. There, and in much of S England, fewer livestock, loss of farm ponds, and an increase in autumn sown cereals, have combined to reduce habitat diversity and feeding opportunities for Pied Wagtails. In the north and west, Pied Wagtails commonly occur in very open, treeless habitats, often breeding in dry stone walls or roadside cliffs, and foraging on roads.

Davies (1977a) found that Diptera (mainly Chironomid midges) comprised about 97% of the prey of Pied Wagtails feeding in Oxfordshire pastures in March. Pairs breeding along roadsides in mid Wales often take insects damaged by vehicles. Caterpillars fallen from overhanging roadside trees are also much exploited.

Clutches are laid from early April to the end of July, with a peak of laying in the last week of April and the first week of May (Mason and Lyczynski 1980). The most frequent clutch size is five (mean 5.1 eggs), with two to three broods attempted by a pair each season. Eggs have a 52.7% success rate with 63.8% hatching, and 82% of hatched eggs producing fledged young (Mason and Lyczynski 1980).

Pied Wagtails *M. alba yarelli* are native to Britain and Ireland, but small numbers breed on the adjacent coasts of NW Europe, especially in the Netherlands (Teixeira 1979). Occasional White Wagtails *M. alba alba* breed, either in pure *alba × alba* or in mixed *alba × yarelli* pairings. The three maps shown here include all possible pairings of Pied and White

Wagtails, but those involving White Wagtails are shown separately on a map in Appendix E. White Wagtails breed only rarely in mainland Britain, and then mostly in Scotland (*Birds in Scotland*), but, as the appendixed map shows, they do so not infrequently in Shetland, where all possible pairing combinations were recorded during 1988–91. In the Channel Islands, however, Pied Wagtails do not breed, but pure *alba × alba* breeding records have become increasingly common, after a gap of 30 years since breeding was first confirmed in Jersey in 1952. Two or three pairs of White Wagtails have bred annually since 1980 on Alderney, and following the second breeding record on Jersey, in 1985, single pairs bred in 1988 and 1989, with two pairs in 1990, and five in 1991. There is thus clear evidence of a range expansion of the White Wagtail from the Continent into the Channel Islands (I.J. Buxton).

During 1988–91 there were an estimated 300,000 Pied Wagtail territories in Britain with a further 130,000 in Ireland.

STEPHANIE J. TYLER and STEVE J. ORMEROD

68–72 Atlas p 400

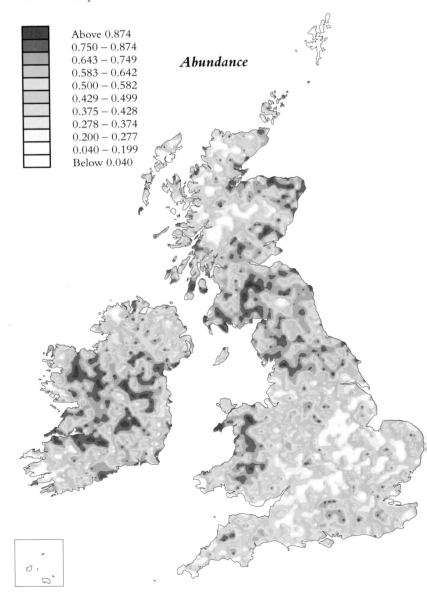

Abundance

Above 0.874
0.750 – 0.874
0.643 – 0.749
0.583 – 0.642
0.500 – 0.582
0.429 – 0.499
0.375 – 0.428
0.278 – 0.374
0.200 – 0.277
0.040 – 0.199
Below 0.040

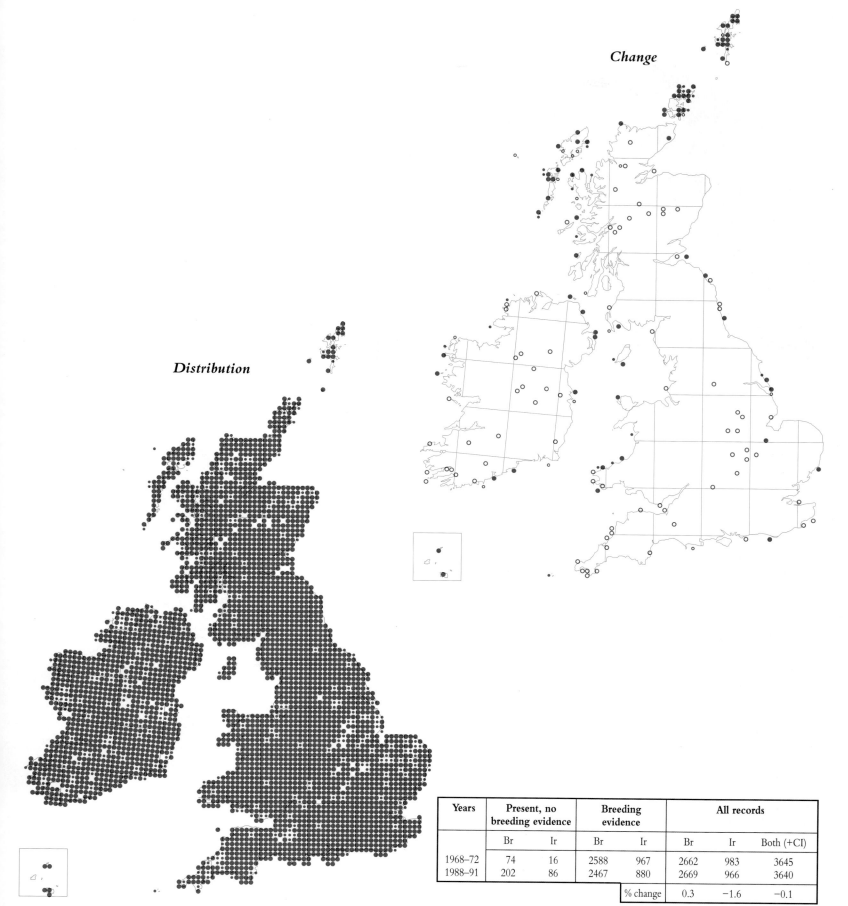

Change

Distribution

Years	Present, no breeding evidence		Breeding evidence		All records		
	Br	Ir	Br	Ir	Br	Ir	Both (+CI)
1968–72	74	16	2588	967	2662	983	3645
1988–91	202	86	2467	880	2669	966	3640
			% change		0.3	−1.6	−0.1

Dipper

Cinclus cinclus

The only truly aquatic species of passerine breeding in Britain, the Dipper is remarkable, too, for its bobbing and blinking behaviour. Traditionally associated with rocks and waterfalls in mountain streams, it breeds also on some lowland rivers in association with weirs and bridges.

Breeding Dippers occur throughout upland areas of N and W Britain, extending into the more lowland counties of Gloucester, Avon, Wiltshire and occasionally Hampshire. Their range includes much of Ireland, although they are absent from the centre, which has few suitable watercourses, and from parts of the southeast and west.

Dippers occur in greatest numbers in mid and SE Wales, much of N England and the Southern Uplands of Scotland, and in scattered areas in the Highlands. Northern and SW Ireland also support high numbers. These are generally areas with base-rich, productive streams. Conversely, acidic watercourses in W and N Wales, parts of the Lake District, and parts of Galloway and the W Highlands of Scotland support no Dippers or only small numbers. This is consistent with results of studies in Wales (Ormerod and Tyler 1987a) and in Scotland (Vickery 1991), which have shown that Dippers breed sparsely on acidic streams. These have poor invertebrate and fish communities compared with those streams flowing over more base-rich substrates, with abundant mayfly nymphs and caddis larvae. These two groups of invertebrates are the preferred prey of Dippers during the breeding season (Ormerod and Tyler 1987a). Ormerod and Tyler (1987a) have also drawn attention to the effect of conifers, which trap acidifying air pollutants on their foliage, and consequently have more acidic water draining from them than from unafforested catchments.

The patterns of distribution in the *68–72 Atlas* and the *88–91 Atlas* are similar, although there is an indication that Dippers may have disappeared from parts of Ireland, W Wales, SW and NE England, and various parts

of Scotland such as Galloway, the NE and W Highlands. All these areas have acidic streams, some draining from catchments extensively planted with conifers. However, poorer coverage during the 1988–91 survey in Ireland may partly explain the apparent decline there. By contrast, Dippers have returned to the Isle of Man.

Following mild winters, Dippers start breeding in early February, but in Wales peak laying occurs during the middle two weeks of April. Irish birds tend to start nesting earlier than those in Britain. Dippers breeding at higher altitudes and/or on acidic streams lay later than birds at lower or better quality sites. The former rarely attempt a second clutch. Pairs on acidic streams also lay significantly smaller clutches and rear smaller broods than those on neutral streams. Difficulties in supporting their broods, because of low invertebrate abundances, and low calcium availability at acidic sites, have both been implicated in this reduced breeding performance (Tyler and Ormerod 1992).

Besides being indicators of acidity, by their absence or scarcity, Dippers may likewise be indicators of river contaminants. Analyses in 1988 and 1990 have shown organochlorine residues and PCBs in most Dipper eggs from Wales, Scotland and Ireland. Eggs from Scotland in 1990 contained

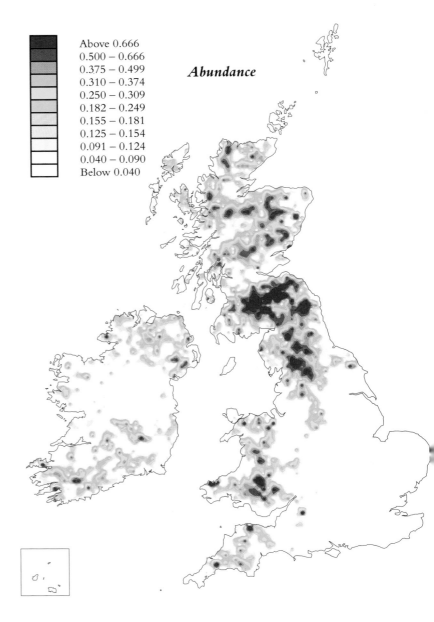

Abundance

■	Above 0.666
	0.500 – 0.666
	0.375 – 0.499
	0.310 – 0.374
	0.250 – 0.309
	0.182 – 0.249
	0.155 – 0.181
	0.125 – 0.154
	0.091 – 0.124
	0.040 – 0.090
	Below 0.040

higher organochlorine residues than those from Wales or Ireland, whereas eggs from Ireland contained higher levels of mercury (Ormerod and Tyler 1990, Ormerod and Tyler 1992). Scottish eggs even contained DDT, indicating possible recent, and illicit, use.

There is some evidence from mid and SE Wales that breeding success and the survival of nestlings and juveniles were reduced in the drought years of 1989 and 1990. Low flows, silting, reduced invertebrate faunas and high predator populations have all been implicated (Tyler and Ormerod).

Two races of Dipper breed in Britain and Ireland; *cinclus* over most of Britain, and *hibernicus* in Ireland and NW Scotland. Various other races of the Dipper breed in suitable upland areas throughout the Palearctic region. Some populations as in Germany and Poland (*BWP*) are declining due to pollution, others due to modifications of watercourses by man for irrigation schemes, hydro-electric schemes and reservoir construction as in Sicily (Maurizio Sara) and others, as in Hungary, for unknown reasons (Horvarth 1988). Although there have been local declines of Dippers in upland Britain, there have also been increases on some rivers, as in the S Wales valleys, where water quality has improved following closure of coal mines and other industries.

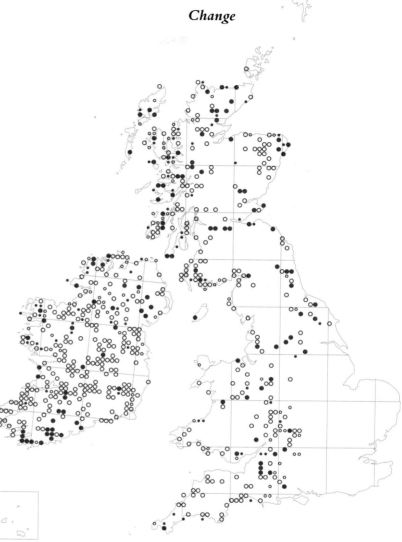

Change

Distribution

The population in the *68–72 Atlas* was put at 30,000 pairs, assuming an average of 15 pairs in an occupied 10-km square. Welsh data suggest that 15 pairs per 10-km square is realistic where there are productive streams, but elsewhere there may only be 5–10 pairs per 10-km square. This would give a population in Britain of *ca* 7,000–21,000 pairs and in Ireland of *ca* 1,750–5,000 pairs.

STEPHANIE J. TYLER and STEVE J. ORMEROD

One breeding record has been moved by up to two 10-km squares on the Distribution and Change maps.

68–72 Atlas p 326

Years	Present, no breeding evidence		Breeding evidence		All records		
	Br	Ir	Br	Ir	Br	Ir	Both (+CI)
1968–72	135	66	1299	551	1434	617	2051
1988–91	212	144	1097	285	1309	429	1738
			% change		−8.7	−30.4	−15.2

In most of Britain the main nesting habitat is the undergrowth of deciduous and mixed woods, and especially along streamsides where the vegetation is densest. When breeding numbers are high, Wrens overflow from their preferred habitats, firstly into gardens and orchards, and then into farmland and other hedges. Some may extend up scrubby valleys into moorland. On the coast Wrens are common on sea cliffs, and several of the Scottish islands have sedentary populations which are subspecifically distinct.

The nest sites recorded by Garson (1980a) in his analysis of BTO Nest Record Cards suggested that within or against a tree trunk was the commonest site, with scrub/herbs and buildings as next preference. This would clearly not be the case on, for example, outlying islands, where the majority of nests would be in fissures of cliffs or equivalent holes.

Males set up their territories in early spring and usually build several nests. When a female is attracted, she may inspect a number of these before choosing one in which to lay. The chosen sites have been shown to be the more cryptic and therefore less prone to subsequent predation (Garson 1980b). Males with large and good quality territories sometimes attract more than one female: such polygamy appears to be more common in woodland than in the less preferred habitats.

Wren

Troglodytes troglodytes

If they were not so lustily and obsessively vocal, Wrens would be easily overlooked, for they prefer to haunt the undergrowth. Thus, isolated pairs, which tend to be relatively quieter, could quite easily be missed.

It is clear from the maps that the Wren remains one of the most widespread breeding birds in Britain and Ireland, and it is probably one of the most numerous, too, although its abundance varies considerably over a period of years.

In Britain it is most numerous south of a line from the Mersey to the Humber rivers, although there are fewer on higher ground in Wales, and especially in the fenland areas around the Wash. Overall S and W Britain seem to hold larger numbers than elsewhere. The south and west were also shown by O'Connor and Shrubb (1986a) to be areas of highest density on farm CBC plots.

In the north, the Wren is mainly a lowland bird, with almost all areas of highest abundance near the coast. In Scotland, for example, Fife, the Moray Firth, and the coastal areas in Strathclyde Region seem especially favoured, and these are areas which have milder winters than much of the rest of Scotland. In Ireland, Wrens seem to be common almost everywhere. The *Trends Guide* makes it clear that winter temperature is the most important determinant of breeding numbers the following summer. Decreases of 20–25% in breeding numbers followed some of the cold winters in the early 1980s but, equally, numbers can recover very quickly after one or two mild winters. Snow, on its own, is relatively unimportant for Wrens can – and often do – feed, and even roost, under it. The areas of highest breeding abundance in any one year depend on which areas escaped the lower temperatures in the preceding winter.

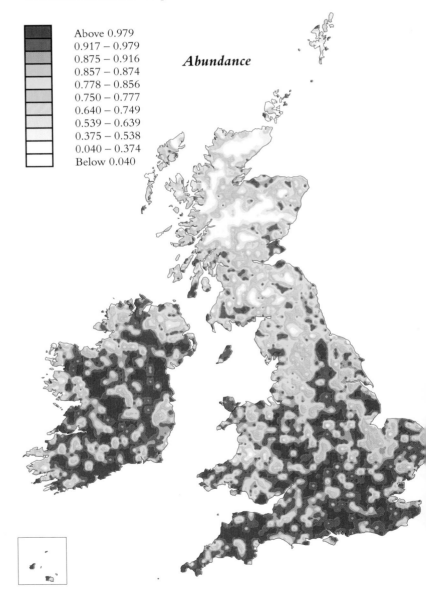

Abundance

Above 0.979
0.917 – 0.979
0.875 – 0.916
0.857 – 0.874
0.778 – 0.856
0.750 – 0.777
0.640 – 0.749
0.539 – 0.639
0.375 – 0.538
0.040 – 0.374
Below 0.040

Wrens are nearly always double brooded in Britain and Ireland, and there is a marked decline in clutch size through the season, which may run into July (Garson 1980a). Nesting attempts early on in the season are also subject to much higher predation than later ones, probably due to the relative lack of vegetation for concealment compared with later (Garson 1980a).

During 1988–91 there were an estimated 7,100,000 Wren territories in Britain, and 2,800,000 in Ireland.

PETER LACK

68–72 Atlas p 324

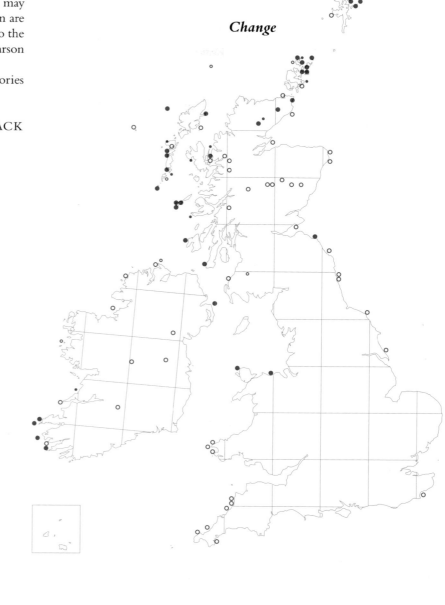

Change

Distribution

Years	Present, no breeding evidence		Breeding evidence		All records		
	Br	Ir	Br	Ir	Br	Ir	Both (+CI)
1968–72	27	5	2730	988	2757	993	3755
1988–91	97	59	2650	928	2747	987	3748
% change					−0.3	−0.5	−0.3

flicking display, often performed by several birds together, accompanies aggressive interactions between neighbouring males, often involving both alpha and beta males from more than one territory.

The ubiquity of Dunnocks in Britain and Ireland is not typical of the species over much of continental Europe. Like its relatives, originally it was almost certainly a montane bird, inhabiting scrubby vegetation not far below the tree line (still its preferred habitat in central Europe). It has adapted to similar habitats at lower levels, especially scrubby coniferous growth in young plantations, but has only locally become a common garden bird, especially in Germany and the Low Countries (*BWP*). During 1988–91 there were an estimated 2,000,000 Dunnock territories in Britain, and 810,000 in Ireland.

DAVID SNOW

Dunnock

Prunella modularis

The Dunnock is one of the more familiar of British breeding birds, but one whose behaviour and social organisation for long remained something of a mystery. When nesting it is rather secretive, the parents often giving little clue to the whereabouts of the nest, by approaching it through thick cover. The thin but distinctive song, a mixture of high-pitched notes and short trills, may be given from anywhere in a male's territory.

In Britain, in contrast to many parts of the Continent, Dunnocks are catholic in their choice of habitat, but low, thick growth of some kind – whether of brambles, mixed herbaceous growth, bushes or young trees – is necessary for their nest site. Over much of England they are the main host of the Cuckoo, making it possible for Cuckoos to breed in many areas of S England where there is no habitat suitable for the other two important hosts, the Reed Warbler and Meadow Pipit (Glue and Morgan 1972).

The Dunnock's distribution has remained essentially unchanged since the *68–72 Atlas* survey, except perhaps in the Outer Hebrides, where it is at best very local and difficult to detect (*Birds in Scotland*). During 1988–91, breeding was recorded in fewer squares in highland areas of Scotland, where the population is also sparse. The CBC has shown a decline in numbers since the mid 1970s, the cause of which is not known (*Trends Guide*). There is a N-S increase in population density on farmland, presumably reflecting the extent and thickness of hedgerow vegetation.

Breeding Dunnocks are very difficult to census accurately, chiefly because of their unusual social system. This is perhaps the most complex known in British birds, and the details of it have been revealed only since the *68–72 Atlas* survey (Snow and Snow 1982, Davies and Lundberg 1984, Davies 1985, Davies 1992). Although many birds are monogamous, polyandry (one female with two males) is common, polygyny (one male with two or three females) is regular, and 'polygynandry' (two or three males with two to four females) occasional. Male and female territories are largely independent of one another, male territories usually being larger, so that one male may encompass the territories of more than one female, leading to polygyny. Male territories, however, are often occupied by two males, one (the alpha male) dominant to the other (the beta male), and both may mate with the same female (polyandry). Beta males feed the young only if they have mated with the female, which they do not always succeed in doing, as the alpha male guards the female closely around the time of egg laying. Both males sing within the territory, leading to underestimation of numbers by conventional census methods. The wing-

68–72 Atlas p 392

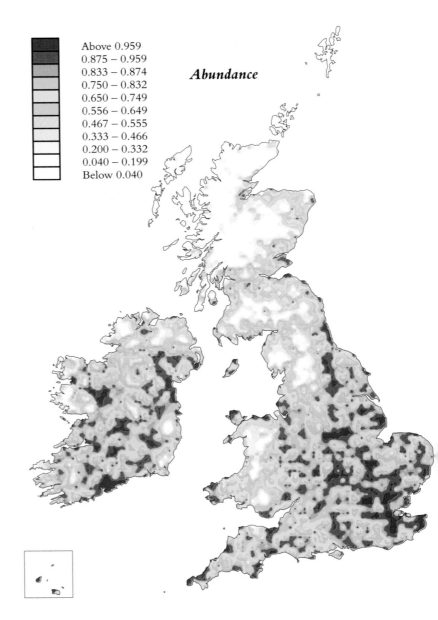

Above 0.959
0.875 – 0.959
0.833 – 0.874
0.750 – 0.832
0.650 – 0.749
0.556 – 0.649
0.467 – 0.555
0.333 – 0.466
0.200 – 0.332
0.040 – 0.199
Below 0.040

Abundance

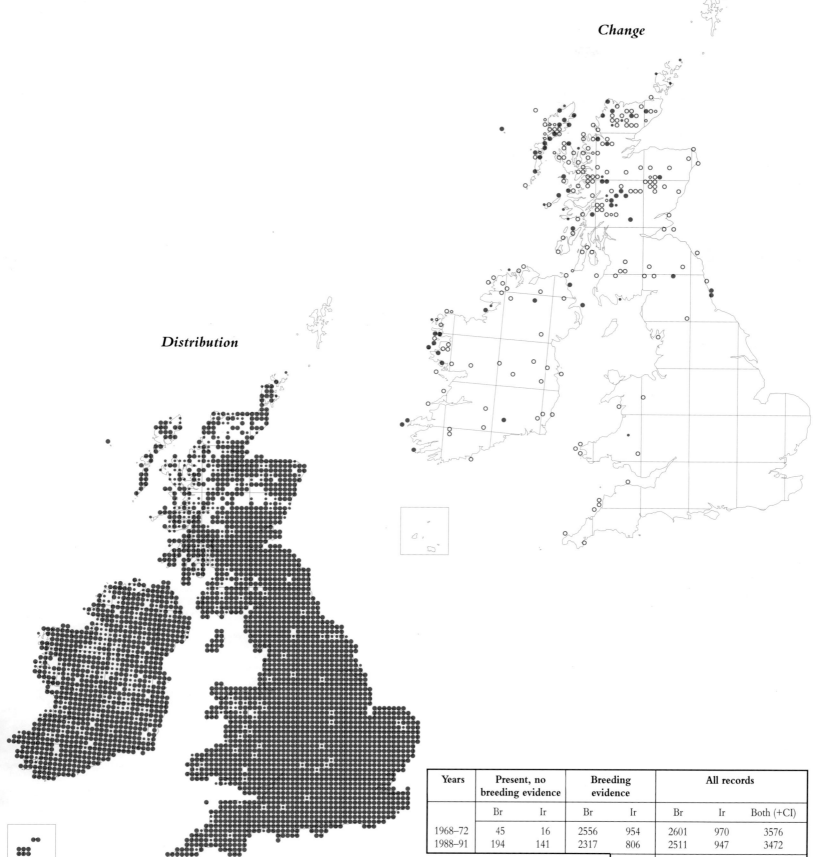

Change

Distribution

Years	Present, no breeding evidence		Breeding evidence		All records		
	Br	Ir	Br	Ir	Br	Ir	Both (+CI)
1968–72	45	16	2556	954	2601	970	3576
1988–91	194	141	2317	806	2511	947	3472
			% change		−3.2	−2.3	−2.9

For example, the well-wooded Weald of SE England shows up on the Abundance map as an area of high overall density, even though farmland densities there are relatively low (*Trends Guide*). The general agreement between the two methods, increases the confidence with which Ireland, where there are very few CBC sites, can be identified as an important area for breeding Robins (*Birds in Ireland*).

Ringing recoveries suggest that few Robins move far from their birth-place (Mead 1984). Therefore it is no surprise that the Abundance map is so similar to the *Winter Atlas* map. There are two obvious differences. First, there is a suggestion of altitudinal movement around, for example, the Great Glen and central lowlands of Scotland, with more birds breeding at upland sites than can winter there. Second, Robins are less widespread as breeders in Shetland than as winter visitors. The claim in the *Winter Atlas* that this is also true for the Outer Hebrides and Orkney appears to be incorrect. The differences between the *68–72 Atlas* and the *Winter Atlas* are probably better explained by limited range expansion, rather than by migration. Closer comparison of the Abundance map with the *Winter Atlas* suggests scope for many local studies. What, for instance, is really happening to Robin numbers in Ireland around the Bog of Allen and Co. Cork?

Robin

Erithacus rubecula

Robins can breed almost anywhere, so long as there is a shrub layer and the ground is not waterlogged. These adaptable songsters are therefore widespread. But why are they so much more common in some regions than others?

A confiding nature and a familiar warbling song make Robins hard to overlook. When, however, they are living at low densities, Robins can be virtually mute and surprisingly furtive. Therefore, while the Distribution map is probably unusually accurate, the Abundance map is likely to underestimate abundance in areas where Robins are sparse.

The Distribution map emphasises that a visit to Orkney or Shetland is the only real hope for a birdwatcher anxious to avoid Robins. Continuing afforestation may eventually remove even this escape (*Birds in Scotland*). For example, nearly three times as many 10-km squares in Orkney and Shetland are occupied on the 1988–91 map compared with that in the *68–72 Atlas*. The Change map shows that Robins are also now more widespread in the Outer Hebrides.

The Abundance map shows that Robins are most common in Britain south of a line from Chester northeast to York and in Ireland. Within these areas, notable pockets of low density occur wherever trees are sparse, ranging from the fens of East Anglia to uplands such as Macgillicuddy's Reeks in Co. Kerry. In N Britain the avoidance of high ground is emphasised by the relatively high densities in the central lowlands of Scotland and around the Moray Firth.

The CBC suggests that there are N–S and E–W gradients in Robin density on British farmland, so that Robins are most common in Wales and SW England (O'Connor and Shrubb 1986a). The trends for woodland are much less distinct, with the highest densities in Wales, but the lowest in SW England (*Trends Guide*). Taking into account differences in land usage, there is broad agreement between the CBC and Atlas results.

Above 0.979
0.917 – 0.979
0.875 – 0.916
0.833 – 0.874
0.750 – 0.832
0.714 – 0.749
0.625 – 0.713
0.500 – 0.624
0.333 – 0.499
0.040 – 0.332
Below 0.040

Abundance

Robins are notoriously territorial. There is considerable scope for studies on the habitat features which influence breeding success. Hoelzel (1986) has shown how important dense vegetation is to Robins, although it is unclear exactly how they benefit. Perhaps dense cover provides more food, or better protection from bad weather and predators. Although they can breed in gorse scrub, the presence of even a few taller song posts increases the chances that a site will be used for breeding (D. Harper). Since Robins capture most of their prey on the ground (*BWP*), open areas with short vegetation may be another important feature of their territories.

The breeding biology of Robins has been described by Lack (1943), East (1981), Mead (1984) and Harper (1985). Although most pairs do not start laying until late March or early April, providing the weather is mild a minority living in S England begin as early as January. Since the last clutches are not laid until July, and the next clutch is often started before the previous brood fledges, many pairs rear two or three broods in a season, and some manage five! The decreases in numbers caused by cold winters are therefore short lived and the population is basically stable (*Trends Guide*).

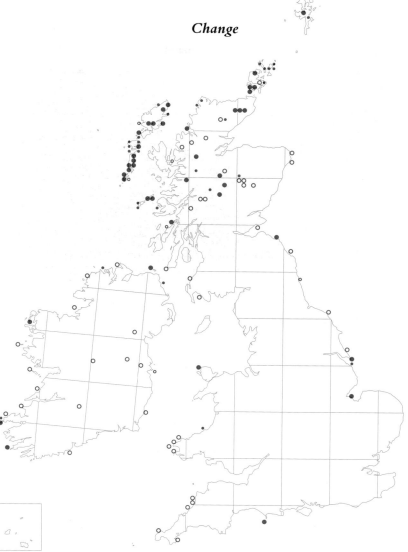

Change

Distribution

The mean density of Robins on farmland can reach 33 males per km^2 and the mean density in woodland is often twice as high (*Trends Guide*). During 1988–91 there were an estimated 4,200,000 Robin territories in Britain, and a further 1,900,000 in Ireland.

DAVID HARPER

68–72 Atlas p 354

Years	Present, no breeding evidence		Breeding evidence		All records		
	Br	Ir	Br	Ir	Br	Ir	Both (+CI)
1968–72	34	2	2575	975	2609	977	3591
1988–91	93	23	2536	944	2629	967	3610
				% change	1.0	−0.9	0.5

Nightingale

Luscinia megarhynchos

'A vision or a waking dream?' The song of the Nightingale has been a recurring inspiration to poets and musicians through the ages. That so inconspicuous a bird should have such a rich – and nocturnal – song has no doubt added to the aura of mystery which surrounds the species. There is still much which we do not know about the Nightingale, even if we, unlike the early poets, now realise that it is the male and not the female which sings.

The Nightingale has a southeast distribution in Britain. Most are found southeast of a line from the Humber to the Severn, with few in Devon and none in Cornwall. Even within this area it is now sporadic in some parts, notably within a broad band extending northwest from London through the English Midlands. This pattern reflects to a degree the species' preference for coppice woodland, which has become increasingly concentrated in SE England, but must also be determined by climatic factors, since plenty of apparently suitable dense cover exists outside the species' range.

The areas where Nightingales are most abundant are mainly in the extreme southeast, in Suffolk, Kent and Sussex. In a national census in 1980, these counties held 45% of the total of 4,770 singing males recorded (Davis 1982).

The contraction in range noted in the *68–72 Atlas* has continued. It has been most marked at the northern and western limits of the distribution – from Lincolnshire to W Dorset – but is also evident in the counties to the north and northwest of London.

Egg laying in Britain extends from late April to mid June, with a peak in mid May. The clutch is most commonly 4 or 5 eggs, and incubation and fledging times are respectively 13 and 11 days (Morgan 1982). Second broods are unknown in Britain, but replacement clutches may follow early failure. Migration to the wintering grounds, assumed to be in West Africa, is most marked during August and the first half of September.

Nightingales feed mainly on the ground, generally under tree canopy but sometimes in open woodland rides. Prey includes ground-dwelling invertebrates such as beetles, especially weevils, wood ants where present, and spiders (A. Henderson). Nestlings receive similar food, plus a higher proportion of lepidopteran larvae than the adults.

Nightingales occupy habitat types from hedgerows and pioneer scrub, through coppice and young conifer plantations to mature deciduous woodland. Very high densities can be found in scrub associated with wetlands such as gravel pits. Although, nationally, other vegetation types hold most Nightingales (Hudson 1979), coppice is very important where actively managed. The key factor is the presence of dense undergrowth close to the ground, and this can be present for several years, typically 3–9 years after coppicing (Bayes and Henderson 1988, Fuller *et al.* 1989, Fuller and Henderson 1992). Coppice, which passes through this stage on each cycle, is more suitable in the long term than scrub or conifer plantations which, although used for a time, may then become unsuitable for many years. Not all coppice is suitable, however, and pure chestnut with its rapid growth and open structure seldom holds Nightingales unless there is also dense bramble.

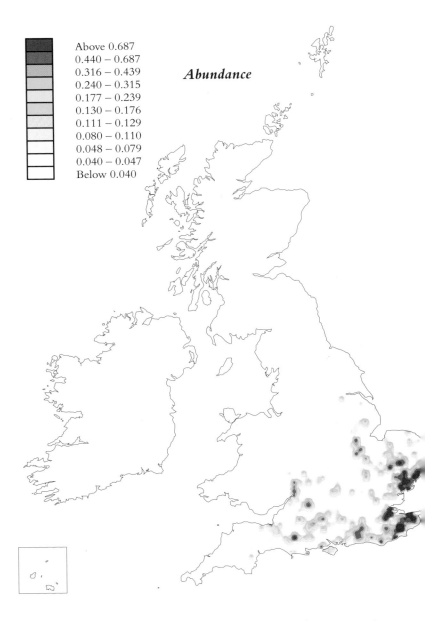

Abundance
Above 0.687
0.440 – 0.687
0.316 – 0.439
0.240 – 0.315
0.177 – 0.239
0.130 – 0.176
0.111 – 0.129
0.080 – 0.110
0.048 – 0.079
0.040 – 0.047
Below 0.040

In Britain, Nightingale song is most noticeable from the birds' arrival in the first half of April until the end of May. It then becomes more sporadic, petering out completely during July. It can be heard throughout the night, but tends to be strongest around first light. There is much individual variation, and some males sing surprisingly less often than one would expect. Comparison of the results of single night counts at the peak of the song period with those from detailed studies in the same areas show that single visits may underestimate numbers by as much as a half (Hudson 1979, A. Henderson). The best time to count Nightingales is in mid May from midnight to dawn, when song is relatively frequent and when other song birds are silent.

There have been two national Nightingale surveys, in 1976 and 1980 (Hudson 1979, Davis 1982), which revealed respectively 3,230 and 4,770 singing males. Fluctuations certainly occur towards the edge of the range but probably are not so great as suggested by these results, which may reflect variation in the extent of coverage and song frequency. Making allowance for sites not visited, and for the numbers of single night visits (which were high in both surveys), the national population between 1988

Change

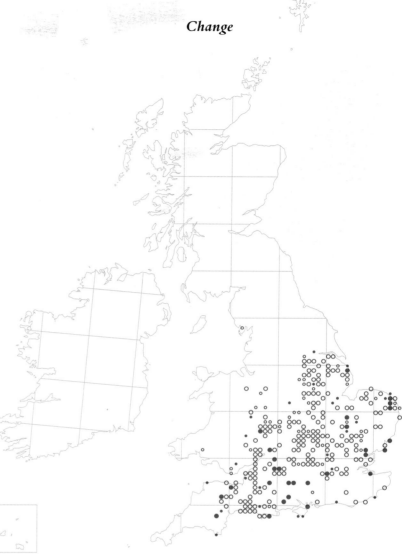

Distribution

and 1991 was probably in the range 5,000–6,000 pairs. The observed decline may mean, however, that the total is now towards the lower end of the range but southeast counties from Kent to Hampshire, which held almost 50% of the total in 1980, have been least affected so far.

ANDREW HENDERSON

68–72 Atlas p 352

Years	Present, no breeding evidence		Breeding evidence		All records		
	Br	Ir	Br	Ir	Br	Ir	Both (+CI)
1968–72	52	0	587	0	639	0	639
1988–91	154	0	303	0	457	0	457
	% change				−28.5	0	−28.5

Black Redstart

Phoenicurus ochruros

The flash of a bright orange-chestnut quivering tail announces the welcome spring arrival of a full-plumage male Black Redstart to brighten a scattering of built-up environments. Others betray their presence through a scolding 'tucc-tucc' alarm call, still more by the short warbling song with characteristic metallic terminal flourish. Even so, some may well go undetected amidst the noise and clutter of industrial and residential waste ground favoured by Black Redstarts today; sites sometimes inaccessible and often of little appeal to birdwatchers.

In Britain, from 1923 onwards, colonising Black Redstarts occupied mainly natural cliff sites in Sussex and Cornwall, but during the 1940s pairs readily took to bombed areas in London and towns in the southeast (notably Dover). In the late 1950s and 1960s pairs clung tenaciously to the few remaining sites as war damage was repaired. Thereafter, the species spread into built-up areas elsewhere, extending thinly north to Lancashire and Northumberland. A surge in numbers began in 1969 and was maintained generally throughout the 1970s, with fewer birds using coastal cliffs and war damaged sites, more often preferring modern industrial complexes and old houses, in rural as well as coastal settings. In 1977 a BTO Black Redstart breeding survey, employing greater field effort than hitherto to detect this elusive species, revealed a minimum of 104 territory-holding males and demonstrated a broad spectrum of nesting places ranging from derelict industrial sites to gas works, sewage farms, railway sheds, power stations and town centres (Morgan and Glue 1981).

Fieldwork during 1988–91 shows the continuing importance of urban sites in central and Greater London and the Home Counties, but reveals a second major concentration embracing towns in the West Midlands, and there are key coastal populations in Norfolk/Suffolk, Lincolnshire and South Humberside. Elsewhere, scattered singing males or pairs were detected west to Dorset and Gwynedd, and in urban localities north to the Lothians.

Since the *68–72 Atlas* there has been an apparent thinning of the nucleus population in London and the southeast counties, possibly reflecting the continued replacement of wartime damaged buildings. This coincides with a strengthening of Black Redstart numbers in the Midlands and East Anglia.

In Scotland the Black Redstart remains primarily a passage migrant, occasional individuals staying the winter, chiefly in the west from Ayr to Galloway. In 1973 a presumed unmated female was discovered incubating a clutch of four infertile eggs on Copinsay (Orkney), and in 1976 nesting was suspected in Aberdeen (*Birds in Scotland*). Similarly, in Ireland it remains a regular spring and autumn migrant, though in variable numbers annually. Few overwinter and just the occasional summer bird is recorded in suitable breeding habitat, e.g. Wicklow in 1971, though nesting has never been suspected (*Birds in Ireland*).

The process of colonisation over almost 70 years has been painfully slow, and it is perhaps a little premature to claim that the Black Redstart is now a firmly established widespread breeding bird in Britain. It has failed to move extensively into towns, village gardens, farms or mountainous rocky areas, all used by Continental birds. Equally, it has not favoured coastal sites used by the earliest colonists.

Indications suggest that the breeding biology of the Black Redstart in Britain, like that in Sweden (colonised in 1910), differs from that of S European populations (Andersson 1990). In Sweden, fledging success is lower, the proportion of second broods fewer, the season later, and the range of nesting habitats more limited (Andersson 1990). In Britain, some site-faithful males may be paired and egg laying started by mid April, but the late arrival of most recruits from winter quarters (late May to mid June) suggests many may be birds displaced from elsewhere (Langslow 1977).

Pairs often occupy holes and ledges in industrial premises, wasteland, and housing under construction, all, by their nature vulnerable to change and disturbance, with known clutches lost to children at play, working men and cats. Inexperienced first-year males are common in Britain, a study in coastal Lowestoft, Suffolk, showing these to be less efficient parents than full-plumage males (Beecroft 1986).

The number of male Black Redstarts located in Britain each year exceeds the total later proved to breed and the population shows sharp year-to-year fluctuations. This is typical of a scarce migrant at the northern limit of its breeding range (Fitter 1965, *Trends Guide*). The RBBP's long-term assessment of Black Redstart numbers peaked in 1988, at 118 reported pairs and singing males, but fell back to a maximum of 82 at 56 localities in 1989, and 74 at 50 localities in 1990 (RBBP). Allowing for known clusters of several pairs at key city and coastal sites, and knowing that some 10-km squares were occupied in single years only, the current British population is likely to fluctuate around 80–120 territory-holding males/pairs, virtually all of them in England.

DAVID E. GLUE

68–72 Atlas p 350

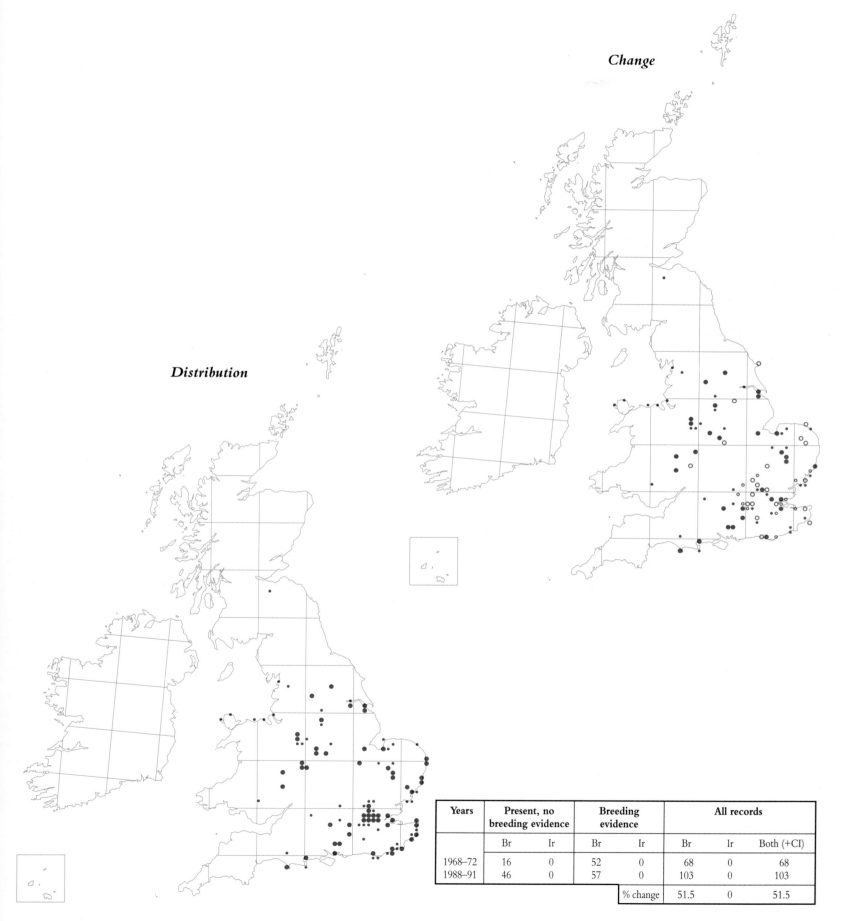

Change

Distribution

Years	Present, no breeding evidence		Breeding evidence		All records		
	Br	Ir	Br	Ir	Br	Ir	Both (+CI)
1968–72	16	0	52	0	68	0	68
1988–91	46	0	57	0	103	0	103
				% change	51.5	0	51.5

Redstart

Phoenicurus phoenicurus

The sweet but slightly hoarse song of the Redstart, often delivered from a tree top, is a sure sign of spring. The Redstart is found in a wide variety of habitats, ranging from old hedgerows and parks to oak and Scots pine woods, and even to tree-fringed streams penetrating open moorland.

The Distribution map shows that Redstarts are widespread, particularly in upland areas, but uncommon or absent from the lowlands and from most islands, from Anglesey to Shetland. This distribution is similar to that of the Wood Warbler and, less so, the Pied Flycatcher, indicating the same preference for western oak woods, where the trio of species forms one of the most distinctive woodland bird communities in Britain. All three species are curiously rare in Ireland.

The Abundance map shows that Wales is perhaps the Redstart's main stronghold, followed by the N Pennines and the Lake District. In Scotland, Redstarts are more patchily distributed, occurring in higher densities in drier, taller oak, birch and Scots pine woodlands.

Since the *68–72 Atlas* there have clearly been large losses from central, E and S England and, given the good coverage achieved in these areas during the 1988–91 period, these are almost certainly real. The reduction is further supported by local studies (e.g. around Banbury) which show an almost total loss of Redstarts between 1968–72 and the mid 1980s (Brownett 1989).

At first sight there is an apparent conflict between these changes and those of the CBC. The *Trends Guide*, drawing on CBC data, shows Redstarts to be more numerous in 1988 than during the *68–72 Atlas*. But regional CBC studies have indicated that numbers in E England have not shown the increases seen elsewhere in CBC work. There is thus the implication that the species is increasing in those parts of its range where it is more abundant, but decreasing at the periphery of its range. The

Abundance map indicates that even where Redstarts do occur in E England, they are at low densities. It is further apparent that individual populations have become more isolated from one another since the *68–72 Atlas*.

Redstarts are one of the Africa-wintering species widely thought to have been affected by the Sahel drought. However, the CBC index has shown a consistent increase during the 1970s and 1980s. The Redstart's recovery has been faster than that of other 'Sahel' species, and was unaffected by the one severe drought year (1984). This suggests that circumstances in Africa are not the only factor controlling the population. The recovery has not been complete, however, as is indicated by evidence from E Britain, Devon (Sitters 1988) and NE Scotland (Buckland *et al.* 1990).

Redstarts take less readily to nest boxes than do some of their potential competitors, such as tits or Pied Flycatchers, though numbers are increased by the provision of boxes. It has been suggested that the increase in nest box schemes may have attracted competitors away from natural holes, allowing Redstarts to breed more freely, though this seems unlikely as many nest box schemes were present long before the increase started (*Trends Guide*).

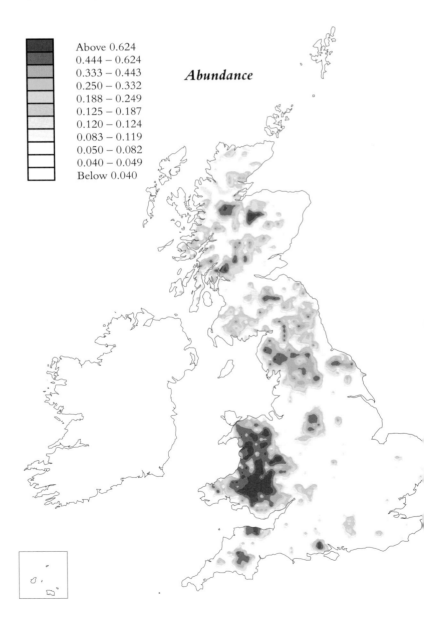

Above 0.624
0.444 – 0.624
0.333 – 0.443
0.250 – 0.332
0.188 – 0.249
0.125 – 0.187
0.120 – 0.124
0.083 – 0.119
0.050 – 0.082
0.040 – 0.049
Below 0.040

Abundance

Redstart populations have declined extensively in much of Europe (*BWP*). Adverse conditions in the wintering grounds have been blamed, but there is also evidence from Switzerland of a deterioration of conditions on the breeding grounds, as a result of habitat loss or reduction in quality, and changing agricultural practices (Bruderer and Hirschi 1984). In Finland a decline in numbers between 1927 and 1977 was attributed to a decrease in the extent of large old forests (Jarvinen and Vaisenen 1978), although there has been a recent increase in S Finland and in Denmark (*Trends Guide*).

The breeding biology of the Redstart has been little studied in Britain, in contrast to several other hole nesting species. Redstarts across Europe appear to breed earlier than Pied Flycatchers, and show a typical decline in clutch size as the season progresses. The clutch size averages about six. Although most nests in natural cavities in Wales were in oak, Redstarts used other species of tree (birch, alder, rowan) in greater proportion than their abundance (Stowe 1987). In dry stone wall areas, holes in walls, including buildings, are much used. The tendency of many nest box schemes in Britain to use nest hole sizes too small for Redstarts has undoubtedly thwarted the existence of a large, box breeding population.

Change

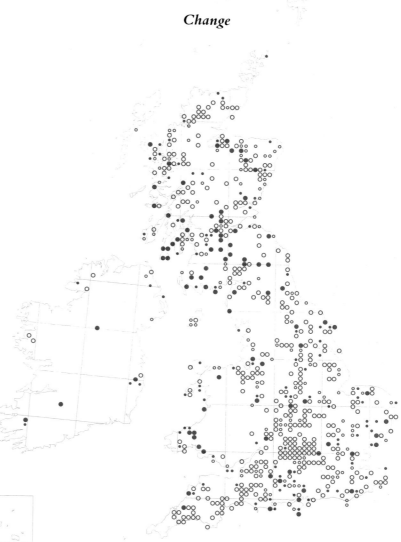

Distribution

The mean CBC index for all habitats during 1968–72 was 65, and rose to 100 during 1988–91. Assuming mean densities increased in the same manner, then the estimate of 45 pairs per occupied 10-km square, used in the *68–72 Atlas*, would have risen to 70 pairs by 1988–91, yielding a total British population of 90,000 pairs. This estimate, however, contrasts markedly with that of 330,000 territories calculated from mean CBC densities. It is likely that the true population lies towards the lower end of these extremes. The Redstart is a very rare breeding bird in Ireland with a maximum of four breeding pairs recorded in any one year (1988, O'Sullivan and Smiddy 1989) during 1988–91.

TIM J. STOWE

68–72 Atlas p 348

Years	Present, no breeding evidence		Breeding evidence		All records		
	Br	Ir	Br	Ir	Br	Ir	Both (+CI)
1968–72	174	2	1487	7	1661	9	1670
1988–91	308	7	1019	4	1327	11	1338
				% change	−20.1	22.2	−19.9

Whinchat

Saxicola rubetra

Often heard before it is seen, the Whinchat scolds with its sharp 'tic tic' any intruder entering its domain: otherwise, the male's clear, crisp song may be the first indication of an occupied territory. A rather attractive summer visitor, which prefers low to medium height perches, the Whinchat is often a colourful addition to many a featureless upland area.

The Distribution map illustrates that although in the preferred upland areas of Britain, the range is not dissimilar to that in the *68–72 Atlas*, there are now many lowland squares, especially in the English Midlands, the lower Thames valley and the coastal outposts of Suffolk and Kent where the species has all but disappeared as a breeding bird. South of a line drawn from the Humber to the Severn, only 89 10-km squares revealed evidence of breeding during the 1988–91 period, whereas in the *68–72 Atlas* breeding (probable and confirmed) was recorded in 238 10-km squares, a decrease of nearly two-thirds. Given that S Britain may be more easy to survey, with ease of access and more flat land, the decline is alarming. In Ireland the trend is similar, with 124 occupied 10-km squares compared with 192 in 1968–72. Only in Wales, NW England and much of Scotland is the distribution apparently little changed, while in the Outer Hebrides the Whinchat may even have gained some ground. In both surveys it seems unlikely that any but the smallest pockets will have been overlooked, for the Whinchat draws attention to itself at all stages of its breeding cycle.

Some causes of the decline in numbers between the two Atlases may be suggested. During the 1970s and early 1980s generous farming subsidies became available to encourage land drainage and promote improved soil conditions. These, together with the advent of monoculture systems over much of lowland Britain, especially in England, threatened those species that favour moist, uncultivated habitats, the Whinchat among them. Even in Ireland, where changes in land use were resisted longer, new farming practices encouraged by the Common Agricultural Policy resulted in severe loss of suitable habitat. Many of the reasons for loss suggested in the

68–72 Atlas are still relevant today, and are likely to have hastened the decline in areas where the species was already thinly spread. The general 'tidying up' of landscape, including urban waste ground, will have done the Whinchat few favours.

Towards the end of the *68–72 Atlas* period much planting of conifers took place, especially in Northumbria, Dumfries and Galloway and the Borders, covering large tracts of land. Whinchats exploited much of this new, temporary, habitat, often colonising on a large scale. But as the crops matured simultaneously, widespread areas became unsuitable again, a pattern which will doubtless be repeated with the cycle of planting and felling.

An analysis of 476 nest sites in Britain (Fuller and Glue 1977) revealed the main vegetation types to be open grass (58%), bracken (14.3%), mixed low vegetation (11%), heather (6.9%) and gorse (4.8%). A study in Ayrshire (Gray 1974), however, showed that there, low gorse was the preferred habitat for arriving males, and that it subsequently contained all first nests. Loss or deterioration of roadside verges reduces the supply of suitable habitat. Gray found that if verge mowing was delayed from mid June to early August, breeding success improved from 46% to 64%, and that where

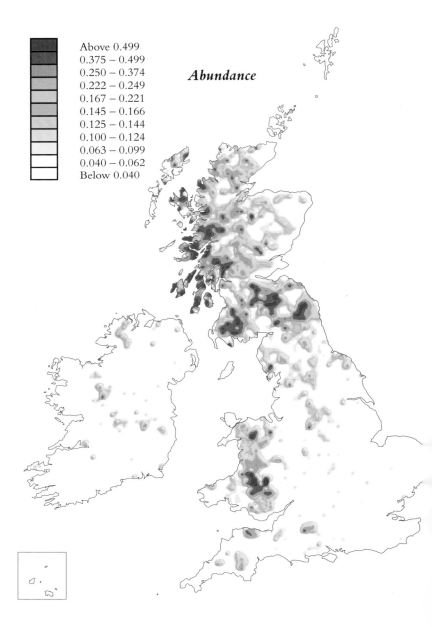

Abundance

Above 0.499
0.375 – 0.499
0.250 – 0.374
0.222 – 0.249
0.167 – 0.221
0.145 – 0.166
0.125 – 0.144
0.100 – 0.124
0.063 – 0.099
0.040 – 0.062
Below 0.040

Whinchats nested in habitats relatively free from human influence, they were usually more successful.

Some habitats, especially on coasts and uplands, have Whinchats and Stonechats nesting as neighbours, leading to interaction between the two. Observations (Greig-Smith 1982, Callion) showed that Whinchats were never the aggressors, always the aggrieved. Presumably resident Stonechats have the advantage over newly arrived migrant Whinchats and are able to establish early dominance. It may be relevant that Whinchats forage and feed more on the ground than do Stonechats, making less use of a strong territorial perch, and are thus easier to dislodge.

Although the Whinchat is often stated to be double brooded in some parts of its range, there has been little evidence to substantiate this. In Cumbria, however, an isolated group of three pairs, occupying a SW-facing fellside of bracken and gorse, all produced two successful broods.

Most W European countries have reported declines, ranging from 'modest' in France to 'catastrophic' in S Finland. Little is known about the E European populations, though it seems likely that, with the recent opening of borders, the inevitable inflow of different farming methods will adversely affect Whinchat habitat.

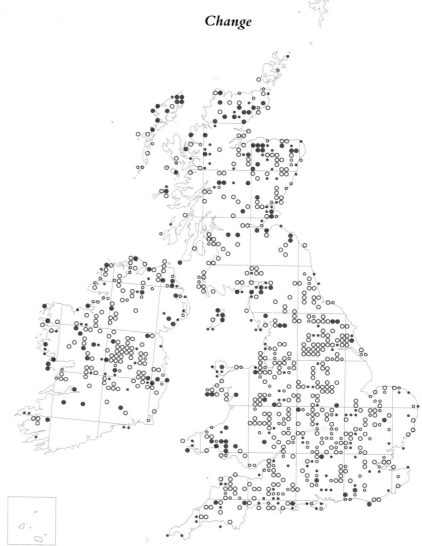

Change

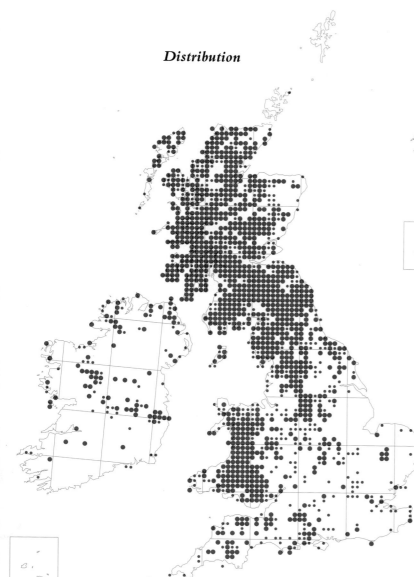

Distribution

Still numerous in some western uplands of Britain, with densities of up to 50 pairs per 10-km square in particularly choice habitats in Wales and Cumbria (Williams, Day), Whinchat numbers may locally be as high as ever. Assuming that the reduction in range since the *68–72 Atlas* has not been accompanied by a reduction in density within occupied squares, then the population in Britain is estimated at 14,000–28,000 pairs, and that in Ireland at 1,250–2,500 pairs.

JOHN CALLION

68–72 Atlas p 346

Years	Present, no breeding evidence		Breeding evidence		All records		
	Br	Ir	Br	Ir	Br	Ir	Both (+CI)
1968–72	192	32	1484	160	1676	192	1868
1988–91	342	48	1062	76	1404	124	1528
			% change		−16.2	−35.1	−18.2

Stonechat

Saxicola torquata

There can be few early springtime pleasures greater than the sight of a male Stonechat atop coastal gorse, or the sound of its captivating song from fellside heather or lowland heath; often unnoticed, the unobtrusive female can be much closer to the observer. Stonechats can be fairly adaptable, though three essential features must be contained in all territories: a regular spread of perches for songposts and from which to drop on to prey, some reasonably short vegetation for ground feeding and a good measure of thick cover for nesting.

With distribution biased towards the more temperate W of Britain, some Stonechats survive even the most severe winters. Following the extremely harsh winter of 1963 a period of recovery and expansion continued throughout the *68–72 Atlas* period and for some years after, resulting in the occupation of many inland and east coast localities, notably in Northumberland (Macfarlane 1980) and much of E Scotland. Like several other species, including the closely related Whinchat, Stonechats at this time also exploited many young forestry plantations. The decline in distribution now evident is not easy to explain. The run of cold winters prior to 1986 will have caused some losses nationwide, as it did in Devon (Sitters 1988), and probably the recovery expected as a response to the recent milder winters was not enough to be reflected in the Distribution map. Since few Stonechats winter inland or on the E coast, they presumably either move towards the W coast or even leave Britain. By 1987 almost 17,000 Stonechats had been ringed, of which 34 have been recovered in W Mediterranean countries including two that had reached Spain as early as October.

The Change map documents a dramatic retreat from the E coast of Britain, with few breeding now between Aberdeen and Norfolk. On the W coast of Britain the almost continuous line of breeding of the *68–72 Atlas* now shows many gaps, especially in Ayrshire, Galloway, Devon and Cornwall. In Scotland populations in the north and west remain healthy, as do those in Wales. Many unoccupied 10-km squares in Ireland, especially in the east and inland, indicate a serious decline in distribution.

Several studies of breeding Stonechats (Phillips 1968, Fuller and Glue 1977, Callion and Holloway) give an average fledging success of over 80%: more than adequate to sustain a stable population. If not affected by man, records show that, given a run of reasonably mild winters, numbers can return to normal within six years (*Irish Birds*). Although too few ringed birds have been recovered to identify migration strategies, a large proportion of the population remains in Britain throughout the year, even

some first-winter birds in Scotland. Maturing young conifers will have had an effect in reversing the expansive trend of the early 1970s. Habitat loss due to intensive agriculture, urban and recreational development and ever-increasing human disturbance are likely to be factors that have eliminated Stonechats from marginally suitable habitats and reduced numbers in former strongholds.

It is possible that nest site choice is inherited; why else should some inland expanses of fellside heather support good numbers of nesting birds whereas adjacent sweeps of gorse (a much favoured habitat) remain vacant? Even as high as 180m, egg laying often begins in March in N England and birds breeding for the first time often start laying in March. Stonechats are highly productive, with most resident pairs attempting three broods (migrants have two). They average 5.06 eggs per clutch (*BWP*) with the second clutch being larger to take maximum advantage of the security of thickening cover and the abundance of food. The nest site is either on or very close to the ground; of 523 nests analysed (*BWP*) 33.5% were in gorse, 21.6% open grass and 20.8% heather. Once fledged, the chicks are ignored by the female after about five days. While she builds the next nest, the male tends the chicks for up to another ten days prior to them becoming independent.

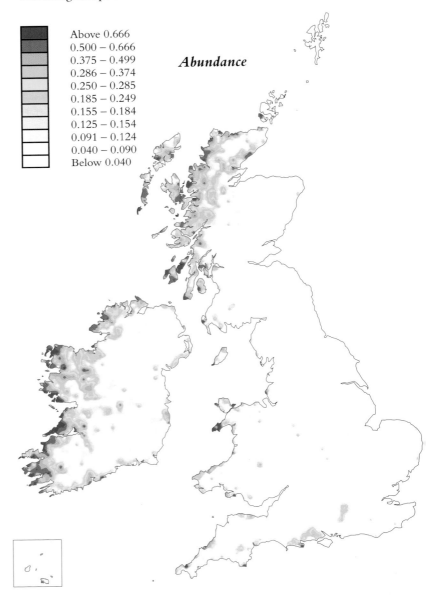

Abundance

Above 0.666
0.500 – 0.666
0.375 – 0.499
0.286 – 0.374
0.250 – 0.285
0.185 – 0.249
0.155 – 0.184
0.125 – 0.154
0.091 – 0.124
0.040 – 0.090
Below 0.040

Those W European countries with data have reported various stages of population decline, while the colonisation of Norway (Munkejord 1981), probably by overshooting British migrants, has ceased, perhaps symptomatic of the retreat in E Britain.

From an estimated population of 30,000–60,000 pairs in Britain during the *68–72 Atlas* period the trend has been steadily downward, though it may have been halted in the late 1980s following two of the warmest winters for 20 years. Sharrock (*68–72 Atlas*) reached his figure by assuming an average of 15–30 pairs per occupied square. That value is likely to be too high for the early 1990s, and most occupied 10-km squares probably contained only 10–25 pairs. On this assumption, there would be an estimated 8,500–21,500 pairs of Stonechats in Britain, with nearly as many (7,500–18,750) in Ireland.

JOHN CALLION

68–72 Atlas p 344

Change

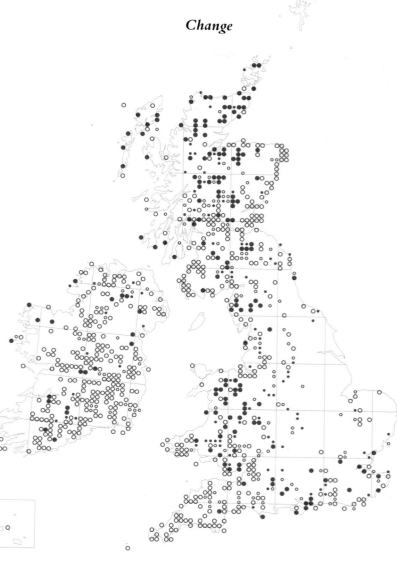

Distribution

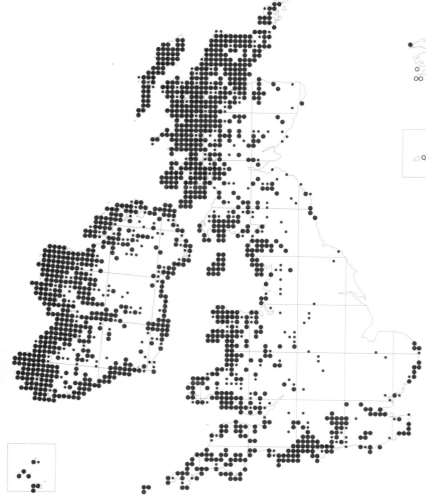

Years	Present, no breeding evidence		Breeding evidence		All records		
	Br	Ir	Br	Ir	Br	Ir	Both (+CI)
1968–72	162	38	1055	756	1217	794	2016
1988–91	184	102	850	467	1034	569	1611
			% change		−14.6	−28.2	−20.0

Wheatear

Oenanthe oenanthe

Whilst the male's plumage makes the Wheatear strikingly conspicuous when first seen in spring against the sheep or rabbit grazed swards of lowland Britain, against the rock screes of its high altitude, primary, habitat it disappears, even though it may be calling aggressively – the blacks, the whites and the greys become the highlights and shadows of the stones and rocks amongst which it is nesting.

The Distribution map shows the chiefly north and west distribution, with the majority of the 10-km squares with breeding Wheatears above the 300m (1,000 feet) contour. Indeed, in W Ireland, SW England, Wales, NW England and most of Scotland, the correspondence between squares occupied by Wheatears and land above 300m is remarkable. In Scotland, only in the outer islands and on the coastal plains of Caithness do Wheatears breed commonly below 300m.

In the southern half of England, Wheatears have a scattered distribution, with concentrations in Breckland and Dartmoor. They prefer rabbit- or sheep-grazed low hills or clifftops, with a close-grazed sward. Large areas of shingle with sparse vegetation, such as on Dungeness, Kent, may be inhabited, provided nest sites are available in nest boxes or other discarded artifacts (Axell 1954). However, nesting success in such habitats may be lower than in more vegetated sites where grass root caterpillars are more abundant (Conder 1989).

In lowland Britain, Wheatears chiefly nest in holes in the ground and prefer rabbit nurseries. Holes in stone walls are widely used where available, but in upland areas most nests are under rocks and in mountain screes. The same hole may be re-used in subsequent years, particularly in territories where there is a shortage of sites.

Lowland Wheatear habitats must have been rare in Britain until Neolithic man cleared the forests for sheep, or after rabbits were introduced by the Normans. Only then were the lowlands increasingly grazed and burrows produced, which expanded the Wheatear's potential breeding range substantially. Prior to this the species presumably occupied the bare, rocky, high altitude habitats, where they still nest up to at least 1,200m in Scotland (*Birds in Scotland*).

Several factors have reduced the population, the most general of which have been agriculture and afforestation, particularly in Scotland (*Birds in Scotland*). Dresser (1871) wrote that the number of Wheatears caught on the South Downs, Sussex, had been greatly reduced since Pennant wrote 100 years previously, because so much of their habitat was being ploughed up. Ploughing-up has continued since, particularly in war years, although

the rate may have declined a little since 1989. In the last 40 years, myxomatosis has reduced rabbit numbers, resulting in a growth in vegetation and a loss of burrows, thus rendering former Wheatear habitats no longer available.

The *Trends Guide* concluded that the population was continuing its decline in the southern half of England, but that in Wales and Scotland there was uncertainty as to whether there had been any substantial change. It doubts the effect of the Sahel drought on Wheatear breeding numbers although, on the Welsh island of Skokholm, the Wheatear population dropped from 20 pairs in 1971, to eight pairs in 1973 – the lowest figure ever – presumably as a result of the Sahelian drought. Since 1976, the population there, as elsewhere, has been increasing slowly, but with a slight drop in 1984 when, according to *Trends Guide*, the effect of a complete failure of rain in the Sahel in 1983 affected other species wintering in the area. The Change map shows an overall drop in the number of occupied 10-km squares, this being most marked in Ireland and central, S and E England, and least marked in the uplands of N and W Britain.

The *68–72 Atlas* used a conservative estimate of 40 pairs per 10-km square with probable or confirmed breeding, to calculate a total British and Irish population of 80,000 pairs. Using the same density, there are

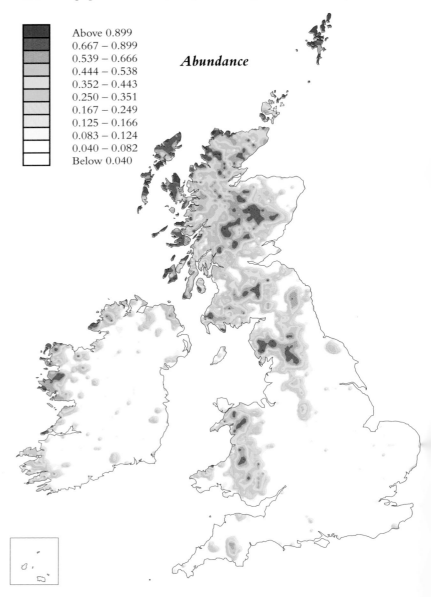

Above 0.899
0.667 – 0.899
0.539 – 0.666
0.444 – 0.538
0.352 – 0.443
0.250 – 0.351
0.167 – 0.249
0.125 – 0.166
0.083 – 0.124
0.040 – 0.082
Below 0.040

Abundance

now an estimated 55,000 pairs of Wheatear in Britain, and 12,000 in Ireland. However, as the density estimate was conservative, these estimates should be considered minimal.

PETER CONDER

68–72 Atlas p 342

Change

Distribution

Years	Present, no breeding evidence		Breeding evidence		All records		
	Br	Ir	Br	Ir	Br	Ir	Both (+CI)
1968–72	274	96	1590	445	1864	541	2409
1988–91	397	105	1341	328	1738	433	2175
				% change	−6.8	−19.8	−9.8

Ring Ouzel

Turdus torquatus

The Ring Ouzel has almost a score of folk names, many of them incorporating the words 'mountain' or 'moor'. It is an upland and montane species, frequenting the broken terrain of crags, gullies, ghylls and cloughs. Especially when feeding young, it will fly some distance onto open moor, or down to pasture, but it prefers more sheltered spots along valley sides, where small bushes and trees such as ash and rowan grow.

The Distribution map shows its association with high ground, most of the land occupied being above the 250m contour. Extensive populated areas include the Pennines, the Lake District, the southern parts of the Grampian mountains, the western parts of the Highland Region of Scotland and throughout the high ground of N and mid Wales.

Comparison of the Distribution map with that in the *68–72 Atlas* reveals that significant attrition and fragmentation of range has occurred during the intervening years. In particular, the Ring Ouzel has disappeared from a large number of 10-km squares in Wales, S Scotland, and Highland and Grampian Regions. The small cluster of records on Dartmoor, which occupied nine 10-km squares in the *68–72 Atlas*, has been reduced to only five squares. Although Ireland was never a stronghold, the species is now less widespread there than at the time of the *68–72 Atlas*, but there has been an interesting colonisation of Co. Kerry.

Even at the time of the *68–72 Atlas*, it was clear that the species was in trouble and the *88–91 Atlas* is registering the continuation of a long-term decline. Up to the end of the 19th century, there were a number of scattered records in lowland England. In this century Baxter and Rintoul (1953) reported serious decreases from many districts in the Highland and Southern Upland districts of Scotland. The trend was confirmed more recently in *Birds in Scotland*. Ussher and Warren (1900) considered that the species bred in all but five of the 32 Irish counties, so the decline there has been most marked.

The areas of rock outcrop and steep-sided valleys favoured by Ring Ouzels (Poxton 1986) are not best suited to afforestation, whereas the uniform tracts of moorland favoured by foresters are marginal nesting habitat for Ring Ouzels, although used by them for feeding. Recent work by Haworth and Thompson (1990) suggest that, at least in the S Pennines, heather and bracken abundance, in providing nesting cover, and the presence of nearby pasture for feeding, are of overriding importance. Disturbance arising from the increased number of visitors, suggested as a possible cause of decline, was not judged to have played a role in the S Pennine study areas. Bracken, identified as valuable to the Ring Ouzel,

has in fact increased in extent in recent years, so that its role in the species ecology is unclear.

It has been suggested that warmer conditions, perhaps aiding an increase in Blackbirds and causing interspecific competition, has been partly responsible for the decline of the Ring Ouzel in Britain and Ireland. Certainly there is usually an altitudinal separation of the two species, but higher temperatures might be expected to benefit both species equally. The role of afforestation, already referred to, remains uncertain.

Most nest sites are in natural positions, frequently on a ledge or a rocky outcrop. Other nest sites include walls, derelict buildings, quarries, mine shafts and pot-holes. Infrequently, nests are built in grass on heather slopes (Flegg and Glue 1975).

The population sizes of Ring Ouzels in Britain and Ireland are difficult to estimate but approximate figures of 5,500–11,000 pairs in Britain and 180–360 in Ireland are arrived at by assuming the same range of density per occupied 10-km square (10–20 pairs) as used in the *68–72 Atlas*, but taking into account the differences in density in Britain and Ireland shown by the timed tetrad counts.

DAVID HILL

68–72 Atlas p 338

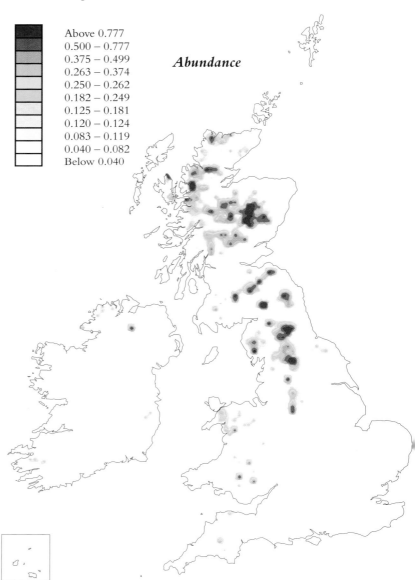

| Above 0.777 |
| 0.500 – 0.777 |
| 0.375 – 0.499 |
| 0.263 – 0.374 |
| 0.250 – 0.262 |
| 0.182 – 0.249 |
| 0.125 – 0.181 |
| 0.120 – 0.124 |
| 0.083 – 0.119 |
| 0.040 – 0.082 |
| Below 0.040 |

Abundance

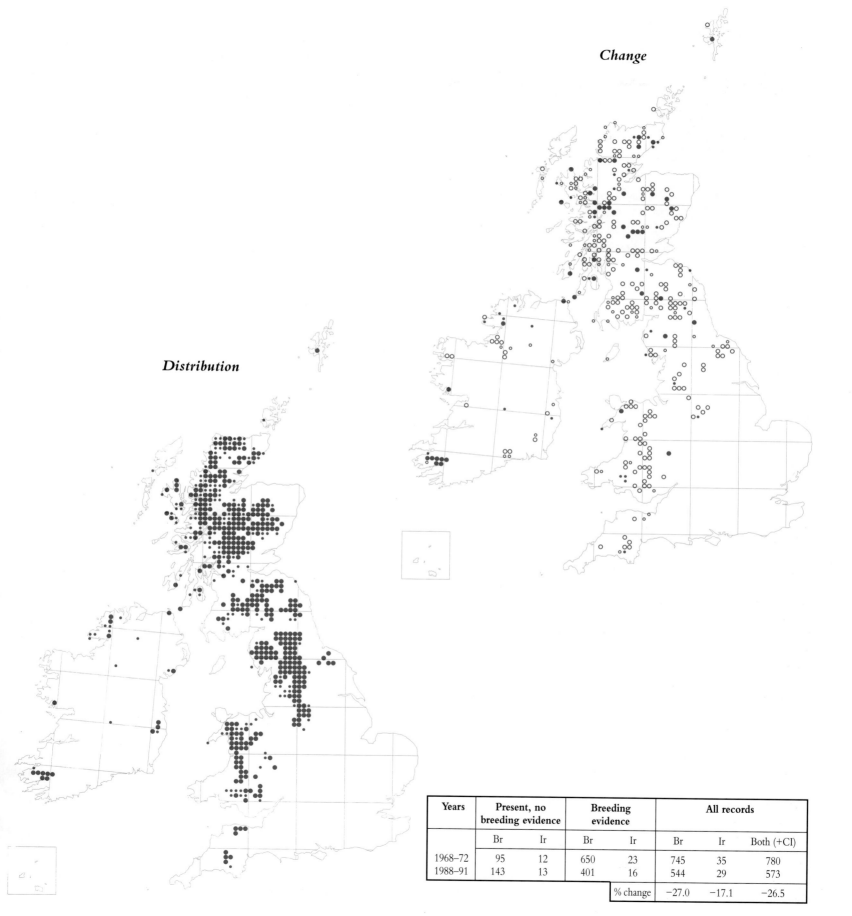

Change

Distribution

Years	Present, no breeding evidence		Breeding evidence		All records		
	Br	Ir	Br	Ir	Br	Ir	Both (+CI)
1968–72	95	12	650	23	745	35	780
1988–91	143	13	401	16	544	29	573
			% change		−27.0	−17.1	−26.5

Blackbird

Turdus merula

One of the few birds known to nearly everyone, the Blackbird is rivalled by few other species in the completeness of its occupancy of Britain and Ireland. As would be expected, densities are low in highland and moorland areas, and on windswept islands, where only buildings, walls and gardens provide some shelter and improved feeding conditions. On farmland, there is a clear N-S increase in abundance (O'Connor and Shrub 1986a), which is presumably related to the abundance of the food supply.

The Blackbird is the most adaptable of the thrushes, as is reflected in its W Palearctic range. It is the only one to have colonised the Atlantic islands as far west as the Azores. It first bred in the Faeroe Islands in 1947 and is now established there; and has bred, but has not yet become established, in Iceland (*BWP*). On the Continent it is still extending its range to the north and east, as man creates suitable, mainly suburban, habitats (Spencer 1975).

This ability to profit from man's modification of the environment has been attributed to the Blackbird's supposed original adaptation to woodland edges rather than high forest (*68–72 Atlas*), thereby pre-adapting it to semi-open habitats. But recent studies in the primeval Bialowieza forest of Poland (Tomialojc) found it to be a bird of the high forest, with no preference for forest edge. This suggests that high forest was, in fact, its original habitat before man came on the scene. The all-black male and dull brown female plumage is paralleled by some tropical forest thrushes, while the low-pitched song is characteristic of birds of forest habitats, and related to the better transmission through forest vegetation of low-frequency rather than high-frequency sounds. Both these features suggest adaptation to forest rather than to semi-open country. However that may

be, in many parts of its range the Blackbird has for long been a common bird in non-forest habitats; and even the primeval forest population of N Europe must have wintered in more varied and open habitats to the south.

In Britain, a gradual spread into man-made habitats began in the 19th century, leading to an expansion of range and great increase in numbers. Indices from ringing and Nest Record Cards indicate that the Blackbird population stopped increasing by the early 1950s, and that there was no marked change thereafter. Numbers fell after the exceptional 1962/63 winter, but soon increased again and remained at a high level during the period of the *68–72 Atlas*. Since then there has been a slight downward trend, presumably related to the colder winters from the mid 1970s to mid 1980s (*Trends Guide*).

Breeding densities of Blackbirds are extremely varied, from as high as 250 pairs per km^2 in suburbia (Batten 1973), through an average of about 26 in farmland, to single figures in largely unsuitable upland areas. During 1988–91 there were an estimated 4,400,000 Blackbird territories in Britain, and 1,800,000 in Ireland.

DAVID SNOW

68–72 Atlas p 340

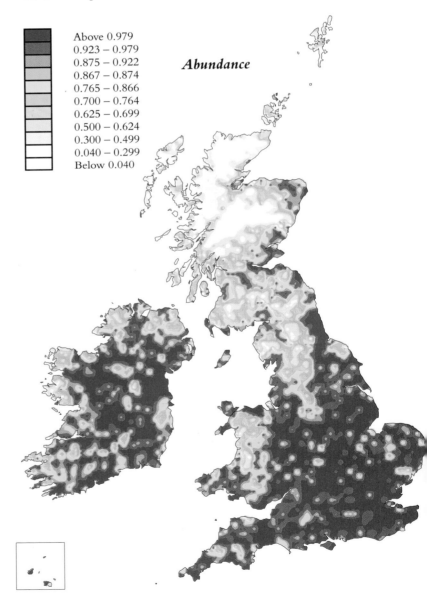

Abundance
Above 0.979
0.923 – 0.979
0.875 – 0.922
0.867 – 0.874
0.765 – 0.866
0.700 – 0.764
0.625 – 0.699
0.500 – 0.624
0.300 – 0.499
0.040 – 0.299
Below 0.040

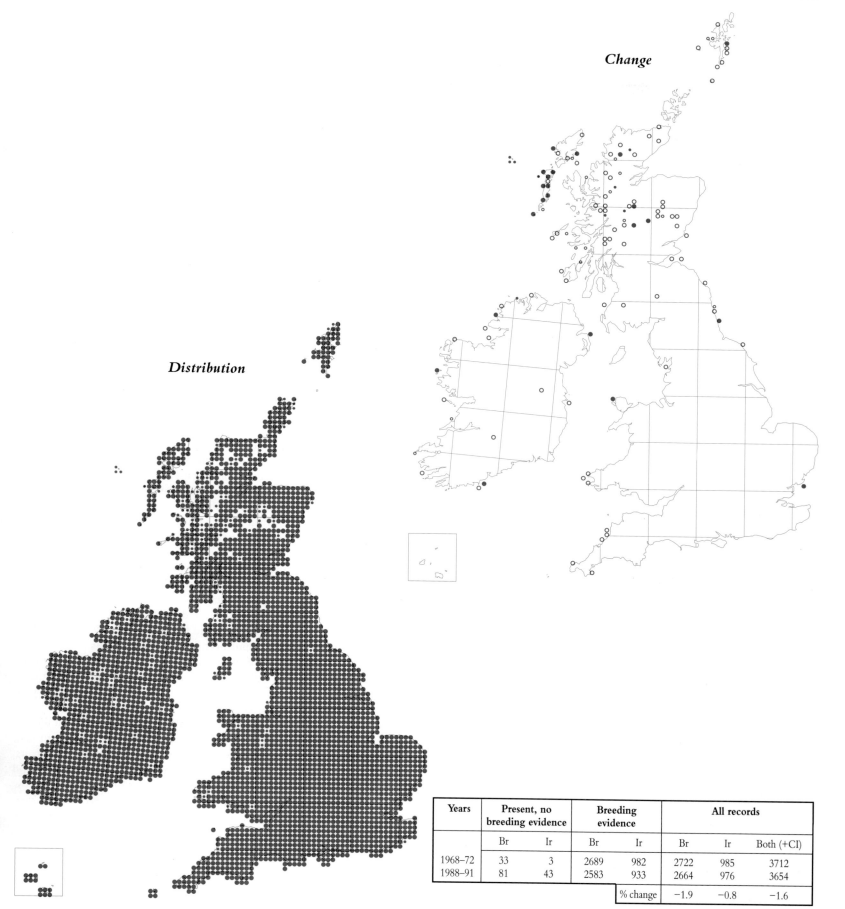

Change

Distribution

Years	Present, no breeding evidence		Breeding evidence		All records		
	Br	Ir	Br	Ir	Br	Ir	Both (+CI)
1968–72	33	3	2689	982	2722	985	3712
1988–91	81	43	2583	933	2664	976	3654
% change					−1.9	−0.8	−1.6

Fieldfare

Turdus pilaris

In Scandinavia and other areas where Fieldfares breed abundantly, their nests are easy to locate, for they are noisy, mainly colonial breeders. Isolated breeding pairs, as all those recorded so far in Britain have been, are much less easy to find. The song, 'a feeble warble interrupted by wheezes and chuckles' (*BWP*), is little help for it is not persistent and not audible from afar. Also, in late spring pairs tend to wander, presumably in search of a suitable nesting area. Hence while it is possible that some breeding pairs were overlooked, it is equally likely that some individual pairs were recorded in more than one 10-km square.

For these reasons the map cannot be taken as presenting a precise picture of Fieldfare breeding distribution in 1988–91. Nevertheless, it shows that the process of colonisation of Britain, which began in the 1960s and resulted in 15 confirmed breeding records in 1968–72, mainly in Shetland, Orkney and mainland Scotland south to between 56°N and 57°N, has not progressed as might have been expected. In NE Scotland, where there was a concentration of breeding records in 1968–72, there were none in 1981–84 (Buckland *et al.* 1990), but two in 1988–91. The few breeding records in 1988–91 were again mainly in N Britain, although none in Shetland and Orkney. Some records, but with no evidence of breeding, extend nearly to the S coast of England. Although the record was not submitted to the Atlas, a pair of Fieldfares successfully fledged young in Kent in 1991. These, together with the lack of records from Shetland and Orkney, indicate a southern shift in the centre of gravity of the small British breeding population, but the distribution still suggests, as it did in 1968–72, a link with the Scandinavian rather than with the central European breeding stock.

The wide range of breeding habitats used by Fieldfares in Scandinavia, from sub-Arctic scrub, and even open tundra, through woodland of various kinds to city parks, suggests that they should easily find suitable breeding habitats in Britain. Most records have in fact been from moorland valleys,

hillside birch woods or the edges of plantations. Possibly a more important factor constraining the spread of the species in Britain is the lack of opportunity for colonial breeding when pairs are so sparsely distributed and in such small numbers. In Scandinavia, where most Fieldfares nest in groups of 5–20 pairs, it has been found that colonial pairs have significantly higher breeding success than solitary pairs, mainly because they suffer less nest predation (Wiklund and Andersson 1980). Nesting Fieldfares are well known for the bold defence of their nests against avian predators, with persistent swooping attacks accompanied by loud rattling calls and well-aimed defecation. These attacks are extremely effective and are continued until the intruder retreats, its plumage soiled and matted. There have even been reports of predators being grounded and dying as a result of the Fieldfare's defensive tactics (Lübke and Furrer 1985, *BWP*).

The British breeding population is clearly very small. With records from all four years, most of which refer to one year only, and with no evidence whether breeding took place in the other years, the map must exaggerate the number of breeding localities recorded in any one year. 1989 and 1990 were the best years on record for this species, with a maximum of 12–13 breeding pairs (RBBP). On the other hand there is the likelihood that a good many isolated pairs in remote areas were missed. One can tentatively suggest a breeding population of fewer than 25 pairs. Probably the Fieldfare's further occupation of Britain will be slow and irregular, unless the central European breeding population, which has now reached N France (Siblet *et al.* 1991) and the Low Countries, spills over into SE England, in which case a rapid, spectacular colonisation may well take place.

DAVID SNOW

68–72 Atlas p 332

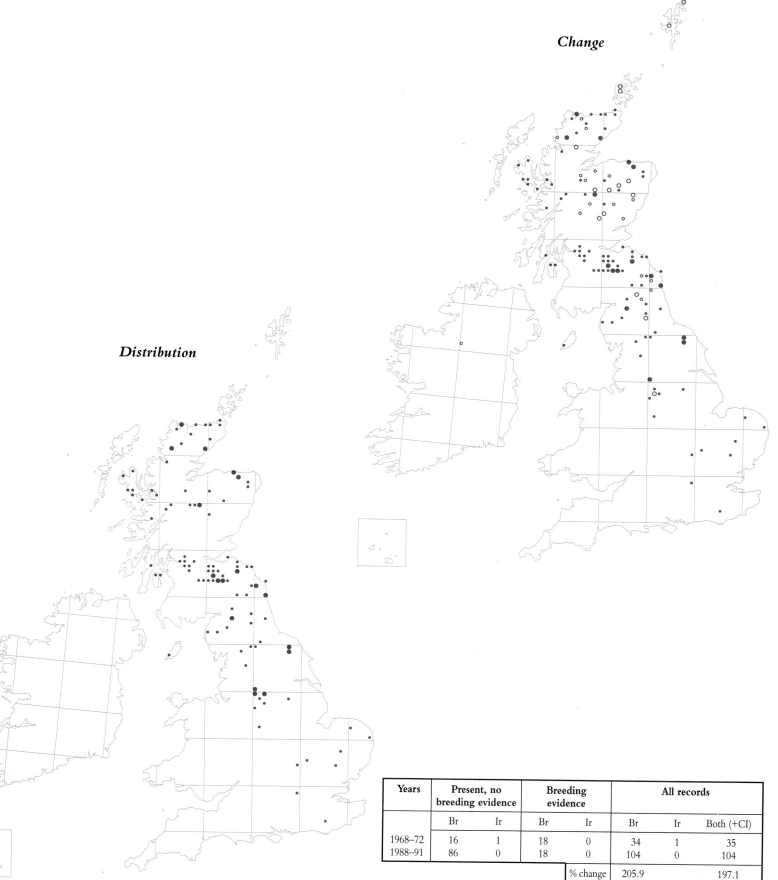

Change

Distribution

Years	Present, no breeding evidence		Breeding evidence		All records		
	Br	Ir	Br	Ir	Br	Ir	Both (+CI)
1968–72	16	1	18	0	34	1	35
1988–91	86	0	18	0	104	0	104
			% change		205.9		197.1

Song Thrush

Turdus philomelos

The Song Thrush's distinctive song, with its repeated phrases, is well known to many who would not regard themselves as birdwatchers. Yet the songs of this species are much more complex than those of our other thrushes, individual birds usually having a repertoire of over 100 different phrases.

Song Thrushes occupy most terrestrial habitats except for completely open areas, and were recorded breeding in nearly all 10-km squares in Britain and Ireland. They are virtually absent from Shetland, where the small population collapsed after the severe winter of 1947, and from a few areas in the N of Scotland. This distribution pattern is unchanged since the *68–72 Atlas*. Densities are generally highest in SE England and East Anglia, decreasing as one goes further north and west. Similar patterns are shown by the *Winter Atlas* and the farmland CBC (O'Connor and Shrubb 1986a). Lower proportions of suitable habitat in areas with high ground or open country, such as central Wales, the Lake District and much of N Scotland, appear to explain at least part of this variation in density.

British Song Thrushes are partial migrants, the proportion of birds which migrate declining as one moves south. Many birds from Scotland and N England winter in Ireland, whereas those leaving S England move to France and Spain. A study in Oxford found that winter territories were held by individual birds, usually males (Davies and Snow 1965). Females arrived in March and departed in July. Pairs defended territories during the breeding season, but sometimes fed up to half a mile outside them. Laying starts in the first half of March and continues until July, giving the Song Thrush the longest breeding season of our thrushes (Myres 1955).

Song Thrushes were more abundant than Blackbirds during the early decades of this century but, since the 1940s, Blackbirds have been the commoner species (Ginn 1969, *Trends Guide*). Blackbird numbers certainly increased during the first part of the century. Song Thrushes suffered a

number of short-term decreases after severe winters during this period, but it is unclear whether they underwent a sustained decline. Their numbers did not show any clear trend between the late 1940s and the mid 1960s (Ginn 1969).

The population declined after the 1962/63 winter but recovered in 3–4 years, reaching a stable level by the *68–72 Atlas* period (*Trends Guide*). Numbers remained stable up to the mid 1970s but declined steadily thereafter. Overall declines, of 54% on farmland and 27% in woodland, occurred between 1970 and 1989. A similar decline has been recorded by the BTO's Garden Bird Feeding Survey. Nesting success has increased slightly since the early 1960s, eliminating the possibility that the decline has resulted from nest predation, due to increasing numbers of Magpies and other corvids. Some of our Song Thrushes winter in France and S Europe, where they are quarry species. However, the proportion of the total population which is at risk from such hunting is small, and the levels of hunting to which they are exposed have not increased (Baillie 1990).

Song Thrushes are particularly vulnerable to cold winters, and the early stages of the decline are explained by the increasing amount of cold weather in the late 1970s and early 1980s, following a run of very mild

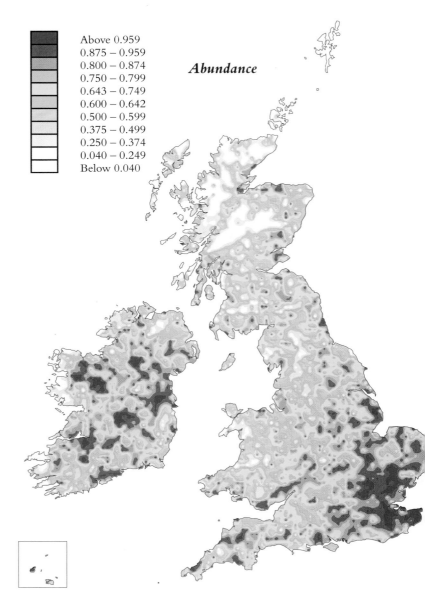

Abundance

	Above 0.959
	0.875 – 0.959
	0.800 – 0.874
	0.750 – 0.799
	0.643 – 0.749
	0.600 – 0.642
	0.500 – 0.599
	0.375 – 0.499
	0.250 – 0.374
	0.040 – 0.249
	Below 0.040

winters which had lasted since 1963. The numbers of freezing days in January and February, and density dependence, together explained 58% of the variation in the population changes of farmland Song Thrushes between 1962 and 1987. However, weather conditions alone were not sufficient to explain the continuing decline of this species (Baillie 1990). A number of possible explanations have been put forward and now require testing. The reduction in the area of spring tillage which has resulted from the switch to autumn sown cereals may have reduced the feeding areas available to Song Thrushes during the early part of the breeding season (O'Connor and Shrubb 1986a). Other aspects of land-use change, such as losses of hedgerows and permanent pasture, might also be involved, but their timing is not obviously linked to that of the population decline.

Pesticides might be affecting Song Thrush populations, either directly by poisoning them, or indirectly by reducing food supplies. During the first three months of the breeding season they feed mainly on earthworms, but in late May and June pupating caterpillars are preferred food, if available. Snails become an important food item only in late summer, when other foods are difficult to obtain, particularly if the weather is dry. They are also important in winter when freezing weather limits access to

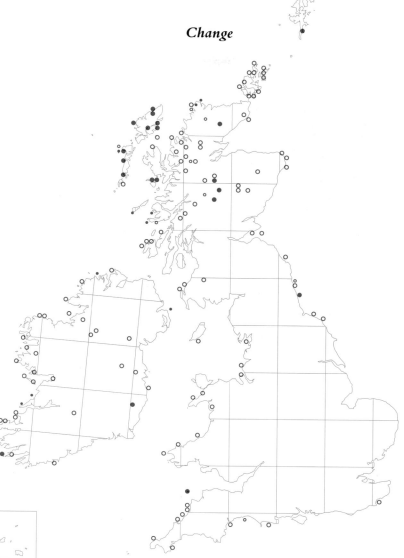

Change

Distribution

earthworms (Davies and Snow 1965). Snails are thus a critical food at times when preferred prey are not available. The possibility that Song Thrush populations have been adversely affected by molluscicide-induced declines in snail populations needs careful evaluation.

During 1988–91 there were an estimated 990,000 Song Thrush territories in Britain, and 390,000 in Ireland.

STEPHEN R. BAILLIE

68–72 Atlas p 334

Years	Present, no breeding evidence		Breeding evidence		All records		
	Br	Ir	Br	Ir	Br	Ir	Both (+CI)
1968–72	32	3	2648	967	2680	970	3655
1988–91	129	112	2491	835	2620	947	3581
				% change	−2.1	−2.3	−2.1

Redwing

Turdus iliacus

The Redwing is a common, almost commonplace, bird of the British and Irish countryside in winter. But a deeper and more lasting impression is created when one hears its unmistakable song piercing the early morning stillness of a Highland glen in May.

In Scotland a range of habitats, frequently close to water, is used. These include the mature, mixed woodland surrounding estate houses, scrubby wooded areas along cultivated river valleys, and even conifer plantations. The important feature appears to be the presence of scrub with easy access to damp patches for feeding (Williamson 1973). The breeding habitat in higher latitudes, Scandinavia for example, is equally diverse - from forest bog with well developed birch and willow scrub to suburban gardens, again typified by well developed scrub.

Breeding Redwings are currently confined mainly to 10-km squares around and north of the Great Glen, especially non-coastal ones. Records of birds not suspected of breeding are more widespread – those on passage may still be present as late as May, even in Ireland (*Birds in Ireland*). It is worthy of note, however, that in recent years Redwings have bred in a few English counties; particularly in Kent (Taylor 1981a) where at least 5 pairs bred between 1975 and 1991. Evidence of breeding was recorded in many fewer squares in the *88–91 Atlas* than in the *68–72 Atlas*, but such an apparently large decline is probably not cause for great concern because the species frequents areas with low coverage by observers.

The Change map indicates a contraction and slight shift of the Scottish breeding range to the south and west. The Abundance map also suggests a western bias to the species' range. This indicates a possible preference for a wetter, more oceanic climate, which would accord well with the birds' presumed preference for nesting habitat containing damp feeding areas.

The Redwing's migratory habits have received much attention but are not fully understood. Certainly, the species appears to be of a somewhat eruptive nature and this may have some ultimate bearing on its breeding ecology. Movements, both within and between winters, are not entirely predictable, and cold weather movements within the winter range give the appearance of nomadism. It may have been similar cold weather dispersal which accounted for the original Redwing colonisation and spread within Scotland – the species being displaced southwestwards from Scandinavia by cold springs (*Birds in Scotland*). Populations of the other small W Palearctic thrush, the Song Thrush, certainly seem to be sensitive to cold weather (Greenwood and Baillie 1991). Alternatively, the increase in the size of the Scottish breeding population may be due to colonisation by birds of Scandinavian origin, on spring migration, being diverted by easterly winds (Williamson 1975).

The history of colonisation of Britain since 1925 is well recorded (see *68–72 Atlas*). Numbers breeding (actual and possible) since 1973 are shown in the Table. Even allowing for uneven fieldwork and reporting rates from year to year, there is clearly much annual variation in the size of the summering population. The reasons for such fluctuations are not known but the species' curious migratory behaviour, which may result in

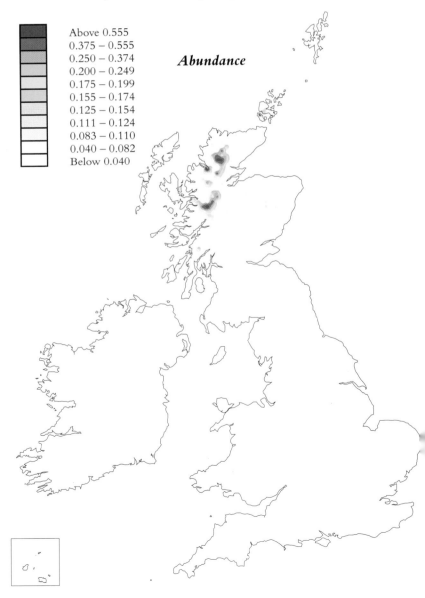

Above 0.555	
0.375 – 0.555	
0.250 – 0.374	*Abundance*
0.200 – 0.249	
0.175 – 0.199	
0.155 – 0.174	
0.125 – 0.154	
0.111 – 0.124	
0.083 – 0.110	
0.040 – 0.082	
Below 0.040	

rather low breeding site fidelity, especially among young birds (Bjerke and Bjerke 1981), could account for it.

The pattern which emerges is of a bird which is very sensitive to inimical changes in the weather, and one which is able to respond to them by a remarkable ability to disperse to regions where more favourable conditions prevail. This may impose a flexible approach to breeding location, and begs the question of whether the Scottish population has resulted from a process of limited colonisation and healthy endemic population growth (sustained by high breeding success and subsequent high breeding site fidelity), or one of annual recolonisation from abroad. Redwings normally breed solitarily, but loose colonial clusters have been recorded on the Continent, and pairs may even nest close together in Scotland (*BWP*, RBBP 1990). The nest, containing 4–6 eggs, is usually a few feet above, or even on, the ground in scrub. Surprisingly little is known of the breeding biology of British Redwings, most of which are probably of the N Eurasian race *T.i. iliacus*.

Despite the apparent decrease in numbers and contraction of breeding range since the *68–72 Atlas*, there is probably no conservation action which could promote population growth (*Red Data Birds*).

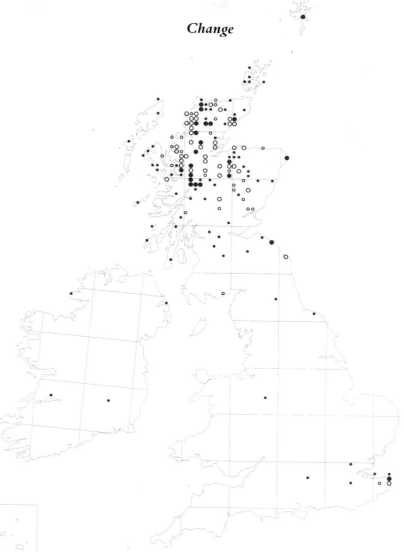

Change

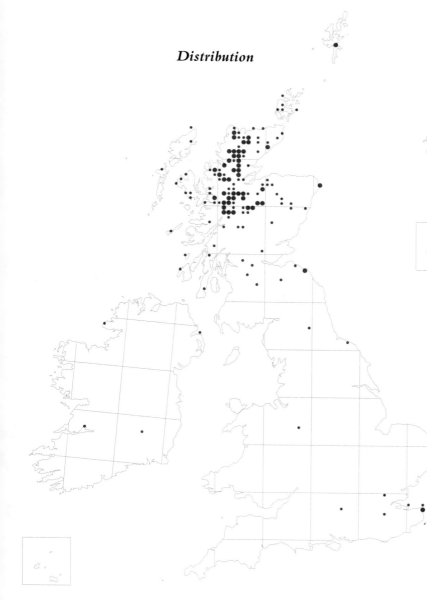

Distribution

The likely population size in any one year is in the region of 40–80 pairs (*Trends Guide*). This figure would allow for inconsistent reporting and non-detection of nesting pairs.

J. B. REID

Two-year averages for breeding Redwings in Britain, 1973–88

	73–74	75–76	77–78	79–80	81–82	83–84	85–86	87–88
Confirmed	5	8	3	4	22	24	21	9
Max total	26	34	22	20	49	73	54	45

Based on data pooled from RBBP and Scottish Bird Reports: the figures refer to number of breeding pairs.

68–72 Atlas p 336

Years	Present, no breeding evidence		Breeding evidence		All records		
	Br	Ir	Br	Ir	Br	Ir	Both (+CI)
1968–72	32	0	79	0	111	0	111
1988–91	90	4	46	0	136	4	140
			% change		22.5		26.1

Mistle Thrush

Turdus viscivorus

Mistle Thrushes are familiar but enigmatic. They are conspicuous while delivering their loud fluting song from an exposed perch, foraging in the open or noisily defending their nest. Nevertheless, it is easy to cross their large territories without noticing them. This may be one reason why their natural history is poorly known.

Mistle Thrushes are widespread in Britain and Ireland, although they are absent from Shetland and Orkney and very local in the Outer Hebrides. They are most frequent in SE England, with pockets of high density occurring as far north as the Moray Firth. As birds of open woodland they are least frequent where trees are scarce: the fens of East Anglia, exposed coasts and upland areas. Yet a few moorland Mistle Thrushes, especially in Ireland, forsake trees entirely, nesting on stone walls or on the ground (*BWP*).

The Change map shows that the explosive spread by Mistle Thrushes in the last two centuries is over, although apparently suitable habitat remains unoccupied in the Hebrides (*Birds in Scotland*). Before 1800, Mistle Thrushes were chiefly associated with the montane forests of continental Europe. They were unknown in Ireland and rare in Britain north of a line from the Gower Peninsula to the Wash. Even in S Britain, they were usually regarded as uncommon. The reasons for their subsequent increase in range and numbers, especially along the Atlantic coast of Europe, remain unknown (*68–72 Atlas*, Simms 1978, *BWP*). The increase in Ireland was dramatic. From the first sighting in 1800, the whole island had been colonised by 1850 (*Birds in Ireland*). Meanwhile, Mistle Thrushes had become regular in Britain as far north as the Central Lowlands of Scotland, and had even bred on Orkney. Since 1850, expansion into the Highlands and Hebrides has been slow (*Birds in Scotland*). Further south, populations continued to increase as suburban and urban habitats were

colonised (Simms 1978). At present, range expansion appears to be continuing in the south from Madeira (A. Swash) to Iraq (P. Ctyroky), but some contraction is occurring in the north from Orkney (*Birds in Scotland*) to Scandinavia (*BWP*).

Mistle Thrushes defend larger territories than other *Turdus* species nesting in the same habitat (Simms 1978): in E Sussex, few were less than one ha in area and some exceeded 10ha (D. Harper). Most territories contain mature trees, which provide song posts and nest sites, and extensive areas of grassland where the birds can forage (*BWP*). Spring tillage can be a partial substitute for grassland and a switch to autumn tillage may explain some local declines in numbers (O'Connor and Shrubb 1986a, *Trends Guide*).

Young nestlings are fed small prey, including flies and caterpillars, collected from the lower canopies of trees and bushes or from the ground under them. This may explain why Mistle Thrushes rarely breed in really open habitats and why they appear to shun trees which support relatively few invertebrates, such as ash (Simms 1971).

Fruit is often fed to older chicks and suitable bushes may be defended by their parents. In Hertfordshire, bushes (mainly holly) were defended

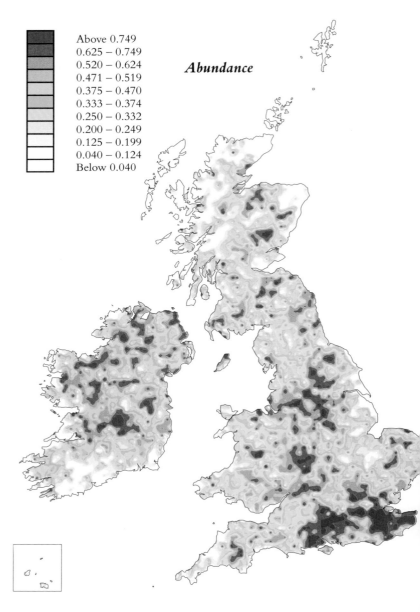

	Above 0.749
	0.625 – 0.749
	0.520 – 0.624
	0.471 – 0.519
	0.375 – 0.470
	0.333 – 0.374
	0.250 – 0.332
	0.200 – 0.249
	0.125 – 0.199
	0.040 – 0.124
	Below 0.040

Abundance

during the winter (Snow and Snow 1984), but in East Sussex fruit defence continued all year, with the birds defending different plants at different seasons. Most defence was by males; those with defended fruit reared more fledglings than average, and some became polygamous (D. Harper). Since winter survival is probably improved by fruit defence (Snow and Snow 1984), and many Mistle Thrushes are sedentary (Snow 1969), local distribution may be influenced by the availability of defendable fruiting plants.

Not all Mistle Thrushes are sedentary. In Britain, virtually all yearling females, many yearling males and a few adults of both sexes are migrants (*Winter Atlas*, D. Harper). Comparison of the Abundance map with that in the *Winter Atlas* supports the conclusion that Scottish birds are highly migratory (Snow 1969). However, there is little sign of the movement to and from Ireland detected by ringing, and the numbers involved may be low (Snow 1969). Intriguingly, pockets of high breeding density around Sligo and the Bog of Allen, not detected by the *Winter Atlas*, suggest that Irish Mistle Thrushes may be more migratory than previously thought.

During 1988–91 there were estimated to have been 230,000 Mistle Thrush territories in Britain and 90,000 in Ireland. Although this is far

Change

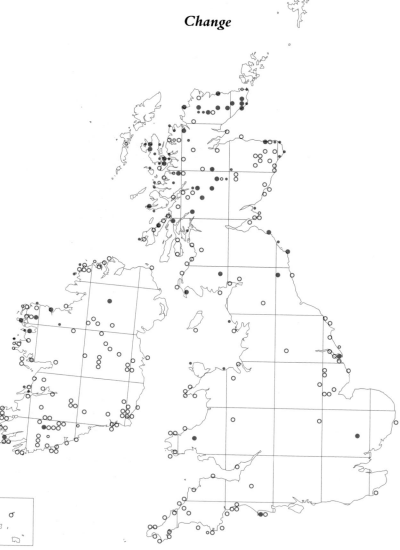

Distribution

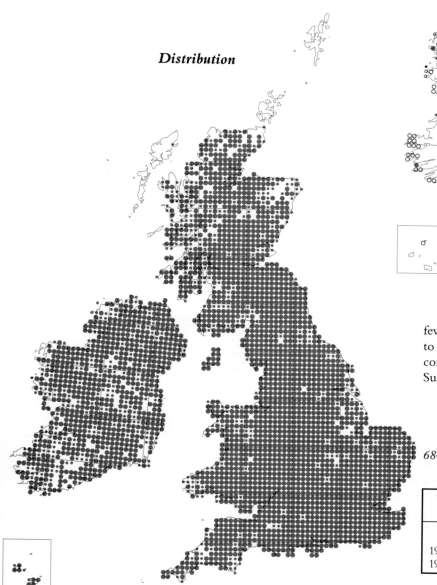

fewer than the estimate for Blackbirds, O'Connor (1986) used CBC data to infer that much of Britain is saturated with Mistle Thrushes. His conclusion is supported by the existence of non-breeding floaters in East Sussex (D. Harper). So why does each pair need so much room?

DAVID HARPER

68–72 Atlas p 330

Years	Present, no breeding evidence		Breeding evidence		All records		
	Br	Ir	Br	Ir	Br	Ir	Both (+CI)
1968–72	67	26	2382	915	2449	941	3394
1988–91	244	152	2153	705	2397	857	3264
				% change	−2.0	−8.8	−4.0

Cetti's Warbler

Cettia cetti

The Cetti's Warbler skulks like a mouse but has an extraordinarily far-carrying song which it uses virtually throughout the year. The song conjures up images of S European marshes and riversides but the species has greatly extended its range to the north in recent years. A northward spread in France starting in the 1920s and accelerating after 1950 is described by Bonham and Robertson (1975).

The *68–72 Atlas* caught the Cetti's Warbler at its moment of colonisation of Britain. The first six records of individuals in Britain and Ireland came in the 1960s. Several birds arrived in 1971. Up to three survived the winter in Kent and probably bred near Canterbury in 1972 (Harvey 1977).

The spread and increase in numbers after 1972 was spectacular, reaching 100 males in five years and a peak of over 300 by 1984. Since then numbers have declined slightly (see Table). Cetti's Warblers are sedentary, with males occupying territories and singing from the autumn onwards. Not surprisingly, they are vulnerable to cold winters, which occurred in 78/79, 81/82, 84/85 and 85/86 leading to a national population decrease of up to 30%. Numbers in Kent fell surprisingly sharply over these winters and not one was left by 1988 although they have been recorded there since. The Suffolk population also declined, from over 30 in 1984 to two in 1989. Several more counties such as Cambridgeshire, Hertfordshire and Essex have lost breeding populations which had reached about five males each, but Hertfordshire was recolonised in 1989.

The Distribution map shows that Cetti's Warblers have got as far west as is possible in Britain. Populations in the more westerly counties, especially Hampshire, Dorset and Devon, seem to have been less badly affected by cold winters, and by 1989 Hampshire had become the most populated county, with 93 reported males. To date, the most northerly singing bird was in Shropshire in 1983.

Cetti's Warblers frequent dense scrub in damp places. Bramble, reeds and willows were the dominant plants in Kentish territories (Harvey 1977). They nest in thick vegetation and forage on damp bare ground beneath the scrub, and in thick cover. Many such places adjoin reed beds but reeds themselves are not much occupied until the winter, when both sexes, but especially the females, move in (Bibby and Thomas 1984). Although suitable breeding habitat occurs along stream and canal sides, most of the population is on sites with major reed beds. Apart from their scarcity in Suffolk and Kent, the present distribution accords closely with the distribution of large reed bed sites shown by Bibby and Lunn (1982).

Cetti's Warblers are rather unusual in several respects. They have only ten tail feathers (most passerines have 12) and lay bright red eggs. Much more interesting is the fact that males are about 30% heavier than females (Bibby and Thomas 1984). The male spends much time patrolling and singing in large territories and makes very little contribution to parental care. Several females may breed in the territory of one male (Bibby 1982). It seems that they can do this because nest predation rates are low compared with those of Reed and Sedge Warblers. They have a relatively prolonged incubation and nestling period, but in the rich foraging conditions of marshland margins, the females rear the young to fledging with very little help. Some rear second broods.

Females are remarkably unobtrusive, which makes nests difficult to find and pairs impossible to count without much effort. The only sensible counting unit is the singing male. Care is needed here because they have a habit of flying between successive song bursts and it is often hard to believe that songs from different directions may have come from the same individual. The songs of many individuals are distinctive and recognisable to the human ear. Counts of singing birds submitted to the RBBP are likely to be quite accurate for most areas where the Cetti's Warbler is still rare enough to attract attention. Peak numbers of over 300 males were

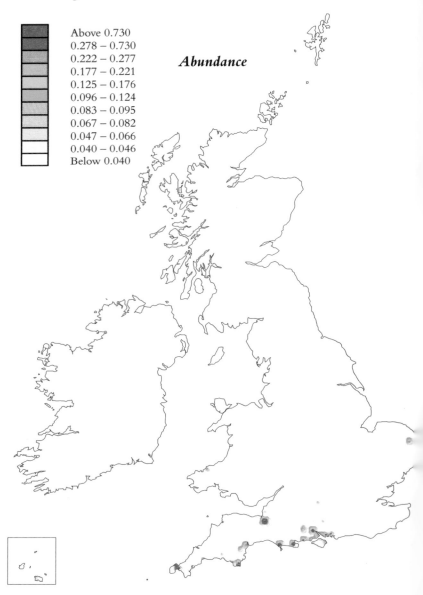

Above 0.730
0.278 – 0.730
0.222 – 0.277
0.177 – 0.221
0.125 – 0.176
0.096 – 0.124
0.083 – 0.095
0.067 – 0.082
0.047 – 0.066
0.040 – 0.046
Below 0.040

Abundance

reached in 1984, and again in 1990 (350 males) after two mild winters. This would correspond to about 450 pairs. Such numbers are of no international significance.

It will be most interesting to see how the story of this dramatic colonisation has developed by the time of the next breeding atlas. There must still be room for further expansion of numbers within the present range. Perhaps Ireland will be colonised and Kent regain some of its former numbers. Much will depend on the severity of future winters.

COLIN BIBBY

68–72 Atlas p 356

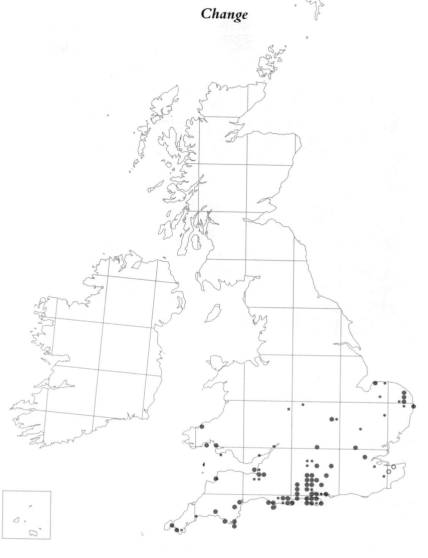

Change

Distribution

Numbers (as reported to the RBBP) of singing male Cetti's Warblers in Britain as two-year means after their first occurrence in 1972.

	72	73–74	75–76	77–78	79–80	81–82	83–84	85–86	87–88
Kent	2	13	61	100	88	65	55	13	1
Norfolk	0	3	16	26	28	28	55	32	21
Dorset	0	0	0	12	20	26	65	43	41
Devon	0	0	3	11	18	38	38	39	33
Hants	0	0	1	4	9	19	27	35	56
Suffolk	0	1	1	6	5	19	27	3	3
Cornwall	0	0	1	3	2	6	12	13	13
Others	0	0	1	6	12	18	21	34	29
Totals	2	17	84	168	182	219	300	212	197

Years	Present, no breeding evidence		Breeding evidence		All records		
	Br	Ir	Br	Ir	Br	Ir	Both (+CI)
1968–72	2	0	3	0	5	0	6
1988–91	29	0	57	0	86	0	89
				% change	1620.0	0	1350.0

Grasshopper Warbler

Locustella naevia

The secretive mouse-like Grasshopper Warbler may go undetected if the high pitched mechanical reeling song is not heard. Freshly arrived migrant males sing strongly from exposed positions, and undertake regular flights during territorial skirmishes and sexual chases. Thereafter, song output diminishes and birds spend much of their time amongst thick herbage close to, or on the ground, where roaming families go undetected to an unknown degree (Hulten 1959, Glue 1990b).

The Grasshopper Warbler requires three basic components within its breeding territory: firstly thick ground cover (normally provided by coarse grasses, bramble, sedges or rushes), secondly several suitable song posts (usually stooks of dead herbage, short bushes or sapling trees), thirdly a rich source of invertebrate foods, preferably within 50m, but sometimes up to 200m from the nest site (Glue 1990b). Such conditions are found in a broad spectrum of 'wet' and 'dry' habitats, hence the widespread if somewhat sparse distribution throughout most of Britain and Ireland, though almost exclusively below 300m (1,000 feet).

The Distribution map indicates concentrations for England and Wales in scrub, marsh, farmland and dune slacks of coastal counties. Inland, areas of young forestry plantations, ill-drained farmland, freshwater marsh and overgrown mineral extraction sites support pockets of populations in valley systems such as the Thames, Stour, Trent and Humber. Grasshopper Warblers are absent from the conurbations of Greater London and Essex, and breed thinly on intensively farmed parts of the Weald and Lincolnshire, the uplands of Devon, the Pennines, Yorkshire Moors and Cambrian Mountains.

In Scotland breeding was confirmed in most of the mainland counties, with concentrations, notably from Dumfries and Galloway north to Clyde and Argyll. Nesting was more sparse over much of Tayside, Grampian and Highland. Pairs were located on Rum and Skye, but were absent from the

Outer Hebrides and Northern Isles. The distribution in Ireland, like that in Britain, shows concentrations in many W coast counties associated, at least in part, with the afforestation programme there.

The *88–91 Atlas* shows a general thinning throughout the entire range of the Grasshopper Warbler since 1968–72, with virtually no region unaffected. Most alarming are the apparent losses over extensive parts of the Home Counties, SW England, the Welsh Borders, lowland Lancashire and E Ireland.

This decline needs to interpreted with care since the Grasshopper Warbler is more prone than most summer visitor passerines to local year-to-year fluctuations in numbers and is erratic in its occupation of some breeding sites (Parslow 1973). CBC indices show sharp fluctuations up to 1971, followed by a crash over three years, so marked that the 1980s value was just one-sixth that of the 1960s. Indices constructed from daily log figures pooled from nine British observatories during 1964–81 show an increase up to 1970, followed by a substantial decline in passage numbers over four years, falling to lower levels than recorded in the 1970s (Riddiford and Findley 1981, Riddiford 1983). Similarly, ringing totals corrected for ringer effort increased fivefold between 1963 and 1970, halving suddenly

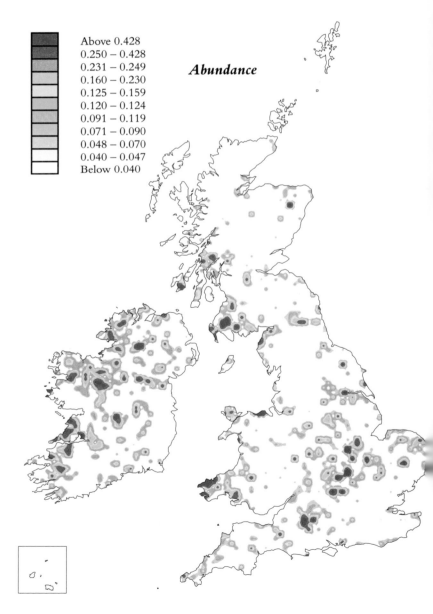

■	Above 0.428
	0.250 – 0.428
	0.231 – 0.249
	0.160 – 0.230
	0.125 – 0.159
	0.120 – 0.124
	0.091 – 0.119
	0.071 – 0.090
	0.048 – 0.070
	0.040 – 0.047
	Below 0.040

Abundance

in 1971, before falling to a level comparable with the early 1960s (*Trends Guide*).

Clearly the Grasshopper Warbler has failed to maintain the surge in numbers it achieved in the 1960s. Some losses can be attributed directly to habitat deterioration and loss. Foremost have been the maturation of much conifer afforestation beyond the preferred thicket stage, and clearance of downland scrub. Similarly, the drainage of damp sedgy fields, carr and fen, the progressive grubbing of suitable thick broad-based hedges, and the clearance of choked ditches, all features of increased mechanisation on farmland during the 1970s and 1980s, will all have removed many potential territories.

Nevertheless, the magnitude of the decline in Grasshopper Warbler numbers since 1971 suggests that habitat loss cannot be the only reason, and that adverse factors may be operating outside the breeding season. Indeed, BTO Nest Records show that the Grasshopper Warbler has a protracted laying season, from late April to early August, and high reproductive potential with an average brood size of 4.87 young, a fledging success rate of 65.2% and the ability to rear two, or even three broods in favourable warm summers (Glue 1990b, Callion *et al.* 1990).

Change

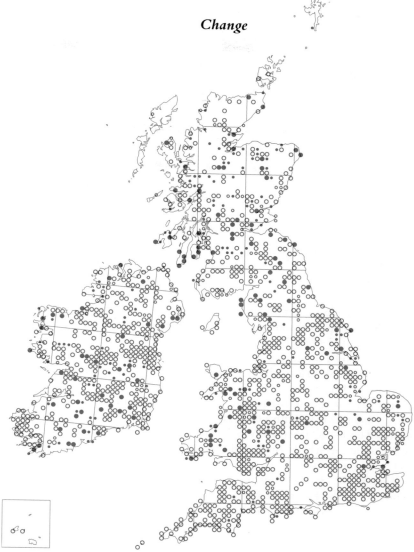

Distribution

The winter quarters are not precisely known but are likely to be similar to those of the nominate race breeding in Europe, which travels to the tropical zone of W Africa (Moreau 1972). Like other trans-Saharan migrants, such as the Whitethroat, the Grasshopper Warbler may have incurred losses there, though a 3–4 year difference in the drought's impact may indicate that they do not share winter quarters.

Using the *68–72 Atlas* assumption of an optimistic 10 pairs per occupied 10-km square, there are estimated to be 10,500 pairs of Grasshopper Warblers in Britain and a further 5,500 in Ireland.

DAVID E. GLUE

68–72 Atlas p 358

Years	Present, no breeding evidence		Breeding evidence		All records		
	Br	Ir	Br	Ir	Br	Ir	Both (+CI)
1968–72	102	29	1780	664	1882	693	2577
1988–91	516	185	673	224	1189	409	1598
	% change				−36.7	−40.9	−37.8

Savi's Warbler

Locustella luscinioides

Savi's Warblers are about the size of a Reed Warbler, but of richer colour. They begin singing as soon as they arrive from their sub-Saharan winter quarters, in mid April. Even where present, they are rarely seen, and are most commonly located by their reeling song. This, although similar to that of their close relative, the Grasshopper Warbler, is higher pitched, and more insect-like.

It is a rare species in Britain and Ireland, and is found in wet reed bed areas. *Phragmites* reeds with a thick ground layer of sedges, rushes and with scattered bushes are favoured (Van der Hut 1986). They may also be found in dry or tidal reed beds (Bibby and Lunn 1982), but have not been known to breed in such sites.

The *68–72 Atlas* documented a population that was mainly concentrated on the E coast of England, from Kent north to Norfolk. The 88–91 Distribution map shows that over the 20 year period between the two Atlases, the population has moved inland, although some of the coastal sites remain. An apparent decline in range in Kent has been partly compensated for by an increase in records along the S coast. There were two Irish records of singing males (one for a protracted period) but no evidence of nesting there during 1988–91, and breeding has yet to be proven (*Birds in Ireland*). Any annual fluctuation in numbers is probably due to overshooting of Continental birds in spring.

The species was first described in 1824. Newton's revision of Yarrell (1871–74) reported breeding records from Norfolk, Suffolk, Cambridgeshire and Huntingdonshire. As a result of extensive drainage the species seems to have become extinct in England by 1856, and it was not until 1954 that it was recorded again, in Cambridgeshire. Breeding was first proven in Kent in 1960, and at two sites in Suffolk in 1970–71 (Pitt 1967, Axell and Jobson 1972).

The Savi's Warbler nests in reeds over water and in such vulnerable a habitat it is difficult to study in the breeding season. As a result virtually no research has been carried out in Britain, although the species has been studied in other parts of its range where it is more abundant. The most extensive study of Savi's Warblers was made in an area of fish ponds near Milicz in Poland in the late 1970s (Pikulski 1986). Within this 560 ha complex, singing males occurred at a density of nearly one per 10 ha, and in favoured areas of reed bed they were recorded at densities of up to 16 males per 10 ha. In any one year only half of the males attracted a mate,

those that were unsuccessful occupying sub-optimal habitats. Males arrived in mid April and set up territories (mean size: 5,000m^2) approximately 12 days before the females arrived. Pikulski distinguished five separate song types which apparently served different functions during the nesting cycle. These were: a 'basic' song (the longest type) for mate attraction; a 'checking' song, used by males to determine whether or not the female was on the nest; a 'warning' song to alert the incubating female to potential danger; a 'distraction' song to lure predators away from the nest, and a 'threatening' song mainly used in territorial defence. Males sang most at night, with a peak in song output at dawn. These observations could prove useful in future monitoring of this species, particularly when an accurate assessment of numbers and proof of breeding are required. The earliest clutch was recorded in late April and the last chick fledged in late July. The mean clutch size was five eggs and about a third of pairs attempted a second brood if their first was successful. The young were fed mainly on insects, Diptera and Lepidoptera accounting for nearly 90% of their diet. Nearly half of all nests were lost, mainly due to predation during the nesting stage, and the average productivity was 2.5 fledglings per pair per season.

Savi's Warblers breed locally from SE England, the Low Countries, Germany, Poland and central Russia, south to the Mediterranean, the Black Sea and Caspian region. Some expansion has occurred in suitable habitats, although there are reports of a decline in Belgium and, possibly, the Netherlands, due to habitat loss, highlighting the need for protection and careful management of sites in Britain and Ireland (*Red Data Birds*).

Ten to 17 singing males were recorded in 1988–90. Only 13 localities were occupied (RBBP). Bearing in mind that not all records are reported and that some males would have remained unmated, the annual population in Britain during 1988–91 was probably about 10–20 pairs and that of Ireland 0–1 unmated males.

L. P. BEAVEN

68–72 Atlas p 360

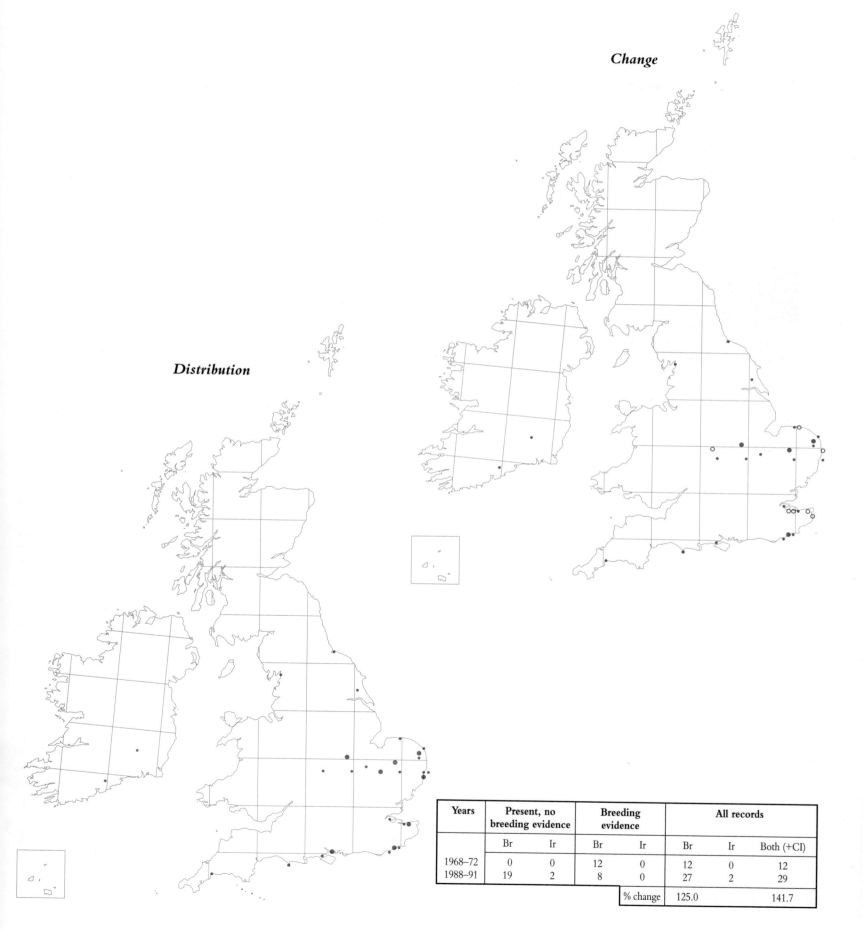

Change

Distribution

Years	Present, no breeding evidence		Breeding evidence		All records		
	Br	Ir	Br	Ir	Br	Ir	Both (+CI)
1968–72	0	0	12	0	12	0	12
1988–91	19	2	8	0	27	2	29
				% change	125.0		141.7

Sedge Warbler

Acrocephalus schoenobaenus

The arrival of the early Sedge Warblers in mid April is announced with vigorous bouts of cheerful chatter, delivered either during flight or from a song post. Characteristic of lowland marsh and waterside habitats, this close relative of the Reed Warbler breeds also in a variety of dry scrub vegetation, including bramble and hawthorn thickets and occasionally young conifers and crops like oil-seed rape.

Although the Sedge Warbler continues to be widely distributed in lowland areas of Britain, the Change map shows that it has lost ground, particularly in parts of S Ireland and central and SW England. Some of the losses in Scotland and N England have been offset by the colonisation of previously unoccupied squares.

Population decline and habitat losses have probably contributed to the disappearance of Sedge Warblers from many areas. The numbers of territorial birds counted on CBC and WBS plots have fluctuated widely since 1962 (*Trends Guide*), and after a severe decline in the mid 1980s numbers at the time of the *88–91 Atlas* fieldwork had recovered to levels typical of the 1970s. Some of the losses shown on the Change map may be a consequence of the occupation of marginal or sub-optimal habitats during 1968, a year of exceptional abundance for British Sedge Warblers. The drainage of wetlands, coupled with the widespread intensification of agricultural methods, may account for the increasingly patchy distribution, even in the present stronghold of E England. Figures documenting the extent of habitat loss are rarely available but, when they are, they are usually worrying. It has been estimated, for example, that 90% of East Anglian fenland was lost between 1934 and 1984 (NCC 1984).

The Abundance map emphasises the importance to Sedge Warblers of counties in E England, particularly Cambridgeshire, Humberside and Essex. Throughout much of W Britain and S Ireland, abundance is greatest in coastal sites or along the lower reaches of major rivers, which to some

extent reflects topography. Relatively high densities of birds exist in many areas of lowland Scotland, central Ireland, and the N plain of the Isle of Man.

Sedge Warblers build nests close to the ground and appear to prefer patches of dense vegetation with a complex structure (Thomas 1984). Nest record cards show that clutches of five eggs are most common and that the peak laying period occurs between 15 and 20 May, just two weeks after the main arrival of birds from Africa (Bibby 1978). Incubation is mainly by the female and lasts approximately 13 days. When feeding chicks the adults forage mainly outside the territory (Catchpole 1972). The young leave the nest 10–14 days after hatching and remain dependent on the parents for a further 1–2 weeks. Although second broods have been reported, little is known about their frequency.

There is no evidence from BTO Nest Record Cards that the decline of the British Sedge Warbler population since the late 1960s has been caused by a decline in breeding success. There is, however, strong evidence that both the population levels and adult survival rates of British Sedge Warblers are determined by conditions on the wintering grounds (Peach *et al.* 1991). In years of normal rainfall in W Africa extensive floodplains

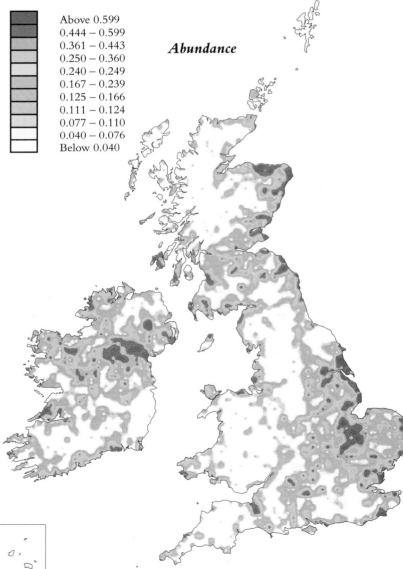

	Above 0.599
	0.444 – 0.599
	0.361 – 0.443
	0.250 – 0.360
	0.240 – 0.249
	0.167 – 0.239
	0.125 – 0.166
	0.111 – 0.124
	0.077 – 0.110
	0.040 – 0.076
	Below 0.040

Abundance

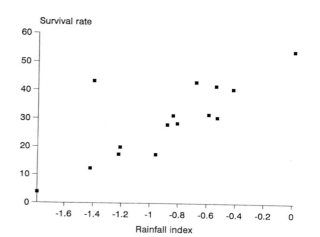

Estimated annual survival rates (%) of adult Sedge Warblers trapped 1969–84 at two sites in S England, plotted against an index of annual rainfall for the previous wet season in the W African winter quarters. For full details see Peach *et al.* 1991.

Change

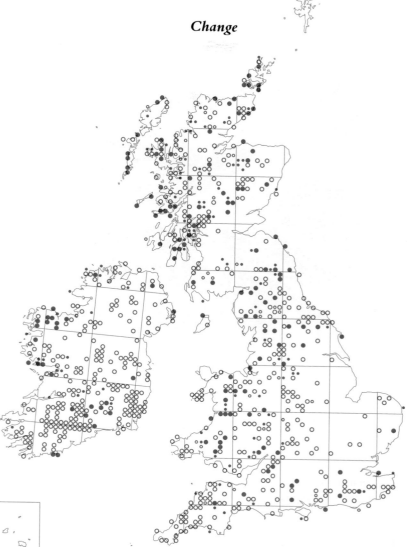

Distribution

develop around the Rivers Niger and Sénégal, creating an abundance of wetland habitat for arriving migrants. In years of drought there is presumably relatively little flooding, and wetland habitat is scarce. Following the most severe failure of the rains, in 1983, fewer than 5% of adult Sedge Warblers were estimated to have survived and returned to their breeding sites in England.

During 1988–91 there were estimated to have been 250,000 Sedge Warbler territories in Britain, and a further 110,000 in Ireland.

WILL PEACH

68–72 Atlas p 366

Years	Present, no breeding evidence		Breeding evidence		All records		
	Br	Ir	Br	Ir	Br	Ir	Both (+CI)
1968–72	81	20	1988	838	2069	858	2929
1988–91	333	175	1554	506	1887	681	2571
			% change		−8.7	−20.5	−12.2

Marsh Warbler

Acrocephalus palustris

The Marsh Warbler is the finest mimic of all British breeding birds, producing, on average, imitations of over 70 different species within a half hour song bout (Dowsett-Lemaire 1979). This song contains faithful renderings of African and European species and upon the male's arrival is delivered for most of the day and night.

Unfortunately, this extraordinary singer is one of the rarest of all breeding birds in Britain, despite the apparent availability over much of S England of suitable breeding habitat. Marsh Warblers require stands of tall, dense vegetation which is generally dominated by willowherbs, meadowsweet, umbellifers or nettles. Adjacent bushes of hawthorn or willow provide additional song perches and foraging areas. Such vegetation may be found along valley bottoms and disused land in many areas. Marsh Warblers are amongst the latest migrants to arrive, corresponding with the late development of their favoured tall vegetation community (Van der Hut 1986).

The Marsh Warbler is common in Europe and its rarity in Britain might be attributable to the phenomenon of island biogeography. As with Golden Orioles and Icterine Warblers, which are also long-distance migrants widespread in W Europe but rare or absent as breeders in Britain, a combination of factors may well be involved. British Marsh Warblers generally arrive later (from late May onwards) than European populations and are less fecund (Kelsey 1989a). Failed breeders attempting to breed again will do so, on average, later than their Continental counterparts and offspring produced will probably be less likely to survive through to the following year.

Until the 1960s, Marsh Warblers bred on average in five English counties each year. However, in the 20 counties in which breeding has occurred, or probably occurred, since 1890, only three have supported populations for over half the period. In two of those, Somerset and Gloucestershire, breeding has not taken place since 1961 and 1984 respectively (Kelsey *et al.* 1989). Most records are of transient breeding, even when more than one pair is involved. However, the West Midlands population, and that of Worcestershire in particular, has been outstanding in its continuity. At its peak there may have been over 100 pairs and surveys showed at least 95 pairs in the 1960s. The population declined during the 1970s to 50–60 singing males.

Why the population showed an enduring preference for this region is not clear, although the sheltered Severn and Avon valleys boast high summer temperatures and the fertile, damp soil there supports a luxuriant vegetation usually well advanced in growth by the time the birds are nest building. The reason for the population decline within its West Midlands heartland is also not clear, but cumulative effects of reduction in optimal habitat during the 1950s and 1960s may have made pairs more susceptible to a succession of poor summers or other external factors. In addition, evidence from the 1980s suggested that the population was isolated from Continental populations, thus making it vulnerable (Kelsey *et al.* 1989).

The pattern of records in the southern counties is typical for any four year period. Sites occupied during some years may not necessarily be favoured in subsequent years. The individuals involved may be overshoots from the Continent, although small populations may be established.

Determination of the size of the breeding population is not easy because the males which pair quickly stop singing, whereas those that remain unpaired may sing for several weeks. Furthermore, some males sing again later in the season, away from their breeding territories (Kelsey 1989b), and are liable to be double counted by survey workers. Although Reed Warblers are almost identical in appearance, they cannot match the quality of the Marsh Warbler's song. Much caution is, however, required in separating the species since they often breed and forage in close proximity. Marsh Warblers are single brooded, building a nest slung between vertical stems in a less compact and tidy fashion than Reed Warblers. Research in Germany has shown that the incidence of successful host parasitism by Cuckoos is significantly less in Marsh Warblers than Reed Warblers, Cuckoo eggs commonly being ejected by their Marsh Warbler hosts (Gärtner 1982).

Although the European population has not shown any general decline, and indeed, is expanding northwards and eastwards (Kelsey *et al.* 1989), the British population has declined markedly in the last 20 years. Compared with the 1968–72 estimate of 50–80 pairs, the population during 1988–91 averaged fewer than 12 pairs a year, although the number of singing males would be higher. It is hoped that the population recently established in SE England may expand in numbers and range.

MARTIN KELSEY

68–72 Atlas p 364

The authenticity of the two confirmed breeding records for Kent shown in the *68–72 Atlas* has been questioned, and it is thought that the first confirmed breeding record for the county was in 1980 when a single pair raised two young (Taylor 1982). However, as it is probable that Marsh Warblers bred in at least one of these two sites, these records were included in the production of the Change map.

Change

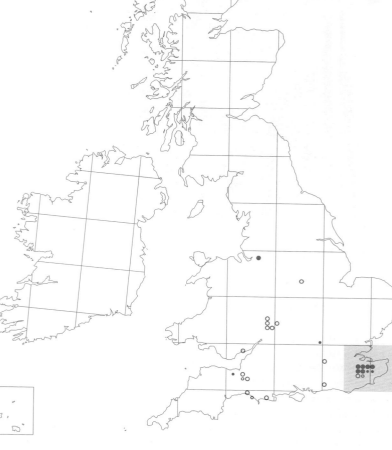

Distribution

Years	Present, no breeding evidence		Breeding evidence		All records		
	Br	Ir	Br	Ir	Br	Ir	Both (+CI)
1968–72	3	0	18	0	21	0	21
1988–91	7	0	8	0	15	0	15
			% change		−28.6	0	−28.6

Reed Warbler

Acrocephalus scirpaceus

The conversational chatter of Reed Warbler song is a conspicuous feature of most reed beds in the summer, but the bird itself is quite often difficult to see, spending much of its time rather low down in the vegetation, even when it is singing.

Reed Warblers are closely associated with *Phragmites* and can be found not only in extensive reed beds, but also where reeds form a fringing vegetation to riverbanks and gravel pits. Indeed, studies of the morphology of *Acrocephalus* warblers suggest that the large foot-span of the Reed Warbler makes it supremely well adapted for a life in the forest of vertical stems (Leisler 1975). The association is not, however, an exclusive one. Reed Warblers do sometimes occupy other vegetation types of similar structure to reeds, such as willowherb and are now being recorded in arable crops such as rape (Sharrock 1986). Likewise, not all British reed beds support Reed Warblers. The Distribution map shows that the species occurs mainly in lowland Britain, south of a line from S Yorkshire to Lancashire, even though *Phragmites* occurs widely elsewhere. The limiting factor for the distribution is not understood, although Reed Warblers may fare better in the drier, warmer south and east where the quality of reed growth may be superior (*68–72 Atlas*).

The Abundance map shows that Reed Warblers are most common in East Anglia, where there are important reed beds and many reed filled dykes in the extensive low-lying country. Similar landscapes in the Humber region, the Somerset Levels, Kent, Romney Marsh and other coastal lowlands also support large populations.

Reed Warblers have extended their range northwards and westwards during the last 20 years, particularly in N England and Wales. Breeding is now regular in Ireland and has been recorded in Scotland (which might be attributed to misplaced birds from the northward expansion of the Scandinavian populations). There has been little change within the main part of the species' range in England. Localised declines arising from the

loss of bankside vegetation may be offset by the availability of new sites in abandoned gravel workings and areas of mining subsidence (*Trends Guide*). Overall, there does not appear to have been any major change to the size of the Reed Warbler population in Britain, although some evidence from CBC studies indicates an increase. On the other hand, the semi-colonial dispersion of breeding Reed Warblers and the nature of their breeding habitat make census techniques based on observation extremely unreliable. The BTO are addressing this drawback with the introduction of Constant Effort Site ringing. Reed bed sites form a strong component of this scheme and future changes in Reed Warbler populations should be detected. In contrast to the Sedge Warbler, the species seems to have been little affected by the Sahel drought because its wintering areas lie further south (*Trends Guide*).

Reeds provide the most important nesting sites, but the density of nests will vary considerably with natural or induced variations in the quality of the reeds. Nesting does take place in adjacent dry-land vegetation, which is also important for feeding (Catchpole 1972). Compared with Sedge Warblers, Reed Warblers have a more prolonged song period. This is partly due to some song resumption with second broods and the persistence

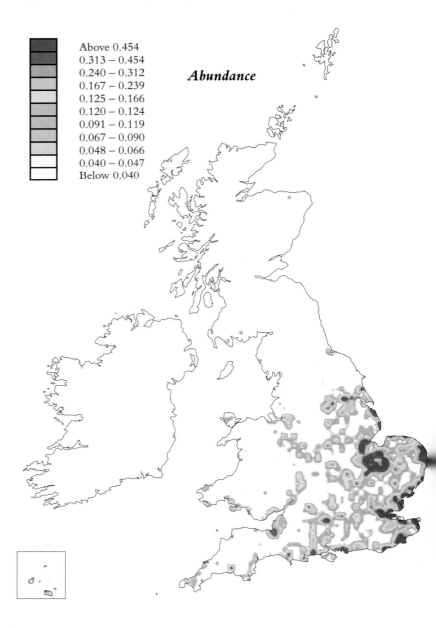

████	Above 0.454
	0.313 – 0.454
	0.240 – 0.312
	0.167 – 0.239
	0.125 – 0.166
	0.120 – 0.124
	0.091 – 0.119
	0.067 – 0.090
	0.048 – 0.066
	0.040 – 0.047
	Below 0.040

Abundance

of song in paired males during dawn and evening choruses (Catchpole 1973). Those individuals singing for long periods, however, are likely to have been unsuccessful at pairing or are failed breeders. The main arrival period is from the second half of April, and individuals may linger in the breeding areas into early October.

There has been considerable recent work on the incidence of brood parasitism by Cuckoos on Reed Warblers. The Reed Warbler is one of the three most important host species for the Cuckoo in Britain (along with Dunnock and Meadow Pipit) and the percentage of Reed Warbler nests parasitised has more than doubled in the last 40 years (Brooke and Davies 1987). In some years, a fifth of all nests may be parasitised at particular sites, although the same study showed that Reed Warblers rejected 19% of Cuckoo eggs laid. The ability of Reed Warblers to recognise unusual eggs in the clutch has probably led to the highly mimetic appearance of the Cuckoo egg, both in colour and size. It has been shown that clutches that are parasitised suffer less subsequent predation than unparasitised clutches, suggesting that the Cuckoo itself is an important egg predator, probably destroying clutches that are too well advanced for its needs, to force Reed Warblers to re-lay (Davies and Brooke 1988).

Change

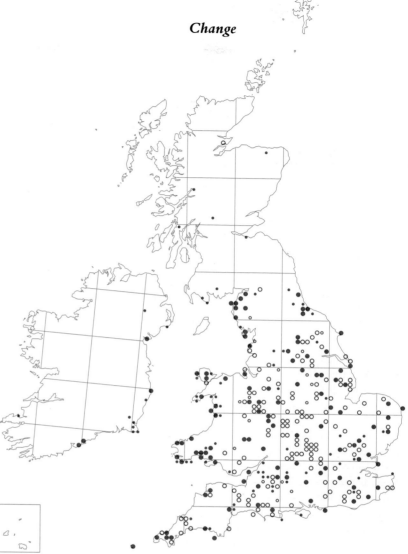

Distribution

Given the similarity in the number of occupied 10-km squares during the two Atlases, it is probable that the population does not differ significantly from the estimate of 40,000–80,000 pairs given in the *68–72 Atlas*. The Irish population, although increasing, is still small with probably no more than 40–50 singing males (and fewer breeding pairs) in any one year.

MARTIN KELSEY

68–72 Atlas p 362

Years	Present, no breeding evidence		Breeding evidence		All records		
	Br	Ir	Br	Ir	Br	Ir	Both (+CI)
1968–72	39	0	736	0	775	0	778
1988–91	152	9	638	4	790	13	812
	% change		1.9				3.6

Dartford Warbler

Sylvia undata

A skulking bird flitting into the cover of gorse on a southern heath may well be the elusive Dartford Warbler. Cold northern winters are not to the taste of such a small and largely resident warbler, which is more at home in the heat of Mediterranean latitudes.

In Britain, Dartford Warblers are symbols of their scarce and much threatened habitats. They are sufficiently studied for numbers and distribution to be quite well understood. The critical habitat is almost exclusively dry heathland with gorse. Their nest may be in deep heather or gorse and these are the two elements of their habitat throughout the year.

Gorse is the predominant feeding place because it is much richer in invertebrates than heather (Bibby 1979a). Especially in winter, food abundance is much lower in heather, and gorse also provides some shelter from drifting snow. The species is rather sedentary, though a good number roam along the S coast of England each autumn and some may winter across the Channel (Bibby 1979b). Like other small insectivorous birds, the Dartford Warbler suffers in cold winters, when numbers can drop by as much as 80%. After the successive cold winters of 1961/62 and 1962/63, only 11 remaining pairs were known. After such collapses, the autumn movement of birds will eventually ensure the recolonisation of remoter sites. In good summers, Dartford Warblers can rear two, or sometimes even three, broods and breeding populations can double in two years. The recent history of numbers consists of periodic crashes between spells of recovery and expansion (Bibby and Tubbs 1975, Robins and Bibby 1985). A run of mild winters up to the mid 1970s produced a population of nearly 600 pairs and at the time it seemed unlikely that this figure would ever be exceeded.

The extensive heaths of the New Forest and Dorset hold the nucleus of the population. The *88–91 Atlas* maps have caught the highest numbers for at least half a century (see Table) and previous predictions have proved to be pessimistic. In the *68–72 Atlas*, Surrey had only just been reoccupied after being depopulated in 1961. In the *88–91 Atlas*, the best heaths in Surrey are well populated. Exceptionally high numbers were counted in the New Forest in 1988 (Westerhoff and Tubbs 1991). In 1980, Dartford Warblers bred in Cornwall for the first time in 40 years and a small population of up to ten pairs persisted for a few years. Numbers in Devon are smaller than their previous high of 30–40 pairs in 1977. Sussex is still occupied though numbers have not recently matched the high point of 15 pairs in 1974. With appropriate management and good gorse cover,

Dartford Warblers can reach densities of 10–15 pairs per km², but most places are less suitable. The remoter populations are all susceptible to complete loss in cold winters, with Surrey probably being colder and its population more vulnerable than those of the coastal counties to the west. Whilst there is a chance that heathland areas in S Wales or East Anglia could be colonised, small populations would not be very likely to persist. In the east, potential habitat is plentiful, but Dartford Warblers would be very vulnerable to cold winters in East Anglia. To the west, the coastal areas of heathland are tiny and fragmented, so populations would be vulnerable through being small and isolated.

The favoured heaths may be damaged by accidental fires but such damage is temporary. After about ten years, burnt heaths have the ideal development of heather and gorse for Dartford Warblers. Farming and forestry have been serious threats in the past, but are now largely prevented by statutory site protection. House and road building pressures on heathland remain enormous, especially in Dorset, but conservation has achieved some notable success in maintaining its claims. Other heaths are being lost through the spread of trees in the absence of grazing. This is an insidious threat which will require a substantial effort to combat.

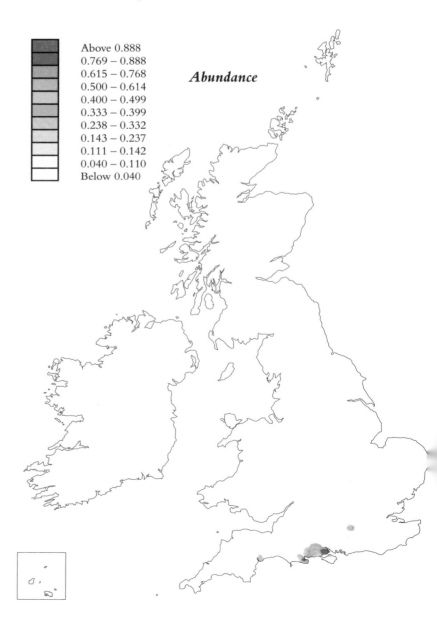

Above 0.888
0.769 – 0.888
0.615 – 0.768
0.500 – 0.614
0.400 – 0.499
0.333 – 0.399
0.238 – 0.332
0.143 – 0.237
0.111 – 0.142
0.040 – 0.110
Below 0.040

Abundance

The population in the *88–91 Atlas* period was probably up to 950 pairs. Although this is a tiny number compared with those in S and W France and Spain, where the species occurs extensively in open scrubby habitats, the Dartford Warbler population is an important emblem in Britain for the conservation of lowland heaths which are very special for a variety of other reasons.

Some recent counts of Dartford Warblers in Britain

	1960–61	1963	1974	1984	1988–89	1990
Hampshire	350	6	255	219	461	441
Dorset	63	4	286	127	132	334
Surrey	40	0	1	69	54	120
Sussex	4	1	15	0	4	8
Devon	?	0	3	2	12	?
Cornwall	0	0	0	6	0	1

COLIN BIBBY

68–72 Atlas p 376

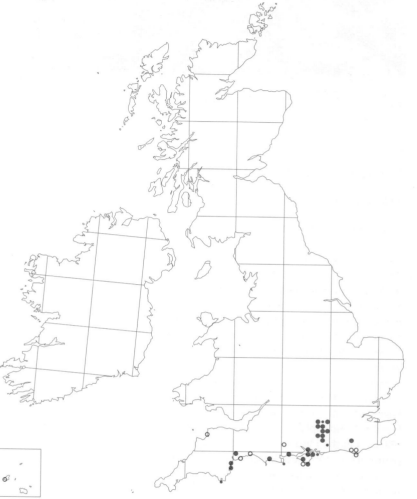

Change

Distribution

Years	Present, no breeding evidence		Breeding evidence		All records		
	Br	Ir	Br	Ir	Br	Ir	Both (+CI)
1968–72	2	0	26	0	28	0	31
1988–91	5	0	40	0	45	0	50
				% change	60.7	0	51.6

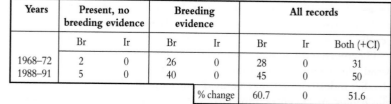

Lesser Whitethroat

Sylvia curruca

The Lesser Whitethroat is the smallest of our four migrant *Sylvia* warblers – and the least conspicuous. It has, to human ears, a rather boring, rattling song, generally given within the deep cover in which it chooses to live. When well seen it is undoubtedly one of our most attractive birds with the broad black stripe from above the bill, through the eye and spreading across the lores, giving it a jaunty, piratical look. It is clearly closely related to the Whitethroat, but the ancestors of the two species became separated during a period when the advancing ice forced the birds southward. Ancestral Whitethroats were isolated in the west and Lessers in the east. To this day, European Lesser Whitethroats migrate through the E Mediterranean to winter in Ethiopian highlands.

Lesser Whitethroats are generally found in scrubby areas, parkland and farmland with mature hedges. They are absent from woodland, open farmland and upland areas. The southeastern origin of the species is reflected in the Distribution map. The species is notably sparse in the SW peninsula – even in the lower areas with apparently very suitable habitat – and from the mainland of NW Wales. There are records from all 10-km squares in Anglesey but none from the Isle of Man. From Ireland there is only one record, and most of the few Scottish records are along, or close to, the E coast, from Fife southwards. Scattered records further north are an indication of the potential for expansion shown by a long-distance migrant that regularly strays north of its normal breeding area in spring. This is a species which has clearly extended its range since the *68–72 Atlas*, especially in S Wales, Anglesey, N England and S Scotland. The breeding record in Ireland is the first ever, and Hutchinson (*Birds in Ireland*) notes only 150 records of migrants, with autumn birds outnumbering spring ones 2:1. Thom (*Birds in Scotland*) notes that there were no records of confirmed breeding in the *68–72 Atlas* but that, since then, breeding has been confirmed in six counties, from Aberdeen southwards. The Caithness and Orkney breeding records shown on the Distribution map are the first for these regions. The Lothians' population was estimated by da Prato (1980) as being 50–100 pairs. Subsequently, Byars *et al.* (1991) reported that 9–12 territories are regularly used in Strathclyde each year, all of them in dense and ungrazed hawthorn scrub.

Hawthorn scrub with dense undergrowth is not a very common habitat. Overgrown railway embankments, disused mineral workings and even derelict industrial sites may be used, but naturally occurring examples of this habitat are largely confined to chalk grassland scrub. Suitable farmland

areas have undoubtedly decreased over the last few decades as hedges have been grubbed out and those that remain have been degraded by mechanical trimming (O'Connor and Shrubb 1986a). Future changes in agricultural management, promoting extensification, may be of considerable help to this species.

In the main part of the Lesser Whitethroat's British range, birds may regularly be found breeding at 200m (or even higher) where suitable habitat exists – as in Hertfordshire (Mead and Smith 1982). In contrast, on the western fringes, in Devon, Sitters (1988) notes that most records are from coastal areas. The percentage of tetrads with records of this species in Hertfordshire was 48%, but only 6% in Devon.

The *Trends Guide* shows marked fluctuations in the CBC index but no consistent change. Certainly the Sahel drought which affected species moving to W Africa, such as the Whitethroat, did not have a measurable effect on this southeasterly migrant. There will still be very few pairs in the newly colonized areas on the fringes of the range, but the extension should be indicative of consolidation within the main part of the range.

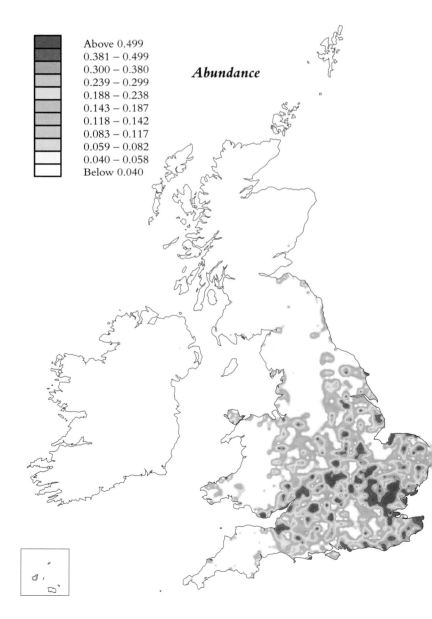

Above 0.499
0.381 – 0.499
0.300 – 0.380
0.239 – 0.299
0.188 – 0.238
0.143 – 0.187
0.118 – 0.142
0.083 – 0.117
0.059 – 0.082
0.040 – 0.058
Below 0.040

Abundance

On farmland, CBC densities are now at about 1.0 pairs per km², compared with 0.65 at the time of the *68–72 Atlas*. During 1988–91 there were estimated to have been 80,000 Lesser Whitethroat territories in Britain.

CHRIS MEAD

68–72 Atlas p 374

Change

Distribution

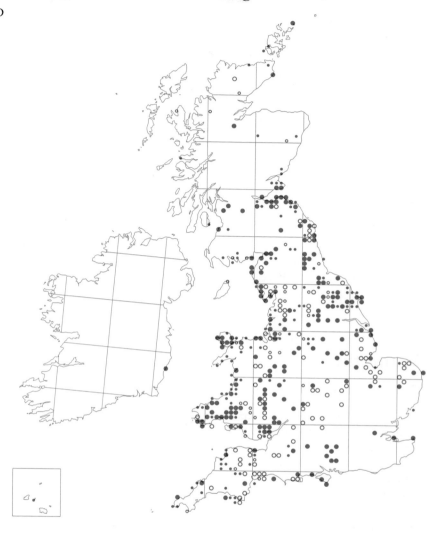

Years	Present, no breeding evidence		Breeding evidence		All records		
	Br	Ir	Br	Ir	Br	Ir	Both (+CI)
1968–72	124	0	970	0	1094	0	1097
1988–91	290	0	981	1	1271	1	1279
				% change	16.2		16.3

Whitethroat

Sylvia communis

The jerky display flight and scratchy song of the Whitethroat are a characteristic sight and sound of scrub, young woodland and hedgerows throughout most of Britain and Ireland. Although sometimes abundant, this species has become only too well known for the dramatic crashes which its populations have made, due to climate related changes in its African wintering grounds.

Whitethroats occur more or less throughout England and Wales, although they are absent from the highest ground of the Pennines, the Lake District, the Welsh mountains, and Dartmoor. In Ireland, some of the patchiness of the distribution may be due to the lower level of coverage there, but there are genuine gaps in parts of the north, in Mayo and Galway, and through the southern third of the island. In Scotland, it is well distributed on lower ground in the south and east, but absent from large areas in the Southern Uplands, the Highlands, most of the Outer Hebrides, Orkney and Shetland.

Whitethroat abundance is typically higher in E England than elsewhere. This is considered to be due to geographical factors, rather than differing agricultural practices (*Trends Guide*). To the north and west, high numbers tend to be in coastal areas, with concentrations in such places as the Burren in Co. Clare and around the Dyfed coast, where scrubby deciduous woodland, hedgerows and cliff tops are especially prevalent. Perhaps these are the only habitats, in these areas of more rigorous climate, with sufficiently dense vegetation and reliable supplies of insect prey, to attract more than occasional pairs.

It was the failure of three-quarters of the Whitethroat population to reappear, in spring 1969, which drew attention to the effect of the Sahelian drought on migratory birds (Winstanley *et al.* 1974). The catastrophe affected other species which winter just south of the Sahara, but perhaps

only for Sand Martin was the decline as steep. Whitethroat numbers reached a low point in 1974, following which they stabilised at a level between a third and a half of the pre-1968 norm. Disaster again overtook the species in the winter of 1983/84, after which CBC indices were similar to those of 1974.

Until 1991, when another sharp decline occurred, the years of the *88–91 Atlas* saw a reversal of the species' fortunes, with a continuous rise in the farmland CBC index from 1985 to 1990, a trend generally mirrored by other BTO monitoring schemes. Yet the Change map indicates a marked retreat from many areas in the north and west since the *68–72 Atlas* (even though only the first year of fieldwork for that Atlas preceded the 1969 population crash). The Pennines, the Southern Uplands and Highlands of Scotland, and Ireland, especially in the southwest, are most obviously affected. CBC densities have remained particularly low at higher altitudes (*Trends Guide*).

Studies of the breeding birds of coppice woodland (eg Fuller and Moreton 1987, Fuller *et al.* 1989) have emphasised the close association of the Whitethroat with the early successional stages of woodland. Like most other woodland summer migrants, it avoids older coppice and in

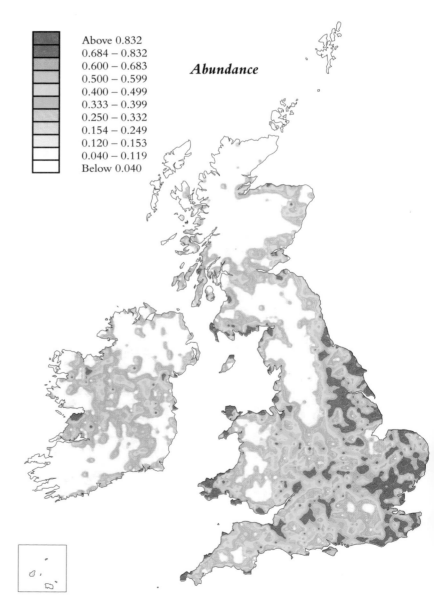

Above 0.832
0.684 – 0.832
0.600 – 0.683
0.500 – 0.599
0.400 – 0.499
0.333 – 0.399
0.250 – 0.332
0.154 – 0.249
0.120 – 0.153
0.040 – 0.119
Below 0.040

Abundance

fact it has one of the narrowest bands of tolerance in terms of coppice age. Densities are highest two or three years after coppicing, and decline quickly once canopy cover passes 50%.

A similar pattern exists in scrub, with Whitethroats disappearing once bushes close over open areas (Fuller 1982), but the pace of development of scrub is generally slower, so that the habitat may be exploited for many years. Scrub supported over half (56%) of Whitethroat nests reported in BTO Nest Records (Mason 1976), whereas woodland held 19% and hedgerows 18%. Scrub can support very high densities, with up to 158 territories per km² recorded (da Prato 1985). The species' ability to use low cover was further illustrated by a BTO study of farmland birds, in which Whitethroat was one of very few hedgerow species not adversely affected by severe hedge cutting (Lack 1987).

During 1988–91 there were estimated to have been 660,000 Whitethroat territories in Britain, and 120,000 in Ireland.

ANDREW HENDERSON

68–72 Atlas p 372

Change

Distribution

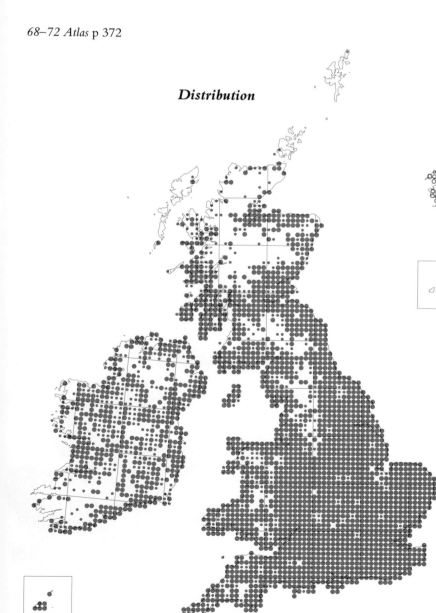

Years	Present, no breeding evidence		Breeding evidence		All records		
	Br	Ir	Br	Ir	Br	Ir	Both (+CI)
1968–72	67	32	2280	812	2347	844	3196
1988–91	252	197	1934	431	2186	628	2824
% change					−6.7	−25.5	−11.6

Garden Warbler

Sylvia borin

The rich contralto song, delivered at a brisk pace from a perch in dense scrub or in woodland canopy, reveals the presence of a Garden Warbler. These subtly attractive, grey-brown birds are long-distance migrants, wintering in central and southern Africa, well to the south of the drought-stricken Sahel region.

Garden Warblers arrive in Britain and Ireland from late April; mainly in May. They select deciduous woodland and scrub, with the dense ground cover which they need for nesting. In farmland areas they are characteristic of small copses and clumps of hawthorn or blackthorn. Conifer plantations are also used, but only when these are relatively young and have a dense ground cover of other plants. Locally the choice of habitat is influenced by the presence of Blackcaps, their close relatives. Blackcaps arrive earlier, and vigorously exclude Garden Warblers, as well as other Blackcaps, from their territories. Removal of established Blackcaps has shown that some of their vacated territories are taken up by Garden Warblers (Garcia 1983). Nevertheless, there is no close relationship between the numbers of each species, suggesting that interspecific territoriality is not a major factor in determining their population sizes.

Garden Warblers are common birds throughout most of England, Wales and S Scotland, being absent only from much of the fenland south of the Wash, where woodland and scrub are largely absent. Probably, most suitable habitat in this area is occupied, but the Abundance map shows highest densities in much of Wales and in S England, from Kent to Hampshire as well as in N Devon, all areas with widespread sessile oak-woods. Elsewhere, Garden Warblers are more thinly spread, and the sessile oakwoods of NW Scotland, W and SE Ireland remain largely unoccupied.

The distribution of Garden Warblers has shown little change since the *68–72 Atlas*. The Change map shows some expansion in Scotland, notably in Kintyre, but the species remains absent from the Highlands and other areas of unsuitable habitat.

Garden Warbler numbers in Britain were relatively high at the time of the *68–72 Atlas*. Subsequently, a marked decline occurred, with a trough in 1975–76 coinciding with drought conditions in the sub-Saharan Sahel region. Garden Warblers cross the Sahel on migration but, unlike White-

throats, they do not winter there. Hence, recovery has been rapid and is continuing. The *Trends Guide* shows that numbers in 1986–88 had returned to their 1968–72 levels in woodlands, but to a lesser extent in farmland habitats. The recovery was largely complete in woodlands by 1983, but is still continuing on farmland.

The relatively stable national distribution shown by the *88–91 Atlas*, and the good evidence for recovery in numbers obtained from CBC census plots, suggest that suitable Garden Warbler habitat remains widely available and that numbers are stable or increasing. The widespread damage to woodlands in SE England, caused by the gales of October 1987, may be expected to produce local increases in this and other species which favour open woodland, particularly where the many newly created clearings are allowed to become overgrown with brambles and scrub. As with many long-distance migrants, the hazards of migration and conditions in winter quarters seem to be the principal determinants of the breeding populations in Britain. During 1988–91 there were estimated to have been 200,000 Garden Warbler territories in Britain and a further 180–300 pairs annually in Ireland (after Herbert 1991).

ERNEST GARCIA

68–72 Atlas p 370

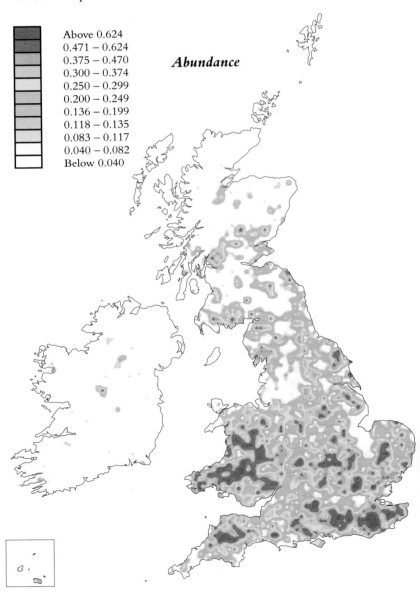

Abundance

	Above 0.624
	0.471 – 0.624
	0.375 – 0.470
	0.300 – 0.374
	0.250 – 0.299
	0.200 – 0.249
	0.136 – 0.199
	0.118 – 0.135
	0.083 – 0.117
	0.040 – 0.082
	Below 0.040

Change

Distribution

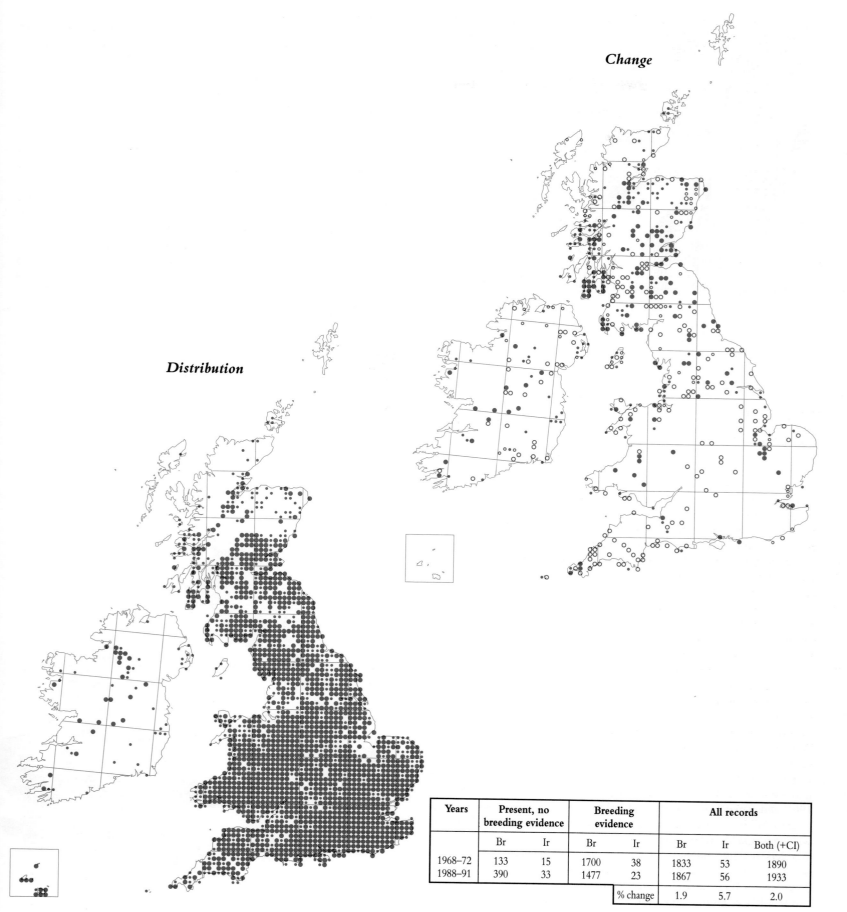

Years	Present, no breeding evidence		Breeding evidence		All records		
	Br	Ir	Br	Ir	Br	Ir	Both (+CI)
1968–72	133	15	1700	38	1833	53	1890
1988–91	390	33	1477	23	1867	56	1933
				% change	1.9	5.7	2.0

The Change map shows clear evidence of a spread northwards, with birds increasingly abundant in Scotland. However, they still remain scarce and probably irregular as breeding birds in most of the Scottish Highland counties (*Birds in Scotland*). Blackcaps are still regarded as local in Ireland (*Birds in Ireland*) but a spread into western and southern regions is obvious, while records in N Ireland and in the Isle of Man, too, have clearly increased.

It is uncertain why Blackcaps are increasing. There is no evidence of any widespread changes in breeding habitat availability, nor of increased reproductive output. Improved survival in winter seems a possibility however. The increase antedates the droughts in the Sahel region in the mid seventies. These would probably have not affected many Blackcaps directly, since good numbers winter north of the Sahara. Ringing has shown that most British and Irish Blackcaps winter around the W Mediterranean, where they feed on fruiting shrubs.

Berthold *et al.* (1990) have demonstrated that migration in Blackcaps has a strong genetic component, and that natural selection can very rapidly favour either sedentary or migratory behaviour in response to

Blackcap

Sylvia atricapilla

Blackcaps are possibly the most successful warblers in Britain and Ireland. Ringing totals, Nest Record Card returns and CBC census data all point to a vigorous and continuing increase dating from at least the mid 1950s. Blackcaps use very similar habitats to those selected by Garden Warblers, with which they compete for space. However, they are less inclined to occur in continuous scrub and are more typical of deciduous woodland. The measured song of Blackcaps, with its clearly spaced and relatively relaxed phrases distinguishing it from that of Garden Warblers, is heard in most British woodlands from March onwards.

The Distribution map shows Blackcaps to be widely distributed throughout England, Wales and S Scotland with absences, as with Garden Warblers, from the higher ground and the fenland region. The distribution in N Scotland is similar to that of Garden Warblers, with Blackcaps being absent from the Highlands. Birds are locally but widely spread in Ireland, where the species is clearly more common than the Garden Warbler.

The Abundance map shows that the density of Blackcaps is uneven, being greatest in S England, south of a line running from the Gower peninsula to the Wash. They are less closely associated than Garden Warblers with sessile oakwoods, a fact shown by the absence of any marked concentration of population in Wales.

The increase of Blackcaps relative to Garden Warblers has been most noticeable in N Britain. In Lancashire (Oakes 1953), Yorkshire (Nelson 1907) and generally in N England, Blackcaps were regarded as scarcer than Garden Warblers in the first half of this century. Baxter and Rintoul, writing in 1953, similarly regarded Blackcaps as less common than Garden Warblers in Scotland. Now both species are at least equally widespread in the northern half of Britain, and more recent accounts for Lancashire (Spencer 1973), Yorkshire (Mather 1986) and Scotland (*Birds in Scotland*) consider Blackcaps to be the more numerous of the two species.

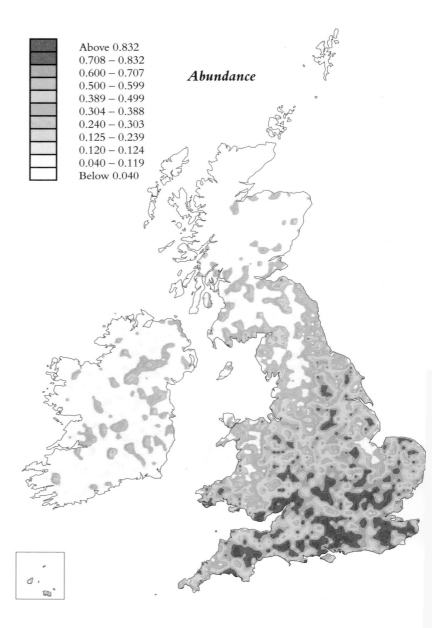

Above 0.832
0.708 – 0.832
0.600 – 0.707
0.500 – 0.599
0.389 – 0.499
0.304 – 0.388
0.240 – 0.303
0.125 – 0.239
0.120 – 0.124
0.040 – 0.119
Below 0.040

Abundance

environmental change. Thus Blackcaps can adapt their migratory behaviour to take advantage of novel winter quarters. An excellent example of this is the rapid increase in Blackcaps wintering in Britain and Ireland (Berthold and Terrill 1988), where they undoubtedly benefit from artificial food sources in gardens. These wintering birds are not British breeders but individuals from central European populations. The increase in Blackcaps is a European trend affecting all those populations which winter in S Europe and N Africa. It is possible that the tendency not to cross the Sahara has become more widespread, and this might account for higher breeding populations as a consequence of improved winter survival.

During 1988–91 there were estimated to have been 580,000 Blackcap territories in Britain and 40,000 in Ireland.

ERNEST GARCIA

68–72 Atlas p 368

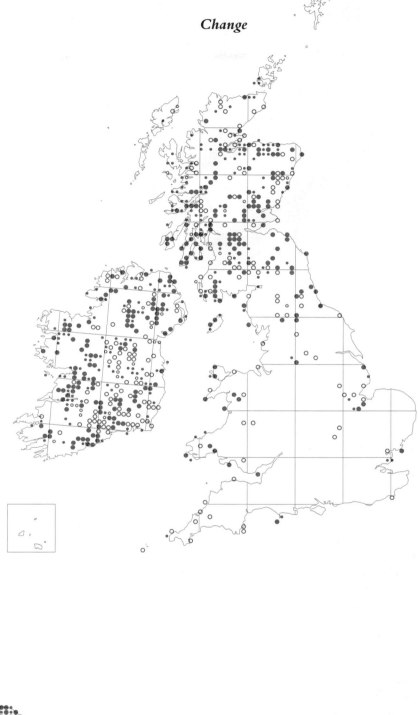

Change

Distribution

Years	Present, no breeding evidence		Breeding evidence		All records		
	Br	Ir	Br	Ir	Br	Ir	Both (+CI)
1968–72	59	35	1864	221	1923	256	2184
1988–91	291	148	1757	209	2048	357	2417
				% change	6.5	39.5	10.3

Wood Warbler

Phylloscopus sibilatrix

There can be few more evocative sounds in a western oakwood in May than the song of the Wood Warbler. Perched on a low branch, head slightly raised and its whole body quivering, it projects its distinctive trilling song far through the wood.

The ease of detecting singing Wood Warblers should mean that the Distribution map is an accurate reflection of their occurrence; which is important, for the species is poorly represented in the CBC plots, and the *Trends Guide* indicates no known recent population trend. The reduction in number of squares with breeding pairs (22%; see Table) is not a genuine range change, but merely reflects the lower emphasis placed on proof of breeding during the *88–91 Atlas*.

There is no clear pattern of change in distribution, although changes vary regionally and locally. For example, Wood Warblers are less common around London and the southeast, continuing the trend reported in the *68–72 Atlas*, and are also less widespread in the southwest. They appear more widespread in SW Wales, where more breeding records were reported in the old county of Carmarthen, and in parts of NW England. In Scotland there have also been gains and losses. For example, Wood Warblers are more widespread on Mull and Arran, and expansion of the range northward, referred to in the *68–72 Atlas*, appears to have continued, albeit slowly: Wood Warblers were reported from more squares in the Highlands including, for the first time, Caithness. There have been local decreases in Easter Ross, Angus, and from Ayrshire and surrounding counties. Although differences in coverage could account for some changes, others are thought to be real, such as in Surrey where declines had already been reported (Herber 1985).

A similar pattern of gains and losses was reported in the 1984–85 Wood Warbler survey, when 149 10-km squares, with probable or confirmed breeding in 1968–72, had no birds in 1984–85; and 68 10-km squares, without Wood Warblers or possible breeding in 1968–72, had gained birds (Bibby 1989).

In Ireland, Wood Warblers are expanding their range, with breeding now occurring in the SW of the Republic and more records from the north. In the *68–72 Atlas*, Wood Warblers were recorded in eight 10-km squares in Ireland. By 1984–85 this had increased to nine squares, involving 16 birds (Bibby 1989); and in the *88–91 Atlas*, to 28 10-km squares.

The Abundance map demonstrates that the Wood Warbler strongholds remain in the upland oakwoods, particularly those of Devon, Wales and the Marches, and parts of central W Scotland and the Great Glen. The map is consistent with that produced in the 1984–85 survey, and also identifies local concentrations such as those in the New Forest, the Forest of Dean and parts of the Lake District. There are some apparent anomalies, however, such as the high numbers reported for a few squares in East Anglia.

The causes of the changes in distribution are not revealed by atlas surveys. In Devon, some losses were attributed to conversion of deciduous woods to conifers (Goodfellow 1986), whilst *Birds in Scotland* mentions both clear-felling and underplanting with conifers as threats to the Wood Warbler's habitat. Curiously, in Surrey, apparently suitable habitat was unoccupied in 1984–85 (Herber 1985). The habitat requirements of Wood

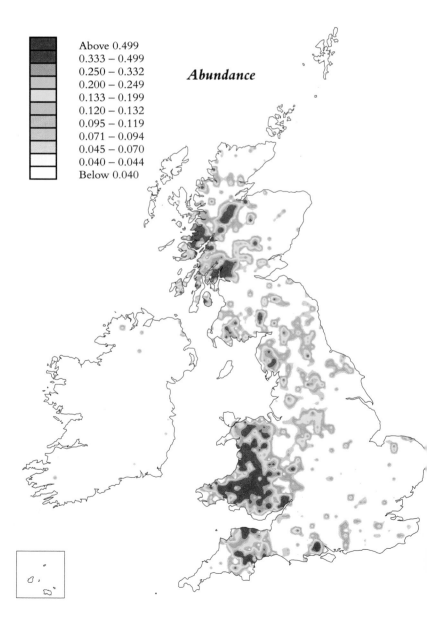

	Above 0.499
	0.333 – 0.499
	0.250 – 0.332
	0.200 – 0.249
	0.133 – 0.199
	0.120 – 0.132
	0.095 – 0.119
	0.071 – 0.094
	0.045 – 0.070
	0.040 – 0.044
	Below 0.040

Abundance

Warblers have been well described elsewhere. Of particular interest is their dependence on closed canopy woods with little or no understorey. Often it is winter grazing, mainly by sheep, which prevents the understorey from developing or becoming too dense. Possible future changes in the agricultural support systems may influence the amount of grazing in, for example, some of the best Welsh woods, and this could have adverse effects on the Wood Warbler population if a dense birch or shrub layer were allowed to develop unchecked (Stowe 1987).

The *68–72 Atlas* estimated the British and Irish population to be in the range 30,000–60,000 singing males. This was based on a supposed 25–50 pairs per 10-km square, a figure that was certainly too high for all but the western oakwood strongholds (Bibby 1989). Bibby's (1989) estimate of 17,200 singing males in Britain, based on the 1984–85 single species survey, remains the best estimate currently available.

Although increasing, the population in Ireland is still small. Given that 18 singing males were recorded in Co. Wicklow in 1989 (O'Sullivan and Smiddy 1990), there could well be up to 30 pairs of Wood Warblers in Ireland.

TIM J. STOWE

68–72 Atlas p 382

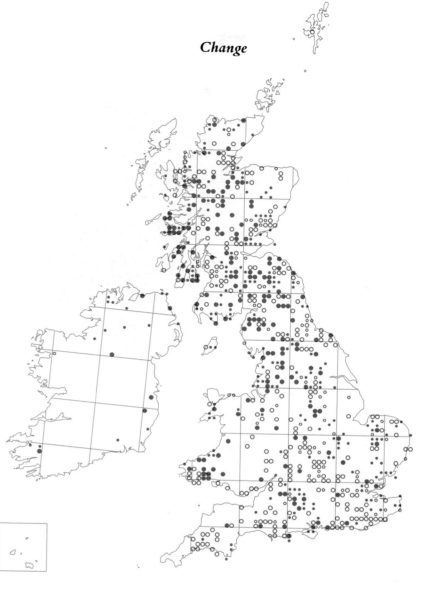

Change

Distribution

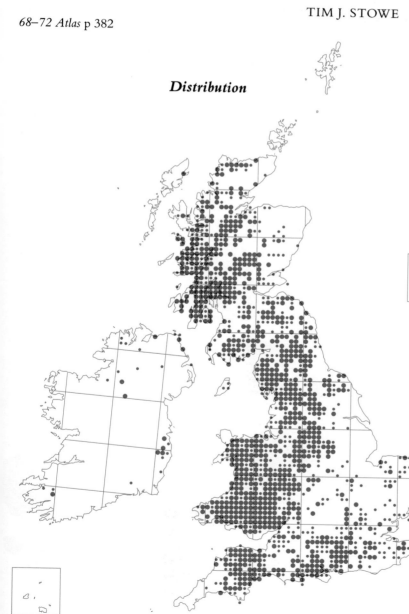

Years	Present, no breeding evidence		Breeding evidence		All records		
	Br	Ir	Br	Ir	Br	Ir	Both (+CI)
1968–72	125	3	1105	5	1230	8	1238
1988–91	411	18	859	10	1270	28	1298
% change					3.3	250.0	4.8

GBB

Chiffchaff

Phylloscopus collybita

In the breeding season it is the Chiffchaff's rather monotonous song that usually announces its presence, and offers the easiest way to distinguish it from its more numerous relative, the Willow Warbler. The two species can interbreed, and the male hybrid's song may contain both Chiffchaff and Willow Warbler phrases (da Prato and da Prato 1986).

Chiffchaffs are more restricted to habitat containing tall deciduous trees than are Willow Warblers (Lack 1971). Survey work in Wales indicates that Chiffchaffs will occur in coniferous plantations, only if some broad-leaves are present (Avery and Leslie 1990). The *68–72 Atlas* and the *88–91 Atlas* both show most of our Chiffchaffs in the southern half of Britain. Unlike several other warblers, Chiffchaffs do not exhibit an east-west bias in distribution and are well represented in the SW England, Wales and Ireland as well as in SE England. The Abundance map emphasises that, within these regions, some areas hold more Chiffchaffs than others, apparently due to the availability of suitable habitat, as in the Severn and other river valleys. Chiffchaffs are much scarcer north of the English Midlands. In Scotland they are sparse, and often associated with policy woodlands, especially where a shrub layer, often of Rhododendron, occurs (*Birds in Scotland*).

BTO ringing suggests that some of the Chiffchaffs that winter in Britain are native here, though most winter around the Mediterranean or in tropical Africa. The fact that many of our Chiffchaffs winter nearer to their breeding grounds may allow them to return earlier in the spring than Willow Warblers. In most parts of Britain, however, Chiffchaffs are much less common than Willow Warblers: CBC data suggest a ratio of 1:3.6 overall. Because their habitat requirements overlap, this raises the question why Chiffchaffs do not 'fill up' territories at the expense of later arriving Willow Warblers. Though Willow Warblers tend to be larger and, pre-sumably, stronger birds, work in Norway indicates that they co-exist with

Chiffchaffs by taking vacant territories, rather than fighting to displace entrenched Chiffchaffs (Saether 1983). However, this study was carried out in an area where both species were at relatively low densities.

It may be that a combination of habitat requirements and winter mortality controls Chiffchaff numbers at levels which reduce direct competition with Willow Warblers. This could explain the fluctuation in the numbers of Chiffchaff territories revealed by the CBC, which shows a decline in the early 1970s, a partial recovery towards the end of that decade, a further decline in the early 1980s, and a recovery since 1985. Chiffchaffs winter further north in Africa than Willow Warblers and may, therefore, be more affected by drought (Lack 1989). Birds that stay in W Europe for the winter could also be affected by unusually severe weather.

Birds in Scotland interpreted the increase in reports received from local recorders as indicative that Chiffchaffs had increased in Scotland, though increase in observer coverage and the difficulty in distinguishing breeding birds from migrants has to be taken into consideration. In NE Scotland, Buckland *et al.* (1990) thought range expansion could be occurring, though they also noted that many of the breeding season records were of singing, but possibly unpaired, males.

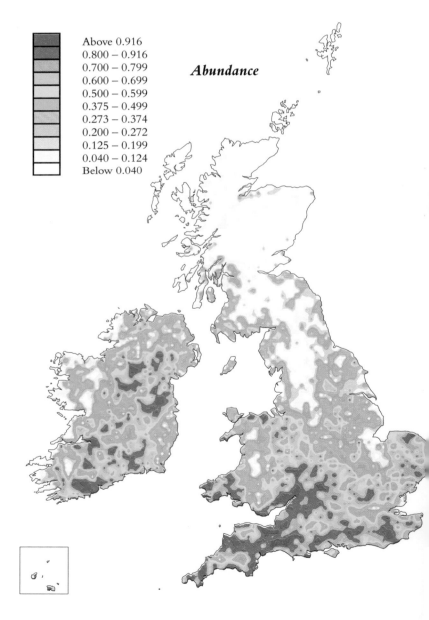

	Above 0.916
	0.800 – 0.916
	0.700 – 0.799
	0.600 – 0.699
	0.500 – 0.599
	0.375 – 0.499
	0.273 – 0.374
	0.200 – 0.272
	0.125 – 0.199
	0.040 – 0.124
	Below 0.040

Abundance

The Change map presents a somewhat contradictory picture for Scotland and N England, albeit with an overall increase in the number of records, though suggesting some recent reduction in Ireland. There, Chiffchaffs had increased since 1850, when they were known in seven out of 32 counties, to their widespread distribution shown in the *68–72 Atlas* which suggested they might be commoner than Willow Warblers in SW Ireland, where many oakwoods lack the shrubby growth that Willow Warblers prefer. For Britain, the total number of Chiffchaff records is similar for both Atlases, but the proportion of breeding records is lower in 1988–91. This may be due to stricter criteria for proof of breeding for a species that is usually located by song.

During 1988–91 there were estimated to have been 640,000 Chiffchaff territories in Britain, and 290,000 in Ireland. Abroad, fluctuations in Chiffchaff populations have been reported from Germany, Czechoslovakia, Denmark and Finland (*Trends Guide*).

S. R. D. DA PRATO

68–72 Atlas p 380

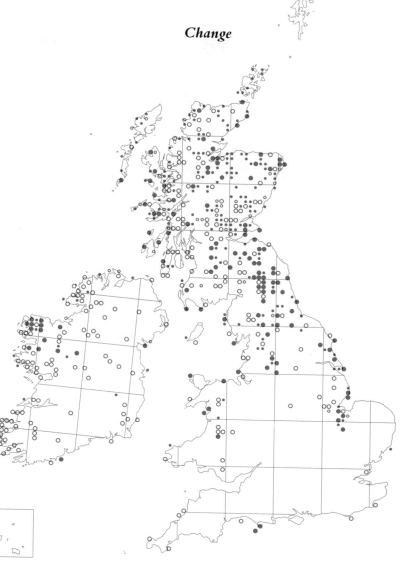

Change

Distribution

Years	Present, no breeding evidence		Breeding evidence		All records		
	Br	Ir	Br	Ir	Br	Ir	Both (+CI)
1968–72	64	11	1939	892	2003	903	2911
1988–91	438	235	1662	601	2100	836	2949
				% change	5.0	−7.3	1.2

Willow Warbler

Phylloscopus trochilus

The Willow Warbler is our commonest summer visitor. To birdwatchers its cascading song is as much a sign of the changing seasons as the sight of the first Swallow. Many male Willow Warblers arrive in Britain in April, though unfavourable weather can delay the arrival of some until May, especially in Scotland. The females are smaller and arrive later than the males. Though they are less obvious than the singing males, experienced observers can deduce the arrival of the first females from the noticeable increase in territorial behaviour among the males.

Willow Warblers normally nest at ground level, though they feed on invertebrates picked from trees, bushes and, especially in late summer, from tall herbs such as nettle, willowherb or reeds (da Prato 1986). This is reflected in their habitats. They are most numerous in young woods and scrub or along woodland edges, rides and clearings. Often they are the commonest breeding species in young conifer plantations, including replanted areas, until the canopy closes over (Avery and Leslie 1990). They are particularly numerous in birch woods, where densities of 300 per km² are not unusual, but scarce in hedgerows and gardens unless these are rather overgrown.

Most British Willow Warblers seem to be single brooded, especially in the north. Some birds make more than one breeding attempt, not always with the same mate. In Surrey, Lawn (1982 and unpublished) found that, between 1977 and 1990, 35 out of 272 (13%) territorial males were polygamous. Other studies have found fewer polygamous males.

Most young Willow Warblers fledge between June and mid July, depending on latitude. As they do so, their parents start to moult, with males normally in advance of females. The parents of late broods can even

be in moult before their young fledge. Moult is rapid and many have finished by the end of July, especially in S England. By the end of August most adults have renewed all their feathers and are on the way to Africa. Juveniles undergo a partial moult which starts as soon as natal growth has finished. Our Willow Warblers seem to fly to W France and Iberia before setting off on the crossing of the Sahara to wintering grounds in the forests of W Africa (Norman and Norman 1985).

Willow Warblers are not long-lived birds, the oldest known from BTO ringing being six years and eight months. Mortality on the breeding grounds, at least in the territorial and nesting phases, is low (Tiainen 1983, da Prato 1986). Mortality may be higher in late summer (M.R. Lawn). It seems likely that many die on migration. Unfortunately, data are lacking on this or on true winter mortality, as opposed to return rates to British breeding grounds, which can be 47.9% in adult males and 41.7% in adult females but much lower for one year old birds (da Prato 1986). CBC data show none of the extreme fluctuations in numbers associated with those migrants that winter in the Sahel region, suggesting that Willow Warbler winter habitat is more stable. The 1988–91 Distribution map shows a similar picture to that in 1968–72, with Willow Warblers distributed

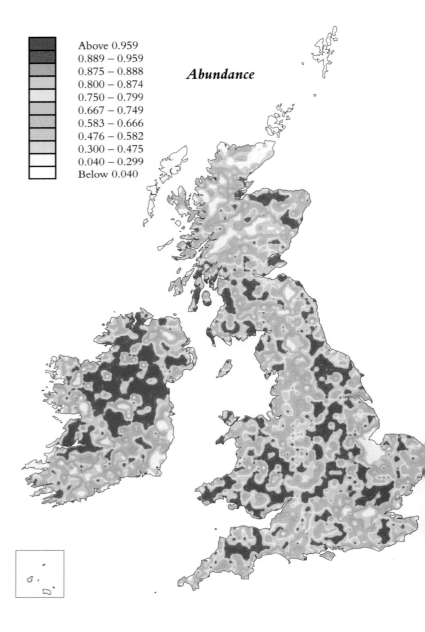

Abundance

	Above 0.959
	0.889 – 0.959
	0.875 – 0.888
	0.800 – 0.874
	0.750 – 0.799
	0.667 – 0.749
	0.583 – 0.666
	0.476 – 0.582
	0.300 – 0.475
	0.040 – 0.299
	Below 0.040

throughout Britain, except in treeless uplands and islands. The Abundance map and local tetrad atlases show lower densities in some urban areas and in some intensively farmed landscapes.

During 1988–91 there were an estimated 2,300,000 Willow Warbler territories in Britain, and 830,000 in Ireland. Although the Willow Warbler is our most numerous summer visitor, it ranks only eighth or ninth in order of abundance among our breeding birds. This fact, and their high numbers in certain habitats such as species-poor Scottish birchwoods (Bibby *et al.* 1989), may bear out O'Connor's (1981) theory that summer visitors exploit niches and habitats not already taken up by residents. Willow Warblers are said to occur at lower densities in Irish woods, where resident insectivores are more numerous, and winters tend to be less severe, than in Britain (*Birds in Ireland*). Abroad, Willow Warblers are common throughout N Europe, from where there is no evidence of long-term trends as opposed to fluctuations.

S. R. D. DA PRATO

68–72 Atlas p 378

Change

Distribution

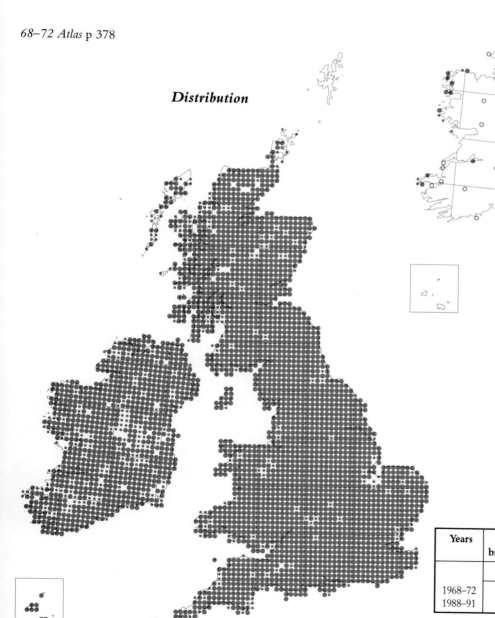

Years	Present, no breeding evidence		Breeding evidence		All records		
	Br	Ir	Br	Ir	Br	Ir	Both (+CI)
1968–72	23	3	2567	938	2590	941	3536
1988–91	156	121	2446	806	2602	927	3539
				% change	0.6	−1.4	0.1

Goldcrest

Regulus regulus

This incessantly busy, tiny green bird is very characteristic of coniferous woodland. As it darts, hops, hangs and hovers among the outermost branches of spruce and pine trees, searching for minute insects, it gives its high-pitched 'tseee' call and in summer sings a surprisingly far-carrying, rhythmic, tinkling song to give away its presence. It is one of the world's smallest song birds and yet is at least partially resident even in the coldest areas of N Scotland.

One of the more widespread birds in Britain and Ireland, the Goldcrest is absent or scarce only in generally treeless areas such as in the Northern Isles, the Outer Hebrides, and the Fens around the Wash. With the exception of a slight expansion in range on the Outer Hebrides and Orkney, probably related to increased afforestation there, the Goldcrests distribution has changed little since 1968–72.

The patchy pattern of the Abundance map reflects well the distribution of the Goldcrest's preferred coniferous woodland in Britain. Afforestation has helped this species and high densities are seen in Kielder (Northumberland), Thetford (Norfolk), New Forest (Hampshire), Wales, Dumfries and Galloway, and in N Scotland. The high densities in Ireland are consistent with Batten's (1976) oft quoted work, which found as many as 400–600 territories per km^2 in spruce plantations, and reasonably high densities in the sessile oak woods of Kerry (150–190 per km^2). These high densities may also reflect the more equable climate in Ireland. It should be noted that these figures greatly exceed values found in Finland (33 per km^2: Tiainen *et al.* 1983) and may be a consequence of small census plot size (*ca* 10ha).

The Goldcrest is the eleventh most abundant breeding passerine in British woods, occurring in 85% of them, although less likely to be found in those smaller than 10ha (Fuller 1982). In coniferous woods, it can form 30–60% of the total bird community, but it also occurs in deciduous woods, especially in Ireland. In areas of forest or plantation, Goldcrests prefer to forage in European larch, pine, spruce and sycamore, and tend to avoid oak, Japanese larch, birch, beech, alder and ash (Peck 1989).

Despite their tiny size – most weigh about 5g – some Goldcrests manage to survive as residents in even the coldest parts of their range. It has been calculated that they can survive 18 hours of darkness at −25°C in Scandinavia, by burning off fat equivalent to about 20% of their body weight, while huddling with other Goldcrests in the depths of a tree's foliage (Reinertsen *et al.* 1988). Their breeding habits seem also geared to minimising heat losses. Their nests are very well-insulated cups of moss, lichens and spiders' webs, lined with a warm layer of feathers. Some of these feathers form a loose umbrella over the nest cup which may help to stop warm air escaping (Haftorn 1978). The 7–12 eggs form several layers in the nest and the female pushes her hot, well-vascularised legs deep into the pile to provide added warmth to those eggs not touching her brood patch (Thaler-Kottek 1988). The nestlings try to minimise heat losses by

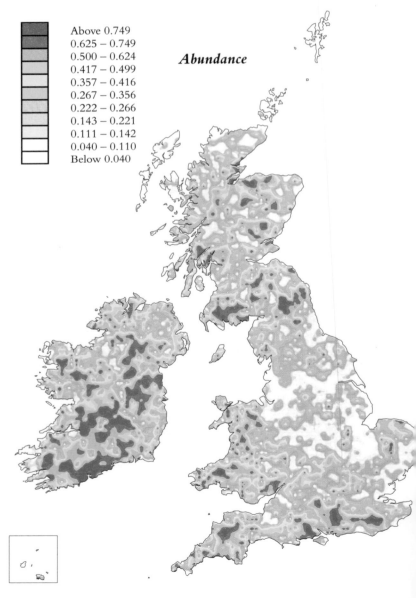

Abundance

Above 0.749
0.625 – 0.749
0.500 – 0.624
0.417 – 0.499
0.357 – 0.416
0.267 – 0.356
0.222 – 0.266
0.143 – 0.221
0.111 – 0.142
0.040 – 0.110
Below 0.040

burrowing to the lower parts of the nest once fed, thereby pushing nestlings that need to be fed to the top of the nest. Despite the fact that egg laying is a large burden to a female (a clutch of 10 eggs represents nearly 1½ times her body weight!), she will start to lay a second clutch in another nest, built earlier by the male, when the first brood is only half grown (Haftorn 1978). In this way, the species has a prodigious ability to produce many young in the summer, thereby making up the large losses they sustain during prolonged cold winters, when food is in short supply.

During 1988–91 there were estimated to have been 560,000 Goldcrest territories in Britain, and a further 300,000 in Ireland.

HUMPHREY Q. P. CRICK

68–72 Atlas p 384

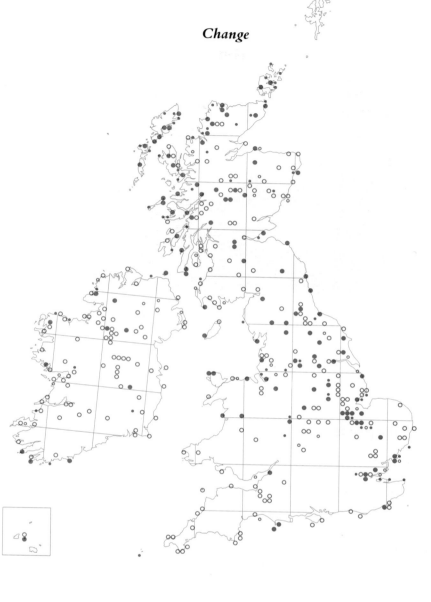

Change

Distribution

Years	Present, no breeding evidence		Breeding evidence		All records		
	Br	Ir	Br	Ir	Br	Ir	Both (+CI)
1968–72	113	26	2230	887	2343	913	3259
1988–91	397	204	1930	648	2327	852	3189
				% change	−0.6	−6.6	−2.3

Firecrest

Regulus ignicapillus

The persistence of the thin but penetrating song, gradually impinging on the consciousness of a birdwatcher, is often the first indication of the presence of this tiny rarity. Tantalising glimpses of a white supercilium, high in the canopy, may be followed by full views on open lower branches as the bird passes close by, seemingly unconcerned at the presence of an enthralled observer. Despite its (at times) confiding behaviour, the Firecrest remains a difficult bird to locate and its presence is almost certainly under-reported.

The map shows a remarkably sparse but widespread scatter of breeding records north to Merseyside and west to Powys, but with none north of Norfolk along the E coast and a dearth of records also from the far west. Taking all records into account, there appear to be seven major concentrations, in the New Forest, the Chilterns, the Hampshire/Surrey border, the Forest of Dean, Epping Forest, the Lyminge area of SE Kent, and coastal Suffolk. The last five are new since the *68–72 Atlas*. Colonisation of Britain, first detected in 1962, is clearly continuing.

In 1968–72 the distribution was very much a southeasterly one, as would be expected for a species newly arrived from the Continent. Now, while there has been some infilling in the southeast, the appearance of isolated pairs and new colonies in scattered localities well to the northwest is particularly noteworthy. This pattern might suggest a species whose habitat requirements are exacting and met only in widely scattered pockets within Britain. Yet Firecrests breed here in various types of woodland and seem most unlikely to be limited by habitat availability. In reality, the mapped distribution must reflect chance discoveries of breeding birds and, perhaps, also regional variations in searching effort. But rather few of

the mapped records relate to birds newly discovered during fieldwork in 1988–91, and there may be few major concentrations still remaining to be found.

On a local scale, the clumping of Firecrest distribution is more extreme. In 1975, which RBBP figures show to have been a peak year, there were 46 singing males in Wendover Woods in Buckinghamshire and 36 in Lyminge Forest, together amounting to two-thirds of the known British total. Throughout the 1980s, just 65ha of spruce plantation in Buckinghamshire accounted for up to 16 territories and 9–23% of the RBBP total (J.H. Marchant). Of course, not all known breeding areas are monitored annually for the Panel.

Comparison with the *Winter Atlas* shows that very few squares were occupied during both seasons. In winter, most Firecrests are found close to the W or S coasts of Britain, with the few inland birds concentrated along river valleys. Summer haunts, which show no aquatic or maritime link, are occupied mainly from April to early August, although a few individuals may be present and singing from late March and in autumn up to early October. Whilst the breeding distribution has spread considerably since colonisation, the winter distribution has not altered since the 1940s. This suggests that most wintering birds may be Continental in origin and as yet there is no direct evidence to indicate whether any of our breeding birds also winter here. Most Firecrests arrive in spring to breed at sites where Goldcrest territories are already established. In Norway spruce plantations in Buckinghamshire, Firecrest territories cover typically 0.5–1.5 ha and overlap considerably with Goldcrest territories, which average a third or a half that size (J.H. Marchant). Females and young are seldom seen, and probably many singing males remain unmated. The Firecrests' larger territories are defended with a louder song which penetrates further through the woodland than that of the Goldcrest, at least to the human ear. Mutual avoidance might be expected in sibling species which are apparently so similar in their feeding and nesting biology, but there are structural and behavioural differences between them which permit coexistence and, except very rarely, prevent interbreeding (e.g. Thaler and Thaler 1982, Thaler-Kottek 1986).

Firecrests are confined to the W Palearctic throughout the year. The current expansion of the breeding range suggests that the total population might be increasing, but *Red Data Birds* confirms that the species is still vulnerable in Britain. The main threat is clearance of occupied forestry plantations, some of which have already been allowed to stand beyond normal felling dates. Estimates of the British population must be approximate, because an unknown proportion of breeding sites remain undiscovered. The RBBP maximum totals varied between 29 and 175 during the 1980s, and wide fluctuations are also evident at individual sites. Perhaps 80–250 singing males hold territory in most years, although half or more of these may fail to attract a mate.

Two-year averages for Firecrests in Britain during 1973–90

	73–74	75–76	77–78	79–80	81–82	83–84	85–86	87–88	89–90
Confirmed	2	4	2	8	10	5	3	10	14
Possible	26	72	20	68	64	124	35	72	85
Max total	28	76	21	76	73	129	38	81	114
Localities	9	24	10	28	28	61	22	40	50

Based on RBBP reports.

JOHN H. MARCHANT

68–72 Atlas p 386

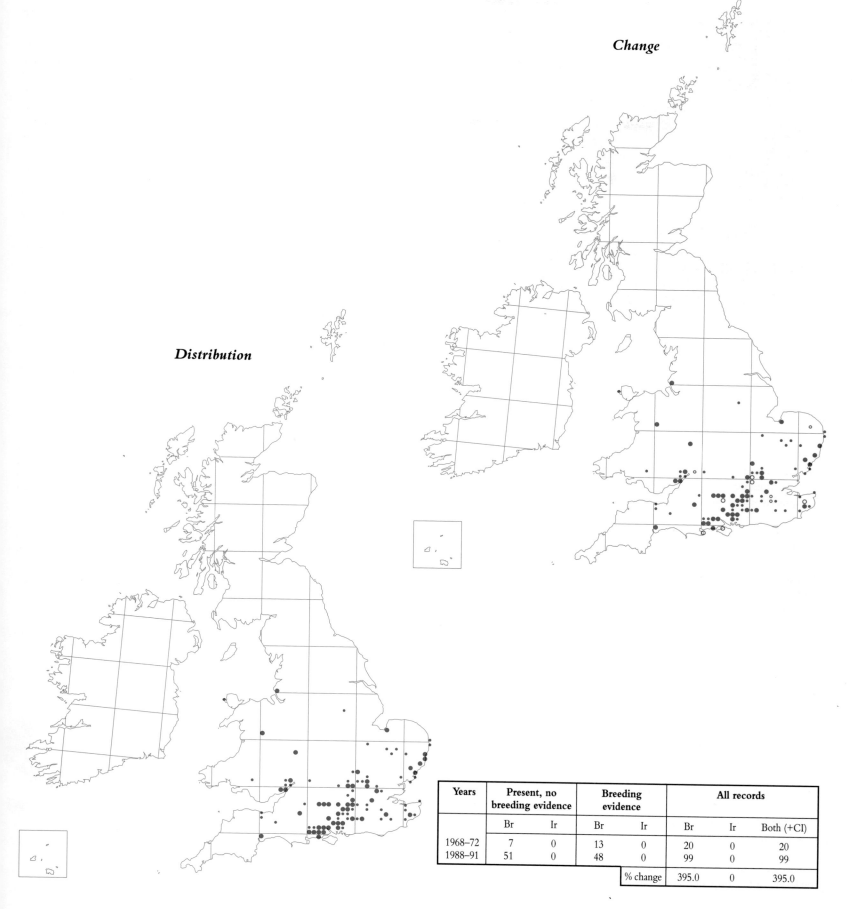

Change

Distribution

Years	Present, no breeding evidence		Breeding evidence		All records		
	Br	Ir	Br	Ir	Br	Ir	Both (+CI)
1968–72	7	0	13	0	20	0	20
1988–91	51	0	48	0	99	0	99
			% change		395.0	0	395.0

The insects they catch by these methods are often small, such as aphids and small flies, and tend not to be fed to their young. After a week of cool, wet conditions, broods often starve to death.

Despite being late arrivers, many Spotted Flycatchers manage to fit in a second clutch in early July (Summers-Smith 1952). The later clutches are smaller than earlier ones, consisting of 3–4 eggs as opposed to 4–5. While forming their eggs, females switch to feeding on some calcium-rich prey such as woodlice or even snails (Davies 1977b). Males provide about a third of the female's food during this period, probably speeding up the process of egg formation, and in this respect it is rather unexpected to find occasional cases of bigamy (Kämpfer and Lederer 1990). Incubation and nesting periods last approximately 13 days each and the young are fed for about two weeks after fledging (Davies 1976). A strategy that may help to promote the growth of the second brood, before their migration to Africa in late August, is the provision of food by members of the first brood (Erard 1991).

During 1988–91 there were estimated to have been 120,000 Spotted Flycatcher territories in Britain and 35,000 in Ireland.

HUMPHREY Q. P. CRICK

68–72 Atlas p 388

Spotted Flycatcher

Muscicapa striata

Despite being a rather dull-brown bird with a very missable, thin, squeaky song, the Spotted Flycatcher is generally thought of with warm affection by birdwatchers. When it appears in May, as one of the last Palearctic-African migrants to arrive, its feeding behaviour and confiding nature make its presence obvious. It sits upright on an exposed twig or branch and quickly sallies forth to catch flying insects, often returning to the same perch.

The *68–72 Atlas* showed that Spotted Flycatchers were found throughout most of Britain and Ireland, thinning out only towards the far north and west. The present Distribution map essentially repeats that picture. Birds are still all but absent from Shetland, Orkney and the Outer Hebrides, with more gaps having appeared in the S and W of Ireland, and sporadically throughout other areas. The Abundance map shows a similarly patchy distribution, with very low densities occurring seemingly at random throughout Britain and Ireland.

Certainly the *Trends Guide* shows that Spotted Flycatchers have been in long-term decline since the early 1960s. The population has dropped to nearly a quarter of what it was. It is unclear what factors have caused this decline, but a major drop occurred when rains failed in the Sahel region, just south of the Sahara, between 1983 and 1984, despite the fact that Spotted Flycatchers use this area only on passage to and from their wintering grounds in southern Africa. Breeding success, also, may have been affected by a series of cooler, wetter summers in Britain and Ireland. The breeding of Spotted Flycatchers seems strongly influenced by the weather in May: more birds breed early if temperatures are warmer, and clutch sizes are bigger with more sunshine (O'Connor and Morgan 1982).

Spotted Flycatchers prefer to feed on flying insects, particularly large flies, but the availability of these is strongly influenced by air temperatures. On cold days, and in the cool of the early morning, they have to resort to searching for insects among foliage of trees, either picking them up from the leaves or flushing them into the air by their movements (Davies 1977b).

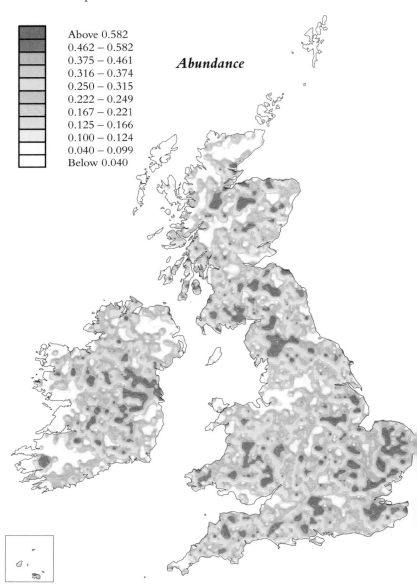

	Abundance
Above 0.582	
0.462 – 0.582	
0.375 – 0.461	
0.316 – 0.374	
0.250 – 0.315	
0.222 – 0.249	
0.167 – 0.221	
0.125 – 0.166	
0.100 – 0.124	
0.040 – 0.099	
Below 0.040	

Change

Distribution

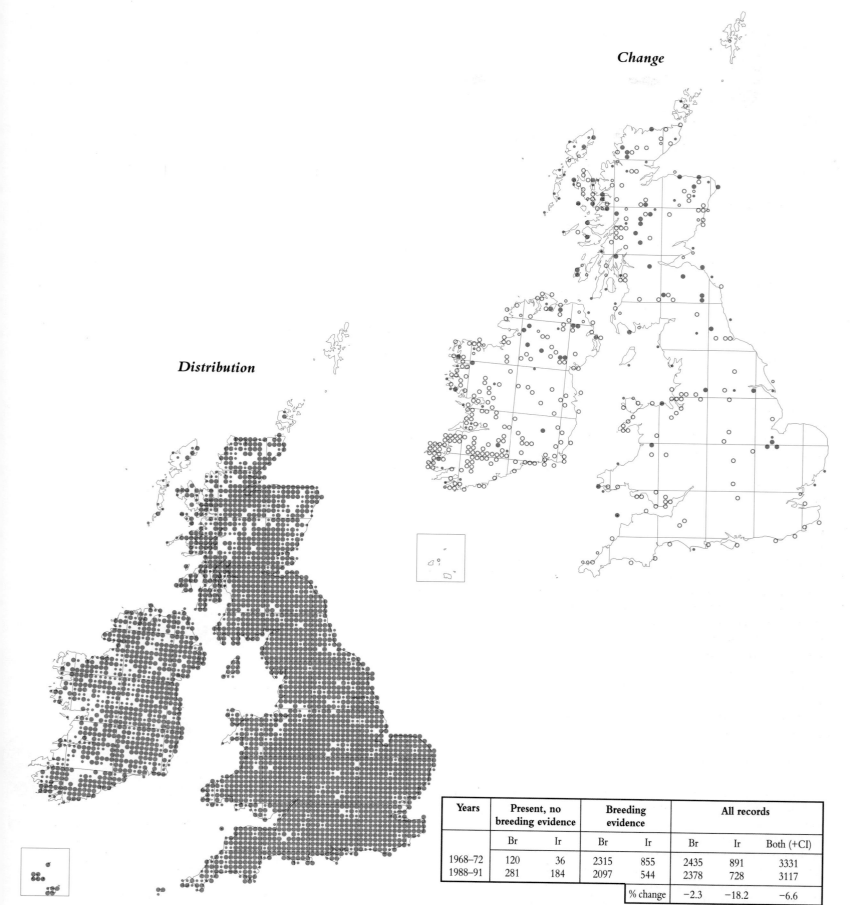

Years	Present, no breeding evidence		Breeding evidence		All records		
	Br	Ir	Br	Ir	Br	Ir	Both (+CI)
1968–72	120	36	2315	855	2435	891	3331
1988–91	281	184	2097	544	2378	728	3117
% change					−2.3	−18.2	−6.6

Pied Flycatcher

Ficedula hypoleuca

Pied Flycatcher males are conspicuously black and white, and sing vigorously from perches in trees during a short period after their arrival from Africa, sometime between mid April and late May. After pairing, males almost completely stop singing, but may later resume in a second territory away from the first. The species breeds in holes in trees in many forested areas of the Palearctic region. It can be found in almost any suitable forest type, but in Britain it is mainly found in oak-dominated deciduous woodlands. The Pied Flycatcher's breeding ecology is widely studied, mainly because it takes readily to nest boxes. Indeed, good quality nest boxes are preferred to natural cavities, making it possible to attract almost the whole breeding population within a woodland to boxes.

The Pied Flycatcher is widespread in forest areas in upland W Britain, and most common in Wales, W England from Devon northwards to Cumbria (with some gaps), and in parts of SW Scotland. In SE England, most of N Scotland, and in Ireland it is a rare inhabitant. On a finer geographical scale the distribution is more patchy, and seems to be concentrated in upland valleys and hillside woods dominated by mature sessile oak (Campbell 1955). It is likely that the spread of the species in Britain is limited by suitable habitat; the most important restricting factor within potential breeding habitats probably being the availability of nesting holes, and oak may offer more natural cavities than do other deciduous trees. Since resident (and thus early breeding) species probably take the better nesting holes, later-arriving migratory species, such as the Pied Flycatcher, may be excluded from habitats with relatively few natural holes. If nest boxes are provided in oak habitats many more of them will be occupied by Pied Flycatchers than by tits. This may be because many natural cavities in oaks are of higher quality than nest boxes: the former will be occupied by tits, leaving many of the boxes for flycatchers.

The Pied Flycatcher leaves Britain during mid August to mid September, and most autumn ringing recoveries are concentrated at stop-over areas in Portugal (Hope Jones et al. 1977). The winter quarters of Pied Flycatchers are presumed to be in tropical W Africa, and two British ringing recoveries would seem to confirm this (Dowsett et al. 1988). In spring a more easterly route is taken than in autumn, and males arrive at the breeding grounds about a week ahead of females; in N England this is in the latter half of April. Egg laying starts in the first week of May and peaks in mid May.

The Change map shows that there has been a slight expansion of range, in particular in SW Wales and NW England. There were, in addition,

more records from SE England, with breeding in Hertfordshire, Sussex and Kent. The species may also be attempting to colonise E Ireland. This expansion of range is possibly due to the great increase in the use of nest boxes, and it seems likely that the majority of British Pied Flycatchers now use nest boxes rather than natural sites. In Scandinavia, where more mature woodlands with natural holes are available, probably more birds breed in natural cavities than in boxes.

With regard to breeding biology and behaviour, Pied Flycatcher females raise only a single brood per season, though a replacement clutch may be laid if the first is depredated at an early stage. Males, on the other hand, are often polygamous, and have two (or sometimes three) females in widely separated territories (this mating system is called polyterritorial polygyny). The polygamous males mainly help their first female in feeding her young, but 'secondary' females are less supported, leading to reduced breeding success among the secondary females. Thus, most Pied Flycatcher nests lacking a male probably belong to females that have paired with an already-mated male. In Cumbria the proportion of 'single parents' is commonly 8–17%, depending on the time-lag between male and female arrival, and habitat and nesting density.

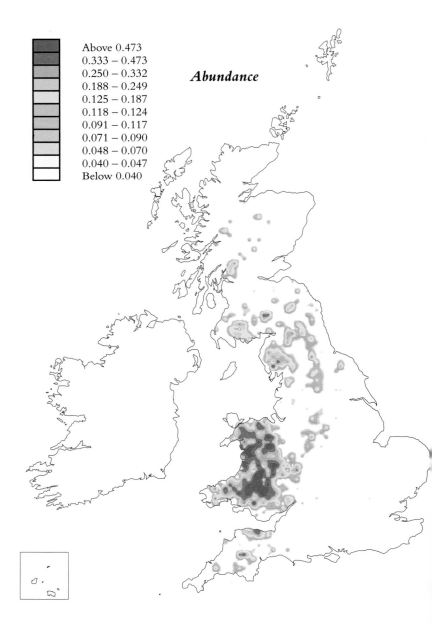

Above 0.473
0.333 – 0.473
0.250 – 0.332
0.188 – 0.249
0.125 – 0.187
0.118 – 0.124
0.091 – 0.117
0.071 – 0.090
0.048 – 0.070
0.040 – 0.047
Below 0.040

Abundance

In the most densely populated parts of its range (e.g. Cumbria and Wales) there may be up to 200 pairs per occupied 10-km square. However, such densities are rare, and the average density of 50 pairs per occupied 10-km square used in the *68–72 Atlas* is probably still reasonable. If this is the case, then there are estimated to have been 35,000–40,000 pairs of Pied Flycatchers, annually, in Britain between 1988 and 1991. The population of Ireland during the same period was 1–2 pairs each year. About 25,000 nestlings are ringed in Britain and Ireland annually (Mead and Clark 1990), which approximately corresponds to 5,000 pairs or 13% of the total estimated population.

ARNE LUNDBERG

68–72 Atlas p 390

Change

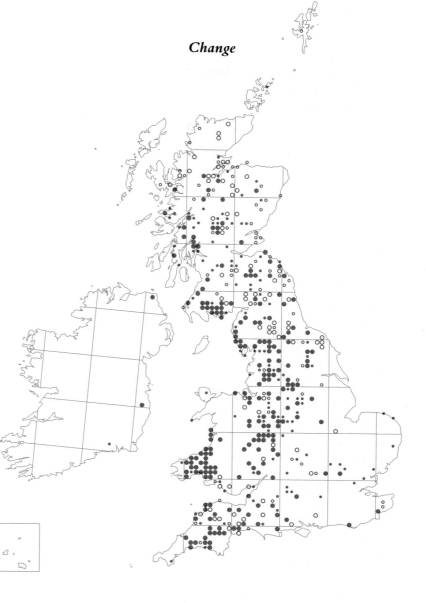

Distribution

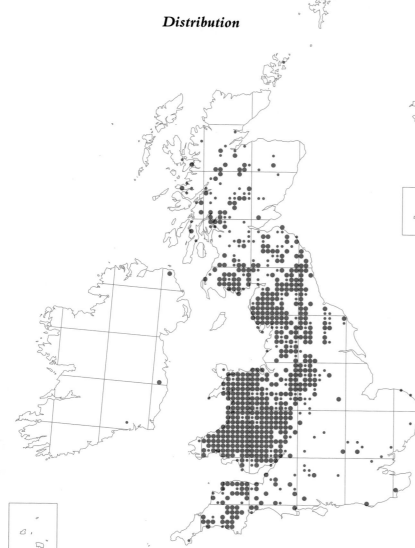

Years	Present, no breeding evidence		Breeding evidence		All records		
	Br	Ir	Br	Ir	Br	Ir	Both (+CI)
1968–72	116	0	430	0	546	0	546
1988–91	185	1	547	2	732	3	735
			% change		34.1		34.6

Bearded Tit

Panurus biarmicus

Bearded Tits are striking birds which clamber acrobatically amongst reed beds and are often detected by their 'pinging' contact calls. They live almost exclusively in or near reeds but make eruptive autumn movements, and are more widely seen in winter than in summer.

The history of Bearded Tits in Britain has been summarised by Axell (1966). They were scarce early in this century and only 2–4 pairs were known in 1947 after a cold winter. A few years later, saltwater flooding of East Anglian reed beds caused much damage to their habitats and recovery of numbers was slow. An impressive expansion began in the mid 1960s. It was driven by immigration from massive populations on newly created polders in the Netherlands (Axell 1966) and, by 1974, the population was at least 590 pairs in 11 counties (O'Sullivan 1976). The *68–72 Atlas* predicted further expansion of what at the time was a very dynamic species in much of W Europe.

The Change map does not indicate as much change as might have been expected. Only a few more 10-km squares were occupied during 1988–91 compared to 1968–72 in the main range on the coasts of East Anglia and Kent, and although there has been a marked range expansion along the S coast of England, breeding numbers there remain low. Away from these areas there are only a few outlying sites. One of these, Leighton Moss in Lancashire, was colonised in 1973. Numbers rose rapidly to about 40 pairs by 1980 but fell to about 15 in more recent years. The other outlying sites have very small populations which may prove to be ephemeral. The overall distribution is very similar to that of large reed beds described by Bibby and Lunn (1982). Bearded Tits have bred in Co. Wicklow (*Birds in Ireland*) but there were no breeding records from Ireland during 1988–91.

Bearded Tits are insectivorous in summer, when they gather midges from wetter areas, or wainscot moth larvae and pupae from reed stems and litter (Bibby 1981b). Nests are built near the ground, in drier areas where sedges or fallen reed stems provide cover. Their long breeding season enables them to rear possibly the largest number of young of any British species (Bibby 1983). After the autumn moult, which is also complete in juveniles, they gather in flocks and may disperse widely (Axell 1966). Most winter in or near reeds but they can also be found in other damp grassy areas. This movement is eruptive but some individuals return to their place of origin. In winter they eat seeds almost exclusively, especially those of the common reed (Bibby 1981b).

Bearded Tits are difficult to count because individuals range very widely between nesting and feeding areas. There is no obvious territorial

behaviour, and a long breeding season during which pairs may move from one nest to another. Most reed beds are not very accessible and are often unvisited in summer to guard against disturbance of nesting Marsh Harriers. Counts of flocks may be made in autumn, but it is never certain that these do not include individuals that have moved from other sites. In spite of these difficulties, the species is rare and interesting enough to ensure that population estimates from many sites are published in county reports.

Some recent numbers are summarised in the Table. The Bearded Tit has not lived up to the prediction of further increase. The 590 pairs reported in 1974 must have been close to a maximum, which was perhaps maintained until the mid 1980s. The 320 pairs shown in the Table are probably too low because of under-reporting from Kent and Norfolk, and it seems likely that present numbers may be about 400 pairs. Population fluctuations are partly caused by cold winters but it is not clear whether these are a sufficient explanation. It is probable that numbers were much enhanced by immigration from the Netherlands between 1966 and 1975. Changes in reed bed habitats, especially the tendency to drier conditions, may come to be important at some sites, if it has not already done so. It will be interesting to see whether populations at the major sites in Kent

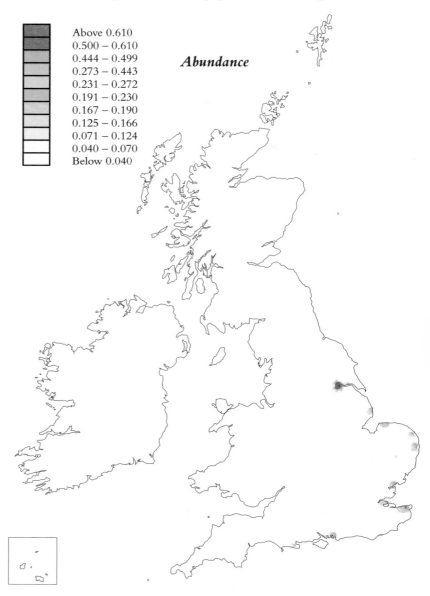

Above 0.610
0.500 – 0.610
0.444 – 0.499
0.273 – 0.443
0.231 – 0.272
0.191 – 0.230
0.167 – 0.190
0.125 – 0.166
0.071 – 0.124
0.040 – 0.070
Below 0.040

Abundance

Recent population estimates of Bearded Tits

County	Sites	Pairs	Status since 1970
Cambridgeshire	1	4–5	1988; highest recent count
Dorset	3	21	Max counts 30–40
Essex	5	22–25	1989; highest recent count
Hampshire	3	9	1989; best season on record
Humberside	1	80	Max 100–120 1977
Kent	3	(30)	Up to 100 pairs late 1970s
Lancashire	1	15	30–40 pairs 1977–85
Lincolnshire	2	4	Max 30–40 pairs 1983
Norfolk	7	(64)	Possible max 200 + pairs about 1980
Suffolk	5	ca 65	Well over 200 pairs 1980
Sussex	1	4–5	1989; highest recent count
Wales	1	1–2	Never well established
Totals (minima)	33	319	

Summarised from county reports mainly for 1987–89. Numbers in brackets may not be complete.

Change

Distribution

and the East Anglian coast reach their previous numbers. The colony in Humberside is presently the only very large one extant, and has maintained its numbers in spite of declines elsewhere. The colony is interestingly different in ecology because it is generally dry, but subject to brackish tides.

COLIN BIBBY

68–72 Atlas p 328

Years	Present, no breeding evidence		Breeding evidence		All records		
	Br	Ir	Br	Ir	Br	Ir	Both (+CI)
1968–72	6	0	39	0	45	0	45
1988–91	8	2	52	0	60	2	63
	% change				33.3		40.0

Long-tailed Tit

Aegithalos caudatus

Weighing 7–9 g, with a long tail and pinkish back, the Long-tailed Tit is one of Britain's most distinctive small birds. These features, together with its habit of spending the year (outside the breeding season) in flocks, make it unmistakable. The flocks roam around winter territories very actively, with noisy chirruping calls. The Long-tailed Tit's diet is almost exclusively small insects, so in winter it must search throughout the daylight hours in order to get enough food to keep itself alive.

The flocks are basically family parties, sometimes together with a few relatives of the breeding pair. Flocking may be a key to the survival of this small bird in areas with winters as cold as those in Britain. The birds in the flock roost together and, in very cold weather, huddle close, keeping each other warm during the night. It seems almost certain that a single bird, without this assistance, would not be able to survive a cold night. Even so, Long-tailed Tits can suffer badly in severe winter weather. It has been estimated that upwards of 80% of the population may perish in particularly harsh winters such as occurred in 1916/17, 1939/40 and 1961/62 and 1962/63 (Dobinson and Richards 1964). For several years after the prolonged cold of 1962/63, the species was locally scarce in many areas. This is not always the case, however. For example, in the winter of 1981/82, many areas of Britain experienced two short periods of extreme cold, with the temperatures dropping to record levels locally, yet the numbers of Long-tailed Tits in the following summer were not particularly low. Other factors, in addition to low temperatures, must play an important role in determining winter survival. Prolonged periods of glazed frost, which prevent access to food, may be more dangerous than the low temperatures themselves. Provided the birds can get sufficient food during the day, they can survive the night.

As winter wanes, the birds pair up. Females tend to move from their winter, family flock to join another group, a process which prevents inbreeding. The birds nest within the winter territory of the male and

start to build quite early in the season – often by mid March in S England. Their nests are elaborately woven, domed structures which may contain over 2,000 feathers, bound together with spiders' webs, and camouflaged with a covering of lichens. They may be placed high up, against the trunk of a tree or in thick cover, such as a gorse bush. So complex a nest takes some time to build, but once the dome is complete, the pair stop roosting with the flock and spend the night, well-insulated, in their nest.

The usual clutch is 7–12 eggs, which are incubated by the female alone; but the chicks, which take some 15–16 days to fledge, are fed by both parents. Many of the nests are found by predators such as Jays. If this happens early in the season the pair may attempt a replacement clutch; later on they may not. When they do give up, they often go to help at another nest. The interesting thing is it seems that only males go to help at another nest within the winter territory area. Since each winter territory was held by a family party, and the females move to other territories to breed, the males within such an area are likely to be brothers. The females, however, do not have relatives nesting within the family territory. Although it needs further confirmation, when they lose their own nest, some females return to their old winter territory and help at a nest of a

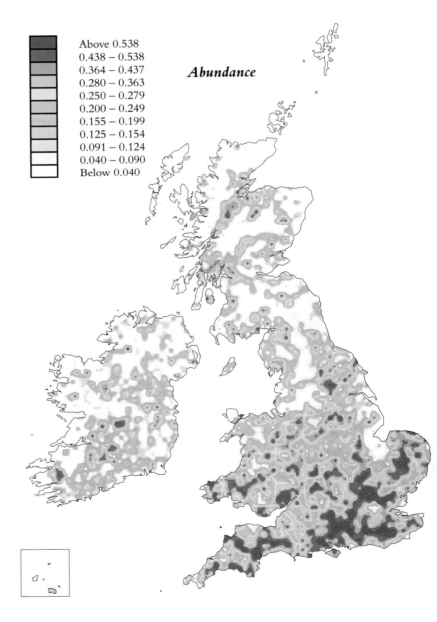

Above 0.538
0.438 – 0.538
0.364 – 0.437
0.280 – 0.363
0.250 – 0.279
0.200 – 0.249
0.155 – 0.199
0.125 – 0.154
0.091 – 0.124
0.040 – 0.090
Below 0.040

Abundance

brother there. In this way it seems that most 'helpers' help to raise the brood of a brother (Glen and Perrins 1988).

The Distribution map is characteristic of many species, with birds being recorded in most areas in S Britain, but thinning out northwards. Areas of high ground such as the Pennines and the Cairngorms show up as being poorly populated, as one would expect. The Abundance map confirms this pattern, but some of the lowland regions within the Highlands – such as the Great Glen – are clearly areas of higher density. The opposite applies to large urban areas, such as London, which are marked by low densities, for Long-tailed Tits are never common in built-up areas.

The Change map shows that the Long-tailed Tit's distribution is similar to that in the *68–72 Atlas*, with the one important exception, that the species seems to be clearly less well-distributed in Ireland than it was at that time. The reason for this is unclear, but may partly reflect less complete cover during the 1988–91 survey.

In the years of the *88–91 Atlas* there were an estimated 210,000 Long-tailed Tit territories in Britain, and a further 40,000 in Ireland.

CHRISTOPHER PERRINS

68–72 Atlas p 318

Change

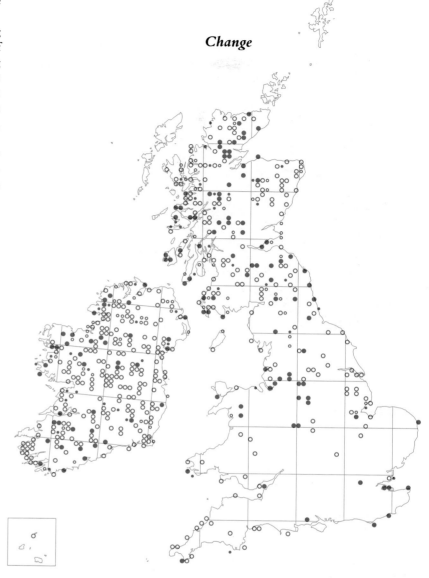

Distribution

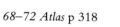

Years	Present, no breeding evidence		Breeding evidence		All records		
	Br	Ir	Br	Ir	Br	Ir	Both (+CI)
1968–72	134	60	2060	659	2194	719	2916
1988–91	238	149	1868	397	2106	546	2660
				% change	−4.0	−24.1	−9.0

Marsh Tit

Parus palustris

The two black-capped tits found in Britain, Marsh and Willow, are a pair of species which presented real problems of identification during Atlas fieldwork. Their distributions are not dissimilar, and so it is fortunate that both possess distinctive calls: the 'pitchou' of the Marsh and the nasal 'tchay-tchay' of the Willow are diagnostic.

In appearance Marsh Tits are generally more upright and a little bigger than Willows, the adults with glossy black caps. Their tails are squarer, their black bibs smaller and their flanks paler than those of the Willow Tit – which tends also to show a pale panel on the closed wing, lacking in the Marsh. It is nevertheless possible that a very small minority of the records published could be misidentifications. The early history of the two species in Britain is uncertain since it was not realised until 1900 that both were present.

Despite their name, Marsh Tits in Britain are typically found in open deciduous woodland, parkland, mature gardens or partly wooded farmland: they normally shun conifers and are seldom encountered in urban areas. They are residents and are extremely territorial throughout the year with pairs defending 2.5–6.0 ha (Perrins 1979).

The Distribution map shows that Marsh Tits are found over most of England and much of Wales, although absent from several quite large areas. They are also present in the very southeastern corner of Scotland. There were no records from the Isle of Man, or Ireland. Marsh Tits are absent from the Fens, from much of E Lincolnshire and Humberside, and also from much of the area between the Humber and the Mersey.

Compared with the *68–72 Atlas*, Marsh Tits seem to be rather more sparse, with local changes of distribution. In SE Scotland, *Birds in Scotland* records breeding from 1921, and a gradual expansion which may be continuing. As one would expect for so sedentary a species, the Dis-

tribution map closely matches that in the *Winter Atlas*. There does, however, appear to have been a progressive decline, from 1,366 occupied 10-km squares in 1968–72, to 1,208 in the *Winter Atlas* (1981/82 to 1983/84), and to 1,133 in 1988–91. This may partly be because the fieldwork undertaken has differed for each survey.

Detailed local atlas work has shown differences in distribution at the tetrad level, which would not show up on the coarser 10-km grid. For instance, Mead and Smith (1982), reporting the results from the 504 tetrads in Hertfordshire, recorded Marsh Tit breeding or probably breeding in 142 tetrads, and Willow in 206. Of the 253 tetrads which had at least one of the species, 95 (38%) had both, 47 (19%) had Marsh only, and 111 (44%) Willow only. The tetrads with only Marsh Tits were along the dry woodland and scrub of the Chiltern scarp, while many with Willow Tits only were situated along the river valleys of SE Hertfordshire. In Devon, Sitters (1988) had records for Marsh Tit in 60% of the county's tetrads, with the species missing only from the high areas of moorland.

The *Trends Guide* records a long-term shallow decline in numbers, with several troughs, mostly following cold winters. As the *Trends Guide* records a very healthy level of breeding numbers of Coal, Blue and Great Tits

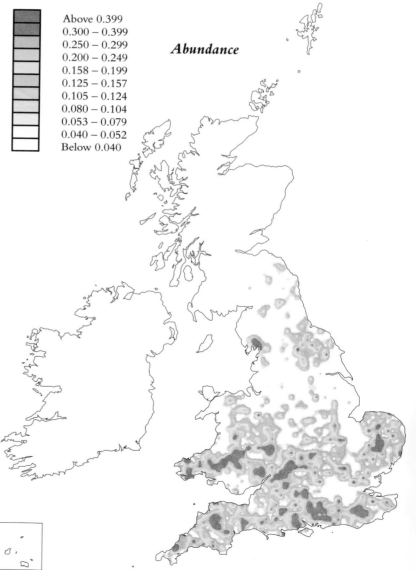

Above 0.399
0.300 – 0.399
0.250 – 0.299
0.200 – 0.249
0.158 – 0.199
0.125 – 0.157
0.105 – 0.124
0.080 – 0.104
0.053 – 0.079
0.040 – 0.052
Below 0.040

Abundance

over the same period, it is possible that Marsh Tits have been losing out in inter-specific competition. This may be both for food and for nesting sites, although Marsh Tits generally nest in holes at lower levels than the three commoner species of tit (Perrins 1979). The decrease recorded in the *Trends Guide*, the species' poor performance in 1989 (Marchant 1991) and the reduction in occupied 10-km squares since the *68–72 Atlas* must all indicate a decline in the British population in recent years.

The Abundance map is patchy, with high levels mainly (but not exclusively) reported from the southern half of England and Wales. During 1988–91 there were estimated to have been 60,000 Marsh Tit territories in Britain.

CHRIS MEAD

68–72 Atlas p 314

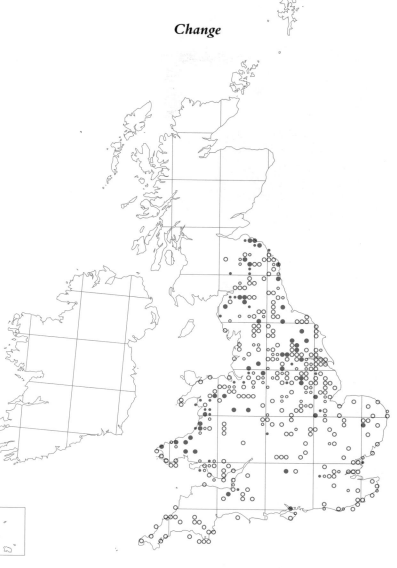

Change

Distribution

Years	Present, no breeding evidence		Breeding evidence		All records		
	Br	Ir	Br	Ir	Br	Ir	Both (+CI)
1968–72	162	0	1204	0	1366	0	1366
1988–91	275	0	858	0	1133	0	1133
	% change		−17.1		0		−17.1

extinction of small populations in the Highlands. Even though food hoarding is common, and populations of other sub-species live in the boreal forests of N Europe with much more severe winter conditions, the species is absent from many Scottish woodlands. Of the common species of tits, Willow Tits are the least likely to join mixed feeding flocks.

The *Winter Atlas* plots no records from N Scotland but otherwise shows a very similar distribution. In the *68–72 Atlas*, records were received from 1,220 10-km squares, for the *Winter Atlas* from 1,152 squares, and during 1988–91 from 1,100 10-km squares. This could be interpreted as a gradual decline; however, recording methods are very different for winter and breeding season atlas work.

In England, some tetrad atlas results have shown this species to be marginally more common than Marsh Tits – e.g. Hertfordshire (Mead and Smith 1982) with records in 53% of tetrads compared with Marsh Tits in 46%. In Devon, however, Willow Tits were much less widespread, occurring in only 12% of tetrads compared with 60% for Marsh Tits (Sitters 1988). Not only were they absent from moorland but also from

Willow Tit

Parus montanus

This small sooty-capped tit was not thought to be a member of the British avifauna until the British race, *kleinschmidti*, was described in 1900, from a bird taken near Finchley in 1897. Until then it had been thought that only Marsh Tits were native to Britain. Happily, identification in the field is aided by its distinctive nasal call, for its other field characters are rather similar to those of the Marsh Tit.

Willow Tits are resident and sedentary, and defend their territories as a pair or a small party, throughout the year (Perrins 1979). Their typical habitats are generally damper than those used by Marsh Tits. Alder carr, streamside wood and scrub, the wooded surroundings of gravel pits and reservoirs, as well as lowland coniferous woodland, are all used. Apart from the Crested Tit, this is the only British *Parus* species which excavates its own nest hole. Generally it is in a well rotted stump and, as with woodpeckers, the debris is characteristically littered below it. Unfortunately the bigger and more aggressive Marsh Tit may supplant the nest excavator to the confusion of the observer.

At first sight the distribution seems to follow that of the Marsh Tit. The northern limits are similar, and both are absent from the relatively treeless area of the Fens. There are, however, significant differences, with the Willow Tit absent from the S Lake District, but present throughout SW Scotland, and very sparse in N Yorkshire, coastal East Anglia and Dorset, Devon and Cornwall.

Compared with the *68–72 Atlas*, the range appears to have contracted in SW Scotland. The single northerly Scottish breeding record for 1988–91 compares with two records of confirmed breeding, one of probable and one of possible, in the five years of Atlas fieldwork from 1968–72. *Birds in Scotland* reports breeding early in this century north to Stirling, Perth and Angus and, sporadically, further north still. The succession of cold winters during the middle part of the century may have led to the

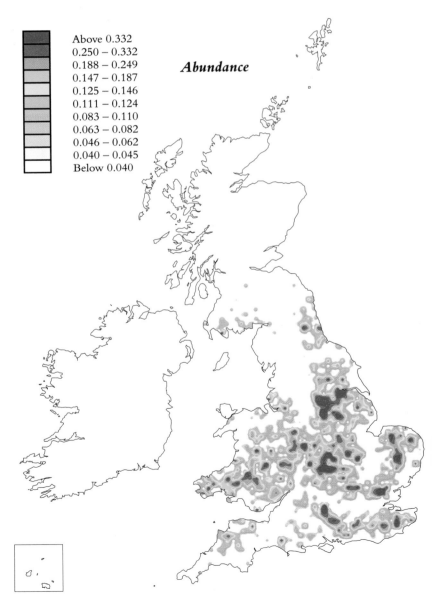

Abundance

Above 0.332
0.250 – 0.332
0.188 – 0.249
0.147 – 0.187
0.125 – 0.146
0.111 – 0.124
0.083 – 0.110
0.063 – 0.082
0.046 – 0.062
0.040 – 0.045
Below 0.040

the southern part of the county, and they were very sparse in the western area extending towards Dorset. Sitters points out the considerable similarity between the Devon distribution maps for birch and for this species – birch is a favourite tree for nest excavation.

The speculation in the *68–72 Atlas*, based on ringing totals, that there was a marked change in fortunes during the early 1960s, with the Marsh Tit losing out and Willow Tit prospering, may have been ill founded. The introduction of mist nets as a means of catching birds in the wider countryside would have shifted the perceived balance from Marsh Tit, readily trapped in gardens, to Willow Tits, the preferred habitat of which is also attractive to many ringers. Ringing totals for 1985 to 1989, inclusive, amount to 4,473 Marsh Tits and 6,597 Willow Tits.

The *Trends Guide* shows a recent decline, although the index is based on small samples. This may be related to the more intensive cultivation of farmland and the loss of small damp areas through drainage. The recent sequence of dry summers, if continued, may further threaten such habitats in the dryer areas of Britain. Although CBC results suggest that Marsh Tits outnumber Willow Tits 3:1 in broadleaved woodland, and 2.5:1 in

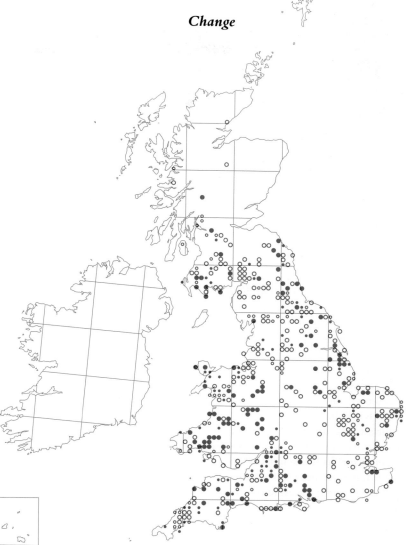

Change

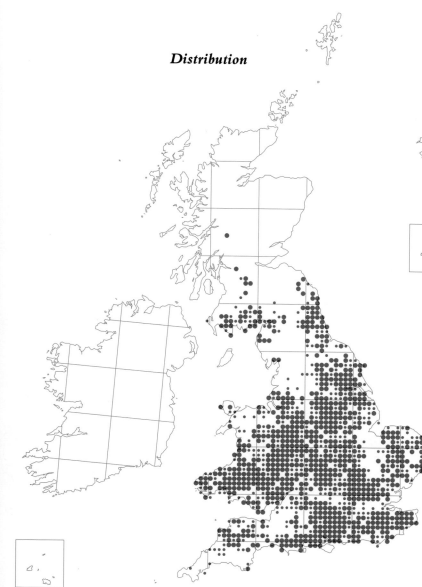

Distribution

farmland, high densities of Willow Tit may be encountered in waterlogged woods and carrs. In many of the occupied 10-km squares, only a small part may be suitable for the Willow Tits to breed. During 1988–91 there were estimated to have been 25,000 Willow Tit territories in Britain.

CHRIS MEAD

68–72 Atlas p 316

Years	Present, no breeding evidence		Breeding evidence		All records		
	Br	Ir	Br	Ir	Br	Ir	Both (+CI)
1968–72	173	0	1047	0	1220	0	1220
1988–91	311	0	789	0	1100	0	1100
				% change	−9.8	0	−9.8

Crested Tit

Parus cristatus

The Crested Tit is distinctive among British birds, its perky crest, persistent indignant trilling and confiding nature all contributing to its unique personality. To experience the excitement of watching Crested Tits in Britain the birdwatcher must travel to the Highlands of Scotland, where they exclusively occupy Scots pine, from the magnificent native pinewoods of Speyside to the mature plantations on the Moray coast.

When compared with the *68–72 Atlas* and 1979–80 Survey (Cook 1982) it would appear that the range of the Crested Tit has changed little. Closer scrutiny, however, reveals some differences. There are fewer reports from Ross-shire and Sutherland, and none from around Loch Laggan. By contrast, a southwesterly extension is evident beyond Glen Affric into Glen Garry and down the Great Glen almost as far as Fort William.

The chief habitat requirements appear to be an extensive needle canopy for summer feeding but with enough light penetration to permit rank heather growth. Access to this field layer is improved by low side-branches on larger trees and/or regeneration. Also essential is the presence of dead stumps in which the birds can excavate their nest holes. One of the aspects of Crested Tit distribution which continues to intrigue is their absence from extensive stands of mature and native pinewood along Deeside in Aberdeenshire. Was the population exterminated by some natural catastrophe or were they never there? The habitat appears to be eminently suitable today but perhaps some unknown subtle ecological factors operate against the species there. It may now be that only the Crested Tit's sedentary nature serves to prevent colonization from Speyside through the barrier of the Cairngorms. During the early and mid 1970s a few were seen on Deeside but since then only one has appeared east of the Cairngorms and so natural expansion through the mountains into these woods seems unlikely. Perhaps colonization via the commercial plantations of Donside would be possible if improved understanding of the birds' habitat requirements were to lead to more sympathetic management of these woods.

Crested Tits breed earlier than other tits in the same woods. The female begins to excavate the nest hole in March, egg laying varies between mid April and early May, and parties of fledged young can be encountered from late May. Of 324 natural nests recorded by Nethersole-Thompson, 74% were in pine stumps, 14% in dead pines, 2.5% in living pines and the remainder in alders, birches, posts and a red squirrel's drey. Most nests were low, 90% of them within 3m of the ground (Nethersole-Thompson and Watson 1974). In the coastal plantations, where rotting stumps are scarcer than in natural pinewoods, nest boxes have been accepted. Twelve pairs bred in boxes in Culbin Forest in 1989.

As the year proceeds, the choice of foraging sites and food changes. Studies in the pinewoods along the northern foothills of the Cairngorms showed that during winter Crested Tits spend over one-third of their foraging time in small pines below 3m in height and one quarter of their time among heather in the field layer (Hartley 1987). During April and May they concentrate less on the twigs and branches and move into the canopy of taller trees to gather moths, caterpillars, aphids and other small invertebrates. From mid summer until autumn, foraging is mostly concentrated among the pine needles high in the canopy. Recent work in Abernethy Forest has highlighted the importance of food storage for Crested Tits, behaviour documented in Norway in the 1950s but not previously studied in Scotland. During April and May seeds are taken from opening pine cones and secreted under lichens on the branches. In autumn, especially October, moth larvae are targeted, with those of the bordered white moth forming the great majority of stored items (H. Young). Crested Tits can be found throughout Europe, with the chief exceptions of most of Italy and much of Britain and Ireland. The Scottish subspecies, *P.c. scoticus*, is by far the most restricted in range. Population

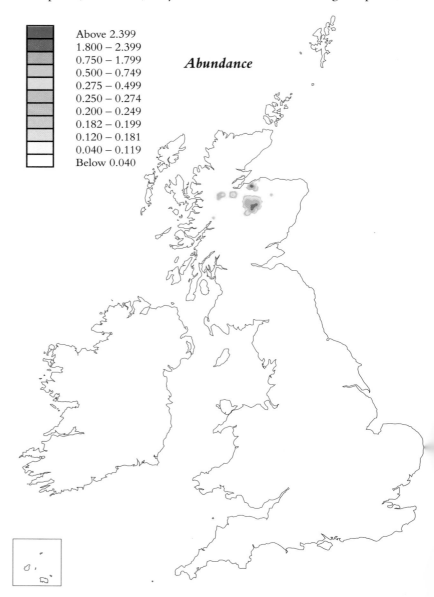

Above 2.399
1.800 – 2.399
0.750 – 1.799
0.500 – 0.749
0.275 – 0.499
0.250 – 0.274
0.200 – 0.249
0.182 – 0.199
0.120 – 0.181
0.040 – 0.119
Below 0.040

Abundance

densities are never high, approximations ranging from 0.15 pairs per ha in the open mature pinewoods to 0.01 pairs per ha in plantations. Application of these densities to the occupied range gave an estimated population, in 1980, of around 900 pairs (Cook 1982). Factors likely to influence this figure are the attainment of an adequate age by Scots pine plantations, the felling of such plantations and their possible replacement with exotic species, and the effects of winter weather. Different bouts of severe winter weather on Speyside have apparently had widely differing effects on the breeding population, ranging from a 70% decrease in 1947 (Nethersole-Thompson and Watson 1974) to a 5.5% increase in 1982 (*Birds in Scotland*).

Current research should throw further light on the Crested Tit's habitat requirements and preferences with a view to improving the attractiveness of commercial plantations to this endearing little bird.

MARTIN J. H. COOK

68–72 *Atlas* p 312

Change

Distribution

Years	Present, no breeding evidence		Breeding evidence		All records		
	Br	Ir	Br	Ir	Br	Ir	Both (+CI)
1968–72	14	0	32	0	46	0	46
1988–91	20	0	31	0	51	0	51
				% change	10.9	0	10.9

Competition with Marsh and Willow Tits has been proposed as one reason why Coal Tits concentrate in coniferous forest, from where those species are largely absent (*68–72 Atlas*). However, Coal Tits are well adapted for feeding amongst pine needles (Snow 1954) with their fine bill structure and acrobatic skill, typical of the tit family, enabling them to exploit this habitat to maximum effect.

Coal Tits are adept at finding food items hidden in crevices, as observers at garden bird feeding stations will be aware. This ability is indicative of their characteristic food storing behaviour, which has been studied in considerable depth (e.g. McNamara *et al.* 1990). Hoarding food is the Coal Tit's insurance policy against food shortage and compensates for its inability to carry large fat reserves, owing to its small body size. The type of food stored depends on seasonal variations in the diet: aphids, lepidopteran larvae and Diptera in summer, tree seeds in winter.

CBC data suggest that Coal Tits are less affected by severe weather than other small birds, and only slight decreases were noted in CBC indices following the hard winters of 1978/79, 1981/82, 1983/84 and 1985/86. Their ability to feed on the underside of snowclad branches (Snow 1954) has survival advantage in severe weather, and their food hoarding behaviour

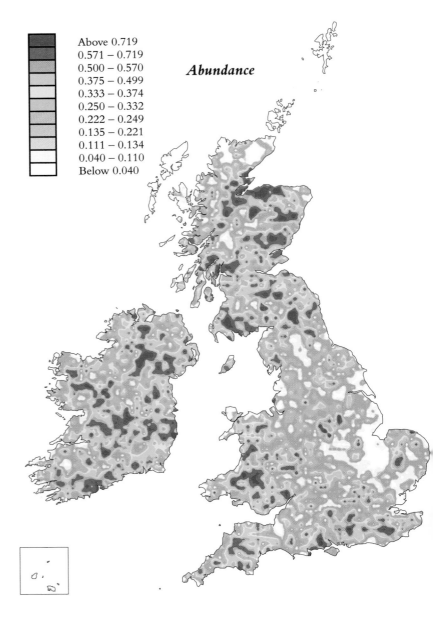

Coal Tit

Parus ater

The Coal Tit, perched at the top of a conifer, singing its persistent 'peechoo-peechoo-peechoo' song, is a familiar feature of approaching summer. This species, more so than its conspecifics – the Blue Tit and Great Tit – is confined to woodlands, particularly coniferous, but also western sessile oak and northern birch woods.

The Distribution map illustrates that Coal Tits are widespread throughout, while the Abundance map shows that they are most common in Wales and parts of mainland Scotland, Ireland, the Isle of Man and in localised pockets of England. They are absent as breeding birds from many outer Scottish isles and from most of the fenland area of E England. Their distribution coincides largely with areas of coniferous forest, hence they are particularly abundant in, for example, the Thetford Forest area of Norfolk, and Kielder Forest in Northumberland. Even single conifers in gardens are attractive to Coal Tits.

As may be seen from the Change map, there is little difference in the overall distribution from that of the *68–72 Atlas*, at which time the species was increasing and expanding its range (*Trends Guide*), presumably due to the rapid expansion of commercial forestry. Breeding numbers, at least on CBC plots, have been stable since the late 1970s, probably as the rate of afforestation has slowed down. However, the Coal Tit's presence as a breeding bird on Lewis, and at more locations along the coast of W Ireland, is a notable development since the *68–72 Atlas*.

The numbers of breeding birds increase markedly during the thicket stage of development in a conifer plantation (*Birds in Scotland*) and breeding density may attain 100 pairs per km² in favoured habitats (*68–72 Atlas*). In sessile oak woods of N and W England and Ireland, and birch woods of NW Scotland, which are also favoured habitats, Coal Tits may be more numerous than Blue Tits or Great Tits (*Birds in Scotland, Birds in Ireland*).

Abundance

Above 0.719
0.571 – 0.719
0.500 – 0.570
0.375 – 0.499
0.333 – 0.374
0.250 – 0.332
0.222 – 0.249
0.135 – 0.221
0.111 – 0.134
0.040 – 0.110
Below 0.040

is another adaptation for times of reduced food availability.

According to Perrins (1979), in years of high breeding productivity and/or peak population levels, inroads on their food supplies cause density-dependent reduction in overwinter survival. The Coal Tit is predominantly a resident, sedentary species, as highlighted by the similarity of the Abundance map to that in the *Winter Atlas*, and most movement recorded away from the home range is in response to food shortage. Thus food supply is likely to be the main determinant of breeding population size.

The nest is in a hole, usually low down, and may be in the ground or amongst tree roots (Campbell and Ferguson-Lees 1972), but Coal Tits will also use nest boxes. The nest is constructed mostly of moss, lined with hair or fur. The species is usually single brooded, and lays early, from the beginning of April. The clutch size is most commonly 10 eggs (Perrins 1979).

During 1988–91 there were estimated to have been 610,000 Coal Tit territories in Britain and 270,000 in Ireland.

ROWENA LANGSTON

68–72 Atlas p 310

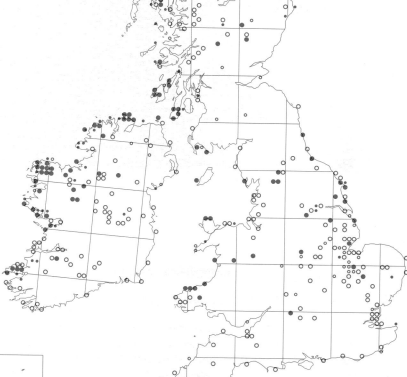

Change

Distribution

Years	Present, no breeding evidence		Breeding evidence		All records		
	Br	Ir	Br	Ir	Br	Ir	Both (+CI)
1968–72	128	44	2264	812	2392	856	3250
1988–91	306	170	2009	684	2315	854	3170
			% change		−3.2	−0.1	−2.4

Blue Tit

Parus caeruleus

The colourful and acrobatic Blue Tit is the commonest member of the tit family in Britain and Ireland and may be found wherever mature trees provide suitable nest holes. Although a familiar visitor to urban gardens, where it readily makes use of bird feeders and nest boxes, it is primarily a bird of deciduous woodland, particularly of oak. Blue Tits also breed, but at lower densities, in parkland, hedgerows and conifers.

Nesting usually begins in mid April with most broods being fledged by the end of June. Spring temperatures strongly influence the timing of breeding, however. The Blue Tit produces one of the largest clutches of any passerine. In Britain most clutches contain 7–13 eggs but clutches of 15 and 16 are not infrequent (Perrins 1979). It is extremely rare for a pair to rear two broods in a single season. Incubation normally takes 12–14 days and the chicks remain in the nest for a further three weeks. The nestlings are fed almost entirely on insects, predominantly caterpillars.

The Distribution map shows the Blue Tit to be almost ubiquitous in England, Wales and Ireland. In Scotland it is absent only from the most mountainous areas, and from Orkney, Shetland and many of the Hebridean islands.

In Britain, Blue Tits are most abundant south of a line from the Tees to the Mersey, though low densities occur in the Welsh mountains and in the relatively treeless fens of East Anglia.

In optimum habitat, Blue Tit densities may reach more than two pairs per ha. CBC data indicate that the highest average density occurs in the woodlands of Wales (*Trends Guide*). In upland areas the species occurs at relatively low densities. Within such regions of generally lower densities, however, there are concentrations of Blue Tits around many major urban areas, where gardens provide an acceptable breeding habitat. This effect is also evident to some extent in Ireland, where Blue Tits are most abundant in the east.

The Distribution map shows very little change from that of the *68–72 Atlas*. The CBC indices show relative stability of numbers during the intervening period, although severe winters have caused short-term fluctuations. The spread of the Blue Tit in N Scotland, which began in the first half of this century and has probably been assisted by the establishment of forestry plantations, has continued, albeit at a slower rate. Thom (*Birds in Scotland*) reports some extension of range in the Hebrides, with nesting in Lewis away from the site originally colonised in 1963, and on the islands of Eigg and Canna, but these subtle changes are not clearly seen on the Change map.

Increasing public interest in birds has resulted in many people providing food for them throughout the year. Consequently, greater numbers of Blue Tits have been attracted to urban areas. Breeding densities in suburban gardens may approach those found in woodland, but reproductive success is relatively low (Cowie and Hinsley 1987). Urban Blue Tits, on average, lay fewer eggs than those in woodland, and a smaller proportion of their chicks survives to fledging. This low productivity is attributed to the relatively poor nutritive value of the food available for both the laying female and the nestlings. Caterpillars, which provide the bulk of the chicks' diet in woodland, are comparatively scarce in the urban environment.

The ecological requirements of Blue Tits and Great Tits overlap considerably, resulting in competition between the species for nest sites and food. It has been shown that when nest sites are in short supply the Great Tit is the superior competitor, being able to exclude the smaller Blue Tits from tree holes and nest boxes (Minot and Perrins 1986). Neither species defends territories against the other, however. Blue Tits appear to be able to take advantage of this in that, when both species occur at high density and there is no shortage of nest sites, the Blue Tit enjoys greater breeding success than the Great Tit. This appears to be a result of the Blue Tit's

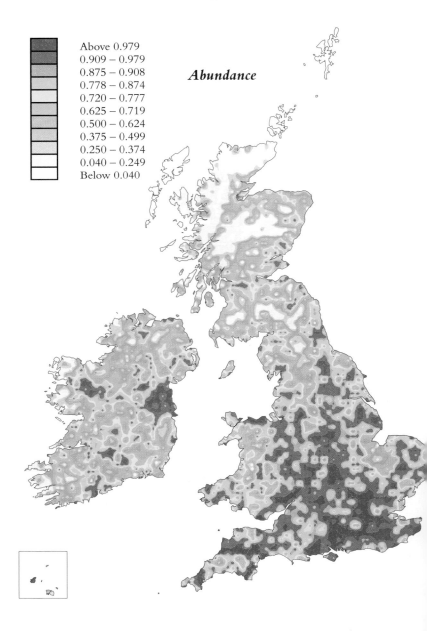

Above 0.979
0.909 – 0.979
0.875 – 0.908
0.778 – 0.874
0.720 – 0.777
0.625 – 0.719
0.500 – 0.624
0.375 – 0.499
0.250 – 0.374
0.040 – 0.249
Below 0.040

Abundance

greater efficiency in exploiting food resources shared by both species (Minot 1981).

The Blue Tit is widespread in continental Europe, being absent only from the most northerly regions of Fennoscandia and the former USSR. The species is also found in North Africa and the Middle East. British Blue Tits are classified as a distinct subspecies and, unlike many Continental populations, tend to be sedentary. Few Blue Tits ringed in Britain are recovered more than 10km from the place of ringing (Perrins 1979). Irruptive autumn movements occasionally bring significant numbers of Continental Blue Tits to the E coast of Britain, but no major influx has occurred since 1957.

During 1988–91 there were estimated to have been 3,300,000 Blue Tit territories in Britain and 1,100,000 in Ireland.

NEIL McCULLOCH

68–72 Atlas p 308

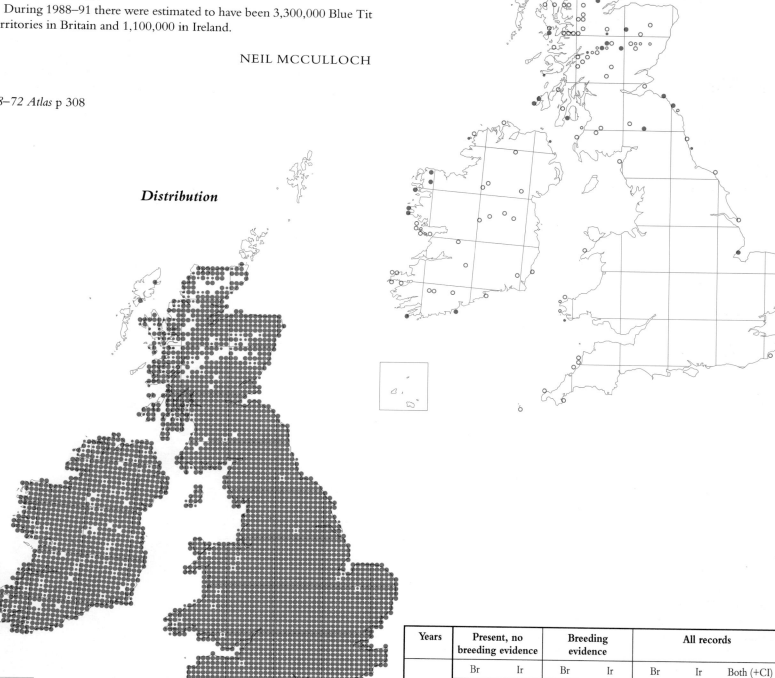

Change

Distribution

Years	Present, no breeding evidence		Breeding evidence		All records		
	Br	Ir	Br	Ir	Br	Ir	Both (+CI)
1968–72	50	9	2468	941	2518	950	3473
1988–91	98	62	2382	868	2480	930	3424
% change					−1.4	−2.0	−1.5

Great Tit

Parus major

This, the largest of the tits, is a distinctive and well-known species. Its black cap, white cheeks and yellow underparts with a black central line make it a bird familiar to most, as does its ringing 'teacher-teacher' song which may be heard in mild weather from January onwards, until it nests in May. It is a common visitor to bird tables in winter and regularly occupies nest boxes in gardens in the summer. As a result of this ready acceptance of nesting boxes, it has become one of the most studied birds in the world. The Great Tit is very catholic in its choice of habitat. Whilst its main habitat is open deciduous woodland, where it may nest at densities of 1–1.5 pairs per ha, it is common in hedgerow trees and readily occupies conifer plantations, albeit at lower densities (of around 0.2 pairs per ha, or even lower). The main prerequisite is that, being hole nesters, a pair requires a sizeable hole in which to raise their young. Young plantations or well tended woodland with few old trees to provide holes do not suit them well. The loss of hedgerow elms due to Dutch elm disease may well have affected Great Tit nesting success on farmland. They do not readily nest in holes in the ground, but will use a wide range of cracks in walls, drainpipes etc, if nothing else suitable is available.

The winter seed crops are important to the Great Tit, especially beech. Breeding numbers tend to increase between one year and the next when there is a beech crop in the intervening winter and to decrease when there is not (Perrins 1967). This correlation may be upset by prolonged snow, apparently because this covers the seeds and prevents the tits from reaching them. There is also a tendency for birds to wander more widely in years without beech crops; the recoveries of BTO ringed birds show that a higher proportion of the recovered Great Tits have moved more than 5km

in winters without a beech crop than in those with a crop. Although these correlations are strongest with beech, the population also fluctuates in the same way in areas without beech; it seems likely that the seed crops of other trees (which tend to vary in parallel with those of beech) are also important to the birds.

The usual clutch is 5–11 eggs, though this varies considerably with conditions, being smaller in gardens than in woodland and smaller at the end than the beginning of the season. Clutch size also varies in relation to the number of caterpillars available during the season. The female alone incubates the eggs, hatching them after about 12–14 days, but both parents feed the chicks, which fledge in about 20 days. Great Tits rarely have a second brood in oak woodland, but do so not uncommonly in pine woods.

The Great Tit is almost exclusively a woodland species and the Distribution map reflects this, with presence recorded throughout Britain and Ireland except in the more exposed parts of the west and north. The Abundance map adds to this by showing that the bird is relatively scarcer in areas of high ground as one goes northwards and westwards. It also clearly shows up areas such as the fens of East Anglia where woodland, and even hedgerows with large trees, are scarce. Nevertheless, it seems likely that Great Tits probably breed in almost every 10-km square in mainland Britain.

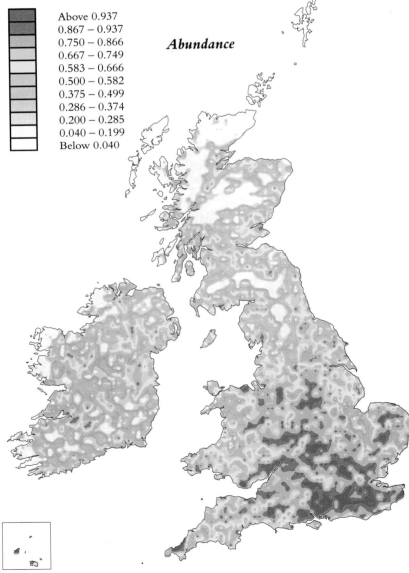

Above 0.937
0.867 – 0.937
0.750 – 0.866
0.667 – 0.749
0.583 – 0.666
0.500 – 0.582
0.375 – 0.499
0.286 – 0.374
0.200 – 0.285
0.040 – 0.199
Below 0.040

Abundance

Compared with the *68–72 Atlas* there is no reason to suppose that numbers have changed significantly. The long-term study in Wytham Woods near Oxford (Perrins 1979) shows no signs of any trend (although numbers may vary markedly from year to year). This is in contrast to a steady increase in the numbers of Blue Tits in Wytham, each year from 1973 onwards, from about 1.34 Blue Tits per Great Tit up to the late 1960s, to 1.87 in the 1970s. There is some sign that the ratio has started to decline again in the late 1980s.

During 1988–91 there were estimated to have been 1,600,000 Great Tit territories in Britain, and 420,000 in Ireland.

CHRISTOPHER PERRINS

68–72 Atlas p 306

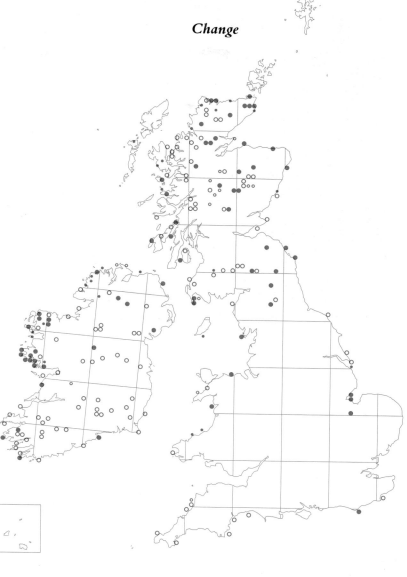

Change

Distribution

Years	Present, no breeding evidence		Breeding evidence		All records		
	Br	Ir	Br	Ir	Br	Ir	Both (+CI)
1968–72	48	27	2409	877	2457	904	3366
1988–91	126	131	2317	752	2443	883	3340
% change					−0.5	−2.2	−1.0

Nuthatch

Sitta europaea

The Nuthatch is best known for its remarkable ability to climb head downwards on tree trunks, a feat not mastered by any other European bird. To do this, it relies on long claws and sturdy legs, instead of gaining support from the tail as woodpeckers do. It is a bird of mature, deciduous woodlands, but is equally at home in parks and large gardens, provided that large trees are present, and is a common visitor to bird tables.

A key feature to understanding the Nuthatch is its strict year-round territoriality (Enoksson 1988). Once settled in a territory, a Nuthatch will usually remain there. Nuthatches are confirmed food hoarders, and their detailed knowledge of the territory and the food hoarded there will tend to encourage sedentariness. Although some moves occur, they are more common during a Nuthatch's first year of life than later, and usually cover short distances.

Nuthatches are strictly monogamous and live in pairs the year round. They breed in early spring and lay only one clutch (of up to nine eggs) with no replacement clutches after failure. Young Nuthatches must establish territories shortly after becoming independent (Matthysen 1988); thus young from late broods are likely to be at a disadvantage. Long distance dispersal is rare, and juveniles often settle very near to their parents, even as nearest neighbours. (Eruptive movements are known only for the Siberian subspecies.)

Nuthatches breed in natural holes in trees, and are less willing than many other hole nesters to use nest boxes. Perhaps this is why they have been less extensively studied than, for example, the Great Tit or the Pied Flycatcher. Consequently, this review is mainly based on data from Swedish and Belgian populations, breeding in natural cavities.

The female alone builds the nest, but the male feeds her during egg laying and incubation. Using mud, the female reduces the entrance hole until she can only just squeeze through, thus decreasing the risk that Starlings will claim the hole. In a nest box, she usually plasters mud along the ceiling. Substantial amounts of clay may be brought into the nest box; the latter should be inspected with care, or the contents could be crushed under the clay. The nest itself is a thick layer of loose pine bark or dry leaves.

Local population density is at its peak when all young have fledged (Matthysen 1988). In autumn, it is limited by food abundance (e.g. hazel nuts or beech mast) (Nilsson 1987, Enoksson 1990). When food is plentiful, Nuthatches defend smaller territories. During winter, territory density declines. Not surprisingly, mortality is higher during colder winters (Nilsson 1987, Enoksson 1988). Furthermore, most Nuthatches breed within their former winter territory (Enoksson 1987). As a result, breeding density depends more on conditions in autumn (food abundance) and winter (weather), than during the breeding season itself. This is reflected in the great similarity between the maps in the *Winter Atlas* and the *88–91 Atlas*.

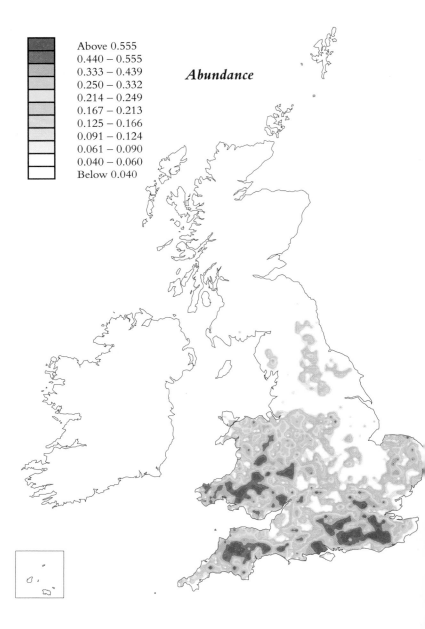

Above 0.555
0.440 – 0.555
0.333 – 0.439
0.250 – 0.332
0.214 – 0.249
0.167 – 0.213
0.125 – 0.166
0.091 – 0.124
0.061 – 0.090
0.040 – 0.060
Below 0.040

Abundance

Although breeding density fluctuates between years, 30 pairs per km² is not uncommon in oak-dominated forests in Sweden. At particularly good sites, territories may be as small as one ha. The population density of Nuthatches varies from site to site, however, and the species may be missing, even, from apparently good sites. In England, the Abundance map reveals that more Nuthatches live south and west of a line drawn from the Mersey to the Thames. The low dispersal propensity may be one reason for the uneven distribution. For example, Nuthatches seem to be less numerous in woods within a predominantly open landscape, than in one where forests – of any kind – are common.

Nuthatches are not shy of humans, and make good use of parks and gardens, not to mention bird feeders. Furthermore, the area requirements of a pair are fairly modest. Thus the *Trends Guide* shows an increase in the number of Nuthatches in Britain; and a comparison between the *88–91 Atlas* and the *68–72 Atlas* shows that it is expanding its range northwards, with the most dramatic extension being in Cumbria. As the species occurs as far north as 60°N in Scandinavia, climate is unlikely to restrict its range in Britain. Many Scottish forests would appear to be excellent Nuthatch habitat, and the increased number of reports from Scotland, with the first

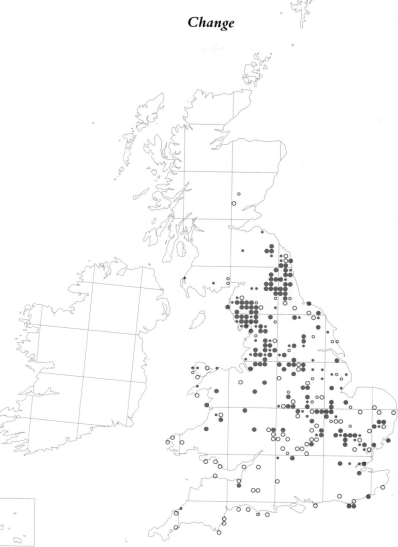

Change

Distribution

successful breeding in 1989 (Murray 1991b), suggests that the species has already found some of them. Probably, we can expect a continued extension of range in Britain, but the colonisation of Ireland seems less likely.

During 1988–91 there were estimated to have been 130,000 Nuthatch territories in Britain.

BODIL ENOKSSON

68–72 Atlas p 320

Years	Present, no breeding evidence		Breeding evidence		All records		
	Br	Ir	Br	Ir	Br	Ir	Both (+CI)
1968–72	114	0	1060	0	1174	0	1174
1988–91	207	0	1063	0	1270	0	1270
				% change	8.2	0	8.2

Treecreeper

Certhia familiaris

Mouse-like and inconspicuous, the Treecreeper is well camouflaged for a life largely spent climbing on tree trunks and branches, propped up by its relatively long, strong tail. Rarely are its silvery-white underparts seen, and even then they may be stained greenish-brown by the lichens on the bark. Nor are its bold white wing bars, for flights from tree to tree tend to be brief and not very frequent.

On the Continent, Treecreepers are usually to be found in upland coniferous forests, and are replaced in lowland deciduous woodlands by their sibling species, the Short-toed Treecreeper *C. brachydactyla*. It is assumed that after the last glaciation, the conifer-loving Treecreeper colonised Britain and Ireland in company with the rapid northward expansion of the Scots pine. The land bridge to the Continent was severed by rising sea levels before the Short-toed Treecreeper, moving more slowly north in association with the broadleaved trees that it favours, reached England. When deciduous woodland appeared in Britain and Ireland, it did so without Short-toed Treecreepers, and was 'taken over' by the Treecreeper in the absence of sibling species competition.

Differences between the *88–91 Atlas* map and its predecessor are generally small, when the coverage irregularities have been taken into account. The Treecreeper remains widely distributed in Britain, the thinnest areas of colonisation being associated with highlands, open moorland, and perhaps with very extensive areas of commercially-managed conifer forestry. The now more or less complete loss of mature elms, particularly from the English countryside, does not seem to have had any marked effect, but perhaps this should not be unexpected as the NRS cards show elm to be a minor tree species so far as Treecreeper nesting is concerned. The boosted insect fauna on dead and decaying elms can only have been a beneficial, if short-lived, bonus.

In Ireland, it does seem that there are several areas where Treecreeper distribution is appreciably more fragmented than in the *68–72 Atlas*,

irrespective of the degree of coverage. Results 20 years ago suggested that, in Ireland, Treecreepers were relatively scarcer than in Britain: the Abundance map for 1988–91 puts Irish abundance on the same level as Scotland and N England, and lower only when compared with S and W England. It may be that the indicated reduction in range in Ireland is a real one, but no clear reason is evident.

Deciduous or mixed woodlands are prime Treecreeper habitat in Britain and Ireland, while conifer forest (especially with older trees), well-treed farmland, parks and gardens are secondary habitats, colonised most extensively when population levels are fairly high. In consequence, current changes both in forestry policy, with less intensive management of conifers and greater plantings of deciduous species, and with changes in agriculture, where farm woodland initiatives are encouraging deciduous tree planting in habitats previously inhospitable, should operate in favour of the Treecreeper.

The song and calls of the Treecreeper, though distinctive, are high-pitched and perhaps frequently unheard. The species is most vocal, and thus most easily located, early in the year between February and April. Nests are most often placed behind a flap of bark or in a slit in a tree, as well as behind ivy, in brick or stone work crevices, and behind loose

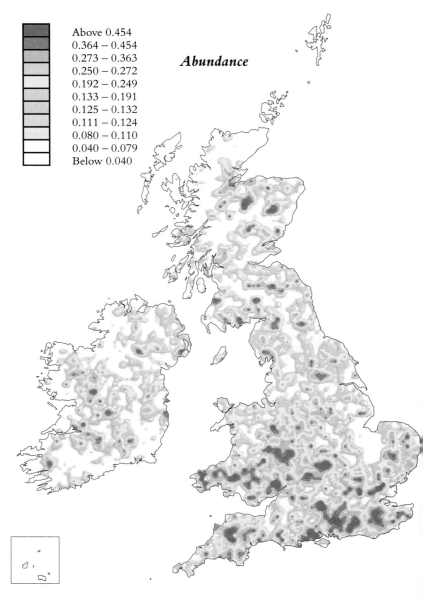

Above 0.454
0.364 – 0.454
0.273 – 0.363
0.250 – 0.272
0.192 – 0.249
0.133 – 0.191
0.125 – 0.132
0.111 – 0.124
0.080 – 0.110
0.040 – 0.079
Below 0.040

Abundance

building cladding. The low level of occupancy of the specially designed nest boxes, though challenging to designers, has not allowed the further colonisation of strictly managed conifer plantations to the degree that could have been hoped. Presumably sufficient natural sites can be found, commensurate with the available food. Nest Record Cards show peak laying in late April and early May (Flegg 1973), with a mean clutch size of about five eggs. Most Treecreepers appear to be single brooded and, as might be deduced from this and from the comparatively small clutch size, losses of eggs and nestlings are small.

Data provided from ringing recoveries show that Treecreepers are among the most sedentary of our birds (Flegg 1973) and that major population changes are associated with severe winters. Extended periods of freezing rain or of glazed frost may be particularly damaging (*Trends Guide*), but overall the Treecreeper seems to be increasing slowly and steadily in numbers.

During 1988–91 there were estimated to have been 200,000 Treecreeper territories in Britain and a further 45,000 in Ireland.

JIM FLEGG

68–72 Atlas p 322

Change

Distribution

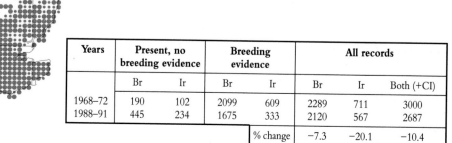

Years	Present, no breeding evidence		Breeding evidence		All records		
	Br	Ir	Br	Ir	Br	Ir	Both (+CI)
1968–72	190	102	2099	609	2289	711	3000
1988–91	445	234	1675	333	2120	567	2687
			% change		−7.3	−20.1	−10.4

Golden Oriole

Oriolus oriolus

In spite of the male's bright yellow and black plumage, the Golden Oriole is a bird more often heard than seen, even where it is known to have a nest. When eventually seen it is arguably one of the most spectacular of British breeding passerines. The female is a much duller green and black, and is far more difficult to see against a leafy background. The cat-like contact squawk or alarm call is the best indication that a female is present.

Breeding in Britain is largely confined to the East Anglian fen basin, where it nests almost exclusively in hybrid poplars. Since 1985, the species has been closely studied by the Golden Oriole Group in co-operation with the RSPB. Because of the felling of 280ha of commercial poplar plantation in which many pairs were concentrated, concern was felt that the species would dramatically decline in the UK. The study has come to the welcome interim conclusion that numbers have not fallen, but that pairs have become less concentrated in one specific locality. In most cases, they are nesting in sites on privately owned and intensively farmed arable land, basically on low lying damp, but drained, peat.

Away from East Anglia, 2–3 breeding pairs were recorded between 1972 and 1987 in a southern county, nesting in oak and sweet chestnut coppice. In a northern county, a pair nested in poplar for three successive years in the early 80s. In Scotland, where the annual average of sight records is 2–3 (but 15 in 1978), some birds do remain in a particular area for 2–3 weeks, and in 1974 a pair bred in Fife, the only confirmed Scottish breeding record (*Birds in Scotland*).

Golden Orioles reach Britain mostly during early May, arriving all along the coasts from Cornwall to Suffolk, and then filter northwards to varying degrees (Dymond *et al.* 1989). Many of these will be non-breeding first-year males, which frequent all sorts of arboreal habitats. Birds reach breeding sites about 10 May, but it can be very difficult to confirm nesting, and it is important to exercise great caution in recording data.

The *68–72 Atlas* confirmed that Golden Orioles were breeding in five 10-km squares. Two of these squares were in the fen basin, where breeding now occurs in ten 10-km squares. Within the two original squares, the birds are now found in more tetrads. Elsewhere, breeding was not recorded in the three outlying 10-km squares with confirmed breeding in 1968–72, but has been recorded in four new ones, suggesting a degree of spasmodic occurrence. The substantial increase in sightings of individuals indicates greater coverage by observers and, probably, more wandering by individual birds.

The northern breeding limit is Finland, where there is a surprising

estimate of 2,000 pairs (Feige 1986). The southern limit is N Africa. The nominate race, with which we are concerned, winters in sub-Saharan Africa, in a broad range from the W to the E coasts but with apparent concentrations from the Gambia to Sierra Leone in the west to Kenya, Uganda and Tanzania in the east.

In the north of its breeding range, the Golden Oriole is more specific in its habitat requirements, i.e. poplar/alder varieties on damp or wet soils, often near rivers or open water. Nearer to S Europe, it is more catholic, breeding in wooded gardens, parkland, orchards and even, in the Iberian peninsula, in coniferous areas.

The *68–72 Atlas* gave no estimate of numbers of British breeding pairs. This was possibly because of insufficiently detailed coverage, even within the ten occupied 10-km squares. Greater coverage, and co-ordination by the GOG, together with records to the RBBP from outside the Group's main study area, enable realistic lower and upper limits to be placed on the numbers currently breeding in Britain.

Now that there is closer monitoring, it can be seen that weather conditions have a great effect on breeding success. 1987, for example, was very wet at the critical hatching and early nestling stage, causing many known failed nests. 1988 and 1989 were good years, but in 1990, despite a very hot summer, for some unascertained reason numbers were down.

Efforts are now under way to promote planting of new hybrid cultivars more suitable for commercial uses, which also happen to be suitable for nesting Orioles. Part of the detailed GOG/RSPB study to determine which of the existing cultivars are suitable has been incorporated into a long-term plan (Prater 1990), a public launch of which took place in April 1991, and it is hoped that over the next two decades the Oriole population will consolidate and expand.

JAKE ALLSOP and PAUL MASON

Recent trends in Golden Oriole numbers.

	1987	1988	1989	1990
Confirmed	11	16	15	10
Possible	20	25	22	32
Max total	31	41	37	42

Based on RBBP reports and those of the Golden Oriole Group: numerals refer to pairs breeding.

68–72 Atlas p 288

Change

Distribution

Years	Present, no breeding evidence		Breeding evidence		All records		
	Br	Ir	Br	Ir	Br	Ir	Both (+CI)
1968–72	4	0	6	0	10	0	10
1988–91	31	0	14	0	45	0	45
				% change	350.0	0	350.0

Red-backed Shrike

Lanius collurio

Shrikes are like small, fierce hawks, and are notable for impaling their prey on thorns, both as larders and as an aid to tearing up and eating larger items. We tend to think that Red-backed Shrikes are most at home in sunny, southern areas, where large insects buzz and chirp, yet over much of their Eurasian range breeding extends well to the north of 60°N.

The Red-backed Shrike was common and widespread in Britain in the 19th century, but has declined ever since (Peakall 1962, Ash 1970, Bibby 1973). The present map offers little scope for interpretation for it has caught the species at the point of breeding extinction. A pair bred in a scrubby clearing in Thetford Forest in 1988. Voluntary wardening of the site was organised by the Forestry Commission, and thousands of visitors enjoyed watching the last of these lovely birds. In 1989 a single male returned to the site but could not find a mate (two females appeared several km away in early July). This was the first year in recorded ornithological history when Red-backed Shrikes are not known to have bred in England.

The species is a trans-Saharan migrant which departs southeastwards in autumn to winter in East Africa. On the return journey some passage migrants, or 'overshoots', arrive each spring, and are likely to account for the scatter of non-breeding records. Future occurrences are likely to become ever more irregular because numbers are declining throughout N and W Europe. The Red-backed Shrike becomes only the third regular breeder to become extinct in England in the present century, and the first for three decades.

When it was common, the Red-backed Shrike occurred in a variety of habitats which combined scrub for perching and nesting with patches of more open ground for hunting. By the time of the *68–72 Atlas* it had become primarily a heathland bird, in the New Forest, Surrey, the East Anglian coast and the Brecks. One by one, these last bastions collapsed, to the final retreat in the Brecks. The history of these last pairs is shown in the Table, which records the erratic but downward nature of the trends. Sporadic breeding over quite a wide area has been a feature of these last years.

The species breeds further to the north of us in Scandinavia, where it particularly favours forest gaps. There was a short-lived hope that it might colonise Scotland, where several pairs bred in the late 1970s. However, only two pairs have bred there since 1980, one of these rearing a single youngster in Shetland in 1990.

The causes of the demise of the Red-backed Shrike are not known. The decline has been of very long duration and has occurred in much of NW Europe. It is rather unlikely that habitat change has been responsible overall, though it might have caused local losses. Growth of taller ground vegetation when myxomatosis made rabbits scarcer would have destroyed some sites, but many previous sites remain and appear suitable.

Change of climate is often cited, but without evidence of any climate factor which has been deteriorating continuously for 150 years. Indeed, the present trend to warmer and drier summers might be expected to suit the shrike, which mainly feeds on large, sun-loving insects. The final retreat from England was by way of the areas with the driest and most continental climate. Perhaps these places were the last areas sufficiently rich in large flying insects which elsewhere may have become more scarce as a result of widespread changes in the countryside.

If conditions in Britain were becoming less suitable, one might have expected a marked decline in breeding success. This does not appear to have been the case. The last pairs continued to be able to rear young even in periods of wet weather.

It is most likely that for unknown reasons, the origin of which is outside Britain, the species is undergoing a range contraction. Individuals may have settled in preferable habitats further south and east in Europe after a slightly shorter migration. If that were the case, the British population would have vanished in spite of seeming to be productive of young right to the end.

The Red-backed Shrike remains very abundant further south and east in its range. Numbers in Britain would never have been a significant part of the total numbers. Although a pair bred in East Anglia in 1992, recovery from extinction in Britain would now be a most unexpected turn round for its fortunes.

COLIN BIBBY

68–72 Atlas p 406

Numbers of pairs of Red-backed Shrikes in Britain as maxima reported to the RBBP in two-year periods.

	73–74	75–76	77–78	79–80	81–82	83–84	85–86	87–88	89–90
Suffolk	33	32	30	30	16	5	6	2	
Norfolk	9	13	11	7	5			1	
Hants	4	3	3	3					
Essex	1	2	1	1	1				
Kent	2	1	1						
Lincs			3	1					
Beds	1	1							
Herts	1								
Berks	1								
Avon		1							
Notts			1						
Wilts			1						
Cambs				1					
Wales					1				
Scotland			7	2	1		1	1	1
Totals	52	53	58	45	24	5	7	4	1

Some records may not have been reported for reasons of secrecy.

Change

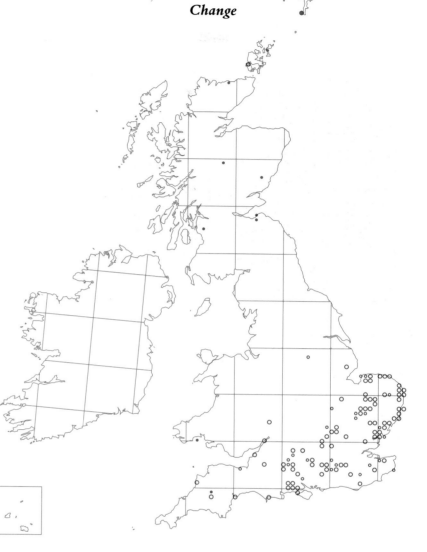

Distribution

Years	Present, no breeding evidence		Breeding evidence		All records		
	Br	Ir	Br	Ir	Br	Ir	Both (+CI)
1968–72	24	0	87	0	111	0	111
1988–91	13	0	2	0	15	0	15
% change					−86.5	0	−86.5

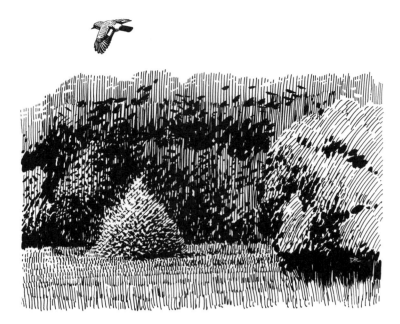

The Jay has always been heavily persecuted. As long ago as the 1870s, gamekeeping was given as a reason for a widespread decline in Scotland (Lumsden 1876). The species was probably in retreat throughout Britain until the First World War, since when it has spread and consolidated its range considerably. These recent gains in N Scotland and W Ireland represent the most recent advances in what is likely to be a continuing expansion. In Scotland the spread has been aided considerably by afforestation, though the rate of expansion remains slow, probably because of the essentially sedentary nature of the bird.

The Jay is reasonably catholic in its use of woodland habitats, occurring in both broadleaved and coniferous. It requires woods where there is adequate cover for nesting and hiding from predators. Thicket stage plantations and any woods with a reasonably dense shrub layer are suitable habitats. The nest is typically placed 2–6m high in the crown or fork of a small tree or bush, sometimes concealed in honeysuckle. Nests can occasionally be clumped in their distribution, with simultaneously occupied nests less than 100m apart (R. J. Fuller). Most studies, however, have found that nests are more widely spaced. Outside the breeding season Jays

Jay

Garrulus glandarius

Although the Jay has adapted to parks and gardens, even in urban areas, it remains essentially a woodland bird. During the breeding season it is shy and secretive, being heard more often than seen. Only in autumn does it become at all conspicuous when it makes many flights over open countryside to gather and hoard acorns. Jays have evolved a close association with oak trees (Bossema 1979). Adults have a mixed diet but acorns are a staple food in most months. This is possible only because individual birds store literally thousands of them, mainly in October. Nestlings are also fed some acorns, though leaf-eating caterpillars from oaks are their main food.

Jays are widespread south of the Southern Uplands but the Abundance map shows that numbers are relatively sparse north of the Humber. The Fens and E Lincolnshire also have few Jays. The greatest concentration occurs in Hampshire, Sussex and Kent, which form the most heavily wooded region of Britain (Locke 1987). The species is also particularly common in East Anglia, despite the comparative dearth of broadleaved woods, and in the W Midlands from Cheshire to the Severn Estuary. Within England and Wales there has been no appreciable change in distribution since the *68–72 Atlas*. But in Scotland, the Jay appears to be pushing out the northern edge of its range into the Great Glen and Grampian. These changes have presumably occurred since the early 1980s, because the *68–72 Atlas* and the *Winter Atlas* show a very similar, more southerly northern limit. By comparison with much of England and Wales, the Jay remains a local bird in Ireland. This may be linked with the relative scarcity of woodland over much of that country. There is evidence of a spread in the N and W of Ireland, including the development of a population in Co. Kerry. This is in contrast to the centre and southeast of the country where the Jay is far less widespread than is shown in the *68–72 Atlas*. This decline is so striking that it is most unlikely to be an artefact of coverage.

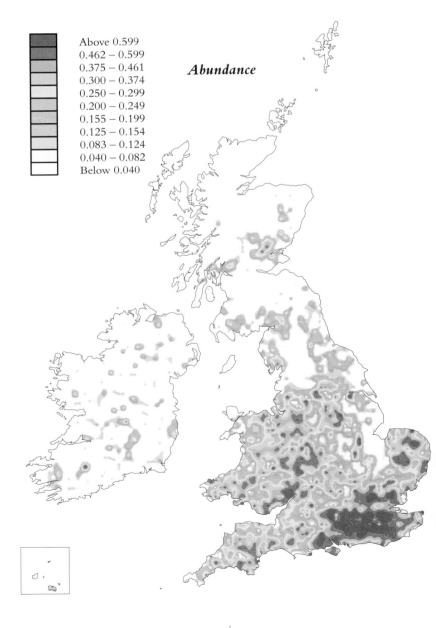

	Above 0.599
	0.462 – 0.599
	0.375 – 0.461
	0.300 – 0.374
	0.250 – 0.299
	0.200 – 0.249
	0.155 – 0.199
	0.125 – 0.154
	0.083 – 0.124
	0.040 – 0.082
	Below 0.040

Abundance

are not territorial, or only weakly so, and the birds appear to remain in the general vicinity of the breeding site. As the breeding season approaches they start to defend territories. Those birds, usually first-years, that do not manage to establish a territory become floaters, obliged to occupy the same habitat as the breeders (Grahn 1991). The normal system of breeding seems to be one of pairs. Despite their territoriality Jays can form 'social gatherings' in early spring (Goodwin 1976). These involve from 3–20 birds, which chase one another through the trees. The function of these gatherings is unknown but Goodwin (1976) considered they may serve to bring unpaired birds together.

The Jay is virtually ubiquitous in mainland Europe, being absent only from the extreme north. Periodically in autumn, Jays irrupt into W Europe, their movements presumably triggered by food shortages to the north and east. It seems that British and Continental Jays are less sedentary in those autumns when the acorn crop fails, such as in 1983 (John and Roskell 1985). In Tuscany, Jays have been reported living at far higher densities than in Scandinavia or W Europe (Patterson *et al.* 1991). The Italian birds ranged over much larger areas than was the case in a detailed

Change

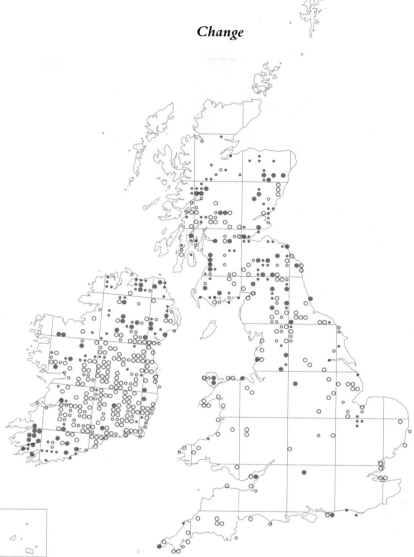

Distribution

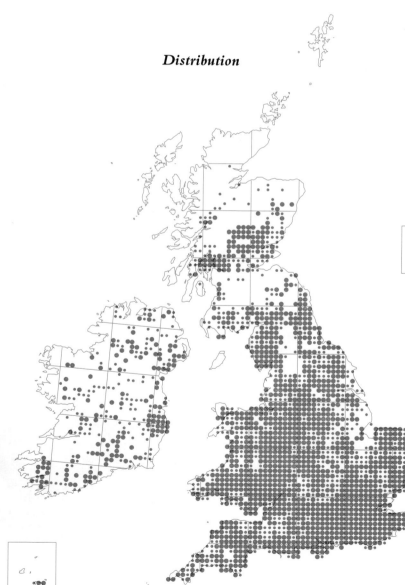

study in Sweden (Grahn 1990). It seems possible that the Jay adjusts its densities and behaviour to match the types and amounts of food that are available in different parts of its range.

During 1988–91 there were estimated to have been 160,000 Jay territories in Britain and 10,000 in Ireland.

ROB FULLER

68–72 Atlas p 302

Years	Present, no breeding evidence		Breeding evidence		All records		
	Br	Ir	Br	Ir	Br	Ir	Both (+CI)
1968–72	199	94	1545	303	1744	397	2142
1988–91	446	134	1267	135	1713	269	1986
			% change		−1.8	−32.2	−7.4

Magpie

Pica pica

The distinctive plumage and conspicuous behaviour of the Magpie make this one of our most easily seen and best known breeding birds. Adults remain bold and noisy throughout the breeding season as they attend their stick nest, which is often barely concealed in an isolated tree, or thorn hedge. These characteristics help to make the Magpie probably one of the most accurately mapped species in this Atlas.

Magpies are found throughout England, Wales and Ireland, but are quite localised in Scotland, being found in the Forth-Clyde valley, the northeast and, very sparsely, in the Southern Uplands. Prior to persecution by gamekeeping interests in the 19th century, they were much more widespread in Scotland. Although over the last 80 years losses due to keepering have been more than made good in the rest of Britain and Ireland, the Scottish breeding range has not recovered to the same extent. Persecution may still play a part in some areas, but the main reasons for this continued restriction of range are unclear (*Birds in Scotland*).

The Abundance map shows high densities of breeding Magpies in Ireland and over most of England and Wales, excepting some of the more treeless areas such as parts of East Anglia and E Yorkshire, where they are quite sparsely distributed. Magpies become generally scarcer in N England, and in Scotland their stronghold is in the urbanised lowlands of the Forth-Clyde valley, with lower densities of birds in the southwest and northeast of the country.

The Magpie has undergone little change in its breeding range between the *68–72* and *88–91 Atlases*. In England, its distribution in the Fens and East Anglia is now more continuous, and there has been a slight spread in Northumberland. In Scotland, there are perhaps fewer breeding now in Tayside. Within the main breeding range, however, Magpies continue to increase in numbers. This increase dates from the First World War, when gamekeeper activity began to wane, and has continued ever since, checked only temporarily in some eastern areas by the agricultural use of organo-chlorine pesticides (*Trends Guide*). Part of this increase is reflected by a spread into urban and suburban areas since the 1940s. There, relative freedom from persecution, and a catholic diet (Tatner 1982), have allowed the population to reach breeding densities of up to 16 pairs per km², exceeding those in either farmland or woodland habitats (Birkhead 1991). Remarkably, this density, one of the highest recorded anywhere, was in Ireland, a country where Magpies were absent until the end of the 17th century.

The marked and continuing increase in Magpie numbers had led to

Abundance

	Above 0.979
	0.909 – 0.979
	0.875 – 0.908
	0.828 – 0.874
	0.750 – 0.827
	0.667 – 0.749
	0.600 – 0.666
	0.462 – 0.599
	0.250 – 0.461
	0.040 – 0.249
	Below 0.040

concern about the possible effect of nest predation by Magpies on urban populations of songbirds. Detailed studies, however, indicate that nestlings and eggs are only a minor component of the diet of Magpies (Tatner 1983), and that the increase of this species has not affected the breeding success of garden passerines (Gooch *et al.* 1991).

Magpies have been the subject of long-term studies in Britain, Sweden, Denmark and the United States. A wealth of information on their behaviour, ecology and breeding biology can be found in Birkhead (1991).

Breeding densities average over 10 pairs per km² in woodland and suburban habitats, and over 5 pairs per km² on farmland (Gooch *et al.* 1991). During 1988–91 there were estimated to have been 590,000 and 320,000 Magpie territories in Britain and Ireland respectively.

JEREMY WILSON

68–72 Atlas p 300

Change

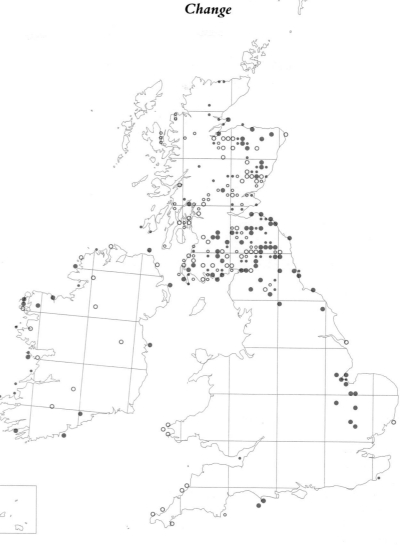

Distribution

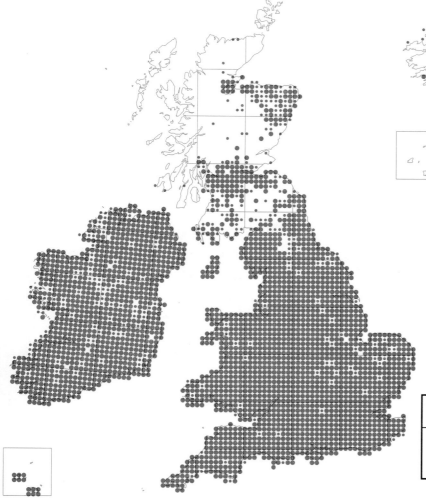

Years	Present, no breeding evidence		Breeding evidence		All records		
	Br	Ir	Br	Ir	Br	Ir	Both (+CI)
1968–72	141	19	1799	935	1940	954	2898
1988–91	183	94	1775	868	1958	962	2932
	% change				1.1	0.9	1.1

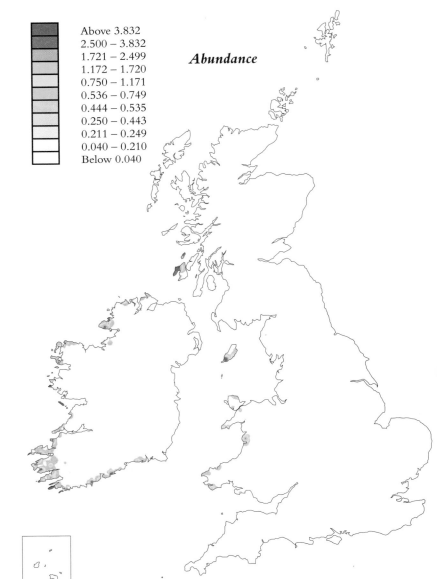

Chough

Pyrrhocorax pyrrhocorax

The Chough is one of the supreme fliers amongst passerines, in its element when riding the updraught on exposed sea cliffs. It is a specialised feeder on soil arthropods, especially beetles, and this restricts it to areas where low-intensity livestock farming occurs close to suitable nesting sites in cliffs, caves and old buildings. Only on the wilder and more remote W coasts of Britain and Ireland is this association found.

Apart from one pair in Scotland, W Wales holds the only breeding Choughs on the British mainland. The remainder breed on the islands of Anglesey, Bardsey, Man, Islay, Colonsay and Jura. In Ireland there is a similar western pattern of distribution, extending from Waterford along the S and W coast to Donegal; in the north a few pairs are found on the Antrim coast and on Rathlin.

The Distribution and Abundance maps are very similar to that in the *Winter Atlas*, and ringing recoveries and long-term colour-ringing studies have not found evidence of interchange between the main breeding areas (Roberts 1989, E. Bignal, A. Sapsford).

Even within its relatively limited British and Irish distribution the Chough has a very low breeding density. There are only three major concentrations, in S and W Ireland, the Isle of Man and Islay. The Abundance map shows no large clusters of breeders or flock birds in Wales and over 70% of the population there are in Pembrokeshire and Caernarvonshire (Bullock *et al.* 1983).

Although the pattern of distribution has altered little over the past 20 years there have been some subtle changes. For instance, in Wales there have been decreases on the W coast but increases on Anglesey. Over 30% of Welsh Choughs breed inland and there is the suggestion of an extension in this part of their range.

On the Isle of Man, losses in the north are thought to be associated with agricultural intensification coupled with the loss of artifact nesting

sites (A. Sapsford). In Scotland, Choughs have been lost from the Mull of Kintyre but there have been increases on Islay, Colonsay and Jura and recolonisation, by a single pair, of the mainland.

The Scottish expansion is attributed to the continuing health of the population on Islay, where land management has remained suitable, and direct conservation of nest sites instigated.

Throughout their world range, wherever a lack of natural nest sites limits breeding opportunities, Choughs readily use old buildings, and contractions of range inland are often associated with losses of such nest sites through dereliction or renovation. It is relatively easy to provide cave-like artificial nest sites in buildings, and this has been particularly successful in the Hebrides and the Isle of Man (Bignal and Bignal 1987). In Scotland 27% of all Chough nests are in man-made structures (Monaghan *et al.* 1989).

Of 198 clutches recorded during long-term monitoring of the Islay population (Bignal *et al.* 1988) the median clutch size was five eggs, with two young fledging per pair. Eggs were laid during April, and there were no repeats following losses after incubation had commenced. The incubation period of four pairs was 20–21 days, beginning with the laying

Above 3.832
2.500 – 3.832
1.721 – 2.499
1.172 – 1.720
0.750 – 1.171
0.536 – 0.749
0.444 – 0.535
0.250 – 0.443
0.211 – 0.249
0.040 – 0.210
Below 0.040

Abundance

of the third egg. Only the female incubated. Young flew at between six and seven weeks old. Mortality was highest in the six months following fledging, and ranged from 24% to 92% in the seasons 1983 to 1990. Of 485 birds fledged in the Hebrides, only 16 males and 20 females were recruited to the breeding population. The median dispersal distance of the males was 4.5 km, and of the females 10.9 km. The most frequent age of first breeding of both sexes was three years and there was a maximum life expectancy of around ten years, with a mean expectancy of three years further life after reaching breeding age (E.Bignal).

A review of literature in 1988 suggested that Choughs are decreasing in all European countries but there is no reliable estimate of the current population (Bignal and Curtis 1989).

In 1982 there were estimated to be up to 274 breeding pairs in Britain and 685 in Ireland (Bullock *et al.* 1983). In 1986 the minimum Scottish population was put at 105 pairs of which 90% were on Islay (Monaghan *et al.* 1989); a survey on the Isle of Man estimated up to 68 breeding pairs in 1991 (A. Sapsford and A. Moore). Assuming the Welsh population has remained constant in size since 1982, the British population during 1986–91 was about 315 breeding pairs. The preliminary results of a survey of

Change

Distribution

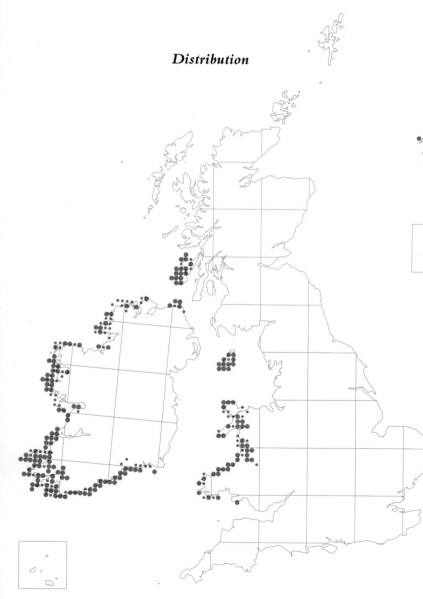

Choughs in Ireland in 1992 suggests a population of 830 pairs there, an increase of 20% since 1982 (O'Sullivan 1992).

ERIC BIGNAL

Two breeding records have been omitted and a third moved by up to two 10-km squares on the Distribution and Change maps. Only two of these three records were obtained during timed visits, and both have been dealt with on the Abundance map in the same manner as the other two maps (i.e. one omitted and one moved).

68–72 Atlas p 304

Years	Present, no breeding evidence		Breeding evidence		All records		
	Br	Ir	Br	Ir	Br	Ir	Both (+CI)
1968–72	15	23	64	142	79	165	244
1988–91	24	52	64	116	88	168	256
				% change	12.8	1.8	5.3

caused by the demolition of old farm buildings could have contributed to these losses. The renovation of old housing stock may have had a similar affect in urban areas, such as London. Sitters (1988) comments on the absence of the Jackdaw in built-up areas of Devon. It is not obvious why the Jackdaw should have become less common on parts of the exposed W coast of Scotland, but one possibility is competition from some other cliff nesting species. Avoidance of exposed coasts is mentioned in *Birds in Ireland*.

Studies in Scotland during the early 1980s showed that Jackdaws exploited a wide range of habitats in the breeding season. During the early nest site guarding phase of March-April, the birds were still dependent on sheepcake in troughs during cold snaps. As the soil warmed, the sheep-grazed land was used heavily and when the nestling period ensued during May-June the adults fed over wider areas. In some years larvae of the antler moth formed the bulk of food items collected for the young, as revealed by neck-collaring nestlings. Newly fledged young were crèched in oak trees, while the adults foraged widely, and older dependent young were in turn led into upland bracken and ripening barley. The pattern, then, is one of successive exploitation of what might be termed micro-habitats in

Jackdaw

Corvus monedula

Many stone ruins are brought to life between April and June by the squabbling of this the smallest crow, when the closeness of nest holes leads to much fighting. The same is true for those individuals nesting in old rabbit warrens and cliff crevices. Less obvious are the quieter pairs occupying the natural tree holes of mature woodland and parkland. Nest sites are at a great premium, the vast majority being cavities, although domed nests of sticks in trees have been reported.

In most of Britain and Ireland this corvid finds places to nest, with adjacent grassland for feeding, but is noticeably absent from the high moorland of NW and W Scotland. In Devon, the number of tetrads where Jackdaws were confirmed breeding declined rapidly above 350m, with none at altitudes above 450m (Sitters 1988). It is suggested that this tendency to become less common on high ground between valleys is a result of the absence of human habitation, and therefore of nest sites, combined with the lack of good grazing pasture. In Cambridgeshire, the Jackdaw is now common in the fenland, nesting in abandoned buildings (Bircham 1988).

During the breeding season, Jackdaws rely mainly on the surface and shallow subsurface invertebrates of grazed grassland, but later in the breeding season they will take to searching rough pasture and bracken, and to foliage feeding in trees (P. Green, Lockie 1955b). The greatest abundance of Jackdaws is in those areas where grazing of sheep and cattle is still common. Areas of mixed farmland provide the favoured short grassland for collecting nestling food, and also livestock feed during periods of cold weather in the early breeding season.

Breeding distribution has not changed markedly since the *68–72 Atlas*. Most losses are from areas adjoining those where they were previously absent, noticeably in upland areas. The disappearance of breeding sites

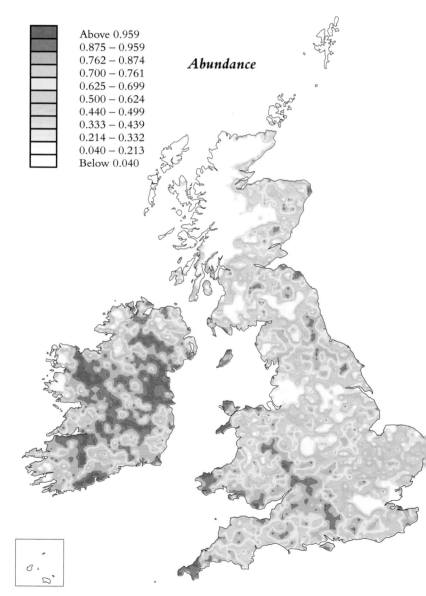

Above 0.959
0.875 – 0.959
0.762 – 0.874
0.700 – 0.761
0.625 – 0.699
0.500 – 0.624
0.440 – 0.499
0.333 – 0.439
0.214 – 0.332
0.040 – 0.213
Below 0.040

Abundance

this area of mixed upland farmland (50–300m). In Scotland, feeding flocks of Jackdaws are regularly disturbed by breeding Lapwings, and by territorial Carrion Crows, and Roell (1978) has shown that larger flocks are less likely to be flushed from the field than smaller ones.

The *68–72 Atlas* estimated the population of Britain and Ireland to be 500,000 pairs, however the CBC has shown that Jackdaw numbers in Britain increased markedly during the inter-Atlas years (*Trends Guide*). In Ireland a similar population increase, which began in the 1950s, is thought to be continuing, and the Abundance map documents how widespread and numerous Jackdaws now are. New colonies have appeared, including one on Cape Clear in 1984 (*Birds in Ireland*). Other sources (e.g. Bircham 1988, Sitters 1988) indicate a continuing population increase. During 1988–91 there are estimated to have been 390,000 Jackdaw territories in Britain, with a further 210,000 in Ireland.

PAUL GREEN

68–72 Atlas p 298

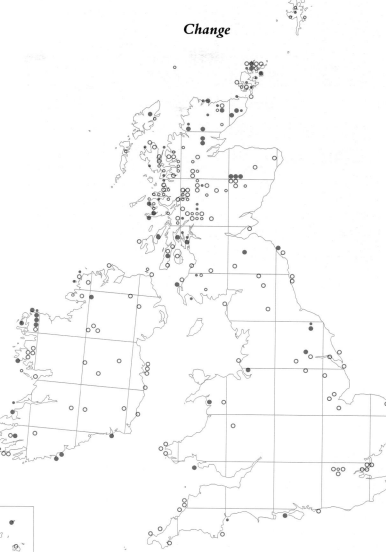

Change

Distribution

Years	Present, no breeding evidence		Breeding evidence		All records		
	Br	Ir	Br	Ir	Br	Ir	Both (+CI)
1968–72	101	13	2319	952	2420	965	3386
1988–91	195	51	2149	893	2344	944	3290
	% change				−3.0	−2.1	−2.7

Rook

Corvus frugilegus

Fewer sounds herald the promise of longer and sunnier days better than the cawing emanating from a busy rookery in late winter. Throughout spring and summer the Rook symbolises the vitality of the agricultural scene, for the bird is an integral part of that landscape.

Rooks are found in all lowland agricultural areas. The only places where they do not breed are in most of the uplands (land higher than enclosed farmland) of Scotland, Wales, N England and W Ireland and the centres of large cities. They prefer farmland with a mixture of field types but in the breeding season depend to a large extent on grass fields (Feare *et al.* 1974). This is illustrated by high Rook abundance wherever there is extensive grassland, particularly in Ireland, the agricultural districts of Scotland and in parts of S England. Generally lower densities prevail in the extensive cereal growing areas of E England. Breeding densities (nests per km²) recorded in 1975 for a selection of counties are shown in the Table.

Comparison of the present breeding distribution with that in the *68–72 Atlas* reveals no noteworthy differences. The winter distribution also is virtually identical, revealing British and Irish Rooks as mainly sedentary birds.

While there is no evidence of distributional change, there have been great changes this century in numbers of Rooks censused in Britain. Many local surveys, two near-complete national surveys in 1943–46 and 1975, and a sample national survey in 1980, have revealed that the Rook population size increased by around 20% between the 1920s and the 1940s (Brenchley 1986, *Trends Guide*). From the 1940s to 1975 there was, however, an overall decrease of over 40% in breeding numbers (Sage and Vernon 1978), but from 1975 to 1980 a small recovery of about 6–7% (Sage and Whittington 1985). The reasons for such a large decline in the British Rook population between the end of the Second World War and the 1970s are not known. However, they are almost certainly related to the intensification of agriculture.

Rooks are sensitive to changes in cropping practices. In NE Fife, for example, variation in breeding Rook numbers in 1945 and 1990 was related to variation in the area of land under grass and cereals. This close relationship did not hold in 1978, however, when the persistent effects of pesticides may have adversely affected population size. This could account partly for the observed decrease of 60% in breeding numbers in this part of Scotland from 1945–1990 (J. B. Reid). The widespread shift away from spring sown corn to autumn sowing also may have adversely affected

invertebrate availability during the breeding season (O'Connor and Shrubb 1986a). Additionally, this effect might be exacerbated by the fact that summer can be the time of greatest shortage of grassland invertebrates (Dunnet and Patterson 1968).

Part of the intensification process in post-war farming was the increased use of pesticides and there is some circumstantial evidence to implicate these in the population decline. The timing of the decline (most likely the late 1950s and early 1960s) coincided with peak usage of organochlorine seed dressings in cereal growing areas, mainly in E Britain. The localities showing the most alarming population crashes were in just such areas, and the recovery and growth of Rook numbers have been greater in northern and western areas – the areas where the chemicals were least used. (The Sparrowhawk, which suffered greatly through organochlorine use, also showed greatest recovery of numbers in northwestern areas, Newton 1979).

Not only did Rook numbers change dramatically between the 1940s and 1970s but the size of rookeries also changed. In 1975 rookeries tended to be smaller than in the 1940s, although typical rookery size in 1980 seems to have been higher than in 1975 (Sage and Vernon 1978, Sage and Whittington 1985). Such a pattern might be expected if, following a

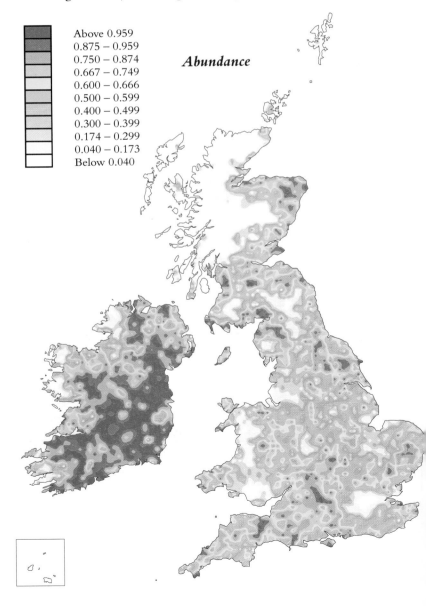

Abundance

	Above 0.959
	0.875 – 0.959
	0.750 – 0.874
	0.667 – 0.749
	0.600 – 0.666
	0.500 – 0.599
	0.400 – 0.499
	0.300 – 0.399
	0.174 – 0.299
	0.040 – 0.173
	Below 0.040

decline up to the 1960s, the population recovered in the 1970s with the foundation of new (and therefore smaller) rookeries which continued to grow in size up to 1980.

The Rook is an integral feature of a habitat which is bound to become less extensive in the future. Unless afforestation of lowland agricultural areas (seen as an antidote to overproductive arable farming) is planned and controlled with care, birds such as the Rook, which are dependent on farmland, may suffer.

Sage and Whittington (1985) estimated there to be 853,000–857,000 pairs in Britain in 1980. By extrapolation, the population in Ireland would be about 520,000 pairs.

J. B. REID

68–72 Atlas p 296

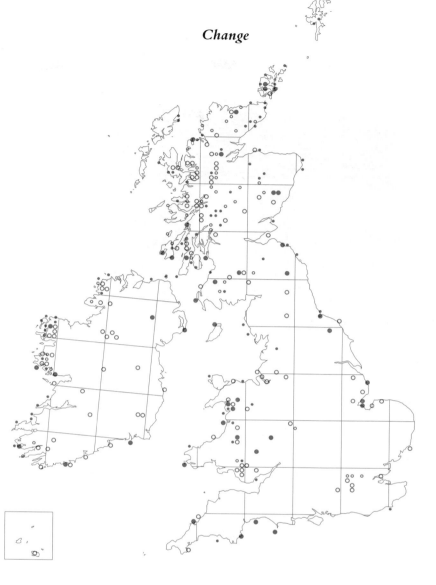

Change

Distribution

Nests per km² in 1975

Aberdeenshire	9.7	Yorkshire	4.0	Londonderry	12.6
Fife	5.7	Caernarvonshire	2.0	Kildare (1981)	47.4
Dumfriesshire	7.8	Hampshire	5.0		

Data taken from Sage and Vernon (1978), Sage and Whittington (1985), and *Birds in Ireland*. The data are from 1975, with the exception of Kildare (1981).

Years	Present, no breeding evidence		Breeding evidence		All records		
	Br	Ir	Br	Ir	Br	Ir	Both (+CI)
1968–72	123	29	2123	902	2246	931	3178
1988–91	283	95	1954	824	2237	919	3156
			% change		−0.4	−1.2	−0.7

Carrion Crow

Carrion and Hooded Crow

Corvus corone corone and *C. c. cornix*

Crows are among the most adaptable of all our birds, being found from lowland open woodlands and farmlands to high heather moor. Their widespread distribution and variety of breeding habitats is partly due to the extremely wide range of foods they take. Their staple diets are usually grain in winter and various insect foods in summer, but they have a highly exploratory and investigative nature and will seek out many alternative sources of food. They can, for example, open milk bottles on doorsteps, but more typically, they are to be found scavenging at rubbish dumps with gulls. They feed extensively on road kills, and take shellfish along the shoreline. In upland areas, in winter, they feed largely on carrion, which they regularly cache, and they are effective predators on amphibians, lizards, and small nesting birds, and have even been reported attacking pigeons and Dunlins and catching Blue Tits and Starlings in flight.

Undoubtedly the most interesting aspect of the breeding distribution of Crows in Britain is the position of the hybrid zone between the Carrion and Hooded Crows, which today runs erratically from the Clyde to Caithness. The grey Hooded and the black Carrion Crow can interbreed and they produce fertile offspring. They are, therefore, regarded as belonging to the same species. The distribution of these two subspecies is, however, difficult to explain and they remain one of the most famous examples of allopatric hybridisation in the animal kingdom. The two forms of Crow are found throughout most of Europe and N Asia. Carrion Crows are found both in the far west and east of this range, one population in W Europe and the other from central Asia right across to the Pacific coast. These two populations are, however, divided by the Hooded Crow,

(continues on p 396)

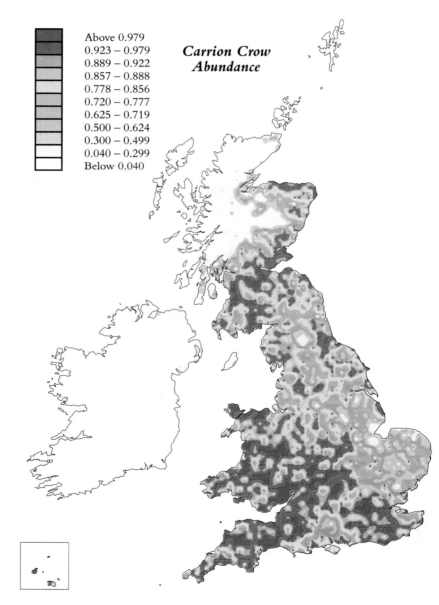

Above 0.979	
0.923 – 0.979	
0.889 – 0.922	***Carrion Crow***
0.857 – 0.888	***Abundance***
0.778 – 0.856	
0.720 – 0.777	
0.625 – 0.719	
0.500 – 0.624	
0.300 – 0.499	
0.040 – 0.299	
Below 0.040	

Note to Carrion Crow Maps

The apparent colonisation of the Isle of Man by Carrion Crows is partly
an artefact of the change in methods between the two Atlases. In 1968–
72 only pure Carrion × Carrion pairings were recorded on the Carrion
map. Although Carrions did breed then on the Isle of Man they did so
only in mixed pairings, as shown on the small map on p 294 of the *68–
72 Atlas*. On the 88–91 Distribution map only one member of a pair
needed to be Carrion for it to be included. The Carrion Crow Change
map for the Isle of Man reflects this.

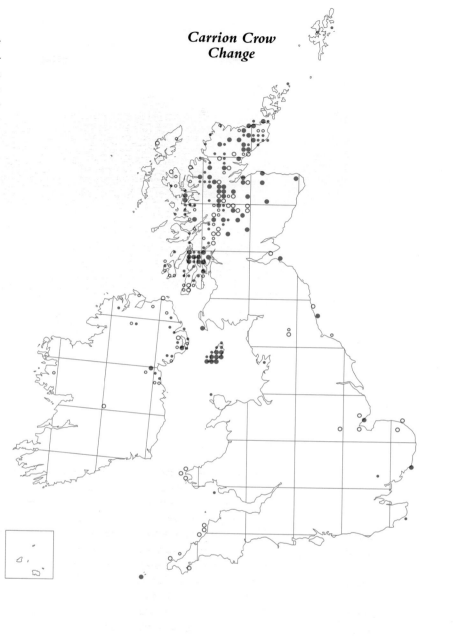

**Carrion Crow
Change**

**Carrion Crow
Distribution**

Years	Present, no breeding evidence		Breeding evidence		All records		
	Br	Ir	Br	Ir	Br	Ir	Both (+CI)
1968–72	136	13	2149	7	2285	20	2310
1988–91	218	16	2124	5	2342	21	2377
				% change	2.8	5.0	2.8

Hooded Crow

(continued from p 394)
which is found throughout virtually all of N and central Europe from the Arctic to the Mediterranean. It is usually assumed that the different colour forms evolved during the last Ice Age, when remnant Crow populations became isolated. In Europe there were ice-free refuges in the Iberian and Balkan Peninsulas, where the birds were separated long enough for them to develop their distinctive plumage differences, but not long enough for them to evolve into full species. With the retreat of the ice these two populations of birds moved north and eventually met in central Europe.

Wherever the two colour forms meet they interbreed, and produce hybrids of intermediate colour (Picozzi 1976). This zone of hybridisation meanders erratically over thousands of km through Europe and Asia, but has been studied in any detail only in Scotland.

One puzzling aspect of this hybrid zone is that it remains so distinct. There is no evidence that Crows deliberately select mates of a similar colour, and it seems that wherever the two colour forms coexist they mate at random. Presumably, therefore, in terms of survival the hybrids are selected against, compared with either pure grey or pure black birds, but it is not known why they show reduced fitness. If this were not the case then the hybrid zone would have progressively widened since the two subspecies met at the end of the Ice Age. As Mayr wrote in 1942, 'Even assuming that adult birds nest year after year in the same tree, and that the young settle down no farther than 500m from the place of birth... the hybrid zone should now cover the whole width of Europe'. With the exception of a few more records of hybrid Crows in SE Scotland, there is no evidence of any widening of the hybrid zone from the *88–91 Atlas* data, and the width of the hybrid zone remains very similar to that recorded in the *68–72 Atlas*.

A second curious feature of Crow breeding distribution is that there are no consistent habitat or climate factors which seem to explain the distinct separate distribution of the Carrion and Hooded Crows throughout their European and Asian range. Cook (1975) showed that within the hybrid zone there was a tendency for Hooded Crows to occupy higher altitude ground. He also showed that there has been, within this century, a northwest movement of the hybrid zone in Scotland. Cook suggested this was due to climatic change. The *88–91 Atlas* provides further evidence of this, for since 1968–72 there has been a clear northwesterly movement of

(continues on p 398)

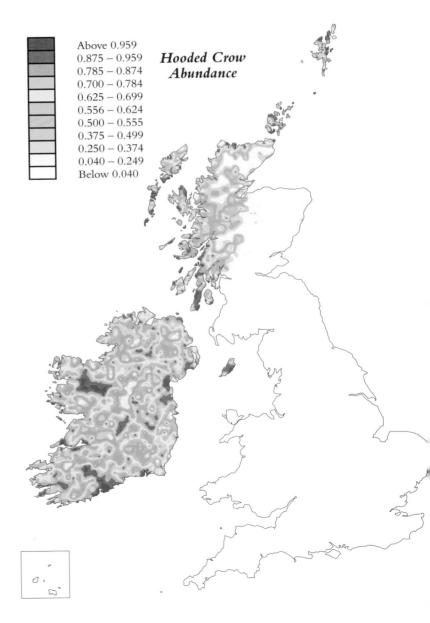

Above 0.959
0.875 – 0.959
0.785 – 0.874
0.700 – 0.784
0.625 – 0.699
0.556 – 0.624
0.500 – 0.555
0.375 – 0.499
0.250 – 0.374
0.040 – 0.249
Below 0.040

Hooded Crow Abundance

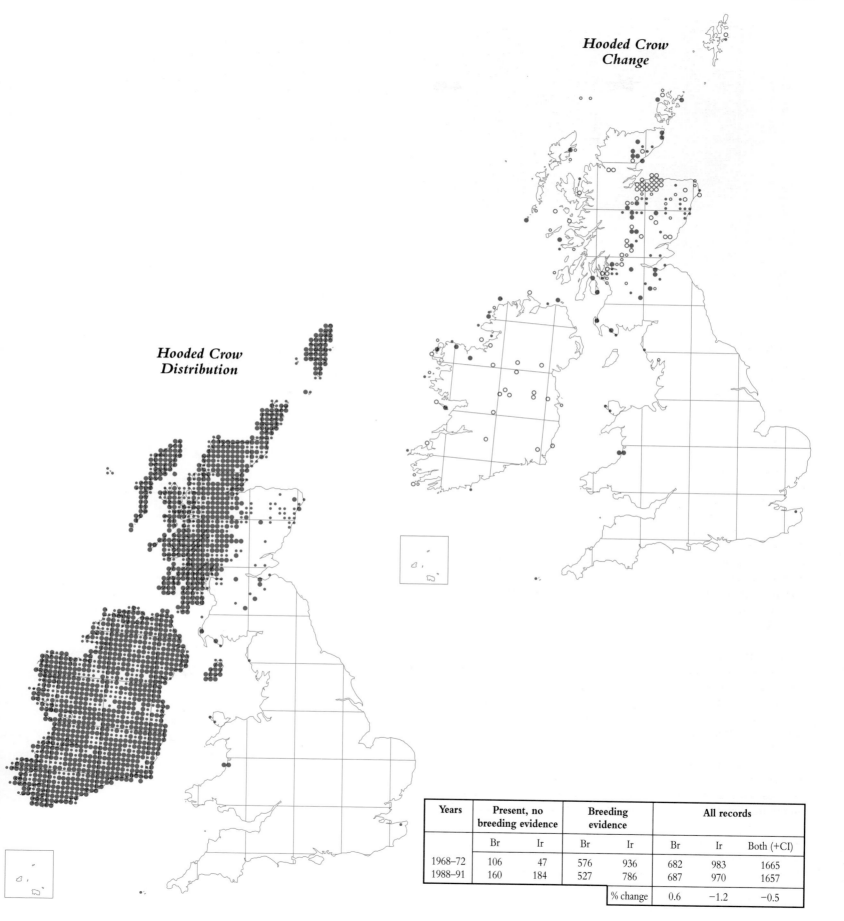

Hooded Crow Change

Hooded Crow Distribution

Years	Present, no breeding evidence		Breeding evidence		All records		
	Br	Ir	Br	Ir	Br	Ir	Both (+CI)
1968–72	106	47	576	936	682	983	1665
1988–91	160	184	527	786	687	970	1657
% change					0.6	−1.2	−0.5

Hybrid Crow

All Crows Combined

Years	Present, no breeding evidence		Breeding evidence		All records		
	Br	Ir	Br	Ir	Br	Ir	Both (+CI)
1968–72	134	47	2641	936	2775	983	3763
1988–91	224	184	2538	786	2762	970	3746
% change					−0.3	−1.2	−0.6

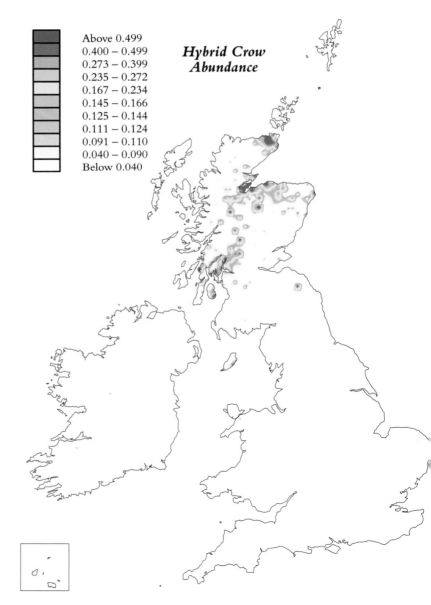

Above 0.499
0.400 – 0.499
0.273 – 0.399
0.235 – 0.272
0.167 – 0.234
0.145 – 0.166
0.125 – 0.144
0.111 – 0.124
0.091 – 0.110
0.040 – 0.090
Below 0.040

Hybrid Crow Abundance

(continued from p 396)
about 30km by Carrion Crows moving into areas previously occupied only by Hooded Crows. As would be expected, this has been accompanied by a westerly shift in the distribution of hybrid birds, which is especially obvious in Caithness, the Black Isle and SW Argyll. Unlike the movement of the hybrid zone up to 1975, which then occurred only at the northern end of the distribution, the shift in the last 15 years has taken place along the whole length of the hybrid zone. Further evidence for continued movement of the hybrid zone comes from Dornoch, in SE Sutherland, where the proportion of black Crows was 15% in 1955/56, had increased to 35% in 1963/4, and had reached 50% by 1974 (Cook 1975). This is a pattern similar to that recorded at Kerloch Moor in Kincardineshire, where Carrion Crows formed 59% of the Crow population in 1966, increasing to 69% in 1967, 72% in 1968 and had reached 100% by 1981.

Crows usually nest in trees. Their large nests are built before the trees come into leaf, which makes them conspicuous early in the season and so breeding can usually be readily confirmed. Where trees are absent, birds will build on rock ledges or low bushes, but this is unusual. Studies in Switzerland have shown that the growth rate of the chicks varies with habitat quality and that poor quality areas give rise to birds that are smaller and of greatly reduced fitness (Richner 1989). A small proportion of crows in that study were also found to have helpers at the nest (Richner 1990), although this has not been reported in Britain.

There are estimated to be 790,000 Carrion Crow, 160,000 Hooded Crow and 20,000 hybrid Crow territories in Britain, with a further 290,000 Hooded Crow territories in Ireland. Although Carrion and hybrid Crows do breed in Ireland, they do so only rarely. Crows continue to be persecuted by gamekeepers and hill shepherds, but it is unlikely that this has any more than a small local effect on their numbers. Their canny behaviour has enabled them to adapt to many changes in the countryside and to steadily increase their numbers in the urban environment.

D. C. HOUSTON

68–72 Atlas p 292

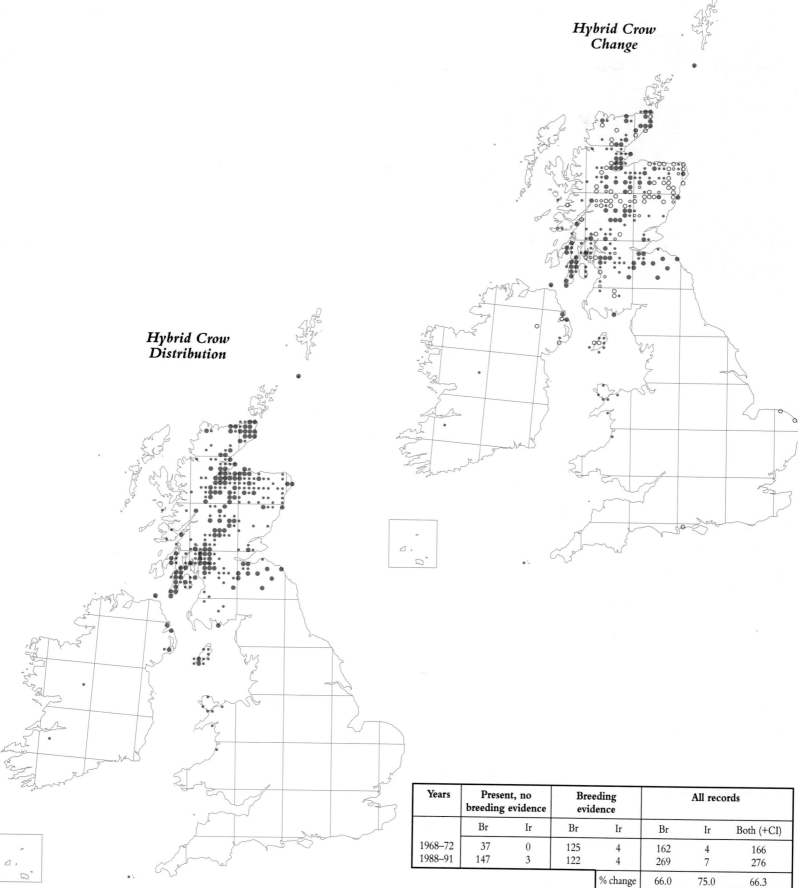

**Hybrid Crow
Change**

**Hybrid Crow
Distribution**

Years	Present, no breeding evidence		Breeding evidence		All records		
	Br	Ir	Br	Ir	Br	Ir	Both (+CI)
1968–72	37	0	125	4	162	4	166
1988–91	147	3	122	4	269	7	276
			% change		66.0	75.0	66.3

Hebridean islands and SW Ireland. The densities in Snowdonia are the highest in Europe, probably because well-stocked sheepwalks and over-wintering of sheep on the hills combine to provide a reliable source of sheep carrion (*Trends Guide*).

A comparison of the 1968–72 and 1988–91 Atlases shows the continuing spread of breeding Ravens across inland Ireland, a trend that is also visible when the *68–72 Atlas* is compared with the *Winter Atlas*. In most of England and Wales there is little evidence of any change, except perhaps for a decline in SW England and along the S coast. In Scotland, there is evidence of a westward contraction of range from the Grampians and Speyside, perhaps a result of poisoning by meat baits (*Birds in Scotland*), but the most striking change is the almost complete loss of breeding Ravens from Northumberland, parts of Galloway and the Scottish Borders since the *68–72 Atlas*. Between the 1950s and 1981, the number of Raven breeding territories fell by over 50% in S Scotland and the Borders. Afforestation, improved sheep husbandry and conversion of pasture to arable are all implicated in these changes (Mearns 1983). *Birds in Scotland* considered that 'if the present rate of decline continues, the Raven may shortly cease to breed inland in southern Scotland'.

Raven

Corvus corax

Resonant calls and acrobatic flight characterise our largest corvid – a bird which even in some of the least hospitable upland and coastal habitats in Britain and Ireland may begin nesting as early as February. From an historical perspective, however, this image of the Raven as a denizen of remote moorland and sea cliffs is misleading.

Ravens were once widespread throughout Britain and Ireland. In the 17th century they could be found scavenging for carrion alongside Red Kites in London's streets (Coombs 1978). Increasing persecution by farmers and gamekeepers in the 19th century, however, caused a rapid contraction of the Raven's breeding range towards the north and west. In lowland England the last breeding pairs were lost by 1895. Shooting by farmers and gamekeepers still occurs, but the scale of persecution has decreased gradually since the First World War as the number of keepers has fallen. In response, the breeding range of the Raven has expanded since the 1920s in SW England, the Welsh Borders and, initially, in the Southern Uplands of Scotland. In Ireland, Ravens had been eliminated from all but some coastal areas by 1900. There too, however, they have increased greatly during the 20th century. They had reoccupied their former range by 1950, and have continued to spread since then (*Winter Atlas*, *Birds in Ireland*).

In lowland Britain, the quite extensive replacement of livestock husbandry by arable farming, and the consequent reduction in the availability of carrion, may mean that the Raven is unlikely to reoccupy much of its former range. In the uplands, blanket afforestation has resulted in loss of habitat that was occupied by breeding Ravens as recently as the *68–72 Atlas*, whilst gamekeeping interests on grouse moors continue to keep densities low in E Scotland and parts of N England. At present, the highest breeding densities are found in Wales, the Northern Isles, some of the

Above 1.249
0.875 – 1.249
0.667 – 0.874
0.500 – 0.666
0.400 – 0.499
0.300 – 0.399
0.250 – 0.299
0.167 – 0.249
0.125 – 0.166
0.040 – 0.124
Below 0.040

Abundance

Dare (1986) studied the breeding population of Ravens in two upland areas of N Wales. In Snowdonia, where sheep stocking densities have increased and there is an abundance of secure rock nest sites, the population has increased by 80% since the 1950s to a density of one pair per 9.5km². The expanding population has not simply filled in the gaps in the 1950s distribution, but has also increased in areas already occupied, thus leading to a contraction in the size of individual territories. On nearby enclosed sheep farms and on moorland, no such increase has occurred, and the population density has remained relatively constant at one pair per 23.9km². Persecution, scarcity of secure nest sites, proximity of lambing flocks to farmsteads, and more rapid removal of carcasses by the farmers there have probably combined to limit population density. Dare considered that future trends in population size and breeding success were likely to be controlled mainly by food availability, which would in turn be dependent on changes in land use and farming practice.

The *68–72 Atlas* estimated a British and Irish breeding population of 5,000 pairs, assuming an average of three pairs per occupied 10-km square, and with approximately three-quarters of these birds in Britain. Extrapolation from tetrad counts during 1988–91 yields a minimum

Change

Distribution

population of 14,000 Ravens in Britain and 7,000 in Ireland. Although these counts may have included some older fledglings, it is equally likely that some adults were overlooked during the two-hour visits. Thus a population of 7,000 pairs is suggested for Britain with a further 3,500 in Ireland.

JEREMY WILSON

68–72 Atlas p 290

Years	Present, no breeding evidence		Breeding evidence		All records		
	Br	Ir	Br	Ir	Br	Ir	Both (+CI)
1968–72	215	131	1031	316	1246	447	1698
1988–91	346	306	785	380	1131	686	1823
	% change				−9.2	53.8	7.4

Along with the availability of nesting holes, similar factors are likely to be the main determinants of Starling abundance throughout the rest of Britain and Ireland.

The preferred foraging habitat of Starlings is grazed permanent pasture, as this provides high densities of the main breeding season food item leatherjackets. Breeding success is generally high amongst first broods with more than 90% of hatchlings surviving to fledging (Feare 1984). A recent Finnish study has, however, shown chick mortality to be much higher in areas of arable and root monoculture than in areas of mixed farming with grazed pasture (Tiainen et al. 1989). It is possible therefore that the continuing loss of permanent pasture, through increased specialisation in cereal production, has resulted in reduced breeding success in many areas. Other factors such as the change from spring sown to autumn sown cereals or subtle climatic changes might also be affecting populations in some regions. Without further research the worrying decline of the Starling across large areas of N Europe will remain unexplained.

Starlings often breed in loose colonies in which only the area immediately surrounding the nest is defended as a territory. Older males begin to occupy territories as early as January although generally eggs are not laid

Starling

Sturnus vulgaris

Being conspicuous and widespread, the Starling has an omnipresence which is arguably unrivalled within the British avifauna. Most familiar as the whistling songster on the chimney pots of suburbia, the Starling is also common on farmland and in woodland habitats. Only remote moorland and mountainous areas seem to be avoided for breeding.

Despite recent reports describing declining populations and local extinctions over large areas of N Europe (Tiainen et al. 1989), there is no indication here of any major contraction in range. The Change map shows the main losses to have occurred in highland areas of Scotland, and to a lesser extent in parts of Wales, Cornwall and Ireland. A high proportion of squares remains occupied, however, including those in Orkney and Shetland where the zetlandicus subspecies occurs.

The apparent decline of the Starling as a breeding species in NW Scotland could be symptomatic of changing land use, particularly the expansion of forestry. This possibility is not, however, supported by the continued presence of breeding birds in Galloway and Northumberland, which have experienced extensive or 'blanket' afforestation. Part of the explanation for the disappearance of Starlings from some of the more remote areas of Scotland may lie in the decline of the lowland British population during the 1980s (Trends Guide). O'Connor and Fuller (1985) have suggested that Starlings breeding at moderate altitudes in Scotland originate as 'overflow' from lowland populations at carrying capacity. However, when numbers in the preferred lowland habitats are depressed, the maintenance of a breeding presence on higher ground may not be possible.

The Abundance map points to interesting regional variations in Starling density. Numbers are generally greatest in central and SE England, with the Central Lowlands and Northern Isles of Scotland showing some of the highest densities in N Britain.

Starlings have two essential requirements for breeding: a hole or crevice in which to nest, and nearby grassland for foraging. In NE Scotland the distribution of breeding Starlings is related mainly to altitude, the availability of permanent grassland and soil quality (Buckland et al. 1990).

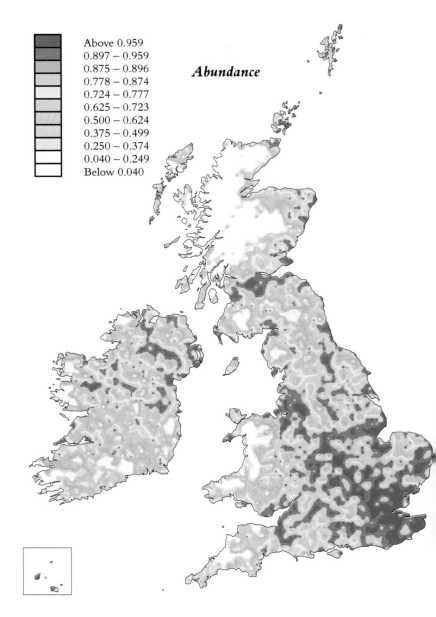

	Above 0.959
	0.897 − 0.959
	0.875 − 0.896
	0.778 − 0.874
	0.724 − 0.777
	0.625 − 0.723
	0.500 − 0.624
	0.375 − 0.499
	0.250 − 0.374
	0.040 − 0.249
	Below 0.040

Abundance

until April. First clutches usually consist of 4–5 eggs, and incubation by the female lasts approximately 12 days. The chicks fledge after 21 days but continue to be fed by both parents for a further 10–12 days. Although second broods are common, clutch size is generally smaller, and fledging weights are lower than in first broods (Feare 1984).

It is not unusual for male Starlings to pair simultaneously with more than one female. Such polygyny usually involves only a second female, but up to five females have been recorded paired to a single male. Brood parasitism, or the laying of eggs by one female in another female's nest, is also common amongst Starlings. In a study of two British nestbox colonies, Evans (1988) found that between 11% and 37% of first clutches were parasitized and that fledging rates in these were lower than those in unparasitized nests.

During 1988–91 there were an estimated 1,100,000 Starling territories in Britain and a further 360,000 in Ireland. However, these estimates should be considered minimal, as the CBC, upon which they are partly based, monitors urban habitats poorly.

WILL PEACH

68–72 Atlas p 408

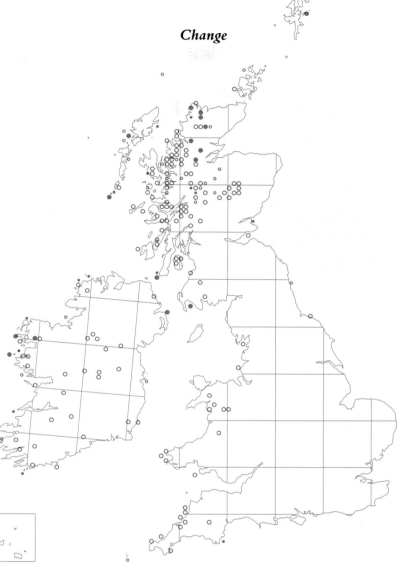

Change

Distribution

Years	Present, no breeding evidence		Breeding evidence		All records		
	Br	Ir	Br	Ir	Br	Ir	Both (+CI)
1968–72	32	17	2688	964	2720	981	3706
1988–91	122	79	2498	878	2620	957	3591
				% change	−3.6	−2.3	−3.3

House Sparrow

Passer domesticus

In Britain there is probably no more familiar bird than the House Sparrow, living as it does in close proximity to man. Even in rural situations it nests predominantly in holes in buildings. It is not, however, restricted to man-made structures for breeding, but uses a variety of other sites, including the nests of House Martins and the holes dug by Sand Martins. In the absence of suitable holes, it will build free-standing nests in thick hedges and conifers, and although the habit is not frequent in Britain, it becomes so further south.

The association with man means that the House Sparrow is widespread throughout Britain, being missing only from high ground in the Scottish Highlands. It is more sparsely distributed in Ireland, particularly in the centre and west, where it is absent from some inhabited areas.

House Sparrow numbers are not merely a function of the human population; the Abundance map shows a clear bias towards an easterly distribution in both Britain and Ireland. This is perhaps related to rainfall, for the sparrows as a genus show a preference for more arid environments.

The Change map shows that significant losses have occurred in N Scotland, the Scottish borders and Ireland, more particularly in the west. Summers-Smith (1988) indicated that there had also been some retraction of range (although at a finer scale than the 10-km square) in England, particularly from some of the more isolated breeding areas.

Prime habitat for the House Sparrow is found in a mosaic of buildings, giving favourable opportunities for nesting, and open green areas that provide a convenient source of the animal food required for rearing the nestlings. These combined requirements mean that the House Sparrow reaches high densities in some urban areas, particularly in central London with its parks. However, the optimum is provided by the more open

suburbs and town edges. Heij and Moeliker (1990), found in the Netherlands that neither their urban nor rural study populations were self-sustaining but depended on a top-up from the neighbouring suburban colonies.

Although there has been a steady increase in the most favoured habitat through the continued development of open suburbs, this has not necessarily resulted in a growth in House Sparrow numbers. Breeding density appears to have fallen in all habitats, though this is difficult to quantify and has not been confirmed by accurate census studies.

Three surveys have given breeding populations in the range 4,000,000–7,000,000 pairs – Summers-Smith (1959) using sample census data for different habitat types, the *68–72 Atlas* and the *Winter Atlas*. The *Trends Guide* suggests that there has been a decline in numbers since then, though this has to be treated with some caution, for the House Sparrow is only marginally covered by the major BTO enquiries. The conclusion is, however, supported by qualitative observations, which suggest a decrease at least in suburban areas. For example, the BTO Garden Bird Feeding Survey indicated a decrease of 15–20% in the numbers of House Sparrows visiting suburban gardens over the decade 1978–88, and there

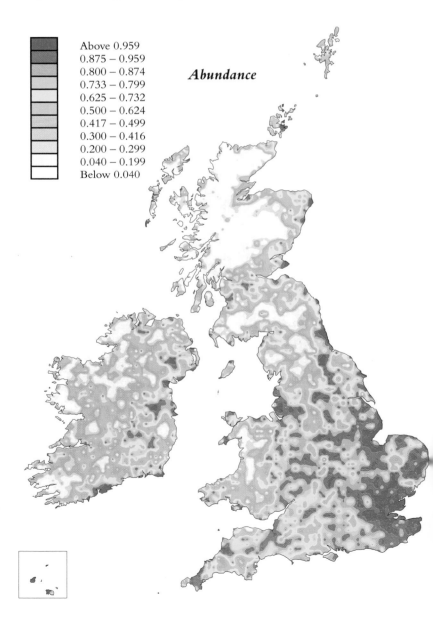

Above 0.959
0.875 – 0.959
0.800 – 0.874
0.733 – 0.799
0.625 – 0.732
0.500 – 0.624
0.417 – 0.499
0.300 – 0.416
0.200 – 0.299
0.040 – 0.199
Below 0.040

Abundance

has also been a noticeable decrease in recent years in the sizes of the post-breeding flocks at the ripening grain fields. Reports from Sweden, Denmark and the Netherlands suggest similar decreases over the same period.

One can only speculate on the possible reasons for the decline. Food is probably a major factor: the supply of invertebrate food required for rearing the young has been reduced by increased crop spraying and the use of garden pesticides; changed farming practices have led to a reduction in weed seeds, and much less grain is spilled in the stubble fields, through the change to autumn sowing of cereals. Increased predation may also have had some influence. Tawny Owls and domestic cats are both significant predators; for example, Churcher and Lawton (1987) found that cats in a Bedfordshire village were responsible for at least 30% of the mortality of the House Sparrow. If these two important predators were a significant factor in controlling House Sparrow populations, then the increase of the Sparrowhawk that took place in the late 1970s (*Trends Guide*), particularly its penetration into suburban areas, may have applied an additional pressure that was sufficient to reduce the House Sparrow population density to a lower level.

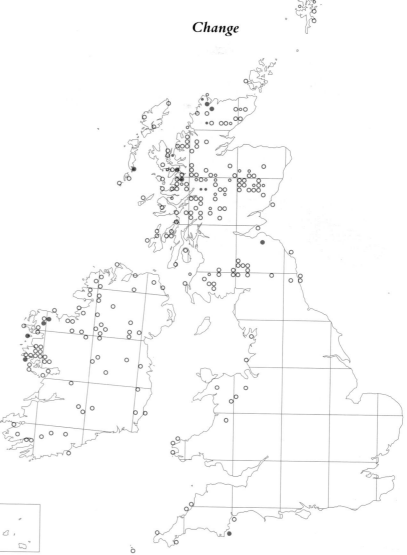

Change

Distribution

Allowing for an overall decline in numbers of 15% since the earlier population figures, there are estimated to have been 2,600,000–4,600,000 pairs of House Sparrow in Britain, and 800,000–1,400,000 in Ireland, annually, during 1988–91.

J. D. SUMMERS-SMITH

68–72 Atlas p 442

Years	Present, no breeding evidence		Breeding evidence		All records		
	Br	Ir	Br	Ir	Br	Ir	Both (+CI)
1968–72	36	5	2633	963	2669	968	3642
1988–91	94	88	2431	813	2525	901	3440
	% change				−5.3	−6.8	−5.7

Tree Sparrow

Passer montanus

Unlike its exceptionally well-known relative, the Tree Sparrow is an inconspicuous bird in Britain and Ireland and is frequently overlooked. As its common name implies, it tends to be associated with trees, but only where these occur in hedgerows or open woodland, not in large or dense forests. In Europe it nests mainly in tree holes, though it also uses holes in buildings where House Sparrows are absent, and very occasionally builds free-standing nests in thick hedges and conifers.

In Britain and Ireland it is most commonly found in farmland and at the edges of built-up areas, seldom penetrating beyond suburbs with large gardens. In Ireland it is mostly restricted to coastal regions. Despite its scientific name, the Tree Sparrow is very much a lowland species, with its stronghold in the central counties of England and adjoining E Wales. As shown in the Distribution map, it is scarce in the southern counties along the Channel coast from Sussex to the extreme southwest. In the north it is largely confined to the E coast, with outliers around the Solway Firth and the Forth-Clyde valley. The Abundance map highlights the preponderance of the Tree Sparrow in the lowlands and its avoidance of high country. This is by no means the case in other parts of its extensive range; for example, it breeds at up to 4,400m in Tibet.

One of the more puzzling features of the Tree Sparrow is the way its numbers fluctuate in an irregular cyclic manner. Summers-Smith (1989) has investigated the fluctuations in Britain and Ireland since 1860, using data from published regional records, supplemented more recently by BTO ringing, Nest Record and census data. This study suggested a high population from the 1880s to the 1930s, and again from 1960–1978, with a low about 1950 when, for a few years, the bird became extinct in Ireland. There has been a steady decline since 1978. These fluctuations in numbers are the result of both a decrease in density, and a retraction of range, particularly in the west and extreme south, as is shown very clearly on the Change map. Between the two Atlas periods the number of 10-km squares in which the species was recorded fell by 18%.

Similar fluctuations also occur in W Europe, but do not necessarily coincide with those in Britain. In Finland, for example, although the minimum in the 1950s (Merikallio 1958 estimated a mere 100 breeding pairs) and the growth in the 1970s (1,000 breeding pairs, Koskimies 1989) parallel the situation in England, there has been no corresponding fall in Finland, and the population has continued to increase through the 1980s.

A characteristic of the Tree Sparrow is that new colonies tend to form, build up rapidly and then disappear over a period of a few years, without any obvious reason. Although the Tree Sparrow is largely sedentary, considerable movements take place from time to time. These tend to be eruptive in character, rather than forming a regular pattern of migration, and movements from Europe appear to coincide with the population build-ups in England. In winter, Tree Sparrows feed largely on the seeds of small weeds. According to Pinowski and Pinowska (1985), winter survival of the Tree Sparrow in Poland is highly dependent on the duration of snow cover, when the seeds become unavailable. They considered the species to be essentially sedentary in Poland and attributed the fall in numbers to increased winter mortality. An alternative explanation could be that it is the result of emigration.

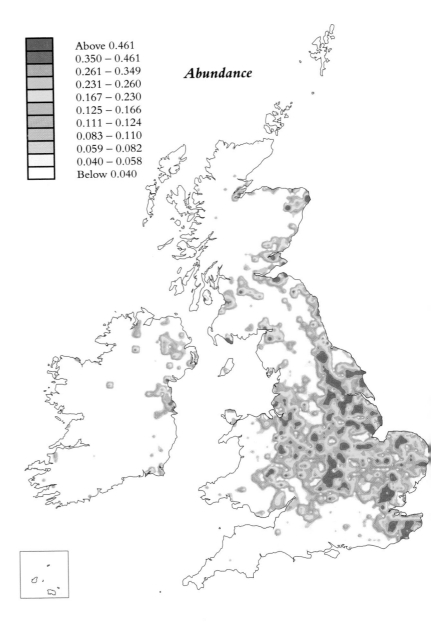

Above 0.461
0.350 – 0.461
0.261 – 0.349
0.231 – 0.260
0.167 – 0.230
0.125 – 0.166
0.111 – 0.124
0.083 – 0.110
0.059 – 0.082
0.040 – 0.058
Below 0.040

Abundance

The formation of new colonies may also depend on similar, though more local, irruptive movements. Such movements could similarly account for the recolonization of Ireland in the 1950s, the separate colonizations of the Faeroes in 1869, 1934 and 1960 (with intervening extinctions) and the spread to Corsica, Sardinia and Malta in this century.

The fluctuations in England are not obviously related to climatic changes. A major factor in the recent decline may have been the change in farming practice to autumn sowing, reducing the availability of weed seeds and scattered grain in stubble fields.

Using a variety of census and sample census data, Summers–Smith (1989) estimated the Tree Sparrow breeding population in Britain and Ireland at a maximum of about 900,000 pairs in 1965, falling to 130,000–140,000 pairs in 1989. From the *88–91 Atlas* data, however, the British population was estimated at 110,000 territories. By extrapolation, there would be a further 9,000 Tree Sparrow territories in Ireland.

J. D. SUMMERS-SMITH

68–72 Atlas p 444

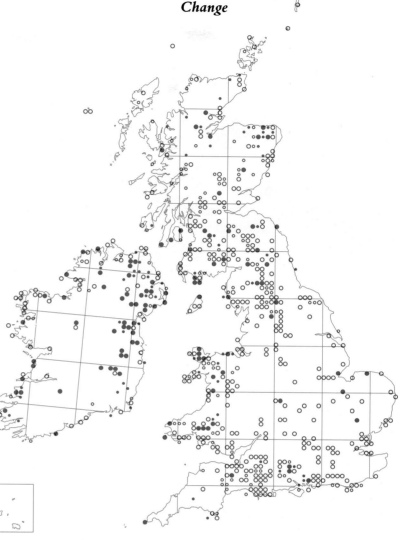

Change

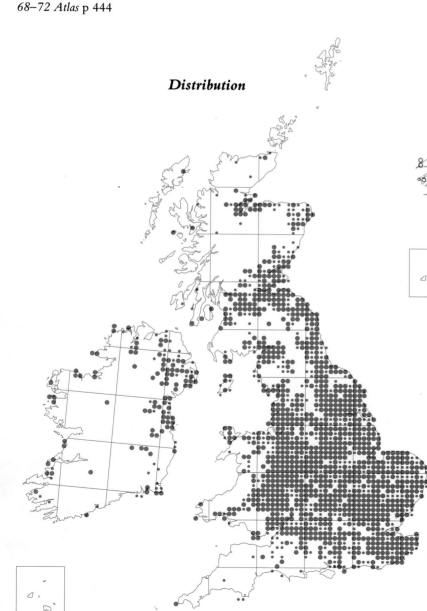

Distribution

Years	Present, no breeding evidence		Breeding evidence		All records		
	Br	Ir	Br	Ir	Br	Ir	Both (+CI)
1968–72	142	18	1533	104	1675	122	1797
1988–91	306	38	1040	92	1346	130	1476
				% change	− 19.6	6.6	− 17.8

Chaffinch

Fringilla coelebs

The Chaffinch is one of the most abundant and widespread birds in Britain, and its cheerful 'spink' call and vigorous song are amongst the most familiar sounds of the countryside. The species breeds wherever there are trees or bushes: in forest and woodland, scattered scrub, hedgerows, parks and gardens, from rural to urban environments. Over the country as a whole its abundance and distribution thus reflect the availability of woody habitats, and the lack of them account for its virtual absence in Shetland, much of the Outer Hebrides and a few other areas.

At the scale of the 10-km square, little change has occurred in the distribution of the Chaffinch since the *68–72 Atlas* survey. Locally its numbers are likely to have increased, however, as planted conifer forests have matured, providing much new nesting habitat, especially in the uplands. The Chaffinch is territorial in the breeding season, but the sizes of territories vary greatly in different types of habitat (Newton 1972). In general, the highest densities occur in rich broadleaved woods, at 49–145 (locally up to 300) pairs per km², and somewhat lower densities in coniferous woods, at 12–102 pairs per km². In most of the woods that have been censused, Chaffinches comprised between one-fifth and two-fifths of the local bird population. In other habitats, such as hedgerows and gardens, densities are generally lower, and depend on the number of trees and large shrubs available. In farmland CBC plots, the Chaffinch averages about 30 pairs per km², and usually forms about one-twelfth of the total bird population.

It is not only its use of a wide range of habitats which makes the Chaffinch one of the commonest birds in Britain, but also its varied diet. For most of the year it feeds mainly on seeds picked from the ground, and

more than 100 types have been recorded in the diet. In woodland it eats seeds of many trees, shrubs and herbaceous plants, and on farmland the seeds of various crop-plants, including cereals and rape, and many common weeds, notably brassicas (*Sinapis*, *Brassica*), persicaria, goosefoots and chickweeds (*Stellaria*, *Cerastium*) (Newton 1967, 1972).

In the breeding season the Chaffinch turns from seed eater to insectivore. It feeds its young on small caterpillars and flies, gleaned from the foliage of trees, and on small beetles and other insects picked from the ground. Pairs forage partly in their territories, and partly outside, on open ground and other undefended areas. Egg laying extends from late April to early June (but mainly in early May), and most clutches contain 3–5 eggs (Newton 1964). The majority of pairs raise no more than one brood in a year, but in the south some manage to raise two. The species is resident in Britain and ringing recoveries indicate that nine-tenths of individuals move no more than 5 km from their birthplace, and the rest no more than 50 km. Breeding individuals tend to return to the same nesting territories year after year. Nonetheless, some of the forests on higher ground are largely abandoned for the winter. At this season, resident Chaffinches occur in small flocks, and feed mostly on farmland and other open ground,

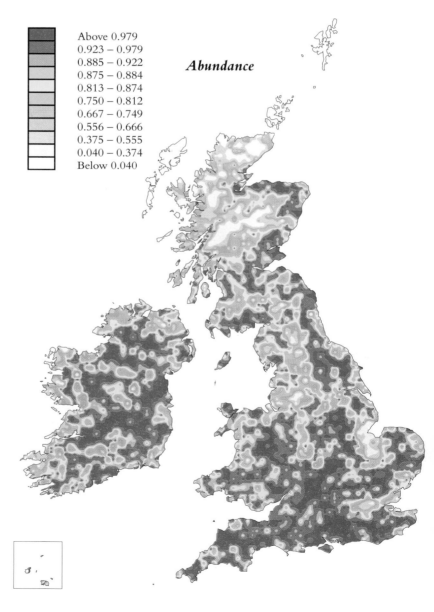

| Above 0.979 |
| 0.923 – 0.979 |
| 0.885 – 0.922 |
| 0.875 – 0.884 |
| 0.813 – 0.874 |
| 0.750 – 0.812 |
| 0.667 – 0.749 |
| 0.556 – 0.666 |
| 0.375 – 0.555 |
| 0.040 – 0.374 |
| Below 0.040 |

Abundance

wherever suitable seeds are available (Newton 1972). Their numbers are augmented each autumn by a large influx of wintering Continental birds, which tend to occur in larger flocks and in more open areas than the residents (Marler 1956).

During 1988–91 there were an estimated 5,400,000 Chaffinch territories in Britain, and a further 2,100,000 in Ireland.

IAN NEWTON

68–72 Atlas p 430

Change

Distribution

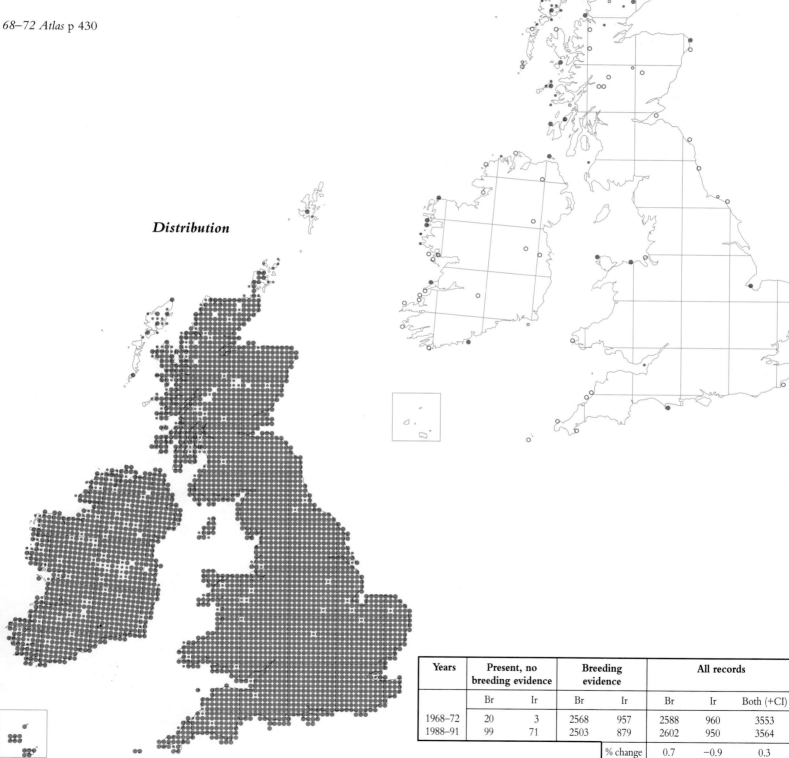

Years	Present, no breeding evidence		Breeding evidence		All records		
	Br	Ir	Br	Ir	Br	Ir	Both (+CI)
1968–72	20	3	2568	957	2588	960	3553
1988–91	99	71	2503	879	2602	950	3564
				% change	0.7	−0.9	0.3

Greenfinch

Carduelis chloris

The Greenfinch is one of the most abundant and widespread of the cardueline finches, familiar to everyone because of its close association with parks and gardens. Its heavy conical bill enables it to take a wide range of seeds, which it obtains directly from trees, shrubs and herbaceous plants, or from the ground (Newton 1967).

It breeds over most of Britain and Ireland, and is absent only from high, largely treeless areas in central Wales and NW Scotland, and from Shetland. It is mostly associated with farmland, villages and towns, so is seldom far from human habitation. It reaches its greatest densities in the S and E of Britain and Ireland. Its distribution has not altered significantly since the *68–72 Atlas* survey, but earlier in the century some expansion was recorded, when the species reached the Isles of Scilly, some W Scottish islands and elsewhere.

Over the past 40 years, however, as food supplies have changed the Greenfinch has become much less abundant on farmland, and more numerous in towns and villages. On farmland, various favoured weed seeds, including charlock and persicaria, have become much less plentiful than formerly because of the increased use of herbicides. This has led to massive declines in the food plants themselves, and to progressive depletion of the seed-bank in the soil, which Greenfinches could exploit in the past whenever tillage brought seeds to the surface.

Over the same period, spilt cereal grains have become less available because of the change from spring ploughing to autumn ploughing of stubble fields, and the replacement of the old binder and threshing machines by less wasteful combine harvesters. Over much of the country the large flocks of Greenfinches, which were once so common on farm-land, are now seldom seen. In towns and villages, on the other hand, more food has become available because of the increased planting of seed-

bearing trees and shrubs, and the deliberate hand-outs of food by house-holders. The Greenfinch is one of the most frequent patronisers of feeding trays when peanuts or sunflower seeds are provided. Indeed, its present numbers probably depend partly on such food sources. Such feeding is likely to be especially important in late winter and early spring, when natural seed supplies reach their lowest level.

Parks and gardens, with their abundance of tall but well-spaced evergreen shrubs, also provide ideal nesting habitat. Like other cardueline finches, Greenfinches tend to nest in loose colonies, but range over large areas to obtain their food, with flocks assembling temporarily wherever suitable seeds are locally available (Newton 1972). Birds may fly more than a km from the nest for food. The breeding season is prolonged. In the south, egg laying extends from late April into August, so that the last young fledge in September. This would give enough time for pairs to raise up to four broods per season, but probably few (if any) breed for the full season, and the output of many pairs is reduced by heavy predation on the eggs and chicks (Monk 1954, Newton 1972). Clutches contain 3–6 eggs and the young are raised mainly on seeds, supplemented with caterpillars and other insects.

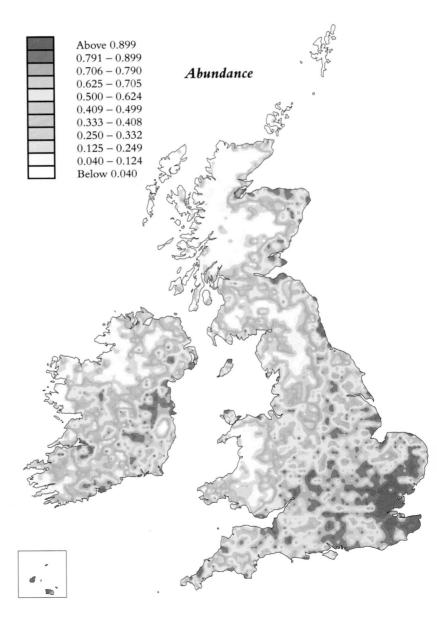

	Above 0.899
	0.791 – 0.899
	0.706 – 0.790
	0.625 – 0.705
	0.500 – 0.624
	0.409 – 0.499
	0.333 – 0.408
	0.250 – 0.332
	0.125 – 0.249
	0.040 – 0.124
	Below 0.040

Abundance

Most Greenfinches remain in Britain throughout the year, but a small proportion leaves the southeast in autumn, to winter on the Continent, while others move southwest into Ireland. At the same time, other Greenfinches arrive in Britain for the winter from N and central Europe (*Winter Atlas*). BTO census data indicate that breeding numbers have remained relatively stable since about the mid 1960s (*Trends Guide*). In the early 1960s, numbers were reduced, probably as a result of widespread poisoning by dieldrin and other organochlorine pesticides, which were used at that time to treat spring sown cereal grains.

During 1988–91 there were an estimated 530,000 Greenfinch territories in Britain, and 160,000 in Ireland. However, there are major difficulties in arriving at these estimates. Not only does a large proportion of pairs nest in private gardens and other sites inaccessible to bird-counters, but the birds also move around a great deal. Individuals may visit several widely separated foraging sites each day, during which time they are absent from their nesting areas. These problems mean that, although Greenfinches are probably the commonest of the cardueline finches, it is difficult to estimate their numbers with any accuracy.

IAN NEWTON

68–72 Atlas p 412

Change

Distribution

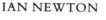

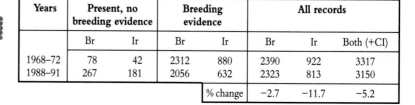

Years	Present, no breeding evidence		Breeding evidence		All records		
	Br	Ir	Br	Ir	Br	Ir	Both (+CI)
1968–72	78	42	2312	880	2390	922	3317
1988–91	267	181	2056	632	2323	813	3150
				% change	−2.7	−11.7	−5.2

Goldfinch

Carduelis carduelis

This colourful and delightful finch, once so popular as a cage bird, is most often seen flitting over the seed heads of thistles and other favoured food plants.

The Goldfinch is now widespread in Britain, and is absent only from open mountains and moorland, especially in NW Scotland and the Northern and Western Isles. For several decades the species has been spreading northward, accounting for the expansion of range that has occurred since the *68–72 Atlas*. In particular, the species now breeds commonly around the Moray Firth in NE Scotland where, in 1968–72, it was almost absent. In contrast, there seems to have been some decline in Ireland, judging by the number of localities abandoned since the *68–72 Atlas*.

Throughout its range, the Goldfinch is commonest on low ground, where its food-plants are most numerous. It specialises on seeds from plants in the family Compositae, such as dandelions, groundsels, ragworts, thistles and knapweeds, and with its long pointed bill, is adept at opening the seed heads. In winter it also takes the seeds of alder, birch and teasel and, in spring, seeds from opening conifer cones (Newton 1967, 1972).

Although Goldfinches may be seen in Britain throughout the year, more than 80% of the population moves south in September and October, to winter in Belgium, France and Spain, returning in April and May (Newton 1972). In much of its Mediterranean wintering range, unlike in Britain, food plants grow throughout the year so that seed supplies are continually replenished. Thus the number of Goldfinches breeding in Britain is probably more dependent on conditions in the Continental wintering areas than on those in Britain and Ireland. This was probably not always so, however, for until the 1930s large scale commercial bird-catching in Britain is thought to have held the numbers of Goldfinches well below the level that could otherwise have occurred. In 1860 it was

alleged that 132,000 were being caught each year near Worthing in Sussex, which lies on the main migration route, and a record exists of twelve dozen being caught in a single morning on the site now occupied by Paddington Station. Such large-scale trapping was so widespread that the newly formed Society for the Protection of Birds (now the RSPB) made 'saving the Goldfinch' one of its first tasks. Legislation against the sale of wild Goldfinches came in 1933, and further legislation against trapping for personal use came in 1954. Although some small-scale trapping has continued illegally to the present day, it seems likely that at least part of the increase and spread of the species observed in recent decades represents a continuing recovery from the impact of earlier exploitation.

Typically, Goldfinches nest in areas of scattered trees and shrubs, including parks and gardens, close to open land where food plants grow. Like other cardueline finches, they tend to nest in loose colonies, but range over wide areas to obtain their food, wherever seeds are available (Newton 1972). The nest is usually placed high on a swaying bough, 4–10 metres above the ground, and has a deep cup to retain the contents in windy weather. Egg laying extends from late April to August, so that the last young fledge in September. Most breeding activity seems to occur in the

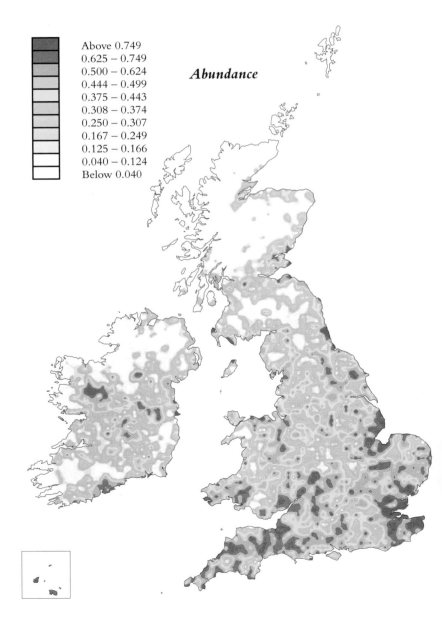

	Abundance
	Above 0.749
	0.625 – 0.749
	0.500 – 0.624
	0.444 – 0.499
	0.375 – 0.443
	0.308 – 0.374
	0.250 – 0.307
	0.167 – 0.249
	0.125 – 0.166
	0.040 – 0.124
	Below 0.040

latter half of this period, when suitable seeds are most plentiful. The eggs number 4–6, and the young are fed mainly on regurgitated seeds, supplemented with some small caterpillars and other insects.

During 1988–91 there were an estimated 220,000 Goldfinch territories in Britain and a further 55,000 in Ireland. However, as pairs are present on their nesting areas for only a small part of each day (except for the female during incubation), and range widely to obtain their food, and as a large part of the population nests in private gardens and other sites not readily accessible for census work, these estimates must be treated with caution.

IAN NEWTON

68–72 Atlas p 414

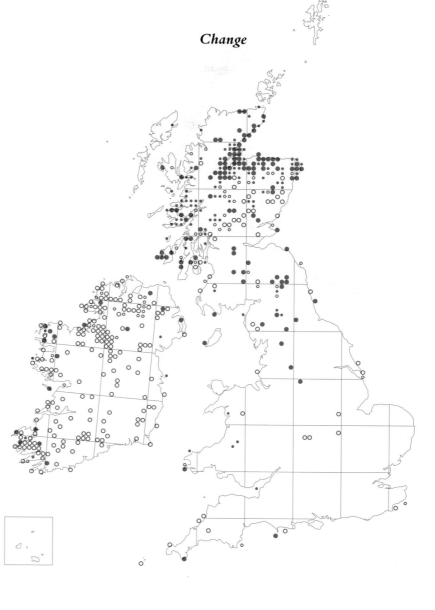

Change

Distribution

Years	Present, no breeding evidence		Breeding evidence		All records		
	Br	Ir	Br	Ir	Br	Ir	Both (+CI)
1968–72	174	42	1925	871	2099	913	3017
1988–91	321	271	1888	478	2209	749	2972
			% change		5.4	−17.9	−1.7

of birds from the northern coniferous forests of Scandinavia to Britain is well recorded. This may have been the source of some of the increased breeding population as birds arriving in Britain found a suitable food source in the maturing plantations.

Shaw (1990) has shown a clear effect of cone production levels in sitka spruce plantation on the timing of breeding in SW Scotland. In poor seed years breeding is later and productivity is consequently reduced because of the limited time for successive breeding attempts.

During the *68–72 Atlas* years there were an estimated 20,000–40,000 pairs in Britain and Ireland. Given the changes in forest age-class structure since 1971, this figure requires revision. Breeding season censuses of Siskins (various sources) suggest population densities of up to one pair per ha, depending on fruiting levels of the crop. Using a conservative estimate of one pair per 5 ha throughout Britain's 1,500,000 ha of coniferous woodland, it is likely that the annual British population during 1988–91 was of the order of 300,000 pairs. By extrapolation, the population in Ireland would have been about 60,000 pairs.

DAVID C. JARDINE

68–72 Atlas p 416

Siskin

Carduelis spinus

At the time of the *68–72 Atlas* the Siskin was a relatively unfamiliar bird to many birdwatchers. It is now a common feeder on peanuts in winter and early spring. The spread of this behaviour, first recorded in 1963, along with other factors has established this colourful finch as a regular member of the birdtable community. It does not, however, breed in many of the gardens of S England, where it was common in the *Winter Atlas* years, but is mainly a bird of the coniferous (spruce) forests of N and W Britain and Ireland. Its rapid spring migration from the gardens to the breeding grounds has been well recorded by ringers during the last decade.

Siskins are principally eaters of tree seeds, and, as most coniferous trees start fruiting once they reach the thicket stage, the population level of this species has been influenced by the age-class structure of the forests present in Britain and Ireland. At the time of the *68–72 Atlas* there were around a third of a million ha of coniferous forest in Britain (excluding Ireland) which were over 25 years old (Forestry Commission Census data). By 1991 the area of this type of habitat had doubled, reflecting the maturing of post-war plantings (see Figure). It is therefore not surprising that the total number of records for the Siskin in the *88–91 Atlas* is significantly greater than that recorded in the *68–72 Atlas*.

In S England, breeding has increased in the New Forest and Dorset and is now recorded in N Devon. Twenty years ago the Siskin was only a localised breeder in Wales, but is now widespread. A similar change from localised to common breeder has also occurred in the Peak and Lake Districts, and in the Pennines. In Scotland there are few areas where breeding is not now recorded. The Siskin is absent only from areas of unsuitable habitat: the treeless Northern and Western Isles, the high mountain tops, the intensive arable areas of the E coast and the industrial central lowlands. In Ireland, where the colonisation has not been as complete as in Scotland, it remains a widespread but localised breeder, and it is now breeding on the Isle of Man.

The production of seed by coniferous trees varies from year to year and this can have profound effects on a Siskin population. Eruptive movement

Above 0.555
0.375 – 0.555
0.273 – 0.374
0.250 – 0.272
0.177 – 0.249
0.155 – 0.176
0.125 – 0.154
0.097 – 0.124
0.063 – 0.096
0.040 – 0.062
Below 0.040

Abundance

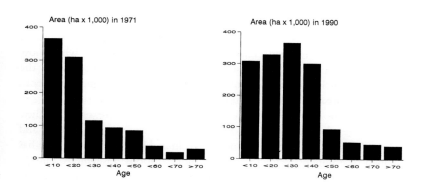

Area (ha x 1,000) in 1971

Area (ha x 1,000) in 1990

Estimates of the age-class distribution of British coniferous forest for 1971 and 1990 (from Forestry Commission data). Note that <10 = less than 10 yrs of age, <20 = 10–19 yrs, <30 = 20–29 yrs etc.

Change

Distribution

Years	Present, no breeding evidence		Breeding evidence		All records		
	Br	Ir	Br	Ir	Br	Ir	Both (+CI)
1968–72	147	67	478	171	625	238	863
1988–91	430	141	728	143	1158	284	1442
				% change	85.3	19.8	67.3

These explanations cannot, however, account fully for the changes apparent in Ireland where, although improvement to pasture and intensification of agriculture have occurred, they are less pronounced than in England. Conifer planting of rough habitats suitable for nesting has affected Linnets at inland sites, once the trees have grown past the thicket stage. Afforestation and improvements to rough grazing are implicated by Thom (*Birds in Scotland*) in the decrease in Scottish breeding Linnets, which now occur at highest breeding densities only in coastal habitats. Indeed, the Linnet is principally a bird of the lowlands, hence its continued scarcity in the upland areas of Scotland and Wales.

Linnets are more dependent on weed seeds than are other finches. They prefer the seeds of *Cruciferae, Chenopodiaciae* - notably fat hen, and *Caryophyllaceae* – particularly chickweed. In arable areas, oilseed rape has helped to compensate for losses to traditional foods. Elsewhere, Linnets have moved into suburban areas, where garden 'weed' seeds are available. This colonisation of formerly little-used habitats has also been documented in Finland (Tast 1968).

Linnets are early nesters, starting in mid April, and are multiple brooded. The species is semi-colonial, nesting in loose colonies in which each pair

Linnet

Carduelis cannabina

Linnets frequent areas of gorse heath, shrubby thickets and hedges, favouring the dense cover which these afford. The bright pink breast and crown of the male in summer contrast well with the yellow flowers of a gorse song perch.

The Distribution map shows the Linnet to be a widespread breeding bird in England, Wales, S Scotland and the Isle of Man, but with a rather restricted often coastal distribution in Ireland and NE Scotland. It is largely absent from most of the central and W Highlands, Hebrides and Shetland.

The Abundance map illustrates the predominance of Linnets in E and S England and around the coastline, the species being uncommon elsewhere. Linnets are only locally common in lowland S, central and NE Scotland (*Birds in Scotland*) and are scarce or local in the Highlands and Inner Hebrides.

The Change map indicates little change in distribution in Scotland since the *68–72 Atlas*, except for losses in S W and central areas and for its demise as a breeding bird in Skye and the Outer Hebrides. In contrast, the Linnet's status as a breeding bird in Ireland has declined substantially. Overall, Linnets are less common than at the time of the *68–72 Atlas*.

These inferences from the Change map are borne out by the CBC which has shown a steep decline in the breeding indices for this species (*Trends Guide*). At the time of the *68–72 Atlas*, Linnets were already in decline in some areas, having earlier recovered after the cessation in legal trade in caged birds. Intensification of agriculture, and the associated application of herbicides, has resulted in arable weeds becoming scarce. The seeds of these plants are particularly important foods for Linnets at all times of year. O'Connor and Shrubb (1986a, b) found that an increase in post-hatching mortality was attributable to starvation, resulting from shortages in seed foods. The removal of scrub hedges has also contributed to their local decline as breeding birds, for these are favoured nest sites.

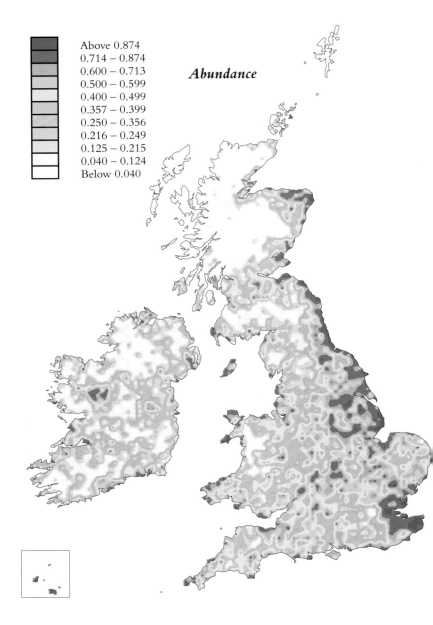

	Above 0.874
	0.714 – 0.874
	0.600 – 0.713
	0.500 – 0.599
	0.400 – 0.499
	0.357 – 0.399
	0.250 – 0.356
	0.216 – 0.249
	0.125 – 0.215
	0.040 – 0.124
	Below 0.040

Abundance

defends a small territory around the nest, but birds forage in flocks away from the colony (Newton 1972). The food, collected during foraging trips, is regurgitated for the nestlings. This strategy has presumably evolved as a result of the dispersion of their preferred food plants and the seasonally changing availability of different seed crops.

Britain is on the northern edge of the Linnets' wintering range, but because they are partial migrants, with many wintering south to Iberia, severe winters have had little impact on the breeding population level in Britain. It is, however, noticeable that the *Winter Atlas* distribution is very similar to that of the Abundance map, which suggests that areas where Linnets are most commonly found in summer are also the most suitable wintering sites.

During 1988–91 there were an estimated 520,000 Linnet territories in Britain and 130,000 in Ireland.

ROWENA LANGSTON

68–72 Atlas p 418

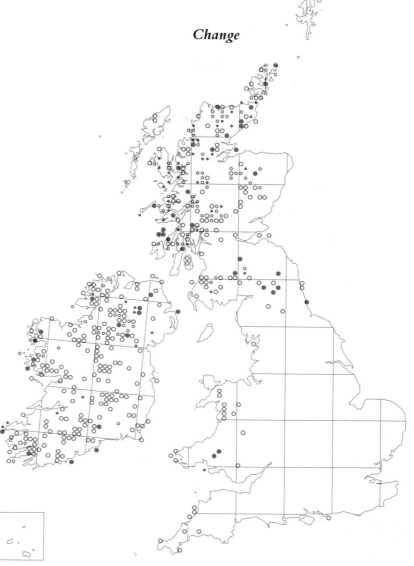

Change

Distribution

Years	Present, no breeding evidence		Breeding evidence		All records		
	Br	Ir	Br	Ir	Br	Ir	Both (+CI)
1968–72	136	58	2247	897	2383	955	3343
1988–91	266	235	2002	548	2268	783	3065
				% change	−4.6	−17.9	−8.4

Twite

Carduelis flavirostris

The Twite is a fascinating little finch whose disjunct world distribution has two centres separated by over 2,500km (Newton 1972). The two regions of occurrence are the uplands of central and SW Asia and the Atlantic coastlines and mountains of NW Europe. This curious distribution may well be attributed to the Twite's response to habitat changes which followed the last major glaciation 18,000–20,000 years ago.

In Britain and Ireland the species is associated with treeless areas, mainly the moorlands of the Highlands and N England and the crofting lands of the N and W seaboards of Scotland and Ireland. It is less widespread as a breeding species in the Southern Uplands and in the Welsh uplands around Snowdonia. Breeding does not occur on the North Yorkshire moors or the moors of SW England.

The Twite is most abundant in the coastal areas of low intensity agriculture and the grouse moors of the S Pennines. Average density on the N Staffordshire moors has been estimated at 7.8 pairs per tetrad (Davies 1988).

The Welsh population shown on the Distribution map was not recorded as breeding in the *68–72 Atlas*, and represents a significant extension in range. The Highland population has also expanded southeastwards. The apparent reduction in range of Twite in Donegal, Mayo, Kerry and other parts of W Ireland during 1988–91 could be due partly to differences in coverage between the two Atlases. It is probable, however, that this reduction in range is real, because overall coverage in Ireland in the *88–91 Atlas* is estimated at slightly less than 90% of that in the *68–72 Atlas*, whereas records of Twite have declined by over 50%. In Britain coverage was similar and yet there has been only a very small decline in the number of records. Hutchinson (*Birds in Ireland*) also describes a long-term decline of the Twite in Ireland during the last 80 years, but does not indicate any

recent trends in breeding. He reports a decline during the last 35 years in the population wintering on the E coast of Ireland but it is not clear whether this is a result of a decline in the breeding population, or whether there has been a change in the wintering range as a result of habitat changes or other factors.

The reasons for a widespread decline of Twite populations in Scotland this century are also unknown (*Birds in Scotland*), but it has been so wide ranging that degradation of moorland habitats through overgrazing and agricultural reclamation may be primary causes. Certainly the latter has been responsible for losses in N England in the last hundred years or so (Orford 1973).

Scottish and Irish Twites appear to be mainly (but not wholly) sedentary (Davies 1988). Those breeding on the English moors, however, move to the coast for the winter. On their return, the S Pennine flocks rely heavily on patches of burnt purple moor grass where they exploit its fallen seeds before dispersing to heather-dominated parts of the moors (Orford 1973). Here the heather and bracken offer the birds safe and concealed nesting places (Haworth and Thompson 1990). Throughout the breeding season they travel to and from agricultural pastures where they feed on weed seeds.

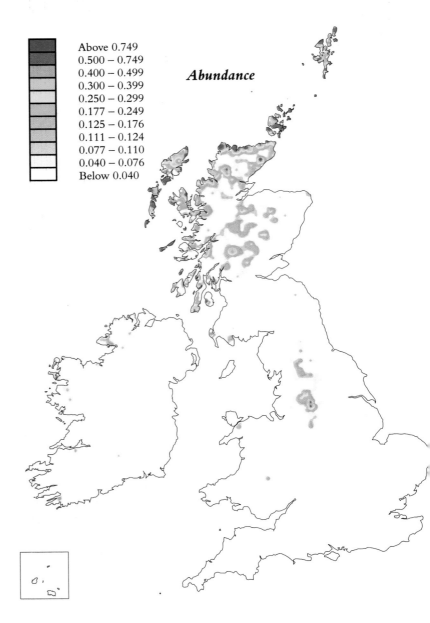

Abundance

	Above 0.749
	0.500 – 0.749
	0.400 – 0.499
	0.300 – 0.399
	0.250 – 0.299
	0.177 – 0.249
	0.125 – 0.176
	0.111 – 0.124
	0.077 – 0.110
	0.040 – 0.076
	Below 0.040

Despite being of international importance little research has been carried out on the subspecies of Twite that occurs in Britain and Ireland (*C.f. pipilans*). *Red Data Birds* advises the safeguarding of heather moorland and crofting land breeding sites but a more thorough investigation of the species' ecology should be a priority for conservation bodies, given its limited distribution and apparent evidence of recent decline, notably in Ireland.

Reported breeding densities in Britain vary between a mean of 2 pairs per 100 ha (Davies 1988) and 9 pairs per 100 ha (B. Campbell from Nest Record Cards). The lower of these figures is four times the maximum estimated density of 50 pairs per occupied 10-km square given for Britain and Ireland in the *68–72 Atlas*. Assuming, perhaps conservatively for inland squares but generously for coastal squares, an average of 100 pairs per occupied 10-km square, the total breeding population in Britain would be 65,000 pairs. By extrapolation, the Irish population would be 3,500 pairs.

DAVID C. JARDINE and J. B. REID

68–72 Atlas p 420

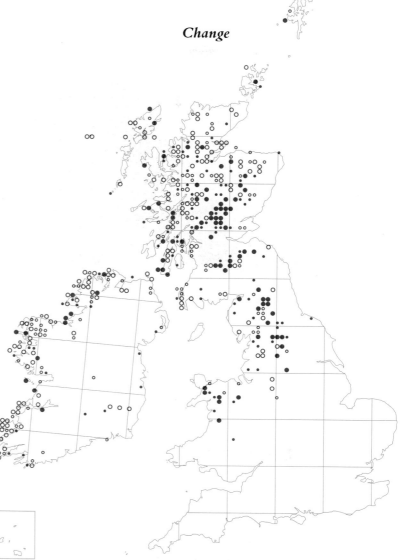

Change

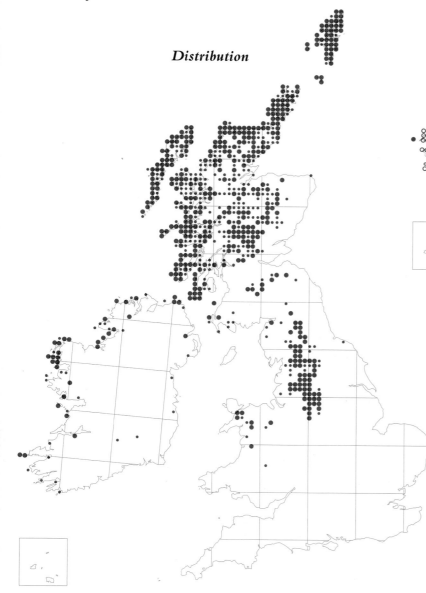

Distribution

Years	Present, no breeding evidence		Breeding evidence		All records		
	Br	Ir	Br	Ir	Br	Ir	Both (+CI)
1968–72	154	42	503	85	657	127	784
1988–91	231	28	420	32	651	60	711
			% change		−1.1	−52.8	−9.5

Redpoll

Carduelis flammea

The distinctive rattling flight call of the Redpoll is usually the first clue to the presence of this woodland finch.

It is a bird of pioneer woodlands – favouring those tree species with small seeds, especially birch. It is also common in the establishment and pre-thicket stages of coniferous plantations (both new planting and restocking) where birch often develops.

As a result, the only places in Britain and Ireland where it was not regularly found breeding were the Northern and Western Isles in Scotland, the Cheshire plain, much of the counties of Hereford, Worcester, Gloucester and Oxford, and an area south of a line from the Severn to Beachy Head.

Sharrock predicted that the striking gap in central S England would be filled quickly following the *68–72 Atlas*. This does not appear to have happened, for changes in distribution have been only localised, and probably reflect changes in the woodland habitat. Contraction of range is apparent around the Vale of York, in the lowlands of Grampian Region, on the Lleyn peninsula and in S and E Ireland. Conversely there has been some expansion on the Island of Lewis, in Dyfed and perhaps more surprisingly, because of the reduction in heathland habitat, in the New Forest and in Dorset. It is now well established on the Isle of Man.

During the last 20 years there has been a slight decline in the area of young conifer forest in Britain, but perhaps more significantly the proportion of birch found in the woodlands of lowland England has declined as a result of woodland succession (Locke 1987). Birch developed rapidly in most woodlands following the wartime fellings of 1939–45 and probably reached its peak in the 1960s and 1970s. Since then other tree species which were slow initial developers have in many places grown enough to compete successfully with birch thereby reducing its seed production. Intensification of agricultural practices during the last 20 years has also

resulted in losses of hedgerow and farmland trees. Taken together these habitat changes have probably been the cause of the marked decline in the number of occupied 10-km squares since the *68–72 Atlas*. The CBC index (*Trends Guide*) shows that the increase in abundance noted during the years of the *68–72 Atlas* peaked in the mid 1970s and has shown a steady decline ever since. Few CBC plots are in coniferous forest habitat because of the well-known successional effects of crop age on bird populations. Thus there are few suitable data on population changes to provide guidance on population estimates.

Breeding densities in various studies have ranged from 0.05 to 0.23 pairs per ha, although some authors, e.g. Thom (*Birds in Scotland*), have suggested that densities up to twice these levels may occur. However, such high densities will be restricted to areas of suitable woodland, and the overall density of 100 pairs per occupied 10-km square used in the *68–72 Atlas* is probably still reasonable. If this is the case, then during 1988–91 there are estimated to have been an annual 160,000 pairs of Redpoll in Britain and 70,000 in Ireland.

DAVID C. JARDINE

68–72 Atlas p 422

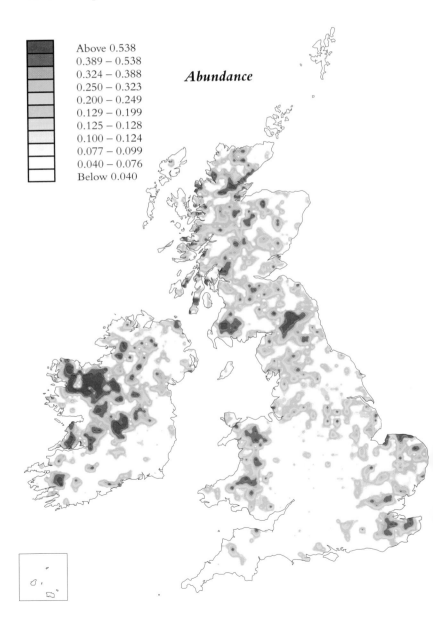

Above 0.538
0.389 – 0.538
0.324 – 0.388
0.250 – 0.323
0.200 – 0.249
0.129 – 0.199
0.125 – 0.128
0.100 – 0.124
0.077 – 0.099
0.040 – 0.076
Below 0.040

Abundance

Change

Distribution

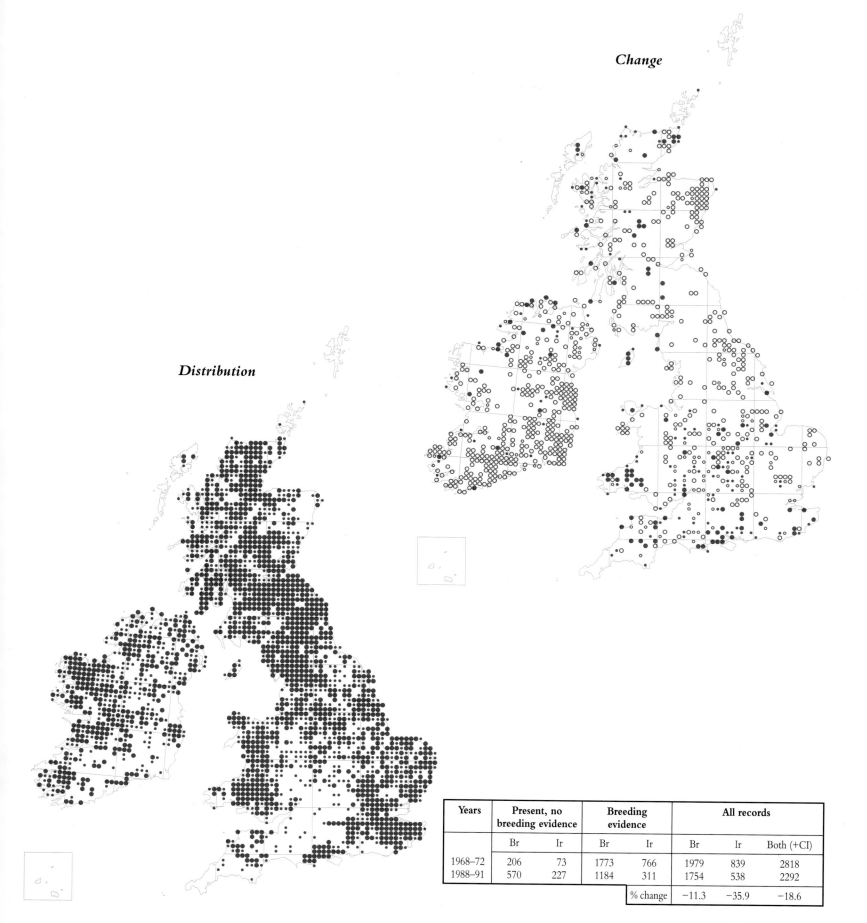

Years	Present, no breeding evidence		Breeding evidence		All records		
	Br	Ir	Br	Ir	Br	Ir	Both (+CI)
1968–72	206	73	1773	766	1979	839	2818
1988–91	570	227	1184	311	1754	538	2292
	% change				−11.3	−35.9	−18.6

Common Crossbill

Common Crossbill and Scottish Crossbill

Loxia curvirostra and *L. scotica*

Often elusive and unpredictable, crossbills are always exciting birds to watch, particularly when they are feeding. With their curved bills and short, stout legs, they look like small parrots, the similarity being emphasised by the way a bird will sidle along a branch and move from twig to twig using its beak. Two species breed regularly in Britain. The endemic Scottish Crossbill is confined to N Scotland. The Common Crossbill may be found throughout Britain and Ireland. Because of the difficulty in separating the species in the field (Knox 1990a), they are treated together here. Parrot Crossbills nest occasionally in Britain (see page 446).

Crossbills live only in conifer woods. They eat conifer seeds and little else, and their dependence on this habitat and an adequate cone crop is almost complete. But the trees only seed well when they reach a certain stage of maturity, and cone crops vary annually. Crossbill numbers change accordingly, as the trees grow, mature and are harvested, and in line with annual coning variations. When cone crops fail over a particular area, the birds leave in search of food elsewhere. Often they travel only short distances but, every few years, large numbers of birds erupt over thousands of km. Crossbills arriving in Britain from the Continent on such occasions may stay with us for only one season, or settle to establish new populations, which survive for many years, the numbers often being supplemented by later irruptions.

During the present survey, the largest numbers of crossbills were seen in Scotland. In the Highlands, birds were widespread along and near the Great Glen, in Strathspey, Deeside and Perthshire. Many will have been in the last remaining fragments of the ancient Caledonian pine forest, or in old planted woods dating from the middle of the last century or earlier. These are the favoured haunts of the Scottish Crossbill. Other Highland crossbills will have been in the younger plantations of pine, larch and spruce, where the Common Crossbill is more frequent. Both species occur in the Highlands in varying numbers, sometimes nesting in the same woods (Knox 1990b).

The Mull of Kintyre is not a traditional crossbill area, but the many

(continues on p 424)

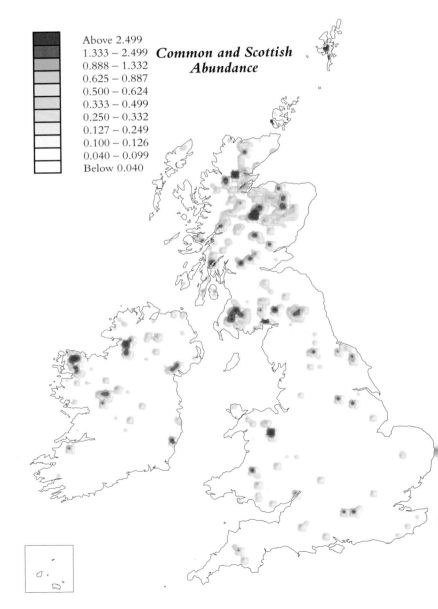

Common and Scottish Abundance

Above 2.499
1.333 – 2.499
0.888 – 1.332
0.625 – 0.887
0.500 – 0.624
0.333 – 0.499
0.250 – 0.332
0.127 – 0.249
0.100 – 0.126
0.040 – 0.099
Below 0.040

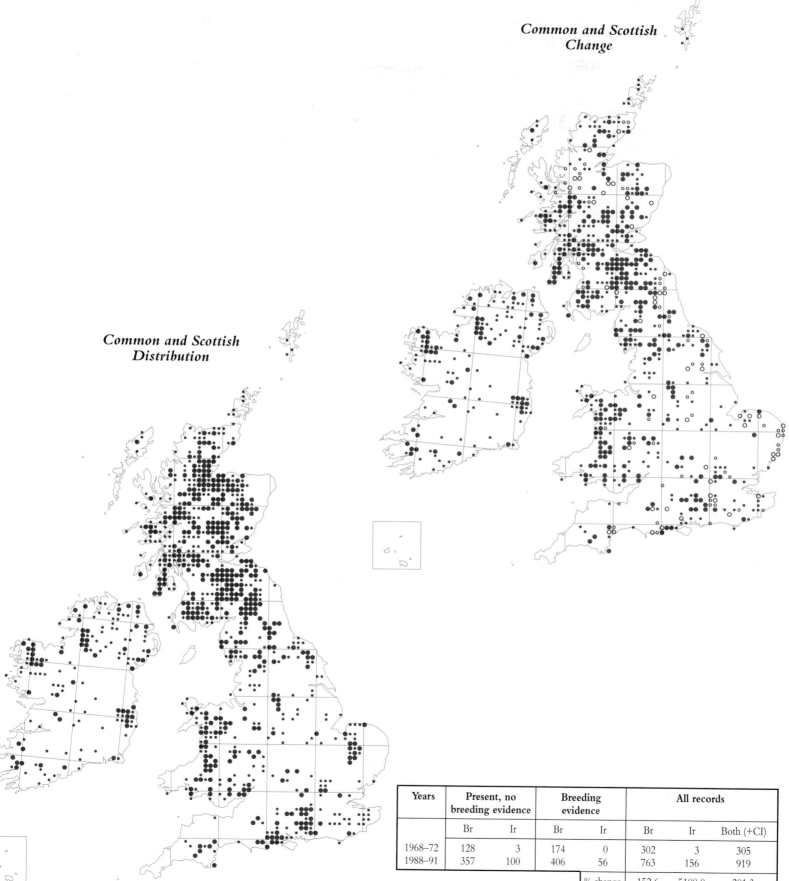

Common and Scottish Change

Common and Scottish Distribution

Years	Present, no breeding evidence		Breeding evidence		All records		
	Br	Ir	Br	Ir	Br	Ir	Both (+CI)
1968–72	128	3	174	0	302	3	305
1988–91	357	100	406	56	763	156	919
				% change	152.6	5100.0	201.3

Scottish Crossbill

(continued from p 422)

occupied squares there probably reflect the maturing forests along the peninsula. There is a clear gap in the distribution in the Central Lowlands of Scotland where agriculture predominates, but the reverse is true for large tracts of S W Scotland, the Southern Uplands and the N Pennines. Huge forests like Kielder have matured in recent decades and become very attractive to crossbills. Some areas, like the North Yorkshire Moors, have held crossbills almost continuously over the last few decades, although not in the same small woods each year. A number of other sites in central and S Britain have been well known for crossbills for some time: the Peak District, the Brecks of East Anglia, the N Norfolk coast, the New Forest, the Forest of Dean and S Devon. Wales and Ireland, like S Scotland, have been heavily planted with conifers, and these are now maturing in increasing number and providing suitable habitat.

The distribution shown is a composite of four years of fieldwork and four years of shifting crossbill populations. In any single season, crossbills were less widely distributed. They were also less common than the map might suggest, with considerable variations from year to year. For example, crossbills were more than ten times commoner during timed visits in 1991 than in 1989 (see Table).

Crossbills tend to breed early in the year, and many broods will have left the nest before the main fieldwork started each spring. They also undertake local movements or begin erupting quite soon after breeding. In 1990 there was a massive post-breeding irruption of Common Crossbills from E Europe, with the first birds arriving in Britain in late May, and further waves of immigrants through to the end of the year.

Most of the forest areas occupied in 1968–72 held crossbills during 1988–91. The main differences were in the number of additional areas which were occupied more recently in the S and S E Highlands, S Scotland, N England and particularly in Wales and Ireland. With the crossbill breeding season starting in late winter, it is no surprise that the broad pattern of the present distribution is similar in many respects to that found during the recent *Winter Atlas* survey.

Although crossbills sometimes feed on other foods when conifer seed is not available during eruptions, their heavy dependence on seeding conifers gives rise to very mobile populations of birds. Pines seed more reliably than spruces and larches and, as a consequence, the pine-feeding Scottish Crossbill moves less than the Common Crossbill. Even so, between years the Scottish Crossbill usually changes wood, whether within the same forest or, sometimes, into a different river valley. It relies on old trees, which are becoming scarcer as ancient forests are felled or die out (Knox 1990b, 1990c). Common Crossbills, on the other hand, are more often found in the younger woods of introduced conifers, and have benefitted considerably from afforestation since the 1914–18 War. The cone crops in these woods generally show greater variations from year to year, and the crossbill populations are consequently less stable.

Crossbills do not often occur on CBC plots. Nevertheless, breeding season populations over the last two decades have shown a clear four year cycle in those woods censused, mainly in the southeast (*Trends Guide*).

The Common Crossbills which occur in Britain and Ireland belong to the nominate race, with a wide range across the N Palearctic, from W Europe to E Siberia. Numerically, their centre of distribution lies in the northern boreal forest of Fenno-Scandia, Russia and Siberia. The Scottish Crossbill is endemic to N Scotland, and is the only species of bird found in Britain and nowhere else.

In the *Winter Atlas* it was estimated that the Common Crossbill population of Britain and Ireland could be less than 1,000 birds between irruptions, and several times that size after an irruption. The 1990 invasions were the largest and most sustained for many years. Point count censusing in Kielder, in November and December 1990, suggested that there might have been as many as 40,000 Common Crossbills in 34,000 ha of forest (Jardine 1991). Extrapolating such numbers to other occupied squares (which may not be realistic) emphasises the magnitude of the differences that can occur between good crossbill years and poor ones. It also reveals the way any population variation arising from birds nesting in Britain and Ireland can be completely swamped by invaders. High numbers occurring here after irruptions are likely to decline over the following years, unless reinforced by further irruptions. Large numbers of crossbills were recorded early in 1991 (see Table), reflecting the invasions which took place after the 1990 field season. Common Crossbills became scarcer through 1991 and into 1992. Because of the lack of good census data and the enormous between year variation in numbers, precise estimates of the British and Irish populations during 1988–91 cannot be calculated.

The population of the Scottish Crossbill is believed to be about 1,500 birds (*Winter Atlas*). There are no more recent estimates. The separate Scottish Crossbill map shown here is based on the unedited records of Scottish Crossbills claimed by observers. It is known to be inaccurate for a number of reasons, mainly due to the difficulty of separating the species in the field, and the relative caution of different fieldworkers.

ALAN G. KNOX

68–72 *Atlas* p 428

Number of crossbills recorded each year during timed fieldwork, 1988–91.

Year	1	2	3	4
1988	39,728	160	1,272	0.032
1989	27,762	63	331	0.012
1990	21,042	108	1,582	0.075
1991	3,798	39	582	0.153

1. Total hours of timed fieldwork.
2. 10-km squares in which crossbills were recorded during timed visits.
3. Crossbills recorded during timed fieldwork.
4. Crossbills per hour of timed fieldwork.

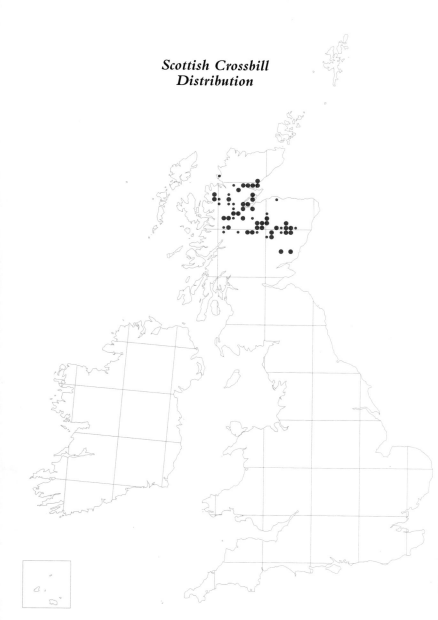

Scottish Crossbill Distribution

Years	Present, no breeding evidence		Breeding evidence		All records		
	Br	Ir	Br	Ir	Br	Ir	Both (+CI)
1968–72	0	0	0	0	0	0	0
1988–91	20	0	39	0	59	0	59
			% change		0		

since the Bullfinch's diet includes a great variety of mainly weed seeds. Recent declines in other seed eaters including Linnet, Tree Sparrow and Corn Bunting have also been attributed to intensification of farming methods. Increases in the abundance of a major predator, the Sparrowhawk, is also suggested as a possible cause of the Bullfinch's decline (*Trends Guide*), but the pattern of decline differs by region and in different habitats, and does not closely match the pattern shown by the spread of the Sparrowhawk.

Due to their habit of eating buds, Bullfinches have been regarded as pests since the 16th century. In a 4ha orchard studied in Herefordshire, about 200 birds were trapped and killed annually for five successive years. Despite this, the number caught did not decline over that period, and the population in the surrounding area of 9km^2 remained at 60–90 pairs (Evans 1984). Most of the damage to orchards is done between February and April, and is more severe in years when natural foods, especially ash seeds, fail. Although a single Bullfinch can eat up to 45 buds a minute, fruit trees can lose at least 50% of their buds without serious loss of fruit in the summer. As with other bird species, however, Bullfinches concentrate their attacks on the edge of orchards. Recent research, therefore, aims to

Bullfinch

Pyrrhula pyrrhula

"Present everywhere, but rarely seen": that is how many birdwatchers might sum up the distribution of the Bullfinch. Although the male is brightly coloured, its subdued song and retiring habits make this species difficult to census accurately. The standardised methods of the *88–91 Atlas* are therefore particularly useful for providing a picture of Bullfinch distribution across Britain and Ireland.

Bullfinches were recorded in almost all regions of Britain and Ireland, except the Northern and Western Isles of Scotland, and the Isle of Man. Gaps in distribution also exist on the mainland, e.g. in the Fens, the extreme western edge of the Irish coast and several parts of Scotland. These blank spots largely mirror the distribution of treeless habitat.

The Abundance map shows high densities of Bullfinches in several regions, from S England to parts of NE Scotland, and especially in Ireland, which forms the western limit of their world distribution. Apart from weed and tree seeds, the major requirement for Bullfinches would seem to be dense bushes or trees for nesting, as suggested by Sitters (1988). The winter distribution closely matches that in the breeding season, which is not surprising, for ringing recoveries show that birds very rarely move more than a few km (*Winter Atlas*).

The Change map shows a small retraction in range over the past 20 years in Scotland, especially in the western half, with less marked decreases in other regions, e.g. around the Fens. Independent evidence of a decline is presented in the *Trends Guide*, where CBC data show a significant fall in breeding numbers between the mid seventies and mid eighties. This downward trend has continued, with the 1990 results for farmland being at the lowest level ever recorded by the CBC (Marchant *et al.* 1991). Hedgerow removal is cited as the most likely factor causing this decline, which was much more severe on farmland than in woodland. Extensive clearance of agricultural 'wasteland' may have been a contributory cause,

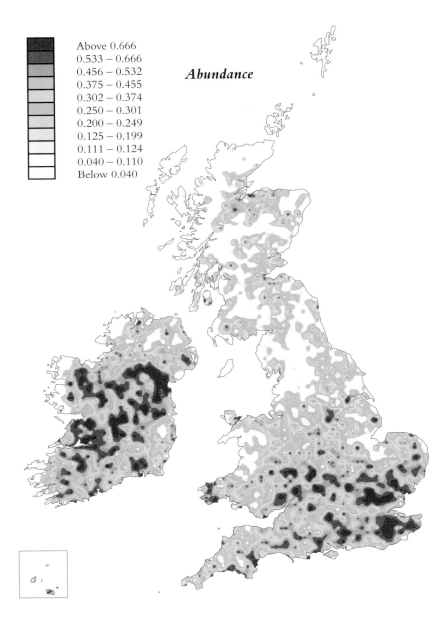

Above 0.666
0.533 – 0.666
0.456 – 0.532
0.375 – 0.455
0.302 – 0.374
0.250 – 0.301
0.200 – 0.249
0.125 – 0.199
0.111 – 0.124
0.040 – 0.110
Below 0.040

Abundance

encourage the birds to disperse their feeding effort so that fewer trees suffer concentrated attacks and less fruit is lost (Greig-Smith 1987).

Unlike the cardueline finches, the Bullfinch does not nest in colonies, but it is certainly not territorial, with neighbouring pairs apparently almost oblivious of one another (Newton 1972). In one region in the Netherlands, Bijlsma (1982) found Bullfinches nesting in loose groups of 3–4 pairs. He found no evidence of territories, and polyandry was noted occasionally. Bullfinches usually nest in thick scrub 2–3m above ground level, and lay 3–6 eggs. Incubation lasts 12–14 days with the young fledging 15–17 days later. In most years, laying in S Britain occurs from May to July, but in some years continues to mid September, so that theoretically even four broods can be raised. Predation of nests, however, is heavy, with the result that pairs raise, on average, 6–10 young per year (Newton 1972).

Although the Bullfinch occurs in a continuous band extending as far east as Japan, the subspecies *pileata* is endemic to Britain and Ireland. During 1988–91 there were an estimated 190,000 Bullfinch territories in Britain and a further 100,000 in Ireland. These figures are very approximate because Bullfinch breeding territories are often large and ill-defined. Standard mapping techniques, upon which these estimates are based, may

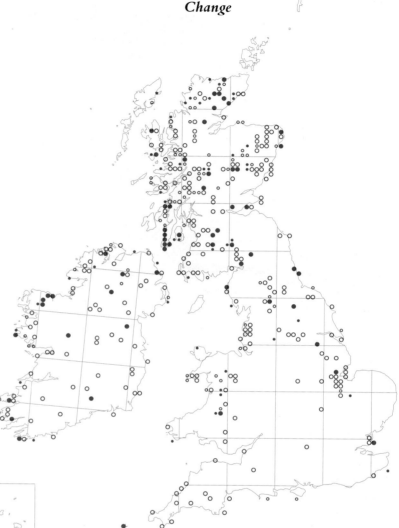

Change

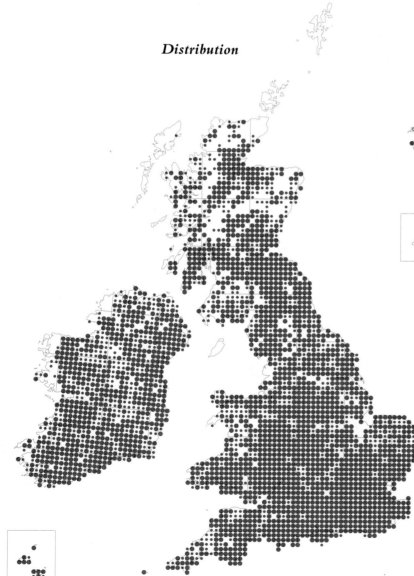

Distribution

seriously under-record breeding density in woodland. In Dutch woodland, Bijlsma (1982) found that on average around a third of pairs was missed. Newton (1972) recommended using ringing and retrapping methods, which revealed densities of over 50 pairs per km^2 around Oxford.

RAY WATERS

68–72 Atlas p 426

Years	Present, no breeding evidence		Breeding evidence		All records		
	Br	Ir	Br	Ir	Br	Ir	Both (+CI)
1968–72	179	55	2147	815	2326	870	3201
1988–91	404	223	1769	602	2173	825	3010
	% change				−6.5	−5.1	−6.1

Hawfinch

Coccothraustes coccothraustes

Any observer seeking Hawfinches, but unfamiliar with their calls, will spend many fruitless hours trying to glimpse these heavy bullnecked finches high in the woodland canopy, their stealthy movements making detection extremely difficult. Once known, however, the woodlands can ring with their explosive 'ticking' calls.

The Distribution map loosely reflects the species' main habitat requirements of mixed broadleaved woodland or extensive orchards. It is widely distributed in the southeast, with local pockets through central and N England. Away from Gwent, Welsh records are few; there is only a scattering of Scottish records in the southeast, and none in Ireland.

Tree seeds (especially hornbeam and the large, hard, stones of cherry) form the staple diet in the autumn and winter, buds (especially beech) and new shoots in spring, and insects (especially the larvae of the oak roller moth) in summer (Mountfort 1957, Newton 1972, S.J. Roberts and J. Lewis).

Where conditions are favourable, Hawfinches may nest in loose colonies where they are easy to detect as the interaction between pairs makes them very vocal. Isolated pairs, however, quickly fall silent after an initial burst of courtship, and are less easily found.

The Hawfinch is generally an elusive species, so coverage is probably incomplete, as it was in the *68–72 Atlas* and *Winter Atlas*. The Change map shows a marked but uneven reduction in overall numbers, with some local populations in the north and west appearing to have been maintained or improved; but the area southeast of a line from the Wash to Dorset still produces more than 50% of all records. The reduction could be due to the loss of broadleaved woodland in the 1970s and 1980s; but (although the species is not indexed by the CBC) the *Trends Guide* suggests overall stability for territory-holding birds, with an indication of increase in the

number of plots which the species visits. The apparent gains in areas such as Gwent, Gloucestershire, E Hampshire and some parts of N England could relate to improved recognition by observers now familiar with the species. The decline has, however, been well recorded in Oxfordshire, where at Blenheim Palace an index of Hawfinch abundance (i.e. 'bird days') declined from 122 in 1984 to 12 in 1990 (D. Doherty). In a second study, at Jarn Mound (again in Oxfordshire), flock sizes on the cherry crop in July and August dropped from 20 in 1970 to none at all since 1986 (J. Brucker).

The catastrophic gale of October 1987, when over 15 million mature trees in the southeast disappeared overnight, could have had a significant (if temporary) effect on Hawfinch feeding and breeding habitat in the subsequent years. Some of the populations formerly around Surrey and West Sussex were not recorded in 1988–91.

Courtship is often involved and elaborate, with males reaching pitches of excitement accompanied by wing shivering, rocking and pivoting towards their mates. Nest building, conversely, is silent, and the only real clue to the nest whereabouts is when the male arrives to feed the incubating female, accompanied by a short burst of feverish calling. Data from the

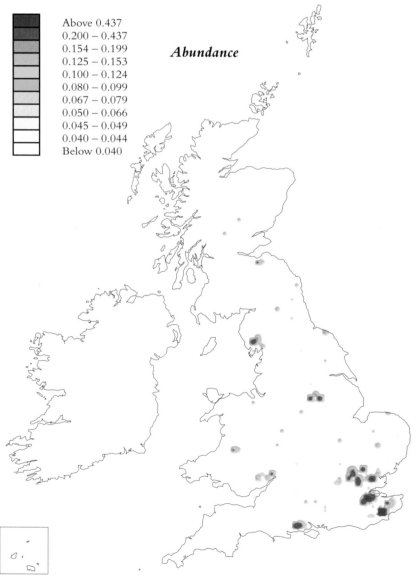

Abundance

	Above 0.437
	0.200 – 0.437
	0.154 – 0.199
	0.125 – 0.153
	0.100 – 0.124
	0.080 – 0.099
	0.067 – 0.079
	0.050 – 0.066
	0.045 – 0.049
	0.040 – 0.044
	Below 0.040

NRS show that 76% of nests are in woodlands (mainly deciduous) with some 15% in orchards, gardens or parkland. The main trees used for nesting are birch, oak and hawthorn, with a wide variety of nest locations: 30% on branches, 26% in forks, 19% against trunks and 18% in the treetop. The height range of nests is considerable (2–18m).

The clutch size is usually 4–5 eggs and there may be several breeding attempts, for as with other cardueline finches the species is heavily predated early in the season (Newton 1972). Jays, in particular, will 'work' Hawfinch colonies and can cause many failures before leaf cover is complete.

"The Hawfinch is notoriously quick to desert its nest at the least disturbance during the early stages of incubation" (Mountfort 1957). This view of high susceptibility has been repeated, "Desertion risk very high at all stages", (Campbell and Ferguson-Lees 1972), but observations in Gwent and Gloucestershire do not support it (Roberts and Lewis 1988). NRS data contain 63 histories where nests with eggs were inspected (sometimes up to five times). Of these, 43 (68%) subsequently had young and 20 failed before hatching, for a variety of reasons. It is likely that the species' high risk of predation and the male's habit of building quick platforms (which are subsequently abandoned) to solicit mates, could have

Change

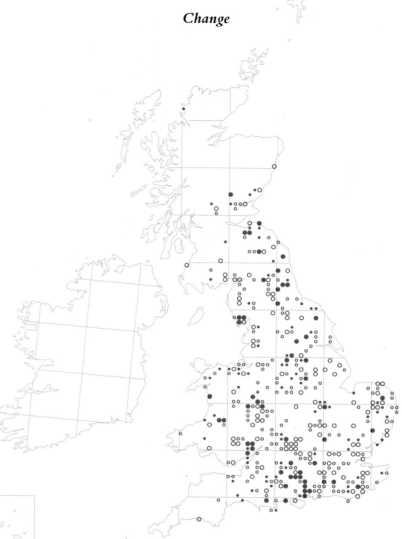

Distribution

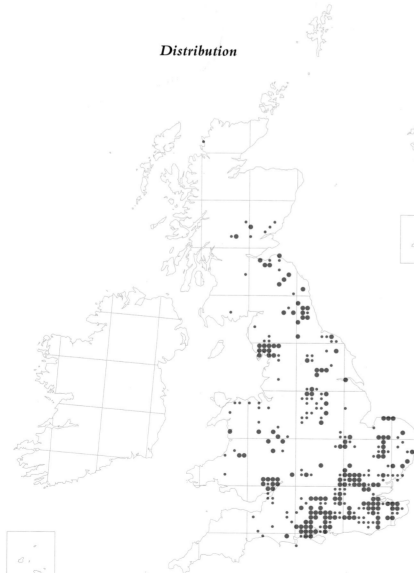

given rise to this early misunderstanding. The Hawfinch does not appear to be more susceptible to desertion than any other cardueline finch.

The *68–72 Atlas* calculated the population on the basis of 10–20 pairs per occupied 10–km square. Using the same formula, the 1988–91 population falls within the range of 3,000–6,500 pairs. On the Continent, Hawfinches are dispersed widely across S and central Europe from the Mediterranean to approximately 60°N, but no population estimates are known.

STEPHEN J. ROBERTS, JERRY LEWIS and CAROLINE DUDLEY

68–72 Atlas p 410

Years	Present, no breeding evidence		Breeding evidence		All records		
	Br	Ir	Br	Ir	Br	Ir	Both (+CI)
1968–72	225	0	234	0	459	0	459
1988–91	145	0	170	0	315	0	315
				% change	−31.4	0	−31.4

Snow Bunting

Plectrophenax nivalis

The familiar shoreline haunts of Snow Buntings in winter contrast starkly with the Arctic quality of their Scottish breeding grounds, the mountain plateaux and corries where few other species are able to exist.

As a consequence of the inaccessibility of many of the Snow Bunting's breeding areas, its scarcity even when present and its inconspicuousness during some parts of the breeding cycle, the Distribution map cannot give more than a general impression of its status and whereabouts. It indicates a thin but widespread occurrence throughout the Scottish Highlands, largely reflecting the availability of montane ground with long-lasting summer snowfields and nearby rocks or boulders suitable for breeding. Occasional summering and breeding attempts are reported at sea-level in the Northern and Western Isles although there were none during 1988–91.

The frequency of records has increased since the *68–72 Atlas*, especially in the W Highlands. This may be a result of increased or more uniform coverage, or of a genuine increase in the Scottish summering population. Watson and Smith (1991) have argued that an increased number of records between 1971 and 1987, compared with earlier in the century, is in part genuine, since it is matched by a similar increase in breeding pairs on a well-studied area of the Cairngorms. The majority of breeding hills, however, are known to hold only one or two sites, and these may be used infrequently or irregularly, and so the recent growth in leisure time and in popularity of outdoor pursuits will undoubtedly have led to increased reporting.

More detailed observations (R. Smith) have shown that the population increase on one massif has been maintained since 1987. Nethersole-Thompson (1966) speculated that the population would increase after cold winters and springs with late snow lie, because these conditions could attract recruits from the wintering populations of Icelandic, Greenland and Scandinavian origin. Milsom and Watson (1984), however, could find no correlation between the extent of spring snow cover and population size. The present breeding population seems to be composed mainly of birds showing plumage characteristics of the Icelandic race, as does the winter immigrant population. Breeding success, survival and site fidelity of native Scottish birds are, however, high and probably capable of sustaining the population without significant influxes of Icelandic birds (Smith 1991).

Scottish Snow Buntings begin breeding from mid May, when late snowfalls can still cause desertion and delays. Often, however, there is still time to rear two broods, although second broods contain fewer young and are more likely to succumb to starvation since the mid summer flush of insect activity is largely over. The vagaries of mountain weather, and its effect on invertebrate availability, is therefore potentially an over-riding factor affecting production. Scraps scavenged from the ever increasing army of hill walkers may modify the effects of weather by supplementing the diet when natural supplies become scarce.

Adults and their young can still be seen on the breeding grounds in the autumn months, although adults become very unobtrusive during the moult in August and September. Indeed, a major loss of adults appears to occur at this vulnerable time. They intermingle freely with flocks of winter visitors, and some breeding birds, especially males, can be seen on or near the breeding grounds throughout the winter, foraging on sedge and grass seeds, or leatherjackets (mainly *Tipula montana*) during thaws.

Snow Buntings exist in Britain at a southern edge of their breeding distribution, in habitat relatively unaffected by man. They may therefore prove to be sensitive indicators of climatic or man-induced change in the uplands, although our present knowledge of their status and breeding parameters is limited. Watson and Smith (1991) report that over 100 sites were used between 1971 and 1987, and estimate that the population may have been up to 50 pairs in some years. Consideration of the increase from the *68–72 Atlas* level to the present one suggests this might be conservative and that an estimate of 70–100 pairs would not be unreasonable.

RIK SMITH

68–72 Atlas p 440

Change

Distribution

Years	Present, no breeding evidence		Breeding evidence		All records		
	Br	Ir	Br	Ir	Br	Ir	Both (+CI)
1968–72	7	0	7	0	14	0	14
1988–91	27	0	15	0	42	0	42
				% change	200.0	0	200.0

Yellowhammer

Emberiza citrinella

When most other birds are moulting and silent in late summer, the song of the Yellowhammer still drifts across the ripening cornfields. Despite his bright colours, however, the singer may be difficult to locate, hidden away among the leaves near the top of a hedgerow tree.

As befits a familiar bird of agricultural environments, the Yellowhammer is widely distributed across lowland Britain. Regional variation in abundance nevertheless occurs, with the population concentrated in E Britain and the Midlands and, except in scattered coastal areas, generally at lower densities in the west. On the higher ground of Wales and N England, and on the uplands and islands of Scotland, the species is considerably less common, and the Distribution map suggests a marked withdrawal from these areas since the *68–72 Atlas*. In Ireland, evidence of a recent decline is much more striking. Around 1970, the Yellowhammer was almost ubiquitous there, but it has now disappeared from large areas of the west and is decreasing where it still occurs, e.g. Co. Galway (Whilde 1990). Indeed away from the southeast of the island, E Donegal is one of the few places where it remains common. Comparison with the *Winter Atlas* shows that a contraction in the Irish range of this essentially sedentary species was already pronounced in the early 1980s.

The pattern of Yellowhammer distribution and abundance across Britain and Ireland supports the findings of Morgan and O'Connor (1980) that the species is more numerous on highly productive farmland. The *Trends Guide* further reveals that while numbers have remained essentially stable on farmland CBC plots, which are mostly concentrated in the productive, lowland, SE Britain, they have declined since 1980 in woodland – a sub-optimal habitat for this species (O'Connor 1980). A map of

agricultural practice in the Republic of Ireland (Gillmor 1979) demonstrates that Yellowhammers have disappeared from areas in which tillage comprises the lowest proportion (<10%) of agricultural land use. Outside the breeding season, the Yellowhammer is heavily dependent for food on cereals and other large grass seeds (Prŷs-Jones 1977). It thus seems not unreasonable to postulate that increasing regional specialisation and efficiency in farming have tended to affect the species, adversely, in the milder, moister areas of W Britain and, in particular, in Ireland, where livestock grazing is most favoured. More generally, the switch since the 1960s from spring to autumn cereal sowing cannot but have had a pronounced impact on the ecology of a bird so characteristically associated with winter stubbles.

Although the literature available for review in the *Trends Guide* provided little evidence of changes in Yellowhammer numbers in continental Europe, a recent Swedish ringing analysis suggests otherwise (Hjort and Pettersson 1990). Most Yellowhammers in Scandinavia are migratory, and autumn ringing results from Ottenby Bird Observatory, spanning the 40 years from 1948 to 1987, point to a steadily declining population from the late 1940s to mid 1960s, followed by a resurgence lasting until 1980,

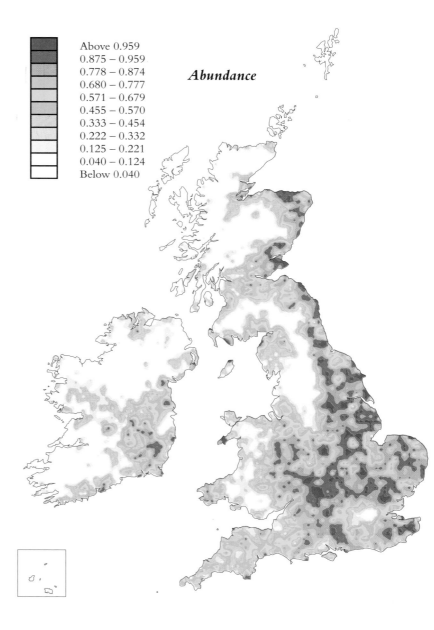

Above 0.959
0.875 – 0.959
0.778 – 0.874
0.680 – 0.777
0.571 – 0.679
0.455 – 0.570
0.333 – 0.454
0.222 – 0.332
0.125 – 0.221
0.040 – 0.124
Below 0.040

Abundance

and then a brief but rapid decline which appeared to have reached its nadir by 1985. The earlier decline seems likely to have been linked to the effects of poisoning from alkyl-mercury seed-dressing compounds, banned in Sweden in 1966; the later decline coincides with a trend towards colder winters in the early 1980s, but might equally reflect broader patterns of change in agricultural environments.

During 1988–91 there were an estimated 1,200,000 Yellowhammer territories in Britain, and 200,000 in Ireland.

ROBERT PRŶS-JONES

68–72 Atlas p 434

Distribution

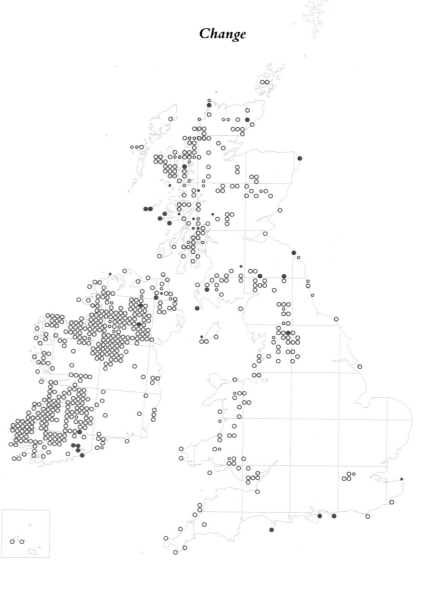

Change

Years	Present, no breeding evidence		Breeding evidence		All records		
	Br	Ir	Br	Ir	Br	Ir	Both (+CI)
1968–72	57	14	2381	925	2438	939	3380
1988–91	262	181	1962	406	2224	587	2814
				% change	−8.6	−37.4	−16.7

Cirl Bunting

Emberiza cirlus

Never abundant like its close relative the Yellowhammer, but equally a bird of lowland agriculture, the Cirl Bunting's population size and range have declined dramatically over the last 50 years so that it is now one of Britain's rarest resident passerines.

The Distribution and Abundance maps show that its main range in Britain is confined to the coastal strip of S Devon between Plymouth and Exeter. Most of the other records relate to isolated occurrences in a single year. The Change map demonstrates the contraction of range that has taken place since 1968–72, with an 83% reduction in the number of 10–km squares occupied.

Comparison with the *Winter Atlas* indicates that the British population is mainly sedentary, a point reinforced by observations of colour-ringed birds in S Devon (A. D. Evans). The sporadic occurrences of birds far from the main population is therefore surprising. They may be spring migrants overshooting, for most of the northern populations on the Continent are migratory (Géroudet 1980).

The Cirl Bunting is principally a bird of the countries surrounding the N and W Mediterranean, but its range extends northwards through France, where it is quite widespread, to S England. Declines have been noted in France, Luxembourg, Belgium and Germany (*Red Data Birds*).

It was not recorded in Britain before the 19th century, when a few birds were discovered in S Devon. Although it is not certain, the species may only then have been colonising Britain for the first time. Throughout the 19th century it spread across the lowlands of S England and, by the 1930s, was common and widespread south of a line between London and Bristol, and also occurred in N Wales (Parslow 1973). The decline first became noticeable in the 1950s, and by the time of the *68–72 Atlas* was well advanced, although the species was still present in 174 10–km squares. The further decline, with only 32 squares occupied during 1988–91, is a clear indication that the British population is in serious jeopardy.

It was estimated in the *68–72 Atlas* that the population was then in the range 350–700 pairs. Sitters (1982) considered that this assessment was optimistic and that, based on county bird report data, 250–300 pairs was nearer the mark. The first census of Cirl Buntings by the BTO took place in 1982 when a maximum of 167 pairs was located in England, mostly in Devon, with another 14 pairs on Jersey (Sitters 1982, 1985). In 1988, the RSPB instigated an ecological study to identify reasons for the decline and to formulate conservation prescriptions. As part of this project, further surveys were carried out in 1989 (in conjunction with the Devon Bird

Watching and Preservation Society) and 1991. The latter survey recorded an increase to a maximum of 229 pairs in S England; 217 of them in Devon and 12 in Cornwall. On Jersey the numbers had fallen to five pairs. The higher numbers in Devon can be attributed partly to increased research effort, partly to the discovery of previously unknown sub-populations and partly to a genuine increase following good breeding success in the fine summers of 1989 and 1990. The decrease on Jersey was attributed to severe winter weather in early 1991 (J. Waldon).

A variety of theories have been advanced for the general decline in the population. The most likely can be grouped under 'habitat change' and 'climatic factors'. The possible significance of climate is indicated by the fact that British Cirl Buntings are at the northern limit of the species' range. It may also be significant that the population has retreated to S Devon, one of the warmest parts of the country, and where the species was first recorded in Britain. Nevertheless, if climatic change is important, the manner in which it acts in the species' life cycle has yet to be identified.

That habitat change and changes in farming practice are the primary cause of the Cirl Bunting's decline are indicated by the provisional results of the RSPB project (Evans 1992). The availability of good winter foraging sites may be important because it has been found that birds show

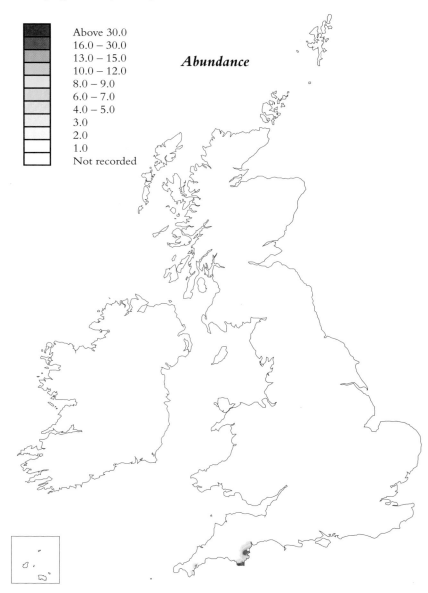

Above 30.0
16.0 – 30.0
13.0 – 15.0
10.0 – 12.0
8.0 – 9.0
6.0 – 7.0
4.0 – 5.0
3.0
2.0
1.0
Not recorded

Abundance

a strong preference for weed-rich stubble fields, a habitat which has declined markedly in recent years with the widespread switch to autumn sown cereals (Evans 1990). Similarly, good breeding habitat may also have declined, for it has been shown that adults habitually collect food for their young in grassland which is rich in invertebrates. Such areas have probably become very scarce in today's farmland, with the widespread use of pesticides (Sitters 1991). These studies not only indicate reasons for the decline but they also offer a basis for promoting possible conservation strategies.

HUMPHREY SITTERS and ANDY EVANS

68–72 Atlas p 436

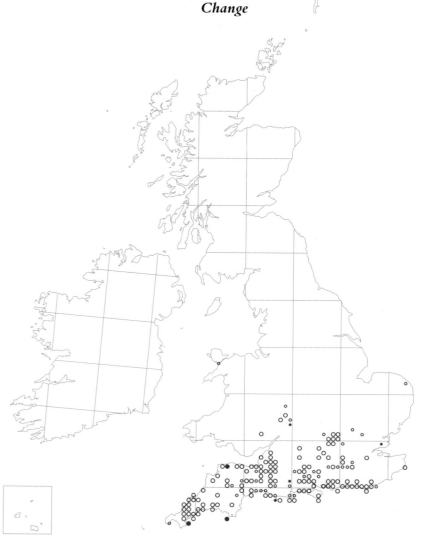

Change

Distribution

Years	Present, no breeding evidence		Breeding evidence		All records		
	Br	Ir	Br	Ir	Br	Ir	Both (+CI)
1968–72	32	0	141	0	173	0	174
1988–91	11	0	18	0	29	0	32
				% change	−83.2	0	−82.8

Reed Bunting

Emberiza schoeniclus

Although typically associated with marshland, gravel pits and riversides, male Reed Buntings can be found voicing their repetitive song alongside Yellowhammers in young forestry plantations, and in company with Corn Buntings in rape fields. The proportionate use made of such less preferred habitats clearly varies with population size, however, and increases when numbers are high.

The overall widespread pattern of Reed Bunting distribution revealed by the *88–91 Atlas* is little changed from that documented by the *68–72 Atlas*, but the species has nevertheless been lost as a breeding bird from numerous predominantly northerly and westerly squares. By contrast, a minor but clear extension of range relates to the N of Shetland, where breeding pairs were formerly absent (*Birds in Scotland*). The species remains generally absent from the higher upland areas, notably in Scotland, but the smaller distributional gaps do not appear to be systematically associated with altitude. The Abundance map reveals striking, relatively fine-grained variation, but with pairs tending to be more numerous in low-land areas. Overall, Reed Buntings occur at higher densities in Ireland than in Britain.

The most recent review of the Reed Bunting's breeding status in Ireland suggested a marked population increase between 1966 and 1986 (*Birds in Ireland*), but its concurrent disappearance from a swathe of squares in the S of the Republic seems at odds with this. In Britain, CBC and WBS results both indicate a recent steep decline in the Reed Bunting breeding population (*Trends Guide*). A relatively stable high level which lasted from the late 1960s into the mid 1970s, was followed by a decrease of more than 50% to a new fairly stable lower level, achieved during the early 1980s. Reed Buntings are year-round residents in Britain and known to be sensitive to cold winters with extensive snow cover (Prŷs-Jones 1984). However, recovery from weather-related declines is typically rapid, and the observed sustained population reduction cannot easily be explained in these terms.

More pertinent evidence is probably provided by the striking coinci-

dence, in the time scale of the Reed Bunting's decline, with decreases recorded for other small passerine species, such as Linnet and Tree Sparrow, which share its non-breeding diet of small grass and herb seeds, in particular of farmland weeds. O'Connor and Shrubb (1986a) have developed a compelling case that the decline of the Linnet has been the result of food shortages, caused by the introduction of efficient herbicides to control an array of these weed species during the 1970s. This is almost certainly true for the Reed Bunting also, notwithstanding O'Connor and Shrubb's suggestion that this species may have been affected more by herbicide-induced changes to vegetation structure, than to effects on seed availability. Circumstantial, corroborative evidence, pointing to food shortages, is provided by results from the BTO's Garden Bird Feeding Survey, which reveal a major and continuing increase in the winter use of gardens by Reed Buntings during the 1980s (Thompson 1988).

During 1988–91 there were an estimated 220,000 Reed Bunting territories in Britain, and 130,000 in Ireland.

ROBERT PRŶS-JONES

68–72 Atlas p 438

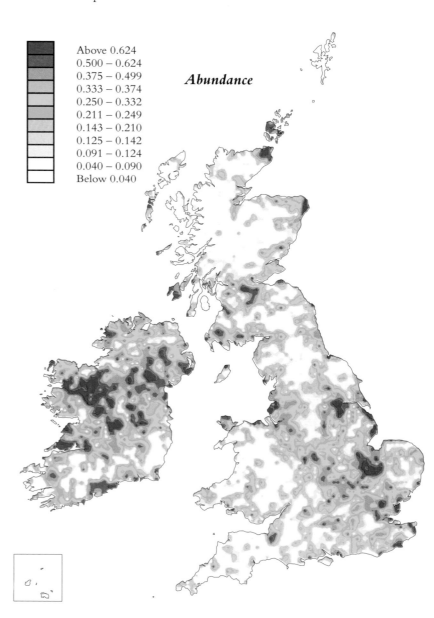

	Above 0.624
	0.500 – 0.624
	0.375 – 0.499
	0.333 – 0.374
	0.250 – 0.332
	0.211 – 0.249
	0.143 – 0.210
	0.125 – 0.142
	0.091 – 0.124
	0.040 – 0.090
	Below 0.040

Abundance

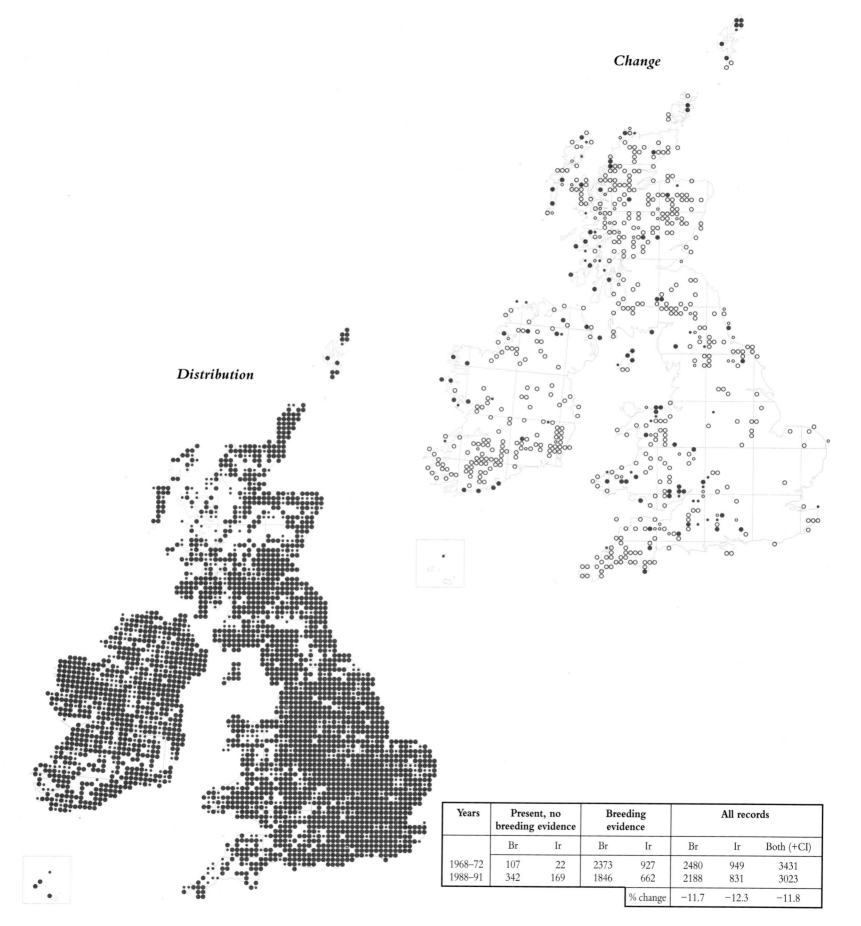

Change

Distribution

Years	Present, no breeding evidence		Breeding evidence		All records		
	Br	Ir	Br	Ir	Br	Ir	Both (+CI)
1968–72	107	22	2373	927	2480	949	3431
1988–91	342	169	1846	662	2188	831	3023
				% change	−11.7	−12.3	−11.8

Corn Bunting

Miliaria calandra

The persistent jangling song of a male Corn Bunting on an exposed perch is a characteristic sound of open farmland, from the flower-rich machair of the Hebrides to the rolling downs of SE England. But for how much longer? Corn Buntings have been in sharp decline over most of NW Europe since at least 1975 (*Trends Guide*, Hustings *et al.* 1990).

The distribution of Corn Buntings has long baffled ornithologists (e.g. Walpole-Bond 1938, *68–72 Atlas*). There is some association with the distribution of cereal farming over much of England and E Scotland (*Winter Atlas*), which ties in well with the results of local studies on habitat selection (e.g. Møller 1983, Thompson and Gribbin 1986, O'Connor and Shrubb 1986a). However there are distinct outliers from the main distribution, notably the birds in the Hebrides, Ireland and SW England. Even within areas of intensive arable farming, some individuals breed very successfully in other habitats (D. Harper), and on the Continent hayfields are often the favoured habitat (Hustings *et al.* 1990). Furthermore, Corn Buntings are virtually absent from some centres of arable production, e.g. East Anglia.

Comparison with the *68–72 Atlas* and *Winter Atlas* highlights the rapid contraction in the Corn Bunting's range, which has been catastrophic in Ireland, much of Scotland, Central England and the SW Peninsula. Even the isolated stronghold population in the Uists and Benbecula, once considered to be one of the densest in Britain (Walpole-Bond 1938), is in decline (*Birds in Scotland*). Retreat was first noted in Wales in the 1920s, but may well have started earlier in Ireland, where at the turn of the century, Corn Buntings were regarded as widespread (*Birds in Ireland*). By the 1940s, declines had been reported from the Isles of Scilly to Fair Isle, concentrated in the west. The *68–72 Atlas* revealed curious gaps in the distribution in E England (e.g. in East Anglia and the Weald) which may be enlarging.

Changes in the Corn Bunting's fortunes correlate with that of cereal farming (*Trends Guide*) and there is good evidence that they are especially linked with barley (e.g. Thompson and Gribbin 1986, O'Connor and Shrubb 1986a). At first sight, the crash of about 50% in Corn Bunting numbers, revealed by the CBC, is hard to explain in terms of a much smaller decrease in barley acreage sown: the figures in O'Connor and Shrubb (1986a) would still leave about 20ha of barley per bunting! If, however, the loss of barley has concentrated on habitats which Corn Buntings prefer for other reasons, the correlation could be cause and effect.

The *Winter Atlas* emphasised that many seed-eating birds are decreasing in numbers on farmland (see *Trends Guide*), and implicated changes in agricultural practices that had restricted seed supplies in the winter. Reduction in invertebrate numbers in the summer may also have been important. Gone are the days when Walpole-Bond (1938) could watch females collecting, with ease, insects for their chicks. Now, in the same area of East Sussex, many chicks starve (D. Harper). Worse, earlier harvesting of the crop (whether cereal or hay) prevents most females there from attempting a second brood, let alone the third broods noted by Walpole-Bond (1938). Access to other habitat types for insect food, and safer nest sites, may explain why territories in East Sussex are associated not only with cereals (especially barley) but with overall habitat diversity (D. Harper). An affinity for sites with an access to water was noted by Walpole-Bond (1938), and has been advanced as a factor in the decline in the Netherlands, due to the in-filling of ditches (Hustings *et al.*. 1990).

Estimating population density is complicated because the mating system can vary from co-operative polyandry (two males share a mate) to polygamy (Ryves and Ryves 1934). Polygamous males are usually bigamists (Thompson and Gribbin 1986), but exceptionally can attract 11 females during a season (D. Harper).

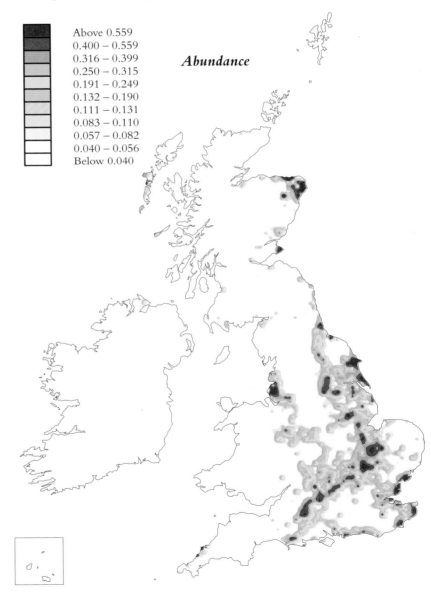

Above 0.559
0.400 – 0.559
0.316 – 0.399
0.250 – 0.315
0.191 – 0.249
0.132 – 0.190
0.111 – 0.131
0.083 – 0.110
0.057 – 0.082
0.040 – 0.056
Below 0.040

Abundance

Since the *68–72 Atlas*, the CBC index for Corn Buntings has fallen by 60%. This suggests that the estimate in the *68–72 Atlas* of 30,000 'pairs' breeding in Britain and Ireland, which was based on a very conservative one-tenth of the mean CBC density, was too low. A simple extrapolation from mean CBC densities in 1989 suggested a population of 160,000 Corn Bunting territories in Britain. However, Corn Buntings were only recorded on a few plots in that year, and this estimate is surely too high. The BTO/JNCC 1992–93 Corn Bunting survey should provide a more reliable estimate. The Irish population is now very small, and may well number less than 30 territories. But how many squares will be occupied in the third breeding atlas? The prospects for this archetypal 'little brown job' are hardly rosy.

DAVID HARPER

68–72 Atlas p 432

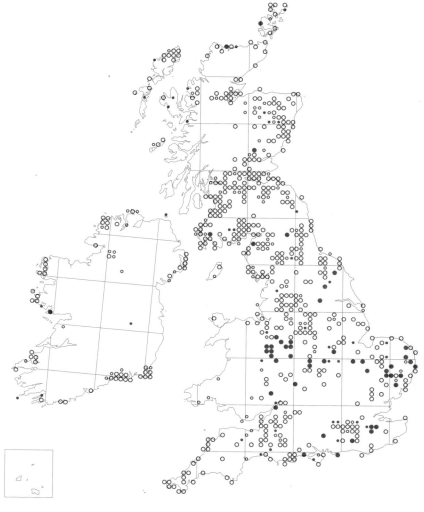

Change

Distribution

Years	Present, no breeding evidence		Breeding evidence		All records		
	Br	Ir	Br	Ir	Br	Ir	Both (+CI)
1968–72	104	17	1254	51	1358	68	1426
1988–91	235	9	686	2	921	11	932
				% change	−32.1	−83.8	−34.6

Red-necked Grebe

Podiceps grisegena

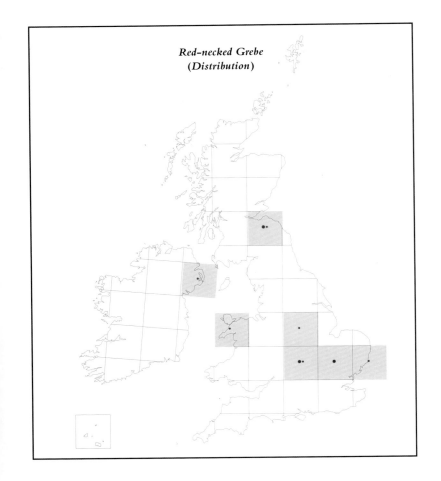

Red-necked Grebe
(Distribution)

Two-year averages for Red-necked Grebes in Britain during 1973–90.

	73–74	75–76	77–78	79–80	81–82	83–84	85–86	87–88	89–90
Individuals	1	4	1	3	2	2	5	11	8
Pairs	0	0	0	1	0	1	1	2	3

Based on RBBP reports; the first line refers to the numbers of summering individuals, the second to pairs. Note: figures are rounded up to nearest whole number.

Years	Present, no breeding evidence		Breeding evidence		All records		
	Br	Ir	Br	Ir	Br	Ir	Both (+CI)
1968–72	0	0	0	0	0	0	0
1988–91	5	1	3	0	8	1	9
Red-necked Grebe			% change				

Barnacle Goose

Branta leucopsis

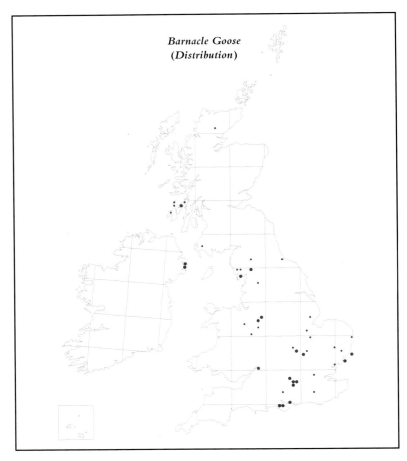

Barnacle Goose
(Distribution)

This species has a Holarctic breeding distribution, with the European breeding population concentrated in Finland (2,000 pairs), Poland (up to 1,000 pairs) and Denmark (up to 400 pairs). It prefers smaller and shallower areas of water than does the Great Crested Grebe, although the two species may overlap. The nest is built on a platform, usually situated further inside emergent vegetation than is the case with other grebes (*BWP*). Despite suggestions that this species has declined in numbers along the edges of its breeding range in recent years (*BWP*), the records of the RBBP have shown an increase in the numbers of individuals summering in Britain (see Table), culminating in the first confirmed breeding attempts in Scotland in 1988 (Anon. 1989) and in England in the same year (Parslow-Otsu and Elliott 1991). Despite the presence of a pair of birds accompanied by a very young juvenile in August 1987, successful breeding in Britain has not yet been proved. Nest failures have been due primarily to egg predation (by Magpies and maybe also by mink), disturbance and flooding.

The abundance of apparently suitable breeding habitat and increasing records of summering, particularly in S Scotland and the E Midlands, suggest that this beautiful grebe may yet establish a firm foothold in Britain. During 1988–90 two to three pairs of Red-necked Grebes were recorded in Britain (RBBP).

PAUL DONALD

In the *68–72 Atlas*, Barnacle Geese were recorded in only two 10-km squares. Both of these records were of confirmed breeding on Strangford Lough, Northern Ireland. In the *88–91 Atlas*, the same two squares remain the only ones occupied in Ireland, but there are an additional 17 10-km squares in Britain with breeding Barnacle Geese, scattered from S England to the Isle of Jura.

Over 40,000 Barnacle Geese winter in Britain and Ireland (*Red Data Birds*). The 30,000 birds breeding in Greenland winter on Islay and in

Ireland, and the 11,000 birds breeding in Spitsbergen winter on the Solway Firth. The distribution of breeding records does not mirror the winter distribution, and all *88–91 Atlas* records almost certainly involve escapes or deliberate releases from captivity. The possible exception is the breeding record from Jura, which may reflect breeding by sick or injured wild birds over-summering. Some of the breeding records may be of pairings between Barnacle and Canada Geese. Lever (1987) makes no mention of any naturalised populations of Barnacle Geese and, although Kear (1990) notes that this species is now 'breeding at large' in Britain, birds are probably still too scattered for there to be a self-sustaining feral population. If the current increase in breeding records continues, however, there seems no reason to doubt that such a population could become established.

The WWT/JNCC survey of introduced geese, in June-July 1991, located 600–700 Barnacle Geese in Britain. It is thus likely that the Distribution map under-represents the extent of the species range in Britain.

JEREMY WILSON

68–72 Atlas p 453

Years	Present, no breeding evidence		Breeding evidence		All records		
	Br	Ir	Br	Ir	Br	Ir	Both (+CI)
1968–72	0	0	0	2	0	2	2
1988–91	26	0	17	2	43	2	45
Barnacle Goose			% change			0	2150.0

Red-crested Pochard

Netta rufina

This handsome species has nested in small numbers in Britain in most years since 1968, but there can be little doubt that these records refer to escapees from waterfowl collections and their progeny.

The headquarters of the Red-crested Pochard in Britain now appears to be the Cotswold Water Park on the Gloucestershire/Wiltshire border where, as anticipated in the *68–72 Atlas*, the population has increased steadily since 1968 to 32 birds in September 1989 (Baatsen 1990). Another well established full-winged population is found in London at St James's Park.

Since they are inexpensive and breed readily in captivity, Red-crested Pochards are found in at least 200 waterfowl collections in Britain, and a small number in Ireland (N. Hewston). The female's habit of egg dumping – laying part of the clutch in the nest of another species – combined with the fact that the ducklings are difficult to catch, mean that this species perhaps more readily evades pinioning by collectors than most ducks, and so may be more likely to escape. Their distribution is therefore largely artificial, reflecting that of collections and the suitability of habitat for nesting and rearing young.

The first records of breeding in the wild were in Lincolnshire in 1937, and in Essex in 1958 (*BWP*). Pyman (1959) suggested that the earlier pair originated from the collection at Woburn, and that the Essex pair were likely to have been from London stock. Sightings at the Cotswold Water Park increased in the 1960s, and from 1972 onward the population rose steadily to present levels (Baatsen 1990).

The bulk of the world population occurs from central Asia west to the Caspian and Black Seas, and the European population is widely and relatively thinly distributed and prone to fluctuations. The British popu-

lation is still extremely small and scattered, and that of Ireland tiny. Adding the known totals for the Cotswold Water Park, the London Parks and sites in E England is likely to produce a total of fewer than 100 individuals, which agrees with the estimate presented in the *Winter Atlas*.

SIMON DELANY

68–72 Atlas p 446

Years	Present, no breeding evidence		Breeding evidence		All records		
	Br	Ir	Br	Ir	Br	Ir	Both (+CI)
1968–72	2	0	2	0	4	0	4
1988–91	9	1	3	0	12	1	13
Red-crested Pochard			% change		200.0		225.0

Scaup

Aythya marila

The Scaup is undoubtedly the rarest of our breeding ducks, with Britain and Ireland lying at the extreme southern edge of an extensive circumpolar breeding range. During the period of the *88–91 Atlas*, breeding occurred at only three sites but pairs or individuals of either sex were present at a further 18 localities, of which three were in Ireland. This general paucity compares with four cases of confirmed breeding in the *68–72 Atlas* and birds present at four other sites, none of which was in Ireland. It is clear when considering all records of breeding, or indeed of summering birds (as collated by the RBBP), that there is no evidence of regular annual breeding attempts anywhere in Britain. Other records are mainly from

Scotland, but birds have been noted during the breeding season as far south as Essex (in 1980), which suggests that breeding could occur virtually anywhere at any time. The current British breeding population must be considered to lie in the range 0–5 pairs.

STEVE CARTER

68–72 Atlas p 76

Years	Present, no breeding evidence		Breeding evidence		All records		
	Br	Ir	Br	Ir	Br	Ir	Both (+CI)
1968–72	4	0	4	0	8	0	8
1988–91	15	3	3	0	18	3	21
Scaup			% change		125.0		162.5

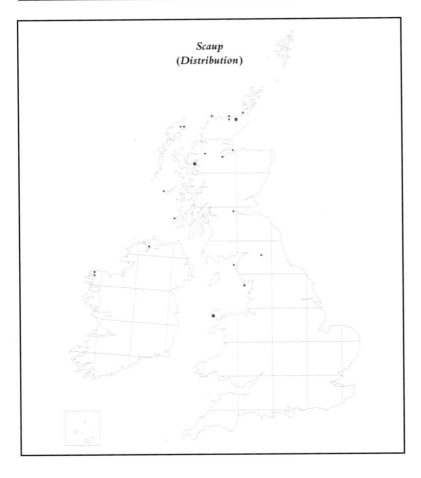

Scaup
(Distribution)

Crane

Grus grus

These magnificent birds commonly bred in East Anglia, probably into the 17th century, and also in Ireland possibly to the 14th (Snow 1971). Ray (1678) records great flocks in the fens of Lincolnshire and Cambridgeshire but, even then, was not certain that they bred. In recent years passage migrants have continued to be recorded regularly, in small but variable numbers. The recolonisation of East Anglia, with at least one pair breeding (or attempting to breed) each year since 1981 came as a surprise. Now this northern species is also attempting to breed in France. The English birds have raised four young in their first eight years (RBBP).

Cranes require large and undisturbed wetlands for breeding. There are many such suitable sites on reserves within Britain and Ireland, should the population start to expand. No records were plotted in the *68–72 Atlas* and there is no suggestion that birds bred in the other 10–km square mapped here, although two Cranes were present for at least a month from late May 1988 (Bell 1989). Care must be taken that individuals of this and related species, escaped from collections, are not erroneously recorded as truly wild birds.

CHRIS MEAD

Years	Present, no breeding evidence		Breeding evidence		All records		
	Br	Ir	Br	Ir	Br	Ir	Both (+CI)
1968–72	0	0	0	0	0	0	0
1988–91	1	0	1	0	2	0	2
Crane			% change			0	

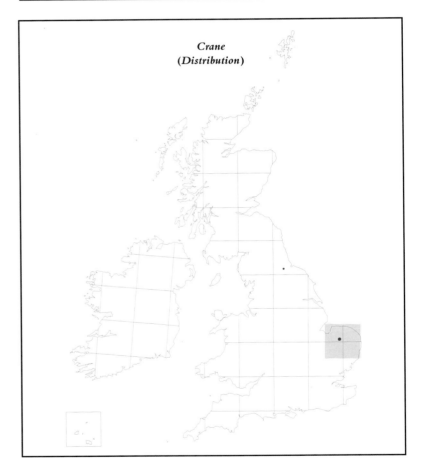

Crane
(Distribution)

Temminck's Stint

Calidris temminckii

These diminutive waders are usually regarded as unobtrusive birds, but when they are newly arrived on the breeding ground they become very active and vocal. Chasing each other across the nesting areas with a beautiful moth-like aerial display flight and a trilling song, they show off the distinctive white outer tail feathers. This is really a bird of the Arctic tundra of Scandinavia and Russia, but small numbers breed further south in N Europe and a tiny population hangs on in Scotland.

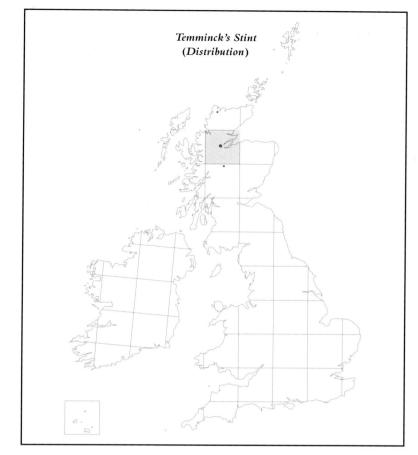

Years	Present, no breeding evidence		Breeding evidence		All records		
	Br	Ir	Br	Ir	Br	Ir	Both (+CI)
1968–72	2	0	2	0	4	0	4
1988–91	2	0	1	0	3	0	3
Temminck's Stint			% change		−25.0	0	−25.0

Purple Sandpiper

Calidris maritima

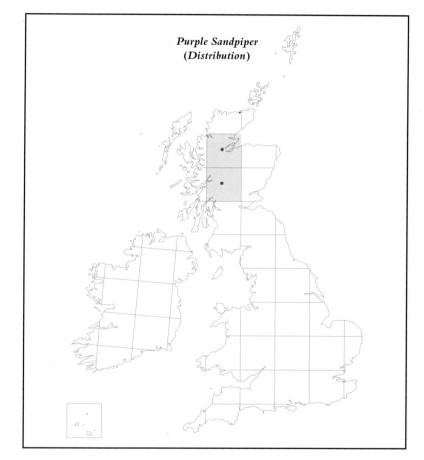

Comparison of the maps for 1968–72 and 1988–91 shows little change. Breeding was confirmed at the same site in each period and the species was recorded in three further 10-km squares in 1968–72 and in two in 1988–91. In the intervening period, however, changes in the fortunes of this bird have taken place in Scotland.

As this species has a complicated breeding strategy whereby both sexes can be successively bigamous (Hildén 1975), the number of adults is a better measure than the number of pairs. Breeding was first proved in 1971, with summering adults having been present at the same site since 1969 (Headlam 1972). Since then, the number of adults recorded on the breeding grounds increased to a maximum of nine in the late 1970s and early 1980s, but has subsequently declined (see Table). Nesting was recorded at five separate sites during the inter-Atlas years, although at never more than four sites in any one year.

The Scottish population is clearly very small and subject to change and may disappear at any time. There is apparently no shortage of habitat similar to the sites already used and the fortunes of this bird as a breeding species in this country are perhaps more likely to be influenced by climatic conditions. Population estimates for Scandinavia are of over 100,000 birds.

Temminck's Stint: the number of adults and nesting sites used between 1971 and 1989 in Scotland

	71	72	73	74	75	76	77	78	79	80	81	82	83	84	85	86	87	88	89
Adults	2	4	4	6	5	6	6	9	9	9	8	3	4	2	3	5	6	6	5
Sites	1	1	1	2	1	1	1	1	3	4	2	2	1	1	2	2	1	1	1

ROY DENNIS

68–72 Atlas p 192

The attractive and confiding Purple Sandpiper has a circumpolar distribution and breeds mainly in the Arctic. It occupies a variety of open habitats from sea level to mountain tops, but the breeding range also extends down through Scandinavia and the Faeroes.

There is much suitable habitat in Scotland but breeding was not recorded until 1978, when one pair bred, hatching three young and fledging at least one of them (Dennis 1983b). The colonists may have originated from Scandinavia, in common with other recent additions to the British breeding list such as Wood Sandpiper and Temminck's Stint.

Purple Sandpipers have probably bred at the original site each year since 1978. Success has not always been proved, however, because local ornithologists have not searched for nests and eggs but have relied on seeing young, thus minimising disturbance and the chance of leading predators to nests. It is very pleasing to report that a second site was added in 1989 when breeding was proved well away from the original site. Up to three pairs have been proved breeding in any one year, but it is likely that the total is at least four pairs. There is every reason to believe that if disturbance can be kept to a minimum, colonisation of Scotland will continue.

JACQUIE CLARK

Years	Present, no breeding evidence		Breeding evidence		All records		
	Br	Ir	Br	Ir	Br	Ir	Both (+CI)
1968–72	0	0	0	0	0	0	0
1988–91	1	0	2	0	3	0	3
Purple Sandpiper			% change		0		

Snowy Owl

Nyctea scandiaca

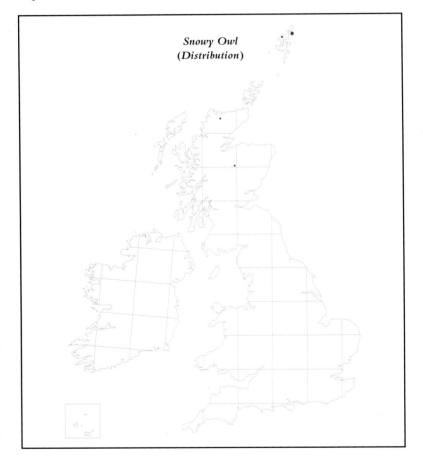

This large white owl with yellow eyes and variable black markings is mainly a circumpolar breeding bird. On the high Arctic tundra, lemmings are its major prey, though it may take other rodents, hares and rabbits. As with other owls, clutch size is dependent on prey availability.

During the winter Snowy Owls disperse southwards, a very few reaching Britain and Ireland. During 1988–91 one or two were usually present on Shetland, and summering birds appeared in April or May. Breeding was first proved on Fetlar in 1967, when eggs were laid in a nest on the ground among grass and ridges of bare rock (*68–72 Atlas*). Breeding continued until 1975, but wardening could not guarantee the presence of a male to partner the two females. A male found exhausted on an oil-rig

was released on Fetlar in 1989, but he remained only two days before leaving. The eggs laid that year were infertile; the Fetlar record shown here refers to this nesting attempt.

Should a pair re-establish themselves on Fetlar, then prevention of disturbance would be vital for the successful raising of a brood.

PHILIP JACKSON

68–72 Atlas p 252

Years	Present, no breeding evidence		Breeding evidence		All records		
	Br	Ir	Br	Ir	Br	Ir	Both (+CI)
1968–72	4	0	1	0	5	0	5
1988–91	3	0	1	0	4	0	4
Snowy Owl			% change		−20.0	0	−20.0

Short-toed Treecreeper

Certhia brachydactyla

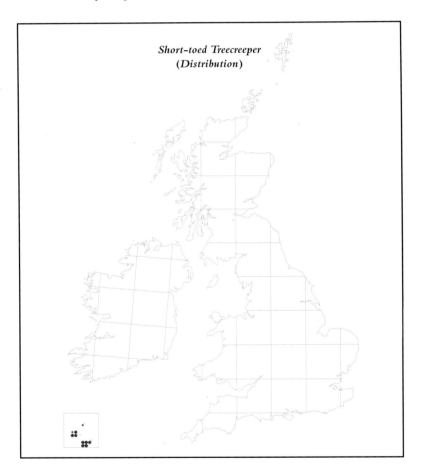

The Short-toed Treecreeper, unlike its sibling species, is absent from Britain and Ireland, but is found in the Channel Islands. There it replaces the common Treecreeper, and mainly occurs in the steep-sided wooded valleys characteristic of many parts of the archipelago. In those parts of Europe where both species are present, the Short-toed Treecreeper favours broadleaved woodland while *C. familiaris* occupies coniferous areas.

The separation of the two species in the field is not easy, and claims that *brachydactyla* has been found holding territory in S England have all been withdrawn. It is probably not under threat in the Channel Islands, but is unlikely to colonise Britain.

CHRIS MEAD

68–72 Atlas p 450

Years	Present, no breeding evidence		Breeding evidence		All records		
	Br	Ir	Br	Ir	Br	Ir	Both (+CI)
1968–72	0	0	0	0	0	0	5
1988–91	0	0	0	0	0	0	10
Short-toed Treecreeper			% change		0	0	−40.0

Brambling

Fringilla montifringilla

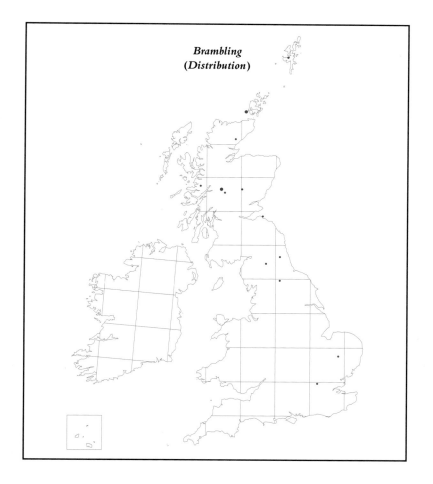

This white-rumped finch, close relative of the abundant Chaffinch, has the merest toehold in Britain as a breeding species. The breeding record for Orkney shown on the Distribution map is the first for the island group. With the exception of the record of a pair feeding fledged young in 1991 in SE Highland, most of the other records scattered through mainly E Britain, are of males recorded singing in summer. Their song is a nasal 'dzweess', not unlike that uttered by a Greenfinch. It is commonly given by winter visitors in April immediately before they return eastwards.

The first accepted breeding record was in Sutherland in 1920. Between then and 1979, when breeding was again confirmed in Scotland (*Birds in*

Scotland), the Brambling did not breed in Britain. Over the next ten years, however, a further seven pairs were confirmed as breeding, with as many as 35 possible breeding records (RBBP). The records were widely scattered in Scotland and E England.

The habitats used in Britain are mainly in birch scrub and plenty of suitable sites are available should the stirrings of colonisation, seen over the last decade, lead to more sustained attempts. Variable numbers are present in the winter, depending on the availability of beech mast in Britain and on the Continent, with 2,000,000 possibly present in some years and as few as 50,000 in others. Annual breeding numbers in Britain are 1–2 pairs.

CHRIS MEAD

68–72 Atlas p 451

Years	Present, no breeding evidence		Breeding evidence		All records		
	Br	Ir	Br	Ir	Br	Ir	Both (+CI)
1968–72	2	0	3	0	5	0	5
1988–91	11	0	2	0	13	0	13
Brambling			% change		160.0	0	160.0

Serin

Serinus serinus

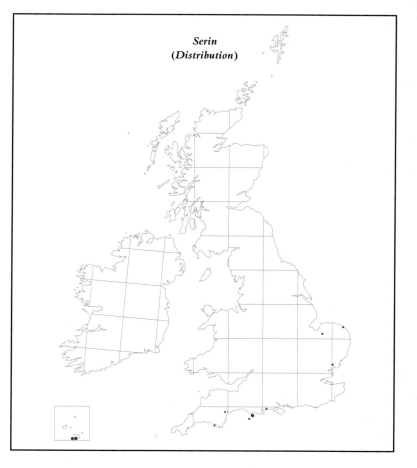

This small yellow and green finch has been expanding its range northwards from the Mediterranean since 1800, and has reached Scandinavia. In

Britain it has become an annual visitor since the 1960s, and is known to have bred twice by 1974 – in Dorset in 1967 and Sussex in 1969. The colonisation then predicted has not occurred, even though there seems to be plenty of suitable habitat. Serins prefer nesting in gardens, parks, orchards and churchyards, and as long as there is plenty of cover will nest in urban as well as rural areas. As they are principally seed eaters, they should be able to find food easily, but the Distribution map shows that the expansion rate has been disappointing. Serins are still reported annually from Dorset, the pioneer county, but from 1978 to 1985 the focus of attention moved to Devon, with nine cases of successful breeding, and other birds present. From time to time there are promising observations in other counties, as far north as Shropshire, and there may be a slight increase in numbers reaching England. During the 1980s there were never more than two confirmed breeding records from Britain in any one year (RBBP). Breeding was first suspected on Jersey in 1972, and by 1980 there were thought to be more than 20 singing males on the island (Long 1981).

The Serin is a small bird and builds a small, well-concealed nest. As with other cardueline finches, where the habitat is suitable it is not unusual for nests to be clustered in small groups. The fledged young remain in family parties, but may move away from the nesting area.

One can only speculate that the water barrier of the English Channel is inhibiting northward breeding expansion by Serins.

PHILIP JACKSON

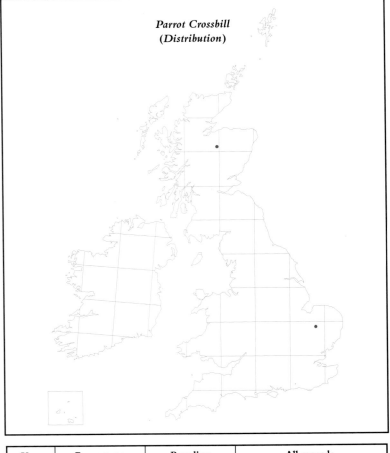

Parrot Crossbill
(Distribution)

68–72 Atlas p 424

Years	Present, no breeding evidence		Breeding evidence		All records		
	Br	Ir	Br	Ir	Br	Ir	Both (+CI)
1968–72	0	0	5	0	5	0	6
1988–91	7	0	1	0	8	0	10
Serin			% change		60.0	0	50.0

Years	Present, no breeding evidence		Breeding evidence		All records		
	Br	Ir	Br	Ir	Br	Ir	Both (+CI)
1968–72	0	0	0	0	0	0	0
1988–91	0	0	2	0	2	0	2
Parrot Crossbill			% change			0	

Parrot Crossbill

Loxia pytyopsittacus

Parrot Crossbills are associated with coniferous trees, and prefer Scots pine. Their main breeding range is from Fenno-Scandia eastwards to the Ural mountains. As with other crossbills, failure of the cone crop may lead them to erupt from their home areas and, following such events, breeding has occurred in Poland, Germany and the Baltic states. Irruptions into Britain occurred in 1962, 1982 and 1990, and although some remained in the Pennines during 1983, the first recorded breeding, in Norfolk, was not until February 1984 (Davidson 1985). During the period of the *88–91 Atlas*, breeding occurred near Loch Garten in Scotland (*Scottish Bird News* 24) and in Breckland, both during 1991.

Perhaps these unusual and attractive birds will retain their tenuous foothold here and expand in range, reinforced by further irruptions. They have a long nesting season, from December to June, and when in the tops of pines may be difficult to detect. The exact relationship between the crossbill species occurring in Britain is complex (Nethersole-Thompson 1975) and it is difficult to separate this species with certainty from Common and Scottish Crossbills.

PHILIP JACKSON

Scarlet Rosefinch

Carpodacus erythrinus

The Scarlet, or Common, Rosefinch breeds in a wide variety of habitats across its N Eurasian range, from copse and scrub to oakwoods, forest and lowland swamp, and even suburban gardens. At the time of the *68–72 Atlas* the breeding of the Scarlet Rosefinch in these islands seemed, at best, a long shot (Ferguson-Lees 1971). Over the following decade, however, its continuing westward range expansion into Scandinavia and N Europe led to a more optimistic attitude, and the first breeding record was eagerly awaited (Sharrock and Hildén 1983). This came in June 1982 when a pair, with nest and eggs, was found in the Highland Region of Scotland (Mullins 1984). Hopes for a more general colonisation were shortlived, for although varying numbers of individuals (mostly singing males) were present in Britain during the breeding season (RBBP), no further breeding records were reported for the rest of the decade.

A second nest was discovered, in coastal Sutherland, during Atlas fieldwork in June 1990. As with the first nest, attention was drawn to the area by the male's distinctive song, although in this case the nest was still being built and the male guarding the female. The nest was in a rough

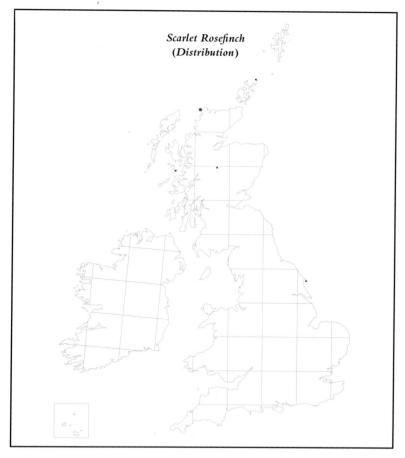

Scarlet Rosefinch
(Distribution)

patch of vegetation (consisting mainly of a gooseberry bush) by a tiny stream. The nest was subsequently monitored and the young fledged successfully.

Although a well watched pair (and second male) were present throughout July 1991 at Flamborough Head and seemed intent on breeding, neither nest nor young were found (P.A. Lassey). The two Scottish nests remained the only proven breeding records for Britain until the unprecedented colonisation of Flamborough Head by up to 15 birds, with at least two nests, in May and June 1992 (Lassey and Wallace 1992). Breeding was also recorded in Suffolk during the same season. Breeding has not yet been recorded in Ireland.

DAVID W. GIBBONS

Years	Present, no breeding evidence		Breeding evidence		All records		
	Br	Ir	Br	Ir	Br	Ir	Both (+CI)
1968–72	0	0	0	0	0	0	0
1988–91	4	0	1	0	5	0	5
Scarlet Rosefinch			% change		0		

Bias and the Key Squares Survey

During the planning of the *88–91 Atlas*, great care was taken to ensure that the measures of abundance obtained were not open to systematic bias (see *Introduction and Methods*). Such bias could have arisen in three ways: regional variation in observer effort, regional variation in the proportion of tetrads visited only once, and by allowing observers to choose which tetrads (out of 25 in each 10-km square) they covered. The first of these biases was overcome by standardising the time spent in each tetrad, as outlined at length in the *Introduction and Methods*. The remaining two sources of bias are examined here, although no corrections have been made for them in the production of the Abundance maps.

Observers were asked to make two one-hour visits to each tetrad wherever possible, but in remote areas it was suggested that a single two-hour visit would be a more efficient use of time. On a single two-hour visit the total number of birds encountered was recorded for 'count' species. On two one-hour visits, one count was performed early and one

late in the season, and the higher of the two counts was recorded. For highly clumped species, the two figures obtained using the alternative visiting regimes would probably be similar, but for evenly distributed species a two-hour count total could be, on average, up to twice the maximum of the two one-hour counts. Frequency indices could also be affected, but to a lesser extent. Fig. 7 shows that there was clear geographical variation in the proportion of tetrads visited only once in each 10-km square.

The final source of bias occurred because observers were allowed to *choose* which tetrads they visited. Observers faced with a choice of eight tetrads out of 25 most likely would have chosen either those tetrads closest to their home, those which covered the widest variety of habitats, those which they thought were best for birds, those which were most accessible, or those based on other criteria. All of these choices could have introduced bias. For example, in areas with good coverage, in which most tetrads were visited in each 10-km square, tetrads which were both good and bad for birds would have been covered. In areas with poor coverage, only the 'best' tetrads may have been visited. This would have led to geographical bias; in particular to biased indices for species in poorly covered areas, compared with more realistic ones in well covered areas.

One solution that was considered was to instruct observers to visit randomly allocated tetrads in every 10-km square, rather than tetrads of their own choice. However, it was felt that there would be strong objections from many volunteer observers if they were not allowed an element of choice in the tetrads they visited. It was crucial that support for the survey was not diminished, because the scale of the project was such that large numbers of enthusiastic volunteers were essential to ensure its successful completion. The methods were already appreciably more complex than those adopted for the *68–72 Atlas*, and to increase this complexity would have dampened enthusiasm. Furthermore, if observers were constrained to visit only certain tetrads many rarities would probably have been missed. So, observers were allowed to visit tetrads of their own choice and the associated potential problems of bias were tackled through a sample survey – the Key Squares Survey (KSS) – which ran concurrently with the main Atlas survey.

There were three fundamental differences between the main Atlas survey and the KSS. Firstly, the former covered all 10-km squares, whilst the latter only covered a sample. Secondly, in the former, observers selected their tetrads, but, in the latter, observers visited randomly allocated ones. Finally, in the former, observers were able to choose between one two-hour visit or two one-hour visits to each tetrad, but, in the latter, had to make two one-hour visits. A comparison of the measures of abundance obtained from the KSS and the main Atlas survey enabled any bias to be estimated.

In addition to the timed tetrad visits, point counts (see Verner 1985 and Bibby *et al.* 1992 for full reviews) were undertaken as part of the KSS. This was because very common species present problems for an index of abundance based on frequency of occurrence. In many 10-km squares these species occur in every tetrad, and thus their index of frequency would always be high, and often 1. Above a threshold density at which a species occurs in every tetrad, increases in density will not be matched by increases in frequency of occurrence for this will already be at its maximum

Fig. 7. The proportion of tetrads in each 10-km square visited only once. The dots of increasing size refer to proportions of: 0–0.040, 0.041–0.500, 0.501–1. These class limits provide similar sample sizes within each grouping.

of 1. Thus the true variation in density of common species can be obscured. To overcome this problem, point counts were carried out within each random tetrad in each key square, as these reflect density more precisely and generally will do so in a linear manner (Gates *et al.* 1993).

KSS timed tetrad visits

In Britain, 1 in 9, and in Ireland, 1 in 36 10-km squares, were designated as key squares. These squares were the same as those used by the BSBI monitoring scheme, with the exception that in Ireland only a 1 in 4 sample of the BSBI's 1 in 9 (and thus a 1 in 36 sample of all squares) was used, because of the lack of observers. Fieldwork was undertaken using the methods already outlined for the main survey, with the following exceptions. Observers (individuals or teams) visited at least 15 specified tetrads, and spent one hour early in the season and one hour late in each. Thus observers were not allowed to choose which tetrads to visit, nor did they have the option of visiting tetrads once only. In each key square the same target tetrads were visited; these were A, B, D, F, H, J, K, M, P, R, T, U, W, X and Y. These tetrads were selected using a randomised statistical design which ensured that each tetrad row (e.g. E to Z) and column (e.g. A to E, see Fig. 1 in the *Introduction and Methods*) contained three target tetrads. Observers were allowed to visit more than the 15 target tetrads if they wished, but if they did so they had to visit either 20 tetrads (those already listed, plus C, G, N, Q and Z) or all 25 tetrads; however, this rarely happened.

Allocation of target tetrads for coastal 10-km squares was a little complex, and the following instructions were given:

(i) If the centres of 15 or fewer tetrads fell on land, all tetrads on land had to be surveyed.

(ii) If the centres of 16–20 tetrads fell on land, the observer(s) had two options: either to survey all of the 15 target tetrads and any of the five extras – C, G, N, Q and Z – whose centre fell on land, or to survey all the tetrads whose centres fell on land.

(iii) If the centres of 21–24 tetrads fell on land, the observer(s) had three options: either to survey all of the 15 target tetrads whose centre fell on land, to survey all of the target tetrads and the five extras – C, G, N, Q and Z - whose centres fell on land, or to survey all of the tetrads whose centres fell on land.

Thus, in some coastal squares the minimum amount of fieldwork could have been more than 15 tetrads.

KSS point counts

Two point counts (one early – April to May – and one late – June to July – in the season) were performed at, or as close to as possible, the centre of each target tetrad. The point counts were of ten minutes' duration, during which time the observer remained stationary and counted all individual birds seen or heard, up to an unlimited distance. Observers were instructed to do the count sometime between 0700 and 1200 (BST), although in remote areas they could count in the afternoon if necessary. The date of the count was recorded, as was the time (either morning or afternoon).

Finally, using the classification in Table 4, the habitat surrounding each counting station up to a distance of 50m was recorded. For most counting stations only a single habitat code was necessary, although if they fell at the boundary of two habitat types, two codes (a major and minor) were recorded. Habitat information was used for estimating population sizes (see *Population Estimates for Breeding Birds in Britain and Ireland*).

Table 4.   *The habitat classification used at key square point count stations.*

Each habitat has a unique two number code (e.g. coniferous woodland 01, and reed bed 22), one number from column A and one number from column B.

	A		B
0	Woodland and Scrub	0	Broadleaved woodland
		1	Coniferous woodland
		2	Mixed woodland (Broadleaved and coniferous)
		3	Scrub (all scrub including downland and coastal scrub)
1	Semi-natural Grassland/Heaths	0	Bracken
		1	Chalk grassland and similar
		2	Damp or unimproved lowland grassland (include flood meadows)
		3	Dry lowland heath
		4	Wet lowland heath
		5	Upland heather moor (unenclosed land; depth of peat less than 0.5m)
		6	Upland grassland (unenclosed and unimproved land; depth of peat less than 0.5m)
		7	High montane heath/grassland (on exposed summits)
			Note: Record reclaimed marsh and other maritime grassland as 42 and 43 respectively. Record improved grasslands as 60, 61 or 62. Record upland heather moor or upland grassland with deep peat (more than 0.5m) as 20.
2	Bog, Fen and Marsh	0	Acid bog (include blanket and raised bog; depth of peat more than 0.5m)
		1	Fen/marsh/swamp
		2	Reed bed (with *Phragmites*)
			Note: Record flood meadow and wet lowland heath as 12 and 14 respectively.
3	Water Bodies (Freshwater)	0	Lowland river/stream (below 800ft = 250m)
		1	Upland river/stream (above 800ft = 250m)
		2	Canal
		3	Standing water body less than 5 ha
		4	Standing water body more than 5 ha
			Note: 5 ha is approximately 12 acres or 8 football pitches.
4	Coastal	0	Intertidal mud/sand (include sandy beaches)
		1	Saltmarsh
		2	Reclaimed marsh
		3	Other maritime grasslands (include machair)
		4	Brackish pools and lagoons
		5	Gravel/pebbles/shells (non-sandy beaches, bar, spit etc)
		6	Sand dunes (include dune slacks, but record scrub as 03)
		7	Intertidal rock
		8	Cliff/small rocky island (record scrub as 03)
5	Exposed and Bare Surfaces	0	Inland cliff/crag/montane rock/scree/boulder slope
		1	Limestone pavement
		2	Quarry surface
		3	Spoil (e.g. slag heap, but record rubbish tip as 77)
6	Improved Farmland	0	Improved lowland grassland
		1	Enclosed, improved upland grassland
		2	Unenclosed, improved upland grassland
		3	Arable (crops)
		4	Mixed farmland (grazing and crops)
		5	Farm buildings

Note: Record chalk grassland, unimproved lowland grassland,

unimproved upland grassland, and montane grassland as 11, 12, 16 and 17 respectively. Hedges are not included as they form part of other habitats (e.g. 60 and 63).

7	Miscellaneous (mostly artificial)	0	Urban/suburban park
		1	Rural park
		2	Golf course
		3	Cemetery/churchyard
		4	Residential housing (including gardens)
		5	Non-residential buildings
		6	Sewage treatment works
		7	Rubbish tip
		8	Waste land (record scrub as 03)

Improved grassland: grass regularly treated with artificial fertilisers, distinguished by its bright colour, lush growth and even texture.

Unimproved grassland: not treated with artificial fertilisers, usually grazed or mown regularly, may be rank and neglected.

Enclosed land: land enclosed within a hedge, stone-wall, fence or equivalent.

Note that a more comprehensive bird habitat coding system now exists (Crick 1992). In future, it is recommended that the Crick system be used instead of that presented here.

As for the main Atlas survey, only birds 'using' tetrads were recorded during KSS timed tetrad visits and point counts; summering, non-breeding birds were included, but late winterers and passage migrants were excluded.

Observers were asked to complete all KSS timed tetrad visits and point counts for a given 10-km square within a single season, wherever possible. In addition, ROs were asked to allocate the main Atlas survey and the KSS of the same 10-km square to different observers, although in practice this was not always possible. This was in an attempt to ensure that the results of the two surveys were independent.

Customised KSS recording forms were provided for tetrad visits and point counts; the former was very similar to the worksheet used in the main Atlas survey (see Fig. 2 in the *Introduction and Methods*), and an example of the latter is shown in Fig. 8. The data were processed and checked in the same manner as for the main Atlas survey, outlined in the *Introduction and Methods*.

Results of the KSS

Timed tetrad visits and point counts were carried out in 304 out of a total of 331 coverable key squares (Fig. 9): 278 out of 304 of these (91%) were in Britain, and 26 out of 27 (96%) in Ireland. In three key squares only timed tetrad visits were undertaken, and in a further three, only point counts were made. In many of the uncovered key squares, fieldwork was commenced but not completed and their data have not been included in these analyses; rather, they were converted to supplementary records and incorporated into the main Atlas survey.

Timed tetrad visits

These data can be used to answer two different questions. Firstly, were the frequency indices biased by allowing observers to choose which tetrads they visited and the number of visits to each? Secondly, was there regional variation in this bias (e.g. no bias in some areas, but strong bias in others)? To test this, the mean frequency index for each species across all key squares has been compared with that obtained from the main survey (in which the observer chose the tetrads and number of visits to each) in the same 10-km squares. The KSS and main survey were both carried out in

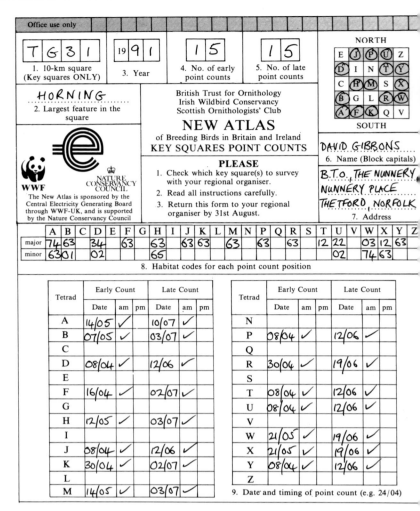

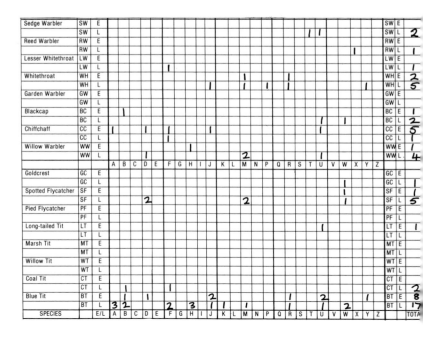

Fig. 8. (a) A completed header section of the KSS point count recording form.

(b) A completed part of the species section of the KSS point count recording form. E = early count, L = late count.

a total of 297 10-km squares, and this comparison can be made within these only. Fig. 10 shows the result of this comparison. If there was no bias due to observer choice, then the slope of the line in Fig. 10 would be 1, with all the data points (each referring to a species) falling upon it. The slope of the line was in fact 0.972 (standard error of estimate 0.0153), but was not significantly different to 1. Thus, for the 'average' species, there was no bias due to observer choice. However, the species are scattered about the line, those falling above it having artificially inflated indices due to observer choice, those falling below it, similarly, having reduced indices.

To test whether or not the indices for each individual species were biased due to observer choice, the indices from the two surveys were compared for each species separately, using a matched pairs signed-rank test. Out of 201 species which could be tested in this manner, significant biases (at the $P < 0.05$ level) were found for only 32 species (see Table 5). In this table, a negative signed-rank means that observers, when given a choice of tetrads, preferred to visit those that did not contain these species; a positive value means that the observer preferred tetrads that did contain these species. As can be seen, some birds of open farmland (Skylark, Rook, Tree Sparrow, Goldfinch, Linnet and Yellowhammer) and birds of mountain and moorland (Merlin, Red Grouse, Meadow Pipit and Raven) had significant negative values (i.e. observers selectively avoided these

habitats), whilst some birds of woodland (Woodcock, Tawny Owl, Nightingale and Wood Warbler) and birds associated with water (Little Grebe, Grey Heron, Mute Swan, Tufted Duck, Coot, Oystercatcher and Kingfisher) had positive values (the observers selectively chose these habitats).

Provided that there was no regional variation in this bias, then these results do not affect the interpretation of the Abundance maps. Thus, for example, the frequency index for the Skylark was biased downwards due to observer choice but, providing it was biased downwards to the same amount throughout its range, then the Abundance map showing regional variation in relative abundance is still a faithful representation of the variation in true, but unknown, breeding density. Such regional bias could have occurred for one of two reasons. Firstly, observers may have gone out of their way to find a species that was rare in one area, but not bothered to search for it in another where it was common. Secondly, biased indices may have occurred in poorly covered areas (those in which few tetrads were covered) because observers may have only covered the most 'interesting' tetrads.

To test whether or not there was any regional variation in bias the analysis presented in Table 5 was extended regionally. For this purpose, the 297 10-km squares which were surveyed, both as part of the main Atlas survey and the KSS, were divided into four separate regions. These were: Ireland, N Britain (mostly Scotland), E Britain (E and central England), and W Britain (W England, Wales, SW Scotland, Isle of Man and Channel Islands) (see Table 5 for further details). These regions were chosen such that, within Britain, sample sizes were similar in each region. The results of the regional analyses are included in Table 5. Although a higher number of species (44) recorded significant bias in at least one of the four regions, for only one of these species (Blue Tit) was the bias in different directions in different regions (positive in Scotland and negative in Ireland). As suggested, the Skylark was negatively biased in all regions.

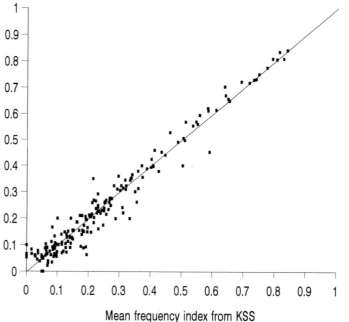

Mean frequency index from main survey

Mean frequency index from KSS

Fig. 9. Coverage of key squares. Open dot = key squares not covered; large filled dot = both timed tetrad visits and point counts undertaken; small filled dot = only point counts undertaken; small filled square = only tetrad visits undertaken.

Fig. 10. The relationship between the mean frequency index of each species from the main survey with that from the KSS. Mean indices calculated only across those squares in which the species was recorded by one or both of the surveys. $n = 201$, $R^2 = 0.953$, $P < 0.0001$. See text for further details.

Table 5. *A comparison of the measures of relative abundance obtained from the main Atlas survey and the KSS. See text and footnotes for explanation.*

	Frequency Indices							Count		
	Britain and Ireland			Direction and significance of the bias in				Britain and Ireland		
	Mean index for:		Direction and significance of the bias	E Britain	W Britain	Scotland	Ireland	Mean count for:		Direction and significance of the bias
	Main Survey	KSS						Main Survey	KSS	
Red-throated Diver	0.2021	0.2123	− NS			− NS		0.5170	0.4449	+ NS
Black-throated Diver	0.1417	0.0905	+ NS			+ NS		0.2613	0.1326	+ ★
Little Grebe	0.0989	0.0741	+ ★★	+ NS	+ ★	+ NS	+ NS	0.2121	0.1667	+ ★
Great Crested Grebe	0.1268	0.1054	+ NS	+ NS	− NS	+ NS	− NS	0.4093	0.3371	+ NS
Slavonian Grebe	0.0909	0.1333	− NS			− NS		0.1818	0.4000	− NS
Black-necked Grebe	0.0000	0.0667	− NS	− NS				0.0000	0.1000	− NS
Fulmar	0.3499	0.3007	+ NS	− NS	− NS	+ NS	+ NS	0.0557	0.2870	− NS
Manx Shearwater	0.1944	0.0639	+ NS			+ NS	+ NS	.	.	
Leach's Petrel	0.0000	0.0625	− NS			− NS		.	.	
Gannet	0.2056	0.2083	− NS		− NS	+ NS	+ NS	.	.	
Cormorant	0.1346	0.1519	− NS	+ NS	− NS	− NS	− NS	0.0000	0.0019	− NS
Shag	0.3098	0.2408	+ NS		+ NS	+ NS	+ NS	.	.	
Bittern	0.0435	0.0385	+ NS	+ NS				0.0435	0.0385	+ NS
Grey Heron	0.2169	0.1840	+ ★	+ NS	+ NS	+ NS	+ NS	0.0749	0.0717	− NS
Mute Swan	0.1783	0.1390	+ ★★★	+ ★	+ ★	+ NS	+ NS	0.8049	0.9018	+ NS
Greylag Goose	0.1142	0.1025	+ NS	+ ★	− NS	− NS	− NS	1.9009	2.0222	− NS
Canada Goose	0.1508	0.1532	− NS	+ NS	− NS	+ NS		2.0173	2.1886	− NS
Egyptian Goose	0.1921	0.0930	+ NS	+ NS	− NS			0.4652	0.4026	− NS
Shelduck	0.2443	0.2298	+ NS	+ NS	− NS	+ NS	+ NS	1.9938	2.1135	+ NS
Wood Duck	0.0000	0.0667	− NS	− NS				0.0000	0.0667	− NS
Mandarin	0.0831	0.0344	+ NS	+ NS	+ NS	− NS		0.1679	0.0711	+ NS
Wigeon	0.0924	0.0903	+ NS	− NS	− NS	+ NS		0.4625	0.3565	+ NS
Gadwall	0.0866	0.0848	− NS	− NS	− NS	− NS		0.4241	0.4167	+ NS
Teal	0.0914	0.0871	+ NS	+ NS	+ NS	− NS	− NS	0.3209	0.2825	+ NS
Mallard	0.4072	0.4015	+ NS	+ NS	− NS	+ NS	+ NS	.	.	
Pintail	0.0492	0.0556	− NS	+ NS	+ NS	− NS		0.1700	0.1111	+ NS
Garganey	0.0289	0.0573	− NS	− NS		− NS		0.0439	0.1291	− NS
Shoveler	0.0651	0.0730	− NS	− NS	− NS	+ NS	− NS	0.2475	0.2302	− NS
Red-crested Pochard	0.0544	0.0000	+ NS	+ NS				0.0822	0.0000	+ NS
Pochard	0.0671	0.0654	− NS	− NS	− NS	+ NS	− NS	0.2845	0.3735	− NS
Tufted Duck	0.1472	0.1251	+ ★	+ NS	+ NS	+ NS	+ NS	1.0984	1.0781	+ NS
Eider	0.2757	0.2747	+ NS	+ NS	− NS	+ NS		4.8402	5.9883	− NS
Common Scoter	0.2083	0.1482	+ NS			+ NS		0.7500	0.3630	+ NS
Goldeneye	0.0621	0.0614	+ NS	− NS	− NS	+ NS		0.1940	0.1074	− NS
Red-breasted Merganser	0.1344	0.1136	+ NS	− NS	− NS	+ NS	+ NS	0.4177	0.3008	+ NS
Goosander	0.0844	0.0921	− NS	+ NS	− NS	− NS		0.2212	0.2247	− NS
Ruddy Duck	0.0752	0.0760	− NS	+ NS	− NS	− NS	+ NS	0.2311	0.2736	− NS
Red Kite	0.2142	0.3500	− NS		− NS			0.2455	0.4667	− NS
Marsh Harrier	0.0978	0.0568	+ NS	+ NS				0.1065	0.0634	+ NS
Hen Harrier	0.1150	0.1096	− NS	− NS	− NS	+ NS	− NS	0.1676	0.1788	− NS
Goshawk	0.0167	0.0667	− NS	− NS	− NS			0.0167	0.0667	− NS
Sparrowhawk	0.1393	0.1314	+ NS	+ NS	− NS	+ NS	− NS	0.6006	0.6457	− NS
Buzzard	0.3709	0.4007	− NS	+ NS	− ★★	+ NS	− NS			
Golden Eagle	0.1755	0.0944	+ ★			+ ★		0.2279	0.1240	+ ★
Osprey	0.0671	0.0905	− NS			− NS		0.0786	0.1095	− NS
Kestrel	0.3214	0.3233	− NS	+ NS	− NS	− NS	− NS	.	.	
Merlin	0.0606	0.0883	− ★	− NS	− NS	− NS	− NS	0.0845	0.1135	− ★
Hobby	0.0513	0.0531	− NS	− NS	+ NS			0.0591	0.0700	− NS
Peregrine	0.0905	0.0972	− NS		− NS	− NS	− NS	0.1307	0.1428	− NS
Red Grouse	0.2167	0.2681	− ★★	− NS	− ★★	− NS	+ NS	.	.	
Ptarmigan	0.0923	0.1078	− NS			− NS		0.1689	0.2495	− NS
Black Grouse	0.0808	0.1125	− NS		− NS	− NS		0.1838	0.2427	− NS
Capercaillie	0.0683	0.0222	+ NS			+ NS		0.0683	0.0444	+ NS
Red-legged Partridge	0.3115	0.3151	− NS	− NS	+ NS	− NS		.	.	
Grey Partridge	0.1911	0.1962	− NS	+ NS	− NS	− NS		.	.	
Quail	0.0489	0.0611	− NS	− NS	− NS	+ NS		0.0618	0.0852	− NS
Pheasant	0.5356	0.5545	− NS	− NS	− NS	+ NS	− NS	.	.	
Golden Pheasant	0.0370	0.0787	− NS		− NS	− NS		0.0370	0.1528	− NS
Water Rail	0.0509	0.0486	− NS	+ NS	+ NS	+ NS	− ★	0.0789	0.0865	− NS
Corncrake	0.2967	0.3603	− NS			+ NS	− NS	0.1894	0.4300	− NS

Table 5 (*continued*).

	Frequency Indices							Count		
	Britain and Ireland			Direction and significance of the bias in				Britain and Ireland		
	Mean index for:		Direction and significance of the bias	E Britain	W Britain	Scotland	Ireland	Mean count for:		Direction and significance of the bias
	Main Survey	KSS						Main Survey	KSS	
Moorhen	0.3736	0.3558	+NS	+NS	−NS	+NS	+NS			
Coot	0.1971	0.1652	+★	+★	−NS	+★★	−NS	1.6123	1.3815	+NS
Oystercatcher	0.4256	0.3790	+★★	+★	+NS	+★	−NS			
Avocet	0.1393	0.1290	+NS	+NS				3.1737	5.2290	−NS
Stone Curlew	0.3333	0.2000	+NS	+NS				1.0000	0.5333	+NS
Little Ringed Plover	0.0542	0.0486	+NS	−NS	+NS			0.1141	0.1238	+NS
Ringed Plover	0.2077	0.2071	−NS	+NS	−NS	−NS	+NS	0.8216	1.1566	−NS
Dotterel	0.0000	0.0667	−NS			−NS		0.0000	0.1334	−NS
Golden Plover	0.2235	0.2652	−NS	−NS	−NS	−NS	+NS	0.9087	1.1622	−NS
Lapwing	0.3873	0.3888	−NS	−NS	−NS	+NS	+NS			
Dunlin	0.1522	0.1555	−NS	+NS	−NS	−NS	+NS	0.8600	0.8712	−NS
Ruff	0.0927	0.0620	+NS	+NS				0.5707	0.4620	+NS
Snipe	0.2083	0.2079	−NS	+NS	+NS	−NS	−NS			
Woodcock	0.0883	0.0559	+★	+NS	+NS	+NS	−NS	0.1185	0.0774	+NS
Black-tailed Godwit	0.0276	0.0620	−NS	−NS	+NS			0.0869	0.2263	−NS
Whimbrel	0.1841	0.0887	+NS			+NS		0.4336	0.1899	+NS
Curlew	0.3588	0.3801	−NS	−NS	−NS	+NS	−NS			
Redshank	0.2065	0.2055	+NS	+NS	−NS	+NS	−NS			
Greenshank	0.1844	0.1504	+NS			+NS		0.3059	0.2333	+NS
Wood Sandpiper	0.0000	0.0667	−NS			−NS		0.0000	0.0667	−NS
Common Sandpiper	0.2353	0.2260	+NS	−NS	−NS	+NS	−NS			
Arctic Skua	0.5033	0.4015	+NS			+NS				
Great Skua	0.5897	0.4534	+NS			+NS				
Mediterranean Gull	0.0500	0.0000	+NS	+NS						
Black-headed Gull	0.3022	0.3060	−NS	−★	−NS	+NS	−NS	1.4271	0.7576	+NS
Common Gull	0.2723	0.2488	+NS	−NS	+NS	+NS	−NS	0.2911	0.4609	−NS
Lesser Black-backed Gull	0.1618	0.1859	−NS	−NS	−NS	+NS	−NS	0.0032	0.0094	−NS
Herring Gull	0.3057	0.3239	−NS	+NS	−NS	+NS	−★	0.1440	0.0528	−NS
Great Black-backed Gull	0.2654	0.2329	+NS	+NS	+NS	+NS	−NS	0.0678	0.0165	+NS
Kittiwake	0.2882	0.1993	+NS	+NS	+NS	+★	−NS			
Sandwich Tern	0.1305	0.1291	−NS	+NS	−NS	+NS	−NS			
Common Tern	0.1699	0.1549	+NS	+NS	−NS	+NS	+NS	0.1396	0.1558	+NS
Arctic Tern	0.3553	0.2618	+NS	−NS	−NS	+★	−NS	1.6401	2.2837	−NS
Little Tern	0.2301	0.2270	+NS	−NS	+NS	−NS	+NS			
Guillemot	0.2508	0.2286	+NS	−NS	+NS	+NS	−NS			
Razorbill	0.3121	0.2209	+NS		+NS	+NS	+NS			
Black Guillemot	0.2320	0.1605	+★		+NS	+★	+NS			
Puffin	0.2483	0.1749	+NS		+NS	+NS	−NS			
Feral Pigeon/Rock Dove	0.2825	0.3233	−★	−NS	−NS	−NS	−NS			
Stock Dove	0.3364	0.3501	−NS	−NS	−NS	−NS	+NS			
Woodpigeon	0.8064	0.8076	−NS	−NS	−NS	+NS	−NS			
Collared Dove	0.4083	0.3976	+NS	+NS	−NS	+NS	−NS			
Turtle Dove	0.3202	0.3094	+NS	−NS	+NS	−NS				
Ring-necked Parakeet	0.0634	0.0828	−NS	−NS				0.2382	0.2422	+NS
Cuckoo	0.3342	0.3447	−NS	+NS	−NS	+NS	−★			
Barn Owl	0.0791	0.0623	+NS	+NS	+NS	−NS	+NS	0.1065	0.0756	+NS
Little Owl	0.0956	0.1032	−NS	−NS	−NS	−NS		0.1322	0.1265	−NS
Tawny Owl	0.1272	0.0897	+★★	+NS	+NS	+NS		0.1604	0.1073	+★★★
Long-eared Owl	0.0350	0.0455	−NS	+NS	−NS	−NS	+NS	0.0396	0.0727	−NS
Short-eared Owl	0.1031	0.1006	−NS	−NS	−NS	−NS		0.1493	0.1309	−NS
Nightjar	0.0698	0.0333	+NS	+NS	+NS			0.1629	0.0917	+NS
Swift	0.5095	0.4981	+NS	+NS	−NS	+NS	+★★			
Kingfisher	0.0971	0.0623	+★★★	+★★	+NS	+NS	+NS	0.1151	0.0767	+★★
Green Woodpecker	0.2392	0.2194	+NS	+NS	−NS	+NS				
Great Spotted Woodpecker	0.2179	0.2029	+NS	+NS	−NS	+NS				
Lesser Spotted Woodpecker	0.0543	0.0534	−NS	−NS	+NS			0.0580	0.0601	−NS
Woodlark	0.1435	0.1067	+NS	+NS	+NS			0.3877	0.2800	+NS
Skylark	0.6377	0.7046	−★★★	−★	−★	−★	−★★			
Sand Martin	0.1466	0.1282	+NS	+NS	+NS	+NS	+NS	2.0114	1.2261	+NS

Table 5 (*continued*).

	Frequency Indices			Direction and significance of the bias in				Count		
	Britain and Ireland							Britain and Ireland		
	Mean index for:							Mean count for:		
	Main Survey	KSS	Direction and significance of the bias	E Britain	W Britain	Scotland	Ireland	Main Survey	KSS	Direction and significance of the bias
Swallow	0.7488	0.7519	− NS	− NS	− NS	+ NS	− ★	.	.	
House Martin	0.4875	0.4915	− NS	− NS	− NS	+ NS	− NS	.	.	
Tree Pipit	0.2107	0.2155	− NS	− NS	− NS	+ NS		.	.	
Meadow Pipit	0.5121	0.5695	− ★★★	− ★★	− NS	− ★★★	− NS	.	.	
Rock Pipit	0.3603	0.3133	+ ★	+ NS	+ NS	+ NS	+ NS	.	.	
Yellow Wagtail	0.2236	0.2369	− NS	− NS	− NS	− NS		.	.	
Grey Wagtail	0.2271	0.2198	+ NS	+ ★★	− NS	+ NS	− ★	.	.	
Pied Wagtail	0.5046	0.5064	− NS	+ NS	− NS	+ NS	+ NS	0.2712	0.2432	+ NS
Dipper	0.1801	0.1555	+ NS	+ NS	− NS	+ NS	+ NS	.	.	
Wren	0.8143	0.8370	− NS	− NS	− NS	− NS	− NS	.	.	
Dunnock	0.6404	0.6712	− ★	+ NS	− ★	− NS	− NS	.	.	
Robin	0.7751	0.7773	− NS	+ NS	− NS	+ NS	+ NS	.	.	
Nightingale	0.1298	0.0729	+ ★★	+ NS	+ ★			.	.	
Black Redstart	0.0000	0.0534	− NS	− NS	− NS			0.0000	0.0734	− NS
Redstart	0.2299	0.2276	+ NS	+ NS	− NS	+ NS		.	.	
Whinchat	0.1635	0.1854	− NS	− NS	− NS	− NS	− NS	.	.	
Stonechat	0.2580	0.2150	+ NS	− NS	+ ★	+ NS	− NS	.	.	
Wheatear	0.3379	0.3531	− NS	+ NS	− NS	− NS	+ NS	.	.	
Ring Ouzel	0.1236	0.1488	− NS	+ NS	− ★	− NS	+ NS	0.2202	0.2894	− NS
Blackbird	0.8285	0.8108	+ ★	+ NS	+ NS	+ ★	− NS	.	.	
Fieldfare	0.0698	0.0392	+ NS	− NS	+ NS			0.1377	0.0392	+ NS
Song Thrush	0.6490	0.6581	− NS	+ NS	− NS	+ NS	− NS	.	.	
Redwing	0.1297	0.1403	+ NS			+ NS		0.2349	0.4312	+ NS
Mistle Thrush	0.4013	0.3948	+ NS	+ NS	− NS	+ NS	− NS	.	.	
Cetti's Warbler	0.1198	0.0764	+ NS	− NS	+ NS			0.2686	0.1285	+ NS
Grasshopper Warbler	0.0861	0.0831	− NS	+ NS	+ NS	− NS	− ★	0.1165	0.0999	+ NS
Savi's Warbler	0.0500	0.0000	+ NS	+ NS				0.0500	0.0000	+ NS
Sedge Warbler	0.2513	0.2380	+ NS	+ ★★	+ NS	− NS	− NS	.	.	
Reed Warbler	0.1960	0.1495	+ ★	+ NS	+ NS	+ NS		.	.	
Dartford Warbler	0.1434	0.2000	− NS	− NS				0.2323	0.4889	− NS
Lesser Whitethroat	0.1899	0.2062	− NS	− NS	− NS	− NS	− NS	.	.	
Whitethroat	0.4345	0.4531	− NS	+ NS	− NS	− NS		.	.	
Garden Warbler	0.2526	0.2589	− NS	+ NS	− NS	− NS		.	.	
Blackcap	0.4022	0.4255	− NS	+ NS	− ★	+ NS	+ NS	.	.	
Wood Warbler	0.1503	0.1225	+ ★	+ NS	+ ★★★	− NS		.	.	
Chiffchaff	0.4466	0.4419	+ NS	+ NS	− NS	+ NS	+ NS	.	.	
Willow Warbler	0.6915	0.7235	− NS	+ NS	− NS	− NS	− ★	.	.	
Goldcrest	0.3405	0.3664	− NS	+ NS	− NS	− NS	− NS	.	.	
Firecrest	0.0833	0.1667	− NS	− NS	− NS			0.1333	0.2334	− NS
Spotted Flycatcher	0.2704	0.2766	− NS	− NS	− NS	+ NS	+ NS	.	.	
Pied Flycatcher	0.1486	0.1559	− NS	− NS	− NS	+ NS		.	.	
Bearded Tit	0.0376	0.0693	− NS	− NS	− NS			0.1055	0.6199	− NS
Long-tailed Tit	0.2576	0.2681	− NS	+ NS	− NS	− NS	− NS	.	.	
Marsh Tit	0.1823	0.1518	+ ★	+ ★★	− NS	+ NS		.	.	
Willow Tit	0.1131	0.1398	− ★	− NS	− ★	+ NS		.	.	
Crested Tit	0.0160	0.0833	− NS			− NS		0.0240	0.1367	− NS
Coal Tit	0.3181	0.3453	− ★	− NS	− ★	+ NS	− ★	.	.	
Blue Tit	0.7319	0.7307	− NS	− NS	− NS	+ ★	− ★	.	.	
Great Tit	0.6104	0.6158	+ NS	+ NS	− NS	+ ★	− NS	.	.	
Nuthatch	0.2308	0.2416	− NS	+ NS	− NS			.	.	
Treecreeper	0.1936	0.1947	− NS	+ NS	− NS	− NS	− NS	.	.	
Golden Oriole	0.0625	0.0889	− NS			− NS		0.0625	0.0889	− NS
Red-backed Shrike	0.0625	0.1034	− NS	+ NS		− NS		0.0625	0.1034	− NS
Jay	0.2536	0.2616	− NS	− NS	− NS	− NS	+ NS	.	.	
Magpie	0.7399	0.7330	+ NS	+ NS	− NS	+ NS	− NS	.	.	
Chough	0.2299	0.2922	− NS		− NS	− NS	+ NS	0.6343	0.6507	− NS
Jackdaw	0.5851	0.6129	− ★	− NS	+ NS	− ★	− NS	.	.	
Rook	0.5574	0.5959	− ★	+ NS	− ★	− NS	− NS	.	.	
Carrion Crow	0.7928	0.8106	− NS	− NS	− ★★★	− NS		.	.	
Hooded Crow	0.5839	0.6226	− NS	+ NS	+ NS	+ NS	− ★★	.	.	

Table 5 (*continued*).

	Frequency Indices							Count		
	Britain and Ireland			Direction and significance of the bias in				Britain and Ireland		
	Mean index for:							Mean count for:		
	Main Survey	KSS	Direction and significance of the bias	E Britain	W Britain	Scotland	Ireland	Main Survey	KSS	Direction and significance of the bias
Hybrid Crow	0.1852	0.1736	+ NS		+ NS	− NS				
Raven	0.2120	0.2589	− ★		− ★	− NS	− NS	0.4608	0.5621	− NS
Starling	0.7173	0.7191	− NS	+ NS	− NS	+ NS	− NS	.	.	
House Sparrow	0.6526	0.6504	+ NS	+ NS	− NS	+ NS	− NS	.	.	
Tree Sparrow	0.1720	0.2107	− ★	− NS	− NS	− NS	− NS	.	.	
Chaffinch	0.8400	0.8428	− NS	− NS	− NS	+ NS	− NS	.	.	
Serin	0.0000	0.0667	− NS	− NS				0.0000	1.3333	− NS
Greenfinch	0.5506	0.5609	− NS	− NS	− NS	+ NS	+ NS	.	.	
Goldfinch	0.4118	0.4602	− ★★	− NS	− ★★	+ NS	− NS	.	.	
Siskin	0.2042	0.2069	+ NS	+ NS	− NS	+ NS	+ NS	.	.	
Linnet	0.4621	0.5282	− ★★★	− ★★	− ★★	− NS	− NS	.	.	
Twite	0.2694	0.2723	+ NS	− NS	− NS	+ NS	+ NS	.	.	
Redpoll	0.1788	0.2096	− NS	− NS	− NS	− NS	− NS	.	.	
Common Crossbill	0.1151	0.0700	+ NS	+ NS	+ NS	+ NS	+ NS	0.7363	0.9500	+ NS
Scottish Crossbill	0.0000	0.1000	− NS			− NS		0.0000	0.2000	− NS
Bullfinch	0.2930	0.3107	− NS	− NS	− NS	+ NS	− NS	.	.	
Hawfinch	0.0902	0.0667	+ NS	+ NS	− NS			0.1399	0.1658	− NS
Snow Bunting	0.0000	0.0667	− NS			− NS		0.0000	0.0667	− NS
Yellowhammer	0.5462	0.5698	− ★	− NS	− NS	− NS	− NS	.	.	
Cirl Bunting	0.1000	0.2000	− NS		− NS			0.3000	0.3333	− NS
Reed Bunting	0.2679	0.2601	+ NS	+ NS	+ NS	+ NS	− ★	.	.	
Corn Bunting	0.2306	0.2476	− NS	− NS	+ NS	+ NS	+ NS	.	.	

The following 100-km squares were included in:

E Britain	NZ, SE, SK, SP, SU, SZ, TA, TF, TG, TL, TM, TQ, TR
W Britain	NX, NY, SC, SD, SH, SJ, SM, SN, SO, SR, SS, ST, SW, SX, SY, WV
Scotland	HP, HU, HY, NB, NC, ND, NF, NG, NH, NJ, NL, NM, NN, NO, NR, NS, NT, NU
Ireland	B, C, D, H, J, L, M, N, O, Q, R, S, T, V, W

The locations of these 100-km squares are shown in Fig. 1 in the *Introduction and Methods*. The total number of squares in which both the main and KSS were undertaken was: E Britain = 87, W Britain = 99, Scotland = 85, Ireland = 26.

For each species the mean frequency index (calculated across all squares included in the analysis in which it was recorded during either the main survey, the KSS or both) is presented for Britain and Ireland combined. Mean counts are similarly shown, but for count species only. Species marked '.' were not count species. Frequency indices were calculated for each 10-km square as the proportion of tetrads visited in which that species was recorded. Mean counts were calculated for each 10-km square as the mean number of adults of that species

recorded per two hours in each visited tetrad (except for Grey Heron, Sand Martin and seabirds for which counts were of apparently occupied nests). Note that seabird, Corncrake and Cirl Bunting data from timed visits were not used in the production of Abundance maps.

−	= negative matched pairs signed-rank
+	= positive matched pairs signed-rank
NS	= not significant
★	= $P < 0.05$
★★	= $P < 0.01$
★★★	= $P < 0.001$

For the individual regions (E Britain, W Britain, Scotland and Ireland), only the direction of the signed-rank, and the significance level are shown, and these only for frequency indices.

Note that no adjustment has been made for the number of tests made simultaneously, so this table exaggerates the number of species with significant bias.

Species for which there are blanks in the table were not recorded during either survey in the region in question.

The results of these analyses show that there was bias in the indices of abundance obtained for a few species, by allowing observers to choose which tetrads to visit and how many visits to make to each, and that this bias occurred particularly among species associated with water, woodland, mountain and moorland, and open farmland. However, there was strong evidence that the bias for each species acted in the same direction in different regions so that, generally, the Abundance map will reflect true patterns of regional variation. Nevertheless, the Abundance maps should be interpreted carefully, and it is advised that reference should be made to Table 5 when critical applications or interpretations are necessary.

Point counts

KSS point count data can be used to produce Abundance maps for common species, particularly those whose frequency indices from the main survey were often high. The 20 species listed in Table 6 had indices

of 1 (i.e. occurred in every visited tetrad) in more than 250 10-km squares. Despite fears to the contrary, the frequency index method censused even common birds quite well, with only six species having indices of 1 in more than a quarter of all 10-km squares with timed tetrad visits. Limiting tetrad visits to only two hours undoubtedly helped ensure this. Nevertheless, point count data can help put detail on to some of the Abundance maps, albeit on a much coarser grid, so colour Abundance maps based on point counts are presented on pp 457–461 for these 20 species. The method of production of these maps is as outlined in the *Introduction and Methods*, although the data used are different. A mean early point count for each species in each key square was calculated as the sum of all early point counts, divided by the number of tetrads in which point counts were performed (generally 15), and a mean late point count calculated in the same manner; the mean of these was used in production of the maps. As for the colour Abundance maps presented with the species accounts, regional variation in mean point count value is represented using ten colours. The class limits were calculated in the same manner, as outlined in *Interpreting the Species Accounts*, and are shown alongside each map. Because point count data were collected on a much coarser grid than the frequency index data from the main survey, these maps inevitably required a much higher level of interpolation. In the most extreme cases this could mean a species being interpolated into an area in which it did not actually occur (as shown by its Distribution map). The most obvious example of this was shown by the Carrion Crow whose high point count values in SW England and Wales were interpolated across to S Ireland, where it did not occur.

For a number of reasons, care is needed when interpreting these point count Abundance maps. First, because of the low density of key squares covered in Ireland compared to that in Britain, geographical variation in abundance in Ireland is represented only crudely. In general, the Irish point count data should only be used when comparing Ireland with Britain, and not in comparing different regions within Ireland. Second, no account was taken of differences in detectability with habitat, and it is likely that some species were more detectable in some habitats than in others. Finally, species which flock (e.g. Starling) can have highly variable counts, and a single very large flock could have influenced the map for such species markedly.

Despite these problems, a comparison between the Abundance maps based on frequency index shown in the species accounts, with those based on point counts shown here, can help highlight areas of high and low density more clearly. Thus, for example, the point count maps show that Robin and Wren densities were much greater in Ireland than in Britain, which was less clear from the frequency index maps. By contrast, the Woodpigeon point count map suggests lower densities in Ireland compared to Britain, and higher densities in E and W Britain, neither of which were so obvious from the frequency index map. Most importantly, perhaps, these maps show how a simple method, involving only a relatively small amount of time in the field, yielded useful information on geographical variation in breeding density of common species.

A summary of the KSS

The KSS was originally undertaken to estimate biases in the methods of the main Atlas survey. However, its uses, which are summarised below, went well beyond this:

1. To estimate bias (and regional bias) in allowing observer choice.
2. To produce point count Abundance maps for the more common species.
3. To provide additional distributional (presence/absence) data for the main survey.
4. To enable national population estimates to be calculated (see *Population Estimates for Breeding Birds in Britain and Ireland*).
5. To collect habitat information that can be used in interpreting patterns of distribution.

Furthermore, it is a highly repeatable survey in its own right – even more so than the main survey – because of its rigorous methods.

Table 6. *Species which occurred in every visited tetrad (i.e. whose frequency index from the main survey was 1) in more than 250 10-km squares.*

Species	No. (and %) of 10-km squares with a frequency index of 1	
Great Tit	262	(7.14)
Yellowhammer	285	(7.76)
Song Thrush	399	(10.87)
Jackdaw	460	(12.53)
Skylark	488	(13.29)
House Sparrow	499	(13.59)
Rook	505	(13.75)
Dunnock	555	(15.11)
Magpie	688	(18.74)
Meadow Pipit	703	(19.14)
Willow Warbler	708	(19.28)
Carrion Crow	732	(19.93)
Blue Tit	739	(20.13)
Starling	740	(20.15)
Swallow	954	(25.98)
Robin	1100	(29.96)
Woodpigeon	1357	(36.96)
Wren	1398	(38.07)
Chaffinch	1417	(38.59)
Blackbird	1671	(45.51)

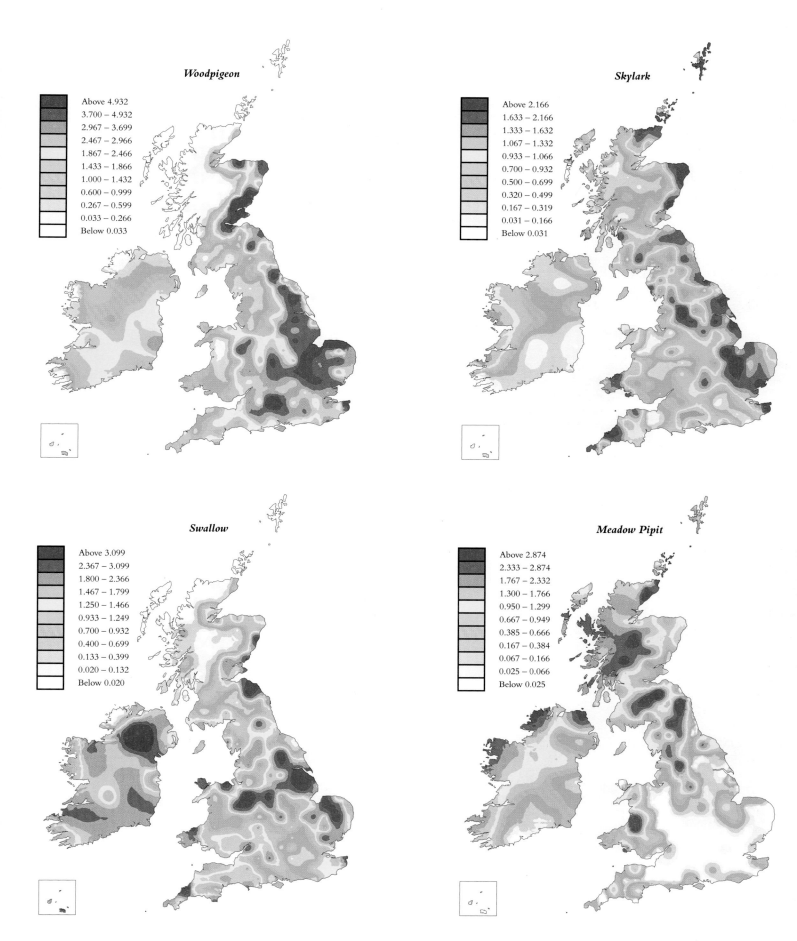

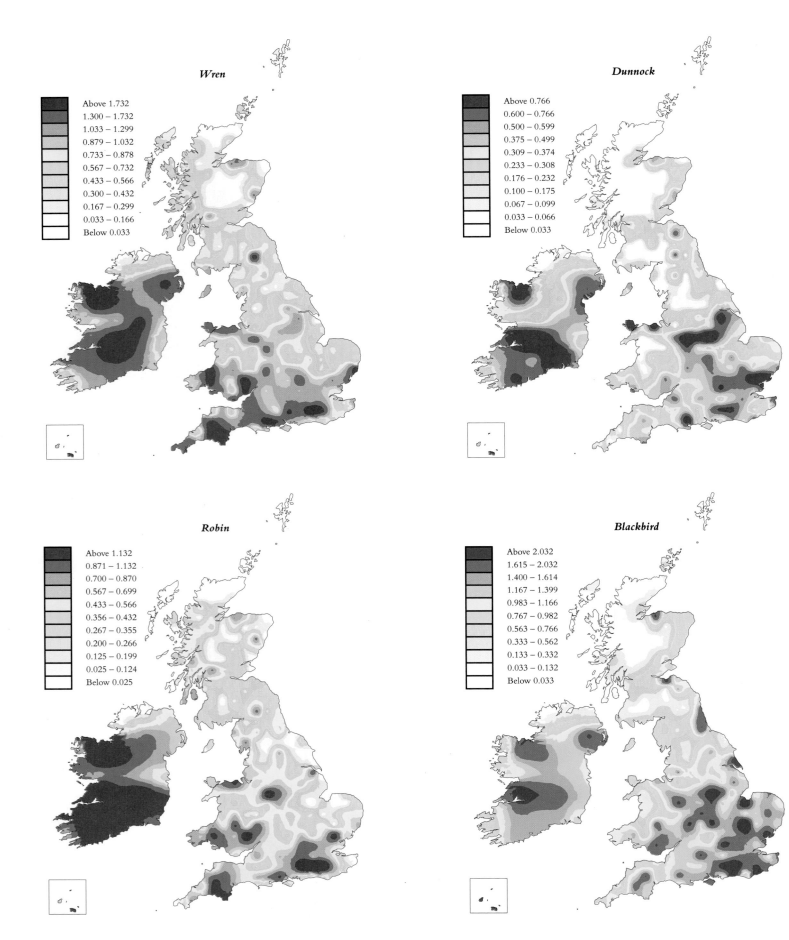

Wren

Above 1.732
1.300 – 1.732
1.033 – 1.299
0.879 – 1.032
0.733 – 0.878
0.567 – 0.732
0.433 – 0.566
0.300 – 0.432
0.167 – 0.299
0.033 – 0.166
Below 0.033

Dunnock

Above 0.766
0.600 – 0.766
0.500 – 0.599
0.375 – 0.499
0.309 – 0.374
0.233 – 0.308
0.176 – 0.232
0.100 – 0.175
0.067 – 0.099
0.033 – 0.066
Below 0.033

Robin

Above 1.132
0.871 – 1.132
0.700 – 0.870
0.567 – 0.699
0.433 – 0.566
0.356 – 0.432
0.267 – 0.355
0.200 – 0.266
0.125 – 0.199
0.025 – 0.124
Below 0.025

Blackbird

Above 2.032
1.615 – 2.032
1.400 – 1.614
1.167 – 1.399
0.983 – 1.166
0.767 – 0.982
0.563 – 0.766
0.333 – 0.562
0.133 – 0.332
0.033 – 0.132
Below 0.033

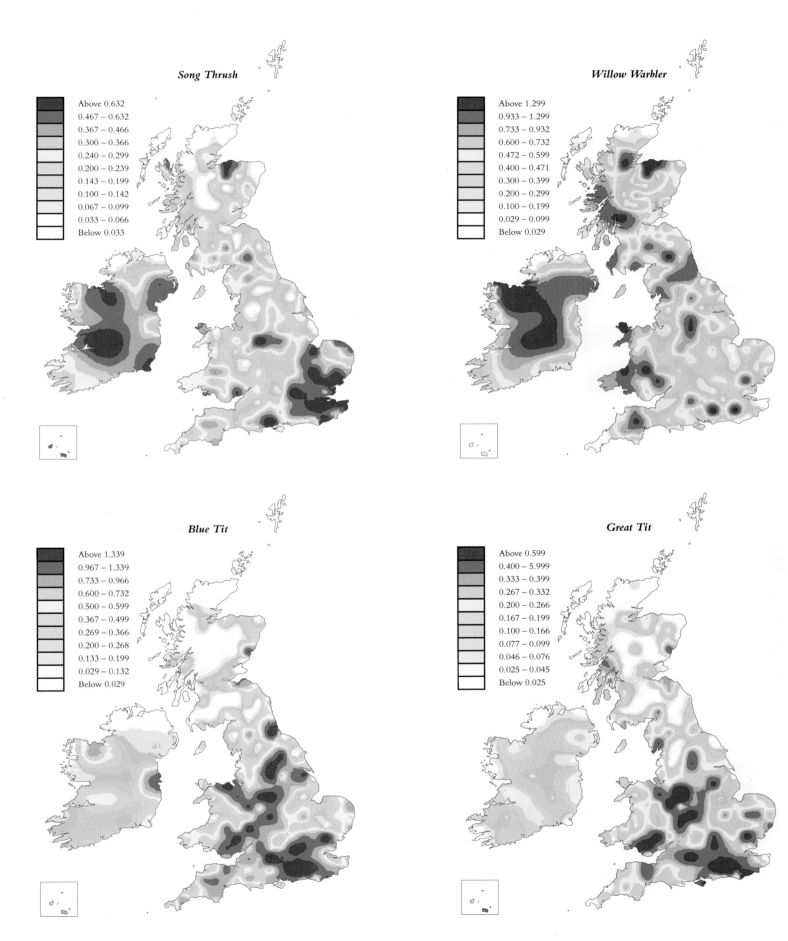

Song Thrush

Above 0.632
0.467 – 0.632
0.367 – 0.466
0.300 – 0.366
0.240 – 0.299
0.200 – 0.239
0.143 – 0.199
0.100 – 0.142
0.067 – 0.099
0.033 – 0.066
Below 0.033

Willow Warbler

Above 1.299
0.933 – 1.299
0.733 – 0.932
0.600 – 0.732
0.472 – 0.599
0.400 – 0.471
0.300 – 0.399
0.200 – 0.299
0.100 – 0.199
0.029 – 0.099
Below 0.029

Blue Tit

Above 1.339
0.967 – 1.339
0.733 – 0.966
0.600 – 0.732
0.500 – 0.599
0.367 – 0.499
0.269 – 0.366
0.200 – 0.268
0.133 – 0.199
0.029 – 0.132
Below 0.029

Great Tit

Above 0.599
0.400 – 5.999
0.333 – 0.399
0.267 – 0.332
0.200 – 0.266
0.167 – 0.199
0.100 – 0.166
0.077 – 0.099
0.046 – 0.076
0.025 – 0.045
Below 0.025

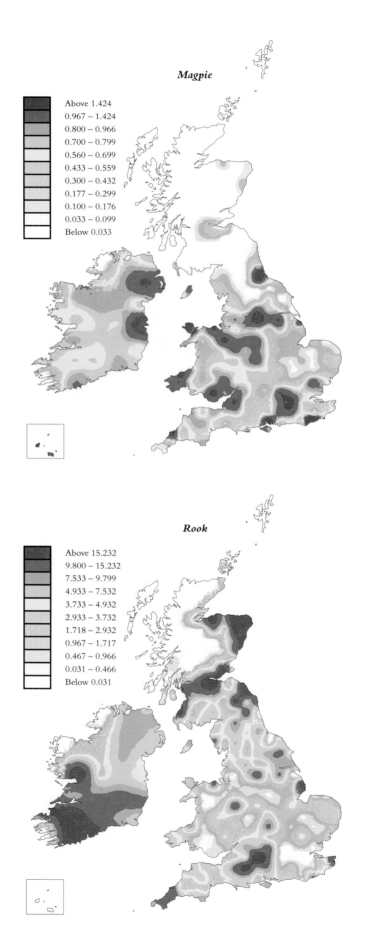

Magpie

	Above 1.424
	0.967 – 1.424
	0.800 – 0.966
	0.700 – 0.799
	0.560 – 0.699
	0.433 – 0.559
	0.300 – 0.432
	0.177 – 0.299
	0.100 – 0.176
	0.033 – 0.099
	Below 0.033

Rook

	Above 15.232
	9.800 – 15.232
	7.533 – 9.799
	4.933 – 7.532
	3.733 – 4.932
	2.933 – 3.732
	1.718 – 2.932
	0.967 – 1.717
	0.467 – 0.966
	0.031 – 0.466
	Below 0.031

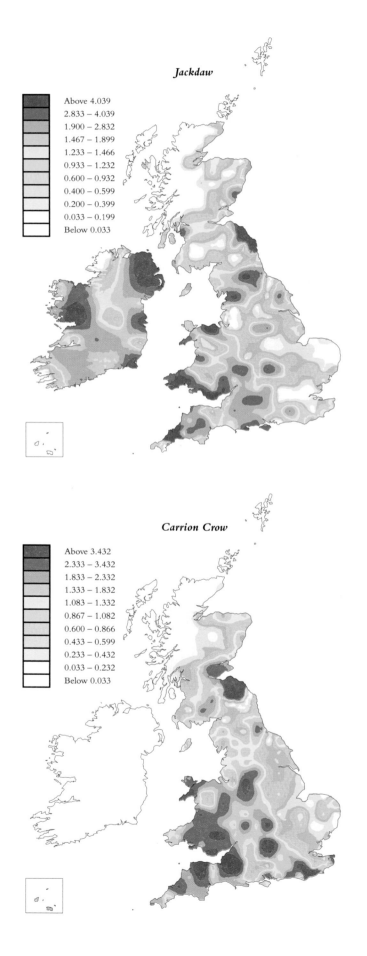

Jackdaw

	Above 4.039
	2.833 – 4.039
	1.900 – 2.832
	1.467 – 1.899
	1.233 – 1.466
	0.933 – 1.232
	0.600 – 0.932
	0.400 – 0.599
	0.200 – 0.399
	0.033 – 0.199
	Below 0.033

Carrion Crow

	Above 3.432
	2.333 – 3.432
	1.833 – 2.332
	1.333 – 1.832
	1.083 – 1.332
	0.867 – 1.082
	0.600 – 0.866
	0.433 – 0.599
	0.233 – 0.432
	0.033 – 0.232
	Below 0.033

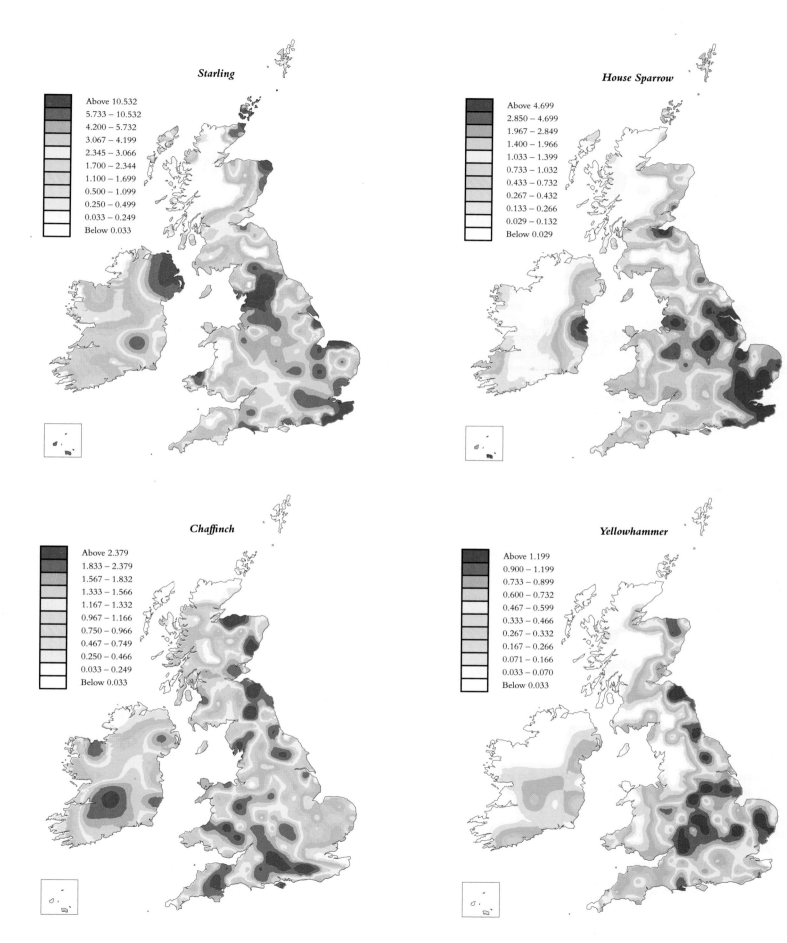

Population Estimates for Breeding Birds in Britain and Ireland

Introduction

The birds of the British Isles are one of the best studied groups of animals in the world, yet for most species the sizes of their populations are poorly known. There has been a number of attempts in the past to estimate national population sizes, for breeding birds (Parslow 1973, *68–72 Atlas*, Hudson and Marchant 1984, *Trends Guide*) and wintering birds (*Winter Atlas*). The figures published in the *68–72 Atlas* have been one of the most widely quoted parts of that book, although they were only intended as a rough guide to the abundance of each species. These population estimates, in common with many others before and since, were based largely on informed guesswork. Typically, they were produced by estimating the number of pairs breeding in a 10-km square, and multiplying this by the number of occupied 10-km squares. The estimates of densities in 10-km squares were taken from a wide variety of sources, and were corrected according to a subjective assessment of how typical the figure was of the species' density over its range. These corrections tended to err on the side of caution. The figures could obviously be improved by basing the estimates of density on quantitative data, and this was done by Hudson and Marchant (1984), whose estimates form the basis of the most recent population figures available (*Trends Guide*). They used densities from CBC plots to estimate national population sizes, by multiplying average densities in a variety of habitats by the extent of those habitats in Britain (they did not produce estimates for Ireland). There were two major drawbacks with Hudson and Marchant's method. First, densities in habitats other than woodland and farmland could not be calculated systematically and had to be guessed. Second, it took no account of the non-random geographical distribution of CBC plots.

The *88–91 Atlas* provides an opportunity both to update and to improve the accuracy of British and Irish population estimates. Various methods have been used to do this, and they are fully described in the following sections. In brief, the methods are:

(1) The *CBC Method*: a modified version of that used by Hudson and Marchant (1984), which derives population estimates by multiplying mean woodland and farmland CBC densities by the area of these habitats in Britain. The modified method calculates densities in habitats other than farmland and woodland in a systematic manner, and takes account of the non-random distribution of CBC plots.

(2) The *Count Method*: for a few conspicuous species, the populations in Britain and Ireland can be estimated directly from counts during timed tetrad visits.

(3) *Extrapolation from Britain to Ireland*: where the population size in Britain was known, that in Ireland could be derived by extrapolation, taking into account the relative abundance of the species in each.

(4) *Species texts*: this includes estimates from a wide variety of censuses, the method or source being outlined in the species account.

(5) The *Number of Squares Method*: estimates based on an informed guess of the species' density in occupied squares, multiplied by the number of occupied squares.

Population estimates produced using methods 1–3 and 5 were generated centrally at BTO headquarters while most of those using method 4 were provided by the authors of the species texts. A detailed account of each method follows.

1. The CBC Method (method CBC in Table 9)

Hudson and Marchant (1984) estimated the population sizes of British breeding bird species from mean densities on CBC plots. Full details of the methods of the CBC are given in the *Trends Guide*. Hudson and Marchant's method assumed that the CBC plots were a random sample of farmland and woodland in the whole of Britain, and they calculated the mean densities of each species across all woodland and farmland CBC plots. These densities were multiplied by the area of woodland and farmland in Britain, to give populations for those habitats. For other habitats, they subjectively estimated densities by comparison with those calculated for farmland and woodland, and again multiplied the estimated densities by the area of other habitats in Britain, to give a population for these habitats. The national population for each species was obtained by summing the populations in all of the different habitats. They used CBC data from 1982 to calculate CBC densities. The accuracy of this method depended on how representative the CBC plots were of each species' range. For a species such as the Robin, which may well occur at its highest densities on CBC plots, this method will probably have overestimated the population, but for a species which occurs at higher densities elsewhere, the population may have been underestimated. Furthermore, populations of species with restricted distributions will have been overestimated if their distribution coincided with that of CBC plots. This was because for a species confined to the south and east, like the CBC plots, the CBC will have sampled most of the areas in which it occurred. The CBC density was then extrapolated to the whole country, including large areas where it did not occur, and the population was therefore overestimated.

The method used to estimate British population sizes here is similar to that of Hudson and Marchant but differs in a number of important respects. The rest of this section outlines the method used, and the manner in which it differed from that of Hudson and Marchant.

Populations in woodland and farmland

The mean density of each species in woodland and farmland was calculated from densities on CBC plots in 1989. All CBC plots were assessed for their suitability for estimating territory densities, and only plots deemed acceptable for this purpose were included. This excluded small plots (less than 40 ha for farmland and 10 ha for woodland) and long, narrow plots, both of which were especially likely to have overestimated territory density

Table 7. *The calculation of British population estimates using the CBC method and their extrapolation to Ireland. (See text for full details).*

Species	No. W plots	No. F plots	Mean W density	Mean F density	Mean O density	British W population	British F population	British O population	Sum of populations	Correction ratio	British population	Irish extrapolation ratio	Irish population
Red-legged Partridge	3	24	0.0016	0.0104	0.0003	3228	120683	2800	126711	0.6965	90000	0.0019	.
Grey Partridge	5	32	0.0058	0.0129	0.0005	11626	150849	4675	167149	0.8084	140000	0.0036	
Pheasant	32	59	0.0691	0.0458	0.0077	138594	533802	71898	744294	0.7817	580000	0.3553	210000
Moorhen	14	43	0.0116	0.0264	0.0015	23184	307821	13898	344904	0.6987	240000	0.3128	75000
Lapwing	2	26	0.0034	0.0223	0.0114	6759	259646	106579	372983	1.1003	410000	0.1270	50000
Stock Dove	26	36	0.0755	0.0180	0.0023	151469	209674	21816	382958	0.6260	240000	0.1275	30000
Woodpigeon	39	43	0.3919	0.1095	0.0625	786660	1277593	586121	2650374	0.7968	2100000	0.3796	800000
Collared Dove	12	40	0.0219	0.0178	0.0046	44044	207912	43399	295355	0.6757	200000	0.1407	30000
Turtle Dove	14	18	0.0268	0.0064	0.0008	53790	74818	7457	136064	0.5448	75000	0.0042	
Great Spotted Woodpecker	48	18	0.0648	0.0047	0.0011	130069	54463	10258	194790	0.5266	100000	0.0000	0
Skylark	4	69	0.0129	0.1156	0.0687	25976	1348888	644149	2019014	0.9892	2000000	0.2860	570000
Swallow	1	53	0.0014	0.0420	0.0156	2842	489313	146206	638361	0.8913	570000	0.4430	250000
Tree Pipit	11	4	0.0236	0.0021	0.0027	47278	23994	25657	96930	1.2120	120000	0.0009	.
Meadow Pipit	3	19	0.0022	0.0230	0.0605	4472	268732	567023	840228	2.2395	1900000	0.4779	900000
Yellow Wagtail	0	15	0.0000	0.0051	0.0003	0	59222	3109	62331	0.7773	5000	0.0006	
Pied Wagtail	3	51	0.0017	0.0177	0.0046	3390	206921	43233	253545	1.1781	300000	0.4259	130000
Wren	66	75	1.3430	0.3307	0.1750	2695763	3857112	1640663	8193538	0.8631	7100000	0.4008	2800000
Dunnock	51	73	0.2664	0.1602	0.0314	534652	1868762	294744	2698158	0.7314	2000000	0.4105	810000
Robin	66	74	1.0438	0.2166	0.0805	2095282	2526957	754973	5377212	0.7874	4200000	0.4477	1900000
Redstart	11	9	0.0420	0.0103	0.0033	84210	120042	31320	235572	1.3892	330000	0.0003	
Blackbird	63	77	0.6688	0.2636	0.1110	1342421	3074221	1040401	5457042	0.8111	4400000	0.4098	1800000
Song Thrush	58	65	0.2456	0.0487	0.0188	492974	568586	176167	1237728	0.7988	990000	0.3896	390000
Mistle Thrush	41	45	0.0543	0.0148	0.0036	108944	172230	33349	314523	0.7446	230000	0.3767	90000
Sedge Warbler	2	17	0.0061	0.0169	0.0022	12307	196586	20579	229471	1.0763	250000	0.4326	110000
Lesser Whitethroat	13	24	0.0184	0.0097	0.0003	36832	113020	2351	152203	0.5212	80000	0.0000	0
Whitethroat	24	57	0.0535	0.0606	0.0049	107429	706335	46390	860154	0.7635	660000	0.1867	120000
Garden Warbler	34	26	0.0813	0.0126	0.0007	163233	146999	6970	317203	0.6187	200000	0.0125	
Blackcap	56	47	0.2747	0.0355	0.0044	551371	414402	41237	1007009	0.5764	580000	0.0655	40000
Chiffchaff	57	35	0.2571	0.0389	0.0084	516108	453444	78606	1048158	0.6113	640000	0.4452	290000
Willow Warbler	61	65	0.4686	0.1046	0.0442	940652	1220447	414323	2575422	0.8794	2300000	0.3686	830000
Goldcrest	47	25	0.2194	0.0144	0.0047	440411	168346	44394	653151	0.8616	560000	0.5384	300000
Spotted Flycatcher	20	24	0.0261	0.0066	0.0011	52306	77162	10511	139979	0.8484	120000	0.2921	35000
Long-tailed Tit	51	44	0.0717	0.0174	0.0009	143829	202477	8303	354609	0.6043	210000	0.1854	40000
Marsh Tit	23	10	0.0403	0.0038	0.0000	80814	44466	0	125280	0.4873	60000	0.0000	0
Willow Tit	10	7	0.0126	0.0015	0.0001	25314	17054	1311	43679	0.5715	25000	0.0000	0
Coal Tit	54	20	0.2665	0.0102	0.0068	534849	118525	64113	717487	0.8561	610000	0.4437	270000
Blue Tit	64	74	0.8241	0.1835	0.0671	1654206	2140901	628789	4423896	0.7456	3300000	0.3341	1100000
Great Tit	63	74	0.4858	0.0966	0.0217	975123	1126581	203781	2305485	0.6967	1600000	0.2591	420000
Nuthatch	37	15	0.0789	0.0068	0.0013	158388	79157	11831	249376	0.5235	130000	0.0000	0
Treecreeper	48	26	0.0920	0.0093	0.0006	184663	108377	5736	298776	0.6586	200000	0.2267	45000
Jay	55	29	0.0883	0.0090	0.0011	177273	104446	10696	292415	0.5389	160000	0.0631	10000
Magpie	50	63	0.0932	0.0528	0.0126	187064	616015	118198	921277	0.6394	590000	0.5499	320000
Jackdaw	24	33	0.0486	0.0224	0.0116	97577	260929	108472	466977	0.8353	390000	0.5463	210000
Carrion Crow	45	63	0.0712	0.0491	0.0270	142883	572669	253400	968952	0.8111	790000	0.0028	
Starling	20	59	0.1246	0.0634	0.0348	250043	739907	326566	1316515	0.8069	1100000	0.3432	360000
House Sparrow	3	38	0.0056	0.0608	0.0213	11209	708960	200041	920210	0.7791	720000	0.2932	210000
Tree Sparrow	4	20	0.0162	0.0090	0.0004	32414	105263	3374	141050	0.7595	110000	0.0858	9000
Chaffinch	65	76	0.7139	0.3224	0.1230	1432904	3761204	1153351	6347459	0.8453	5400000	0.3824	2100000
Greenfinch	25	59	0.0549	0.0483	0.0121	110253	563302	113672	787227	0.6730	530000	0.3100	160000
Goldfinch	11	41	0.0119	0.0209	0.0037	23951	243525	35005	302481	0.7377	220000	0.2504	55000
Linnet	14	56	0.0344	0.0415	0.0082	69031	484157	77166	630354	0.8213	520000	0.2566	130000
Bullfinch	43	34	0.0772	0.0134	0.0009	154889	156098	8472	319459	0.5964	190000	0.5363	100000
Yellowhammer	26	62	0.0759	0.1162	0.0121	152426	1355525	113381	1621333	0.7454	1200000	0.1690	200000
Reed Bunting	4	36	0.0035	0.0173	0.0019	7108	201532	17860	226500	0.9568	220000	0.6002	130000
Corn Bunting	1	14	0.0020	0.0152	0.0002	3960	177596	1777	183333	0.8944	160000	0.0033	.

Footnotes to Table 7 are located on p. 464

Footnotes to Table 7

W: woodland; F: farmland; O: other.

All density estimates are territories/ha

British populations are: [(mean W density × 2,007,298) + (mean F density × 11,664,747) + (mean O density × 9,373,755)] × Correction ratio

The British populations include the Isle of Man but exclude the Channel Islands.

Irish populations are: British population × Irish extrapolation ratio

The method of calculation of mean densities, and of the correction and Irish extrapolation ratios, are detailed in the text. The mean densities presented have been rounded for clarity; less rounded densities were used for population estimation.

Note that British populations have not been extrapolated to Ireland for those species whose extrapolation ratio was 0.02 or less. These species have a full stop in the Irish population column.

The values of all British and Irish populations were rounded in the following manner: values of greater than 1,000,000 were rounded to the nearest 100,000; values of 100,000 to 1,000,000 were rounded to the nearest 10,000; values of 10,000 to 100,000 were rounded to the nearest 5,000; values of less than 10,000 were rounded to the nearest 1,000.

Table 8. *The area of each type of habitat in Britain (including the Isle of Man but excluding the Channel Islands), in ha, with the percentage of the land area that this represents.*

	Farmland[1]	Woodland[2]	Other
Britain	11,664,747 (50.6%)	2,007,298 (8.7%)	9,373,755 (40.7%)

The sources of the figures for farmland and woodland were:
[1] MAFF/DAFS Agricultural census 1988. Note that farmland includes woodland, buildings and roads present on agricultural land.
[2] Forestry facts and figures 1989–1990 (Forestry Commission 1990b)
The figure for other habitats was calculated by subtraction from the total land area.

due to edge effects (Marchant 1981). Following this selection procedure a total of 67 woodland and 77 farmland plots was available for density estimation. The mean density of each species in woodland was then calculated as the total number of territories of each species, across all woodland plots, divided by the total area of all woodland plots. Mean densities on farmland were calculated in the same manner. The number of plots on which each species occurred is shown in Table 7, and mean densities were calculated in this manner only for species which occurred on a total of at least 14 (approximately 10% of) CBC plots. When used thus the CBC estimates territory (rather than pair) density, so all the final population estimates are expressed in terms of numbers of territories.

The total population of each species in British woodland was calculated by multiplying its mean density in woodland by the total extent of woodland in Britain, as documented in Table 8. The total farmland population was calculated in the same manner. Table 7 lists the mean density of each species in woodland and farmland, and their estimated total population in each habitat.

Populations in other habitats

About 40% of the land area of Britain is neither farmland nor woodland, and much of this is either rough grazing, moorland or urban. There are very few CBC plots in these habitats, so it was not possible to calculate densities in the same manner as for farmland and woodland. Rather, the populations in habitats other than woodland and farmland had to be estimated.

The simplest way to explain how this was done is by an example, that of the Chaffinch. Three pieces of information were necessary for the calculation: the ratio of the total area of woodland occupied by the Chaffinch, to the total area of farmland occupied by the Chaffinch, the total area of other habitats occupied, and the mean density of Chaffinches on woodland and farmland CBC plots, where they occurred.

Chaffinches were recorded on 65 out of 67 (= 0.9701) woodland CBC plots in 1989, and 76 out of 77 (= 0.9870) farmland plots (see Table 7). Assuming these proportions were representative of Britain, then Chaffinches occurred in 1,947,280 ha of woodland (= 0.9701 × 2,007,298; see Table 8), and on 11,513,105 ha of farmland (= 0.9870 × 11,664,747). The ratio of the total area of woodland to the total area of farmland occupied by the Chaffinch was thus 1,947,280:11,513,105. Expressed as proportions of 1, this ratio is 0.1447:0.8553.

Unfortunately, the total area of habitats other than woodland and farmland occupied by Chaffinches could not be calculated in the same manner. Instead this was calculated from the habitat codes of the KSS point count stations. As outlined in *Bias and the Key Squares Survey*, KSS observers recorded the habitat at each of 3,837 randomly positioned point count stations, in 301 key squares, using a simple habitat classification (see Table 4 in *Bias and the Key Squares Survey*). This classification was further simplified by separating these habitats into three classes; woodland (classes 00–03), farmland (classes 60–65 and 12) and other habitats (all other classes). Only the major habitat codes were used in this simplification. Of these 3,837 point counts, 1,336 were carried out in other habitats, and Chaffinches were recorded on 426 (= 0.3189) of these. Assuming Chaffinches were always recorded when present, then they occupied 2,989,290 ha of other habitats (= 0.3189 × 9,373,755; see Table 8).

Finally, the mean density in woodland (0.7358 territories/ha) and on farmland (0.3267 territories/ha) CBC plots, where Chaffinches occurred, was calculated. Note that these densities were not the same as those used to calculate total populations in woodland and farmland (as shown in Table 7), which were calculated across all plots including those where Chaffinches did not occur.

To calculate the total population in other habitats, it was assumed that Chaffinches were present at their woodland density in 0.1447 of the area of other habitats occupied, and at their farmland density in the remainder. The total population was thus: [0.7358 (0.1447 × 2,989,290) + 0.3267 (0.8553 × 2,989,290)] = 1,153,556 territories. Although not strictly necessary for the calculation of the final British population estimate, the mean density of Chaffinches in other habitats was then calculated, retrospectively, by dividing the population in other habitats by the total area of other habitats (1,153,556 ÷ 9373755 = 0.1231 territories/ha) (Note that the values quoted here differ very slightly from those in Table 7, as those presented here have been rounded for simplification.)

The method used, here, to calculate the population in other habitats thus has the merit of reflecting each species' habitat preference. For example, a species which mainly occurred in woodland was mainly allocated the woodland density to the area of other habitats occupied. This makes sense because it was likely that a woodland species would have been found in other habitats more similar to woodland than farmland.

The estimated populations and densities for each species in other habitats are shown in Table 7. An overall British population was then calculated as the sum of the populations in woodland, farmland and other habitats.

Correcting for bias in CBC plot distribution

The geographical distribution of CBC plots is not random (see Fig. 2.1

in the *Trends Guide*) as there is a strong bias towards the centres of human population, and particularly towards S and E England. Thus, for a species that is most abundant in the south and east, to extrapolate its mean density, calculated from CBC plots, to the whole of Britain would overestimate its population. Populations of species which are most abundant in the north and west would, by contrast, be underestimated. An attempt has been made to correct for this bias in the following manner. Using the measures of abundance from the main Atlas survey (either frequency index or count – see *Introduction and Methods*), the ratio of the mean abundance over all 10-km squares, to the mean abundance in 10-km squares containing CBC plots, was calculated for each species. The sum of the populations across all habitats was then multiplied by this correction ratio to provide the final British population estimate. Two examples will illustrate the use of these correction ratios. The Meadow Pipit is a species which is most abundant in the north and west, and the ratio of mean abundance across all 10-km squares to that of 10-km squares containing CBC plots was 2.24. Thus the sum of the Meadow Pipit populations across all habitats (840,000 territories) was likely to be a serious underestimate, and was multiplied by this correction ratio to yield the final British population estimate of 1,900,000 territories. At the other extreme, the Nuthatch is a species with a distribution restricted to England and Wales, and thus similar to that of CBC plots. Extrapolation of mean densities from CBC plots to the whole of Britain would inevitably overestimate the size of the Nuthatch population, as the species will be assumed to have been present in Scotland, where it occurs only rarely. The correction ratio for the Nuthatch was 0.52, and the sum of the populations across all habitats (250,000 territories) was multiplied by this ratio to yield the final British population estimate of 130,000 territories. The correction ratios and final British population estimates for all species covered by the CBC method are shown in Table 7. The final estimates are also listed in Table 9.

Comparisons with the method of Hudson and Marchant (1984)

The method presented here differs from that of Hudson and Marchant (1984) in a number of respects. Firstly, they considered four separate habitat types; farmland, woodland, rough grazing and other. Densities in farmland and woodland were calculated from CBC plots, and those on rough grazing and other were based on subjective estimates. Only three habitats have been considered here (farmland, woodland and other), and the density in other habitats has been calculated in an objective and repeatable manner. Secondly, Hudson and Marchant made an assessment of the potential of edge effects for each species, and sometimes downgraded the densities estimated from CBC plots. Thus for a species commonly found at the edge of CBC plots, and whose territories may have overlapped the plot boundaries, densities may have been overestimated, and so the calculated density estimate was reduced. This problem has been addressed here in a more systematic manner by removing all CBC plots with a high ratio of edge to area, particularly small, and long thin plots. Thirdly, Hudson and Marchant were unable to correct for the non-random geographical distribution of CBC plots. Abundance data collected from the main Atlas survey has enabled this correction to be performed here.

Limitations of the CBC Method

Although it was possible to calculate populations in other habitats, the reliability of these estimates is unknown. These populations were ultimately based on woodland and farmland densities, which may not be realistic for some species (for example urban species such as the House Sparrow). However, there was no obvious alternative to this method.

In principle the total areas of woodland and farmland occupied by each species could have been calculated, as for other habitats, using the KSS habitat codes. However, this method assumed that when species were present they were always recorded during point counts, but this may not have been the case. The area of other habitats occupied, and thus the population in other habitats, was likely to have been an underestimate, and this would have been increasingly true for species which were difficult to detect (e.g. Treecreeper). The area of woodland and farmland occupied could be calculated more precisely from the CBC data, and so these were used in preference.

While it was possible to correct for the bias in the geographical distribution of CBC plots, it was not possible to take account of how representative the CBC plots used in the analysis were of the 10-km squares in which they occurred. Whilst Fuller *et al.* (1985) have shown that farmland CBC plots were representative of lowland England, no similar analysis has been conducted for woodland. In addition, it is particularly likely that densities in woodland vary with woodland type, and as CBC woodland plots are mainly deciduous, it is possible that populations of species with a preference for coniferous woodland have been underestimated, and those with a preference for deciduous woodland overestimated.

For some species, it is known that the number of territories estimated by CBC methods is an underestimate of the breeding population (Snow 1965, Tomiałojć and Lontkowski 1989). If it is true that there were generally more breeding pairs on a plot than shown by the CBC data, the population estimates for some species could be too low.

The CBC method was not suitable for non-territorial species (e.g. Mallard and Rook), nocturnal species (e.g. Little and Tawny Owl) and those whose territories were large in relation to CBC plot size (*Trends Guide*). Estimates of territory density for the latter group of species (Curlew, Sparrowhawk, Kestrel, Cuckoo, Tawny Owl and Green Woodpecker) were unreliable as, firstly, the area occupied by each territory was likely to have been underestimated, and, secondly, a plot might have contained parts of several territories; in either case territory density will have been overestimated. In addition, although the estimates shown in Table 7 for Lapwing, Pheasant, Great Spotted Woodpecker and House Sparrow, were produced using the CBC method, these are not presented in Table 9 or the species accounts, because it was thought that alternative methods, outlined in each account, were more suitable.

The results from the CBC method yielded population estimates in terms of the number of territories, corresponding to the 'pairs' of previous estimates. How this figure relates to the number of individuals in the population will depend on the mating system, and the size of the non-breeding population. The normal mating system in birds was for a long time thought to be monogamy, with one male and one female on a single territory, but it is now known that several passerines frequently depart from this system (see Møller 1986 for a recent review). It is possible that such strategies are employed by most species under certain conditions. Davies and Lundberg (1984) found that food availability had a major influence on the mating system, and different levels of food availability in different years may lead to variable levels of non-monogamous mating systems. The number of territories may remain relatively constant, but the number of breeding birds may vary widely.

Table 9. *Annual population estimates for breeding birds in Britain and Ireland.*

Species	Britain			Ireland		
	Estimate	**Date**	**Method**	**Estimate**	**Date**	**Method**
Red-throated Diver *Gavia stellata*	1,200–1,500 p	88–91	TEXT/CNT	< 10 p	89	TEXT
Black-throated Diver *Gavia arctica*	150 p	85–91	TEXT/CNT	0–1 p?	88–91	TEXT
Little Grebe *Tachybaptus ruficollis*	5,000–10,000 p	88–91	NSQ★	3,000–6,000 p	88–91	NSQ
Great Crested Grebe *Podiceps cristatus*	8,000 i	88–91	CNT	4,150 i	88–91	CNT
Red-necked Grebe *Podiceps grisegena*	3 p	88–89	TEXT	0		
Slavonian Grebe *Podiceps auritus*	60 p	88–91	TEXT	0		
Black-necked Grebe *Podiceps nigricollis*	15–40 p	88–89	TEXT	0		
Fulmar *Fulmarus glacialis*	540,000 p	85–87	TEXT★	31,000 p	85–87	TEXT
Manx Shearwater *Puffinus puffinus*	220,000–250,000 p	85–87	TEXT★	30,000–50,000 p	85–87	TEXT
Storm Petrel *Hydrobates pelagicus*	20,000–150,000 p	80s	TEXT★	50,000–100,000 p	80s	TEXT
Leach's Petrel *Oceanodroma leucorhoa*	10,000–100,000 p in Britain and Ireland (TEXT)					
Gannet *Morus bassanus*	162,000 n	84–91	TEXT★	24,500 n	84–90	TEXT
Cormorant *Phalacrocorax carbo*	7,000 p	85–87	TEXT★	4,700 p	85–87	TEXT
Shag *Phalacrocorax aristotelis*	38,500 p	85–87	TEXT★	8,800 p	85–87	TEXT
Bittern *Botaurus stellaris*	16 p	88–91	TEXT	0		
Grey Heron *Ardea cinerea*	10,300 n	91	TEXT	3,650 n	88–91	EXTP
Mute Swan *Cygnus olor*	25,800–27,000 i	90/88–91	TEXT/CNT	7,000 i [19,000–20,000 i]	80s 88–91	TEXT [EXTP/CNT]
Whooper Swan *Cygnus cygnus*	2–5 p	88–90	TEXT	0		
Greylag Goose *Anser anser*	22,000 i	88–91	CNT	700 i	86	TEXT
Canada Goose *Branta canadensis*	59,500 i	88–91	CNT	575– > 700 i	88–91	CNT/TEXT
Barnacle Goose *Branta leucopsis*	600–700 i	91	TEXT	?		
Egyptian Goose *Alopochen aegyptiacus*	750–800 i	88–91	CNT/TEXT	0		
Shelduck *Tadorna tadorna*	44,200 i 10,600 bp	88–91	CNT	4,650 i 1,100 bp	88–91	CNT
Wood Duck *Aix sponsa*	?			0		
Mandarin *Aix galericulata*	> 7,000 i	88–91	TEXT	0		
Wigeon *Anas penelope*	300–500 p	88–91	TEXT	0		
Gadwall *Anas strepera*	770 p	90	TEXT	30 p	90	TEXT

Table 9 (*continued*).

Species	Britain			Ireland		
	Estimate	**Date**	**Method**	**Estimate**	**Date**	**Method**
Teal *Anas crecca*	1,500–2,600 p	88–91	NSQ	400–675 p	88–91	NSQ
Mallard *Anas platyrhynchos*	100,000 p	86	TEXT	23,000 p	86/ 88–91	EXTP
Pintail *Anas acuta*	30–40 p	88–91	TEXT	0–1 p	88–91	TEXT
Garganey *Anas querquedula*	40–100 p	88–89	TEXT	0–1 p	88–91	TEXT
Shoveler *Anas clypeata*	1,000–1,500 p	88–91	TEXT	< 100 p	88–91	TEXT
Red-crested Pochard *Netta rufina*	< 100 i	88–91	TEXT	0		
Pochard *Aythya ferina*	400 p	88–91	TEXT	30 p	88–91	TEXT
Tufted Duck *Aythya fuligula*	7,000–8,000 bp	86	TEXT	1,750–2,000 bp	86–91	EXTP
Scaup *Aythya marila*	0–5 p	88–91	TEXT	0		
Eider *Somateria mollissima*	31,000–32,000 f	88–91	TEXT/CNT	600–1,000 f	88–91	CNT/TEXT
Common Scoter *Melanitta nigra*	100 p	89	TEXT	65–70 p	89	TEXT
Goldeneye *Bucephala clangula*	95 p	90	TEXT	0		
Red-breasted Merganser *Mergus serrator*	2,150 p	88–91	CNT	700 p	88–91	CNT
Goosander *Mergus merganser*	2,700 p	87	TEXT	0–1 p	88–91	TEXT
Ruddy Duck *Oxyura jamaicensis*	570 p	91	TEXT	15–20 p	88–91	EXTP
Honey Buzzard *Pernis apivorus*	< 30 p	88–91	TEXT	0		
Red Kite *Milvus milvus*	92 tp but only 77 bp	91	TEXT	0		
White-tailed Eagle *Haliaeetus albicilla*	11 tp but only 8 bp	88–91	TEXT	0		
Marsh Harrier *Circus aeruginosus*	87 n, 190 i	90	TEXT	0		
Hen Harrier *Circus cyaneus*	630 p	88–89	TEXT	180 p	88–91	EXTP
Montagu's Harrier *Circus pygargus*	12 n ≧ 6 n	90 91	TEXT	0		
Goshawk *Accipiter gentilis*	200 p	88	TEXT	0		
Sparrowhawk *Accipiter nisus*	32,000 p	86	TEXT	11,000 p	86/ 88–91	EXTP
Buzzard *Buteo buteo*	12,000–17,000 tp	83	TEXT	> 150 p	86–91	TEXT
Golden Eagle *Aquila chrysaetos*	420 bp	82–83	TEXT	0		
Osprey *Pandion haliaetus*	72 p	91	TEXT	0		
Kestrel *Falco tinnunculus*	50,000 p	88–91	NSQ★	10,000 p	88–91	EXTP

Table 9 *(continued)*.

Species	Britain			Ireland		
	Estimate	**Date**	**Method**	**Estimate**	**Date**	**Method**
Merlin *Falco columbarius*	550–650 p	83–84	TEXT	110–130 p	83–84/ 88–91	EXTP
Hobby *Falco subbuteo*	500–900 p	88–91	TEXT	0		
Peregrine *Falco peregrinus*	1,100 p	91	TEXT	365 p	88–91	TEXT/EXTP
Red Grouse *Lagopus lagopus*	250,000 p	88–91	TEXT	1,000–5,000 p	88–91	TEXT
Ptarmigan *Lagopus mutus*	> 10,000 p	90	TEXT	0		
Black Grouse *Tetrao tetrix*	10,000–15,000 f	88–91	TEXT	0		
Capercaillie *Tetrao urogallus*	a few 1000 i		TEXT	0		
Red-legged Partridge *Alectoris rufa*	> 90,000 t	88–91	CBC	0		TEXT
Grey Partridge *Perdix perdix*	140,000 t – 150,000 p	88–91/92	CBC/TEXT	500 p	88–91	TEXT
Quail *Coturnix coturnix*	2,600 m, 100–300 p	89 88, 90, 91	TEXT	90 m < 20 m	89 88, 90, 91	TEXT
Pheasant *Phasianus colchicus*	1,500,000 m 1,600,000 f	89	TEXT	530,000 m 570,000 f	88–91	EXTP
Golden Pheasant *Chrysolophus pictus*	1,000–2,000 i	81–84	TEXT	0		
Lady Amherst's Pheasant *Chrysolophus amherstiae*	100–200 i	91	TEXT	0		
Water Rail *Rallus aquaticus*	> 450–900 p	88–91	NSQ★	> 850–1700 p	88–91	NSQ
Spotted Crake *Porzana porzana*	10–21 m	88–90	TEXT	0		
Corncrake *Crex crex*	550–600 m 470 m	88 91	TEXT	903–930 m	88	TEXT
Moorhen *Gallinula chloropus*	240,000 t	88–91	CBC	75,000 t	88–91	EXTP
Coot *Fulica atra*	46,000 i	88–91	CNT	8,600 i	88–91	CNT
Crane *Grus grus*	≧ 1 p	88–91	TEXT	0		
Oystercatcher *Haematopus ostralegus*	33,000–43,000 p	mid 80s	TEXT	3,000–4,000 p	mid 80s	TEXT
Avocet *Recurvirostra avosetta*	400–500 p	88–90	TEXT	0		
Stone Curlew *Burhinus oedicnemus*	150–160 p	88	TEXT	0		
Little Ringed Plover *Charadrius dubius*	825–1,070 p	88–91	NSQ	0		
Ringed Plover *Charadrius hiaticula*	8,480 p	84	TEXT	1,250 p	84/ 88–91	EXTP
Dotterel *Charadrius morinellus*	840–950 p	87–88	TEXT	0		
Golden Plover *Pluvialis apricaria*	22,600 p	88–91	TEXT	400 p	88–91	TEXT
Lapwing *Vanellus vanellus*	185,000–238,000 p	86–87	TEXT	21,500 p	88–91	TEXT

Table 9 (*continued*).

Species	Britain			Ireland		
	Estimate	**Date**	**Method**	**Estimate**	**Date**	**Method**
Temminck's Stint *Calidris temminckii*	5–6 i	88–89	TEXT	0		
Purple Sandpiper *Calidris maritima*	1–4 p	89–91	TEXT	0		
Dunlin *Calidris alpina*	9,150 p	90	TEXT	175 p	mid 80s	TEXT
Ruff *Philomachus pugnax*	< 5 nesting f	88–91	TEXT	0		
Snipe *Gallinago gallinago*	30,000 p	mid 80s	TEXT	> 10,000 p	mid 80s/ 85–87	TEXT
Woodcock *Scolopax rusticola*	> 8,500–21,500 p	88–91	NSQ	> 1,750–4,500 p	88–91	NSQ
Black-tailed Godwit *Limosa limosa*	> 33–36 p	88–90	TEXT	2–3 p	89	TEXT
Whimbrel *Numenius phaeopus*	435–495 p	90	TEXT	0		
Curlew *Numenius arquata*	33,000–38,000 p	85	TEXT	12,000 p	85	TEXT
Redshank *Tringa totanus*	30,600–33,600 p	mid 80s	TEXT	4,400–5,000 p	mid 80s/ /88–91	EXTP
Greenshank *Tringa nebularia*	1,100–1,600 p	88–91	TEXT	0		
Wood Sandpiper *Tringa glareola*	5–6 p	88–91	TEXT	0		
Common Sandpiper *Actitis hypoleucos*	15,800 p	88–91	NSQ	2,500 p	88–91	NSQ
Red-necked Phalarope *Phalaropus lobatus*	20 p	88	TEXT	0–1 p	88–91	TEXT
Arctic Skua *Stercorarius parasiticus*	3,350 p	85–87	TEXT	0		
Great Skua *Catharacta skua*	7,900 p	85–87	TEXT	0		
Mediterranean Gull *Larus melanocephalus*	5–16 p	88–90	TEXT	0		
Black-headed Gull *Larus ridibundus*	147,400 p	85–91	TEXT	53,800 p	85–91	TEXT
Common Gull *Larus canus*	68,000 p	85–87	TEXT	3,600 p	85–87	TEXT
Lesser Black-backed Gull *Larus fuscus*	83,500 p	85–87	TEXT★	5,200 p	85–87	TEXT
Herring Gull *Larus argentatus*	161,000 p	85–87	TEXT★	44,700 p	85–87	TEXT
Great Black-backed Gull *Larus marinus*	19,000 p	85–87	TEXT★	4,500 p	85–87	TEXT
Kittiwake *Rissa tridactyla*	493,400 p	85–87	TEXT★	50,200 p	85–87	TEXT
Sandwich Tern *Sterna sandvicensis*	14,000 p	85–87	TEXT★	4,400 p	85–87	TEXT
Roseate Tern *Sterna dougallii*	< 90 p	90	TEXT	> 400 p	90	TEXT
Common Tern *Sterna hirundo*	12,900 p	84–87	TEXT★	3,100 p	84–87	TEXT
Arctic Tern *Sterna paradisaea*	44,000 p	85–89	TEXT	2,500 p	83–87	TEXT

Table 9   (*continued*).

Species	Britain			Ireland		
	Estimate	Date	Method	Estimate	Date	Method
Little Tern *Sterna albifrons*	2,430 p	85–87	TEXT	390 p	85–87	TEXT
Guillemot *Uria aalge*	1,047,000 i	85–87	TEXT★	153,000 i	85–87	TEXT
Razorbill *Alca torda*	148,000 i	85–87	TEXT★	34,000 i	85–87	TEXT
Black Guillemot *Cepphus grylle*	37,000 i	85–91	TEXT	3,000 i	85–91	TEXT
Puffin *Fratercula arctica*	900,000 i	85–87	TEXT★	41,000 i	85–87	TEXT
Rock Dove/Feral Pigeon *Columba livia*	?			?		
Stock Dove *Columba oenas*	240,000 t	88–91	CBC	30,000 t	88–91	EXTP
Woodpigeon *Columba palumbus*	2,100,000 t – 2,550,000 p	88–91	CBC/TEXT	800,000 t – 970,000 p	88–91	EXTP
Collared Dove *Streptopelia decaocto*	200,000 t	88–91	CBC	30,000 t	88–91	EXTP
Turtle Dove *Streptopelia turtur*	75,000 t	88–91	CBC	0		
Ring-necked Parakeet *Psittacula krameri*	several 1000 i	91	TEXT	0		
Cuckoo *Cuculus canorus*	13,000–26,000 p	88–91	NSQ★	3,000–6,000 p	88–91	NSQ
Barn Owl *Tyto alba*	4,400 p	82–85	TEXT	600–900 p	82–85	TEXT
Snowy Owl *Nyctea scandiaca*	1–2 f	88–89	TEXT	0		
Little Owl *Athene noctua*	6,000–12,000 p	88–91	TEXT	0		
Tawny Owl *Strix aluco*	> 20,000 p	88–91	TEXT	0		
Long-eared Owl *Asio otus*	1,100–3,600 p	88–91	NSQ★	1,100–3,600 p	88–91	NSQ
Short-eared Owl *Asio flammeus*	1,000–3,500 p	88–91	TEXT	0		
Nightjar *Caprimulgus europaeus*	3,000 m	92	TEXT	< 30 p	88–91	TEXT
Swift *Apus apus*	80,000 p	68–72/ 88–91	NSQ★	20,000 p	68–72/ 88–91	NSQ
Kingfisher *Alcedo atthis*	3,300–5,500 p	88–91	NSQ★	1,300–2,100 p	88–91	NSQ
Wryneck *Jynx torquilla*	0–10 p	88–90	TEXT	0		
Green Woodpecker *Picus viridis*	15,000 p	88–91	NSQ	0		
Great Spotted Woodpecker *Dendrocopos major*	25,000–30,000 p	88–91	TEXT	0		
Lesser Spotted Woodpecker *Dendrocopos minor*	3,000–6,000 p	85/88–91	TEXT	0		
Woodlark *Lullula arborea*	350 p	88–91	TEXT	0		
Skylark *Alauda arvensis*	2,000,000 t	88–91	CBC	570,000 t	88–91	EXTP

Table 9 (*continued*).

Species	Britain			Ireland		
	Estimate	Date	Method	Estimate	Date	Method
Sand Martin *Riparia riparia*	77,500–250,000 n	88–91	CNT/TEXT	49,500–150,000 n	88–91	CNT/TEXT
Swallow *Hirundo rustica*	570,000 t	88–91	CBC	250,000 t	88–91	EXTP
House Martin *Delichon urbica*	250,000–500,000 p	88–91	NSQ★	70,000–140,000	88–91	NSQ
Tree Pipit *Anthus trivialis*	120,000 t	88–91	CBC	?0–1 p	88–91	TEXT
Meadow Pipit *Anthus pratensis*	1,900,000 t	88–91	CBC	900,000 t	88–91	EXTP
Rock Pipit *Anthus petrosus*	34,000 p	88–91	NSQ★	12,500 p	88–91	NSQ
Yellow Wagtail *Motacilla flava*	50,000 t	88–91	CBC	0		
Grey Wagtail *Motacilla cinerea*	34,000 p	88–91	NSQ★	22,000 p	88–91	NSQ
Pied Wagtail *Motacilla alba*	300,000 t	88–91	CBC	130,000 t	88–91	EXTP
Dipper *Cinclus cinclus*	7,000–21,000 p	88–91	NSQ	1,750–5,000 p	88–91	NSQ
Wren *Troglodytes troglodytes*	7,100,000 t	88–91	CBC	2,800,000 t	88–91	EXTP
Dunnock *Prunella modularis*	2,000,000 t	88–91	CBC	810,000 t	88–91	EXTP
Robin *Erithacus rubecula*	4,200,000 t	88–91	CBC	1,900,000 t	88–91	CBC
Nightingale *Luscinia megarhynchos*	5,000–6,000 p	88–91	TEXT	0		
Black Redstart *Phoenicurus ochruros*	80–120 p and t	88–91	TEXT	0		
Redstart *Phoenicurus phoenicurus*	90,000 p–330,000 t	88–91	NSQ/CBC	1–4 p	88–91	TEXT
Whinchat *Saxicola rubetra*	14,000–28,000 p	88–91	NSQ	1,250–2,500 p	88–91	NSQ
Stonechat *Saxicola torquata*	8,500–21,500 p	88–91	NSQ★	7,500–18,750 p	88–91	NSQ
Wheatear *Oenanthe oenanthe*	> 55,000 p	88–91	NSQ★	>12,000 p	88–91	NSQ
Ring Ouzel *Turdus torquatus*	5,500–11,000 p	88–91	NSQ	180–360 p	88–91	NSQ
Blackbird *Turdus merula*	4,400,000 t	88–91	CBC	1,800,000 t	88–91	EXTP
Fieldfare *Turdus pilaris*	< 25 p	88–91	TEXT	0		
Song Thrush *Turdus philomelos*	990,000 t	88–91	CBC	390,000 t	88–91	EXTP
Redwing *Turdus iliacus*	40–80 p	88–91	TEXT	0		
Mistle Thrush *Turdus viscivorus*	230,000 t	88–91	CBC	90,000	88–91	EXTP
Cetti's Warbler *Cettia cetti*	450 p	88–91	TEXT	0		
Grasshopper Warbler *Locustella naevia*	10,500 p	88–91	NSQ	5,500 p	88–91	NSQ

Table 9 *(continued)*.

Species	Britain			Ireland		
	Estimate	Date	Method	Estimate	Date	Method
Savi's Warbler *Locustella luscinioides*	10–20 p	88–91	TEXT	0		
Sedge Warbler *Acrocephalus schoenobaenus*	250,000 t	88–91	CBC	110,000 t	88–91	EXTP
Marsh Warbler *Acrocephalus palustris*	< 12 p	88–89	TEXT	0		
Reed Warbler *Acrocephalus scirpaceus*	40,000–80,000 p	68–72/ 88–91	NSQ★	< 40–50 m	88–91	TEXT
Dartford Warbler *Sylvia undata*	< 950 p	88–90	TEXT	0		
Lesser Whitethroat *Sylvia curruca*	80,000 t	88–91	CBC	0–1 p	88–91	TEXT
Whitethroat *Sylvia communis*	660,000 t	88–91	CBC	120,000 t	88–91	EXTP
Garden Warbler *Sylvia borin*	200,000 t	88–91	CBC	180–300 p	88–91	TEXT
Blackcap *Sylvia atricapilla*	580,000 t	88–91	CBC	40,000 t	88–91	EXTP
Wood Warbler *Phylloscopus sibilatrix*	17,200 m	84–85	TEXT	30 t	88–91	TEXT
Chiffchaff *Phylloscopus collybita*	640,000 t	88–91	CBC	290,000 t	88–91	EXTP
Willow Warbler *Phylloscopus trochilus*	2,300,000 t	88–91	CBC	830,000 t	88–91	EXTP
Goldcrest *Regulus regulus*	560,000 t	88–91	CBC	300,000 t	88–91	EXTP
Firecrest *Regulus ignicapillus*	80–250 m	88–91	TEXT	0		
Spotted Flycatcher *Muscicapa striata*	120,000 t	88–91	CBC	35,000 t	88–91	EXTP
Pied Flycatcher *Ficedula hypoleuca*	35,000–40,000 p	88–91	NSQ	1–2 p	88–91	TEXT
Bearded Tit *Panurus biarmicus*	400 p	88–91	TEXT	0		
Long-tailed Tit *Aegithalos caudatus*	210,000 t	88–91	CBC	40,000 t	88–91	EXTP
Marsh Tit *Parus palustris*	60,000 t	88–91	CBC	0		
Willow Tit *Parus montanus*	25,000 t	88–91	CBC	0		
Crested Tit *Parus cristatus*	900 p	80	TEXT	0		
Coal Tit *Parus ater*	610,000 t	88–91	CBC	270,000 t	88–91	EXTP
Blue Tit *Parus caeruleus*	3,300,000 t	88–91	CBC	1,100,000 t	88–91	EXTP
Great Tit *Parus major*	1,600,000 t	88–91	CBC	420,000 t	88–91	CBC
Nuthatch *Sitta europaea*	130,000 t	88–91	CBC	0		
Treecreeper *Certhia familiaris*	200,000 t	88–91	CBC	45,000 t	88–91	EXTP
Short-toed Treecreeper *Certhia brachydactyla*	? (Channel Islands only)			0		

Table 9 (*continued*).

Species	Britain			Ireland		
	Estimate	Date	Method	Estimate	Date	Method
Golden Oriole *Oriolus oriolus*	40 p	88–90	TEXT	0		
Red-backed Shrike *Lanius collurio*	0–1 p	88–91	TEXT	0		
Jay *Garrulus glandarius*	160,000 t	88–91	CBC	10,000 t	88–91	EXTP
Magpie *Pica pica*	590,000 t	88–91	CBC	320,000 t	88–91	EXTP
Chough *Pyrrhocorax pyrrhocorax*	315 bp	86–91	TEXT	830 bp	92	EXTP
Jackdaw *Corvus monedula*	390,000 t	88–91	CBC	210,000 t	88–91	EXTP
Rook *Corvus frugilegus*	853,000–857,000 p	80	TEXT	520,000 p	80/88–91	EXTP
Carrion Crow *Corvus corone corone*	790,000 t	88–91	CBC	?		
Hooded Crow *Corvus corone cornix*	160,000 t	88–91	CBC/ EXTP	290,000 t	88–91	EXTP
Hybrid Crow *C.c. corone × C.c. cornix*	20,000 t	88–91	CBC/ EXTP	?		
Raven *Corvus corax*	7,000 p	88–91	CNT	3,500 p	88–91	CNT
Starling *Sturnus vulgaris*	> 1,100,000 t	88–91	CBC	> 360,000 t	88–91	EXTP
House Sparrow *Passer domesticus*	2,600,000–4,600,000 p	88–91	TEXT	800,000–1,400,000 p	88–91	EXTP
Tree Sparrow *Passer montanus*	110,000 t	88–91	CBC	9,000 t	88–91	EXTP
Chaffinch *Fringilla coelebs*	5,400,000 t	88–91	CBC	2,100,000 t	88–91	EXTP
Brambling *Fringilla montifringilla*	1–2 p	88–89	TEXT	0		
Serin *Serinus serinus*	0–2 p (20 m in Channel Islands)	80–91	TEXT	0		
Greenfinch *Carduelis chloris*	530,000 t	88–91	CBC	160,000 t	88–91	EXTP
Goldfinch *Carduelis carduelis*	220,000 t	88–91	CBC	55,000 t	88–91	EXTP
Siskin *Carduelis spinus*	300,000 p	88–91	TEXT	60,000 p	88–91	EXTP
Linnet *Carduelis cannabina*	520,000 t	88–91	CBC	130,000 t	88–91	EXTP
Twite *Carduelis flavirostris*	65,000 p	88–91	NSQ	3,500 p	88–91	EXTP
Redpoll *Carduelis flammea*	160,000 p	88–91	NSQ	70,000 p	88–91	NSQ
Common Crossbill *Loxia curvirostra*	?			?		
Scottish Crossbill *Loxia scotica*	1,500 i	75 and 81–84	TEXT	0		
Parrot Crossbill *Loxia pytyopsittacus*	0–2 p	88–91	TEXT	0		
Scarlet Rosefinch *Carpodacus erythrinus*	0–1 p	88–91	TEXT	0		

Table 9   (*continued*).

Species	Britain			Ireland		
	Estimate	Date	Method	Estimate	Date	Method
Bullfinch *Pyrrhula pyrrhula*	190,000 t	88–91	CBC	100,000 t	88–91	EXTP
Hawfinch *Coccothraustes coccothraustes*	3,000–6,500 p	88–91	NSQ	0		
Snow Bunting *Plectrophenax nivalis*	70–100 p	88–91	TEXT	0		
Yellowhammer *Emberiza citrinella*	1,200,000 t	88–91	CBC	200,000 t	88–91	EXTP
Cirl Bunting *Emberiza cirlus*	229 p (5 p on Jersey)	89/91	TEXT	0		
Reed Bunting *Emberiza schoeniclus*	220,000 t	88–91	CBC	130,000 t	88–91	EXTP
Corn Bunting *Miliaria calandra*	< 160,000 t	88–91	CBC/TEXT	< 30 t	88–91	TEXT

Notes: Only those species which bred in Britain or Ireland during 1988–91 are included, and an estimate given for Ireland only if breeding occurred during 1988–91.
Only those estimates marked ★ include the Channel Islands.
The British estimates include the Isle of Man.
All population estimates are annual, and the dates to which the estimates refer are given.

Abbreviations used: (a) Estimate: p: pairs; t: territories; i: individual adults; n: nests; tp: territorial pairs; bp: breeding pairs; m: males; f: females; <: less than; >: greater than; ≧: greater than or equal to; ≦: less than or equal to; ?: no estimate available. (b) Method: CBC: CBC Method; CNT: Count Method; EXTP: Extrapolation from Britain to Ireland; TEXT; Method or source explained in species account; NSQ: Number of Squares Method.

It is theoretically possible to produce population estimates for Ireland in the same manner as for Britain. There are, however, extremely few CBC plots in Ireland, and the Irish bird community has major differences from that in Britain, in species composition and the relative abundances of species. Thus, using mean CBC densities from Britain in Ireland is likely to be invalid. Because of this, and because of the lack of accurate data on the areas of woodland and farmland in Ireland, estimates for Ireland for species covered by the CBC method in Britain are generated by extrapolation (see below).

2. The Count Method (method CNT in Table 9)

For a few species, a reasonably accurate estimate of the population can be derived from counts made during the two-hour tetrad visits undertaken as part of the main Atlas survey (see *Introduction and Methods*). Obviously, it was impossible to count every individual of every species present in an area of 4 km² in two hours, so this method only gave a minimum population size; if a certain number of birds was counted, it is clear that there was at least that number present. The degree to which the population was underestimated will have depended on the conspicuousness of the species and the range of habitats that it occupies. For a very cryptic bird that occurred throughout the whole tetrad, little of the population would have been detected in two hours, but it would have been easy to count the whole population of a very conspicuous species that occurred in only one location. Thus, for a few conspicuous species, the British population has been estimated from tetrad counts in the following manner. For each species, a mean count per tetrad over all 10-km squares with timed tetrad visits was calculated and multiplied by the number of tetrads in Britain, or more precisely the total land area of Britain in km², taken from *The Times Atlas of the World* (Times Books 1988), divided by four (= 57,614).

However, this method assumes that the mean count per tetrad was the same for tetrads which were and were not visited, and, as observers were allowed to choose which tetrads to visit, this is unlikely to have been the case. Fortunately, the extent of this bias has been estimated as part of the KSS (see Table 5 in *Bias and the Key Squares Survey*), and can be corrected by multiplying the derived estimate by the ratio of the mean count per tetrad from the KSS, to that from the main survey (i.e. the ratio of the figure contained in the last column of data in Table 5 to that in the column preceding it). An example, that of the Great Crested Grebe, will illustrate this method. The mean count per tetrad, of the Great Crested Grebe, across all 10-km squares in Britain with timed visits was 0.168. From this a total population of 9,700 individuals (= 0.168 × 57614) was calculated. However the KSS suggests that this value was too high, as observers selectively chose tetrads with Great Crested Grebe habitat, and that the true population was only 0.82 (0.3371/0.4093; see Table 5) of this value, i.e. 8,000 individuals. Note that these estimates have been rounded.

This method yielded minimum estimates of the total number of individuals present in Britain and Ireland separately. Unfortunately, only ornithological intuition could determine for which species this method was suitable and, in practice, it was only used for a small number.

3. Extrapolation to Ireland (method EXTP in Table 9)

Many species had good population estimates for Britain but none for Ireland. Examples are species for which the size of the British population was estimated by the CBC method, or those for which a population estimate has been derived from a single species survey covering Britain only (for example Hen Harrier). For these it was necessary to extrapolate the population in Ireland from that in Britain. If the relative sizes of the populations in Britain and Ireland are known, the absolute size of the Irish population can be found by multiplying the British population by the ratio between them. To do this, the relative numbers in Britain and Ireland were calculated from the measures of abundance from the main Atlas survey. For species which were counted, the ratio of the population sizes was found by calculating the sizes of the British and Irish populations as described for the count method above, and then calculating the ratio of the Irish to the British estimates. For species which were not counted, the frequency indices were used (see *Introduction and Methods*). For this, the mean frequency indices for Britain and Ireland were calculated, including all of the 10-km squares that were visited. These were then multiplied by the appropriate land area to estimate the ratio of the Irish to the British population. These ratios are shown in Table 7 for species whose British population sizes were estimated by the CBC method. As an example, the ratio of the Irish to British population of Meadow Pipit was 0.478 and the Irish population was thus 900,000 territories (1,900,000 × 0.478).

One limitation of this method is that when using frequency indices to calculate the ratio, it assumes equal densities in occupied tetrads in Britain and Ireland, and this may not have been the case. No attempt was made to extrapolate populations to Ireland if the ratio calculated from frequency indices was 0.02 or less (i.e. the species was at least 50 times more common in Britain than in Ireland). In addition, Irish estimates calculated in this manner were only as reliable as the British estimates from which they were extrapolated.

4. Estimate provided within the species account (method TEXT in Table 9)

For many species the methods outlined above were unsuitable (e.g. seabirds and rarities), and population estimates were available from a variety of other sources. Thus, for example, for rare breeding birds, RBBP reports provide annual estimates of population sizes; *Status of Seabirds* (and updates provided by JNCC/Seabird Group) give recent estimates of seabird populations and various single species surveys have yielded population sizes. For these species, the method for deriving the population estimate, or the source from which it was obtained, is outlined within the relevant species account. For some species, especially waders, it did not prove possible to provide estimates which were more recent than the mid 1980s. At the time of writing, RBBP reports were available only for 1988, 1989 and 1990.

5. The Number of Squares Method (method NSQ in Table 9)

Finally, for a few species which were neither rare, easy to count, well known from single species surveys, nor suitable for the CBC method, the total British and Irish population has been calculated by multiplying the number of occupied 10-km squares by an informed guess of the number of pairs per occupied 10-km square. Except where stated, the density estimate is the same as that used in the *68–72 Atlas*. The total population was subsequently split into estimates for Britain and Ireland, by taking into account the relative abundance of the species in each, as outlined in the 'Extrapolation to Ireland' method above.

Results

The results of these analyses are presented in Table 9, which lists the

estimate for each species in Britain and Ireland separately, and the method by which it was derived. The majority of British estimates do not include the Channel Islands, as these are geographically and ecologically more similar to continental Europe, but all include the Isle of Man.

SIMON GATES
DAVID W. GIBBONS
JOHN H. MARCHANT

The Breeding Birds of Britain and Ireland: Changing Status and Species Richness

The changing status of breeding birds in Britain and Ireland

Although the species composition of the breeding avifauna of Britain and Ireland changed continually between 1800 and 1949, the total number of regularly breeding species was remarkably constant as extinctions were balanced by colonisations or recolonisations (Parslow 1973, Sharrock 1974). Between 1950 and the period of the *68–72 Atlas*, however, there was a marked gain in the number of breeding species recorded, with Wood Sandpiper, Ruff, Snowy Owl, Fieldfare, Cetti's Warbler, Firecrest and Serin colonising, Savi's Warbler recolonising following an absence of more than a century, and only Kentish Plover being lost as a regularly breeding species, although it continued to breed in the Channel Islands.

One way of determining whether or not this rising trend in the number of breeding species has continued in recent years, is to compare the number of species lost and gained between the *68–72 Atlas* and *88–91 Atlas*. This comparison (Table 10) shows that, when species of introduced, reintroduced or feral origin are excluded, the net gain in number of breeding species was much more modest, amounting to only two species (or three if Scottish Crossbill, recently accorded full species status, is included). In Britain, Great Northern Diver, Black Tern, Snowy Owl and Hoopoe have been lost, while Red-necked Grebe, Purple Sandpiper, Parrot Crossbill and Scarlet Rosefinch have been gained as breeding species, and Whooper Swan, Crane and White-tailed Eagle have recolonised, the latter following a reintroduction programme (Table 11). In Ireland there was a similar number of changes. Black-necked Grebe, Montagu's Harrier, Greenshank, Turtle Dove and Yellow Wagtail were lost as breeding species, while Leach's Petrel, Reed Warbler, Lesser Whitethroat, Pied Flycatcher, and possibly Black-throated Diver were gained as new species, and Garganey and Common Crossbill bred again following a gap of some years.

Table 10. *The number of breeding species in Britain and Ireland during 1968–72 and 1988–91.*

Period	Number of breeding species	
	Britain	Ireland
1968–72	201 (214)	135 (139)
1988–91	204 (219)	137 (142)

The figures in parentheses include species that bred in the wild but which were of introduced or reintroduced origin, or were feral. These are: Barnacle Goose, Canada Goose, Egyptian Goose, Wood Duck, Mandarin, Red-crested Pochard, Ruddy Duck, White-tailed Eagle, Capercaillie, Red-legged Partridge, Pheasant, Golden Pheasant, Lady Amherst's Pheasant, Ring-necked Parakeet and Little Owl. The 1988–91 Snowy Owl record and the 1968–72 Bluethroat record are not included in the figures; although clutches of eggs were laid, no male was present. The Scottish Crossbill is included in the 1988–91, but not the 1968–72 figures. None of the breeding species listed in Appendix F is included in the 1988–91 figures, and none of those included in the list of additional species on p 453 of the *68–72 Atlas*, nor Bob-white Quail or Budgerigar, are included in the 1968–72 figures.

Table 11. *Losses and gains of breeding species in Britain and Ireland since 1968.*

Britain	Ireland

(a) Species that bred in 1968–72 but not 1988–91:

Britain	Ireland
Great Northern Diver (1970)	Black-necked Grebe (1982)
Black Tern (1978)	Montagu's Harrier (1971)
Snowy Owl (1975)[1]	Golden Pheasant★
Hoopoe (1977)	Greenshank (1974)
	Turtle Dove (1977)
	Yellow Wagtail (1982)

(b) Species that bred in 1973–87, but not during either Atlas period:

Britain	Ireland
Little Bittern (1984)	Dotterel (1975)
Black-winged Stilt (1983/1987)[2]	Black Tern (1975)[6]
Kentish Plover (irregular breeder)[3]	Short-eared Owl (irregular breeder)
Spotted Sandpiper (1975)	Bearded Tit (1976/1985)
Little Gull (1975/1987)	
Shore Lark (1973/1977)	
Bluethroat (1985)[4]	
Lapland Bunting (1977/1980)	

(c) Species that bred in 1988–91 but not 1968–72:

Britain	Ireland
Red-necked Grebe (1980)	Black-throated Diver (1990)[7]
Whooper Swan (irregular breeder)	Leach's Petrel (1982)
Barnacle Goose★	Garganey (irregular breeder)
White-tailed Eagle★	Ruddy Duck (1973)★
Crane (recolonised 1981)	Red-legged Partridge★
Purple Sandpiper (1978)	Reed Warbler (1981)[8]
Parrott Crossbill (1984)	Lesser Whitethroat (1990)
[Scottish Crossbill][5]	Pied Flycatcher (1985)
Scarlet Rosefinch (1982)	Common Crossbill (irregular breeder)

General notes: Breeding need not have been successful for a species to be included in this table; attempted breeding by a pair was sufficient. Species marked ★ were of introduced or reintroduced origin, or were feral. Britain includes the Isle of Man, but excludes the Channel Islands. The years in parentheses are, for (a) the last year of recorded breeding, (b) the first and last year of recorded breeding, and (c) the first year of recorded breeding. Species that bred on more than one occasion in the past are considered as irregular breeders.

Footnotes: [1] A number of infertile clutches of Snowy Owl eggs has been laid on Fetlar since 1975, including during 1988–91. [2] Black-winged Stilts also bred in Britain in 1945. [3] Kentish Plovers bred in the Channel Islands during 1968–72. [4] The 1968 Scottish Bluethroat record has been excluded from (a) for although a clutch of eggs was found no male was seen. [5] The Scottish Crossbill was recorded as a breeding species during 1988–91, but not during 1968–72 when it was considered conspecific with the Common Crossbill: it now has the status of a separate species (Voous 1978). [6] Black Terns bred successfully at Lough Erne, Co. Fermanagh in 1967. [7] The 1990 Irish Black-throated Diver record was of a pair with juveniles on the sea off the W coast of Ireland; their origin was unknown. [8] Reed Warblers bred in Ireland in 1935, prior to the recent colonisation. The record of a Tree Pipit 'breeding' in Ireland is excluded for although a male sang on territory for several days during 1988 no nest was ever found.

Sources: *68–72 Atlas*, *88–91 Atlas*, *Red Data Birds*, *Birds in Scotland*, *Birds in Ireland*, RBBP reports, *Irish Birds*.

Such a simple analysis, however, does not take into account the species that bred during the inter-Atlas period only, 1973–87 (Table 11). Five of these species, Little Bittern, Spotted Sandpiper, Little Gull, Shore Lark and Lapland Bunting were additions to the list of British breeding species, as were Dotterel and Bearded Tit to the Irish list. With the possible exception of Bearded Tit in Ireland, none of the species bred regularly, and only time will tell which of those that bred during 1988–91 but not 1968–72 will remain, although White-tailed Eagle, Crane and Purple Sandpiper seem well established, albeit in very small numbers.

Predicting which species may colonise in the future is, as James Ferguson-Lees suggested in his Foreword to the *68–72 Atlas*, 'all a guessing game'. However, two of the species he proposed as potential colonists then, Spoonbill and Penduline Tit, made tentative breeding attempts in Britain during 1988–91, building nests though not laying eggs. The Penduline Tit, in particular, has shown a marked westward range expansion on the Continent in recent years, and the breeding population of Spoonbills in the Netherlands has increased substantially over the same period. Following the unprecedented record of up to seven breeding pairs of Scarlet Rosefinches in Britain in 1992 (previously only two nests had ever been found) there are high hopes that this species may become well established here.

In Ireland several species have been increasingly recorded during the summer months, often following expansion of their breeding range in Britain. Wigeon, Goldeneye, Marsh Harrier, Golden Eagle, Osprey, Hobby, Little Ringed Plover, Savi's Warbler and Firecrest have all been suggested as potential colonists or recolonists (*Birds in Ireland, Irish Birds*). Whooper Swans bred for the first time in Ireland in 1992, in Co. Donegal, and it is possible that a pair of Hawfinches bred in Co. Clare in 1991 although the record (of two adults and a juvenile) was in September and thus outwith the breeding season.

Though interesting, such comparisons are essentially trivial when compared to the much greater changes in status of existing breeding species. An approximate measure of this is the percentage change in the number of 10-km squares occupied by each species during 1968–72 and 1988–91, shown in the tables alongside each species account. When analysing these figures, all the caveats outlined in *Interpreting the Species Accounts*, particularly of differences in coverage and method, should be borne in mind. Although it is tempting to conclude that species with positive percentage change values are those with expanding ranges, and negative values those with contracting ranges, this is not necessarily the case. A species may have maintained the same limits to its range, but simply become more, or less, widespread within those limits. Alternatively the range of a species may have changed completely, yet the species may have been recorded in precisely the same number of 10-km squares, thus giving a zero percentage change value when calculated in this manner. Nevertheless, these values do provide clues to the change in status of each species during the inter-Atlas period. A few of these values are summarised, for Britain and Ireland separately, in Table 12, which lists the species with the ten greatest gains and losses. As outlined in *Interpreting the Species Accounts*, comparisons made for species with only a small number of records can lead to artificially high or low percentage change values, so those species which occurred in only 15 or fewer 10-km squares during each Atlas period are excluded from both this analysis and that of habitat and distributional change (below). Two values are given for each species, the percentage change in all records, and the percentage change in breeding records only ('probable' and 'confirmed' for 1968–72 and 'breeding' for 1988–91), although the ranking is based on the 'all records' column.

With the exception of the Goshawk population, the provenance of

Table 12. *Species that have shown marked range changes.*

Species	Britain percentage change		Species	Ireland percentage change	
	All records	Breeding records		All records	Breeding records
(a) *Top ten species with positive percentage change values (= expanding range)*					
Cetti's Warbler	1620	1800	Common Crossbill	5100	–
Ring-necked Parakeet	1475	400	Buzzard	272	132
Ruddy Duck	1437	1315	Wood Warbler	250	100
Goldeneye	811	160	Canada Goose	217	175
Osprey	572	255	Greylag Goose	188	100
Goshawk	574	194	Peregrine	166	145
Mandarin	459	375	Gadwall	79	117
Firecrest	395	269	Raven	54	20
Egyptian Goose	383	179	Blackcap	40	– 5
Golden Oriole	350	133	Siskin	20	– 16
(b) *Top ten species with negative percentage change values (= contracting range)*					
Wryneck	– 88	– 96	Nightjar	– 88	– 95
Red-backed Shrike	– 87	– 98	Grey Partridge	– 86	– 93
Cirl Bunting	– 83	– 87	Corn Bunting	– 84	– 96
Corncrake	– 76	– 80	Corncrake	– 70	– 93
Capercaillie	– 64	– 75	Red Grouse	– 66	– 83
Bittern	– 63	– 62	Woodcock	– 64	– 74
Woodlark	– 63	– 66	Barn Owl	– 63	– 81
Wood Sandpiper	– 56	– 71	Twite	– 53	– 62
Nightjar	– 51	– 59	Teal	– 52	– 72
Red-necked Phalarope	– 50	– 69	Roseate Tern	– 50	– 50

Notes: Species which were recorded in 15 or fewer 10-km squares (in each country separately) during both Atlas periods are excluded (see text). Only those species which bred in Britain or Ireland during both Atlas periods are included; those which bred in one Atlas period, but not both, are listed in Table 11.

which is uncertain, and the Golden Oriole, whose historical status is poorly documented, all those British species on this list with expanding ranges were either of introduced origin or were feral (Ring-necked Parakeet, Ruddy Duck, Mandarin and Egyptian Goose) or have colonised (Cetti's Warbler, Goldeneye and Firecrest), or recolonised (Osprey) since 1950. In Ireland, the pattern was somewhat different. None of the ten species with expanding ranges were recent natural colonists; most were of established species showing marked range expansions, although two (Canada and Greylag Goose) were feral, and the historical status of breeding Gadwall in Ireland is unclear.

At least two of the ten species with the most extreme range contractions in Britain are likely to become, or may already have become, extinct as regular breeders. The period of the *88–91 Atlas* included the first recorded year (1989) in which Red-backed Shrikes failed to breed in Britain, although breeding has occurred subsequently. The Wryneck seems perilously close to extinction, and may indeed already have failed to breed in recent years; certainly there were no cases of confirmed breeding between 1981 and 1984. However, it is such an unobtrusive species that nesting attempts may have occurred unnoticed. The range of the Corncrake has continued to contract northwestwards and this contraction has been associated with a decline in the overall population size. For three of the species listed, however, despite marked range contractions, the sizes of their breeding populations have stabilised and may even have increased in recent years. Nightjars and Woodlarks have abandoned many of their former haunts, particularly lowland heathland, but their populations have become concentrated on a smaller number of sites, often within lowland forests in areas of young growth or clear-fell. Similarly, despite the marked range contraction of the Cirl Bunting, numbers in its Devon stronghold are now stable, and possibly increasing. Nevertheless, such range contractions are of great concern, even with a stable breeding population, for species become increasingly vulnerable when restricted to fewer sites.

In Ireland, Nightjar, Grey Partridge and Corn Bunting have shown alarming contractions in range and may well be threatened with extinction, while the fate of Corncrake there is now as worrying as it is in Britain and indeed throughout the Palearctic. Although not included in this list (it was recorded in too few 10-km squares) the Red-necked Phalarope retains only the most tenuous foothold as a breeding species in Ireland. The range of the Twite, once a common bird of mountain and rocky coast in all provinces (Ussher and Warren 1900), is now considerably more restricted than at the time of the *68–72 Atlas*, and the decline in distribution and numbers of Red Grouse has continued. The percentage change figures for Barn Owl and Woodcock are perhaps exaggerated by differences in coverage between the two Atlas periods (see *Interpreting the Species Accounts*); but although local declines of the Barn Owl have been reported in Ireland, other evidence suggests that the Irish Woodcock population may be increasing (*Birds in Ireland*).

Habitat and distributional change

Determining the underlying causes of the changes in distribution documented by the Change maps is not straightforward, and although the reason can often be guessed, proof is much harder to obtain. Some general themes do, however, emerge, and one of these is the influence of breeding season habitat. Table 13 lists each species by habitat and whether or not its percentage change value was positive ('expanding range') or negative ('contracting range'). Because of the problem of poorer coverage in Ireland in 1988–91 compared with 1968–72, this analysis only considers Britain.

The habitat classification is very much a personal one, although partially based on published information. Because of the relatively small number of species in total and their categorisation into 'expanding' or 'contracting', it was necessary to limit the classification to as small a number of classes as possible even though additional classes, e.g. heathland, would have made it more comprehensive. There were six classes: coastal, farmland, lowland wetland, urban, upland and woodland. Each species was allocated to one of these classes, although those which could not be easily classified were placed in a seventh (not classified) category. The details of this classification were as follows:

Coastal	Species associated with rocky shores, cliffs, shingle beaches, mudflats, sandy shores and the coastal fringe (e.g. saltmarsh and coastal grassland).
Farmland	Species feeding in open farmland during the breeding season, even though they may nest in woods or hedges. Although Red-backed Shrike was traditionally a species of farmland, during 1968–72 it was restricted mainly to heathland and so has been excluded from this category. The least likely of these species, Reed Bunting, has, however, been included as it was common on farmland in the inter-Atlas years.
Lowland wetland	Species associated with lowland freshwater and fringing habitats such as reed beds – Reed Warbler and Bearded Tit. Marshland – Savi's, Marsh, Cetti's and

Sedge Warblers. Shorelines – Little Ringed Plover. This category also includes those species from more northern latitudes that nest along the valley floors (e.g. Goldeneye and Osprey).

Urban	Species associated mainly with man or the urban environment, and with farm buildings (e.g. Collared Dove). Feral Pigeon/Rock Dove is included in this category, as the bulk of this species' population breeds in the urban environment.
Upland	Species that breed in mountain, upland or hill country. It includes montane species, those of submontane moorland (even down to sea-level in N Scotland, such as Red-necked Phalarope, Whimbrel, Arctic and Great Skua) and those of upland lakes and streams. Species frequenting lakes and rivers within forests or along valley floors were, however, excluded. Dunlin has been included even though a substantial proportion of the population nests on the machair of the Outer Hebrides. This classification closely follows that of Ratcliffe 1990.
Woodland	Species that breed and feed (mainly) in woodland. Some prefer dense scrub or thicket-stage woods rather than mature woods (e.g. Nightingale). This category also includes species which breed in scrub, or woodland with a more open canopy (e.g. Tree Pipit and Lesser Whitethroat). Species which nest in woodland but which feed mainly in surrounding open habitats (e.g. Buzzard, Turtle Dove, Mistle Thrush, Jackdaw and Rook) were excluded. This classification closely follows that of Fuller (1994).

Classifying habitat specialists was generally straightforward, although the same was not true for generalists. Thus, for example, Cuckoo and Crow bred in most habitat types with the possible exception of urban. Others are more restricted than this, but commonly occur in two or more habitat types. Black-headed and Common Gull are both coastal and upland species, Common Tern is a coastal and lowland freshwater species, and Oystercatchers nest on the coast, along the banks of rivers, on open agricultural land in the north of their range, and even on urban roofs. In addition, the habitat occupied by some species has changed in recent years (e.g. Nightjar and Red-backed Shrike). Finally, Swallow and Sand Martin are both aerial insectivores which, although they nest in a restricted range of habitats, feed more widely. Because of these problems, all these species were allocated to the not classified category. Doubtless, different classifications could be drawn up and as the percentage change data are tabulated alongside the species accounts, the analyses presented here could be repeated for any number of alternative classifications.

Despite these problems, Table 13 shows that there were marked differences between habitat types ($\chi^2 = 176$ with 6 *d.f.*, $P < 0.0001$). For all habitat types except farmland, the number of species with positive and negative percentage changes was similar. Out of 174 non-farmland species, 85 showed positive and 89 had negative percentage changes. For farmland species, however, 24 out of 28 species (86%) had negative percentage values.

Table 13. *Habitat and distributional change*

Habitat	Positive Percentage Change	Negative Percentage Change
Coastal	Fulmar 11 Storm Petrel Gannet Shelduck Eider Avocet Ringed Plover Lesser Black-backed Gull Kittiwake Black Guillemot Chough	Manx Shearwater 13 Cormorant Shag Herring Gull Great Black-backed Gull Sandwich Tern Roseate Tern Arctic Tern Little Tern Guillemot Razorbill Puffin Rock Pipit
Farmland	Hobby 4 (3) Red-legged Partridge★ Quail Goldfinch	Montagu's Harrier 24 (23) Kestrel Grey Partridge Corncrake Stone Curlew Lapwing Stock Dove Woodpigeon Turtle Dove Barn Owl Little Owl★ Skylark Yellow Wagtail Whitethroat Jackdaw Rook Starling Tree Sparrow Greenfinch Linnet Yellowhammer Cirl Bunting Reed Bunting Corn Bunting
Woodland	Honey Buzzard 22 (20) Goshawk Sparrowhawk Pheasant★ Golden Pheasant★ Robin Fieldfare Redwing Lesser Whitethroat Garden Warbler Blackcap Wood Warbler Chiffchaff Willow Warbler Firecrest Pied Flycatcher Crested Tit Nuthatch Golden Oriole Chaffinch Siskin Crossbill spp.	Capercaillie★ 28 (27) Woodcock Tawny Owl Long-eared Owl Wryneck Green Woodpecker Great Spotted Woodpecker Lesser Spotted Woodpecker Tree Pipit Wren Dunnock Nightingale Redstart Blackbird Song Thrush Goldcrest Spotted Flycatcher Long-tailed Tit Marsh Tit Willow Tit Coal Tit Blue Tit Great Tit Treecreeper Jay Redpoll Bullfinch Hawfinch
Lowland wetland	Great Crested Grebe 21 (15) Slavonian Grebe Black-necked Grebe	Little Grebe 14 Bittern Mute Swan

Table 13. *(continued)*

Habitat	Positive Percentage Change	Negative Percentage Change
Lowland wetland *(continued)*	Grey Heron Greylag Goose★ Canada Goose★ Egyptian Goose★ Wood Duck★ Mandarin★ Gadwall Garganey Tufted Duck Goldeneye Ruddy Duck★ Marsh Harrier Osprey Little Ringed Plover Cetti's Warbler Savi's Warbler Reed Warbler Bearded Tit	Mallard Pintail Shoveler Pochard Water Rail Spotted Crake Moorhen Coot Kingfisher Sedge Warbler Marsh Warbler
Urban	Feral Pigeon/Rock Dove★ 5 (3) Collared Dove Ring-necked Parakeet★ Swift Black Redstart	House Martin 2 House Sparrow
Upland	Red-throated Diver 19 Wigeon Common Scoter Red-breasted Merganser Goosander Red Kite Hen Harrier Buzzard Golden Eagle Merlin Peregrine Dotterel Dunlin Whimbrel Common Sandpiper Arctic Skua Great Skua Grey Wagtail Snow Bunting	Black-throated Diver 19 Teal Red Grouse Ptarmigan Black Grouse Golden Plover Snipe Curlew Greenshank Wood Sandpiper Red-necked Phalarope Short-eared Owl Meadow Pipit Dipper Whinchat Wheatear Ring Ouzel Raven Twite
Not classified	Oystercatcher 7 Ruff Black-tailed Godwit Swallow Pied Wagtail Dartford Warbler Magpie	Redshank 13 Black-headed Gull Common Gull Common Tern Cuckoo Nightjar Woodlark Sand Martin Stonechat Mistle Thrush Grasshopper Warbler Red-backed Shrike Crow

The percentage change values are based on all British records. Species recorded in 15 or fewer 10-km squares during each Atlas period are excluded (see text). As the percentage of records with evidence of breeding was less in 1988–91 compared with 1968–72 (see Table 2 in *Introduction and Methods*) it is not possible to categorise species based on percentage change in the number of breeding records only, although for a few species, e.g. Dunlin and Goldeneye, this would perhaps be a more realistic statistic. Habitat categories are defined in the text. For species marked ★ either a substantial part or all of the breeding population is of introduced, reintroduced or feral origin (based mainly on BOU categories C and D; British Ornithologists' Union 1992). Goshawk is not included in this category as its provenance is uncertain, nor is Gadwall, even though its range expansion has been assisted by introduction in the past (see species accounts). The figure in the top right hand corner of each cell is the number of species in that cell. Figures in parentheses are after exclusion of species of introduced, reintroduced or feral origin.

These worrying results confirm those of the CBC, summarised in the *Trends Guide*, which has documented long-term declines in populations of many farmland birds. Whilst such simple analyses of changes in distribution and numbers cannot establish cause and effect, there is strong circumstantial evidence that changes in farming practice in recent years, particularly since the mid 1970s, are responsible. O'Connor and Shrubb (1986a) were among the first to establish a link between developments in crop production and bird populations, although it is likely that the scale of the changes in farmland bird populations is now even greater than was reported by them (Fuller 1993). Although for most species the mechanisms underlying their declines are poorly understood, it seems likely that agricultural intensification has been at least partly responsible. The switch from spring to autumn sowing of crops, accompanied by the loss of winter stubbles, and a move away from crop rotation occurred in the 1970s and early 1980s. At the same time inputs of inorganic fertiliser increased, as did the use of insecticide, molluscicide and fungicide, and herbicide usage reached its highest level in the 1980s. Habitat destruction and deterioration, such as the loss of hedgerow and the draining of wet meadows, have also played their part. Taken together, these changes have led to a loss of breeding habitat and to a reduction in the food supply of farmland birds. Some of the best evidence for a link between changing farm practice, a declining food supply and dwindling bird populations comes from the research undertaken by the Game Conservancy Trust on the Grey Partridge (Potts 1986). This work has shown that increased use of herbicide has removed the food plants of the insects eaten by partridges, and in addition a reduction in the undersowing of cereals with clover has led to the disappearance of sawfly larvae, the preferred food of Grey Partridge chicks. Further research on the impact of agricultural change on farmland bird populations is urgently needed.

Species richness and avian diversity

It is tempting to suppose that Fig. 6 in *Introduction and Methods*, which shows the total number of species recorded in each 10-km square, represents geographical variation in bird species richness. Unfortunately this is not the case, because no account was taken of variation in fieldwork effort prior to the production of this map. Fortunately, the methods used in the *88–91 Atlas* allow calculation of a more precise measure of species richness which is corrected for effort, and this is shown in Fig. 11. The data used to produce this map were calculated as follows. The number of species recorded in each of the 42,736 tetrads visited was counted. In the great majority of cases this was the number recorded during two hours but, despite instructions to the contrary, some tetrads were covered more than once (either in different years or by different observers or both). Where this was the case, an average number of species per two hours was calculated across each of the two-hour visits to that tetrad. A measure of species richness for each 10-km square was then calculated as the average of the number of species per two hours in each of its covered tetrads. This value varied from 2.5 to 68 species with an average of 26.6 species per two hours. Even this corrected measure of richness is open to a degree of bias due to observer choice of tetrads (see *Bias and the Key Squares Survey*), but the effect is likely to be slight compared with the great range of species richness across Britain and Ireland. Figure 11 illustrates that there was clear geographical variation in species richness, with SE Britain showing higher species richness than NW Britain, and Ireland lower richness than Britain.

This map is thus a representation of avian diversity in Britain and Ireland, though better diversity measures would take account of the number of individuals of each species in addition to the number of species.

It could be argued, that in order to maximise avian diversity, conservation resources in Britain should not be squandered in the species-poor northwest as this area contains relatively few species. This approach, however, misses an important point, for Fig. 11 includes all recorded species and takes no account of whether or not each species was of conservation concern. The NCC and RSPB listed rare, threatened and important species in Britain in *Red Data Birds*. Breeding species were admitted to this list for one or more of five reasons: (i) the British population was of international significance (Britain held more than 20% of the NW European population), (ii) the species was a rare breeder (less than 300 pairs in Britain), (iii) the species' population was declining (by more than 50% over the previous 25 years), (iv) the species was restricted to vulnerable sites or habitats (more than half of the population in ten or fewer sites), or (v) the species was of special concern (eight species in total). Of the 94 breeding species included, five (Little Bittern, Black-winged Stilt, Kentish Plover, Shore Lark and Lapland Bunting) were not recorded at all for the *88–91 Atlas*. A provisional list of 44 species has similarly been drawn up for Ireland by Tony Whilde. Geographical variation in the numbers of these breeding red data birds is shown in Fig. 12. Because of the rarity of many of the species included on these two lists, and because rare birds were, by their very nature, frequently missed during timed visits, no attempt has been made to correct for fieldwork effort prior to the production of this map. Nevertheless the general pattern is clear: the areas of highest red data bird diversity are in NW Britain, and the least are in the southeast, in complete contrast to the diversity of all species. Close examination of this map reveals a number of areas with higher red data bird diversity than elsewhere. In Britain these are the Ouse and Nene Washes, the Brecks, the Norfolk and Suffolk coasts, the N Kent marshes, the New Forest and surroundings, the Welsh mountains, Anglesey, the Peak District and Pennines, North Yorkshire Moors, parts of Cumbria, the Isle of Man, much of the Southern Uplands, and most of the Highlands and Islands. In Ireland they are Tacumshin and Lady Island Lake in Co. Wexford, the Wicklow Mountains, Strangford Lough, Loughs Neagh and Erne, much of Co. Donegal, the Glens of Antrim, parts of Co. Kerry around Lough Leane, the Killarney Woodlands and Macgillycuddy's Reeks, and the Shannon Callows.

The two maps presented here (Figs. 11 and 12) highlight a conflict of interests for those individuals and organisations involved in the conservation of avian diversity in Britain and Ireland. Should 'all-species richness' be used as a criterion for setting biodiversity conservation priorities, or should the diversity of particular sorts of species, such as endemics with limited distributions, rare species or species with declining populations, be used (ICBP 1992)? The active conservation of species which are in no way threatened would seem to be wasteful of resources, whereas use of the *Red Data Birds* criteria seems eminently more sensible. While biodiversity should be maintained where it exists, or improved where there is scope for doing so, active conservation should be primarily aimed at threatened species and areas of high diversity. The publication of the work of the ICBP on the conservation status of European birds is eagerly awaited as it will put the status of British and Irish species into a wider European and global context and will highlight those species which are of particular European (rather than British or Irish) conservation concern.

Above 11.0
10.0 – 11.0
8.0 – 9.0
7.0
6.0
5.0
4.0
3.0
2.0
1.0
None

Fig. 11   (*below*) Geographical variation in all-species richness corrected for variation in fieldwork effort. Species richness was measured as the mean number of species recorded per two hours in each tetrad visited in each 10-km square.

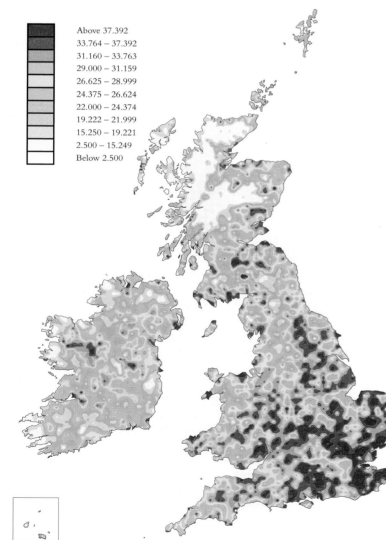

Above 37.392
33.764 – 37.392
31.160 – 33.763
29.000 – 31.159
26.625 – 28.999
24.375 – 26.624
22.000 – 24.374
19.222 – 21.999
15.250 – 19.221
2.500 – 15.249
Below 2.500

Fig. 12   (*above*) Geographical variation in the total number of breeding red data birds in each 10-km square.

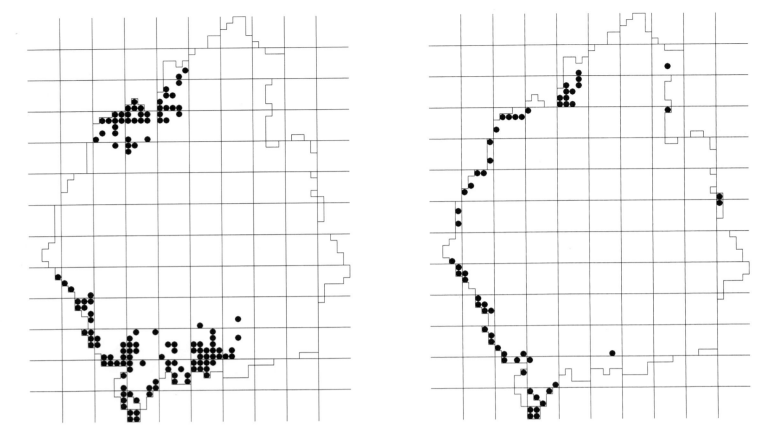

A   B

Fig. 13 Tetrad distributions in Cumbria for (a) Shelduck, (b) Ringed Plover, (c) Ring Ouzel and (d) Blackbird. Each dot refers to presence in each tetrad recorded, mostly, during two hours, although a longer time was spent in some tetrads.

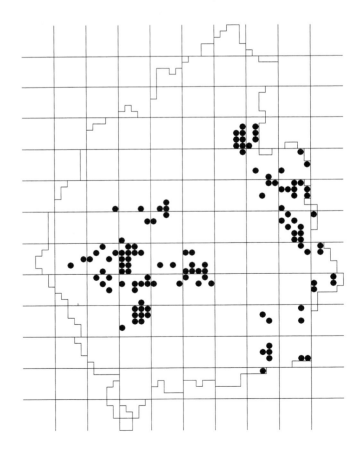

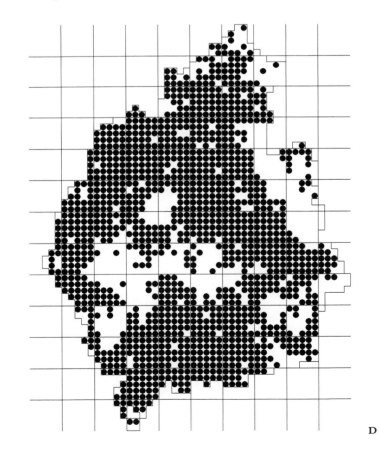

C   D

The importance of timed tetrad visits

The introduction of timed tetrad visits into the *88–91 Atlas* methods was undoubtedly controversial. It is quite likely that many observers, who were used to spending as long as they wished birdwatching in their favoured areas, were put off by the rigour and apparent complexity of timed visits. If this did occur, then it was unfortunate, because standardisation of effort for at least part of the fieldwork was central to the methods. The importance of the timed tetrad visits has been outlined earlier (see *Introduction and Methods*) but is enlarged upon here.

There were three advantages from the adoption of timed tetrad visits, although the first of these was a spin-off, rather than a reason for their adoption. Firstly, although all maps presented so far have been based on 10-km square data, all tetrad data are stored on computer and it is possible to produce maps of approximate distribution by tetrad. Some examples of these are shown for the county of Cumbria (where complete coverage of all tetrads was achieved during 1988–91) in Figs. 13(a)–(d). As only a proportion of all tetrads (more than half in Britain, but less than half in Ireland) was visited and only two hours spent in each, then the tetrad level distribution data are by no means comprehensive, and cannot be considered as valuable as a local tetrad atlas covering the same area. However, for areas without tetrad atlases, the data could be useful for local conservation issues.

Anyone wishing to gain access to these data is invited to write to BTO headquarters.

Secondly, the introduction of a time limit has enabled Abundance maps corrected for variation in fieldwork effort to be produced. For most species, these maps provide the first insight into their geographical variation in abundance during the breeding season.

Finally, if the same methods are adopted for the next breeding Atlas, it will be possible to calculate very much more precise measures of percentage change. Those presented in this volume are, of necessity, imprecise. They are based on the comparison of presence or absence recorded during an unlimited (and unknown) amount of time in a total of 3,875 10-km squares. During fieldwork in 1988–91 two hours was spent in each of 42,736 tetrads. For the next breeding Atlas, the same (or even a sample) of these could be re-visited and two hours spent in each. It would then be possible to calculate very precise estimates of percentage change based on presence or absence in each of those tetrads. Whilst there can be no doubt that the *68–72 Atlas* was an ornithological milestone, it failed to establish a precisely repeatable methodology. Had methods similar to those used in the *88–91 Atlas* been adopted then, it would now be possible to calculate precise change statistics, even for areas with widely differing levels of coverage. It is sincerely hoped that anyone involved in the early stages of planning an Atlas will take these considerations into account.

Appendices

Appendix A: Data used for the Abundance maps

The following table documents the data used for the Abundance maps in the species accounts, Appendix E, and *Bias and the Key Squares Survey*. For each map, one of seven types of data was used. These were:

1. Frequency: the proportion of tetrads visited in which the species was recorded.
2. Count: the mean number of adults recorded per two hours in each visited tetrad.
3. Count (n): the mean number of apparently occupied nests recorded per two hours in each visited tetrad.
4. Count (★): the number of calling male Corncrakes, or Cirl Bunting pairs (data supplied by RSPB).
5. Count (SCR p): the total number of pairs recorded by the SCR.
6. Count (SCR i): the total number of individuals recorded by the SCR.
7. KSS point count: KSS point count data.

Species	Data used for map in:		
	Species account	Appendix E	Bias and the Key Squares Survey
Red-throated Diver	Count		
Black-throated Diver	Count		
Little Grebe	Count		
Great Crested Grebe	Count		
Fulmar	Count (SCR p)		
Manx Shearwater	Count (SCR p)		
Storm Petrel	Count (SCR p)		
Gannet	Count (SCR p)		
Cormorant	Count (SCR p)	Frequency	
Shag	Count (SCR p)		
Grey Heron	Count (n)	Frequency	
Mute Swan	Count		
Greylag Goose	Count		
Canada Goose	Count		
Egyptian Goose	Count		
Shelduck	Count		
Mandarin	Count		
Wigeon	Count		
Gadwall	Count		
Teal	Count		
Mallard	Frequency		
Shoveler	Count		
Pochard	Count		
Tufted Duck	Count		
Eider	Count		
Red-breasted Merganser	Count		
Goosander	Count		
Ruddy Duck	Count		
Marsh Harrier	Count		
Hen Harrier	Count		

Species	Data used for map in:		
	Species account	Appendix E	Bias and the Key Squares Survey
Sparrowhawk	Count		
Buzzard	Count		
Golden Eagle	Count		
Kestrel	Frequency		
Merlin	Count		
Hobby	Count		
Peregrine	Count		
Red Grouse	Frequency		
Ptarmigan	Count		
Black Grouse	Count		
Red-legged Partridge	Frequency		
Grey Partridge	Frequency		
Pheasant	Frequency		
Water Rail	Count		
Corncrake	Count (★)		
Moorhen	Frequency		
Coot	Count		
Oystercatcher	Frequency		
Avocet	Count		
Little Ringed Plover	Count		
Ringed Plover	Count		
Dotterel	Count		
Golden Plover	Count		
Lapwing	Frequency		
Dunlin	Count		
Snipe	Frequency		
Woodcock	Count		
Whimbrel	Count		
Curlew	Frequency		
Redshank	Frequency		
Greenshank	Count		
Common Sandpiper	Frequency		
Arctic Skua	Count (SCR p)		
Great Skua	Count (SCR p)		
Black-headed Gull	Count (SCR p)	Frequency	
Common Gull	Count (SCR p)	Frequency	
Lesser Black-backed Gull	Count (SCR p)	Frequency	
Herring Gull	Count (SCR p)	Frequency	
Great Black-backed Gull	Count (SCR p)		
Kittiwake	Count (SCR p)		
Sandwich Tern	Count (SCR p)		
Roseate Tern	Count (SCR p)		
Common Tern	Count (SCR p)	Frequency	
Arctic Tern	Count (SCR p)		
Little Tern	Count (SCR p)		
Guillemot	Count (SCR i)		
Razorbill	Count (SCR i)		
Black Guillemot	Count (SCR i)		
Puffin	Count (SCR i)		
Feral Pigeon/Rock Dove	Frequency		
Stock Dove	Frequency		
Woodpigeon	Frequency		KSS point count
Collared Dove	Frequency		

Species	Data used for map in:		
	Species account	Appendix E	Bias and the Key Squares Survey
Turtle Dove	Frequency		
Cuckoo	Frequency		
Little Owl	Count		
Tawny Owl	Count		
Short-eared Owl	Count		
Swift	Frequency		
Kingfisher	Count		
Green Woodpecker	Frequency		
Great Spotted Woodpecker	Frequency		
Lesser Spotted Woodpecker	Count		
Skylark	Frequency		KSS point count
Sand Martin	Count (n)	Frequency	
Swallow	Frequency		KSS point count
House Martin	Frequency		
Tree Pipit	Frequency		
Meadow Pipit	Frequency		KSS point count
Rock Pipit	Frequency		
Yellow Wagtail	Frequency		
Grey Wagtail	Frequency		
Pied Wagtail	Frequency		
Dipper	Count		
Wren	Frequency		KSS point count
Dunnock	Frequency		KSS point count
Robin	Frequency		KSS point count
Nightingale	Count		
Redstart	Frequency		
Whinchat	Frequency		
Stonechat	Frequency		
Wheatear	Frequency		
Ring Ouzel	Count		
Blackbird	Frequency		KSS point count
Song Thrush	Frequency		KSS point count
Redwing	Count		
Mistle Thrush	Frequency		
Cetti's Warbler	Count		
Grasshopper Warbler	Count		
Sedge Warbler	Frequency		
Reed Warbler	Frequency		
Dartford Warbler	Count		
Lesser Whitethroat	Frequency		
Whitethroat	Frequency		
Garden Warbler	Frequency		
Blackcap	Frequency		
Wood Warbler	Frequency		
Chiffchaff	Frequency		
Willow Warbler	Frequency		KSS point count
Goldcrest	Frequency		
Spotted Flycatcher	Frequency		
Pied Flycatcher	Frequency		
Bearded Tit	Count		
Long-tailed Tit	Frequency		
Marsh Tit	Frequency		
Willow Tit	Frequency		

Species	Data used for map in:		
	Species account	**Appendix E**	**Bias and the Key Squares Survey**
Crested Tit	Count		
Coal Tit	Frequency		
Blue Tit	Frequency		KSS point count
Great Tit	Frequency		KSS point count
Nuthatch	Frequency		
Treecreeper	Frequency		
Jay	Frequency		
Magpie	Frequency		KSS point count
Chough	Count		
Jackdaw	Frequency		KSS point count
Rook	Frequency		KSS point count
Carrion Crow	Frequency		KSS point count
Hooded Crow	Frequency		
Hybrid Crow	Frequency		
Raven	Count		
Starling	Frequency		KSS point count
House Sparrow	Frequency		KSS point count
Tree Sparrow	Frequency		
Chaffinch	Frequency		KSS point count
Greenfinch	Frequency		
Goldfinch	Frequency		
Siskin	Frequency		
Linnet	Frequency		
Twite	Frequency		
Redpoll	Frequency		
Crossbill spp.	Count		
Bullfinch	Frequency		
Hawfinch	Count		
Yellowhammer	Frequency		KSS point count
Cirl Bunting	Count (★)		
Reed Bunting	Frequency		
Corn Bunting	Frequency		

Appendix B: Names of plants mentioned in the text

Vernacular names are listed alphabetically; both scientific and vernacular names follow current usage.

ALDER   *Alnus glutinosa*
AMPHIBIOUS BISTORT   *Polygonum amphibium*
ASH   *Fraxinus excelsior*

BAMBOO   Bambusoideae
BARLEY   *Hordeum* spp.
BEECH   *Fagus sylvatica*
BILBERRY/BLAEBERRY   *Vaccinium myrtillus*
BIRCH   *Betula pendula/pubescens*
BLACKTHORN   *Prunus spinosa*
BRACKEN   *Pteridium aquilinum*
BRAMBLE   *Rubus fruticosus* agg.

CARROT   *Daucus carota*
CHARLOCK   *Sinapis arvensis*
CHERRY   *Prunus padus/avium*
CHESTNUT, SWEET   *Castanea sativa*
CHICKWEED   *Stellaria* spp.
CLOVER   *Trifolium* spp.
COTTONGRASS   *Eriophorum* spp.
CROWBERRY   *Empetrum nigrum*

DANDELION   *Taraxacum* spp.

EEL GRASS   *Zostera marina*

ELDER   *Sambucus nigra*
ELM   *Ulmus glabra/procera*

FAT HEN   *Chenopodium album*
FUMITORY   *Fumaria* spp.

GOOSEBERRY   *Ribes uva-crispa*
GOOSEFOOT   *Chenopodium* spp.
GORSE   *Ulex europaeus/gallii*
GRASS   Gramineae
  CANARY   *Phalaris* spp.
  HAIR/TUFTED   *Deschampsia* spp.
  MAT   *Nardus stricta*
  PURPLE MOOR   *Molinia caerulea*
  REED   *Glyceria*
GROUNDSEL   *Senecio vulgaris*

HAWTHORN   *Crataegus monogyna*
HAZEL   *Corylus avellana*
HEATHER   Ericaceae/*Calluna vulgaris*
HOLLY   *Ilex aquifolium*
HORNBEAM   *Carpinus betulus*

KNAPWEED   *Centaurea* spp.

LARCH   *Larix* spp.
  EUROPEAN   *L. decidua*
  JAPANESE   *L. kaempferi*

MEADOWSWEET   *Filipendula ulmaria*
MOSS, WOOLLY FRINGE   *Rhacomitrium lanuginosum*

NETTLE   *Urtica dioica*

OAK   *Quercus* spp.
  SESSILE   *Q. petraea*
OIL SEED RAPE   *Brassica napus ssp. oleifera*

PEANUT   *Arachis hypogea*
PERSICARIA   *Polygonum persicaria/lapathifolium*
POPLAR   *Populus* spp.

RAGWORT   *Senecio jacobea*
REED, COMMON   *Phragmites australis*
REEDMACE   *Typha* spp.
RHODODENDRON   *Rhododendron* spp.
ROWAN   *Sorbus aucuparia*
RUSH   *Juncus* spp.

SEA KALE   *Crambe maritima*
SEDGE   *Carex* spp.
  BOTTLE   *C. rostrata*
SPRUCE   *Picea* spp.
  NORWAY   *P. abies*
SUNFLOWER   *Helianthus annus*
SYCAMORE   *Acer pseudoplatanus*

TEASEL   *Dipsacus fullonum*
THISTLE   *Carduus/Cirsium* spp.

WATERLILY   Nymphaceae
WILLOW   *Salix* spp.
WILLOWHERB   *Epilobium* spp.

YEW   *Taxus baccata*

Appendix C: Names of birds mentioned in the text

Vernacular names are listed alphabetically and follow current usage. Scientific names were taken from British Ornithologists' Union (1992).

ARCTIC SKUA   *Stercorarius parasiticus*
ARCTIC TERN   *Sterna paradisaea*
AVOCET   *Recurvirostra avosetta*

BARNACLE GOOSE   *Branta leucopsis*
BARN OWL   *Tyto alba*
BEARDED TIT   *Panurus biarmicus*
BITTERN   *Botaurus stellaris*
BLACKBIRD   *Turdus merula*
BLACKCAP   *Sylvia atricapilla*
BLACK GROUSE   *Tetrao tetrix*
BLACK GUILLEMOT   *Cepphus grylle*
BLACK-HEADED GULL   *Larus ridibundus*
BLACK-NECKED GREBE   *Podiceps nigricollis*
BLACK REDSTART   *Phoenicurus ochruros*
BLACK-TAILED GODWIT   *Limosa limosa*
BLACK TERN   *Chlidonias niger*

BLACK-THROATED DIVER   *Gavia arctica*
BLACK-WINGED STILT   *Himantopus himantopus*
BLACK WOODPECKER   *Dryocopus martius*
BLUE-HEADED WAGTAIL   *Motacilla flava flava*
BLUETHROAT   *Luscinia svecica*
BLUE TIT   *Parus caeruleus*
BRAMBLING   *Fringilla montifringilla*
BULLFINCH   *Pyrrhula pyrrhula*
BUZZARD   *Buteo buteo*

CANADA GOOSE   *Branta canadensis*
CAPERCAILLIE   *Tetrao urogallus*
CARRION CROW   *Corvus corone corone*
CETTI'S WARBLER   *Cettia cetti*
CHAFFINCH   *Fringilla coelebs*
CHIFFCHAFF   *Phylloscopus collybita*
CHOUGH   *Pyrrhocorax pyrrhocorax*

CIRL BUNTING   *Emberiza cirlus*
COAL TIT   *Parus ater*
COLLARED DOVE   *Streptopelia decaocto*
COMMON CROSSBILL   *Loxia curvirostra*
COMMON GULL   *Larus canus*
COMMON SANDPIPER   *Actitis hypoleucos*
COMMON SCOTER   *Melanitta nigra*
COMMON TERN   *Sterna hirundo*
COOT   *Fulica atra*
CORMORANT   *Phalacrocorax carbo*
CORN BUNTING   *Miliaria calandra*
CORNCRAKE   *Crex crex*
CRANE   *Grus grus*
CRESTED TIT   *Parus cristatus*
CUCKOO   *Cuculus canorus*
CURLEW   *Numenius arquata*

DARTFORD WARBLER   *Sylvia undata*
DIPPER   *Cinclus cinclus*
DOTTEREL   *Charadrius morinellus*
DUNLIN   *Calidris alpina*
DUNNOCK   *Prunella modularis*

EGYPTIAN GOOSE   *Alopochen aegyptiacus*
EIDER   *Somateria mollissima*
ELEONORA'S FALCON   *Falco eleonorae*

FIELDFARE   *Turdus pilaris*
FIRECREST   *Regulus ignicapillus*
FULMAR   *Fulmarus glacialis*

GADWALL   *Anas strepera*
GANNET   *Morus bassanus*
GARDEN WARBLER   *Sylvia borin*
GARGANEY   *Anas querquedula*
GOLDCREST   *Regulus regulus*
GOLDEN EAGLE   *Aquila chrysaetos*
GOLDENEYE   *Bucephala clangula*
GOLDEN ORIOLE   *Oriolus oriolus*
GOLDEN PHEASANT   *Chrysolophus pictus*
GOLDEN PLOVER   *Pluvialis apricaria*
GOLDFINCH   *Carduelis carduelis*
GOOSANDER   *Mergus merganser*
GOSHAWK   *Accipiter gentilis*
GRASSHOPPER WARBLER   *Locustella naevia*
GREAT BLACK-BACKED GULL   *Larus marinus*
GREAT CRESTED GREBE   *Podiceps cristatus*
GREAT NORTHERN DIVER   *Gavia immer*
GREAT REED WARBLER   *Acrocephalus arundinaceus*
GREAT SKUA   *Catharacta skua*
GREAT SPOTTED WOODPECKER   *Dendrocopos major*
GREAT TIT   *Parus major*
GREENFINCH   *Carduelis chloris*
GREENSHANK   *Tringa nebularia*
GREEN WOODPECKER   *Picus viridis*
GREY HERON   *Ardea cinerea*
GREYLAG GOOSE   *Anser anser*
GREY PARTRIDGE   *Perdix perdix*

GREY WAGTAIL   *Motacilla cinerea*
GUILLEMOT   *Uria aalge*

HAWFINCH   *Coccothraustes coccothraustes*
HEN HARRIER   *Circus cyaneus*
HERRING GULL   *Larus argentatus*
HOBBY   *Falco subbuteo*
HONEY BUZZARD   *Pernis apivorus*
HOODED CROW   *Corvus c. cornix*
HOOPOE   *Upupa epops*
HOUSE MARTIN   *Delichon urbica*
HOUSE SPARROW   *Passer domesticus*
HYBRID CROW   *Corvus c. corone x C.c. cornix*

JACKDAW   *Corvus monedula*
JAY   *Garrulus glandarius*

KENTISH PLOVER   *Charadrius alexandrinus*
KESTREL   *Falco tinnunculus*
KINGFISHER   *Alcedo atthis*
KITTIWAKE   *Rissa tridactyla*

LADY AMHERST'S PHEASANT   *Chrysolophus amherstiae*
LAPLAND BUNTING   *Calcarius lapponicus*
LAPWING   *Vanellus vanellus*
LEACH'S PETREL   *Oceanodroma leucorhoa*
LESSER BLACK-BACKED GULL   *Larus fuscus*
LESSER SPOTTED WOODPECKER   *Dendrocopos minor*
LESSER WHITETHROAT   *Sylvia curruca*
LINNET   *Carduelis cannabina*
LITTLE BITTERN   *Ixobrychus minutus*
LITTLE GREBE   *Tachybaptus ruficollis*
LITTLE GULL   *Larus minutus*
LITTLE OWL   *Athene noctua*
LITTLE RINGED PLOVER   *Charadrius dubius*
LITTLE TERN   *Sterna albifrons*
LONG-EARED OWL   *Asio otus*
LONG-TAILED TIT   *Aegithalos caudatus*

MAGPIE   *Pica pica*
MALLARD   *Anas platyrhynchos*
MANDARIN DUCK   *Aix galericulata*
MANX SHEARWATER   *Puffinus puffinus*
MARSH HARRIER   *Circus aeruginosus*
MARSH TIT   *Parus palustris*
MARSH WARBLER   *Acrocephalus palustris*
MEADOW PIPIT   *Anthus pratensis*
MEDITERRANEAN GULL   *Larus melanocephalus*
MERLIN   *Falco columbarius*
MISTLE THRUSH   *Turdus viscivorus*
MONTAGU'S HARRIER   *Circus pygargus*
MOORHEN   *Gallinula chloropus*
MUTE SWAN   *Cygnus olor*

NIGHTINGALE   *Luscinia megarhynchos*
NIGHTJAR   *Caprimulgus europaeus*
NUTHATCH   *Sitta europaea*

OSPREY   *Pandion haliaetus*

OYSTERCATCHER   *Haematopus ostralegus*

PARROT CROSSBILL   *Loxia pytyopsittacus*
PEREGRINE   *Falco peregrinus*
PHEASANT   *Phasianus colchicus*
PIED FLYCATCHER   *Ficedula hypoleuca*
PIED WAGTAIL   *Motacilla alba yarrellii*
PINTAIL   *Anas acuta*
POCHARD   *Aythya ferina*
PTARMIGAN   *Lagopus mutus*
PUFFIN   *Fratercula arctica*
PURPLE SANDPIPER   *Calidris maritima*

QUAIL   *Coturnix coturnix*

RAVEN   *Corvus corax*
RAZORBILL   *Alca torda*
RED-BACKED SHRIKE   *Lanius collurio*
RED-BREASTED MERGANSER   *Mergus serrator*
RED-CRESTED POCHARD   *Netta rufina*
RED GROUSE   *Lagopus lagopus*
RED KITE   *Milvus milvus*
RED-LEGGED PARTRIDGE   *Alectoris rufa*
RED-NECKED GREBE   *Podiceps grisegena*
RED-NECKED PHALAROPE   *Phalaropus lobatus*
REDPOLL   *Carduelis flammea*
REDSHANK   *Tringa totanus*
REDSTART   *Phoenicurus phoenicurus*
RED-THROATED DIVER   *Gavia stellata*
REDWING   *Turdus iliacus*
REED BUNTING   *Emberiza schoeniclus*
REED WARBLER   *Acrocephalus scirpaceus*
RINGED PLOVER   *Charadrius hiaticula*
RING-NECKED PARAKEET   *Psittacula krameri*
RING OUZEL   *Turdus torquatus*
ROBIN   *Erithacus rubecula*
ROCK DOVE/FERAL PIGEON   *Columba livia*
ROCK PIPIT   *Anthus petrosus*
ROOK   *Corvus frugilegus*
ROSEATE TERN   *Sterna dougallii*
RUDDY DUCK   *Oxyura jamaicensis*
RUFF   *Philomachus pugnax*

SAND MARTIN   *Riparia riparia*
SANDWICH TERN   *Sterna sandvicensis*
SAVI'S WARBLER   *Locustella luscinioides*
SCARLET ROSEFINCH   *Carpodacus erythrinus*
SCAUP   *Aythya marila*
SCOTTISH CROSSBILL   *Loxia scotica*
SEDGE WARBLER   *Acrocephalus schoenobaenus*
SEA EAGLE   *Haliaeetus albicilla*
SERIN   *Serinus serinus*
SHAG   *Phalacrocorax aristotelis*
SHELDUCK   *Tadorna tadorna*
SHORE LARK   *Eremophila alpestris*
SHORT-EARED OWL   *Asio flammeus*

SHORT-TOED TREECREEPER   *Certhia brachydactyla*
SHOVELER   *Anas clypeata*
SISKIN   *Carduelis spinus*
SKYLARK   *Alauda arvensis*
SLAVONIAN GREBE   *Podiceps auritus*
SNIPE   *Gallinago gallinago*
SNOW BUNTING   *Plectrophenax nivalis*
SNOWY OWL   *Nyctea scandiaca*
SONG THRUSH   *Turdus philomelos*
SOUTH POLAR SKUA   *Catharacta maccormicki*
SPARROWHAWK   *Accipiter nisus*
SPOTTED CRAKE   *Porzana porzana*
SPOTTED FLYCATCHER   *Muscicapa striata*
SPOTTED SANDPIPER   *Actitis macularia*
STARLING   *Sturnus vulgaris*
STOCK DOVE   *Columba oenas*
STONECHAT   *Saxicola torquata*
STONE CURLEW   *Burhinus oedicnemus*
STORM PETREL   *Hydrobates pelagicus*
SWALLOW   *Hirundo rustica*
SWIFT   *Apus apus*

TAWNY OWL   *Strix aluco*
TEAL   *Anas crecca*
TEMMINCK'S STINT   *Calidris temminckii*
TREECREEPER   *Certhia familiaris*
TREE PIPIT   *Anthus trivialis*
TREE SPARROW   *Passer montanus*
TUFTED DUCK   *Aythya fuligula*
TURTLE DOVE   *Streptopelia turtur*
TWITE   *Carduelis flavirostris*

WATER RAIL   *Rallus aquaticus*
WHEATEAR   *Oenanthe oenanthe*
WHIMBREL   *Numenius phaeopus*
WHINCHAT   *Saxicola rubetra*
WHITE-HEADED DUCK   *Oxyura leucocephala*
WHITE-TAILED EAGLE   *Haliaeetus albicilla*
WHITETHROAT   *Sylvia communis*
WHITE WAGTAIL   *Motacilla alba alba*
WHOOPER SWAN   *Cygnus cygnus*
WIGEON   *Anas penelope*
WILLOW TIT   *Parus montanus*
WILLOW WARBLER   *Phylloscopus trochilus*
WOODCOCK   *Scolopax rusticola*
WOOD DUCK   *Aix sponsa*
WOODLARK   *Lullula arborea*
WOODPIGEON   *Columba palumbus*
WOOD SANDPIPER   *Tringa glareola*
WOOD WARBLER   *Phylloscopus sibilatrix*
WREN   *Troglodytes troglodytes*
WRYNECK   *Jynx torquilla*

YELLOWHAMMER   *Emberiza citrinella*
YELLOW WAGTAIL   *Motacilla flava flavissima*

Appendix D: Names of animals mentioned in the text

Vernacular names are listed alphabetically; both scientific and vernacular names follow current usage.

AMPHIPOD   Amphipoda
ANTLER MOTH   *Cerapteryx graminis*
ANT   Formicidae
  WOOD   *Formica rufa*
APHID   Aphidae
ARTHROPOD   Arthropoda

BEETLE   Coleoptera
BLUEBOTTLE   *Calliphora* spp.
BORDERED WHITE MOTH   *Bupalus piniaria*

CADDIS   Trichoptera
CAECAL THREADWORM   Nematoda
CAT, FERAL   *Felis catus*
CATERPILLAR   Lepidoptera (larva)
CHIRONOMID   Chironomidae including
              Chironomus salinarius
CRAB   Decapoda
CRANEFLY   Tipulidae

DAMSELFLY   Zygoptera
DEER   Cervidae
DIPTERAN   Diptera
DOG   *Canis familiaris*
DUNGFLY   *Scathophaga* spp.

EARTHWORM   Oligochaeta
EEL   *Anguilla anguilla*

FLY   Diptera
FOX   *Vulpes vulpes*
FROG   *Rana* spp.
FRUITFLY   Drosophilidae

HARE, COMMON   *Lepus europaeus*
  MOUNTAIN   *L. timidus*
HERRING   *Clupea harengus*
HORSEFLY   Tabanidae
HOVERFLY   Syrphidae

LEATHERJACKET   Tipulidae (larva)
LEMMING   *Lemmus lemmus*
LEPIDOPTERAN LARVA   see Caterpillar
LIZARD   *Lacerta* spp.

MACKEREL   *Scomber scombrus*
MARTEN, PINE   *Martes martes*
MAYFLY   Ephemeroptera
MIDGE   Nematocera

MINK, AMERICAN   *Mustela vison*
MOLE   *Talpa europaea*
MOLLUSC   Mollusca
MOUSE, WOOD   *Apodemus sylvaticus*
MUSSEL   Lammellibranchia
  ZEBRA   *Dreissena polymorpha*

NEMATOCERA   Diptera

OAK ROLLER MOTH   *Tortrix viridiana*

PIKE   *Esox lucius*
PRAWN   *Palaemonetes varians*

RABBIT   *Oryctolagus cuniculus*
RAGWORM, COMMON   *Nereis diversicolor*
RAT, COMMON (BROWN)   *Rattus norvegicus*
ROACH   *Rutilus rutilus*
ROBBERFLY   Asilidae

SALMON   *Salmo salar*
SANDEEL, COMMON   *Hyperophus lanceolatus*
  LESSER   *Ammodytes tobianus*
SAWFLY   Symphyta
SHEEP   *Ovis ammon*
SHREW, PYGMY   *Sorex minutus*
SHRIMP   *Corophium volutator*
SNAIL   Gastropoda
  COMMON BITHYNIA   *Bithynia tentaculata*
SPIDER   Araneae
SPRAT   *Sprattus sprattus*
SQUIRREL, RED   *Sciurus vulgaris*
  GREY   *S. carolinensis*
STOAT   *Mustela erminea*

TICK   Acarina
TIPULID LARVA   see Leatherjacket
TROUT, BROWN   *Salmo trutta*
  RAINBOW   *Salmo gairdneri*

VOLE, BANK   *Clethrionomys glareolus*
  SKOMER   *C. glareolus ssp. skomerensis*
  FIELD   *Microtus agrestis*
  ORKNEY   *M. arvalis*

WASP   Vespidae
WEEVIL   Curculionidae
WOODLOUSE   Malacostraca

Appendix E: Additional maps

Thirteen maps are included in this Appendix. The eight colour Abundance maps are based on frequency of occurrence data, rather than counts of pairs or nests at colonies (maps of which are included with the species accounts). Records of summering individuals are included, so these maps represent geographical variation in the abundance of individuals of each species during the breeding season. As these species are often recorded away from their nesting colonies during the breeding season, these maps show different patterns of abundance when compared to those based on colony counts alone. Explanations of the five dot Distribution maps are given in the relevant species accounts: for Blue-headed Wagtail see the Yellow Wagtail account, and for White Wagtail the Pied/White Wagtail account.

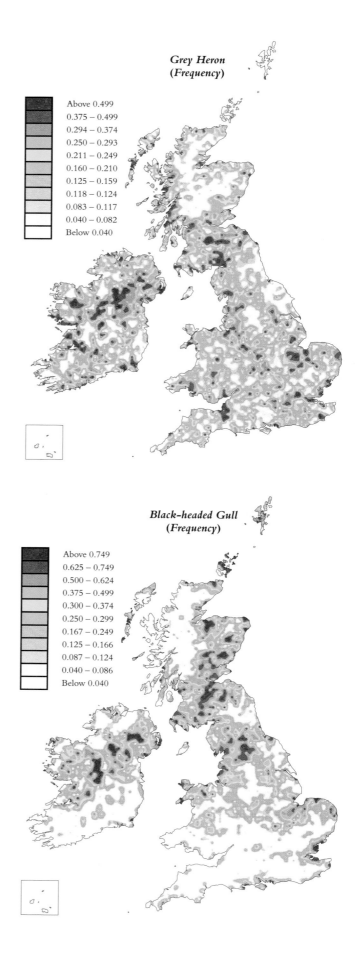

Grey Heron
(Frequency)

Above 0.499
0.375 – 0.499
0.294 – 0.374
0.250 – 0.293
0.211 – 0.249
0.160 – 0.210
0.125 – 0.159
0.118 – 0.124
0.083 – 0.117
0.040 – 0.082
Below 0.040

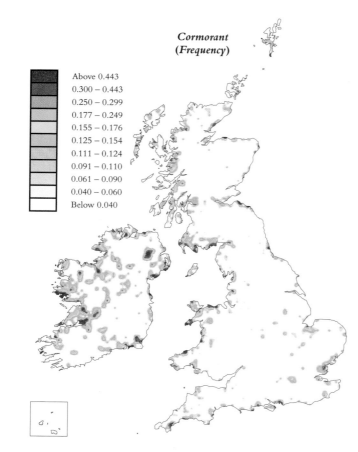

Cormorant
(Frequency)

Above 0.443
0.300 – 0.443
0.250 – 0.299
0.177 – 0.249
0.155 – 0.176
0.125 – 0.154
0.111 – 0.124
0.091 – 0.110
0.061 – 0.090
0.040 – 0.060
Below 0.040

Black-headed Gull
(Frequency)

Above 0.749
0.625 – 0.749
0.500 – 0.624
0.375 – 0.499
0.300 – 0.374
0.250 – 0.299
0.167 – 0.249
0.125 – 0.166
0.087 – 0.124
0.040 – 0.086
Below 0.040

Common Gull
(Frequency)

Above 0.749
0.546 – 0.749
0.400 – 0.545
0.333 – 0.399
0.250 – 0.332
0.200 – 0.249
0.125 – 0.199
0.111 – 0.124
0.077 – 0.110
0.040 – 0.076
Below 0.040

Herring Gull
(Frequency)

Above 0.946
0.667 – 0.946
0.500 – 0.666
0.375 – 0.499
0.286 – 0.374
0.222 – 0.285
0.143 – 0.221
0.125 – 0.142
0.080 – 0.124
0.040 – 0.079
Below 0.040

Lesser Black-backed Gull
(Frequency)

Above 0.624
0.400 – 0.624
0.300 – 0.399
0.250 – 0.299
0.167 – 0.249
0.125 – 0.166
0.120 – 0.124
0.091 – 0.119
0.063 – 0.090
0.040 – 0.062
Below 0.040

Common Tern
(Frequency)

Above 0.499
0.333 – 0.499
0.250 – 0.332
0.200 – 0.249
0.155 – 0.199
0.125 – 0.154
0.111 – 0.124
0.087 – 0.110
0.059 – 0.086
0.040 – 0.058
Below 0.040

Sand Martin
(Frequency)

Above 0.374
0.280 – 0.374
0.250 – 0.279
0.200 – 0.249
0.133 – 0.199
0.125 – 0.132
0.118 – 0.124
0.091 – 0.117
0.063 – 0.090
0.040 – 0.062
Below 0.040

Quail
(1989 Distribution)

Quail
(1988 Distribution)

Quail
(1990 Distribution)

Blue-headed Wagtail
(Distribution)

White Wagtail
(Distribution)

Appendix F: List of other species recorded

Species for which only records of occurrence, but without evidence of breeding, were submitted (or for which the only breeding records were thought to be of feral or introduced birds) are included in the following list. For each species the 10-km square(s) in which it was recorded are listed, along with the status of the record (S = present, but no evidence of breeding, B = breeding evidence). The species list is annotated where relevant.

Great Northern Diver
Gavia immer

The great majority of records were of summering birds around the NW coast of Scotland, but a few were located on inland lochs, particularly in Ireland. There have been no confirmed breeding records of the Great Northern Diver since 1970 (*68–72 Atlas*).

HU32 S	ND16 S	NG92 S	NR63 S
C01 S	ND26 S	NG94 S	NR64 S
C03 S	ND27 S	NG99 S	NR65 S
G10 S	ND49 S	NH56 S	NR69 S
G11 S	NF72 S	NH86 S	NR71 S
V66 S	NF86 S	NM43 S	NR73 S
NB24 S	NF99 S	NM46 S	NR78 S
NB43 S	NG08 S	NM55 S	NR79 S
NC01 S	NG09 S	NM62 S	NR83 S
NC02 S	NG73 S	NM69 S	NR85 S
NC14 S	NG77 S	NR25 S	NR95 S
NC37 S	NG78 S	NR47 S	NR96 S
NC55 S	NG80 S	NR56 S	NS05 S
NC90 S	NG82 S	NR62 S	

Black-browed Albatross
Diomedea melanophris

HP51 S

Little Egret
Egretta garzetta
V47 S   NR78 S   SU70 S

Spoonbill
Platalea leucorodia

One further record of a pair that started to build a nest in early July 1989 in E England (RBBP) was not submitted to the Atlas. The Spoonbill is now well established in the Netherlands, with 150 breeding pairs in 1968–69 rising to 320 in 1984 (SOVON 1987), so the colonisation of Britain is a realistic possibility.

SH27 S   TF84 S   TG50 S

Black Swan
Cygnus atratus

TG31 B

Swan Goose
Anser cygnoides

TQ59 S

Pink-footed Goose
Anser brachyrhynchus

L97 S	NN80 S	NZ05 S	SK68 B
NJ92 B	NS89 S	SD29 S	TF25 S
NN31 S	NY30 S	SD75 S	TQ08 S
NN75 S	NY52 S	SE24 S	TQ59 S

White-fronted Goose
Anser albifrons

NX25 S   TM13 S

Bar-headed Goose
Anser indicus

ND06 S   NX25 S   SH37 S   TL81 B

Snow Goose
Anser caerulescens

NM26 S	NY51 S	SD17 S	SK80 S
NX25 S	NY52 S	SE26 B	SU86 S
NY41 S	NZ71 S	SH37 S	

Emperor Goose
Anser canagicus

TL81 S   TL91 S

Brent Goose
Branta bernicla

HU35 B   S70 S   SU70 S   TF44 S   TF45 S   TL91 S

Ruddy Shelduck
Tadorna ferruginea

NZ52 S   SH37 S   SO70 B   SP93 S   SU86 S

Muscovy Duck
Cairina moschata
Since the early 1980s a feral population of Muscovy Ducks has been present on the River Ouse at Ely. Broods were first noticed in 1987, and in November 1991 the population was estimated at approximately 130 individuals, about 60% of which were that year's juveniles (R. Milwright).

TL15 S   TL57 B   TL58 B   TM59 B   TQ40 B

Ring-necked Duck
Aythya collaris

NZ36 S   SD84 S   SE43 S   SS42 S

King Eider
Somateria spectabilis

A male King Eider paired with a female Eider.

NK02 S

Long-tailed Duck
Clangula hyemalis

NC96 S	NG08 S	NH98 S	SD26 S
ND27 S	NG77 S	NJ36 S	SH37 S
ND35 S	NG82 S	SD17 S	SH38 S
ND49 S			

Velvet Scoter
Melanitta fusca

H15 S   SD17 S   TF64 S

Smew
Mergus albellus

ND27 S

Rough-legged Buzzard
Buteo lagopus

SS14 S

Red-footed Falcon
Falco vespertinus

J54 S   SK46 S

Bobwhite Quail
Colinus virginianus

TG04 S

Chukar
Alectoris chukar

This species was unfortunately not included on the recording forms and it is likely that many records of this species were not reported to the Atlas.

NH70 S	SD31 B	SU47 S	TL14 B
NH80 B	SU26 S	SU48 S	TL88 B
NJ01 B	SU27 S	SU52 B	TM47 S
NJ92 S	SU33 S	SU56 S	TM57 S
NT86 B	SU35 B	SU57 S	TQ33 S
NT90 S	SU36 B	SU58 S	TQ64 B
NT91 B	SU37 B	SU66 S	TQ73 S
NT92 S	SU40 B	TF62 B	TR35 S
NT96 B	SU42 B	TG04 S	
NZ09 S	SU46 S	TL13 B	

Reeves' Pheasant
Syrmaticus reevesii

There have been a number of attempts at introducing this species to Britain although none has proved successful (*68–72 Atlas*). It is possible that the species was overlooked in some areas.

SU36 S   SU56 S   SU57 S

Helmeted Guineafowl
Numida meleagris

TQ40 B

Grey Plover
Pluvialis squatarola

NC11 S A single bird summered on Cul Mor (Inverpolly NNR) in 1990. SU70 S SZ39 S TF44 S TF52 S TF71 S TL91 S

Knot
Calidris canutus

SJ39 S TL91 S

Sanderling
Calidris alba

NC26 S NG99 S NS98 S SN20 S SN30 S SU70 S TA16 S

Little Stint
Calidris minuta

NS98 S TL91 S

Jack Snipe
Lymnocryptes minimus

NM66 S SP51 S

Bar-tailed Godwit
Limosa lapponica

J56 S NC90 S NF74 S NH96 S NS98 S NZ52 S
SJ39 S SZ39 S TF84 S

Spotted Redshank
Tringa erythropus

NC54 S SZ57 S SZ68 S TF44 S

Green Sandpiper
Tringa ochropus

SU56 S SU67 S SU77 S SZ58 S TM28 S TM35 S TQ78 S

Turnstone
Arenaria interpres

HU16 S	J48 S	ND47 S	NZ52 S
HU30 S	J53 S	NF74 S	NZ53 S
HU31 S	J55 S	NF76 S	ST36 S
HU32 S	J56 S	NH19 S	SV80 S
HU33 S	J58 S	NH98 S	SV81 S
HU34 S	NA91 S	NJ66 S	SV90 S
HU35 S	NB45 S	NM15 S	SV91 S
HU36 S	NC96 S	NR25 S	SX96 S
HU41 S	ND16 S	NR88 S	SX98 S
HU58 S	ND27 S	NR96 S	TA16 S
HU68 S	ND35 S	NS05 S	TF74 S
J21 S	ND37 S	NX19 S	TL91 S
J31 S			

Little Gull
Larus minutus

NO66 S NY37 S NZ52 S SD47 S SD57 S SJ39 S
SJ58 S SJ68 S SK80 S TA14 S TF84 S

Ring-billed Gull
Larus delawarensis

NR25 S SX05 S

Iceland Gull
Larus glaucoides

NB13 S NC10 S

Glaucous Gull
Larus hyperboreus

NM69 S NR26 S NR86 S NS98 S

Lesser Crested Tern
Sterna bengalensis

A Lesser Crested Tern (paired with a male Sandwich Tern) successfully raised a chick in Northumberland in 1989 (RBBP).

NU23 B

Black Tern
Chlidonias niger

IF71 S ND27 S NZ13 S TF30 S TL19 S TL29 S
TL39 S TM28 S

White-winged Black Tern
Chlidonias leucopterus

ND27 S

Budgerigar
Melopsittacus undulatus

TL27 S TL37 S

Peach-faced Lovebird
Agapornis roseicollis

TL94 S WV65 S

Monk Parakeet
Myiopsitta monachus

SJ84 B SJ85 B

Eagle Owl
Bubo bubo

NM81 S

Hoopoe
Upupa epops

NZ04 S SH12 S SK47 S SN96 S SS62 S ST14 S SU94 S

Waxwing
Bombycilla garrulus

SJ88 S

Bluethroat
Luscinia svecica

SY58 S

Icterine Warbler
Hippolais icterina

S61 S   NM77 S

A single male sang on territory for at least two weeks in 1989.

Subalpine Warbler
Sylvia cantillans

J53 S

Penduline Tit
Remiz pendulinus

A single male built two nests during the Atlas years in a county in S England (RBBP).

Great Grey Shrike
Lanius excubitor

NO25 S   TM48 S

Two-barred Crossbill
Loxia leucoptera

NX47 S

Black-headed Weaver
Ploceus melanocephalus

TG51 B

Ortolan Bunting
Emberiza hortulana

IG50 S

References

AEBISCHER, N.J. 1986. Retrospective investigation of an ecological disaster in the Shag *Phalacrocorax aristotelis*: a general method based on long-term marking. *J. Anim. Ecol.* **55**: 613–629.

AEBISCHER, N.J. 1990. Assessing pesticide effects on non-target invertebrates using long-term monitoring and time-series modelling. *J. Functional Ecol.* **4**: 369–373.

AINSLIE, J.A. and R. ATKINSON. 1937. On the breeding habits of Leach's Fork-tailed Petrel. *Brit. Birds* **30**: 234–248.

ALEXANDER, I. and B. CRESSWELL. 1990. Foraging by Nightjars *Caprimulgus europaeus* away from their nesting areas. *Ibis* **132**: 568–574.

ALLISON, A., I. NEWTON and C. CAMPBELL. 1974. *Loch Leven National Nature Reserve.* WAGBI, Chester.

ANDERSSON, R. 1990. Svarta rödstjärtens *Phoenicurus ochruros* häckningsbiologi i Västsverige. *Var Fagelvärld* **49**: 201–210.

ANON. 1989. Goshawk breeding habitat in lowland Britain. *Brit. Birds* **82**: 56–67.

ANON. 1989. Red-necked Grebe breeding in Scotland. *Scott. Birds* **15**: 133.

ANON. 1990. Breeding biology of Goshawks in lowland Britain. *Brit. Birds* **83**: 527–540.

ASBIRK, S. and T. DYBBRO. 1978. Population size and habitat selection of the Great Crested Grebe *Podiceps cristatus* in Denmark, 1975. *Dansk Orn. Foren. Tidsskr.* **72**: 1–13 (in Danish with English summary).

ASH, J.S. 1970. Observations on a decreasing population of Red-backed Shrikes. *Brit. Birds* **63**: 185–205.

ASPINALL, S. and M.L. TASKER. 1992. *Birds of the Solent.* JNCC, Aberdeen.

ATKINSON-WILLES, G.L. 1963. *Wildfowl in Great Britain.* Nature Conservancy Monograph No. 3, London.

ATKINSON-WILLES, G.L. 1970. Wildfowl situation in England, Scotland and Wales. *Proc. Int. Reg. Meet. Conserv. Wildfowl Res. Leningrad 1968*: 101–107. IWRB, Slimbridge.

AVERY, M.I. 1987. *Emergency Action to Protect Endangered Bird Species in the European Community. Article 6616. Protection of* Sterna dougallii. RSPB and IWC Contract No. 12.05.87★003832.

AVERY, M.I. 1990. Seabirds, fisheries and politics. *RSPB Conserv. Rev.* **4**: 36–39.

AVERY, M.I. and A. DEL NEVO. 1991. Action for Roseate Terns. *RSPB Conserv. Rev.* **5**: 54–59.

AVERY, M.I. and R.E. GREEN. 1989. Not enough fish in the sea. *New Scientist* 22 July, pp 28–29.

AVERY, M.I. and R.H. HAINES-YOUNG. 1990. Population estimates for the Dunlin *Calidris alpina* derived from remotely sensed satellite imagery of the Flow Country of northern Scotland. *Nature* **344**: 860–862.

AVERY, M.I. and R. LESLIE. 1990. *Birds and Forestry.* Poyser, London.

AXELL, H.E. 1954. The Wheatear on Dungeness. *Bird Notes* **36**: 38–41.

AXELL, H.E. 1966. Eruptions of Bearded Tits during 1959–65. *Brit. Birds* **59**: 513–543.

AXELL, H.E. and G.J. JOBSON. 1972. Savi's Warblers breeding in Suffolk. *Brit. Birds* **65**: 229–232.

BAATSEN, R.G. 1990. Red-crested Pochard *Netta rufina* in the Cotswold Water Park. *Hobby* **16**: 64–67, Wiltshire Ornithological Society.

BAILLIE, S.R. 1990. Integrated population monitoring of breeding birds in Britain and Ireland. *Ibis* **132**: 151–166.

BAILLIE, S.R. and H. MILNE. 1982. The influence of female age on breeding in the Eider *Somateria mollissima*. *Bird Study* **29**: 55–66.

BAILLIE, S.R. and H. MILNE. 1989. Movements of Eiders *Somateria mollissima* on the east coast of Britain. *Ibis* **131**: 321–335.

BAINES, D. 1988. The effects of improvement of upland grassland on the distribution and density of breeding wading birds (*Charadriiformes*) in Northern England. *Biol. Conserv.* **45**: 221–236.

BAINES, D. 1989. The effects of improvement of upland, marginal grasslands on the breeding success of Lapwings *Vanellus vanellus* and other waders. *Ibis* **131**: 497–506.

BAINES, D. 1991. Factors contributing to local and regional variations in Black Grouse breeding success in northern Britain. *Ornis Scand.* **22**: 264–270.

BAKER, E.C.S. 1942. *Cuckoo Problems.* Witherby, London.

BARNARD, C.J. and D.B.A. THOMPSON. 1985. *Gulls and Plovers: the Ecology and Behaviour of Mixed-species Feeding Groups.* Croom Helm, London.

BARRETT, R.T. and W. VADER. 1984. The status and conservation of breeding seabirds in Norway. Pp 323–334 in Croxall, J.P, P.G. Evans and R.W. Schreiber (eds), *Status and Conservation of the World's Seabirds.* ICBP Technical Publication No. 2, Cambridge.

BARTONEK, J.C. 1989. Status of waterfowl, cranes and other water birds in North America. Pp 64–75 in Boyd, H. and J.-Y. Pirot *Flyways and Reserve Networks for Water Birds.* IWRB Special Publication No. 9. IWRB, Slimbridge.

BATTEN, L.A. 1973. Population dynamics of suburban Blackbirds. *Bird Study* **20**: 251–258.

BATTEN, L.A. 1976. Bird communities of some Killarney woodlands. *Proc. Roy. Irish Acad.* **76 (B)**: 285–313.

BATTEN, L.A., C.J. BIBBY, P. CLEMENT, G.D. ELLIOTT and R.F. PORTER. 1990. *Red Data Birds in Britain.* Poyser, London.

BAXTER, E.V. and L.J. RINTOUL. 1953. *The Birds of Scotland.* Oliver and Boyd, Edinburgh.

BAYES, K. and A. HENDERSON. 1988. Nightingales and coppiced woodland. *RSPB Conserv. Rev.* **2**: 47–49.

BAYLISS, M. 1985. Much grunting in the marsh. *BTO News* **138**: 8–9.

BEECROFT, R. 1986. The Suffolk Black Redstart breeding survey 1986. *Suffolk Ornithologists' Group Bulletin* **75**: 16–19.

BEINTEMA, A.J. and G.J.D.M. MUSKENS. 1987. Nesting success of birds breeding in Dutch agricultural grasslands. *J. Appl. Ecol.* **24**: 743–758.

BEKHUIS, J. 1991. The European Ornithological Atlas Project. *Sitta* **5**: 1–10.

BELL, D.G. 1989. *County of Cleveland Bird Report 1988.* Teesmouth Bird Club.

BENGTSON, S.-A. 1972. Reproduction and fluctuations in the size of the duck populations at Lake Myvatn, Iceland. *Oikos* **23**: 35–38.

BERTHEMY, B., P. DABIN and M. TERRASSE. 1983. Recensement et protection d'une espéce protégéé: Le Buzard Cendré. *Le Courrier de la Nature*: 10–16.

BERTHOLD, P., G. MOHR and U. QUERNER. 1990. Sterung und potentielle Evolutionsgeschwindigkeit des obligaten Teilzieherverhaltens: Ergebnisse eines Zweiweg-Selektionsexperiments mit der Monchsgrasmucke *Sylvia atricapilla. J. Orn.* **131**: 33–45.

BERTHOLD, P. and S.B. TERRILL. 1988. Migratory behaviour and population growth of Blackcaps wintering in Britain and Ireland: some hypotheses. *Ring. and Migr.* **9**: 153–159.

BIBBY, C.J. 1973. The Red-backed Shrike: a vanishing British species. *Bird Study* **20**: 103–110.

BIBBY, C.J. 1978. Some breeding statistics of Reed and Sedge Warblers. *Bird Study* **25**: 207–222.

BIBBY, C.J. 1979a. Foods of the Dartford Warbler *Sylvia undata* on southern English heathland (Aves: Sylviidae). *J. Zool. Lond.* **188**: 557–576.

BIBBY, C.J. 1979b. Mortality and movements of Dartford Warblers in England. *Brit. Birds* **72**: 10–22.

BIBBY, C.J. 1981a. Wintering Bitterns in Britain. *Brit. Birds.* **74**: 1–10.

BIBBY, C.J. 1981b. Food supply and diet of the Bearded Tit. *Bird Study* **28**: 201–210.

BIBBY, C.J. 1982. Polygyny and the breeding ecology of the Cetti's Warbler *Cettia cetti. Ibis* **124**: 288–301.

BIBBY, C.J. 1983. Studies of west Palearctic birds 186. Bearded Tit. *Brit. Birds* **76**: 549–563.

BIBBY, C.J. 1986. Merlins in Wales: site occupancy and breeding in relation to vegetation. *J. Appl. Ecol.* **23**: 1–12.

BIBBY, C.J. 1987. Foods of breeding Merlins *Falco columbarius* in Wales. *Bird Study* **34**: 64–70.

BIBBY, C.J. 1989. A survey of breeding Wood Warblers *Phylloscopus sibilatrix* in Britain, 1984–1985. *Bird Study* **36**: 56–72.

BIBBY, C.J., C.G. BAIN and D.J. BURGESS. 1989. Bird communities of highland birchwoods. *Bird Study* **36**: 123–133.

BIBBY, C.J., N.D. BURGESS and D.A. HILL. 1992. *Bird Census Techniques*. Academic Press, London.

BIBBY, C.J. and B. ETHERIDGE. 1993. Status of the Hen Harrier *Circus cyaneus* in Scotland in 1988–1989. *Bird Study* **40**: 1–11.

BIBBY, C.J. and J. LUNN. 1982. Conservation of reed-beds and their avifauna in England and Wales. *Biol. Conserv.* **23**: 167–186.

BIBBY, C.J. and M. NATTRASS. 1986. Breeding status of the Merlin in Britain. *Brit. Birds* **79**: 170–185.

BIBBY, C.J. and D.K. THOMAS. 1984. Sexual dimorphism in size, moult and movements of Cetti's Warbler *Cettia cetti*. *Bird Study* **31**: 28–34.

BIBBY, C.J. and C.R. TUBBS. 1975. Status, habitats and conservation of the Dartford Warbler in England. *Brit. Birds* **68**: 177–195.

BIGNAL, E. and S. BIGNAL. 1987. The provision of nesting sites for Choughs. *NCC Chief Scientist Directorate Report* No. 765. Peterborough.

BIGNAL, E., S. BIGNAL and N. EASTERBEE. 1988. The recent status and distribution of the Chough *Pyrrhocorax pyrrhocorax* in Scotland. *NCC Chief Scientist Directorate Report* No. 843. Peterborough.

BIGNAL, E. and D.J. CURTIS. 1989. *Choughs and land-use in Europe*. Proceedings of an International Workshop on the Conservation of the Chough, *Pyrrhocorax pyrrhocorax*, in the EC. 11–14 November 1988. Scottish Chough Study Group.

BIJLSMA, R.G. 1982. Census problems with Bullfinches *Pyrrhula pyrrhula* as breeding birds. *Limosa* **55**: 9–16. (Dutch, with English summary.)

BIJLSMA, R.G. 1991. Trends in European Goshawks *Accipiter gentilis*: an overview. *Bird Census News* **4**: 3–47.

BIRCHAM, P.M.M. 1988. *The Birds of Cambridgeshire*. Cambridge University Press, Cambridge.

BIRKHEAD, T.R. 1991. *The Magpies. The Ecology and Behaviour of Black-billed and Yellow-billed Magpies*. Poyser, London.

BIRKHEAD, T.R. and D.N. NETTLESHIP. 1980. Census methods for murres *Uria* species – a unified approach. *Canad. Wildl. Serv. Occ. Paper No. 43*: 1–25.

BIRKS, J. 1990. Feral Mink and nature conservation. *Brit. Wildlife* **1**: 313–323.

BJERKE, T.K. and T.H. BJERKE. 1981. Song dialects in the Redwing. *Ornis Scand.* **12**: 40–50.

BLAKER, G.B. 1934. *The Barn Owl in England and Wales*. RSPB, London.

BLONDEL, J. 1975. L'analyse des peuplements d'oiseaux, élément d'un diagnostic écologique. 1. La méthod des échantillonnages fréquentiels progressifs (E.F.P.). *Terre et Vie* **29**: 533–589.

BLONDEL, J., C. FERRY and B. FROCHOT. 1981. Point counts with unlimited distance. *Studies in Avian Biology* **6**: 414–420.

BLUME, D. 1977. *Die Buntspechte*. Wittenberg, Lutherstadt.

BONHAM, P.F. and J.C.M. ROBERTSON. 1975. The spread of the Cetti's Warbler in north-west Europe. *Brit. Birds* **68**: 393–408.

BOOTH, C.J., M. CUTHBERT and P. REYNOLDS. 1984. The birds of Orkney. *The Orkney Press*.

BOSSEMA, L. 1979. Jays and oaks: an eco-ethological study of a symbiosis. *Behaviour* **70**: 1–117.

BOWDEN, C.G.R. 1990. Selection of foraging habitats used by Woodlarks nesting in pine plantations. *J. Appl. Ecol.* **27**: 410–419.

BOWDEN, C. and R. HOBLYN. 1990. The increasing importance of restocked conifer plantations for Woodlarks in Britain: implications and consequences. *RSPB Conserv. Rev.* **4**: 26–31.

BRENCHLEY, A. 1986. The breeding distribution and abundance of the Rook *Corvus frugilegus* L. in Great Britain since the 1920s. *J. Zool., Lond. (A)* **210**: 261–278.

BRITISH ORNITHOLOGISTS' UNION. 1992. *Checklist of Birds of Britain and Ireland*. 6th edn. British Ornithologists' Union, Tring.

BRITTAS, R. and T. WILLEBRAND. 1991. Nesting habitat and egg predation in Swedish Black Grouse. *Ornis Scand.* **22**: 261–263.

BROAD, R.A., A.J.E. SEDDON and D.A. STROUD. 1986. The waterfowl of freshwater lochs in Argyll. *The Third Argyll Bird Report* 77–84.

BROMHALL, D. 1980. *Devil Birds*. Hutchinson, London.

BROOKE, M. 1990. *The Manx Shearwater*. Poyser, London.

BROOKE, M. de L. and N.B. DAVIES. 1987. Recent changes in host usage by Cuckoos *Cuculus canorus* in Britain. *J. Anim. Ecol.* **56**: 873–883.

BROWN, L. 1976. *British Birds of Prey*. Collins, Glasgow.

BROWNETT, A. 1989. Redstart survey 1989. Unpublished Report. Banbury Ornithological Society.

BRUDERER, B. and W. HIRSCHI. 1984. Langfristige bestandsentwicklung vor *Phoenicurus phoenicurus* und *Ficedula hypoleuca* nach Schweizerischen Beringungszahlen und Nisthohlenkontrollen. *Der Ornith. Beobachter*. **81**: 285–302.

BRYANT, D.M. 1975. Breeding biology of House Martins *Delichon urbica* in relation to aerial insect abundance. *Ibis* **117**: 180–216.

BRYANT, D.M. 1978. Moulting Shelducks on the Forth estuary. *Bird Study* **25**: 103–108.

BRYANT, D.M. 1981. Moulting Shelducks on the Wash. *Bird Study* **28**: 157–158.

BUCKLAND, S.T., M.V. BELL and N. PICOZZI. 1990. *The Birds of North-East Scotland*. North-East Scotland Bird Club, Aberdeen.

BUCKLEY, P.A. and F.G. BUCKLEY. 1984. Seabirds of the north and middle Atlantic coasts of the United States: their status and conservation. Pp 101–133 in Croxall, J.P., P.G.H. Evans and R.W. Schreiber (eds), *Status and Conservation of the World's Seabirds*. ICPB Tech. Pub. No. 2, Cambridge.

BULLOCK, I.D., D.R. DREWITT and S.P. MICKLEBURGH. 1983. The Chough in Britain and Ireland. *Brit. Birds* **76**: 377–401.

BULLOCK, I.D. and C.H. GOMERSALL. 1981. The breeding populations of terns in Orkney and Shetland in 1980. *Bird Study* **28**: 187–200.

BURGESS, N., C. EVANS and J. SORENSEN. 1990. Heathland management for Nightjars. *RSPB Conserv. Rev.* **4**: 32–35.

BURTON, H., T.L. EVANS and D.N. WEIR. 1970. Wrynecks breeding in Scotland. *Scott. Birds* **6**: 154–156.

BUXTON, E.J.M. 1962. The inland breeding of the Oystercatcher in Great Britain, 1958–59. *Bird Study* **8**: 194–209.

BYARS, T., D.J. CURTIS and I. McDONALD. 1991. The breeding distribution and habitat requirements of the Lesser Whitethroat in Strathclyde. *Scott. Birds* **16**: 66–76.

BYRKJEDAL, I. 1985. Time budget and parental labour division in breeding Black-tailed Godwits *Limosa limosa*. *Fauna norv. Ser. C, Cinclus* **8**: 24–34.

CADBURY, C.J. 1980. The status and habitats of the Corncrake in Britain 1978–79. *Bird Study* **27**: 203–218.

CADBURY, C.J. 1991. *Death by Design: the Persecution of Birds of Prey and Owls in the UK 1979–1989*. RSPB/NCC, Sandy.

CADBURY, C.J., R.E. GREEN and G. ALLPORT. 1987. Redshanks and other breeding waders of British saltmarshes. *RSPB Conserv. Rev.* **1**: 37–40.

CADBURY, C.J., D. HILL., J. PARTRIDGE. and J. SORENSEN. 1989. The history of the Avocet population and its management in England since recolonisation. *RSPB Conserv. Rev.* **3**: 9–13.

CADBURY, C.J. and P.J.S. OLNEY. 1978. Avocet population dynamics in England. *Brit. Birds* **78**: 102–121.

CADBURY, C.J. and P.A. RICHARDS. 1978. The breeding and feeding ecology of the Avocet *Recurvirostra avosetta* in Suffolk, England. Unpublished RSPB report.

CADE, T.J. 1982. *The Falcons of the World*. Collins, London.

CADE, T.J., J.H. ENDERSON, C.G. THELANDER and C.M. WHITE. (eds) 1988. *Peregrine Falcon populations: their Management and Recovery*. The Peregrine Fund, Inc., Boise, Idaho.

CALDOW, R. W. G. and R. W. FURNESS. 1991. The relationship between kleptoparasitism and plumage polymorphism in the Arctic Skua *Stercorarius parasiticus* L. *J. Functional Ecol.* **5**: 331–339.

CALLION, J., N. WHITE and D. HOLLOWAY. 1990. Grasshopper Warblers raising two and three broods in Cumbria. *Brit. Birds* **83**: 506–508.

CAMPBELL, B. 1955. The breeding distribution and habitats of the Pied Flycatcher *Muscicapa hypoleuca* in Britain. Part II. *Bird Study* **2**: 24–32.

CAMPBELL, B. 1960. The Mute Swan census in England and Wales, 1955–56. *Bird Study* **7**: 208–223.

CAMPBELL, B. and I.J. FERGUSON-LEES. 1972. *A Field Guide to Birds' Nests.* Constable, London.

CAMPBELL, L. H. 1988. Loon conservation in the British Isles. Pp 78–85 in *Papers from the 1987 Conference on Loon Research and Management.* North American Loon Fund, Meredith, NH.

CAMPBELL, L.H. and G.P. MUDGE. 1989. Conservation of Black-throated Divers in Scotland. *RSPB Conserv. Rev.* **3**: 72–74.

CAMPBELL, L.H. and T.R. TALBOT. 1987. Breeding status of Black-throated Divers in Scotland. *Brit. Birds* **80**: 1–8.

CAMPHUYSEN, C.J. 1989. *Beached Bird Survey in the Netherlands, 1915–1988.* Techn. Rapport Vogelbescherming 1, Amsterdam.

CARSS, D.N. 1989. Sawbill ducks at fish-farms in Argyll, western Scotland. *Scott. Birds* **15**: 145–150.

CARSS, D.N. and M. MARQUISS. 1992. Avian predation at farmed and natural fisheries. *Proc. 22nd Inst. Fisheries Management, Annual Study Course:* 179–196.

CARTER, S.P. 1990. *Studies of the Goosander Mergus merganser.* PhD thesis, University of Durham.

CARTER, S.P. and P.R. EVANS. 1986. Goosander and Merganser studies in Scotland, 1986. Report to NCC.

CATCHPOLE, C.K. 1972. A comparative study of territory in the Reed Warbler *Acrocephalus scirpaceus* and Sedge Warbler *A. schoenobaenus. J. Zool. Lond.* **166**: 213–231.

CATCHPOLE, C.K. 1973. The functions of advertising song in the Sedge Warbler *Acrocephalus schoenobaenus* and the Reed Warbler *A. scirpaceus. Behaviour* **46**: 300–320.

CAWTHORNE, R.A. and J.H. MARCHANT. 1980. The effects of the 1978/79 winter on British bird populations. *Bird Study* **27**: 163–172.

CHABRZYK, G. and J.C. COULSON. 1976. Survival and recruitment in the Herring Gull *Larus argentatus. J. Anim. Ecol.* **45**: 205–214.

CHANCE, E.P. 1940. *The Truth about the Cuckoo.* Country Life, London.

CHENG, TSO-HSIN. 1963. *China's Economic Fauna: Birds.* English translation 1964. Washington.

CHURCHER, P.B. and J.H. LAWTON. 1987. Predation by domestic cats in an English village. *J. Zool. Lond.* **212**: 439–455.

CLARK, F. and D.A.C. McNEIL. 1980. Cliff-nesting colonies of House Martins *Delichon urbica* in Great Britain. *Ibis* **122**: 27–42.

CLARKE, E.R. 1974. *Woodcock Status Report.* US Fish and Wildlife Service. Washington D.C.

CONDER, P. 1989. *The Wheatear.* Christopher Helm, London.

COOK, A. 1975. Changes in the Carrion/Hooded Crow hybrid zone and the possible importance of climate. *Bird Study* **22**: 165–168.

COOK, M.J.H. 1982. Breeding status of the Crested Tit. *Scott. Birds* **12**: 97–106.

COOKE, A.S., A.A. BELL and M.B. HAAS. 1982. *Predatory Birds, Pesticides and Pollution.* Institute of Terrestrial Ecology, Cambridge.

COOMBS, C.F.B., A.J. ISAACSON, R.K. MURTON, R.J.P. THEARLE and N.J.WESTWOOD. 1981. Collared Doves *Streptopelia decaocto* in urban habitats. *J. Appl. Ecol.* **18**: 41–62.

COOMBS, F. 1978. *The Crows. A Study of the Corvids of Europe.* B.T. Batsford, London.

COTTIER, E.J. and D. LEA. 1969. Black-tailed Godwits, Ruffs and Black Terns breeding on the Ouse Washes. *Brit. Birds* **62**: 259–270.

COULSON, J.C. 1956. Mortality and egg production of the Meadow Pipit with special reference to altitude. *Bird Study* **3**: 119–132.

COULSON, J.C. 1983. The changing status of the Kittiwake in the British Isles, 1969–1979. *Bird Study* **30**: 9–16.

COULSON, J.C. 1984. The population dynamics of the Eider Duck *Somateria mollissima* and evidence of extensive non-breeding by adult ducks. *Ibis* **126**: 525–543.

COULSON, J.C. 1991. The population dynamics of culling Herring Gulls *Larus argentatus* and Lesser Black-backed Gulls *L. fuscus.* In Perrins C.M., J.P. Lebreton and G.M. Hirons (eds), *Bird Population Studies. Relevance to Conservation and Management.* Oxford University Press, Oxford.

COULSON, J.C. and C.S. THOMAS. 1985. Changes in the biology of the Kittiwake *Rissa tridactyla*: a 31–year study of a breeding colony. *J. Anim. Ecol.* **54**: 9–26.

COWIE, R.J. and S.A. HINSLEY. 1987. Breeding success of Blue Tits and Great Tits in suburban gardens. *Ardea* **75**: 81–90.

COWLEY, E. 1979. Sand Martin population trends in Britain, 1965–1978. *Bird Study* **26**: 113–116.

CRAMP, S. (ed) 1977–93. *Handbook of the Birds of Europe, the Middle East and North Africa: the birds of the Western Palearctic.* Oxford University Press, Oxford. 7 Vols.

CRAMP, S., W.R.P. BOURNE and D. SAUNDERS. 1974. *The Seabirds of Britain and Ireland.* Collins, London.

CRICK, H.Q.P. 1992. A bird-habitat coding system for use in Britain and Ireland incorporating aspects of land-management and human activity. *Bird Study* **39**: 1–12.

CROOKE, C.H., R.H. DENNIS and G.P. MUDGE. 1987. *Breeding Slavonian Grebes in Great Britain.* RSPB Highland Office Report.

DA PRATO, S.R.D. 1980. How many Lesser Whitethroats breed in the Lothians? *Scott. Birds* **13**: 108–111.

DA PRATO, S.R.D. 1985. The breeding birds of agricultural land in southeast Scotland. *Scott. Birds* **13**: 203–216.

DA PRATO, S.R.D. 1986. *The Ecology of Three Species of Warblers, Sylviidae, in Scrubland in South East Scotland.* PhD thesis, University of Edinburgh.

DA PRATO, S.R.D. and E.S. DA PRATO. 1986. Appearance and song of possible Chiffchaff × Willow Warbler hybrid. *Brit. Birds* **79**: 341–342.

DAGLEY, J.R. 1987. *Golden Oriole Oriolus oriolus.* Unpublished RSPB Report.

DARE, P.J. 1966. The breeding and wintering populations of the Oystercatcher *Haematopus ostralegus* L. in the British Isles. *Fishery Invest. London,* Ser. II, **25** (5): 169.

DARE, P.J. 1986. Raven *Corvus corax* populations in two upland regions of North Wales. *Bird Study* **33**: 179–189.

DAVIDSON, C. 1985. Parrot Crossbills: A new breeding species. *Norfolk Bird Report 1984*: 99–102.

DAVIDSON, R. 1987. Breeding birds of Lough Neagh, 1987. Unpublished Report, Craigavon.

DAVIES, A.K. 1988. The distribution and status of the Mandarin Duck *Aix galericulata* in Britain. *Bird Study* **35**: 203–208.

DAVIES, A.K. and G.K. BAGGOTT. 1989a. Egg-laying, incubation and intraspecific nest-parasitism by the Mandarin Duck *Aix galericulata. Bird Study* **36**: 115–122.

DAVIES, A.K. and G.K. BAGGOTT. 1989b. Clutch size and nesting sites of the Mandarin Duck *Aix galericulata. Bird Study* **36**: 32–36.

DAVIES, M. 1988. The importance of Britain's Twites. *RSPB Conserv. Rev.* **2**: 91–94.

DAVIES, N.B. 1976. Parental care and the transition to independent feeding in the young Spotted Flycatcher *Muscicapa striata. Behaviour* **59**: 280–295.

DAVIES, N.B. 1977a. Prey selection and social behaviour in wagtails (Aves: Motacillidae). *J. Anim. Ecol.* **46**: 37–57.

DAVIES, N.B. 1977b. Prey selection and the search strategy of the Spotted Flycatcher. *Anim. Behav.* **25**: 1016–1033.

DAVIES, N.B. 1985. Cooperation and conflict among Dunnocks, *Prunella modularis*, in a variable breeding system. *Anim. Behav.* **33**: 628–648.

DAVIES, N.B. 1992. *Dunnock Behaviour and Social Evolution.* Oxford University Press, Oxford.

DAVIES, N.B. and M. BROOKE. 1991. Coevolution of the Cuckoo and its hosts. *Sci. Am.* **264**: 92–98.

DAVIES, N.B. and M. de L. BROOKE. 1988. Cuckoos versus Reed Warblers: adaptations and counteradaptations. *Anim. Behav.* **36:** 262–284.

DAVIES, N.B. and A. LUNDBERG. 1984. Food distribution and a variable mating system in the Dunnock, *Prunella modularis. J. Anim. Ecol.* **53:** 895–912.

DAVIES, P.W. and D.W. SNOW. 1965. Territory and food of the Song Thrush. *Brit. Birds* **58:** 161–174.

DAVIS, P.E. and J.E. DAVIS. 1981. The food of the Red Kite in Wales. *Bird Study* **28:** 33–40.

DAVIS, P.E. and I. NEWTON. 1981. Population and breeding of Red Kites in Wales over a 30-year period. *J. Anim. Ecol.* **50:** 759–772.

DAVIS, P.G. 1982. Nightingales in Britain in 1980. *Bird Study* **29:** 73–79.

DAWSON, D.G. 1981. Experimental design when counting birds. *Studies in Avian Biology* **6:** 392–398.

DAY, J.C.U. 1981. Status of Bitterns in Europe since 1976. *Brit. Birds.* **74:** 10–16.

DAY, J.C.U. and J. WILSON. 1978. Breeding Bitterns in Britain. *Brit. Birds.* **71:** 285–300.

DELANY, S., J.J.D. GREENWOOD and J. KIRBY. 1992. *National Mute Swan Survey 1990.* Unpublished Report to JNCC.

DEMENT'EV, G.P. and N.A. GLADKOV. 1967. *Birds of the Soviet Union* Vol IV. Israel Program for Scientific Translations, Jerusalem (translated from 1952 original).

DENNIS, R.H. 1968. Sea Eagles. *Fair Isle Bird Observatory Report* **21:** 17–21.

DENNIS, R.H. 1969. Sea Eagles. *Fair Isle Bird Observatory Report* **22:** 23–29.

DENNIS, R.H. 1983a. Population studies and conservation of Ospreys in Scotland. Pp 207–214 in *Biology and management of Ospreys and Bald Eagles.* Montreal.

DENNIS, R.H. 1983b. Purple Sandpipers breeding in Scotland. *Brit. Birds* **76:** 563–566.

DENNIS, R.H. 1984. *Birds of Badenoch and Strathspey.* Roy Dennis Enterprises, Inverness.

DENNIS, R.H. 1987a. Boxes for Goldeneyes: a success story. *RSPB Conserv. Rev.* **1:** 85–87.

DENNIS, R.H. 1987b. Osprey recolonisation. *RSPB Conserv. Rev.* **1:** 88–90.

DENNIS, R.H. 1991. *Ospreys.* Colin Baxter, Lanark.

DENNIS, R.H. and H. DOW. 1974. The establishment of a population of Goldeneye breeding in Scotland. *Bird Study* **34:** 217–222.

DENNIS, R.H., P.M. ELLIS, R.A. BROAD and D.R. LANGSLOW. 1984. The status of the Golden Eagle in Britain. *Brit. Birds* **77:** 592–607.

DEVILLERS, P., W. ROGGEMAN, J. TRICOT, P. DEL MARMOL, C. KERWIJN, J-P. JACOB and A. ANSELIN. 1988. *Atlas des oiseaux nicheurs de Belgique.* Institut Royal des Sciences Naturelles de Belgique, Bruxelles.

DIX, T. and P. CUNNINGHAM. 1991. *Outer Hebrides Bird Report, 1989 and 1990.* Outer Hebrides Ornithological Group.

DOBINSON, H.M. and A.J. RICHARDS. 1964. The effects of the severe winter of 1962/63 on birds in Britain. *Brit. Birds* **57:** 373–434.

DOBSON, A.P and P.J. HUDSON. 1992. Regulation and stability of a free-living host-parasite system *Trichostrongylus tenuis* in Red Grouse: II Population models. *J. Anim. Ecol.* **61:** 487–500.

DOUSE, A.F.G. 1981. *The Use of Agricultural Land by Common Gulls* Larus canus *L. Wintering in North-east Scotland.* PhD thesis, University of Aberdeen.

DOWSETT, R.J., G.C. BACKHURST and T.B. OATLEY. 1988. Afro-tropical ringing recoveries of Palaearctic migrants 1. Passerines (Turdidae to Oriolodae). *Tauraco* **1:** 29–63.

DOWSETT-LEMAIRE, F. 1979. The imitative range of the song of the Marsh Warbler *Acrocephalus palustris,* with special reference to imitations of African birds. *Ibis* **121:** 453–467.

DRESSER, H.E. 1871–1881. *A History of the Birds of Europe.* London.

DUNNET, G.M., J.C. OLLASON and A. ANDERSON. 1979. A 28-year study of breeding Fulmars *Fulmarus glacialis* in Orkney. *Ibis* **121:** 293–300.

DUNNET, G.M. and I.J. PATTERSON. 1968. The Rook problem in northeast Scotland. Pp 119–139 in Murton, R.K. and E.N. Wright (eds), *The Problems of Birds as Pests.* Academic Press, London.

DYMOND, J.N., P.A. FRASER and S.J.M. GANTLETT. 1989. *Rare Birds in Britain and Ireland.* Poyser, Calton.

EAST, M. 1981. Aspects of courtship feeding and parental care of the European Robin *Erithacus rubecula. Ornis Scand.* **12:** 230–239.

ELLIOTT, G. 1988. Montagu's Harrier conservation. *RSPB Conserv. Rev.* **2:** 20–21.

ELLIOTT, G.D. and M.I. AVERY. 1991. A review of reports of Buzzard persecutions 1975–1989. *Bird Study* **38:** 52–56.

ELLIOTT, G., R. DENNIS, J. LOVE, M. PIENKOWSKI and R. BROAD. 1991. A future for the White-tailed Eagle in Britain. *RSPB Conserv. Rev.* **5:** 41–46.

ENOKSSON, B. 1987. Local movements in the Nuthatch *Sitta europaea. Acta Reg. Soc. Sci. Litt. Gothoburgensis Zoologica* **14:** 36–47.

ENOKSSON, B. 1988. *Prospective Resource Defence and its Consequences in the Nuthatch* Sitta europaea L. PhD thesis, Uppsala University, Uppsala.

ENOKSSON, B. 1990. Autumn territories and population regulation in the Nuthatch *Sitta europaea:* An experimental study. *J. Anim. Ecol.* **59:** 1047–1062.

ERARD, C. 1991. Aide au nourrissage chez le Gobe-mouche gris *Muscicapa striata. L'Oiseau* **61:** 154–155.

ETHERIDGE, B. 1982. Distribution of Dunlin *Calidris alpina* nests on an area of South Uist machair. *Bird Study* **29:** 239–243.

EVANS, A. 1990. Save the village bunting. *Birds* **13:** 29–31.

EVANS, A.D. 1992. The numbers and distribution of Cirl Buntings *Emberiza cirlus* breeding in Britain in 1989. *Bird Study* **39:** 17–22.

EVANS, I.B. 1984. A thousand Bullfinches. *Herefordshire Ornithological Club – Annual Report 1984:* 171–174.

EVANS, I.M. and M.W. PIENKOWSKI. 1991. World status of the Red Kite. *Brit. Birds* **84:** 171–187.

EVANS, P.G.H. 1988. Intraspecific nest parasitism in the European Starling *Sturnus vulgaris. Anim. Behav.* **36:** 1282–1294.

EVERETT, M.J. 1971. Breeding status of Red-necked Phalaropes in Britain and Ireland. *Brit. Birds* **64:** 293–302.

EWINS, P.J. 1985. Colony attendance and censusing of Black Guillemots *Cepphus grylle* in Shetland. *Bird Study* **32:** 176–185.

EWINS, P.J. 1989. The breeding biology of Black Guillemots *Cepphus grylle* in Shetland. *Ibis* **131:** 507–520.

EWINS, P.J., P.M. ELLIS, D.R. BIRD and A. PRIOR. 1988. The Distribution and Status of Arctic and Great Skuas in Shetland, 1985–86. *Scot. Birds* **15:** 9–20.

EWINS, P.J. and M.L. TASKER. 1985. Breeding distribution of Black Guillemots *Cepphus grylle* in Orkney and Shetland, 1982–84. *Bird Study* **32:** 186–193.

FALLODON, Lord GREY of. 1927. *The Charm of Birds.* Hodder and Stoughton, London.

FEA, J. 1775. *The Present State of the Orkney Islands Considered.* Brown, Edinburgh (1884).

FEARE, C.J. 1984. *The Starling.* Oxford University Press, Oxford.

FEARE, C.J. 1990. Pigeon control: towards a humane alternative. *Env. Health* **98:** 155–156.

FEARE, C.J., G.M. DUNNET and I.J. PATTERSON. 1974. Ecological studies of the Rook *Corvus frugilegus* L. in North-east Scotland. Food intake and feeding behaviour. *J. Appl. Ecol.* **11:** 867–896.

FEIGE, K.D. 1986. *Der Pirol.* Ziemsen Verlag.

FELTHAM, M.J. 1990. The diet of the Red-breasted Merganser *Mergus serrator* during the smolt run in N.E. Scotland; the importance of salmon *Salmo salar* smolts and parr. *J. Zool., Lond.* **222:** 285–292.

FERGUSON-LEES, I.J. 1971. Coming to breed? *BTO News* **44:** 6.

FERGUSON-LEES, I.J. *et al.* 1983. British Ornithologists' Union Records Committee 11th Report. *Ibis* **126:** 441.

FERGUSON-LEES, I.J. and J.T.R. SHARROCK. 1970. A comparison between basic atlas and tetrad techniques. *Bird Study* **18:** 227–228.

FERRY, C. and B. FROCHOT. 1970. L'avifaune nidificatrice d'une foret de chenes pedoncules en Bourgogne: etude de deux successions ecologiques. *Terre et Vie* **24:** 153–250.

FISHER, J. 1952. *The Fulmar*. Collins, London.

FISHER, J. 1953. The Collared Turtle Dove in Europe. *Brit. Birds* **46**: 153–181.

FITTER, R.S.R. 1965. The breeding status of the Black Redstart in Britain. *Brit. Birds* **58**: 481–492.

FIUCZYNSKI, D. 1987. *Der Baumfalke*. Neue Brehm-Bucherei, Lutherstadt.

FIUCZYNSKI, D. and D. NETHERSOLE-THOMPSON. 1980. Hobby studies in England and Germany. *Brit. Birds* **73**: 275–295.

FLEGG, J.J.M. 1973. A study of Treecreepers. *Bird Study* **21**: 287–302.

FLEGG, J.J.M. and T.J. BENNETT. 1974. The birds of oak woodlands. Pp 324–340 in Morris, M.G. and F.H. Perring (eds), *The British Oak*, Faringdon.

FLEGG, J.J.M. and D.E. GLUE. 1973. A Water Rail Study. *Bird Study* **20**: 69–79.

FLEGG, J.J.M. and D.E. GLUE. 1975. The nesting of the Ring Ouzel. *Bird Study* **22**: 1–8.

FORESTRY COMMISSION. 1984. *Census of Woodland and Trees, Great Britain*. Forestry Commission, Edinburgh.

FORESTRY COMMISSION. 1990a. *Forestry Commission Annual Report for 1990*. HMSO, London.

FORESTRY COMMISSION 1990b. Forestry facts and figures 1989–1990.

FORSHAW, J.M. 1980. *Parrots of the World*. Lansdowne, Melbourne.

FOWLER, J.A. 1982. Leach's Petrels present on Ramna Stacks, Shetland. *Seabird Report* **6**: 93.

FOX, A.D. 1986. The breeding Teal *Anas crecca* of a coastal raised mire in central West Wales. *Bird Study* **33**: 18–23.

FOX, A.D. 1988. Breeding status of the Gadwall in Britain and Ireland. *Brit. Birds* **81**: 51–66.

FOX, A.D. 1991. History of the Pochard breeding in Britain. *Brit. Birds* **84**: 83–97.

FOX, A.D., N. JARRETT, H. GITAY and D. PAYNTER. 1989. Late summer habitat selection by breeding waterfowl in northern Scotland. *Wildfowl* **40**: 106–114.

FOX, A.D. and D.G. SALMON. 1989a. The winter status and distribution of Gadwall in Britain and Ireland. *Bird Study* **36**: 37–44.

FOX, A.D. and D.G. SALMON. 1989b. Changes in the non-breeding distribution and habitat of Pochard *Aythya ferina* in Britain. *Biol. Conserv.* **46**: 303–316.

FULLER, R.J. 1982. *Bird Habitats in Britain*. Poyser, Calton.

FULLER, R.J. 1993. Farmland birds in trouble. *BTO News* **184**: 1.

FULLER, R.J. 1994. *Bird Life of Woodland and Forest*. Cambridge University Press, Cambridge.

FULLER, R.J., J.K. BAKER, R.A. MORGAN, R. SCROGGS and M. WRIGHT. 1985. Breeding populations of the Hobby *Falco subbuteo* on farmland in the southern Midlands of England. *Ibis* **127**: 510–516.

FULLER, R.J., J.H. MARCHANT and R.A. MORGAN. 1985. How representative of agricultural practice in Britain are Common Birds Census farmland plots? *Bird Study* **32**: 56–70.

FULLER, R.J. and D.E. GLUE. 1977. The breeding biology of the Stonechat and Whinchat. *Bird Study* **24**: 215–228.

FULLER, R.J. and A.C.B. HENDERSON. 1992. Distribution of breeding songbirds in Bradfield Woods, Suffolk, in relation to vegetation and coppice management. *Bird Study* **39**: 73–88.

FULLER, R.J. and B.D. MORETON. 1987. Breeding bird populations of Kentish Sweet Chestnut *Castanea sativa* coppice in relation to age and structure of the coppice. *J. Appl. Ecol.* **39**: 73–88.

FULLER, R.J., T.M. REED, N.E. BUXTON, A. WEBB, T.D. WILLIAMS and M.W. PIENKOWSKI. 1986. Populations of breeding waders *Charadrii* and their habitats on the crofting lands of the Outer Hebrides, Scotland. *Biol. Conserv.* **37**: 333–361.

FULLER, R.J., P. STUTTARD and C.M. RAY. 1989. The distribution of breeding songbirds within mixed coppice woodland in Kent, England, in relation to vegetation age and structure. *Ann. Zool. Fennici* **26**: 265–275.

FURNESS, R.W. 1987. *The Skuas*. Poyser, Calton.

FURNESS, R.W. 1990. Evolutionary and ecological constraints on the breeding distributions and behaviour of skuas. *Proc. Int. 100 DO-G Meeting, Topics Avian Biol., Bonn 1988*: 153–158.

FURNESS, R.W. and C.M. TODD. 1984. Diets and feeding of Fulmars *Fulmarus glacialis* during the breeding season: a comparison between St Kilda and Shetland colonies. *Ibis* **126**: 379–387.

GALBRAITH, H. 1988. The breeding ecology of Lapwings *V. vanellus* on Scottish agricultural land. *J. Appl. Ecol.* **25**: 487–504.

GALBRAITH, H., R.W. FURNESS and R.J. FULLER. 1984. Habitats and distribution of waders breeding on Scottish agricultural land. *Scott. Birds* **13**: 98–107.

GALBRAITH, H., S. MURRAY, K. DUNCAN, R. SMITH, D.P. WHITFIELD and D.B.A. THOMPSON. 1993. Diet and habitat use in the Dotterel *Charadrius morinellus*. *Ibis*. **135**: 148–155.

GALBRAITH, H., S. MURRAY., S. RAE., D.P. WHITFIELD and D.B.A. THOMPSON. 1993. Numbers and breeding distribution of Dotterel *Charadrius morinellus* in Great Britain. *Bird Study*. In press.

GAMMELL, A. 1979. The world situation of Montagu's Harrier. *Proc. World Conference on Birds of Prey*. ICBP, Macedonia.

GARCIA, E.F.J. 1983. An experimental test of competition for space between Blackcaps *Sylvia atricapilla* and Garden Warblers *Sylvia borin* in the breeding season. *J. Anim. Ecol.* **52**: 795–805.

GARCIA, E.F.J. 1989. *The Blackcap and the Garden Warbler. Shire Natural History No. 43*. Shire Publications Ltd.

GARDARSSON, A. 1979. Waterfowl populations of Lake Myvatn and recent changes in numbers and food habits. *Oikos* **32**: 250–270.

GARSON, P.J. 1980a. The breeding ecology of the Wren in Britain. *Bird Study* **27**: 63–72.

GARSON, P.J. 1980b. Male behaviour and female choice: mate selection in the Wren? *Anim. Behav.* **28**: 491–502.

GÄRTNER, K. 1982. Zur Ablehnung von Eiern und Jungen des Kuckucks *Cuculus canorus* durch die Wirtsvöge-Beobachtungen und experimentelle Untersuchungen am Sumpfrohrsänger *Acrocephalus palustris*. *Die Vogelwelt* **103**: 210–224.

GASTON, A.J. 1973. The ecology and behaviour of the Long-tailed Tit. *Ibis* **115**: 330–351.

GATES, S., D.W. GIBBONS and J.H. MARCHANT. 1993. Methodology for producing national population estimates for breeding birds. *BTO Research Report*.

GENSBØL, B. 1984. *Birds of Prey of Britain and Europe, North Africa and the Middle East*. Collins, London.

GENTZ, K. 1965. *Die Grosse Dommel*. Wittenberg, Lutherstadt.

GERELL, R. 1985. Habitat selection and nest predation in a Common Eider population in Southern Sweden. *Ornis Scand.* **16**: 129–39.

GÉROUDET, P. 1980. *Les Passereaux 1. Du Coucou aux Corvidés*. Delachaux et Niestlé, Neuchâtel.

GIBB, J. 1956. Food, feeding habits and territory of the Rock Pipit *Anthus spinoletta*. *Ibis* **98**: 506–530.

GIBBONS, D.W. 1986. Brood parasitism and cooperative nesting in the Moorhen *Gallinula chloropus*. *Behav. Ecol. Sociobiol.* **19**: 221–232.

GIBBONS, D.W. 1987. The New Atlas Pilot Fieldwork. *BTO Research Report* No. 30.

GILES, N. 1990. Effects of increasing larval chironomid densities on the underwater feeding success of downy Tufted ducklings *Aythya fuligula*. *Wildfowl* **41**: 99–106.

GILLMOR, D.A. 1979. Agriculture. Pp 109–136 in Gillmor, D.A. (ed), *Irish Resources and Land Use*. Institute of Public Administration, Dublin.

GINN, H.B. 1969. The use of annual ringing and nest record card totals as indicators of bird population levels. *Bird Study* **16**: 210–248.

GITAY, H., A.D. FOX and S.C. RIDGILL. 1990. Survival estimates of Teal *Anas crecca* ringed at three stations in Britain. *The Ring* **13**: 45–58.

GLEN, N.W. and C.M. PERRINS. 1988. Co-operative breeding by Long-tailed Tits. *Brit. Birds* **81**: 630–641.

GLUE, D.E. 1973. Seasonal mortality of four small birds of prey. *Ornis. Scand.* **4:** 97–102.

GLUE, D.E. 1977a. Breeding biology of Long-eared Owls. *Brit. Birds* **70:** 318–331.

GLUE, D.E. 1977b. Feeding ecology of the Short-eared Owl in Britain and Ireland. *Bird Study* **24:** 70–78.

GLUE, D.E. 1990a. Little Grebes rearing three broods in Buckinghamshire. *Brit. Birds* **83:** 278.

GLUE D.E. 1990b. Breeding biology of the Grasshopper Warbler in Britain. *Brit. Birds* **83:** 131–145.

GLUE, D.E. and G.J. HAMMOND. 1974. Feeding ecology of the Long-eared Owl in Britain and Ireland. *Brit. Birds* **67:** 361–369.

GLUE, D.E. and R. MORGAN. 1972. Cuckoo hosts in British habitats. *Bird Study* **19:** 187–192.

GLUE, D.E. and D. SCOTT. 1980. Breeding biology of the Little Owl. *Brit. Birds* **73:** 167–180.

GLUTZ VON BLOTZHEIM, U.N. and K.M. BAUER. 1980. *Handbuch der Vogel Mitteleuropas* **9**. Akademische Verlagsgellschaft, Wiesbaden.

GLUTZ VON BLOTZHEIM, U.N. and K.M. BAUER. 1985. *Handbuch der Vogel Mitteleuropas* **10**. AULA-Verlag, Wiesbaden.

GOMERSALL, C.H. 1986. Breeding performance of the Red-throated Diver *Gavia stellata* in Shetland. *Holarct. Ecol.* **9:** 277–284.

GOMERSALL, C.H., J.S. MORTON and R.M. WYNDE. 1984. Status of breeding Red-throated Divers in Shetland, 1983. *Bird Study* **31:** 223–229.

GOOCH, S., S. BAILLIE and T.R. BIRKHEAD. 1991. The impact of Magpies *Pica pica* on songbird populations. Retrospective investigation of trends in population density and breeding success. *J. Appl. Ecol.* **28:** 1068–1086.

GOODE, D. 1981. *Lead Poisoning in Swans*. Report of the Nature Conservancy Council's Working Group.

GOODFELLOW, P.F. 1986. The Wood Warbler survey 1984–85. *Devon Birds* **39:** 87–92.

GOODWIN, D. 1948. Incubation habits of the Golden Pheasant. *Ibis* **90:** 280–284.

GOODWIN, D. 1976. *Crows of the World*. British Museum (Natural History), London.

GRAHN, M. 1990. Seasonal changes in ranging behaviour and territoriality in the European Jay *Garrulus g.glandarius*. *Ornis Scand.* **21:** 195–201.

GRANT, M.C. 1991. Nesting densities, productivity and survival of breeding Whimbrel *Numenius phaeopus* in Shetland. *Bird Study* **38:** 160–169.

GRAY, D.B. 1974. Breeding behaviour of Whinchats. *Bird Study* **21:** 280–282.

GREEN, R.E. 1978. Factors affecting the diet of farmland Skylarks *Alauda arvensis*. *J. Anim. Ecol.* **47:** 913–928.

GREEN, R.E. 1984. Double nesting of the Red-legged Partridge *Alectoris rufa*. *Ibis* **126:** 332–346.

GREEN, R.E. 1986. *The Management of Lowland Wet Grassland for Breeding Waders*. Unpublished RSPB report.

GREEN, R.E. 1988a. Stone Curlew conservation. *RSPB Conserv. Rev.* **2:** 30–33.

GREEN, R.E. 1988b. Effects of environmental factors on the timing and success of breeding of Common Snipe *Gallinago gallinago* (Aves: Scolopacidae). *J. Appl. Ecol.* **25:** 79–93.

GREEN, R.E., S.R. BAILLIE and M.I. AVERY. 1990. Can ringing recoveries help to explain the population dynamics of British terns? *The Ring.* **13:** 133–138.

GREEN, R.E., C.J. CADBURY and G. WILLIAMS. 1987. Floods threaten Black-tailed Godwits breeding at the Ouse Washes. *RSPB Conserv. Rev.* **1:** 14–16.

GREEN, R.E., G.J.M. HIRONS and J.S. KIRBY. 1990. The effectiveness of nest defence by Black-tailed Godwits *Limosa limosa*. *Ardea* **78** (3): 405–413.

GREEN, R.E. and G.J.M. HIRONS. 1991. The relevance of population studies to the conservation of threatened birds. In: Perrins, C.M., J.D. Lebreton and G.J.M. Hirons (eds) *Bird Population Studies: Relevance to Conservation and Management*. Oxford University Press, Oxford.

GREENWOOD, J.J.D. and S.R. BAILLIE. 1991. Effects of density-depen-

dence and weather on population changes of English passerines using a non-experimental paradigm. *Ibis* **133** suppl. 1: 121–133.

GREIG-SMITH, P.W. 1982. Interspecific aggression between chats. *Bird Study* **29:** 162.

GREIG-SMITH, P.W. 1987. Bud-feeding by Bullfinches: methods for spreading damage evenly within orchards. *J. Appl. Ecol.* **24:** 49–62.

GREY, Lord, of FALLODON. 1927. *The Charm of Birds*. Hodder and Stoughton, London.

GRIBBLE, F.C. 1976. A census of Black-headed Gull colonies. *Bird Study* **23:** 135–145.

GRIBBLE, F.C. 1983. Nightjars in Britain and Ireland in 1981. *Bird Study* **30:** 165–176.

GRIFFIN, B. 1990. Breeding sawbills on Welsh rivers 1990. Unpublished RSPB Report.

HAAG, D. 1987. Regulationsmechanismen bei der Strassentaube *Columba livia forma domestica* (Gmelin 1789). *Verhandl. Naturf. Ges. Basel* **97:** 31–42.

HAAPANEN, A. and L. NILSSON. 1979. Breeding waterfowl populations in northern Fennoscandia. *Ornis Scand.* **10:** 145–219.

HADDON, P.C. and R.C. KNIGHT. 1983. *A Guide to Little Tern Conservation*. RSPB, Sandy.

HAFTORN, S. 1978. Energetics of incubation by the Goldcrest *Regulus regulus* in relation to ambient air temperatures and the geographical distribution of the species. *Ornis Scand.* **9:** 22–30.

HÅGVAR, S., G. HÅGVAR and E. MØNNESS. 1990. Nest site selection in Norwegian woodpeckers. *Holarct. Ecol.* **13:** 156–165.

HAMER, K.C., R.W. FURNESS and R.W.G. CALDOW. 1991. The effects of changes in food availability on the breeding ecology of Great Skuas *Catharacta skua* in Shetland. *J. Zool., Lond.* **223:** 175–188.

HAN LINXIAN, YANG LAN and ZHENG BAOLAI. 1990. Observations of wild breeding ecology of Lady Amherst's Pheasant. In: Hill D.A., P.J. Garson, and D. Jenkins (eds), *Pheasants in Asia 1990*. Proceedings of 4th International Symposium of the World Pheasant Association. Beijing, China 1989. WPA, Reading, UK.

HARPER, D.G.C. 1985. Pairing strategies and mate choice in female Robins *Erithacus rubecula*. *Anim. Behav.* **33:** 862–875.

HARRIS, M.P. 1984. *The Puffin*. Poyser, Calton.

HARRIS, M.P. 1991. Population changes in British Common Murres and Common Puffins, 1969–1988. Pp 52–58 in Easton, A.J., and R.D. Elliot (eds), *Studies of High Latitude Seabirds, 2. Conservation Biology of the Thick-billed Murre in the Northwest Atlantic*. Canadian Wildlife Service, Ottawa.

HARRIS, M.P. and R.S. BAILEY. 1992. Mortality rates of Puffin *Fratercula arctica* and Guillemot *Uria aalge* and fish abundance in the North Sea. *Biol. Conserv.* **60:** 39–46.

HARRIS, M.P. and P. ROTHERY. 1988. Monitoring of Puffin burrows on Dun, St Kilda, 1977–1987. *Bird Study* **35:** 97–99.

HARRIS, M.P. and S. WANLESS. 1988. The breeding biology of Guillemots *Uria aalge* on the Isle of May over a six year period. *Ibis* **130:** 172–192.

HARRIS, M.P. and S. WANLESS. 1990. Breeding success of British Kittiwakes *Rissa tridactyla* in 1986–88: evidence for changing conditions in the northern North Sea. *J. Appl. Ecol.* **27:** 172–187.

HARRIS, M.P. and S. WANLESS. 1991. Population studies and conservation of Puffins, *Fratercula arctica*. Pp 230–248 in Perrins, C.M., J.-D. Lebreton and G.J.M. Hirons (eds), *Bird Population Studies: Relevance to Conservation and Management*. Oxford University Press, Oxford.

HARRIS, M.P. and S. WANLESS. 1992. The importance of the Lesser Sandeel *Ammodytes marinus* in the diet of the Shag *Phalacrocorax aristotelis*. *Ornis Scand.* **22:** 375–382.

HARRISSON, T.H. and P.A.D. HOLLOM. 1932. The Great Crested Grebe enquiry, 1931. *Brit. Birds* **26:** 62–92, 102–131, 142–155, 174–195.

HARTLEY, P.H.T. 1987. Ecological aspects of the foraging behaviour of Crested Tits *Parus cristatus*. *Bird Study* **34:** 107–111.

HARVEY, H.J. 1979. Great Crested Grebe breeding on rivers. *Brit. Birds* **71:** 385–386.

HARVEY, P. 1983. Breeding seabird populations, Isles of Scilly. *NCC Report*, SW England Region, Taunton.

HARVEY, W. G. 1977. Cetti's Warblers in east Kent in 1975. *Brit. Birds* **70**: 89–96.

HAVERSCHMIDT, F. 1946. Observations on the breeding habits of the Little Owl. *Ardea* **34**: 214–246.

HAWORTH, P. F. and D. B. A. THOMPSON. 1990. Factors associated with the breeding distribution of upland birds in the south Pennines, England. *J. Appl. Ecol.* **27**: 562–577.

HEADLAM, C. G. 1972. Temminck's Stints breeding in Scotland. *Scott. Birds* **7**: 94.

HEIJ, C. J. and C. W. MOELIKER. 1990. Population dynamics of Dutch House Sparrows in urban, suburban and rural habitats. In Pinowski, J. and J. D. Summers-Smith (eds), *Granivorous Birds in the Agricultural Landscape.* P W N, Warsaw.

HELLE, P. 1985. Effects of forest regeneration on the structure of bird communities in northern Finland. *Holarct. Ecol.* **8**: 120–132.

HENRY, J. and J. Y. MONNAT. 1981. *Oiseaux Marins de la Facade Atlantique Francaise.* Report for SEPNB.

HEPPLESTON, P. B. 1972. The comparative breeding ecology of Oystercatchers *Haematopus ostralegus* L. in inland and coastal habitats. *J. Anim. Ecol.* **41**: 23–51.

HERBER, K. J. 1985. BTO Wood Warbler survey, 1984. *Surrey Bird Report (1984)*: 51–53.

HERBERT, I. J. 1991. The status and habitat of the Garden Warbler at Crom Estate, Co. Fermanagh, and a review of its status in Ireland. *Irish Birds* **4**: 369–376.

HEUBECK, M. (ed) 1989. *Seabirds and Sandeels: Proceedings of a Seminar held in Lerwick, Shetland, 15–16th October 1988.* Shetland Bird Club, Shetland.

HIBBERT-WARE, A. 1937–38. Report of the Little Owl food inquiry, 1936–37. *Brit. Birds* **31**: 162–187, 205–229, 249–264.

HICKEY, J. J. (ed) 1969. *Peregrine Falcon Populations: their Biology and Decline.* Univ. Wisconsin Press, Madison and London.

HILDÉN, O. 1975. Breeding system of Temminck's Stint *Calidris temminckii*. *Ornis Fenn.* **52**: 117–146.

HILDÉN, O. and S. VUOLANTO. 1972. Breeding biology of the Red-necked Phalarope *Phalaropus lobatus* in Finland. *Ornis Fenn.* **49**: 57–85.

HILL, D. 1984a. Clutch predation in relation to nest density in Mallard and Tufted Duck. *Wildfowl* **35**: 151–156.

HILL, D. 1984b. Factors affecting nest success in the Mallard and Tufted Duck. *Ornis Scand.* **15**: 115–122.

HILL, D. 1984c. Population regulation in the Mallard *Anas platyrhynchos* L. *J. Anim. Ecol.* **53**: 192–202.

HILL, D. 1988. Population dynamics of the Avocet *Recurvirostra avosetta* breeding in Britain. *J. Anim. Ecol.* **57**: 638–669.

HILL, D. and N. CARTER. 1991. An empirical simulation model of an Avocet *Recurvirostra avosetta* population. *Ornis Scand.* **22**: 65–72.

HILL, D. and N. ELLIS. 1984. Survival and age-related changes in the foraging behaviour and time budgets of Tufted ducklings *Aythya fuligula*. *Ibis* **126**: 544–550.

HILL, D. A. and P. A. ROBERTSON. 1988. *The Pheasant: Ecology Management and Conservation.* Blackwell Scientific Publications, London.

HILL, D., R. WRIGHT and M. STREET. 1987. Survival of Mallard ducklings and competition with fish for invertebrates on a flooded gravel quarry in England. *Ibis* **129**: 159–167.

HILLIS, J. P. and D. C. F. COTTON. 1989. Black-necked Grebes breeding in Ireland. *Irish Birds* **4**: 72.

HIOM, L., M. BOLTON, P. MONAGHAN and D. WORRALL. 1991. Experimental evidence for food limitation of egg production in gulls. *Ornis Scand.* **22**: 94–97.

HJORT, C. and J. PETTERSSON. 1990. Changing numbers of migrant birds versus the changing environments. *Calidris* **19**: 13–23. (Swedish, with English summary and figure captions).

HOCKEY, P. A. R., L. G. UNDERHILL, M. NEATHERWAY and P. G. RYAN. 1989. *Atlas of the Birds of the Southwestern Cape.* Cape Bird Club, Cape Town.

HOELZEL, A. R. 1989. Territorial behaviour of the European Robin *Erithacus rubecula*: the importance of vegetation density. *Ibis* **131**: 432–436.

HOLLAND, P. K., J. E. ROBSON and D. W. YALDEN. 1982. The status and distribution of the Common Sandpiper *Actitis hypoleucos* in the Peak District. *Naturalist* **107**: 77–86.

HOLLAND, P. K. and D. W. YALDEN. 1991. Population dynamics of Common Sandpipers *Actitis hypoleucos* breeding along an upland river system. *Bird Study* **38**: 151–159

HOLLOM, P. A. D. 1936, 1951, 1959. (Great Crested Grebe sample counts). *Brit. Birds* **30**: 138–158, **44**: 361–369; *Bird Study* **6**: 1–7.

HOPE JONES, P., C. J. MEAD and R. F. DURMAN. 1977. The migration of the Pied Flycatcher from and through Britain. *Bird Study* **24**: 2–14.

HORI, J. 1966. Observations on Pochard and Tufted Duck breeding biology with particular reference to colonisation of a home range. *Bird Study* **13**: 297–305.

HORVARTH, R. 1988. Angaben uber die Wasseramsel *Cinclus cinclus* in Ungarn. *Egretta* **31**: 12–17.

HUDSON, A. V. 1982. Great Black-backed Gulls on Great Saltee Island, 1980. *Irish Birds* **2**: 167–175.

HUDSON, A. V. and R. W. FURNESS. 1989. The behaviour of seabirds foraging at fishing boats around Shetland. *Ibis* **131**: 225–237.

HUDSON, A. V., T. J. STOWE and S. J. ASPINALL. 1990. Status and distribution of Corncrakes in Britain in 1988. *Brit. Birds* **83**: 173–187.

HUDSON, P. J. 1986. *The Red Grouse: Biology and Management of a Wild Gamebird.* Game Conservancy, Fordingbridge.

HUDSON, P. J. 1989. Black Grouse in Britain. *Game Cons. Ann. Rev.* **20**: 119–124.

HUDSON, P. J. 1992. *Scottish Red Grouse Populations: The Biology of Low Density Populations.* Game Conservancy, Fordingbridge.

HUDSON, P. J., D. NEWBORN and A. P. DOBSON. 1992. Regulation and stability of a free-living host-parasite system, *Trichostrongylus tenuis* in Red Grouse: I Monitoring and Experiments. *J. Anim. Ecol.* **61**: 437–486.

HUDSON, R. 1965. The spread of the Collared Dove in Britain and Ireland. *Brit. Birds* **58**: 105–139.

HUDSON, R. 1972. Collared Doves in Britain and Ireland during 1965–1970. *Brit. Birds* **65**: 139–155.

HUDSON, R. 1974. Feral parakeets near London. *Brit. Birds* **67**: 33,174.

HUDSON, R. 1979. Nightingales in Britain in 1976. *Bird Study* **26**: 204–212.

HUDSON, R. and J. H. MARCHANT. 1984. Population estimates for British breeding birds. *BTO Research Report* 13.

HUGHES, B. 1990. The ecology and behaviour of the North American Ruddy-Duck *Oxyura jamaicensis* in Great Britain and its interaction with native waterbirds: a progress report. *Wildfowl* **41**: 133–138.

HUGHES, S. W. M., P. BACON and J. J. M. FLEGG. 1979. The 1975 census of the Great Crested Grebe in Britain. *Bird Study* **26**: 213–226.

HULTEN, M. 1959. Beitrag zur Kenntnis des Feldschwirls *Locustella naevia*. *Regulus* **39**: 95–117.

HUSTINGS, F. 1991. Explosive increase of breeding Black-necked Grebes in the Netherlands in 1983–89. *Limosa* **64**: 17–24

HUSTINGS, F., F. POST and F. SCHEPERS. 1990. Are Corn Buntings disappearing as breeding birds in the Netherlands? *Limosa* **63**: 103–111.

HUTCHINSON, C. 1979. *Ireland's Wetlands and their Birds.* Irish Wildlife Conservancy, Dublin.

HUTCHINSON, C. D. 1989. *Birds in Ireland.* Poyser, Calton.

IBANEZ, F. and B. TROLLIET. 1990. Le Canard Souchet *Anas clypeata* nicheur dans le Marais Breton: Effectif, repartition et liaison avec les limicoles. *Gibier Faune Sauvage* **7**: 95–106.

ICBP. 1992. *Putting biodiversity on the map: priority areas for global conservation.* ICBP, Cambridge, UK.

IMAGE, R. A. 1991. Montagu's and Marsh Harriers in West Norfolk 1987–1991. *Norfolk Bird and Mammal Report 1991*: 270–271.

INGLIS, I. R., A. J. ISAACSON, R. J. P. THEARLE and N. J. WESTWOOD. 1990. The effects of changing agricultural practice on Woodpigeon *Columba palumbus* numbers. *Ibis* **132:** 262–272.

IRONS, A. 1980. *Breeding of the Honey Buzzard in Nottinghamshire.* Trent Valley Bird Watchers and Nottinghamshire Ornithological Society.

JARDINE, D. 1991. 1990 – The year of the Crossbill. *Birds in Northumbria 1990*: 103–106.

JARVINEN, O. and R. A. VAISENEN. 1978. Long term population changes of the most abundant south Finnish forest birds during the past 50 years. *J. Orn.* **119:** 441–449.

JOHN, A. W. G. and J. ROSKELL. 1985. Jay movements in autumn 1983. *Brit. Birds* **78:** 611–637.

JOHNSGARD, P. A. 1986. *Pheasants of the World.* Oxford University Press, Oxford.

JOHNSON, R. F. and S. G. JOHNSON. 1990. Reproductive ecology of Feral Pigeons. Pp 237–252 in Pinowski, J. and J. D. Summers-Smith (eds), *Granivorous Birds in the Agricultural Landscape.* INTECOL, Warszawa.

JONES, G. 1986. The distribution and abundance of Sand Martins breeding in central Scotland. *Scott. Birds* **14:** 33–38.

JONES, G. 1987. Selection against large size in the Sand Martin *Riparia riparia* during a dramatic population crash. *Ibis* **129:** 274–280.

JORGENSEN, H. E. 1985. Bestandsudrikling, habitatvalg og ungeproduktion hos Rorhog *Circus aeruginosus* 1971–83. *Dansk Ornithologisk Forenings Tidsskrift* **79:** 81–102.

JOYNER, D. E. 1975. *Nest Parasitism and Brood-related Behaviour of the Ruddy Duck* Oxyura jamaicensis rubida. PhD Thesis, Univ. Nebraska, Lincoln.

KALCHREUTER, H. (translated P. D. K. HESSEL). 1982. *The Woodcock.* Verlag Dieter Hoffman. Mainz.

KÄMPFER, A. and W. LEDERER. 1990. Nachweis von Bigynie bein Grauschnäpper *Muscicapa striata. Vogelwelt* **111:** 189–196.

KAVANAGH, B. 1992. Irish Grey Partridge census 1991: preliminary report. In: Birkan M., G. R. Potts, N. J. Aebischer and S. D. Dowell (eds), *Perdix VI: The First International Partridge, Quail and Francolin Symposium. Gibier Faune Sauvage* **9**.

KEAR, J. 1990. *Man and Wildfowl.* Poyser, London.

KELLY, G. 1986. *The Norfolk Bird Atlas.* Norfolk and Norwich Naturalists' Society Occ. Publ. No. 1.

KELSEY, M. G. 1989a. Breeding biology of Marsh Warblers *Acrocephalus palustris* in Worcestershire: a comparison with European populations. *Bird Study* **36:** 205–210.

KELSEY, M. G. 1989b. A comparison of the song and territorial behaviour of a long distance migrant, the Marsh Warbler *Acrocephalus palustris*, in summer and winter. *Ibis* **131:** 403–414.

KELSEY, M. G., G. H. GREEN, M. C. GARNETT and P. V. HAYMAN. 1989. Marsh Warblers in Britain. *Brit. Birds* **82:** 239–256.

KENNEDY, P. G., R. F. RUTTLEDGE and C. F. SCROOPE. 1954. *Birds of Ireland.* Oliver and Boyd, Edinburgh.

KIRBY, J. S. and C. MITCHELL. 1993. Distribution and status of wintering Shoveler in Great Britain. *Bird Study.* In press.

KLOMP, N. I. and R. W. FURNESS. 1990. Variations in numbers of non-breeding Great Skuas attending a colony. *Ornis Scand.* **21:** 270–276.

KNOX, A. G. 1990a. The identification of Crossbill and Scottish Crossbill. *Brit. Birds* **83:** 89–94.

KNOX, A. G. 1990b. The sympatric breeding of *Loxia curvirostra* and *L. scotica* and the evolution of crossbills. *Ibis* **132:** 452–66.

KNOX, A. G. 1990c. The native pinewoods. Pp 453–6 in Buckland, S. T., M. V. Bell, and N. Picozzi (eds), *The Birds of North-East Scotland.* North-east Scotland Bird Club, Aberdeen.

KOSKIMIES, P. 1989. Distribution and numbers of Finnish breeding birds. Appendix to *Suomen Lintuatlas.* SLY, Lintuieto Oy: Helsinki.

KOSTRZEWA, A. 1989. The effects of weather on density and reproduction in Honey Buzzards *Pernis apivorus.* In: Meyburg, B-U. and R. D. Chancellor (eds), *Raptors in the Modern World.* Proceedings of the Third World Conference on Birds of Prey and Owls, 1987. World Working Group on Birds of Prey.

KRUJIT, J. P. and J. A. HOGAN. 1967. Social behaviour on the lek in Black Grouse, *Lyrurus tetrix tetrix. Ardea* **55:** 203–240.

KUHNEN, K. 1975. Bestandsentwicklung, Verbreitung, Biotop und Siedlungsdichte der Uferscwalbe (*Riparia riparia*) 1966–1973 am Niederrhein. *Charadrius* **11:** 1–24.

LACK, D. 1943. *The Life of the Robin.* Witherby, London.

LACK, D. 1956. *Swifts in a Tower.* Methuen, London.

LACK, D. 1966. *Population Studies of Birds.* Clarendon Press, Oxford.

LACK, D. 1971. *Ecological Isolation in Birds.* Blackwell, Oxford.

LACK, P. C. 1986. *The Atlas of Wintering Birds in Britain and Ireland.* Poyser, Calton.

LACK, P. C. 1987. The effects of severe hedge cutting on a breeding bird population. *Bird Study* **34:** 139–146.

LACK, P. C. 1989. Overall and regional trends in warbler populations of British farmland over 25 years. *Ann. Zool. Fennici* **26:** 219–225.

LANGSLOW, D. R. 1977. Movements of Black Redstarts between Britain and Europe as related to occurrences at observatories. *Bird Study* **24:** 169–178.

LASSEY, P. and D. I. M. WALLACE. 1992. Breeding breakthrough by Scarlet Rosefinches. *Bird Watching*, September 1992: 84–85

LAWN, M. R. 1982. Pairing systems and site tenacity of the Willow Warbler, *Phylloscopus trochilus*, in southern England. *Ornis Scand.* **13:** 193–199.

LAWTON, J. H. 1990. Red Grouse populations and management. British Ecological Society, *Ecological Issues No. 2.* Field Studies Council.

LAWTON ROBERTS, J. and N. BOWMAN. 1986. Diet and ecology of Short-eared Owls *Asio flammeus* breeding on heather moor. *Bird Study* **33:** 12–17.

LEISLER, B. 1975. Die Bedeutung der Fußmorphologie für die ökologische Sonderung mitteleuropäischer Rohrsänger *Acrocephalus* und Schwirle *Locustella. J. Orn.* **116:** 117–153.

LESLIE, R. 1985. The populations and distribution of Nightjars *Caprimulgus europaeus* on the North York Moors. *Naturalist* **110:** 23–28.

LESLIE, R. C. B. and E. G. PEDLER. 1938. Nesting of the Little Ringed Plover in Hertfordshire. *Brit. Birds* **32:** 90–102.

LEVER, C. 1977. *The Naturalised Animals of the British Isles.* Hutchinson, London.

LEVER, C. 1987. *Naturalised Birds of the World.* Longman, London.

LLOYD, C., M. L. TASKER and K. PARTRIDGE. 1991. *The Status of Seabirds in Britain and Ireland.* Poyser, London.

LOCKE, G. M. L. 1987. *Census of Woodlands and Trees 1979–82.* Forestry Commission Bulletin 63. HMSO, London.

LOCKIE, J. D. 1955a. The breeding habits and food of Short-eared Owls after a vole plague. *Bird Study* **2:** 53–69.

LOCKIE, J. D. 1955b. The breeding and feeding of Jackdaws and Rooks with notes on Carrion Crows and other corvidae. *Ibis* **97:** 341–369.

LONG, J. L. 1981. *Introduced Birds of the World.* David and Charles, London.

LONG, R. 1981. Review of birds in the Channel Islands, 1951–80. *Brit. Birds* **74:** 327–344.

LOVE, J. A. 1983. *The Return of the Sea Eagle.* Cambridge University Press.

LOVE, J. A. 1989. The re-introduction of the White-tailed Sea Eagle to Scotland: 1975–1987. *Research and Survey in Nature Conservation* No 12. Nature Conservancy Council, Peterborough.

LÜBKE, W. and R. FURRER. 1985. *Die Wacholderdrossel.* Neue Brehm-Bücherei, Lutherstadt.

LUMSDEN, J. 1876. Notes on the distribution of the Common Jay in Scotland. *Scottish Naturalist* **3:** 233–240.

MACFARLANE, G. 1980. The Stonechat, its status as a breeding species in Northumberland. Private Publication.

MACMILLAN, A. T. 1970. Goldeneye breeding in east Invernessshire. *Scott. Birds* **6:** 197–198.

MADDEN, B. 1987. The Mediterranean Gull in Ireland, 1956–1985. *Irish Birds* **3:** 363–376.

MARCHANT, J. H. 1981. Residual edge effects with the mapping bird census method. *Studies in Avian Biology* **6:** 488–491.

MARCHANT, J. 1991. Common Bird Census Report 1988–89. In Stroud, D. A. and D. Glue (eds), *Britain's Birds in 1989/90: the Conservation and Monitoring Review.* BTO/NCC, Thetford.

MARCHANT, J.H. and P.A. HYDE. 1980. Aspects of the distribution of riparian birds on waterways in Britain and Ireland. *Bird Study* **26**: 183–202.

MARCHANT, J.H., R. HUDSON, S.P. CARTER and P. WHITTINGTON. 1990. *Population Trends in British Breeding Birds.* BTO, Tring.

MARCHANT, J.H., L.J. MUSTY and S.P. CARTER. 1991. Common Bird Census: 1989–90 index report. *BTO News.* **177**: 11–14.

MARCHANT, J.H., R. HUDSON and P. WHITTINGTON. 1990. CBC Index report 1988–89. *BTO News* **171**.

MARLER, P. 1956. Behaviour of the Chaffinch *Fringilla coelebs. Behaviour, supplement* **5**: 1–184.

MARQUISS, M. 1989. Grey Herons *Ardea cinerea* breeding in Scotland: numbers, distribution, and census techniques. *Bird Study* **36**: 181–191.

MARQUISS, M. 1981. The Goshawk in Britain – its provenance and current status. Pp 43–55 in Kenwood, R.E. and M. Lindsay (eds), *Understanding the Goshawk.* International Association for Falconry, Oxford.

MARQUISS, M. and A.F. LEITCH. 1990. The diet of Grey Herons *Ardea cinerea* breeding at Loch Leven, Scotland, and the importance of their predation on ducklings. *Ibis* **132**: 535–549.

MARQUISS, M. and I. NEWTON. 1982. The Goshawk in Britain. *Brit. Birds* **75**: 243–260.

MARQUISS, M., D.A. RATCLIFFE and R. ROXBURGH. 1985. The numbers, breeding success and diet of Golden Eagles in southern Scotland in relation to changes in land use. *Biol. Conserv.* **33**: 1–17.

MARTIN, M.W.A. 1988. Colonial nesting Red-breasted Mergansers. *Scott. Bird News* **9**: 8.

MASON, C.F. 1976. Breeding biology of the *Sylvia* warblers. *Bird Study* **23**: 213–232.

MASON, C.F. 1986. Invertebrate populations and biomass over four years in a coastal, saline lagoon. *Hydrobiol.* **133**: 1–29.

MASON, C.F. and F. LYCZYNSKI. 1980. Breeding biology of the Pied and Yellow Wagtail. *Bird Study* **27**: 1–10.

MATHER, J.R. 1986. *The Birds of Yorkshire.* Croom Helm, London.

MATTHYSEN, E. 1988. *Populationsdynamiek, Sociale Organisatie en Habitat-kwaliteit de Boomklever* Sitta europaea L. PhD thesis, University of Antwerpen, Wilrijk.

MAYES, E. and T.J. STOWE. 1989. The status and distribution of Corncrakes in Ireland, 1988. *Irish Birds* **4**: 1–12.

MAYHEW, P.W. 1988. The daily energy intake of European Wigeon in Winter. *Ornis Scand.* **19**: 217–233.

MAYR, E. 1942. *Systematics and the Origin of Species.* New York.

McNAMARA, J.M., A.I. HOUSTON and J.R. KREBS. 1990. Why hoard? The economics of food storing in tits, *Parus* species. *Behav. Ecol.* **1**: 12–23.

MEAD, C.J. 1984. *Robins.* Whittet Books, London.

MEAD, C.J. 1989. Mono-kill. *BTO News* **163**: 1.

MEAD, C.J. and J.A. CLARK. 1990. Report on bird ringing for Britain and Ireland for 1989. *Ring. and Migr.* **11**: 137–176.

MEAD, C.J., P.M. NORTH and B.R. WATMOUGH. 1979. The mortality of British Grey Herons. *Bird Study* **26**: 13–22.

MEAD, C.J. and K. SMITH. 1982. *Hertfordshire Breeding Bird Atlas.* HBBA, Tring, Herts.

MEARNS, R. 1983. The status of the Raven in southern Scotland and Northumbria. *Scott. Birds* **12**: 211–218.

MEARNS, R. and I. NEWTON. 1984. Turnover and dispersal in a Peregrine *Falco peregrinus* population. *Ibis* **126**: 347–355.

MEEK, E.R. and B. LITTLE. 1977. The spread of the Goosander in Britain and Ireland. *Brit. Birds* **70**: 229–237.

MEININGER, P.L. and J.F. BEKHUIS. 1990. De Zwartkopmeeuw *Larus melanocephalus* als broedvogel in Nederland en Europa. *Limosa* **63**: 121–134.

MENDENHALL, V.M. and H. MILNE. 1985. Factors affecting duckling survival of Eiders *Somateria mollissima* in northeast Scotland. *Ibis* **127**: 148–158.

MERIKALLIO, E. 1958. Finnish birds: their distribution and numbers. Helsinki Societas pro Fauna et Flora. *Fauna Fennica* **5**.

MEURY, R. 1989. Population density and home-range of the Tree Pipit *Anthus trivialis* in the Lower Reuss valley, in a landscape with a patchy habitat distribution. *Orn. Beobachter* **86**: 105–135.

MIKKOLA, H. 1976. Owls killing and killed by other owls and raptors in Europe. *Brit. Birds* **69**: 144–154.

MIKKOLA, H. 1983. *Owls of Europe.* Poyser, Calton.

MILSOM, T.P and A. WATSON. 1984. Numbers and spacing of summering Snow Buntings and snow cover in the Cairngorms. *Scott. Birds* **13**: 19–23.

MINISTRY OF AGRICULTURE, FISHERIES AND FOOD and DEPARTMENT OF AGRICULTURE AND FISHERIES FOR SCOTLAND. Annual returns from the June Census of Agriculture.

MINOT, E.O. 1981. Effects of interspecific competition for food in breeding Blue and Great Tits. *J. Anim. Ecol.* **50**: 375–385.

MINOT, E.O. and C.M. PERRINS. 1986. Interspecific interference competition – nest sites for Blue and Great Tits. *J. Anim. Ecol.* **55**: 331–350.

MITCHELL, C. 1990. The movements of British ringed Shoveler. Unpublished Report. Wildfowl and Wetlands Trust, Slimbridge.

MITCHELL, J. 1977. Observations on the Common Scoter on Loch Lomond. *Loch Lomond Bird Report* **5**: 8–13 (see also Report **6**: 9).

MOERBEEK, D.J., W.H. VAN DOBBEN, E.R. OSIECK, G.C. BOERE and C.M. BUNGENBERG DE JONG. 1987. Cormorant damage prevention at a fish farm in the Netherlands. *Biol. Conserv.* **39**: 23–38.

MØLLER, A.P. 1983. Song activity and territory quality in the Corn Bunting *Miliaria calandra* with comments on mate selection. *Ornis Scand.* **14**: 81–89.

MØLLER, A.P. 1986. Mating systems among European passerines: a review. *Ibis* **128**: 234–250.

MØLLER, A.P. 1987. Advantages and disadvantages of coloniality in the Swallow, *Hirundo rustica. Anim. Behav.* **35**: 819–832.

MØLLER, A.P. 1989. Population dynamics of a declining Swallow *Hirundo rustica* population. *J. Anim. Ecol.* **58**: 1051–1063.

MØLLER, A.P. 1991. Density-dependent extra-pair copulations in the Swallow *Hirundo rustica. Ethology* **87**: 316–329.

MONAGHAN, P. 1983. Gulls: populations and problems. Pp 232–237 in Hickling, R. (ed), *Enjoying Ornithology.* Poyser, Calton.

MONAGHAN, P., E. BIGNAL, S. BIGNAL, N. EASTERBEE and C.R. MCKAY. 1989. The distribution and status of the Chough in Scotland in 1986. *Scott. Birds* **15**: 114–118.

MONAGHAN, P. and J.C. COULSON. 1977. The status of large gulls nesting on buildings. *Bird Study* **24**: 89–104.

MONAGHAN, P., J.D. UTTLEY, M.D. BURNS, C. THAINE and J. BLACKWOOD. 1989. The relationship between food supply, reproductive effort and breeding success in Arctic Terns *Sterna paradisaea. J. Anim. Ecol.* **58**: 261–274.

MONK, J.F. 1954. The breeding biology of the Greenfinch. *Bird Study* **1**: 2–14.

MONK, J.F. 1955. Wryneck survey. *Bird Study* **2**: 87–89.

MONK, J.F. 1963. The past and present status of the Wryneck in the British Isles. *Bird Study* **10**: 112–132.

MONVAL, J.-Y., and J.-Y. PIROT. 1989. *Results of the IWRB International Waterfowl Census 1967–1986. IWRB Spec. Publ. No. 8.* IWRB, Slimbridge.

MOORE, N.W. 1957. The past and present status of the Buzzard in the British Isles. *Brit. Birds* **50**: 173–197.

MOORE, N.W. and M.D. HOOPER. 1975. On the number of bird species in British Woods. *Biol. Conserv.* **8**: 239–250.

MOREAU, R.E. 1951. The British status of the Quail and some problems of its biology. *Brit. Birds.* **44**: 257–276.

MOREAU, R.E. 1972. *The Palearctic-African Bird Migration Systems.* Academic Press, London.

MORGAN, R. 1982. The breeding biology of the Nightingale *Luscinia megarhynchos* in Britain. *Bird Study* **29**: 67–72.

MORGAN, R. A. and D. E. GLUE. 1981. Breeding survey of Black Redstarts in Britain, 1977. *Bird Study* **28**: 163–168.

MORGAN, R. A. and R. J. O'CONNOR. 1980. Farmland habitat and Yellowhammer distribution in Britain. *Bird Study* **27**: 155–162.

MORGAN, S. and M. THORP. 1990. Interpolation with UNIRAS. *UNIRAS UK User Group Newsletter* **6**: 11–20.

MORRIS, A. 1993. The 1992 Nightjar survey – a light at the end of the tunnel for this threatened species? *BTO News* **185**: 8–9.

MOSS, D. and G. M. MOSS. 1993. Breeding biology of the Little Grebe *Tachybaptus ruficollis* in Britain and Ireland. *Bird Study* **40**: 107–114.

MOSS, R. 1985. Rain, breeding success and distribution of Capercaillie *Tetrao urogallus* and Black Grouse *Tetrao tetrix* in Scotland. *Ibis* **128**: 65–72.

MOSS, R. and J. OSWALD. 1985. Population dynamics of Capercaillie in a Northeast Scottish glen. *Ornis Scand.* **16**: 229–238.

MOSS, R. and A. WATSON. 1984. Maternal nutrition, egg quality and breeding success of Scottish Ptarmigan *Lagopus mutus*. *Ibis* **126**: 212–220.

MOSS, R. and A. WATSON. 1991. Population cycles and kin selection in Red Grouse, *Lagopus lagopus scoticus*. *Ibis* **113**: 113–120.

MOSS, R. and D. N. WEIR. 1987. Demography of Capercaillie *Tetrao urogallus* in north-east Scotland. III. Production and recruitment of young. *Ornis Scand.* **18**: 141–145.

MOSS, R., D. N. WEIR and A. M. JONES. 1979. Capercaillie management in Scotland. Pp 140–155 in Lovel, T. W. I. (ed) *Woodland Grouse Symposium*. World Pheasant Association, Bures, Suffolk.

MOUNTFORT, G. 1957. *The Hawfinch*. Collins, London.

MUDGE, G. P., R. H. DENNIS, T. R. TALBOT and R. A. BROAD. 1991. Changes in the breeding status of Black-throated Divers in Scotland. *Scott. Birds.* **16**: 77–84.

MULLER, Y. 1988. Recherches sur l'ecologie des oiseaux forestiers des Vosges du Nord IV. Etude de l'avifaune nicheuse de la succession du Pin sylvestre. *L'Oiseau et R. F. O.* **58**: 89–112.

MULLINS, J. R. 1984. Scarlet Rosefinch breeding in Scotland. *Brit. Birds* **77**: 133–135.

MUNKEJORD, A. 1981. The Stonechat in western Norway south of 62°N 1973–1980. *Fauna Norv. Ser C. Cinclus* **4**: 69–75.

MURRAY, R. D. 1991a. Quail in Scotland. *Scottish Bird Report 1989* **22**: 45–50.

MURRAY, R. D. 1991b. The first successful breeding of Nuthatch in Scotland. *Scottish Bird Report 1989* **22**: 51–55.

MURRAY, S. and S. WANLESS. 1986. The status of the Gannet in Scotland 1984–85. *Scott. Birds* **14**: 74–85.

MURTON, R. K. 1965. *The Woodpigeon*. Collins, London.

MURTON, R. K. 1968. Breeding, migration and survival of Turtle Doves. *Brit. Birds.* **61**: 193–212.

MURTON, R. K., C. F. B. COOMBES and R. J. P. THEARLE. 1972. Ecological studies of the Feral Pigeon *Columba livia* var. II. Flock behaviour and social organisation. *J. Appl. Ecol.* **9**: 875–889.

MURTON, R. K. and A. J. ISAACSON. 1964. Production and egg predation in the Woodpigeon. *Ardea* **52**: 30–47.

MURTON, R. K. and J. KEAR. 1978. Photoperiodism in waterfowl: phasing of breeding cycles and zoogeography. *J. Zool., Lond.* **186**: 243–283.

MURTON, R. K., N. J. WESTWOOD and A. J. ISAACSON. 1964. The feeding habits of the Woodpigeon *Columba palumbus*, Stock Dove *C. oenas* and Turtle Dove *Streptopelia turtur*. *Ibis* **106**: 174–188.

MYRES, M. T. 1955. The breeding of the Blackbird, Song Thrush and Mistle Thrush in Great Britain. Part I. Breeding seasons. *Bird Study* **2**: 2–24.

NCC. 1984. *Nature Conservation in Great Britain*. NCC, Peterborough.

NEILL, P. 1806. *A Tour of Orkney*. London.

NELSON, B. 1978. *The Gannet*. Poyser, Berkhamsted.

NELSON, T. H. 1907. *The Birds of Yorkshire*. Vol. 1. Brown and Sons, London.

NETHERSOLE-THOMPSON, D. 1966. *The Snow Bunting*. Oliver and Boyd, Edinburgh.

NETHERSOLE-THOMPSON, D. 1973. *The Dotterel*. Collins, London.

NETHERSOLE-THOMPSON, D. 1975. *Pine Crossbills*. Poyser, Berkhamsted.

NETHERSOLE-THOMPSON, D. and M. NETHERSOLE-THOMPSON. 1979. *Greenshanks*. Poyser, Berkhamsted.

NETHERSOLE-THOMPSON, D. and M. NETHERSOLE-THOMPSON. 1986. *Waders, their Breeding, Haunts and Watchers*. Poyser, Calton.

NETHERSOLE-THOMPSON, D. and A. WATSON. 1974. *The Cairngorms*. Collins, London.

NETHERSOLE-THOMPSON D. and A. WATSON. 1981. *The Cairngorms*. Melven Press, Perth.

NEWTON, I. 1964. The breeding biology of the Chaffinch. *Bird Study* **11**: 47–68.

NEWTON, I. 1967. The adaptive radiation and feeding ecology of some British finches. *Ibis* **109**: 33–98.

NEWTON, I. 1972. *Finches*. Collins, London

NEWTON, I. 1979. *Population Ecology of Raptors*. Poyser, Berkhamsted.

NEWTON, I. 1986. *The Sparrowhawk*. Poyser, Calton.

NEWTON, I., J. A. BOGAN and M. B. HAAS. 1989. Organochlorines and mercury in the eggs of British Peregrines. *Ibis* **131**: 355–376.

NEWTON, I. and C. R. G. CAMPBELL. 1975. Breeding of ducks at Loch Leven, Kinross. *Wildfowl* **26**: 83–102.

NEWTON, I., P. E. DAVIS and J. E. DAVIS. 1989. Age of first breeding, dispersal, and survival of Red Kites *Milvus milvus* in Wales. *Ibis* **131**: 16–21.

NEWTON, I., P. E. DAVIS and D. MOSS. 1981. Distribution and breeding of Red Kites in relation to land-use in Wales. *J. Appl. Ecol.* **18**: 173–186.

NEWTON, I. and M. B. HASS. 1984. The return of the Sparrowhawk. *Brit. Birds* **77**: 47–70.

NEWTON, I. and M. B. HAAS. 1988. Pollutants in Merlin eggs and their effects on breeding. *Brit. Birds* **81**: 258–269.

NEWTON, I., E. R. MEEK and B. LITTLE. 1986. Population and breeding of Northumbrian Merlins. *Brit. Birds* **79**: 155–170.

NEWTON, I., I. WYLLIE and A. ASHER. 1991. Mortality causes in British Barn Owls with a discussion of aldrin-dieldrin poisoning. *Ibis* **133**: 162–169.

NEWTON, I., I. WYLLIE and R. M. MEARNS. 1986. Spacing of Sparrowhawks in relation to food-supply. *J. Anim. Ecol.* **55**: 361–370.

NILSSON, S. G. 1987. Limitation and regulation of population density in the Nuthatch *Sitta europaea* (Aves) breeding in natural cavities. *J. Anim. Ecol.* **56**: 921–937.

NORMAN, S. C. and W. NORMAN. 1985. Autumn movements of Willow Warblers ringed in the British Isles. *Ring. and Migr.* **3**: 165–172.

NORRISS, D. W. 1991. The status of the Buzzard as a breeding species in the Republic of Ireland, 1977–1991. *Irish Birds* **4**: 291–298.

NORTH, P. M. 1979. Relating Grey Heron survival rates to winter weather conditions. *Bird Study* **26**: 23–28.

O'BRIAIN, M. and P. FARRELLY. 1990. Breeding biology of Little Terns at Newcastle, Co. Wicklow and the impact of conservation action, 1985–1990. *Irish Birds* **4**: 149–168.

O'CONNOR, R. J. 1980. Population regulation in the Yellowhammer *Emberiza citrinella* in Britain. Pp 190–200 in Oelke, H (ed), *Bird Census Work and Nature Conservation*. Dachverband Deutscher Avifaunisten, West Germany.

O'CONNOR, R. J. 1981. Comparisons between migrant and non migrant birds in Britain. Pp 167–195 in Aidley D.J. (ed), *Animal Migration*. Soc. for Experimental Biology, Seminar 13. Cambridge University Press, Cambridge.

O'CONNOR, R. J. 1986. Dynamical aspects of avian habitat use. Pp 235–240 in Verner, J., M. L. Morrison and C. J. Ralph (eds), *Modeling Habitat Relationships of Terrestrial Vertebrates*. University of Wisconsin Press, Wisconsin.

O'CONNOR, R. J. and R. J. FULLER. 1985. Bird population responses to habitat. Pp 197–211 in Taylor, K., R. J. Fuller and P. C. Lack (eds), *Bird Census and Atlas Studies*. Proceedings of the 8th International Bird Census Conference. BTO, Tring.

O'CONNOR, R. J. and C. J. MEAD. 1984. The Stock Dove in Britain, 1930–80. *Brit. Birds* **77**: 181–201.

O'CONNOR, R.J. and R.A. MORGAN. 1982. Some effects of weather conditions on the breeding of the Spotted Flycatcher *Muscicapa striata* in Britain. *Bird Study* **29:** 41–48.

O'CONNOR, R.J. and M. SHRUBB. 1986a. *Farming and Birds.* Cambridge University Press, Cambridge.

O'CONNOR, R.J. and M. SHRUBB. 1986b. Recent changes in bird populations in relation to farming practices in England and Wales. *J. Roy. Agric. Soc.* (England) **147:** 132–141.

O'DONALD, P. 1983. *The Arctic Skua: a Study of the Ecology and Evolution of a Seabird.* Cambridge University Press, Cambridge.

O'FLYNN, W.J. 1983. Population changes of the Hen Harrier in Ireland. *Irish Birds* **2:** 337–343.

O'HALLORAN, J., A.A. MYERS and P.F. DUGGAN. 1991. Pp 389–395 in Sears, J. and P.J. Bacon (eds), *Proc. Third IWRB International Swan Symposium. Oxford, 1989.* Wildfowl Supplement No. 1.

O'MEARA, M. 1979. Distribution and numbers of Corncrakes in Ireland in 1978. *Irish Birds* **1:** 381–405.

O'SULLIVAN, J.M. 1976. Bearded Tits in Britain and Ireland 1966–74. *Brit. Birds* **69:** 473–489.

O'SULLIVAN, O. 1992. Chough numbers up. *IWC News* **72:** 6.

O'SULLIVAN, O. and P. SMIDDY. 1989. Thirty-sixth Irish Bird Report, 1988. *Irish Birds* **4:** 79–114.

O'SULLIVAN, O. and P. SMIDDY. 1990. Thirty-seventh Irish Bird Report, 1989. *Irish Birds* **4:** 231–257.

OAKES, C. 1953. *The Birds of Lancashire.* Oliver and Boyd, Edinburgh.

OGILVIE, M.A. 1981. The Mute Swan in Britain, 1978. *Bird Study* **28:** 87–106.

OGILVIE, M.A. 1986. The Mute Swan in Britain, 1983. *Bird Study* **33:** 121–137.

OKILL, J.D. and S. WANLESS. 1990. Breeding success and chick growth of Red-throated Divers *Gavia stellata* in Shetland 1979–88. *Ring. and Migr.* **11:** 65–72.

OLNEY, P.J.S. 1963. The food and feeding habits of Tufted Duck *Aythya fuligula*. *Ibis* **105:** 55–62.

OPDAM, P., G. RIJSDIJK and F. HUSTINGS. 1985. Bird communities in small woods in an agricultural landscape: Effects of area and isolation. *Biol. Conserv.* **34:** 333–352.

ORFORD, N. 1973. Breeding distribution of the Twite in central Britain. *Bird Study* **20:** 50–62, 121–126.

ORMEROD, S.J. and S.J. TYLER. 1987a. Dippers *Cinclus cinclus* and Grey Wagtails *Motacilla cinerea* as indicators of stream acidity in upland Wales. *ICBP Technical Publication No. 6.*

ORMEROD, S.J. and S.J. TYLER. 1987b. Aspects of the breeding ecology of Welsh Grey Wagtails *Motacilla cinerea*. *Bird Study* **34:** 43–51.

ORMEROD, S.J. and S.J. TYLER. 1990. Environmental pollutants in the eggs of Welsh Dippers *Cinclus cinclus*: a potential monitor of organochlorine and mercury contamination in upland rivers. *Bird Study* **37:** 171–176.

ORMEROD, S.J. and S.J. TYLER. 1991. The influence of stream acidification and riparian land use on the feeding ecology of Grey Wagtails *Motacilla cinerea* in Wales. *Ibis* **133:** 53–61.

ORMEROD, S.J. and S.J. TYLER. 1992. Patterns of contamination by organochlorines and mercury in the eggs of two river passerines in Britain and Ireland with reference to individual PCB congeners. *Env. Pollution* **76:** 233–244.

OSBORNE, P. 1982. Some effects of Dutch elm disease on nesting farmland birds. *Bird Study* **29:** 2–16.

OWEN, M., G.L. ATKINSON-WILLES and D.G. SALMON. 1986. *Wildfowl in Great Britain.* Cambridge University Press, Cambridge. 2nd edition.

OWEN, M. and D.G. SALMON. 1988. Feral Greylag Geese *Anser anser* in Britain and Ireland, 1960–1986. *Bird Study* **35:** 37–45.

OWENS, I.P.F. 1992. *Sexual Selection in the Sex-role Reversed Dotterel, Charadrius morinellus.* Unpublished PhD thesis, Univ. of Durham.

PALMER, R.S. (ed). 1976. *Handbook of North American Birds.* Vol. III. Yale Univ. Press, New Haven, Connecticut.

PARR, R. 1980. Population study of Golden Plover *Pluvialis apricaria*, using marked birds. *Ornis Scand.* **11:** 179–189.

PARR, R. 1992. The decline to extinction of a population of Golden Plover in north-east Scotland. *Ornis Scand.* **23:** 152–158.

PARR, S.J. 1985. The breeding ecology and diet of the Hobby *Falco subbuteo* in southern England. *Ibis* **127:** 60–73.

PARRINDER, E.D. 1989. Little Ringed Plovers *Charadrius dubius* in Britain in 1984. *Bird Study* **36:** 147–153.

PARSLOW, J.L.F. 1973. *Breeding Birds of Britain and Ireland.* Poyser, Berkhamsted.

PARSLOW-OTSU, M. and G.D. ELLIOTT. 1991. Red-necked Grebe breeding in England. *Brit. Birds* **84:** 188–191.

PARTRIDGE, J.K. 1988. *The Northern Ireland Breeding Wader Survey.* Unpublished RSPB report. RSPB, Sandy.

PARTRIDGE, J.K. 1989. Lower Lough Erne's Common Scoters. *RSPB Conserv. Rev.* **3:** 25–28.

PARTRIDGE, J.K. and K.W. SMITH. 1992. Breeding wader populations in Northern Ireland, 1985–87. *Irish Birds* **4:** 497–518.

PATERSON, I.W. 1987. The status and distribution of Greylag Geese *Anser anser* in the Uists, Scotland. *Bird Study* **34:** 235–238.

PATERSON, I.W. 1991. The status and distribution of Greylag Geese *Anser anser* in the Uists (Scotland) and their impact upon crofting agriculture. *Ardea* **79:** 243–252.

PATERSON, I.W., P.R. BOYER and D.D. MASSEN. 1990. Variations in clutch size and breeding success of Greylag Geese *Anser anser* in the Uists, Scotland. *Wildfowl* **14:** 18–22.

PATTERSON, I.J. 1982. *The Shelduck.* Cambridge University Press, Cambridge.

PATTERSON, I.J., P. CAVALLINI and A. ROLANDO. 1991. Density, range size and diet of the European Jay *Garrulus glandarius* in the Maremma Natural Park, Tuscany, Italy, in summer and autumn. *Ornis Scand.* **22:** 79–87.

PEACH, W.J., S.R. BAILLIE and L. UNDERHILL. 1991. Survival of British Sedge Warblers *Acrocephalus schoenobaenus* in relation to west African rainfall. *Ibis* **133:** 300–305.

PEAKALL, D.B. 1962. The past and present status of the Red-backed Shrike. *Bird Study* **9:** 198–216.

PEAL, R.E.F. 1968. The distribution of the Wryneck in the British Isles, 1964–66. *Bird Study* **15:** 111–126.

PECK, K.M. 1989. Tree species preferences shown by foraging birds in forest plantations in northern England. *Biol. Conserv.* **48:** 41–57.

PERCIVAL, S.M. 1990. Recent trends in Barn and Tawny Owl populations in Britain. *BTO Research Report No. 57.* BTO, Tring.

PERCIVAL, S.M. 1991. Population trends in British Barn Owls – a review of some possible causes. *Brit. Wildlife* **2:** 131–140.

PERRING, F.H. and S.M. WALTERS. 1962. *Atlas of the British Flora.* London and Edinburgh.

PERRINS, C.M. 1967. The effect of beech crops on Great Tit populations and movements. *Brit Birds* **59:** 419–432.

PERRINS, C.M. 1979. *British Tits.* Collins, London.

PERRINS, C.M. 1990. Factors affecting clutch-size in Great and Blue Tits. Pp 121–130 in Blondel, J., A. Gosler, J-D Lebreton and R.H. McCleery. *Population Biology of Passerine birds.* Springer-Verlag, Berlin.

PERSSON, C. 1987. Sand Martin *Riparia riparia* populations in south-west Scandia, Sweden, 1964–1968. *J. Zool., Lond.* (B) **1:** 619–637.

PETTY, S. 1987. Breeding of Tawny Owls in relation to their food supply in an upland forest. *Proceedings of the Breeding and Management of Birds of Prey Conference, University of Bristol.*

PETTY, S.J. 1985. A study of Tawny Owls in commercial spruce forests in the uplands. *Argyll Bird Report* **2:** 70–71.

PETTY, S.J. 1989. Goshawks: their status, requirements and management. *Forestry Commission Bulletin 81.* HMSO, London.

PETTY, S.J. (ed). 1991. *The Seventh Argyll Bird Report.* Argyll Bird Club.

PHILLIPS, J.S. 1968. Stonechat breeding statistics. *Bird Study* **15:** 104.

PICOZZI, N. 1976. Hybridization of Carrion and Hooded Crows *Corvus c. corone* and *Corvus c. cornix* in Northeastern Scotland. *Ibis* **118:** 254–257.

PICOZZI, N. and D. WEIR. 1976. Dispersal and causes of death of Buzzards. *Brit. Birds* **69:** 193–201.

PIENKOWSKI, M.W. 1984. Breeding biology and population dynamics of Ringed Plovers *Charadrius hiaticula* in Britain and Greenland: nest predation as a possible factor limiting distribution and timing of breeding. *J. Zool., Lond.* **202:** 83–114.

PIENKOWSKI, M.W. and P.R. EVANS. 1982. Breeding behaviour, productivity and survival of colonial and non-colonial Shelducks *Tadorna tadorna* L. *Ornis Scand.* **13:** 101–116.

PIERSMA, T. 1986. Breeding waders in Europe: A review of population size estimates and a bibliography of information sources. *Wader Study Group Bulletin* **48,** Supplement.

PIKULSKI, A. 1986. Breeding biology and ecology of Savi's Warbler at Milicz fish-ponds (preliminary report). *Birds of Silesia* **4:** 2–39.

PINOWSKI, J. and B. PINOWSKA. 1985. The effect of the snow cover on the Tree Sparrow *Passer montanus* survival. *The Ring* **124–125:** 51–56.

PINTOS MARTIN, R. and M. RODRIGUEZ de los SANTOS. 1992. Presencia de la Malvasia Cariblanca en Andalucia: sus efectos sobre *Oxyura leucocephala. IWRB Threatened Waterfowl Research Group Newsletter* **2.**

PIROT, J.-Y., K. LAURSEN, J. MADSEN and J.-Y. MONVAL. 1989. Population estimates of swans, geese, ducks and Eurasian Coot *Fulica atra* in the Western Palearctic and Sahelian Africa. Pp 14–23 in Boyd, H. and J.-Y. Pirot (eds), *Flyways and Reserve Networks for Water Birds. IWRB Spec. Publ. 9.* Slimbridge, UK. 109pp.

PITT, R.G. 1967. Savi's Warblers breeding in Kent. *Brit. Birds* **60:** 349–355.

POOLE, A.F. 1989. *Ospreys, A Natural and Unnatural History.* Cambridge University Press, Cambridge.

PORTER, R.F. and M.A.S. BEAMAN. 1985. Resumé of raptor migration in Europe and the Middle East. In Newton, I. and R.D. Chancellor, (eds), *Conservation Studies on Raptors.* ICBP, Cambridge.

POTTS, G.R. 1969. The influence of eruptive movements, age, population size and other factors on the survival of the Shag *Phalacrocorax aristotelis. J. Anim. Ecol.* **38:** 53–102.

POTTS, G.R. 1970. Recent changes in the farmland fauna with special reference to the decline of the Grey Partridge *Perdix perdix. Bird Study* **17:** 145–166.

POTTS, G.R. 1980. The effects of modern agriculture, nest predation and game management on the population ecology of partridges *Perdix perdix* and *Alectoris rufa. Adv. Ecol. Res.* **11:** 2–79.

POTTS, G.R. 1984. Monitoring changes in the cereal ecosystem. Pp 128–134 in D. Jenkins (ed), *Agriculture and the Environment.* ITE, Cambridge.

POTTS, G.R. 1986. *The Partridge: Pesticides, Predation and Conservation.* Collins, London.

POTTS, G.R. 1989. The impact of releasing hybrid partridges on wild Red-legged Partridge populations. *Game Cons. Ann. Rev.* **20:** 81–85.

POXTON, I.R. 1986. Breeding Ring Ouzels in the Pentland Hills. *Scott. Birds* **14:** 44–48.

PRATER, A.J. 1989. Ringed Plover *Charadrius hiaticula* breeding population of the United Kingdom in 1984. *Bird Study* **36:** 154–159.

PRATER, A.J. (ed) 1990. *Orioles and Poplars in East Anglia: a Golden Opportunity.* RSPB.

PRESTT, I. and D.H. MILLS. 1966. A census of the Great Crested Grebe in Britain, 1965. *Bird Study* **13:** 163–203.

PRŶS-JONES, R.P. 1977. *Aspects of Reed Bunting Ecology, with Comparisons with the Yellowhammer.* DPhil. thesis, Oxford University.

PRŶS-JONES, R.P. 1984. Migration patterns of the Reed Bunting, *Emberiza schoeniclus schoeniclus,* and the dependence of wintering distribution on environmental conditions. *Le Gerfaut* **74:** 15–37.

PYMAN, G.A. 1959. The status of the Red-crested Pochard in the British Isles. *Brit. Birds* **52:** 42–56.

RATCLIFFE, D.A. (ed). 1977. *A Nature Conservation Review.* Vol 1. Cambridge University Press, Cambridge.

RATCLIFFE, D.A. 1976. Observations on the breeding of the Golden Plover in Great Britain. *Bird Study* **23:** 63–116.

RATCLIFFE, D.A. 1980. *The Peregrine Falcon.* Poyser, Calton.

RATCLIFFE, D.A. 1990. *Bird Life of Mountain and Upland.* Cambridge University Press, Cambridge.

RAY, J. 1678. *The Ornithology of Francis Willoughby.* John Martyn, London.

REED, T. 1985. Estimates of British breeding wader populations. *Wader Study Group Bulletin* **45:** 11–12.

REES, E.C., J.M. BLACK, C.J. SPRAY and S. THORISSON. 1991. Comparative study of the breeding success of Whooper Swans *Cygnus cygnus* nesting in upland and lowland regions of Iceland. *Ibis* **133:** 365–373.

REINERTSEN, R.E., S. HAFTORN and E. THALER. 1988. Is hypothermia necessary for the winter survival of the Goldcrest *Regulus regulus? J. Orn.* **129:** 433–437.

REYNOLDS, C.M. 1979. The heronries census: 1972–1977 population changes and a review. *Bird Study* **26:** 7–12.

REYNOLDS, J.V. 1990. The breeding gulls and terns of the islands of Lough Derg. *Irish Birds* **4:** 217–226.

RICHARDSON, M.G. 1990. The distribution and status of Whimbrel *Numenius p. phaeopus* in Shetland and Britain. *Bird Study* **37:** 61–68.

RICHNER, H. 1989. Habitat-specific growth and fitness in Carrion Crows *Corvus corone corone. J. Anim. Ecol.* **58:** 427–440.

RICHNER, H. 1990. Helpers at the nest in Carrion Crows *Corvus corone corone. Ibis* **132:** 105–108.

RIDDIFORD, N. 1983. Recent declines of Grasshopper Warblers *Locustella naevia* at British bird observatories. *Bird Study* **30:** 143–148.

RIDDIFORD, N. and P. FINDLEY. 1981. *Seasonal Movements of Summer Migrants.* BTO, Tring.

RIDGILL, S.C. and A.D. FOX. 1990. *Cold Weather Movements of Waterfowl in Western Europe. IWRB Spec. Publ. No. 13.* IWRB, Slimbridge.

RIEHM, H. 1970. Okologie and Verhalten der Schwanzmeise *Aegithalos caudatus. Zool. Jahrb. Syst.* **97:** 338–400.

ROBERTS, K.A. 1991. A report on Thames Water's tern breeding raft project. Unpublished RSPB Report.

ROBERTS, P. 1989. The numbers, distribution and movements of Choughs in Wales. Pp 9–11 in Bignal, E. and D.J. Curtis, *Choughs and Land-use in Europe.* Scottish Chough Study Group.

ROBERTS, S.J. and J.M.S. LEWIS. 1988. Observations on the sensitivity of nesting Hawfinch. *Gwent Bird Report* **23:** 7–10.

ROBERTSON, H.A. 1990. Breeding of Collared Doves *Streptopelia decaocto* in rural Oxfordshire, England. *Bird Study* **37:** 73–83.

ROBERTSON, P.A. and S.D. DOWELL. 1990. The effects of handrearing on wild gamebird populations. Pp 58–171 in Lumeij, J.T. and Y.R. Hoogeveen (eds), *The Future of Wild Galliformes in the Netherlands.* Organisatiecommissie Nederlandse Wilde Hoenders, Amersfoot, Netherlands.

ROBERTSON, P.A., M.I.A. WOODBURN, S.C. TAPPER and C. STOATE. 1989. Estimating game densities in Britain from land-use maps. Report to the Institute of Terrestrial Ecology, December 1989.

ROBINS, M. 1987. Somerset Moors breeding birds 1987. Unpublished RSPB Report, Sandy.

ROBINS, M. and C.J. BIBBY. 1985. Dartford Warblers in 1984 in Britain. *Brit. Birds* **78:** 269–280.

ROBINSON, A.H., R.D. SALE, J.L. MORRISON and P.C. MUEHRCKE. 1984. *Elements of Cartography.* 5th Edn. John Wiley and Sons, New York.

ROELL, A. 1978. The social behaviour of the Jackdaw in relation to its niche. *Behaviour* **64:** 1–125.

RUFINO, R., A. ARAÚJO and M.V. ABREU. 1985. Breeding raptors in Portugal: distribution and population estimates. Pp 15–28 in Newton, I., and R.D. Chancellor (eds), *Conservation Studies on Raptors.* ICBP, Cambridge.

RUTTLEDGE, R.F. 1987. The breeding distribution of the Common Scoter in Ireland. *Irish Birds* **3:** 417–426.

RYVES, I.N. and B.H. RYVES. 1934. The breeding habits of the Corn Bunting as observed in north Cornwall: with special reference to its polygynous habit. *Brit. Birds* **28:** 2–26, 154–164.

SAETHER, B. 1983. Mechanism of interspecific spacing out in a territorial system of the Chiffchaff, *Phylloscopus collybita*, and the Willow Warbler, *Phylloscopus trochilus*. *Ornis Scand.* **14**: 154–160.

SAGE, B.L. and J.D.R. VERNON. 1978. The 1975 national survey of rookeries. *Bird Study* **25**: 64–86.

SAGE, B.L. and P.A. WHITTINGTON. 1985. The 1980 sample survey of rookeries. *Bird Study* **32**: 77–81.

SANDEMAN, P. 1965. Attempted reintroduction of White-tailed Eagle to Scotland. *Scott. Birds* **3**: 411–412.

SAVAGE, C. 1952. *The Mandarin Duck.* A. and C. Black, London.

SCHIPPER, W.J.A. 1973. A comparison of prey selection in sympatric harriers, *Circus*, in western Europe. *Gerfaut* **63**: 17–120.

SCHLAPFER, A. 1988. Populationsokologie der Feldlerche *Alauda arvensis* in der intensiv genutzen Agrarlandschaft. *Orn. Beob.* **85**: 309–371.

SCHÖNN, S., W. SCHERZINGER, K.M. EXO and R. ILLE. 1991. *Der Steinkauz.* Athene noctua. Die Neue Brehm-Bücherei. Wittenberg Lutherstadt.

SEARS, J. and A. HUNT. 1991. Lead poisoning in Mute Swans, *Cygnus olor*, in England. Pp 383–388 in Sears, J. and P.J. Bacon (eds), *Proc. Third IWRB International Swan Symposium, Oxford, 1989. Wildfowl* Supplement No. 1.

SELLERS, R.M. 1991. Breeding and wintering status of the Cormorant in the British Isles. *Proc. workshop 1989 on Cormorants* Phalacrocorax carbo. Rijkswaterstaat Directorate Flevoland, Lelystad: 30–35.

SELLERS, R.M. 1993. Racial identity of Cormorants *Phalacrocorax carbo* breeding at the Abberton Reservoir colony, Essex. *Seabird* **15**: 45–52.

SHARROCK, J.T.R. 1974. The changing status of breeding birds in Britain and Ireland. Pp 203–220 in Hawksworth, D.L. (ed), *The Changing Flora and Fauna of Britain.* Academic Press, London and New York.

SHARROCK, J.T.R. 1976. *The Atlas of Breeding Birds in Britain and Ireland.* Poyser, Berkhamsted.

SHARROCK, J.T.R. 1986. Reed Warbler singing in oil-seed rape field. *Brit. Birds* **79**: 432.

SHARROCK, J.T.R. and O. HILDÉN. 1983. Survey of some of Europe's breeding birds. *Brit. Birds* **76**: 118–123.

SHAW, G. 1990. Timing and fidelity of breeding of Siskin *Carduelis spinus* in Scottish conifer plantations. *Bird Study* **37**: 30–35.

SHAWYER, C. 1987. *The Barn Owl in the British Isles; Its Past, Present and Future.* Hawk Trust, London.

SHRUBB, M. 1990. Effects of agricultural change on nesting Lapwings *Vanellus vanellus* in England and Wales. *Bird Study* **37**: 115–127.

SHRUBB, M. and P.C. LACK. 1991. The numbers and distribution of Lapwings *V. vanellus* nesting in England and Wales in 1987. *Bird Study* **38**: 20–37.

SIBLET, J.-Ph., O. TOSTAIN and J. Chr KOVACS. 1991. Nouveaux cas de reproduction de la Grive Litorne *Turdus pilaris* en Ile-de-France. Statut actuel dans le bassin parisien. *L. 'Oiseau et R.F.O.* **61**: 85–90.

SIMMONS, K.E.L. 1974. Adaptions in the reproductive biology of the Great Crested Grebe. *Brit. Birds* **67**: 413–437.

SIMMS, E. 1971. *Woodland Birds.* Collins, London.

SIMMS, E. 1978. *British Thrushes.* Collins, London.

SIOKHIN, V., I. CHERNICHKO and T. ARDAMATSKAYA. 1988. *Colonial-nesting waterbirds of the Ukraine.* Akademia Nauk Ukrainiskoj SSR, Kiew.

SITTERS, H.P. 1982. The decline of the Cirl Bunting in Britain, 1968–80. *Brit. Birds* **75**: 105–108.

SITTERS, H.P. 1985. Cirl Buntings in Britain in 1982. *Bird Study* **32**: 1–10.

SITTERS, H.P. 1986. Woodlarks in Britain, 1968–83. *Brit. Birds* **79**: 105–116.

SITTERS, H.P. 1988. *Tetrad Atlas of the Breeding Birds of Devon.* Devon Birdwatching and Preservation Society, Yelverton.

SITTERS, H.P. 1991. *A study of Foraging Behaviour and Parental Care in the Cirl Bunting* Emberiza cirlus. Unpublished MSc thesis, University of Aberdeen.

SMITH, K.W. 1983. The status and distribution of waders breeding on wet lowland grasslands in England and Wales. *Bird Study* **30**: 177–192.

SMITH, K.W. 1987. The ecology of the Great Spotted Woodpecker. *RSPB Conserv. Rev.* **1**: 74–77.

SMITH, R. 1991. Monitoring of breeding Snow Buntings in 1988 and 1989. Pp

112–113 in Stroud, D. and D. Glue (eds), *Britain's Birds in 1989–90: The Conservation and Monitoring Review.* BTO/NCC, Thetford.

SNOW, B.K. and D.W. SNOW. 1982. Territory and social organisation of Dunnocks *Prunella modularis*. *J. Yamashina Inst. Orn.* **14**: 281–292.

SNOW, B.K. and D.W. SNOW. 1984. Long-term defence of fruit by Mistle Thrushes *Turdus viscivorus*. *Ibis* **126**: 39–49.

SNOW, D.W. 1954. The habitats of Eurasian Tits (*Parus* species). *Ibis* **96**: 565–585.

SNOW, D.W. 1965. The relationship between census results and the breeding population of birds on farmland. *Bird Study* **12**: 287–304.

SNOW, D.W. 1969. Some vital statistics of British Mistle Thrushes. *Bird Study* **16**: 34–44.

SNOW, D.W. 1971. *The Status of Birds in Britain and Ireland.* Blackwell, Oxford for *BOU*.

SOUTHERN, N. 1970. The natural control of a population of Tawny Owls. *J. Zool., Lond.* **162**: 197–285.

SOVON 1987. *Atlas van de Nederlandse Vogels.*

SPENCER, K.G. 1973. *The Status and Distribution of Birds in Lancashire.* Turner and Earnshaw, Burnley.

SPENCER, R. 1975. Changes in the distribution of recoveries of ringed Blackbirds. *Bird Study* **22**: 177–190.

SPRAY, C.J. 1981. An isolated population of *Cygnus olor* in Scotland. *Proc. 2nd Int. Swan Symp.* Sapporo, 1980. IWRB.

STAFFORD, J. 1962. Nightjar enquiry 1957–58. *Bird Study* **9**: 104–115.

STIEFEL, A. and H. SCHEUFLER. 1984. *Der Rotschenkel.* A Ziemsen Verlag, Wittenberg, Lutherstadt.

STOWE, T.J. 1987. *The Habitat Requirements of Some Insectivorous Birds and the Management of Sessile Oakwoods.* Unpublished PhD thesis, CNAA.

STOWE, T.J. and A.V. HUDSON. 1991. Radio telemetry studies of Corncrake in Great Britain. *Die Vogelwelt* **112**: 10–16.

STROUD, D.A. 1989. Breeding waterfowl on Coll and Tiree. Pp 43–47 in Stroud, D.A. (ed), *The Birds of Coll and Tiree: Status, Habitats and Conservation.* NCC/SOC, Edinburgh.

STROUD, D.A., T.M REED and N.J. HARDING. 1990. Do moorland waders avoid plantation edges? *Bird Study* **37**: 177–186.

STROUD, D.A, T.M. REED, M.W. PIENKOWSKI and R.A. LINDSAY. 1987. *Birds, bogs and forestry. The peatlands of Caithness and Sutherland.* NCC, Peterborough.

SUGDEN, L.G. 1973. Feeding ecology of Pintail, Gadwall, American Wigeon and Lesser Scaup ducklings. *Canadian Wildlife Service Report Services No. 24.* CWS, Ottawa.

SUMMERS-SMITH, D. 1952. Breeding biology of the Spotted Flycatcher. *Brit. Birds* **45**: 153–167.

SUMMERS-SMITH, D. 1959. The House Sparrow *Passer domesticus*: population problems. *Ibis* **101**: 449–454.

SUMMERS-SMITH, J.D. 1988. *The Sparrows.* Poyser, Calton.

SUMMERS-SMITH, J.D. 1989. A history of the status of the Tree Sparrow *Passer montanus* in the British Isles. *Bird Study* **36**: 23–31.

SUTHERLAND, W.J. and G. ALLPORT. 1991. The distribution and ecology of naturalised Egyptian geese *Alopochen aegyptiacus* in Britain. *Bird Study* **38**: 128–134.

SWENNEN, C. 1989. Gull predation upon Eider *Somateria mollissima* ducklings: destruction or elimination of the unfit? *Ardea* **77**: 21–45.

SYKES, J.M., V.P.W. LOWE and D.R. BRIGGS. 1989. Some effects of afforestation on the flora and fauna of an upland sheepwalk during 12 years after planting. *J. Appl. Ecol.* **26**: 299–320.

TASKER, M.L., P.R. MOORE and R.A. SCHOFIELD. 1988. The seabirds of St Kilda, 1987. *Scott. Birds* **15**: 21–29.

TASKER, M.L., A. WEBB and J.M. MATTHEWS. 1991. A census of the large inland Common Gull colonies of Grampian. *Scott. Birds* **16**: 106–112.

TAST, J. 1968. Changes in the distribution, habitat requirements and nest sites of the Linnet, *Carduelis cannabina* in Finland. *Ann. Zool. Fennici* **5**: 159–178.

TATNER, P. 1978. A review of House Martins *Delichon urbica* in part of South Manchester, 1975. *Naturalist* **103**: 59–68.

TATNER, P. 1982. Factors influencing the distribution of Magpies *Pica pica* in an urban environment. *Bird Study* **29**: 227–234.

TATNER, P. 1983. The diet of urban Magpies *Pica pica*. *Ibis* **125**: 90–107.

TAVERNER, J.H. 1970. Mediterranean Gulls nesting in Hampshire. *Brit. Birds* **63**: 67–79.

TAYLOR, D.W. 1981a. *Birds of Kent*. Kent Ornithological Society.

TAYLOR, D.W. (ed) 1981b. *The (1979) Kent Bird Report* **28**: 78.

TAYLOR, D.W. (ed) 1982. *The (1980) Kent Bird Report* **29**: 55.

TAYLOR, I.R., A. DOWELL, T. IRVING, I.K. LANGFORD and G. SHAW. 1988. The distribution and abundance of the Barn Owl in southwest Scotland. *Scott. Birds* **15**: 40–43.

TAYLOR, K. 1984. The influence of watercourse management on Moorhen breeding biology. *Brit. Birds* **77**: 183–202.

TAYLOR, K. 1985. Great Black-backed Gull *Larus marinus* predation of seabird chicks on three Scottish islands. *Seabird* **8**: 45–52.

TAYLOR, K., R. HUDSON and G. HORNE. 1988. Buzzard breeding distribution and abundance in Britain and Northern Ireland in 1983. *Bird Study* **35**: 109–118.

TEIXEIRA, R.M. (ed) 1979. *Atlas van de Nederlandse Broedvogels*. SOVON.

TEMPERLEY, G.W. and E. BLEZZARD. 1951. Status of the Green Wood-pecker in northern England. *Brit. Birds* **44**: 24–26.

THALER, E. and K. THALER. 1982. Feeding biology of Goldcrest and Firecrest and their segregation by choice of food. *Ökol. Vögel.* **4**: 191–204.

THALER-KOTTEK, E. 1986. The behaviour of Goldcrest *Regulus regulus* and Firecrest *R. ignicapillus* etho-ecological differentiation and adaptation to habitat. *Orn. Beob.* **83**: 281–289.

THALER-KOTTEK, E. 1988. The genus *Regulus* as an example of different survival strategies: adaptation to habitat and etho-ecological differentiation. *Proc. Int. Orn. Congr.* **XIX**: 2007–2020.

THISSEN, J., G. MÜSKENS and P. OPDAM. 1981. Trends in the Dutch Goshawk *Accipiter gentilis* population and their causes. Pp 28–43 in Kenwood, R.E. and M. Lindsay (eds), *Understanding the Goshawk*. International Association for Falconry, Oxford.

THOM, V.M. 1986. *Birds in Scotland*. Poyser, Calton.

THOMAS, D.K. 1984. Aspects of habitat selection in the Sedge Warbler *Acrocephalus schoenobaenus*. *Bird Study* **31**: 187–194.

THOMPSON, D.B.A. and J. BADDELEY. 1991. Some effects of acidic deposition on montane *Rhacomitrium lanuginosum* heaths. Pp 17–28 in Woodin, S.J. and A. Farmer (eds), *The effects of acid deposition on nature conservation in Great Britain*. NCC, Peterborough.

THOMPSON, D.B.A. and A. BROWN 1992. Biodiversity in montane Britain: habitat variation, vegetation diversity and objectives for conservation. *Biodiv. and Conserv.* **1**: 179–208.

THOMPSON, D.B.A., H. GALBRAITH and D. HORSFIELD. 1987. Ecology and resources of Britain's montane plateux: land-use conflicts and issues. Pp 22–31 in Bell, M. and R.G.H. Bunce (eds), *Agriculture and Conservation in the Hills and Uplands*. NERC, Cambridge.

THOMPSON, D.B.A. and S. GRIBBIN. 1986. Ecology of Corn Buntings *Miliaria calandra* in NW England. *Bull. Brit. Ecol. Soc.* **17**: 69–75.

THOMPSON, D.B.A., J.H. MARSDEN, A.M. MACDONALD and C.G. GALBRAITH. 1993. Upland heather moorland: international importance, vegetation change and some objectives for nature conservation. *Biol. Conserv.* In press.

THOMPSON, D.B.A., D.A. STROUD and M.W. PIENKOWSKI. 1988. Effects of afforestation on upland birds: consequences for population ecology. Pp 237–259 in Usher, M.B. and D.B.A. Thompson, *Ecological change in the uplands*. Blackwell Scientific Publications, Oxford.

THOMPSON, D.B.A., P.S. THOMPSON and D. NETHERSOLE-THOMPSON. 1986. Timing of breeding and breeding performance in a population of Greenshanks *Tringa nebularia*. *J. Anim. Ecol.* **55**: 181–199.

THOMPSON, D.R., K.C. HAMER and R.W. FURNESS. 1991. Mercury accumulation in Great Skuas *Catharacta skua* of known age and sex and its effects upon breeding and survival. *J. Appl. Ecol.* **28**: 672–684.

THOMPSON, K.R. 1987. *The Ecology of the Manx Shearwater* Puffinus puffinus *on Rhum, West Scotland*. PhD thesis, University of Glasgow.

THOMPSON, P.S. 1988. Long-term trends in the use of gardens by birds. *BTO Research Report 32*, Tring.

THOMPSON, P.S. and D.B.A. THOMPSON. 1991. Greenshanks *Tringa nebularia* and long-term studies of breeding waders. *Ibis* **133**: suppl. 1: 99–112.

THOMPSON, P. and W.G. HALE. 1989. Breeding site fidelity and natal philopatry in the Redshank *Tringa totanus*. *Ibis* **131**: 214–224.

THOMPSON, P.S. and W.G. HALE. 1991. Age-related reproductive variation in the Redshank *Tringa totanus*. *Ornis Scand.* **22**: 353–359.

THOMSON, D.L. and T.W. DOUGALL. 1988. The status of breeding Wigeon in Ettrick Forest. *Scott. Birds* **15**: 61–64.

TIAINEN, J. 1983. Dynamics of a local population of the Willow Warbler, *Phylloscopus trochilus*, in southern Finland. *Ornis Scand.* **14**: 1–15.

TIAINEN, J., I.K. HANSKI, T. PAKKALA, J. PIIRROINEN and R. YRJOLA. 1989. Clutch size, nestling growth and nestling mortality of the Starling *Sturnus vulgaris* in south Finnish agroenvironments. *Ornis Fenn.* **66**: 41–48.

TIAINEN, J., M. VICKHOLM, T. PAKKALA, J. PIIROINEN, and E. VIROLAINEN. 1983. The habitat and spacial relations of breeding *Phylloscopus* warblers and the Goldcrest *Regulus regulus* in southern Finland. *Ann. Zool. Fenn.* **20**: 1–12.

TIMES BOOKS 1988. *The Times Atlas of the World* (Comprehensive Edition).

TOMIAŁOJĆ, L. (MS). Breeding ecology of Blackbirds *Turdus merula* studied in the primeval forest of Bialowieza, Poland.

TOMIAŁOJĆ, L. and J. LONTKOWSKI 1989. A technique for censusing territorial song thrushes, *Turdus philomelos*. *Ann. Zool. Fenn.* **26**: 235–243.

TRODD, P. and D. KRAMER. 1991. *Birds of Bedfordshire*. Castlemead Publications, Welwyn Garden City.

TUBBS, C.R. 1974. *The Buzzard*. David and Charles, Newton Abbot.

TURNER, A. and C. ROSE. 1989. *Swallows and Martins of the World*. Christopher Helm, London.

TURNER, A.K. 1982. Counts of aerial feeding birds in relation to pollution levels. *Bird Study* **29**: 221–226.

TYE, A. 1980. The breeding biology and population size of the Wheatear *Oenanthe oenanthe* on the Breckland of East Anglia, with implications for its conservation. *Bull. Ecol. t.* **11.3**: 559–569.

TYLER, S.J. 1972. Breeding biology of the Grey Wagtails. *Bird Study* **19**: 69–80.

TYLER, S.J. 1979. Mortality and movements of the Grey Wagtail. *Ring. and Migr.* **2**: 122–131.

TYLER, S.J. 1985. *The Wintering and Breeding Status of Goosanders in Wales*. Unpublished RSPB Report.

TYLER, S.J., J. LEWIS, A. VENABLES and J. WALTON. 1987. *The Gwent Atlas of Breeding Birds*. Gwent Ornithological Society.

TYLER, S.J. and S.J. ORMEROD. 1991. The influence of stream acidification and riparian land use on the breeding biology of Grey Wagtails *Motacilla cinerea* in Wales. *Ibis* **133**: 286–292.

TYLER, S.J. and S.J. ORMEROD. 1992. A review of the likely causal pathways relating the reduced density of breeding Dippers *Cinclus cinclus* to the acidification of upland streams. *Env. Pollution* **78**: 49–56.

UNDERHILL-DAY, J.C. 1984. Population and breeding biology of Marsh Harriers in Britain since 1900. *J. Appl. Ecol.* **21**: 773–787.

UNDERHILL-DAY, J.C. 1985. The food of breeding Marsh Harriers *Circus aeruginosus* in East Anglia. *Bird Study* **32**: 199–206.

UNDERHILL-DAY, J.C. 1989. The effect of predation by Marsh Harriers *Circus aeruginosus* on the survival of ducklings and gamebird chicks. *Ardea* **77**: 47–56.

UNDERHILL-DAY, J.C. 1990. *The Status and Breeding Biology of Marsh Harrier and Montagu's Harrier in Britain since 1900*. PhD Thesis, CNAA, London.

USSHER, R.J. and R. WARREN. 1900. *The Birds of Ireland*. London.

VAN DER HUT, R.M.G. 1986. Habitat choice and temporal differentiation in reed passerines of a Dutch marsh. *Ardea* **74**: 159–176.

VAN DIJK, J. 1975. The English Yellow Wagtail as a breeding bird in the Netherlands. *Limosa* **48**: 86–99 (In Dutch, with English summary).

VAN RHIJN, J.G. 1983. On the maintenance and origin of alternative strategies in the Ruff *Philomachus pugnax*. *Ibis* **115**: 482–498.

VAN RHIJN, J.G. 1985. Scenario for the evolution of social organisation in Ruffs *Philomachus pugnax* and other *Charadriiform* species. *Ardea* **73**: 25–37.

VEEN, J. 1977. *The Sandwich Tern: functional and causal aspects of nest distribution.* Leiden, E.J. Brill.

VENABLES, L.S.V. and U.M. VENABLES. 1955. *Birds and Mammals of Shetland.* Oliver and Boyd, Edinburgh.

VERNER, J. 1985. Assessment of counting techniques. Pp 247–295 in Johnston, R.F. (ed) *Current Ornithology* 2. Plenum Press, New York and London.

VICKERY, J.A. 1991. Breeding density of Dippers *Cinclus cinclus*, Grey Wagtails *Motacilla cinerea* and Common Sandpipers *Actitis hypoleucos* in relation to the acidity of streams in south-west Scotland. *Ibis* **133**: 178–185.

VILLAGE, A. 1981. The diet and breeding of Long-eared Owls in relation to vole numbers. *Bird Study* **28**: 215–224.

VILLAGE, A. 1990. *The Kestrel.* Poyser, London.

VOOUS, K.H. 1960. *Atlas of European Birds.* Nelson, Amsterdam and London.

VOOUS, K.H. 1977. Three lines of thought for conservation and eventual action. Pp 343–347 in Chancellor, R.D. (ed), *World Conference on Birds of Prey: Report of Proceedings, Vienna, 1975.* ICBP, Cambridge.

VOOUS, K.H. 1978. The Scottish Crossbill: *Loxia scotia. Brit. Birds* **71**: 3–10.

WALPOLE-BOND, J. 1938. *A History of Sussex Birds.* Witherby, London.

WALSH, P.M., M. AVERY and M. HEUBECK. 1990. Seabird numbers and breeding success in 1989. *NCC Chief Scientist Directorate Report* No. 1071.

WALSH, P.M., J. SEARS and M. HEUBECK. 1991. Seabird numbers and breeding success in 1990. *NCC Chief Scientist Directorate Report* No. 1235.

WALTERS DAVIES, P. and P.E. DAVIS. 1973. The ecology and conservation of the Red Kite in Wales. *Brit. Birds* **66**: 183–224, 241–270.

WANLESS, S. 1987. A survey of the numbers and breeding distribution of the North Atlantic Gannet *Sula bassana* and an assessment of the changes which have occurred since Operation Seafarer 1969/70. *Research and Survey in Nature Conservation No. 4.* NCC, Peterborough.

WANLESS, S., M.P. HARRIS and J.A. MORRIS. 1991. Foraging range and feeding locations of Shags *Phalacrocorax aristotelis* during chick rearing. *Ibis* **133**: 30–36.

WATSON, A. 1979. Bird and mammal numbers in relation to human impact at ski lifts on Scottish hills. *J. Appl. Ecol.* **16**: 753–764.

WATSON, A. 1982. Effects of human impact on Ptarmigan and Red Grouse near ski lifts in Scotland, *Ann. Rep. Inst. of Terrestrial Ecol.* for 1981, 51.

WATSON, A. 1987a. Weight and sexual cycle of Scottish Rock Ptarmigan. *Ornis Scand.* **18**: 231–232.

WATSON, A. 1987b. Review: The Partridge: pesticides, predation and conservation by G.R. Potts. *J. Appl. Ecol.* **24**: 719–720.

WATSON, A. 1988. Dotterel, *Charadrius morinellus*, numbers in relation to human impact in Scotland. *Biol. Conserv.* **43**: 254–256.

WATSON, A., S. PAYNE and R. RAE. 1989. Golden Eagles. *Aquila chrysaetos*: land use and food in northeast Scotland. *Ibis* **131**: 336–348.

WATSON, A. and R. RAE. 1987. Dotterel numbers, habitats and breeding success in Scotland. *Scott. Birds* **14**: 191–198.

WATSON, A. and R.D. SMITH. 1991. Scottish Snow Bunting numbers in summer 1970–87. *Scott. Birds* **16**: 53–56.

WATSON, D. 1977. *The Hen Harrier.* Poyser, Berkhamsted.

WATSON, J. 1992. Review of the status of the Golden Eagle in Europe. *Bird Cons. Int.* **2**: 175–183.

WATSON, J., A.F. LEITCH and R.A. BROAD. 1992. The diet of the Sea Eagle *Haliaeetus albicilla* and Golden Eagle *Aquila chrysaetos* in western Scotland. *Ibis* **134**: 27–31.

WATSON, J., S.R. RAE and R. STILLMAN. 1992. Nesting density and breeding success of Golden Eagles in relation to food supply in Scotland. *J. Anim. Ecol.* **61**: 543–550.

WAUGH, D.R. 1979. The diet of Sand Martins during the breeding season. *Bird Study* **26**: 123–128.

WEBB, A., N.M. HARRISON, G.M. LEAPER, R.D. STEELE, M.L. TASKER and M.W. PIENKOWSKI. 1990. *Seabird distribution west of Britain.* NCC, Peterborough.

WESTERHOFF, D. and C.R. TUBBS. 1991. Dartford Warblers *Sylvia undata*, their habitat and conservation in the New Forest, Hampshire, England in 1988. *Biol. Conserv.* **56**: 89–100.

WHILDE, A. 1979. Auks trapped in salmon drift-nets. *Irish Birds* **1**: 370–376.

WHILDE, A. 1984. Some aspects of the ecology of a colony of Common Gulls. *Irish Birds* **2**: 466–481.

WHILDE, A. 1985. The 1984 all Ireland tern survey. *Irish Birds* **3**: 1–32.

WHILDE, A. 1990. *Birds of Galway: a review of recent records and field studies.* Galway Branch, Irish Wildbird Conservancy, Galway.

WIKLUND, C.G. and M. ANDERSSON. 1980. Nest predation selects for colonial breeding among Fieldfares *Turdus pilaris. Ibis* **122**: 363–366.

WILLIAMSON, K. 1964. A census of breeding land birds on Hirta, St Kilda, in summer 1963. *Bird Study* **11**: 153–167.

WILLIAMSON, K. 1967. The bird community of farmland. *Bird Study* **14**: 210–226.

WILLIAMSON, K. 1968. Bird communities in the Malham Tarn region of the Pennines. *Field Studies* **68**: 651–668.

WILLIAMSON, K. 1973. Habitat of Redwings in Wester Ross. *Scott. Birds* **7**: 268–269.

WILLIAMSON, K. 1975. Birds and climatic change. *Bird Study* **22**: 143–164.

WINSTANLEY, D., R. SPENCER and K. WILLIAMSON. 1974. Where have all the Whitethroats gone? *Bird Study* **21**: 1–14.

WITHERBY, H.F., F.C.R. JOURDAIN, N.F. TICEHURST and B.W. TUCKER. 1938–41. *The Handbook of British Birds.* London. 5 vols.

WOODBURN, M.I.A. and P.A. ROBERTSON. 1990. Woodland Management for pheasants: economics and conservation effects. Pp 185–197 in Lumeij, J.T. and Y.R. Hoogeveen (eds), *The Future of Wild Galliformes in The Netherlands.* Organisatiecommissie Nederlandse Wilde Hoenders, Amersfoort, Netherlands.

WRIGHT, G.A. 1986. Breeding wildfowl of Loch Leven National Nature Reserve. *Scott. Birds* **14**: 39–43.

WRIGHT, R. and N. GILES. 1988. Breeding success of Canada and Greylag Geese *Branta canadensis* and *Anser anser* on gravel pits. *Bird Study* **35**: 31–36.

WRIGHT, R.M. and V.E. PHILLIPS. 1991. Reducing the breeding success of Canada and Greylag Geese, *Branta canadensis* and *Anser anser*, on gravel pits. *Wildfowl* **42**: 42–44.

WYLLIE, I. 1981. *The Cuckoo.* Batsford, London.

YALDEN, D.W. 1986. The habitat and activity of Common Sandpipers *Actitis hypoleucos* breeding by upland rivers. *Bird Study* **33**: 214–222

YALDEN, D.W. 1992. The influence of recreational disturbance on Common Sandpipers *Actitis hypoleucos* breeding by an upland reservoir. *Biol. Conserv.* **61**: 41–49.

YALDEN, D.W. and P.E. YALDEN. 1988. Golden Plovers and recreational disturbance. Report to NCC, Peterborough.

YAPP, W.B. 1956. The theory of line transects. *Bird Study* **3**: 93–104.

YARKER, B. and G.L. ATKINSON-WILLES. 1972. The numerical distribution of some British breeding ducks. *Wildfowl* **22**: 63–70.

YARRELL, W. 1871–74. *A History of British Birds.* 3rd edition, revised by A. Newton. London.

YARRELL, W. 1882–84. *A History of British Birds.* Revised by H. Saunders. John van Voorst, London.

YATES, B., K. HENDERSON and N. DYMOND. 1983. Red-necked Phalaropes in Britain and Ireland 1983. Unpublished RSPB Report.

ZHANG, Z.W. 1991. Pheasant densities in China. *Game Cons. Ann. Rev.* **22**: 116.

ZUCKERBOT, Y.D., U.N. SAFRIEL and U. PAZ. 1980. Autumn migration of the Quail *Coturnix coturnix* at the north coast of the Sinai Peninsula. *Ibis* **122**: 1–14.

Acknowledgements

Although only three individuals are credited on the title page of this volume, it is in reality the outcome of the efforts, both in the field and at the desk, of thousands of individuals, the great majority of them volunteers.

Undoubtedly the greatest debt of gratitude is owed to the army of amateur ornithologists, mainly members of the BTO, SOC and IWC, who between them visited a grand total of 42,736 tetrads in 3,858 10-km squares throughout Britain and Ireland. Unfortunately, the records of many individual fieldworkers were often collated on to a single recording form under one name only, so it is not possible to produce a complete listing of all those who helped in the field. Because of this, and because of the premium on space in the book, they remain, as for the *68–72 Atlas* and *Winter Atlas*, anonymous. We sincerely hope that seeing the results of their labours presented here will, at least in part, be a reward for all their efforts.

Whilst it would be invidious to single out any particular individual, a number of groups and organisations undertook more than their fair share of the fieldwork. In particular we would like to thank: the National Parks and Wildlife Service (Ireland) for allowing their staff to cover many squares in Ireland as part of their work; the Royal Air Force Ornithological Society (and particularly Squadron Leader Peter Montgomery) for field-work expeditions in 1989 and 1990 to parts of Scotland which, because of their remoteness may not otherwise have been fully surveyed; and finally the 86 members of fieldwork expeditions and 13 paid field staff who helped out in poorly covered areas during 1990 and 1991.

A number of individuals and organisations provided data from their own work, and these are listed on p 10. A glance at this list will show the debt that is owed to the RSPB and NCC/JNCC. Paul Walsh and Mark Tasker, in particular, undertook a large amount of work in merging the SCR and Atlas seabird data. Paul also wrote the first draft of the *Counts of seabird colonies* section of the *Introduction and Methods*.

Coordination of the efforts of all the fieldworkers was crucial to the success of the project, and this rôle fell to the Regional Organisers listed on pp 7–9. The ROs were the backbone of the project; their local knowledge of both birds and birdwatchers put them in a unique position to ensure that coverage was as complete as possible. In addition to those listed, Ian Herbert oversaw the first season of fieldwork in the Republic of Ireland, and Hugh Brazier helped drum up enthusiasm in the Republic during the 1988/89 winter.

Secretarial support at BTO headquarters was very capably supplied by Liz Murray, Janice Riddle, and Sophie Foulger, while Tracey Brookes was responsible for word-processing (and re-processing) the entire text through numerous drafts. Susan MacKenzie, Pat Webster and Sylvia Laing at the SOC, and Maggie Keogh and the IWC staff and volunteers at Ruttledge House provided similar support. Mike Moser, Jim Wolf, Stuart Dorward and Andy Elvin at BTO headquarters, John Davies, Alistair Peirse-Duncombe and Michael Murphy at the SOC, and Richard Nairn and Micheál O'Briain at the IWC, provided the necessary expertise to ensure the Atlas did not founder through poor administration.

The Atlas was initially sponsored by the Central Electricity Generating Board, and subsequently by its four successor companies, National Grid Company, National Power, PowerGen and Nuclear Electric. Not only did they part fund the year of pilot fieldwork, they also fully funded the next four years of work and, quite unasked, provided additional money in 1990 and 1991 to help cover the expenses of fieldwork expeditions and to employ field staff. The late Debbie Bruce (CEGB) was involved in arranging the sponsorship, and Sarah Hiscock (National Power PLC), George Barrett (PowerGen plc), Tony Robson and Tony Free (Nuclear Electric plc) were subsequently responsible for managing it, as was Stephen Hill (The National Grid Company plc) who was the link between the Atlas and its sponsors.

The NCC, and subsequently JNCC, part funded both the year of pilot fieldwork and the work involved in the estimation of population sizes, and fully funded the data processing. The NCC additionally provided substantial financial support to the SOC to ensure coverage of Scotland in 1988–90, and the Nature Conservancy Council for Scotland honoured this commitment in 1991. Mike Pienkowski, and latterly Colin Galbraith, were instrumental in the allocation of these funds.

The SOC Atlas fund-raising committee (chaired by Ian Darling) raised further funds for the organisation of the Atlas in Scotland, and the Carnegie Fund for the Universities of Scotland contributed towards the fieldwork expenses of the Scottish Organiser (JBR) in 1989 and 1990. Mike Harris, Bryan Nelson and Peter Slater assisted the Scottish Organiser raise additional personal funding.

WWF UK (World Wide Fund for Nature) part funded the year of pilot fieldwork, and subsequently negotiated the sponsorship with the CEGB and its successor companies. Sally Jones, Joy Grover, Kay Tulley, Julie Wade, Denise Laver and Chris Robertson at WWF UK were all involved at one stage or another with the negotiation and management of the sponsorship.

The IWC financed the first season of fieldwork in the Irish Republic, and subsequently covered administrative costs there. No full-time organiser was ever appointed in the Republic, and although the organiser there (RAC) was funded for the last two field seasons, nearly all of the work he undertook was in his spare time.

All data were processed by Host Data Services Ltd and, given that the processing required about 26,000,000 key depressions, and that there was great variation in the legibility of each observers handwriting, the rapidity and efficiency with which this task was undertaken was wholly remarkable. Steve Unsworth and Tony Middleton at Host Data oversaw this part of the work.

The acquisition of DMAP (the software used to produce the dot maps) revolutionised the whole project, and its developer, Alan Morton, kindly modified the program to suit the needs of the Atlas. Ian Woiwod at the Institute of Arable Crops Research at Rothamsted introduced colour mapping to the project and he, Lynda Alderson, Chris Thomas and Adrian Ball helped overcome innumerable technical problems during the production of the colour maps.

The artists – listed on p vi – and the authors of species accounts – who are credited at the bottom of the relevant texts – provided their skills and expertise free of charge. Although commissioning texts to be written by a wide variety of experts ensured that the most up-to-date and comprehensive information was incorporated into the species accounts, this inevitably led to an equally wide variety of writing styles. Bob Spencer undertook the daunting task of editing the texts into a standard format, drawing not only on his experience as a text editor, but also as an ornithologist. Trevor Poyser aided Bob in this task and performed the final edit of the entire text, particularly trimming any species accounts that were too long. Robert Gillmor provided the list of potential artists for the species vignettes, and produced the front cover and title page artwork.

The members of the Atlas Working Group (listed on pp v–vi) provided a very stimulating environment for the discussion of ideas, both of methodology and presentation, and five particular members of the group played central rôles. Humphrey Sitters (as Chairman) ably steered the group through most of its 14 meetings, whilst James Ferguson-Lees not only sat in as Chairman in Humphrey's absence, but throughout the project was able to draw on the experience from his close involvement with the two previous national Atlases. Both within and without the Working Group meetings Jeremy Greenwood and Rob Fuller were a constant source of advice and support, particularly during the final stages. Peter Lack helped both in his role as BTO's Computer Officer and from the experience gained as organiser of the *Winter Atlas*.

Tim Sharrock, David Scott and Paul Hillis helped locate a few lost confidential records from the *68–72 Atlas*, and the members of the RBBP (listed on p 12) advised on the best manner of presentation of maps of sensitive species. Paul Hillis further helped here in providing an Irish perspective, and Dick Roxburgh, Ray Murray and John Callion gave advice on the treatment of confidential records of particular species.

The estimation of the sizes of breeding populations in Britain and Ireland led to more sleepless nights than any other stage of the project and was a minefield of technical problems which were very capably solved by Simon Gates. The territory density data on which the estimates were based were provided by CBC fieldworkers, and collated by members of the BTO's Census Unit, particularly John Marchant. Su Gough produced Appendices B and D, Fig. 1 and all the graphs in the text, in addition to editing the Reference list.

All maps and text presented here have been reproduced electronically, and Chris Gibson at the publishers provided the necessary technical support to overcome the inevitable complexities. Sutapas Bhattacharya acted as Production Editor, Debra Kruse as Manufacturing Manager, and Andy Richford (as Senior Editor and subsequently Publisher) oversaw the publication process from beginning to end. Andy is to be thanked particularly for his patience during the unavoidably protracted final stages of manuscript preparation.

Finally, we should like to thank all those members of the BTO, SOC and IWC who provided us with such warm hospitality during our travels throughout Britain and Ireland.

David Gibbons
Jim Reid
Bob Chapman

Index of birds mentioned in the text

Vernacular names are indexed under the last word, and scientific nomenclature under the generic name. The principal references to all the main species can easily be identified because they are given adjacent page numbers hyphenated; other references are indexed only where a significant point is stated or mapped.

Accipiter gentilis, 108–109, 477
   nisus, 110–111, 246, 250, 392, 426, 465
Acrocephalus arundinaceus, 245
   palustris, 332–333, 478
   schoenobaenus, 244, 324, 330–331, 334, 478
   scirpaceus, 244, 245, 296, 324, 328, 330, 332, 334–335, 476, 478
Actitis hypoleucos, 192–193
   macularia, 476, 477
Aegithalos caudatus, 362–363
Agapornis roseicollis, 497
Aix galericulata, 62, 64–65, 476, 477
   sponsa, 62–63, 476
Alauda arvensis, 106, 120, 122, 272–273, 282, 451, 457
Albatross, Black-browed, 495
Alca torda, 10, 11, 226–227
Alcedo atthis, 260–261, 451
Alectoris chukar, 134, 135, 496
   graeca, 134
   rufa, 134–135, 476
Alopochen aegyptiacus, 58–59, 476, 477
Anas acuta, 74–75
   clypeata, 78–79
   crecca, 70–71, 477
   penelope, 66–67, 477
   platyrhynchos, 72–73, 85, 465
   querquedula, 76–77, 476
   strepera, 68–69, 477
Anser albifrons, 496
   anser, 54–55, 477
   brachyrhynchus, 496
   caerulescens, 496
   canagicus, 496
   cygnoides, 496
   indicus, 496
Anthus petrosus, 284–285
   pratensis, 106, 120, 244, 245, 282–283, 284, 296, 335, 451, 457, 465, 475
   trivialis, 244, 280–281, 476, 478
Apus apus, 2, 122, 258–259, 278
Aquila chrysaetos, 12, 18, 22, 100, 114–115, 477
Ardea cinerea, 2, 17, 46, 48–49, 451, 492
Arenaria interpres, 497
Asio flammeus, 254–255, 476
   otus, 252–253
Athene noctua, 248–249, 465, 476
Avocet, 158–159
Aythya collaris, 496

   ferina, 80–81
   fuligula, 66, 80, 82–83, 451
   marila, 441–442

Bittern, 46–47, 477
   Little, 476, 477, 480
Blackbird, 290, 312, 314–315, 318, 323, 458, 482
Blackcap, 342, 344–345, 477
Bluethroat, 476, 498
Bombycilla garrulus, 497
Botaurus stellaris, 46–47, 477
Brambling, 445
Branta bernicla, 496
   canadensis, 56–57, 441, 476, 477
   leucopsis, 440–441, 476
Bubo bubo, 497
Bucephala clangula, 18, 88–89, 477, 478
Budgerigar, 476, 497
Bullfinch, 426–427
Bunting, Cirl, 10, 12, 17, 434–435, 477
   Corn, 426, 436, 438–439, 477, 478
   Lapland, 476, 477, 480
   Ortolan, 498
   Reed, 436–437, 478
   Snow, 430–431
Burhinus oedicnemus, 10, 160–161
Buteo buteo, 18, 58, 98, 112–113, 477, 478
   lagopus, 496
Buzzard, 18, 58, 98, 112–113, 477, 478
   Honey, 96–97
   Rough-legged, 496

Cairina moschata, 496
Calcarius lapponicus, 476, 477, 480
Calidris alba, 497
   alpina, 5, 18, 172–173, 394, 478
   canutus, 497
   maritima, 443–444, 476, 477
   minuta, 497
   temminckii, 442–443, 443
Capercaillie, 132–133, 476, 477
Caprimulgus europaeus, 256–257, 477, 478
Carduelis cannabina, 416–417, 426, 436, 451
   carduelis, 12, 412–413, 451
   chloris, 410–411, 445
   flammea, 420–421
   flavirostris, 418–419, 477, 478

   spinus, 414–415, 477
Carpodacus erythrinus, 446–447, 476, 477
Catharacta maccormicki, 198
   skua, 10, 183, 198–199, 478
Cepphus grylle, 10, 11, 228–229
Certhia brachydactyla, 18, 378, 444–445
   familiaris, 378–379, 444, 465
Cettia cetti, 324–325, 476, 477, 478
Chaffinch, 11, 408–409, 445, 461, 464
Charadrius alexandrinus, 476, 480
   dubius, 162–163, 477, 478
   hiaticula, 164–165, 482
   morinellus, 10, 12, 166–167, 194, 476, 477
Chiffchaff, 348–349
Chlidonias leucopterus, 497
   niger, 476, 497
Chough, 18, 388–389
Chrysolophus amherstiae, 142, 144–145, 476
   pictus, 142–143, 144, 476
Chukar, 134, 135, 496
Cinclus cinclus, 288, 292–293
Circus aeruginosus, 4, 102–103, 360, 477
   cyaneus, 10, 104–105, 474
   pygargus, 106–107, 476
Clangula hyemalis, 496
Coccothraustes coccothraustes, 428–429, 477
Colinus virginianus, 476, 496
Columba livia, 232–233, 478
   oenas, 234–235
   palumbus, 108, 236–237, 456, 457
Coot, 154–155, 451
Cormorant, 2, 10, 11, 42–43, 44, 230, 492
Corncrake, 10, 12, 17, 136, 150–151, 477, 478
Corvus corvus cornix, 185, 394–399
   corvus corone × corvus cornix, 394–399
   corax, 124, 185, 400–401, 451, 477
   corone corone, 391, 394–399, 456, 460
   frugilegus, 2, 392–393, 451, 460, 465, 478
   monedula, 65, 88, 242, 390–391, 460, 478
Coturnix coturnix, 10, 138–139, 151, 494
Crake, Spotted, 148–149
Crane, 442, 476, 477
Crex crex, 10, 12, 17, 136, 150–151, 477, 478
Crossbill, Common, 2, 422–425, 446, 476, 477
   Parrot, 422, 446, 476
   Scottish, 422–425, 446, 476
   Two-barred, 498
Crow, Carrion, 391, 394–399, 456, 460

Hooded, 185, 394–399
Hybrid, 394–399
Cuckoo, 244–245, 282, 284, 296, 332, 335, 465, 478
Cuculus canorus, 244–245, 282, 284, 296, 332, 335, 465, 478
Curlew, 156, 182, 184–185, 465
Stone, 10, 160–161
Cygnus atratus, 495
columbianus, 52
cygnus, 52–53, 198, 476, 477
olor, 10, 50–51, 52, 451

Delichon urbica, 2, 278–279, 404
Dendrocopos major, 266–267, 268, 465
minor, 268–269
Diomedea melanophris, 495
Dipper, 288, 292–293
Diver, Black-throated, 10, 20, 22–23, 476
Great Northern, 476, 495
Red-throated, 18, 20–21, 22
Dotterel, 10, 12, 166–167, 194, 476, 477
Dove, Collared, 238–239, 478
Rock, 232–233, 478
Stock, 234–235
Turtle, 106, 238, 240–241, 476, 478
Dryocopus martius, 88
Duck, Long-tailed, 496
Mandarin, 62, 64–65, 476, 477
Muscovy, 496
Ring-necked, 496
Ruddy, 18, 94–95, 476, 477
Tufted, 66, 80, 82–83, 451
White-headed, 94
Wood, 62–63, 476
Dunlin, 5, 18, 172–173, 394, 478
Dunnock, 244, 245, 296–297, 335, 458

Eagle, Golden, 12, 18, 22, 100, 114–115, 477
White-tailed, 100–101, 476, 477
Egret, Little, 495
Egretta garzetta, 495
Eider, 84–85, 86, 496
King, 496
Emberiza cirlus, 10, 12, 17, 434–435, 477
citrinella, 432–433, 434, 436, 451, 461
hortulana, 498
schoeniclus, 436–437, 478
Eremophila alpestris, 476, 477, 480
Erithacus rubecula, 1, 245, 298–299, 456, 458, 462

Falco columbarius, 120–121, 451
eleonorae, 122
peregrinus, 10, 18, 124–125, 477
subbuteo, 122–123, 477
tinnunculus, 4, 110, 118–119, 180, 222, 235, 248, 250, 465
vespertinus, 496

Falcon, Eleonora's, 122
Red-footed, 496
Ficedula hypoleuca, 304, 305, 358–359, 376, 476
Fieldfare, 316–317, 476
Firecrest, 354–355, 476, 477
Flycatcher, Pied, 304, 305, 358–359, 376, 476
Spotted, 2, 356–357
Fratercula arctica, 10, 11, 196, 197, 226, 230–231
Fringilla coelebs, 11, 408–409, 445, 461, 464
montifringilla, 445
Fulica atra, 154–155, 451
Fulmar, 2, 10, 32–33, 100
Fulmarus glacialis, 2, 10, 32–33, 100

Gadwall, 68–69, 477
Gallinago gallinago, 176–177
Gallinula chloropus, 152–153, 154
Gannet, 10, 40–41, 198
Garganey, 76–77, 476
Garrulus glandarius, 362, 384–385, 429
Gavia arctica, 10, 20, 22–23, 476
immer, 476, 495
stellata, 18, 20–21, 22
Godwit, Bar-tailed, 497
Black-tailed, 180–181
Goldcrest, 352–353, 354
Goldeneye, 18, 88–89, 477, 478
Goldfinch, 12, 412–413, 451
Goosander, 12, 88, 90, 92–93
Goose, Bar-headed, 496
Barnacle, 440–441, 476
Brent, 496
Canada, 56–57, 441, 476, 477
Egyptian, 58–59, 476, 477
Emperor, 496
Greylag, 54–55, 477
Pink-footed, 496
Snow, 496
Swan, 496
White-fronted, 496
Goshawk, 108–109, 477
Grebe, Black-necked, 30–31, 476
Great Crested, 12, 26–27, 440, 474
Little, 12, 24–25, 451
Red-necked, 440, 476
Slavonian, 28–29
Greenfinch, 410–411, 445
Greenshank, 188–189, 476
Grouse, Black, 130–131
Red, 114, 120, 126–127, 129, 451, 477, 478
Grus grus, 442, 476, 477
Guillemot, 10, 11, 212, 224–225, 226, 227
Black, 10, 11, 228–229
Guineafowl, Helmeted, 496
Gull, Black-headed, 2, 10, 11, 30, 159, 200, 202–203, 214, 478, 492
Common, 2, 10, 183, 200, 204–205, 478, 493
Glaucous, 497

Great Black-backed, 2, 10, 198, 210–211
Herring, 2, 10, 84, 202, 206, 207, 208–209, 230, 493
Iceland, 497
Lesser Black-backed, 2, 10, 206–207, 493
Little, 476, 477, 497
Mediterranean, 200–201
Ring-billed, 497

Haematopus ostralegus, 156–157, 451, 478
Haliaeetus albicilla, 100–101, 476, 477
Harrier, Hen, 10, 104–105, 474
Marsh, 4, 102–103, 360, 477
Montagu's, 106–107, 476
Hawfinch, 428–429, 477
Heron, Grey, 2, 17, 46, 48–49, 451, 492
Himantopus himantopus, 476, 480
Hippolais icterina, 332, 498
Hirundo rustica, 4, 258, 276–277, 278, 350, 457, 478
Hobby, 122–123, 477
Hoopoe, 476, 497
Hydrobates pelagicus, 10, 36–37, 38, 198

Ixobrychus minutus, 476, 477, 480

Jackdaw, 65, 88, 242, 390–391, 460, 478
Jay, 362, 384–385, 429
Jynx torquilla, 16, 262–263, 477

Kestrel, 4, 110, 118–119, 180, 222, 235, 248, 250, 465
Kingfisher, 260–261, 451
Kite, Red, 4, 98–99, 400
Kittiwake, 10, 11, 196, 197, 212–213
Knot, 497

Lagopus lagopus, 126
lagopus scoticus, 114, 120, 126–127, 129, 451, 477, 478
mutus, 128–129
Lanius collurio, 16, 382–383, 477, 478
excubitor, 498
Lapwing, 156, 170–171, 391, 465
Lark, Shore, 476, 477, 480
Larus argentatus, 2, 10, 84, 202, 206, 207, 208–209, 230, 493
canus, 2, 10, 183, 200, 204–205, 478, 493
delawarensis, 497
fuscus, 2, 10, 206–207, 493
glaucoides, 497
hyperboreus, 497
marinus, 2, 10, 198, 210–211
melanocephalus, 200–201
minutus, 476, 477, 497
ridibundus, 2, 10, 11, 30, 159, 200, 202–203, 214, 478, 492

Limosa lapponica, 497
 limosa, 180–181
Linnet, 416–417, 426, 436, 451
Locustella luscinioides, 328–329, 476, 477, 478
 naevia, 326–327, 328
Lovebird, Peach-faced, 497
Loxia curvirostra, 2, 422–425, 446, 476, 477
 leucoptera, 498
 pytyopsittacus, 422, 446, 476
 scotica, 422–425, 446, 476
Lullula arborea, 270–271, 477
Luscinia megarhynchos, 300–301, 451, 478
 svecica, 476, 498
Lymnocryptes minimus, 497

Magpie, 82, 236, 318, 386–387, 440, 460
Mallard, 72–73, 85, 465
Mandarin, 62, 64–65, 476, 477
Martin, House, 2, 278–279, 404
 Sand, 2, 17, 274–275, 278, 340, 404, 478, 494
Melanitta fusca, 496
 nigra, 5, 86–87
Melopsittacus undulatus, 476, 497
Merganser, Red-breasted, 90–91
Mergus albellus, 496
 merganser, 12, 88, 90, 92–93
 serrator, 90–91
Merlin, 120–121, 451
Miliaria calandra, 426, 436, 438–439, 477, 478
Milvus milvus, 4, 98–99, 400
Moorhen, 152–153, 154
Morus bassanus, 10, 40–41, 198
Motacilla alba alba, 290–291, 492, 495
 alba yarrellii, 290–291, 492
 cinerea, 288–289
 flava flava, 492, 495
 flava flavissima, 286–287, 476, 492
Muscicapa striata, 2, 356–357
Myiopsitta monachus, 497

Netta rufina, 441, 476
Nightingale, 300–301, 451, 478
Nightjar, 256–257, 477, 478
Numenius arquata, 156, 182, 184–185, 465
 phaeopus, 182–183, 478
Numida meleagris, 496
Nuthatch, 376–377, 465
Nyctea scandiaca, 444, 476

Oceanodroma leucorhoa, 10, 38–39, 476
Oenanthe oenanthe, 310–311
Oriole, Golden, 10, 332, 380–381, 477
Oriolus oriolus, 10, 332, 380–381, 477
Osprey, 116–117, 477, 478
Ouzel, Ring, 312–313, 482
Owl, Barn, 17, 235, 246–247, 248, 250, 477, 478
 Eagle, 497

Little, 248–249, 465, 476
Long-eared, 252–253
Short-eared, 254–255, 476
Snowy, 444, 476
Tawny, 88, 235, 248, 250–251, 252, 405, 451, 465
Oxyura jamaicensis, 18, 94–95, 476, 477
 leucocephala, 94
Oystercatcher, 156–157, 451, 478

Pandion haliaetus, 116–117, 477, 478
Panurus biarmicus, 360–361, 476, 477, 478
Parakeet, Monk, 497
 Ring-necked, 242–243, 476, 477
Partridge, Grey, 134, 135, 136–137, 240, 477, 478, 480
 Red-legged, 134–135, 476
 Rock, 134
Parus ater, 364, 370–371
 caeruleus, 5, 364, 370, 372–373, 375, 394, 451, 459
 cristatus, 10, 366, 368–369
 major, 5, 364, 370, 372, 374–375, 376, 459
 montanus, 364, 366–367, 370
 palustris, 364–365, 366, 367, 370
Passer domesticus, 248, 268, 404–405, 406, 461, 465
 montanus, 406–407, 426, 436, 451
Perdix perdix, 134, 135, 136–137, 240, 477, 478, 480
Peregrine, 10, 18, 124–125, 477
Pernis apivorus, 96–97
Petrel, Leach's, 10, 38–39, 476
 Storm, 10, 36–37, 38, 198
Phalacrocorax aristotelis, 10, 44–45, 230
 carbo, 2, 10, 11, 42–43, 44, 230, 492
Phalarope, Red-necked, 18, 194–195, 477, 478
Phalaropus lobatus, 18, 194–195, 477, 478
Phasianus colchicus, 103, 140–141, 142, 465, 476
Pheasant, 103, 140–141, 142, 465, 476
 Golden, 142–143, 144, 476
 Lady Amherst's, 142, 144–145, 476
 Reeves', 496
Philomachus pugnax, 174–175, 476
Phoenicurus ochruros, 302–303
 phoenicurus, 245, 304–305
Phylloscopus collybita, 348–349
 sibilatrix, 281, 304, 346–347, 451, 477
 trochilus, 348, 349, 350–351, 459
Pica pica, 82, 236, 318, 386–387, 440, 460
Picus viridis, 264–265, 465
Pigeon, Feral, 232–233, 478
Pintail, 74–75
Pipit, Meadow, 106, 120, 244, 245, 282–283, 284, 296, 335, 451, 457, 465, 475
 Rock, 284–285
 Tree, 244, 280–281, 476, 478
Platalea leucorodia, 477, 495
Plectrophenax nivalis, 430–431

Ploceus melanocephalus, 498
Plover, Golden, 168–169, 173
 Grey, 497
 Kentish, 476, 480
 Little Ringed, 162–163, 477, 478
 Ringed, 164–165, 482
Pluvialis apricaria, 168–169, 173
 squatarola, 497
Pochard, 80–81
 Red-crested, 441, 476
Podiceps auritus, 28–29
 cristatus, 12, 26–27, 440, 474
 grisegena, 440, 476
 nigricollis, 30–31, 476
Porzana porzana, 148–149
Prunella modularis, 244, 245, 296–297, 335, 458
Psittacula krameri, 242–243, 476, 477
Ptarmigan, 128–129
 Willow, 126
Puffin, 10, 11, 196, 197, 226, 230–231
Puffinus puffinus, 10, 34–35
Pyrrhocorax pyrrhocorax, 18, 388–389
Pyrrhula pyrrhula, 426–427

Quail, 10, 138–139, 151, 494
 Bobwhite, 476, 496

Rail, Water, 17, 146–147
Rallus aquaticus, 17, 146–147
Raven, 124, 185, 400–401, 451, 477
Razorbill, 10, 11, 226–227
Recurvirostra avosetta, 158–159
Redpoll, 420–421
Redshank, 186–187
 Spotted, 497
Redstart, 245, 304–305
 Black, 302–303
Redwing, 320–321
Reeve, 174–175
Regulus ignicapillus, 354–355, 476, 477
 regulus, 352–353, 354
Remiz pendulinus, 477, 498
Riparia riparia, 2, 17, 274–275, 278, 340, 404, 478, 494
Rissa tridactyla, 10, 11, 196, 197, 212–213
Robin, 1, 245, 298–299, 456, 458, 462
Rook, 2, 392–393, 451, 460, 465, 478
Rosefinch, Scarlet, 446–447, 476, 477
Ruff, 174–175, 476

Sanderling, 497
Sandpiper, Common, 192–193
 Green, 497
 Purple, 443–444, 476, 477
 Spotted, 476, 477
 Wood, 190–191, 443, 476, 477
Saxicola rubetra, 306–307, 308
 torquata, 307, 308–309
Scaup, 441–442

Scolopax rusticola, 17, 178–179, 451, 477, 478
Scoter, Common, 5, 86–87
   Velvet, 496
Serin, 445–446, 476
Serinus serinus, 445–446, 476
Shag, 10, 44–45, 230
Shearwater, Manx, 10, 34–35
Shelduck, 58, 60–61, 482
   Ruddy, 496
Shoveler, 78–79
Shrike, Great Grey, 498
   Red-backed, 16, 382–383, 477, 478
Siskin, 414–415, 477
Sitta europaea, 376–377, 465
Skua, Arctic, 10, 11, 183, 196–197, 198, 478
   Great, 10, 183, 198–199, 478
   South Polar, 198
Skylark, 106, 120, 122, 272–273, 282, 451, 457
Smew, 496
Snipe, 176–177
   Jack, 497
Somateria mollissima, 84–85, 86, 496
   spectabilis, 496
Sparrow, House, 248, 268, 404–405, 406, 461, 465
   Tree, 406–407, 426, 436, 451
Sparrowhawk, 110–111, 246, 250, 392, 426, 465
Spoonbill, 477, 495
Starling, 248, 376, 394, 402–403, 456, 461
Stercorarius parasiticus, 10, 11, 183, 196–197, 198, 478
Sterna albifrons, 10, 18, 214, 218, 222–223
   bengalensis, 497
   dougallii, 10, 216–217, 477
   hirundo, 2, 10, 214, 218–219, 220, 478, 493
   paradisaea, 2, 10, 183, 197, 214, 218, 220–221
   sandvicensis, 10, 214–215, 218, 497
Stilt, Black-winged, 476, 480
Stint, Little, 497
   Temminck's, 442–443, 443
Stonechat, 307, 308–309
Streptopelia decaocto, 238–239, 478
   turtur, 106, 238, 240–241, 476, 478
Strix aluco, 88, 235, 248, 250–251, 252, 405, 451, 465
Sturnus vulgaris, 248, 376, 394, 402–403, 456, 461
Swallow, 4, 258, 276–277, 278, 350, 457, 478
Swan, Bewick's, 52
   Black, 495

Mute, 10, 50–51, 52, 451
   Whooper, 52–53, 198, 476, 477
Swift, 2, 122, 258–259, 278
Sylvia atricapilla, 342, 344–345, 477
   borin, 245, 342–343, 344
   cantillans, 498
   communis, 327, 338, 340–341, 342
   curruca, 338–339, 476, 478
   undata, 336–337
Syrmaticus reevesii, 496

Tachybaptus ruficollis, 12, 24–25, 451
Tadorna ferruginea, 496
   tadorna, 58, 60–61, 482
Teal, 70–71, 477
Tern, Arctic, 2, 10, 183, 197, 214, 218, 220–221
   Black, 476, 497
   Common, 2, 10, 214, 218–219, 220, 478, 493
   Lesser Crested, 497
   Little, 10, 18, 214, 218, 222–223
   Roseate, 10, 216–217, 477
   Sandwich, 10, 214–215, 218, 497
   White-winged Black, 497
Tetrao tetrix, 130–131
   urogallus, 132–133, 476, 477
Thrush, Mistle, 2, 322–323, 478
   Song, 318–319, 320, 459
Tit, Bearded, 360–361, 476, 477, 478
   Blue, 5, 364, 370, 372–373, 375, 394, 451, 459
   Coal, 364, 370–371
   Crested, 10, 366, 368–369
   Great, 5, 364, 370, 372, 374–375, 376, 459
   Long-tailed, 362–363
   Marsh, 364–365, 366, 367, 370
   Penduline, 477, 498
   Willow, 364, 366–367, 370
Treecreeper, 378–379, 444, 465
   Short-toed, 18, 378, 444–445
Tringa erythropus, 497
   glareola, 190–191, 443, 476, 477
   nebularia, 188–189, 476
   ochropus, 497
   totanus, 186–187
Troglodytes troglodytes, 294–295, 456, 458
Turdus iliacus, 320–321
   merula, 290, 312, 314–315, 318, 323, 458, 482
   philomelos, 318–319, 320, 459
   pilaris, 316–317, 476

   torquatus, 312–313, 482
   viscivorus, 2, 322–323, 478
Turnstone, 497
Twite, 418–419, 477, 478
Tyto alba, 17, 235, 246–247, 248, 250, 477, 478

Upupa epops, 476, 497
Uria aalge, 10, 11, 212, 224–225, 226, 227

Vanellus vanellus, 156, 170–171, 391, 465

Wagtail, Blue-headed, 492, 495
   Grey, 288–289
   Pied, 290–291, 492
   White, 290–291, 492, 495
   Yellow, 286–287, 476, 492
Warbler, Cetti's, 324–325, 476, 477, 478
   Dartford, 336–337
   Garden, 245, 342–343, 344
   Grasshopper, 326–327, 328
   Great Reed, 245
   Icterine, 332, 498
   Marsh, 332–333, 478
   Reed, 244, 245, 296, 324, 328, 330, 332, 334–335, 476, 478
   Savi's, 328–329, 476, 477, 478
   Sedge, 244, 324, 330–331, 334, 478
   Subalpine, 498
   Willow, 348, 349, 350–351, 459
   Wood, 281, 304, 346–347, 451, 477
Waxwing, 497
Weaver, Black-headed, 498
Wheatear, 310–311
Whimbrel, 182–183, 478
Whinchat, 306–307, 308
Whitethroat, 327, 338, 340–341, 342
   Lesser, 338–339, 476, 478
Wigeon, 66–67, 477
Woodcock, 17, 178–179, 451, 477, 478
Woodlark, 270–271, 477
Woodpecker, Black, 88
   Great Spotted, 266–267, 268, 465
   Green, 264–265, 465
   Lesser Spotted, 268–269
Woodpigeon, 108, 236–237, 456, 457
Wren, 294–295, 456, 458
Wryneck, 16, 262–263, 477

Yellowhammer, 432–433, 434, 436, 451, 461

The British Trust for Ornithology

Founded in 1933, the British Trust for Ornithology has throughout its history been deeply committed to the idea of membership participation in the many surveys it organises. Often it is difficult, or even impossible, to evaluate the findings of a single observer, but let 500 or 1,000 pool their results and the broad picture starts to emerge. There could be no more convincing demonstration of this than the maps which form this Atlas.

Among its many functions, the BTO administers the national Bird Ringing Scheme and organises the Common Birds Census, designed to detect changes in the population levels of commoner species breeding in Britain. For similar reasons, the BTO organises the Heronries Census and, with the Wildfowl and Wetlands Trust, regularly censuses wetland and estuary birds, while the Nest Record Scheme measures, year by year, the breeding success of Britain's birds.

Thus, the theme common to all BTO studies is that the observations of individual members, guided and co-ordinated by a small research staff, supply information vital to the conservation of Britain's resident and migrant bird populations. A measure of this usefulness is that much of the research is done at the request of, and with financial support from, the Joint Nature Conservation Committee.

The BTO also undertakes many surveys funded by British conservation bodies in order to provide these groups with the exact data needed to conserve birds and their habitats. Studies are also carried out under contract to government or other bodies, often in the form of impact assessments for a proposed industrial or recreational development.

The members of the BTO come from all walks of life, and the keen beginner is as welcome as the expert. Members enjoy access to the postal library operated by the BTO, which is unparalleled in any country, while the many national and local conferences it organises are a happy amalgam of information and enthusiasm.

For details of membership please write to: Membership Secretary, BTO, National Centre for Ornithology, The Nunnery, Thetford, Norfolk IP24 2PU (telephone 0842 750050).

The Scottish Ornithologists' Club

Founded in 1936, the Scottish Ornithologists' Club has served as a focus for ornithology in Scotland for well over 50 years. Its national headquarters in Edinburgh serves 14 separate branches located throughout Scotland from Orkney to Stranraer. Members can avail themselves of a full programme of evening talks at branches during the winter and outings and other events throughout the year. Those members not living in Scotland – of whom there are many all over the world – are kept in touch with developments in the Scottish bird scene through the quarterly newsletter *Scottish Bird News*, the internationally known bi-annual journal *Scottish Birds*, and the annual *Scottish Bird Report*; many travel considerable distances to attend the annual weekend conference which brings together members for a couple of days of talks, outings and general social relaxation in congenial surroundings.

The Club also encourages serious research by its members, its endowment fund being continuously used to support various projects which widen our knowledge of Scottish birds. The contribution of data for this Atlas and its two predecessors is but one aspect of this work, which also illustrates the close cooperation the Club maintains with the BTO, a relationship underlined by the annual one-day Scottish Birdwatchers' Conference jointly hosted by the two organisations.

The Edinburgh headquarters also houses the Waterston Library, the largest ornithological library in Scotland and one of the finest in the United Kingdom, of which extensive use is made both by amateur birdwatchers and professional biologists. Both are equally welcome as members: one of the traditional strengths of the Club is the way in which it brings together the keen beginner and the professional expert. For membership details contact the Scottish Ornithologists' Club, 21 Regent Terrace, Edinburgh EH7 5BT (telephone 031 556 6042).

The Irish Wildbird Conservancy

The Irish Wildbird Conservancy is the largest independent conservation organisation in Ireland. Established in 1968, following the merger of the Irish Society for the Protection of Birds, the Irish Ornithologists' Club and the Irish Wildfowl Conservancy, the IWC is a recognised charity supported almost entirely by membership subscriptions and donations. It has over 4,500 members – from the most active field ornithologists to those who simply enjoy the beauty of birds in their gardens.

The IWC owns or manages an increasing number of reserves which protect threatened habitats for birds. Among the best known are the Wexford Wildfowl Reserve and Little Skellig. Species which have special conservation requirements, such as Corncrake, Chough and Roseate Tern, have been the subject of recent surveys and acquisitions. In 1992, the IWC acquired a Corncrake reserve on the Shannon Callows. It also helps manage Rockabill Island, the home of Europe's largest Roseate Tern colony.

Research and surveys of birds and their habitats form the basis of the IWC's conservation policies. Surveys are carried out on a voluntary basis by IWC members or, on a professional basis, with government agencies and other conservation organisations. Cooperation with the BTO is close, especially for joint projects such as the Breeding and Winter Atlases.

The IWC has a network of over 20 branches nationwide, running annual programmes of outings, film shows and lectures, in addition to various local and nationwide campaigns for conservation, recruitment and fundraising. In addition, the IWC's journal *Irish Birds*, including the Irish Bird Report, is produced annually and its magazine, the *IWC News*, is circulated to all members quarterly.

For more details on how to support or join the IWC, contact the Irish Wildbird Conservancy at Ruttledge House, 8 Longford Place, Monkstown, Co. Dublin. Telephone (-353-1-) 2804322/2808237. Fax (-353-1-) 2844407.

Life Cycle of a

Frog

Angela Royston

Heinemann
LIBRARY

First published in Great Britain by Heinemann Library
Halley Court, Jordan Hill, Oxford OX2 8EJ
a division of Reed Educational and Professional Publishing Ltd
Heinemann is a registered trademark of Reed Educational and Professional
Publishing Limited.

Oxford   Florence   Prague   Madrid   Athens   Melbourne
Auckland   Kuala Lumpur   Singapore   Tokyo   Ibadan
Nairobi   Kampala   Johannesburg   Gaborone   Portsmouth NH
Chicago   Mexico City   São Paulo

Designed by Celia Floyd
Illustrations by Alan Fraser
Printed in Hong Kong / China

02 01 00
10 9 8 7 6 5 4 3

ISBN 0 431 08362 2

British Library Cataloguing in Publication Data

Royston, Angela
    Life cycle of a frog
    1.Frogs - Juvenile literature
    I.Title II.Frog
    597.8'9

Acknowledgements
The Publisher would like to thank the following for permission to reproduce
photographs:
Bruce Coleman Ltd/Hans Reinhard p21; Bruce Coleman Ltd/Jane Burton p14; Bruce
Coleman Ltd/Kim Taylor pp20, 23; Bruce Coleman Ltd/William
S Paton p22; Natural Science Photos/O C Roura p13; Natural Science
Photos/Richard Revels pp6, 11; Natural Science Photos/Ward p.18; NHPA/David
Woodfall p27; NHPA/G I Bernard p15; NHPA/Melvin Grey p26; NHPA/Stephen
Dalton p4; OSF pp5, 10; OSF/David Thompson p9; OSF/G I Bernard pp7, 16, 24;
OSF/Paul Franklin pp8, 12, 17; OSF/Stephen Dalton p19; OSF/Terry Heathcote p25.

Cover photograph: Colin Varndell/Bruce Coleman Ltd.

Contents

	page
Meet the frogs	4
A mass of eggs	6
Hatching	8
1–4 weeks	10
5 weeks	12
5–12 weeks	14
12 weeks	16
3 months	18
3–6 months	20
6–12 months	22
2 years	24
Pond life	26
Life cycle	28
Fact file	30
Glossary	31
Index	32

Meet the frogs

There are many different kinds of frogs all over the world. This frog lives in trees. The reed frog lives in swamps. All frogs live near water.

1 day   1 week   2 weeks   5 weeks

Frogs are amphibians. This means they spend part of their life in water and part on land. The frog in this book is a common frog.

12 weeks   14 weeks   6–12 months   2 years

A mass of eggs

A frog begins life as a tiny egg
laid in water. These frogs have
just laid masses of eggs in a pond.

| 1 day | 1 week | 2 weeks | 5 weeks |

The eggs stick together in a big blob of slimy jelly called **spawn**. The black dot inside each egg is a tiny tadpole.

12 weeks   14 weeks   6–12 months   2 years

Hatching

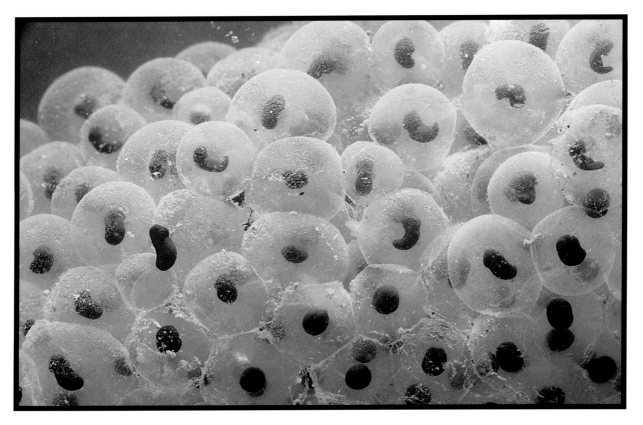

Fish and other animals nibble at the **spawn**, but hundreds of eggs survive. The tadpoles grow bigger and bigger.

1 day 1 week 2 weeks 5 weeks

One day the tadpoles push their way out of the eggs. They hang onto the spawn until their tails grow longer and stronger.

12 weeks
14 weeks
6–12 months
2 years

1–4 weeks

gills

Can you see this tadpole's feathery **gills**? The gills take in **oxygen** from the water. Soon the feathery gills will disappear.

1 day   1 week   2 weeks   5 weeks

Many tadpoles are eaten by water insects and other animals. But the other tadpoles nibble tiny plants and grow bigger and stronger.

12 weeks 14 weeks 6–12 months 2 years

5 weeks

back leg

The tadpole is beginning to change into a frog. The back legs grow first.

| 1 day | 1 week | 2 weeks | 5 weeks |

Lungs grow inside the tadpole's body. Now the tadpole swims up to the surface of the water to breathe in air.

13

12 weeks

14 weeks

6–12 months

2 years

5–12 weeks

gill pouches

Front legs are growing inside the
tadpole's **gill** pouches. The gill
pouches are now bulging.
Soon her legs will push through them.

I day

I week

2 weeks

5 weeks

water fleas

The tadpole still uses her long tail to swim among the plants. She catches some of these water fleas in her wide mouth.

12 weeks   14 weeks   6–12 months   2 years

12 weeks

The tadpole is almost a froglet!
Her tail is getting shorter. Now she
swims by pushing back her long
back legs and **webbed feet**.

| 1 day | 1 week | 2 weeks | 5 weeks |

The tiny froglet now has no tail.
She climbs out of the water onto a
leaf. She looks around and listens
for danger.

12 weeks
14 weeks
6–12 months
2 years

3 months

One day lots of tiny froglets leave the pond and scramble onto the bank. They hide under leaves and stones.

1 day	1 week	2 weeks	5 weeks

The froglet's skin is very thin and she mustn't let it dry out. She dives back into a pond to get wet again.

12 weeks

14 weeks

6–12 months

2 years

3–6 months

The little frog is hungry. She sits very still and waits. Then her long, sticky tongue flicks out and she catches an insect.

| 1 day | 1 week | 2 weeks | 5 weeks |

This grass snake is hungry too. The frog hears it slithering towards her. With a warning croak, she dives quickly into the pond.

12 weeks 14 weeks 6–12 months 2 years

6–12 months

Winter comes and the weather is very cold. The frog looks for a safe hole in the bank. She will stay here and **hibernate**.

1 day

1 week

2 weeks

5 weeks

One warm spring day, she wakes up, feeling very hungry. She crawls out of the hole and leaps off to find food.

12 weeks   14 weeks   6–12 months   2 years

2 years

Another year has passed and the frog's body is fat with eggs. The male frogs are calling from the pond. She hops towards them.

A male frog holds on to her and they **mate**. After a while the eggs leave her body and a new mass of **spawn** floats away.

12 weeks

14 weeks

6–12 months

2 years

Pond life

A frog's life is dangerous. Birds, fish and many animals feed on **spawn**, tadpoles and frogs. Only a few eggs will live to become fully grown frogs. Some frogs may live for up to 10 years.

1 day   1 week   2 weeks   5 weeks

Every spring, adult frogs return to the pond where they were born. They lay thousands of new eggs in the pond.

12 weeks   14 weeks   6–12 months   2 years

Life cycle

Frogspawn

Eggs hatching

2 weeks

5 weeks

12 weeks

5 months

1 year

2 years

29

Fact file

A frog can jump over 3 metres, so it could jump from the foot of your bed to right over your head!

Frogs can breathe through their skins as well as through their mouths.

When a frog first leaves the water it is about the same size as your thumb nail.

The largest frog is the goliath. It lives in Africa. It is large enough to eat small birds and mice.

Glossary

gills part of the body used to breathe in water

hibernate to rest or sleep all winter

lungs part of the body used to breathe in air

mate a male and female come together to produce young

oxygen a gas which living things need to breathe into their bodies to stay alive

spawn a mass of eggs surrounded by jelly

swamp wet, marshy ground

webbed feet feet which have a layer of skin stretched between the toes

Index

amphibian   5

back legs   12, 16

gill pouches   14

gills   10, 31

grass snake   21

insects   11, 15, 20

lungs   13, 31

oxygen   10, 31

skin   19, 30

spawn   6–9, 25–27, 31

tadpole   7–16, 26

tree frog   4

tongue   20

webbed feet   16, 31

winter   22